# *USA in Space*

## *Third Edition*

# USA in Space

## Third Edition

### Volume 1

Air Traffic Control Satellites—ITOS and NOAA
Meteorological Satellites

1–678

*Edited by*

Russell R. Tobias
David G. Fisher

SALEM PRESS, INC.
Pasadena, California • Hackensack, New Jersey

*Editor in Chief:* Dawn P. Dawson
*Editorial Director:* Christina J. Moose
*Acquisitions Editor:* Mark Rehn
*Research Supervisor:* Jeffry Jensen
*Manuscript Editor:* Anna A. Moore
*Production Editor:* Kathy Hix
*Design and Graphics:* James Hutson
*Editorial Assistant:* Dana Garey
*Layout:* Eddie Murillo
*Photograph Editor:* Cynthia Beres

∞ The paper used in these volumes conforms to the American National Standard for Permanence of Paper for Printed Library Materials, Z39.48-1992 (R1997).

**Library of Congress Cataloging-in-Publication Data**

USA in space / edited by Russell R. Tobias and David G. Fisher.— 3rd ed.
  p. cm.
Includes bibliographical references and index.
ISBN-10: 1-58765-259-5 (set : alk. paper)
ISBN-10: 1-58765-260-9 (vol. 1 : alk. paper)
ISBN-13: 978-1-58765-259-2 (set : alk. paper)
ISBN-13: 978-1-58765-260-8 (vol. 1 : alk. paper)
[etc.]
  1. Astronautics—United States.  I. Title: United States of America in space. II. Tobias, Russell R. III. Fisher, David G.
  TL789.8.U5U83 2006
  629.40973—dc22
                                                2005030756

First Printing

PRINTED IN THE UNITED STATES OF AMERICA

# Table of Contents

# Publisher's Note

*USA in Space, Third Edition* updates the Second Edition (3 vols., 2001) with 42 new essays plus 5 replaced essays for a total of 280 essays on the major space programs, piloted and robotic missions, satellites, space centers, space planes, and issues from the earliest missions to the present. The many changes in the space program over the past half decade made a careful reading of every page necessary to ensure the content's accuracy. The editors—longtime historian of the space program Russell R. Tobias, joined in this edition by David G. Fisher, Associate Professor of Physics and Astronomy, Lycoming College, Williamsport, Pennsylvania—carefully pored over all the old essays, revising and often adding paragraphs of material to bring the text up to date. All bibliographies have been expanded with annotated citations to the latest sources for further reading.

## Scope and Coverage

The scope of space exploration history has been expanded at both ends: Two essays on the history of rocket science have been added, as well as the latest missions—both piloted missions, such as those to the International Space Station, and science missions such as the Mars Rovers and Deep Impact.

The new essay on Space Shuttle Mission STS-107, in which seven astronauts tragically lost their lives aboard the orbiter *Columbia* upon reentry into Earth's atmosphere in February of 2003, is accompanied by candid discussion of the institutional conditions that may have contributed to the accident as well as consideration of the direction in which space exploration is headed. An essay has been added to cover the various types of escape mechanisms used for piloted missions. The latest essay covers the return to space, Space Shuttle Mission STS-114, in summer of 2005. Several essays address missions that have yet to be launched but have already undergone years of planning and development.

Despite the title *USA in Space*, the increasingly international nature of space exploration is acknowledged here as well. The International Space Station, Russia's Soyuz and Progress spacecraft, and the contributions of Canada, Japan, China, India, and many European nations in spacecraft, science expriments, equipment, and personnel are well represented among these pages.

## Organization and Format

Articles are arranged in alphabetical order by key word, and by chronological order within groupings. Hence, the reader will find articles on the Apollo missions under "A" in volume 1; space shuttle missions are covered under "Space Shuttle." Within these groups the missions are ordered chronologically. Explorer satellites from mission 1 through Solar Explorers are covered under "Explorers," although those Explorer missions with distinctive names, such as the *International* Ultraviolet Explorer, are alphabetized under those names.

Each article retains the familiar Magill approach to formatting information in a predictable manner to facilitate access to the information. Essays are substantial in length, ranging from 2,000 to 5,000 words. All articles begin with ready-reference top matter, including dates of mission launch and the type of mission, program, satellite, technology, or issue addressed. The old "Principal personages" listings from the first edition, eliminated from the second, have been reinstated in all appropriate essays as "Key Figures." A "Summary of the Mission" (or "Facility," "Program," "Satellite," or "Spacecraft," depending on the topic) follows and introduces the text of the essay, providing an overview of events surrounding the topic, typically from launch to return. The subhead for "Knowledge Gained" has been replaced by a more all-embracing subhead, "Contributions," which describes the section of each essay that encapsulates for the reader the essential discoveries engendered by the mission or program, or the achievements of the essay's subject—such as a space center. "Context," the concluding section, places the topic in perspective in relation to associated events and scientific discoveries, pointing to the importance of the topic within the broader history of space exploration. Finally, each essay ends with a lengthy, annotated list of sources for "Further Reading."

### Finding Aids and Special Features

At the beginning of each volume we have placed a List of Abbreviations, now expanded by more than one-third its previous length; it includes common abbreviations (and their pronunciations if uttered as words) and what they stand for. The List of Illustrations arranges photographs alphabetically by topic, accompanied by page numbers. A Complete List of Contents appears in each volume's front matter in addition to the volume's Table of Contents, to assist those looking for a related topic without access to its volume. Finally, the "Category List" in each volume's front matter arranges the essays by type, from Aerospace Agencies to Spaceports.

To link related essays, we have added a new feature to each essay: "See also" cross-references to other essays in the set. Those who consult the essay on SpaceShipOne, for example, should be aware of the fact that the Ansari X Prize is covered separately.

At the end of Volume 3 are the following appendices, substantially expanded over the second edition:

- Chronology of the U.S. Space Program (expanded 51%)
- Key Figures in the History of Space Exploration (expanded 69%)
- Glossary (expanded 25%)
- Space Shuttle Missions (new)
- World Wide Web Pages (expanded 43%)

Finally, a substantial number of the older images have been replaced with crisper and more informative photos, and many new images have been added. In all, more than 450 black-and-white photographs provide readers with images of what is described in accompanying essays.

### Usage Notes

Units of measurement adhere to the Système International d'Unités (SI units): Launch thrust is measured in newtons rather than pounds, distances in kilometers rather than miles, temperature in Kelvins or degrees Celsius rather than degrees Fahrenheit.

Mission duration for all crewed flights begins at liftoff and ends at touchdown. "Clock times" are rendered, NASA style, in numerals for hours, minutes, and seconds, spelled-out numbers for days, months, and years. Times of day, which in the former editions were rendered in a variety of styles (eastern, Pacific, Central, etc.) have been standardized to that accepted by scientists and engineers as Coordinated Universal Time, abbreviated UTC (for its common alternative, "Universal Time Coordinated").

For our nonspecialist audience, we have generally refrained from abbreviating scientific units of measure and instead spell them out. Likewise, we tend to spell most of the abbreviations and acronyms upon first mention in any essay—and the List of Abbreviations in the front matter to volume 1 provides a ready reference. For those using this publication who are most familiar with the abbreviation rather than its spelled-out version, abbreviations were routinely added to the Subject Index as cross-references.

### Acknowledgments

Grateful acknowledgment is made to the many dedicated contributors who worked to make the language of space science accessible to the general reader. A list of those scientists, academicians, and other space experts (among them some of the foremost historians of space exploration) is found in the front matter to volume 1. Special thanks go to Editors Russell R. Tobias and David G. Fisher for content development, manuscript editing, revisions, additions, and research. This publication benefits incalculably from their scientific expertise and insights.

# Contributors

Rajkumar Ambrose
*Monmouth College*

Michael S. Ameigh
*Saint Bonaventure University*

Joseph A. Angelo, Jr.
*Florida Institute of Technology*

Thomas W. Becker
*Webster University*

Raymond D. Benge, Jr.
*Richland College*

Alvin K. Benson
*Brigham Young University*

Alan F. Bentley
*Eastern Montana College*

John L. Berkley
*State University of New York
    College at Fredonia*

Massimo D. Bezoari
*Huntingdon College*

Terry D. Bilhartz
*Sam Houston State University*

T. Parker Bishop
*Georgia Southern College*

Penelope J. Boston
*Complex Systems Research, Inc.*

Walter E. Bressette
*National Aeronautics and Space
    Administration
Langley Research Center*

Larry M. Browning
*South Dakota State University*

Michael L. Broyles
*Collin County Community College*

Martin Burkey
*The Huntsville Times*

Dennis Chamberland
*Independent Scholar*

Monish R. Chatterjee
*State University of New York at
    Binghamton*

D. K. Chowdhury
*Indiana University*

Eric Christensen
*Independent Scholar*

W. David Compton
*Independent Scholar*

Robert E. Davis
*National Aeronautics and Space
    Administration*

Ronald W. Davis
*Western Michigan University*

Christopher F. Dickens
*Independent Scholar*

Beth Dickey
*Independent Scholar*

Dave Dooling
*The Space and Rocket Center*

John J. Dykla
*Loyola University*

James C. Elliott
*National Aeronautics and Space
    Administration*

Judith Belsky Farrin
*Independent Scholar*

Ronald J. Ferrara
*Middle Tennessee State University*

David G. Fisher
*Lycoming College*

Dennis R. Flentge
*Cedarville College*

George J. Flynn
*State University of New York
    College at Plattsburgh*

Stephanie Gallegos
*Jet Propulsion Laboratory*

Douglas Gomery
*University of Maryland at College
    Park*

Noreen A. Grice
*Boston Museum of Science*

Nicole E. Gugliucci
*Independent Scholar*

Paul A. Heckert
*Western Carolina University*

Rod Heelis
*University of Texas at Dallas*

T. J. Herczeg
*University of Oklahoma*

Niles R. Holt
*Illinois State University*

Linda J. Horn
*Jet Propulsion Laboratory*

Thomas D. Inman
*Northwestern State University*

Patricia Jackson
*Davenport College*

Rebecca B. Jervey
*Independent Scholar*

David W. Jex
*Independent Scholar*

Karen N. Kähler
*California Institute of Technology*

Christopher Keating
*University of Texas at Dallas*

Gregory P. Kennedy
*The Space Center*

John Kenny
*Bradley University*

Robert Klose
*University College of Bangor*

Narayanan M. Komerath
*Georgia Institute of Technology*

William J. Kosmann
*Jet Propulsion Laboratory*

Joseph J. Kosmo
*National Aeronautics and Space Administration*

Adam Lee
*Huntingdon College*

Josué Njock Libii
*Purdue University, Fort Wayne*

M. A. K. Lodhi
*Texas Tech University*

Clarice Lolich
*Oklahoma State University*

Reinhart Lutz
*University of California, Santa Barbara*

Dana P. McDermott
*Independent Scholar*

Kerrie L. MacPherson
*University of Hong Kong*

Paul Madden
*Hardin-Simmons University*

David W. Maguire
*C. S. Mott Community College*

Joseph T. Malloy
*Hamilton College*

Nancy Farm Mannikko
*Independent Scholar*

Robert G. Melton
*Pennsylvania State University, University Park*

Charles Merguerian
*Hofstra University*

J. D. Mihalov
*National Aeronautics and Space Administration*

Randall L. Milstein
*Oregon State University*

Ellen F. Mitchum
*The Space Center*

Christina M. Nestlerode
*Lycoming College*

Peter Neushul
*California Institute of Technology*

John Newman
*University of Illinois at Urbana-Champaign*

Henry W. Norris
*Independent Scholar*

Divonna Ogier
*Oregon Museum of Science and Industry*

Glenn S. Orton
*Jet Propulsion Laboratory*

Robert J. Paradowski
*Rochester Institute of Technology*

Gordon A. Parker
*University of Michigan, Dearborn*

Curtis Peebles
*Independent Scholar*

Donna Pivirotto
*Jet Propulsion Laboratory*

Raj Rani
*Queens College, City University of New York*

Marc D. Rayman
*New Millennium Program*

Robert V. Reeves
*Independent Scholar*

Nicholas A. Renzetti
*California Institute of Technology*

Mike D. Reynolds
*University of North Florida*

Rex Ridenoure
*Jet Propulsion Laboratory*

Jacqueline J. Robinson
*Huntingdon College*

Charles W. Rogers
*Southwestern Oklahoma State University*

Philip J. Sakimoto
*Whitman College*

John M. Shaw
*Educational Systems, Inc.*

David J. Shayler
*Astro Info Service*

R. Baird Shuman
*University of Illinois at Urbana-Champaign*

Paul P. Sipiera
*William Rainey Harper College*

Clyde Curry Smith
*University of Wisconsin, River Falls*

Roger Smith
*Independent Scholar*

Michael Smithwick
*Independent Scholar*

Barry M. Stentiford
*Grambling State University*

Alan Stern
*University of Colorado*

Lulynne Streeter
*Independent Scholar*

Locke Stuart
*Goddard Space Flight Center*

Richard A. Sweetsir
*Independent Scholar*

Loyd S. Swenson, Jr.
*University of Houston*

F. W. Taylor
*University of Oxford*

James Thornton
*Huntingdon College*

Russell R. Tobias
*Independent Scholar*

Greg Tomko-Pavia
*Independent Scholar*

Marsha R. Torr
*National Aeronautics and Space Administration*

Jeffery S. Underwood
*Louisiana State University*

Robert Veiga
*National Aeronautics and Space Administration*

Mary Walsh
*National Aeronautics and Space Administration*

Randii R. Wessen
*Jet Propulsion Laboratory*

Manfred N. Wirth
*National Aeronautics and Space Administration*

Clifton K. Yearley
*State University of New York at Buffalo*

# List of Abbreviations

*Pronunciations are provided for those abbreviations that are pronounced as words. Where no pronunciation is given, the abbreviation is referred to simply by its letters uttered in sequence.*

## Pronunciation Key

| | | | | |
|---|---|---|---|---|
| a | *as in* answer = AN-sihr | | o | *as in* cotton = CO-tuhn |
| ah | *as in* father = FAH-thur | | oh | *as in* below = bee-LOH |
| aw | *as in* awful = AW-ful | | oo | *as in* good = good |
| ay | *as in* blaze = blayz | | ow | *as in* couch = kowch |
| ch | *as in* beach = beech | | oy | *as in* boy = boy |
| eh | *as in* bed = behd | | s | *as in* cellar = SEL-ur |
| ee | *as in* believe = bee-LEEV | | sh | *as in* champagne = sham-PAYN |
| ew | *as in* boot = bewt | | uh | *as in* about = uh-BOWT |
| g | *as in* beg = behg | | ur | *as in* birth = burth |
| i | *as in* buy = bi | | y | *as in* useful = YEWS-ful |
| ih | *as in* bitter = BIH-tur | | z | *as in* business = BIHZ-ness |
| j | *as in* digit = DIH-jiht | | zh | *as in* vision = VIH-zhuhn |
| k | *as in* cat = kat | | | |

**A/L**: air lock

**AAP**: Apollo Applications Program

**ABM**: antiballistic missile

**ABMA**: Army Ballistic Missile Agency

**ACCESS** (AK-sehs): Assembly Concept for Construction of Erectable Space Structures

**ACDWS**: Aircraft Dropwindsonde System

**ACES** (AYS-ehz): Acoustic Containerless Experiment System

**ACP**: Aerosol Collector and Pyrolyser

**ACRC**: assured crew return capability

**ACRIM** (AK-rihm): Active Cavity Radiometer Irradiance Monitor

**ACRV**: Assured Crew Return Vehicle

**ACTS** (akts): Advanced Communications Technology Satellite

**AEM**: Animal Enclosure Module; *also,* Applications Explorer Mission

**AEPI**: Atmospheric Emissions Photometric Imaging spectrometer

**AFB**: Air Force Base

**AGC**: Apollo Guidance Computer

**AGN**: active galactic nucleus

**AGS** (agz): Abort Guidance System

**AIAA**: American Institute of Aeronautics and Astronautics

**AID**: Agency for International Development

**AIRS** (ayrz): Atmospheric Infrared Sounder

**ALAE**: Atmospheric Lyman-Alpha Emission ultraviolet spectrometer

**ALSEP** (ahl-sehp): Apollo Lunar Surface Experiments Package

**ALT**: approach and landing test

**AM**: Airlock Module

**AMOS** (AY-muhs): Air Force Maui Optical System

**AMPTE** (AMP-tee): Active Magnetospheric Particle Tracer Explorer

**AMS**: Alpha Magnetic Spectrometer investigation

**AMSU**: Advanced Microwave Sounding Unit

**AMTEX** (AM-tehks): Air-Mass Transformation Experiment

**AMU**: Astronaut Maneuvering Unit

**APEX** (AY-pehks): Advanced Photovoltaic and Electronic Experiments satellite

**APS** (aps): Ascent Propulsion System (Apollo Lunar Module)

**APT**: automatic picture transmission

**APU**: auxiliary power unit

**APXS**: Alpha Proton X-Ray Spectrometer

**ARPA**: Advanced Research Projects Agency

**AS**: Apollo-Saturn

**ASA**: Astronaut Science Advisor

**ASAT** (AY-sat): antisatellite

**ASC**: L'Agence Spatiale Canadienne (Canadian Space Agency)

**ASI**: Agenzia Spaziale Italiana (Italian Space Agency)

**ASTER** (AS-tuhr): Advanced Spaceborne Thermal Emission and Reflection Radiometer

**ASTP**: Apollo-Soyuz Test Project

**ATDA**: Augmented Target Docking Adapter

**ATLAS** (AT-lehs): Atmospheric Laboratory for Applications and Science

**ATM**: Apollo Telescope Mount

**ATMOS** (AT-mohz): Atmospheric Trace Molecules Spectroscopy experiment

**ATN**: Advanced TIROS-N satellite

**ATP**: Advanced Turboprop program

**ATS**: Applications Technology Satellites

**ATV**: Automated Transfer Vehicle

**AU**: astronomical unit

**AVCS**: advanced vidicon camera system

**AVHRR**: Advanced Very High-Resolution Radiometer

**AXAF**: Advanced X-Ray Astrophysics Facility (renamed Chandra X-Ray Observatory)

**BAT** (bat): Burst Alert Telescope

**BATSE**: Burst and Transient Source Experiment

**BECO** (BEE-koh): Booster Engine Cutoff

**BMD**: ballistic missile defense

**BMEWS** (BEE-mewz): Ballistic Missile Early-Warning System

**CAENEX**: Complex Atmospheric Energetics Experiment

**CAM** (kam): Centrifuge Accommodations Module

**CapCom** (CAP-cahm): capsule communicator

**CAS**: Committee on Atmospheric Sciences

**CCAFS**: Cape Canaveral Air Force Station

**CCD**: charge-coupled device

**CDR**: commander

**CERES** (SIHR-eez): Clouds and the Earth's Radiant Energy System

**CERV**: Crew Emergency Return Vehicle

**CFES**: Continuous Flow Electrophoresis System

**CGRO**: Compton Gamma Ray Observatory

**CHASE** (chays): Coronal Helium Abundance Experiment

**CIRRIS** (SIHR-ihs): Cryogenic Infrared Radiance Instrument for Shuttle

**CLAES**: Cryogenic Limb Array Etalon Spectrometer

**CLIVAR**: Climate Variability project

**CM**: Command Module

**CMB**: cosmic microwave background

**CMP**: Command Module pilot

**CNES**: Centre National d'Études Spatiales (French Space Agency)

**COBE** (KOH-bee): Cosmic Background Explorer

**COMPTEL** (KOM-tehl): Imaging Compton Telescope

**COMSAT** (KOM-sat): Communications Satellite Corporation

**CONTOUR** (KON-tewr): Comet Nucleus Tour mission

**COSPAR** (KOH-spahr): Committee on Space Research

**COSPAS** (KOH-spas): Soviet Space System for Search of Vessels in Distress

**COSPAS/SARSAT** (KOH-spas SAHR-sat): Cooperative Soviet Space System for Search of Vessels in Distress/U.S. Search and Rescue Satellite-Aided Tracking

**COSTAR** (KOH-stahr): Corrective Optics Space Telescope Axial Replacement

**CRISTA-SPAS** (KRIH-stah-spahs): Cryogenic Infrared Spectrometers and Telescopes for the Atmosphere-Shuttle Pallet Satellite-2

**CRN**: Cosmic-Ray Nuclei instrument

**CRV**: Crew Return Vehicle

**CSA**: Canadian Space Agency

**CSM**: Command and Service module

**CTS**: Communications Technology Satellite

**CTV**: Crew Transfer Vehicle

**CXO**: Chandra X-Ray Observatory

**CXRO**: Chandra X-Ray Observatory

**DAX**: Dutch Additional Experiment

**DBS**: direct broadcast satellite

**DE**: Dynamics Explorer

**DGE**: Doppler Gravity Experiment

**DGPS**: Differential Global Positioning System

**DISR**: Descent Imager and Spectral Radiometer

**DLR**: Deutsches Zentrum für Luft- und Raumfahrt (German Aerospace Center)

**DM**: Docking Module

**DMOS** (DEE-mohs): Diffusion Mixing of Organic Solution experiment

**DMSP**: Defense Meteorological Satellite Program

**DNSS**: Defense Navigation Satellite System

**DoD**: Department of Defense

**DPS** (dips): Descent Propulsion System (Apollo Lunar Module)

**DSKY** (DIHS-kee): Display Keyboard for the Apollo onboard computer

**DSN**: Deep Space Network

**DWE**: Doppler Wind Experiment

**Dyna-Soar**: a contraction of "dynamic soaring"

**EASE** (eez): Experimental Assembly of Structures in Extravehicular Activity

**EASEP** (ee-sep): Early Apollo Scientific Experiments Package

**ECLSS**: Environmental Control and Life Support System

**EE**: Electrodynamics Explorer

**EGO**: Eccentric Geophysical Observatory

**EGRET** (EE-greht): Energetic Gamma Ray Experiment Telescope

**ELINT** (EE-lihnt): Electronic Intelligence (satellite)

**ELT**: emergency locator-transmitter

**ELV**: expendable launch vehicle

**EMI**: electromagnetic interference

**EMU**: extravehicular mobility unit

**EOPAP**: Earth and Ocean Physics Applications Program

**EOS**: Earth Observing System

**EOSAT** (EE-oh-sat): Earth Observation Satellite Company

**EPIRB**: emergency pointing/indicating radio beacon

**ERBE** (EHR-bee): Earth Radiation Budget Experiment

**ERIS**: Exoatmospheric Reentry Interceptor System

**EROS** (EHR-ahs): Earth Resources Observation Satellite

**ERTS** (ehrtz): Earth Resources Technology Satellite

**ESA**: European Space Agency

**ESP**: Extravehicular Support Package; External Stowage Package

**ESRO**: European Space Research Organization

**ESSA**: Environmental Science Services Administration

**ETM+**: Enhanced Thematic Mapper Plus

**EURECA** (yew-REE-kah): European Retrievable Carrier

**EUVE** (YEW-vee): Extreme Ultraviolet Explorer

**EVA**: extravehicular activity (spacewalk)

**FAA**: Federal Aviation Administration

**FAI**: Fédération Aéronautique Internationale

**FARE** (fayr): Fluid Acquisition and Resupply Experiment

**FAST** (fast): Fast Auroral Snapshot Explorer

**FAUST** (fowst): Far Ultraviolet Space Telescope

**FBC**: "faster, better, cheaper" policy

**FCSS**: Flight Crew Support System

**FDF**: Flight Dynamics Facility

**FGB**: Functional Cargo Block (*funktsionalya-gruzovod blokor*); also known as the Zarya Control Module

**FGGE**: First GARP Global Experiment

**FILE** (feyl): Feature Identification and Location Experiment

**FOBS** (fobz): Fractional Orbit Bombardment System

**FORTE** (fohrt): Fast On-Orbit Recording of Transient Events satellite

**FRC**: Flight Research Center

**FREESTAR** (FREE-stahr): Fast Reaction Experiments Enabling Science, Technology, Applications and Research

**FUSE** (fewz): Far Ultraviolet Spectroscopic Explorer

**GALCIT**: Guggenheim Aeronautical Laboratory

**GALEX** (GAY-lehks): Galaxy Evolution Explorer

**GARP** (gahrp): Global Atmospheric Research Program

**GAS** (gas): Get-Away Special

**GATE** (gayt): GARP Atlantic Tropical Experiment

**GATV**: Gemini-Agena Target Vehicle

**GCMS**: Gas Chromatograph Mass Spectrometer

**GEO** (GEE-oh): geosynchronous satellite

**GEOS** (GEE-oz): Geodetic Earth-Orbiting Satellite; Geodynamics Experimental Ocean Satellite

**Geosat** (GEE-oh-sat): Geodynamic Earth and Oceans Satellite

**GET**: Ground-elapsed time

**GEWEX**: Global Energy and Water Cycle Experiment

**GFFC**: Geophysical Fluid Flow Cell experiment

**GLAS** (glas): Geoscience Laser Altimeter System

**GLAST** (glast): Gamma-ray Large Area Space Telescope

**GLOMR** (GLOH-mahr): Global Low-Orbiting Message Relay

**GLV**: Gemini Launch Vehicle

**GOES** (gohz): Geostationary Operational Environmental Satellite

**GONG** (gawng): Global Oscillations Network Group

**GPS**: Global Positioning System

**GRC**: Glenn Research Center

**GRC** LeRC: John H. Glenn Research Center at Lewis Field

**GRO**: Compton Gamma Ray Observatory

**GRT**: general relativity theory

**GSFC**: Goddard Space Flight Center

**GT**: Gemini-Titan

**GTO**: geosynchronous transfer orbit

**HALOE** (HAY-loh): Halogen Occultation Experiment

**HASI**: Huygens Atmospheric Structure Instrument

**Hazcam** (HAZ-cam): Hazard Avoidance Camera

**HCMM**: Heat Capacity Mapping Mission

**HEAO** (HEE-oh): High-Energy Astronomical Observatory

**HEAO 2** (HEE-oh-tew): Einstein Observatory

**HERCULES** (HUHR-kew-leez): Handheld Earth-oriented, Real-time, Cooperative, User-friendly, Location-targeting and Environmental System

**HESSI** (HEHS-see): High Energy Solar Spectroscopic Imager

**HHMU**: Handheld Maneuvering Unit

**HiMAT** (HEY-mat): Highly Maneuverable Aircraft Technology

**HPTE**: High-Precision Tracking Experiment

**HRTS**: High-Resolution Telescope and Spectrograph

**HST**: Hubble Space Telescope

**HTPB**: hydroxyl-terminated polybutadiene

**ICBM**: intercontinental ballistic missile

**ICE** (EYS): International Cometary Explorer

**ICESat** (EYS-sat): Ice, Cloud, and Land Elevation Satellite

**ICSU**: International Council of Scientific Unions

**IDCSP**: Interim (or Initial) Defense Communications Satellite Project

**IEEE** (eye-trihpuhl-ee): Institute of Electrical and Electronics Engineers

**IEF**: Isoelectric Focusing

**IEH**: International Extreme Ultraviolet Hitchhiker

**IFA**: in-flight anomaly

**IGBP**: International Geosphere-Biosphere Program

**IGY**: International Geophysical Year, 1957-1958

**IJPS**: Initial Joint Polar-orbiting Operational Satellite System

**IMCO**: Intergovernmental Maritime Consultative Organization

**IMEWS** (eye-mewz): Integrated Missile Early-Warning Satellite

**IMF**: interplanetary magnetic field

**IML**: International Microgravity Laboratory

**IMP** (ihmp): Interplanetary Monitoring Platform; Imager for Mars Pathfinder

**INF**: Intermediate Nuclear Forces

**INMS**: Ion and Neutral Mass Spectrometer

**INP**: Instituto Nacional de Pesquisas Espacias (Brazilian National Institute for Space Research)

**INS**: inertial navigation system

**INTA**: Instituto Nacional de Tecnica Aeroespacial (Spanish Space Agency)

**Intelsat**: International Telecommunications Consortium

**IONS** (EY-awnz): Ionization of Solar and Galactic Cosmic-Ray Heavy Nuclei experiment

**IPS**: Instrument Pointing System

**IRAS** (EY-ras): Infrared Astronomical Satellite

**IRBM**: intermediate range ballistic missile

**IRIS** (EY-rihs): Infrared Imagery of Shuttle; Infrared Interferometer Spectrometer

**ISAMS** (EY-samz): Improved Stratospheric and Mesospheric Sounder

**ISEE** (EY-see): International Sun-Earth Explorer

**ISO** (EY-soh): Imaging Spectrometric Observatory

**ISPM**: International Solar-Polar Mission

**ISPR**: International Standard Payload Rack

**ISS**: International Space Station

**ISTP**: International Solar-Terrestrial Physics Science Initiative; Integrated Space Transportation Plan

**ITOS**: Improved TIROS Operational System

**ITS**: Integrated Truss Structure; Impactor Target Sensor

**IUE**: International Ultraviolet Explorer

**IUGG**: International Union of Geodesy and Geophysics

**IUS**: Inertial Upper Stage

**JAM** (jam): (Quest) Joint Airlock Module

**JASPIC** (JAS-pihc): Joint American-Soviet Particle Intercalibration

**JATO** (JAY-toh): jet-assisted takeoff

**JAWSAT** (JAW-sat): Joint Air Force Academy-Weber State University Satellite

**JAXA**: Japan Aerospace Exploration Agency

**JEM** (jehm): Japanese Experiment Module, *or* Kibo (Japanese for "hope")

**JGOFS**: Joint Global Ocean Flux Study

**JMR**: Jason Microwave Radiometer

**JPL**: Jet Propulsion Laboratory

**JSC**: Johnson Space Center

**J2000.0**: The Julian date reference for January 1, 2000, 11:58:55.816 Coordinated Universal Time

**KBO**: Kuiper Belt object

**KH**: Keyhole

**KSC**: Kennedy Space Center

**LAGEOS** (LAY-gee-ohz): Laser Geodynamics Satellite

**LaRC**: Langley Research Center

**LDD**: Lunar Dust Detector

**LDEF** (EHL-dehf): Long Duration Exposure Facility

**LEAM**: Lunar Ejecta and Meteorites experiment

**LEO** (LEE-oh): low-Earth orbit *or* low-Earth-orbiting

**LEO sat** (LEE-oh sat): Low-Earth-orbiting satellite

**LeRC**: Lewis Research Center

**LES** (lehs): Launch Escape System; land Earth station

**LHe**: liquid helium

**LH$_2$**: liquid hydrogen

**LIDAR** (LEY-dahr): Light Detection and Ranging Instrument

**LJ**: Little Joc

**LLRV**: Lunar Landing Research Vehicle

**LLTV**: Lunar Landing Test Vehicle

**LM** (lehm): Lunar Module; Laboratory Module

**LMLV**: Lockheed Martin Launch Vehicle

**LMP**: Lunar Module pilot

**LMS**: Life and Microgravity Spacelab

**LN₂**: liquid nitrogen

**LOR**: Lunar Orbit Rendezvous

**LORE**: Limb Ozone Retrieval Experiment

**LOX** (lawks): liquid oxygen

**LRC**: Langley Research Center

**LRICBM**: long-range intercontinental ballistic missile

**LRO**: Lunar Reconnaissance Orbiter

**LRV**: Lunar Roving Vehicle

**LSP**: Laminar Soot Processes experiment

**MA**: Mercury-Atlas

**MAD** (mad): mutual assured destruction

**Magsat** (MAG-sat): Magnetic Field Satellite

**MAHRSI**: Middle Atmosphere High Resolution Spectrograph Investigation

**MAPS** (maps): Measurement of Air Pollution from Satellites experiment

**MARCI** (MAHR-see): Mars Color Imager

**MARDI** (MAHR-dee): Mars Descent Imager

**MARIE** (mah-REE): Martian Radiation Environment Experiment

**MAS**: Millimeter-Wave Atmospheric Sounder

**mascons** (MAS-conz): mass concentrations

**MAST**: Multi-Application Survivable Tether

**MDA**: Multiple Docking Adapter

**MEEP** (meep): Mir Environmental Effects Payload

**MEIDEX** (MEY-dehks): Mediterranean Israeli Dust Experiment

**MEO** (MEE-oh): medium-Earth orbit, *or* medium-Earth-orbiting

**MER**: Mars Exploration Rover

**MESA** (MAY-sah): Modularized Equipment Stowage Assembly

**MESSENGER**: Mercury Surface, Space Environment, Geochemistry, and Ranging

**M.E.T.**: Mission Elapsed Time

**MET** (meht): Modular Equipment Transporter

**MGS**: Mars Global Surveyor

**MHV**: Miniature Homing Vehicle

**MiDAS** (MEY-dahs): Missile Defense Alarm System

**MIDEX** (MEY-dehks): Medium Explorer Program

**MILA**: Merritt Island Launch Area

**Mini-TES** (mini-tess): Miniature Thermal Emission Spectrometer

**MISR** (MEY-zuhr): Multi-Angle Imaging Spectroradiometer

**MISS** (mihs): Man-in-Space-Soonest

**MISSE** (MIHS-see): Materials International Space Station Experiment

**MKS**: Russian abbreviation for the International Space Station

**MLR**: Monodisperse Latex Reactor

**MLS**: Microwave Limb Sounder

**MMH**: monomethylhydrazine (a liquid rocket propellant)

**MMS**: multimission modular spacecraft

**MMU**: Manned Maneuvering Unit

**MOC**: Mars Orbiter Camera

**MODIS**: Moderate-Resolution Imaging Spectroradiometer

**MOL**: Manned Orbiting Laboratory

**MOLA**: Mars Orbiter Laser Altimeter

**MONEX**: Monsoon Experiment

**MOPITT**: Measurements of Pollution in the Troposphere

**MPLM**: Multi-Purpose Logistics Module

**MR**: Mercury-Redstone

**MRO**: Mars Reconnaissance Orbiter

**MSC**: Manned Spacecraft Center (later Johnson Space Center Manned Spacecraft Center)

**MSFC**: Marshall Space Flight Center

**MSL**: Microgravity Science Laboratory

**MSS**: Multispectral Scanner Subsystem

**MTPE**: Mission to Planet Earth

**MVACS**: Mars Volatiles and Climate Surveyor

**NACA**: National Advisory Committee for Aeronautics (always pronounced as individual letters, never as "naca")

**NASA** (NA-sah): National Aeronautics and Space Administration

**NASCOM** (NAS-com): NASA Communications Division

**NASP**: National AeroSpace Plane

**NAVAID** (NAV-ayd): Navigational Aid

**Navcam** (NAV-cam): Navigation Camera

**NBS**: Neutral Buoyancy Simulator

**NCOS** (NIHK-ohz): National Commission on Space

**NEAR** (near): Near Earth Asteroid Rendezvous

**NEO** (NEE-oh): Near-Earth object

**NEP**: Nuclear Electric Propulsion

**NERVA** (NEHR-vah): Nuclear Engine for Rocket Vehicle Application

**NICMOS** (NIHK-mohz): Near Infrared Camera and Multi-Object Spectrometer

**NIMA**: National Imagery and Mapping Agency

**NM**: Neuron Spectrometer

**NMP**: New Millennium Program

**NNSS**: Navy's Navigation Satellite System

**NOAA** (NOH-ah): National Oceanic and Atmospheric Administration

**NOMSS**: National Operational Meteorological Satellite System

**NORAD** (NOHR-ad): North American Air Defense Command

**NOSL**: Night/Day Observations from Space of Lightning experiment

**NOSS** (nos): Naval Ocean Surveillance System

**NPO**: Nauchno-Proizvodstvennoe Obedinenie (Soviet/Russian Organization for Scientific Production or Research)

**NPO Molniya** (mohl-NEE-yah): Soviet/Russian Research and Industrial Corporation (founded to create the first Russian piloted reusable spacecraft, *Buran*, in 1976; *molniya* is Russian for "lightning")

**NPOESS**: National Polar-orbiting Operational Environmental Satellite System

**NSBF**: National Scientific Balloon Facility

**NSSS**: Naval Space Surveillance System

**NTP**: Nuclear Thermal Propulsion

**NTS**: Navigation Technology Satellite

**N$_2$O$_4$**: nitrogen tetroxide

**OAMS** (ohmz): Orbital Attitude Maneuvering System

**OAO**: Orbiting Astronomical Observatory

**OAST** (ohst): Office of Aeronautics and Space Technology

**OFO**: Orbiting Frog Otolith

**OFT**: Orbital Flight Test Program

**OGO**: Orbiting Geophysical Observatory

**OKB**: Opytnoe Konstructorskoe Buro (Soviet/ Russian Experimental Design Bureau)

**OKB-1**: Opytnoe Konstructorskoe Buro 1 (classified Soviet name of Sergei Korolev's design bureau, later Energia Rocket and Space Corporation, or RSC Energia)

**OKB-51**: Opytnoe Konstructorskoe Buro 51 (Soviet Special Design Bureau, designed the Soviets' first pilotless aircraft)

**OKB-52**: Opytnoe Konstructorskoe Buro 52 (classified Soviet name of Vladimir Chelomei's design bureau in Moscow)

**OMB**: Office of Management and Budget

**OMS** (ohmz): orbital maneuvering system

**OMSF**: Office of Manned Space Flight

**OPF**: Orbiter Processing Facility

**ORFEUS** (OHR-fee-uhs): Orbiting Retrievable Far and Extreme Ultraviolet Spectrograph

**OSCAR** (OS-kahr): Orbiting Satellites Carrying Amateur Radio

**OSO**: Orbiting Solar Observatory

**OSSA**: Office of Space Science and Applications

**OSSE**: Oriented Scintillation Spectroscopy Experiment

**OSTA** (OH-stah): Office of Space and Terrestrial Applications

**OWS**: Orbital Workshop (Skylab)

**PACS** (paks): Particle-Analysis Camera System

**PAGEOS** (PAY-gee-ohz): Passive Geodetic Earth-Orbiting Satellite

**PAM** (pam): Payload Assist Module

**Pancam** (PAN-cam): Panoramic Camera

**PANSAT** (PAN-sat): Petite Amateur Naval Satellite

**PARD** (pahrd): Pilotless Aircraft Research Division

**PDP**: Plasma Diagnostics Package

**PERSI** (PURH-see): Pluto Exploration Remote Sensing Investigation

**PGNS** (pihngz): Primary Guidance and Navigation System

**PLS**: Personnel Launch System

**PMIRR**: Pressure Modulator Infrared Radiometer

**POES** (pohz): Polar Operational Environmental Satellite

**POIC**: Payload Operation Integration Center

**PRIME** (preym): Precision Recovery Including Maneuvering Entry project

**PSEP**: passive seismograph experiment

**PVTOS**: Physical Vapor Transport of Organic Solids

**RAT** (rat): Rock Abrasion Tool

**RCC**: reinforced carbon-carbon

**RCS**: Reaction Control System; Reentry Control System

**RDM**: Research Double Module (SPACEHAB)

**REX** (rehks): Radio Science Experiment

**RKA**: Russian Aviation and Space Agency (now Roscosmos)

**RKK** Energia: S. P. Korolev Rocket and Space Corporation Energia

**RMS**: Remote Manipulator System, *or* Canadarm

**ROSAT** (ROH-sat): Roentgen X-Ray Satellite

**Roscosmos** (ros-cos-mohz): Russian Federal Space Agency

**RTG**: radioisotope (or radioisotopic) thermal (or thermoelectric) generator

**SA**: Saturn-Apollo

**SAFER** (SAY-fuhr): Simplified Aid for Extravehicular Activity Rescue

**SAGE** (sage): Stratospheric Aerosol and Gas Experiment

**SAINT** (saynt): Satellite Interceptor

**SALT** (sahlt): Strategic Arms Limitation Talks and Treaties

**SAM** (sam): Stratospheric Aerosol Measurement instrument

**SAMOS**: Satellite and Missile Observation System

**SAMPEX** (SAMP-ehks): Solar Anomalous and Magnetospheric Particle Explorer

**SAMS**: Stratospheric and Mesospheric Sounder, Space Acceleration Measurement System

**SAMTO** (SAMP-toh): Space and Missile Test Organization

**SAREX** (SAHR-eks): Shuttle Amateur Radio Experiment

**SARSAT** (SAHR-sat): Search and Rescue Satellite-Aided Tracking

**SAS**: Small Astronomy Satellite

**SBI**: space-based interceptor

**SBUV**: Solar Backscattered Ultraviolet instrument

**SB-WASS**: Space Based Wide Area Surveillance System

**SCA**: Shuttle Carrier Aircraft

**SDI**: Strategic Defense Initiative

**SDIO**: Strategic Defense Initiative Organization

**SDRN**: Satellite Data-Relay Network

**SDS**: Satellite Data System

**SEDS**: Small Expendable Deployer System

**SEM**: Space Environment Monitor; Space Experiment Module

**SEPAC**: Space Experiments with Particle Accelerators

**SEPS**: Space Experiments with Plasmas in Space

**SETII** (SEH-tee): Search for Extraterrestrial Intelligence Institute

**SEWS** (sewz): Satellite Early-Warning System

**SFRB**: solid-fueled rocket booster

**SHARAD**: Shallow Radar

**SIDE** (seyd): Suprathermal Ion Detector Experiment

**SigInt** (SIHG-ihnt): signal intelligence (satellite)

**SIM**: Scientific Instrument Module; Space Interferometry Mission

**SINS** (sihnz): Shipboard Inertial Navigation System

**SIR** (suhr): Shuttle Imaging Radar

**SIR-C/X-SAR** (SURH-see EHKS-sahr): Spaceborne Imaging Radar-C/X-Band Synthetic Aperture Radar

**SIRTF**: Space Infrared Telescope Facility, renamed the Spitzer Space Telescope (originally the Shuttle Infrared Telescope Facility)

**SITE** (seyt): Satellite Instructional Television Experiment

**SLA** (slah): Spacecraft Lunar Module Adapter

**SLC** (slihk): Space Launch Complex

**SLF**: Shuttle Landing Facility

**SLS**: Spacelab Life Sciences

**SLV**: satellite launch vehicle

**SLWT**: Super Lightweight External Tank

**SM**: Service Module

**SMEAT** (smeet): Skylab Medical Experiments Altitude Test

**SMEX**: Small Explorer program

**SMM**: Solar Maximum Mission

**SMS**: Synchronous Meteorological Satellite; Shuttle Mission Simulator

**SNAP** (snap): System for Nuclear Auxiliary Power

**SNOE** (snoh): Student Nitric Oxide Explorer

**SOFAR** (SOH-fahr): sound fixing and ranging bomb

**SOFBALL** (SOF-bahl): Structure of Flame Balls at Low Lewis Number experiment

**SOHO** (SOH-hoh): Solar and Heliospheric Observatory

**SOLCON** (SOHL-con): Measurement of Solar Constant spectrometer

**SOLSE**: Shuttle Ozone Limb Sounding Experiment

**SOLSPEC** (SOHL-spehk): Solar Spectrum Measurement

**SOLSTICE** (SOL-stihs): Solar/Stellar Irradiance Comparison Experiment

**SORCE** (sohrs): Solar Radiation and Climate Experiment

**SOUP** (sewp): Solar Optical Universal Polarimeter

**Spacecom** (SPAYS-com): Space Communications Company

**SPACEHAB** (SPAYS-hab): Space Habitat Module

**Spadoc** (SPAY-doc): Space Defense Operations Center

**SPAS** (spas): Shuttle Pallet Satellite

**SPS**: Service Propulsion System; Standard Positioning Service

**SQUID** (skwihd): superconducting quantum interference device

**SRB**: solid-fueled rocket booster, *or* solid rocket booster

**SRC**: Science and Engineering Research Council (British group)

**SRL**: Space Radar Laboratory

**SRTM**: Shuttle Radar Topography Mission

**SSBUV**: Shuttle Solar Backscatter Ultraviolet spectrometer

**SSCC**: Space Station Control Center

**SSIP**: Shuttle Student Involvement Project

**SSME**: space shuttle main engine

**SSP**: Surface Science Package

**SSRMS**: Space Station Remote Manipulator System

**SSTO**: single-stage-to-orbit

**STADAN**: Space Tracking and Data Acquisition Network

**S*T*A*R*S** (stahrs): Space Technology and Research Students Program

**START** (stahrt): Spacecraft Technology and Advanced Reentry Test

**STDN**: Spaceflight Tracking and Data Network

**STEDI**: Student Explorer Demonstration Initiative

**STG**: Space Task Group

**STS**: Space Transportation System (shuttle); Structural Transition Section (Skylab)

**SUSIM**: Solar Ultraviolet Spectral Irradiance Monitor

**SWAP** (swahp): Solar Wind Around Pluto

**SWAS**: Submillimeter Wave Astronomy Satellite

**SWC**: Sheath and Wake Charging experiment; Solar Wind Composition by Foil Entrapment experiment

**SWUIS**: Southwest Ultraviolet Imaging System

**Syncom** (SIHN-com): Synchronous Communication Satellite

**TACS** (taks): Tracking and Communication System

**TAD** (tad): Thrust-Augmented Delta (launch vehicle)

**TAI**: Temps Atomique International (International Atomic Time)

**TAV**: Transatmospheric Vehicle

**TDOA**: time difference of arrival technique

**TDRS** (TEE-drehs): Tracking and Data-Relay Satellite

**TDRSS** (TEE-drehs SIHS-tehm): Tracking and Data-Relay Satellite System

**THEMIS** (THEE-mihs): Thermal Emission Imaging System

**TIM**: Total Irradiance Monitor

**TIP**: Transit Improvement Program

**TLI**: translunar injection

**TOMS** (tomz): Total Ozone Mapping Spectrometer

**TOPS** (tops): Thermoelectric Outer Planet Spacecraft group

**TOS** (tos): TIROS Operational System

**TOVS**: TIROS Operational Vertical Sounder

**TRACE** (trays): Transition Region and Coronal Explorer

**TRMM**: Tropical Rainfall Measuring Mission

**TsPK**: Tsentr Podgotovka Kosmonavtov (Russian for Center for Preparation of Cosmonauts)

**TSS**: tethered satellite system

**TT**: Terrestrial Time

**UARS** (YEW-ahrs): Upper Atmosphere Research Satellite

**UNEX** (YEW-nehks): University-class Explorer Program

**USA**: United Space Alliance

**USGCRP**: U.S. Global Change Research Program

**USML**: United States Microgravity Laboratory

**USMP**: United States Microgravity Payload

**USSAS**: United States Standard Atmosphere Supplements

**USSC**: United States Space Command

**UTC**: Coordinated Universal Time (sometimes "Universal Time Coordinated")

**UVOT**: Ultra-Violet/Optical Telescope

**VAB**: Vehicle Assembly Building

**VCAP** (VEE-kap): Vehicle-Charging And Potential experiment

**VIM**: Voyager Interstellar Mission

**VKS**: Russian Military Space Forces

**VLA**: Very Large Array

**VLBI**: very long-baseline interferometry

**VRM**: Venus Radar Mapper Project

**VTOL**: vertical takeoff and landing

**WAAS**: Wide Area Augmentation System

**WCRP**: World Climate Research Program

**WFPC**: Wide Field/Planetary Camera

**WIMP**: weakly interacting massive particle

**WIRE** (weyr): Wide-Field Infrared Explorer

**WMAP**: Wilkinson Microwave Anisotropy Probe

**WMO**: World Meteorological Organization

**WSF**: Wake Shield Facility

**WWW**: World Weather Watch

**X**: experimental (for example, X-15, X-20)

**XRT**: X-Ray Telescope

# List of Illustrations

# Complete List of Contents

## Volume 1

# Volume 2

# Volume 3

# Category List

## List of Categories

### Aerospace Agencies

National Aeronautics and Space Administration, 973
National Commission on Space, 989
Space Task Group, 1620
United Space Alliance, 1760
United States Space Command, 1765

### Communications Satellites

Amateur Radio Satellites, 6
Applications Technology Satellites, 174
Intelsat Communications Satellites, 569
Mobile Satellite System, 967
Telecommunications Satellites: Maritime, 1692
Telecommunications Satellites: Military, 1698
Telecommunications Satellites: Passive Relay, 1704
Telecommunications Satellites: Private and Commercial, 1710
Tracking and Data-Relay Communications Satellites, 1748

### Earth Observation

Dynamics Explorers, 292
Earth Observing System Satellites, 303
Explorers: Air Density, 332
Explorers: Atmosphere, 346
Explorers: Ionosphere, 353
Geodetic Satellites, 481
Global Atmospheric Research Program, 497
Heat Capacity Mapping Mission, 519
Landsat 1, 2, and 3, 697
Landsat 4 and 5, 704
Landsat 7, 709
Mission to Planet Earth, 961
Seasat, 1159

### Expendable Launch Vehicles

Atlas Launch Vehicles, 202-208
Delta Launch Vehicles, 284
Launch Vehicles, 719
Saturn Launch Vehicles, 1142
Soyuz Launch Vehicle, 1230
Titan Launch Vehicles, 1741

# Preface to the Third Edition

Science deals with facts—often inspired by fiction. Without science-fiction writers to dream about what is believed to be impossible, no one would accomplish the seemingly unattainable goals that have been reached in the natural sciences and engineering over the past few centuries. Nowhere has this been truer than in the space program. In 1865 Jules Verne, in *De la terre à la lune* (*From the Earth to the Moon*, 1873), envisioned a lunar landing more than a century before Neil Armstrong and Buzz Aldrin set foot on the lunar surface. An American, George Tucker, produced the first reasonable design concept for a spacecraft in his 1827 book *A Voyage to the Moon*.

In 1945, Sir Arthur C. Clarke developed the notion of placing communications satellites into geostationary orbit 35,800 kilometers above the equator. At this altitude, a satellite travels through its flight path at an angular speed equivalent to the rotation rate of the Earth and therefore appears to hover over a single spot on the surface (hence the term "geostationary"). Clarke also described the advantages of placing a piloted space station into a geostationary orbit. "The Sentinel," a short story Clarke published in 1951, was the basis for the screenplay he wrote with filmmaker Stanley Kubrick that in 1968 became the classic science-fiction film *2001: A Space Odyssey*.

Still very early in the twenty-first century, we can only wonder what happened to the dreams of those early visionaries. The permanently occupied orbiting International Space Station is expected to be completed by 2010. However, there will be no piloted spacecraft on its way to Jupiter, as had been projected in Clarke's book *2010: The Year We Make Contact*. The space shuttle fleet is to be retired in 2010 by a nonimaginative administrative directive. A proposed replacement for the shuttle, the Crew Exploration Vehicle, is expected to begin test flights perhaps as early as 2009. Whether it will be available by the time shuttle orbiters are cast off to honored positions in museums is a matter of serious debate among scientists and engineers. Some anticipate that there could be a four-year period between shuttle retirement and flight operations of its replacement vehicle, resulting in the requirement to rely exclusively upon Russian launch services to take American astronauts to the International Space Station. In any event, American space exploration is at a crossroads. The next several decades could see a vigorous and exciting expansion of the human presence within the solar system, including a return to the Moon, this time to stay, and preliminary missions of exploration to Mars. We could also see an American retreat from the strides made over the past fifty years, just as other nations begin to push the limits of their own efforts in space.

Space exploration is a dynamic endeavor and much has happened in the years since the second edition of *USA in Space* was published in 2001. Many projects being planned in 2000 have been completed, some are currently under way, and others remain in development. For *USA in Space, Third Edition*, we have updated the essays in previous editions: Several essays underwent extensive changes or additions to accommodate significant changes in the project, vehicle, or spacecraft during the intervening years. The *Columbia* STS-107 accident on February 1, 2003, radically altered space shuttle operational support to the International Space Station (ISS) and its construction plans. Some of the Second Edition essays addressed promising technologies since abandoned—the National AeroSpace Plane and X-33/X-34 are examples—although we did not abandon the essays. In addition, a full read of the previous editions has identified errors that have been corrected. The intervening years have also seen the appearance of additional print and Internet resources, which have been added to update each essay's bibliography.

Of the 238 essays from the second edition, 25 required extensive revisions, 5 essays were replaced entirely, and 42 essays were added. A few of the new topics were chosen for their notoriety, while others were raised from the depths of obscurity. Some were the result of triumphs, such as the Mars Exploration Rovers and Huygens probe. At least one, covering the ill-fated STS-107 mission, was added to remind us that space travel is still a dangerous business. We did not try to avoid controversy, and in doing so have included essays on topics that continue to engender serious debate, such as the search for extraterrestrial life and the use of nuclear power in space.

The completion of some programs and new developments in ongoing programs—such as the space shuttle, the Hubble Space Telescope, and the ISS—needed new essays as well as updates of old ones. Essays providing a more in-depth look at some of the craft involved in space travel, as well as a couple of the ancestors of the shuttle and ISS, have been added.

Although the focus of this work is on the U.S. space program, we would have been remiss to omit coverage of the Russian Mir Space Station, the Soyuz spacecraft and launch vehicle, and certain European Space Agency activities in our coverage. Research conducted on the now discarded Mir advanced the development and deployment of the International Space Station. It served as an experimental platform for research on the effects of long-term spaceflight and the uses of the microgravity environment. Had it not been for the aged but reliable Soyuz spacecraft and launch vehicle, the ISS would have laid dormant during the more than two-year hiatus from shuttle flights following the STS-107 accident. Several essays related to NASA's Return-to-Flight STS-114 mission were updated at the conclusion of the flight.

Readers consulting essays on the newer missions and programs will notice a distinct shift over the years toward increased international cooperation in space. It is for this reason that there is an essay on the Huygens Titan atmospheric/lander probe in addition to the updated essay on the Cassini orbiter. Huygens is a cooperative project of the European Space Agency and the Italian Space Agency that traveled piggyback to the Saturn system aboard NASA's Cassini spacecraft. International cooperation will be required to carry out the human exploration of Mars and the return to the Moon. Continued robotic studies of the solar system and neighboring galaxial region will require us to pool the talents of several nations. No single national budget can pay for these ambitious goals in the exploration of space and investigation of astrophysics and the space sciences.

One person could not undertake an effort of this magnitude alone. For this edition, two authoritative editors were necessary to check facts and fantasies. Editing this and the previous editions of *USA in Space* has been both a pleasure and an arduous task. With the updating of the old essays and the reading of the new ones came hours of time spent in front of the computer screen, consulting the most trustworthy Web sites for information often changing by the hour. Dozens of books were perused and billions of cyber kilometers traversed to check for new information and to verify old. Many hundreds of e-mail messages were required to transcend time and space. Without the Internet, none of this would have been accomplished in a timely manner.

The editors took special care to identify outdated language as well as substance, and to ensure that the information presented here is the latest as of publication. Although information on launch and project dates often slips with changes in political will as well as technological developments, we strove to eliminate "predictive" language where possible, and to allow references to the future that are reasonable.

The essayists for this collection come from varied backgrounds and academic disciplines. They approached the topics from wide-ranging points of view, writing with a variety of styles, while staying within a set of guidelines designed to make the information in each essay formatted in a predictable way and accessible to the general reference audience. Pulling all this together to make a cohesive collection took the skills of experienced professionals. The editors at Salem Press collected and organized all the material from the first two editions and coordinated the gathering of new essays. They read every essay, helped us identify areas in need of updating, and checked every corrective update we sent.

Events of the past five years have continued to demonstrate that spaceflight is not routine and that the realm beyond the atmosphere is a very dangerous place in which to live and work. Nearly fifty years after getting into the business of space exploration, NASA has not learned to put safety and reliability ahead of business. A rush to meet the goals of the Apollo Program cost the lives of three astronauts in the winter of 1967. Overambitious launch schedules and underfunded projects cost the lives of seven astronauts in the winter of 1986. A rush to construct and supply the International Space Station cost the lives of seven astronauts in the winter of 2003. The problems that haunt the aging space shuttle must be fixed. We must not make plans for lunar habitation and human Martian exploration when we cannot assure the safety of the crews leaving from and returning to Earth.

What will the next five years, the next two decades bring us? What worlds will we humans explore? Whom or what will we send to those distant landscapes and spacescapes? Will we still have an Earth to call home? The future belongs to the dreamers and the visionaries who make those dreams come true. We know that today's science fiction can become tomorrow's science fact. However, history has often shown that we tend to overestimate the near term (a decade) and grossly underestimate the longer term (a century). We must believe in the dreams of the visionaries, and in ourselves. The century-old words of Robert H. Goddard, the American grandfather of rocketry, are as true today as they were when he uttered them at his 1904 high school graduation ceremony:

It is difficult to say what is impossible, for the dream of yesterday is the hope of today and the reality of tomorrow.

*Russell R. Tobias*
*David G. Fisher, Ph.D.*
*September 9, 2005*

# Introduction to the Second Edition

An odd-looking craft hovers above the horizon, shooting out puffs of fire from small engines around its periphery. A larger engine at the base of the craft belches out flame in greater quantity. An apparently futile battle between craft and gravity is being fought gallantly by the craft. It appears to be alive, moving about, as if searching for something. A little forward, a little back, this way and that way it flitters until it seems to have decided on a place to land. Long, spearlike tentacles jut from the bases of three of its four legs. The spears touch the soil and the large engine suddenly goes silent. The craft drops just as suddenly to the ground. What is this unworldly eagle that has set down in the quiet valley? Why has it come? Are there creatures inside waiting to conquer this newly discovered territory? Lifelessly the craft sits for several hours.

Then, a small hatch opens and an unearthly looking creature emerges. It has a bubble for a head and two arms and legs. There is a large hump on its back and many tubes around its body. A small black box is on its chest. Could this be a weapon of some sort? The creature moves down a ladder to the base of the craft. Awkwardly it tries to acclimate itself to the alien environment. Shortly, a second creature joins it. They move about the landscape—cautiously at first—then, with near reckless abandon. They take several devices from the mother craft and place them around the landing site. Scientific probes, no doubt.

After several hours of investigation, they apparently decide that this is no place to invade. They pack up some of their equipment and clamber back up the ladder to the safety of their ship. Suddenly, with a large explosion of fire, the upper part of the craft blasts away from the base, carrying its crew and discoveries back to the dark sky above. Who were these creatures and why did they come to our world? Are they a scouting expedition for a larger invasion force from another world? Will they return to destroy our quiet home?

While this may sound like the opening of a science-fiction novel, it is how an observer might have described the first lunar landing. Many of us who were around to hear the words of Neil Armstrong as he took the first step on another world consider July 20, 1969, to be the most significant date in space exploration. Politically, it marked the completion of a lofty goal set by a dead president eight years earlier. Technically, it was the culmination of the lifelong goal of a former German rocket scientist. Socially, it was the symbol of the cooperation of thousands of Americans working toward a common objective. Not in the twenty-plus years since World War II had Americans labored so ardently on a single project. Not in the twenty-plus years since have we come together for one single, peaceful event.

There are other events and projects that punctuate our quest for new territories on the frontier of space—some of them uplifting, some routine, and some heartbreaking. This collection of essays is our humble attempt to chronicle the American space program during its first forty years. The essayists come from a variety of backgrounds, and each has his or her own perspective of the events in the collection—as do you, the reader. We have attempted to present the most significant projects, places, and events that shaped the American space program. We hope that we have not offended anyone by an oversight.

Many of the essays have retained their original text from their inclusion in our ancestral work, *Magill's Survey of Science: Space Exploration Series.* Salem Press published this compendium of essays on the history of world space exploration in 1989. Some of the essays have been updated because a mission or project was ongoing in 1989. We have added many more essays to include those events that made an important contribution to the continuing quest for space.

What have we been doing since 1989? A lot! Much of the work in the crewed and uncrewed sectors of spaceflight has been the routine. Send up some people; send up some machines. Some missions fail; most succeed. We have pressed on during the past eight years with reckless overcaution. We seem to have forgotten that our goal was to "boldly go where no one has gone before."

Our most ambitious projects have been in the field of robotic exploration. We have explored the depths of Jupiter's atmosphere and thoroughly mapped the unseen face of Venus. We have taken a long, close-up look at the Sun and its polar regions. We have studied the fragile atmosphere of our own planet, looking for a way to keep it habitable for our descendants. Perhaps our most notable achievement in space has been the deployment of the Hubble Space Telescope. Criticized at first for an error in its optics, it has given us a glimpse into the history of our universe, almost back to the beginning. It has revealed some of the mysteries of our solar system and has shown us the birthplace of stars. We have seen planets orbiting around a star other than our own Sun. If there are planets, why not inhabitants? Perhaps Hubble will focus on a tiny world and see a welcome mat. Who is to say what this orbiting observatory will reveal.

Other observatories have been placed in the heavens looking not in visible light, but in the infrared, ultraviolet, and x-ray regions of the spectrum. The data from these telescopes have presented us with as many new questions as they have answers.

Concern over what we humans have been doing to tear down our own planet has led to an increase in Earth observation from space. A series of space shuttle flights have studied the Earth with monitors and cameras as part of NASA's Mission to Planet Earth. Of particular interest has been the atmosphere, that tenuous blanket of life-giving air, which seems to be developing bald spots. Will we kill it with pollution first or simply let it all leak into space? We continue the mapping of our globe with the Landsat and other geodetic satellites. We save thousands of lives and billions of dollars in property with our weather satellites. We study the oceans and clouds, landmasses and ice floes, temperature variations and deforestation.

Our search for answers to Earthly problems has led us to send out more extraterrestrial probes. Magellan went to Venus with its radar mapping equipment piercing the thick atmosphere. It has shown us the scars of a continuous, global acid storm and what pollution could do to our own atmosphere. No signs of Venusians or ancient visitors from a far-off galaxy were found, however, much to the disappointment of many.

Galileo has gone to Jupiter and is surveying the gas giant and its nearby moons. On the way, it snapped some pictures of an asteroid and sent a probe plunging into the Jovian atmosphere. Once again planetary scientists have been sent scrambling for answers to violations of their long-standing theories of planetary construction and activity.

Put together with spare parts and chewing gum, Mars Observer provided an inexpensive means for exploration. The probe successfully traversed the distance from Earth to Mars, only to fall silent as it prepared to enter orbit around the Red Planet. Was it an internal malfunction that caused it to fail, or did it run into a Martian invasion fleet?

A by-product of the Strategic Defense Initiative, Clementine was also put together from parts of other spacecraft. Its short-term mission was to study the Moon. We had returned, not to establish colonies, but to take more pictures. After a thorough job of lunar photography, Clementine headed for a rendezvous with an asteroid. A minor mishap caused the probe to expend all of its control fuel and the rendezvous was lost.

The Ulysses solar-polar spacecraft has begun to return data about the polar regions of the Sun. Its five-year journey got a gravitational assist from Jupiter, which increased the speed of Ulysses as well as altering its flight path so that it would fly above and below the plane of the ecliptic. Ulysses is studying the solar corona, the solar wind, the Sun's magnetic field and galactic cosmic rays.

The space shuttle remains America's only avenue for crewed spaceflight. NASA has recovered from the *Challenger* accident, having completed more than fifty missions since STS-26 in 1988—the first space shuttle mission following the accident. Commercial and military payloads are a thing of the past. Most of the flights have been dedicated to studying the effects of the microgravity environment of low-Earth orbit. Several communications satellites have been deployed, mainly newer members of the Tracking and Data-Relay Satellite System, which permits greater coverage of the scientific studies being conducted. An errant Syncom satellite was even repaired on-site in a dramatic multiperson spacewalk.

Magellan, Galileo, Ulysses, the Hubble Space Telescope, the Upper Atmosphere Research Satellite, and the Compton Gamma Ray Observatory are a few of the more notable payloads successfully deployed from the shuttle. The Tethered Satellite System demonstrated the fact that the "yo-yo" theory applies to space vehicles on a string, too. Its all-too-brief exploration of electrical conduction has given scientists a chance to revise their ideas about generating electricity in space.

Where the shuttle missions have really excelled has been in the study of the effect of spaceflight on people, animals, and other living organisms. While we have only begun to examine the effects of prolonged spaceflight, a number of Spacelab missions have revealed the causes of space maladies. If we can overcome these stumbling blocks, we will be able to develop long-term orbital flights and interplanetary missions. Earth observation and astronomy research are also being conducted from a vantage point beyond the atmosphere. The commercialization of space is being developed through materials processing experiments aboard the shuttle. Testing various pieces of equipment and processing principles is demonstrating the economical advantage of building orbiting factories. Once private industry gets involved in space exploration, the accomplishments will steadily increase.

The Space Station has become a victim of budget cuts and lackluster support from the American public. An American space station is a great idea, they feel, as long as it does not take money away from social reform programs. The Russians have had a crewed permanent presence in space for many years, but the breakup of the Soviet Union has left them strapped for funds. What would have been impossible during most of the twentieth century has become a reality. The United States and its former sworn enemy, Russia, have joined to build an international space station. For the first time in twenty years Americans and Russians have shaken hands in space. In addition, cosmonauts have been ferried to and from the Russian Mir Complex aboard the space shuttle, and an American has been launched aboard a Soyuz spacecraft. This exchange of space travelers is the first step toward that space station. By the end of the next decade, explorers, scientists, and engineers will be conducting long-term research on the Space Station.

As we approach the fortieth anniversary of the world's first small step into the universe, we can look back with amazement at what we have accomplished. For better or worse, technology is bringing us into the new millennium. We have had visions of a Stanley Kubrick twenty-first century. We have the means to continue to investigate the vast regions of space. Will we have the courage to take the giant leap to other planets in our twenty-first century?

This collection of essays is dedicated to the courage and conviction of Americans to explore the unknown. This is not just for those who have given their lives in the pursuit of the quest, but also for those who continue it—astronauts, scientists, engineers, mechanics, technicians, planners and dreamers.

*Russell R. Tobias*
*August 4, 1996*

# USA in Space
## Third Edition

# Air Traffic Control Satellites

*Date:* Beginning February 22, 1978
*Type of spacecraft:* Navigational satellites

*Beginning with the earliest satellite programs, the United States has conducted experiments using artificial satellites to aid in air traffic control operations. These efforts have led to Navstar, a network of navigational satellites, and to Nusat, the first satellite designed exclusively for air traffic control studies.*

## Summary of the Satellites

Since the earliest days of the Space Age, space scientists and aviation specialists have envisioned applications for satellite technology in the area of air traffic control. Shortly after the launch of Sputnik 1 in 1957, the U.S. Navy orbited a series of satellites called Transit that allowed Polaris submarines to get navigational fixes accurate to within 161 meters. In continuous operation since 1964, the Transit system was used by more than one thousand stations worldwide, providing users with about twenty-four position fixes per day from an altitude of 1,100 kilometers. The U.S. Air Force, using identical technology, developed the Navstar satellites, which were designed to give navigational aid to aircraft, spacecraft, marine vessels, and land-based vehicles. Originally, twenty-four such satellites were proposed to form a network that would beam continuous signals such that anyone with a receiver, anywhere on Earth, could plot his or her geographical position to within tens of meters. Eventually, budget cuts would result in a reduction in the original proposal to eighteen satellites, each in a polar orbit on one of three different planes at 17,703 kilometers above Earth. This reduction in the number of satellites in the network also resulted in a slight decrease in the system's accuracy.

The Navstar system was divided into three segments: the space segment, which transmitted satellite position coordinates; the user segment, which received the information transmitted by the satellites; and the ground control segment, which tracked and corrected the position coordinates and timing devices of each satellite daily.

Each user had a radio receiver with an omnidirectional antenna, a signal processor, and a monitor for data readout. Because this equipment did not include a transmitter, an infinite number of users could access radio signals from the Navstar satellites at any one time without revealing their positions. The set was designed to lock automatically onto the four most favorably positioned satellites and automatically compute the approximate range of each, providing data for a small computer programmed to equate mathematically the values of four variables. From these values, the position and velocity of the user craft could be determined.

The Navstar Global Positioning System became fully operational in the mid-1980's. The first Navstar satellite was launched on February 22, 1978, and was used to conduct a series of tests in preparation for the full deployment of the system.

Before Navstar, during the mid-1960's and the early 1970's, the versatile and long-lived Applications Technology Satellites (ATS) system was employed to relay communications from aircraft to terrestrial receiving stations. Although the ATS's were used strictly for communications, they provided impetus for the development of air traffic control programs.

Perhaps the best-known air traffic control experiment with satellites was the Northern Utah Satellite, or Nusat, an extremely small satellite designed to help in the calibration of air traffic control radar equipment around the world. The Nusat project began in 1979 and was conducted primarily by university and corporate scientists as part of a volunteer effort at Weber State College in Ogden, Utah. It is estimated that $1.5 million in parts and labor was donated to the project by a variety of corporate and educational entities. The satellite was launched by the space shuttle *Challenger* during the Spacelab 3 mission on April 29, 1985.

Nusat weighed 52 kilograms and was 75 centimeters in diameter. It contained twelve groups of solar panels that provided power to the onboard systems. The satellite also contained six fixed antennae for tracking air traffic control radar and four antennae for telemetry and ground control communications.

The satellite's mission was to track L-band radar antenna patterns around Earth and provide data on their elevation angles. Ground control facilities were located on the Weber State campus in Ogden, and the satellite was available to anyone wanting access to it.

Nusat contained six L-band radar receivers and a timer. It was designed to distinguish among radar signals, store data in its memory, and return the data to ground control on command. A radar station authorized to work with the Nusat satellite would be instructed to transmit a unique pulse-position code whenever Nusat was on the horizon. That would enable the satellite to distinguish between the radar being calibrated and all others that might be operating within range of the satellite's receivers at the time.

The spacecraft's attitude would ensure that more than one onboard receiver would acquire the signal, thus storing additional data to be relayed on command and analyzed for the radar operator. The data would be adjusted to compensate for varying distances between Nusat and the ground-based radar installation and for known variations and deviations in the antenna patterns of the radar and the satellite itself.

Unfortunately, Nusat never fulfilled its mission. Ground controllers were unable to activate its onboard systems after it had been in orbit for only two and a half months. The last date of contact was July 15, 1985. This development followed a two-week interruption in communications between ground controllers and the satellite during June, 1985. Nevertheless, Nusat stimulated the Federal Aviation Administration to begin development of a follow-up program that would be designed to accomplish the same tasks envisioned for Nusat but would use space-tested hardware.

Nusat also attracted the attention of the International Civil Aviation Organization (ICAO), the International Telecommunications Union (ITU) of the United Nations, and other agencies that worked in concert to ensure that the aeronautics industry would continue to enjoy exclusive use of the aeronautical satellite band on the electromagnetic frequency spectrum. Their efforts were successful, although other frequency users began to pressure ITU for access to that band, known as the L band, during the second half of the 1980's. That band was restricted to aeronautical operations communications and air traffic control communications.

ICAO set up the Future Air Navigation System committee to represent the interests of the air transport industry when the U.S. government proposed that all mobile satellite users, not only aircraft, be allowed to use the aeronautical satellite band until such time as that space was more fully utilized by the aeronautical sector. These other users included drivers of trucks and public safety vehicles.

In the summer of 1986, the Federal Communications Commission ruled that land-based and aeronautical satellite users must share frequencies. Technical discussions immediately ensued to determine the feasibility of assuring certain user priorities, such as public safety operations and uninterrupted air traffic control. It was subsequently determined that a uniform system design could be constructed that would uphold the necessary priorities and yet allow fuller use of the L band. At the time, the ITU also agreed to allow the L band to be used for administrative and private telephone communications between aircraft and ground stations.

### Contributions

Nusat, Navstar, the U.S. Navy Transits, and other space-based systems engaged in the study and implementation of space technology for air traffic control represented a significant step forward in navigation procedures for aircraft, marine traffic, and land-based vehicles. The accuracy of such systems enhanced the safety of air travel by providing precise location coordinates for pilots and air traffic control monitors on a global scale. Thus, an increase in air travel resulting in heavier air traffic was possible. Some satellite systems have included receivers that track coded radio signals emitted continually by transmitters aboard aircraft, thus providing ground control stations with constant, real-time location coordinates for most aircraft aloft. The coded signal provides immediate identification of the aircraft, its operator, its destination, its point of origin, and its flight plan. Occasionally, these signals are tracked for search-and-rescue operations when an aircraft is presumed to be lost.

Satellites have also been used to provide accurate airport ground control information at some of the busiest airports around the world. Landing hundreds of aircraft daily and moving them to and from gates scattered over a wide area are daunting tasks, and equipment similar to that aboard weather satellites is occasionally used to locate taxiing aircraft and to monitor their movement. This technology was first used at London's Heathrow Airport in Great Britain during the 1970's.

The Nusat program was designed to increase the efficiency of ground-based radar, which must be recalibrated regularly to ensure its reliability. Before the development of the Nusat technology, such recalibration required that a radar system be taken out of service for several hours. Moreover, the procedure for recalibration was complicated. With Nusat, the radar operator, working with ground controllers, could recalibrate a system quickly and without significantly interrupting normal operations.

### Context

During the early years of space development, the amount of international air traffic grew significantly. The jet airliner ushered in an era of fast, economical air travel for the public. Aircraft could fly from continent to continent, without refueling, faster than ever before; the result was a greater demand for commercial and private air services.

That increase in demand brought with it a need for a dependable system of air navigation to ensure that air lanes would operate efficiently and safely around the world. In addition, the increasing power and longer range of military land- and sea-based missiles demanded that a global system for instantaneous location and positioning of aircraft be established. The satellites developed for that purpose contributed to breakthroughs in the evolution of safe and dependable air travel.

The effectiveness of early-warning radar systems, a key element in defense against long-range missile attack, was enhanced by the technology employed in Nusat and in related satellite systems. The success of the Navy's Transit tests in the early 1960's was quickly answered by the Soviets, who established a virtually identical system using similar procedures and the exact same frequencies beginning in November of 1967 with the launch of Kosmos 192. Five such satellites were launched by the Soviets every year from 1968 to 1971, when the system became fully operational.

Air transport systems have been proposed that will require more sophisticated satellites for communications and telemetry between aircraft or spacecraft and ground control centers. The United States, for example, has experimented with aircraft that can travel at very high speeds and very high altitudes and that may eventually allow intercontinental air travel in a fraction of the time it now takes to travel from one side of Earth to the other. Such aircraft may travel at the upper limit of the atmosphere, in effect becoming spacecraft for part of the trip, even though they take off from and land at airports in the same way that regular airliners do. Tracking such craft and communicating with them will require satellite networks of great sophistication and efficiency.

In addition, radar systems, which send and receive electromagnetic signals, are technically inferior to satellite-based systems in that they are subject to electromagnetic interference and vulnerable to a variety of ailments, not the least of which is the changing magnetic character of the atmosphere. (Atmospheric changes are one reason that radar instruments require frequent recalibration.) Changes in weather patterns also affect the efficiency of radar, as does the inaccessibility of vast stretches of ocean. Air traffic control satellite networks are a dependable alternative to radar and will be used to track all manner of air-, sea-, and land-based craft with greater efficiency and reliability in years to come.

**See also:** Applications Technology Satellites; Global Positioning System; Navigation Satellites; Space Shuttle: Radar Imaging Laboratories; Telecommunications Satellites: Private and Commercial.

### Further Reading

Andrade, Alessandra A. L. *The Global Navigation Satellite System: Navigating into the New Millennium.* Montreal: Ashgate, 2001. Provides an international view of issues of availability, cooperation, and reliability of air navigation services. Attention is specifically paid to the American GPS and Russian GLONASS systems, although the development of the Galileo civilian system in Europe is also presented.

Butrica, Andrew J. *Beyond the Ionosphere: Fifty Years of Satellite Communications.* NASA SP-4217. Washington, D.C.: Government Printing Office, 1997. Part of the NASA History series, this book looks into the realm of satellite communications. It also delves into the technology that enabled the growth of satellite communications. The book includes many tables, charts, photographs, and illustrations, a detailed bibliography, and reference notes.

Gatland, Kenneth. *The Illustrated Encyclopedia of Space Technology: A Comprehensive History of Space Exploration.* New York: Salamander, 1989. Perhaps the most comprehensive overview of international space programs prior to 1980, this book contains a section on navigational satellites and a completely illustrated description of the technical configuration of the Navstar satellites of the late 1970's.

Gavaghan, Helen. *Something New Under the Sun: Satellites and the Beginning of the Space Age.* New York: Copernicus Books, 1998. This book focuses on the history and development of artificial satellites. It centers on three major areas of development—navigational satellites, communications satellites, and weather observation and forecasting satellites.

Lee, Wayne. *To Rise from Earth: An Easy to Understand Guide to Spaceflight.* New York: Checkmark Books, 1996. This is a good introduction to the science of spaceflight. Although written by an engineer with the NASA Jet Propulsion Laboratory, it is presented in easy-to-understand language. In addition to the theory of spaceflight, it gives some of the history of the human endeavor to explore space.

Ley, Willy. *Events in Space.* New York: David McKay, 1969. Contains summaries of all satellite programs, national and international, that were carried out during the 1960's. Particularly useful is a series of tables and glossaries that define space jargon and describe many of the satellites and rockets of the decade. Also included are lists of satellite launches, complete with launch dates and other information about the satellites and the programs with which they were associated.

National Aeronautics and Space Administration. *NASA: The First Twenty-Five Years, 1958-1983.* NASA EP-182. Washington, D.C.: Government Printing Office, 1983. Lavishly illustrated with color photographs, this NASA publication traces the space agency's history from its beginnings. Included are chapters on space science, technology utilization, tracking and data systems, and applications satellites.

Paul, Günter. *The Satellite Spin-Off: The Achievements of Space Flight.* Translated by Alan Lacy and Barbara Lacy. Washington, D.C.: Robert B. Luce, 1975. A survey of the commercial, scientific, and communications applications that developed from the space research of the 1960's and early 1970's. This book is written from the perspective of the European community, which makes it necessary reading for those desiring an international perspective on the U.S. space program. In addition to navigational applications, space medicine, meteorology, cartography, agriculture, and oceanography are discussed.

Shelton, William Roy. *American Space Exploration: The First Decade.* Boston: Little, Brown, 1967. A historical account of the U.S. space program, with comprehensive listings of every American spaceflight launched between 1957 and 1967.

Zimmerman, Robert. *The Chronological Encyclopedia of Discoveries in Space.* Westport, Conn.: Oryx Press, 2000. Provides a complete chronological history of piloted and robotic spacecraft and explains flight events and scientific results. Suitable for all levels of research.

*Michael S. Ameigh*

# Amateur Radio Satellites

*Date:* Beginning December 12, 1961
*Type of spacecraft:* Communications satellites

*The first amateur radio satellite, launched in 1961, was also the first privately owned, nongovernmental satellite launched. Today, amateur radio satellites allow residents of different countries to communicate with one another, bring space exploration into the classroom, and assist in emergency relief projects.*

## Summary of the Satellites

Amateur, or "ham," radio is a hobby enjoyed by approximately one million people throughout the world. Radio amateurs talk with one another over short or long distances. They communicate by voice; in international Morse code; via slow-scan television, which transmits one picture every eight seconds; via fast-scan television, which is very similar to broadcast television; via radio teletype; or through a repeater or translator that is either ground-based or on board a satellite orbiting Earth. To obtain a license to operate an amateur radio station, one must pass certain examinations on radio theory and government regulations. The U.S. government agency that administers these regulations is the Federal Communications Commission (FCC). The FCC assigns bands of the radio spectrum to the Amateur Radio Service. These bands are denoted by their frequency (number of vibrations per second) or wavelength (distance from one wave crest to the next). An example is the 2-meter band, which has a frequency of 144 megahertz.

Radio amateurs make international contacts using the reflection of their signals back to Earth by the ionosphere, a layer of charged particles approximately 300 kilometers above Earth's surface. If the frequency is very high, however, the radio wave is no longer reflected back to Earth.

These frequencies are very high frequencies (VHF), ultrahigh frequencies (UHF), or frequencies higher than UHF. In the 1950's, radio amateurs hoped for intercontinental contacts using either VHF or UHF. The fulfillment of these hopes began in 1959, when several amateurs began to discuss the possibility of putting a satellite in orbit. A group of amateurs formed the Orbiting Satellite Carrying Amateur Radio (OSCAR) Association in October, 1959. The name OSCAR was suggested by Don Stoner of W6TNS. The OSCAR Association, with the help of many others, built the satellite and obtained permission to operate a transmitter in space without positive control. They also obtained permission for their satellite to act as dead weight, or ballast, in a launch vehicle; most rockets of that time had more payload capacity than they used, so the OSCAR Association was able to persuade the Air Force to "piggyback" the OSCAR satellite on the launch of another spacecraft.

The first amateur radio satellite, OSCAR 1, was launched December 12, 1961, at 20:42 Coordinated Universal Time (UTC). It was carried aloft by an Agena rocket, aboard a Discoverer reconnaissance satellite from Vandenberg Air Force Base, California. All subsequent OSCARs were launched as ballast by the National Aeronautics and Space Administration (NASA) or other appropriate space agencies at no cost to the radio amateurs.

OSCAR 1 was not very sophisticated. It consisted of a transmitter that broadcast the message "Hi" in Morse code. It was inserted into a low north-south orbit. More than 570 stations in twenty-eight countries and forty-four states heard OSCAR 1 before it ceased functioning on January 3, 1962.

OSCAR 2 was launched June 2, 1962, at 00:32 UTC from Vandenberg. OSCAR 2 was very similar to OSCAR 1. Its orbit was lower than OSCAR 1's, and it stopped operating June 20, 1962.

OSCAR 3 was launched March 9, 1965, at approximately 19:30 UTC from Vandenberg. It was the first OSCAR satellite to contain a battery-powered translator. A translator, or repeater, is a device that receives a radio signal on one frequency, amplifies it, and retransmits it on another frequency. Its uplink frequency, or the frequency sent from Earth to the satellite, was in one part of the 2-meter band, and its downlink frequency, or the frequency sent from the satellite to Earth, was in a different part of the 2-meter band. Stations within 3,200 kilometers of each other could communicate through the translator, which functioned for 250 orbits (eighteen days).

OSCAR 4 was launched December 21, 1965, by a Titan III-C rocket from Cape Kennedy, Florida. Its period (the time required to orbit Earth once) was approximately 10 hours; previous OSCARs had had periods of about 90 minutes. OSCAR 4 was built by Herb Gleed of W6ZPX, Dave Moore, and other members of the TRW Amateur Radio Club. The satellite contained a translator and a beacon.

Amsat (Amateur Radio Satellite Corporation) was incorporated March 3, 1969, by a group of radio amateurs on the East Coast. It took over the launching of OSCAR satellites beginning with OSCAR 5.

OSCAR 5 was launched by a Delta 76 rocket from NASA's Western Test Range at Vandenberg Air Force Base in California at 11:31 UTC on January 23, 1970. OSCAR 5 is sometimes called Australis-OSCAR 5, because it was constructed mainly by Australian amateurs. Paul Dunn of VK3ZPD, Owen Mace, and Richard Torhim were the leaders. This OSCAR did not have a translator but did have several beacons. OSCAR 6 was launched at 18:34 UTC on October 15, 1972, by a Thor-Delta rocket from the Western Test Range. OSCAR 6 had two repeaters or translators, one of which was built by German radio amateurs. OSCAR 7 was launched November 15, 1974, at 17:11 UTC, also from the Western Test Range. At the time of OSCAR 7's launch, OSCAR 6 was still active, so radio amateurs had two satellites available. OSCAR 7 had three active repeaters and two beacons. Launched March 5, 1978, from Vandenberg Air Force Base at 17:54 UTC, OSCAR 8 contained two translators. OSCAR 8 was a relatively long-lived satellite. It ceased transmission on June 24, 1983, after more than five years of service.

Amateur radio satellites were also launched by the Soviet Union. On October 26, 1978, at approximately 06:00 UTC, satellites Radio 1 and Radio 2 were launched, together with a Kosmos 1045 satel-

*Astronaut Michel Tognini, mission specialist, uses the Shuttle Amateur Radio Experiment II (SAREX-II) on* Columbia's *flight deck.* (NASA)

lite, into low-Earth orbits from Plesetsk, 750 kilometers north of Moscow. The radio satellites were designed and built by students of Moscow's higher schools and by amateur radio enthusiasts of the Soviet Union's Voluntary Society for Assistance of the Army, Air Force, and Navy (DOSAAF). The satellites' translators differed in size, equipment, antenna systems, and solar batteries.

A launch of a new amateur radio satellite was attempted on May 23, 1980. It was launched by a European Space Agency (ESA) Ariane rocket from Kourou, French Guiana, at 14:29 UTC. The satellite did not achieve orbit but instead splashed down into the Atlantic about 27 kilometers downrange of the launch site. The loss of this satellite was a devastating blow to the amateur space program. Much money and hardware and many hours of volunteer labor had been lost. To ensure that the amateur space program would continue, the American Radio Relay League (ARRL), a national organization for amateur radio operators, pledged its support to help raise funds to replace the satellite.

The satellite that was lost was of a new type called phase 3. Phase 1 satellites were low-orbiting satellites without translators. Phase 1 included OSCAR 1 and OSCAR 2. Phase 2 satellites were also low orbiting, but they contained translators and used solar cells to recharge batteries. OSCAR 6 and OSCAR 7 were phase 2 satellites. Phase 3 satellites contained translators and had highly elliptical orbits with apogees (the apogee is the most distant point from Earth) of about 29,000 kilometers. Such satellites move slowly near the apogee to facilitate signal reception on Earth.

Britain's first amateur radio satellite was launched October 6, 1981, from Vandenberg Air Force Base. This satellite was named OSCAR 9, or Uosat, for the University of Surrey, Guildford, England. Uosat contained no translators but had many beacons and a CCD (charge-coupled device) camera imaging experiment that transmitted images of Earth's surface. Also on board was a speech synthesizer, which relayed various data in English via some of the beacons.

Six new Soviet satellites were placed in orbit December 17, 1981. They were launched from Plesetsk, as Radio 1 and Radio 2 had been, and were designated Radios 3 through 8. All the satellites had translators, and Radio 3 and Radio 5 had robots, or automatic operators. These robots communicated in Morse code. The six satellites were placed in low orbits, permitting radio contact at maximum distances of approximately 8,000 kilometers.

On May 17, 1982, Iskra 2 was launched from the Soviet Salyut 7 space station. It had one translator and one beacon; its translator, however, never functioned. It was placed in a low orbit and it operated for only a few weeks. Iskra 3 was launched in November, 1982, from Salyut 7. It lasted a shorter time than did Iskra 2. Both satellites were used by Soviet students for experiments.

Amsat-OSCAR 10 was launched at 11:59 UTC on June 16, 1983, with the ESA Ariane 16 mission from Kourou, French Guiana. This satellite contained two translators. It had an apogee of 38,400 kilometers, a perigee (the closest point to Earth) of 1,600 kilometers, and a 10-hour orbital period. In order to achieve this orbit, the satellite had first to be placed in a low, circular orbit; subsequently, a kick motor was fired for 170 seconds, and the satellite achieved its final orbit.

On March 1, 1984, OSCAR 11 (Uosat 2) was launched at 17:59 UTC from Vandenberg by a Delta 3920 rocket. OSCAR 11 was very similar to OSCAR 9. OSCAR 11's digital communication experiment, built by Amsat and Volunteers in Technical Assistance, consisted of a digitized "voice" that would transmit certain pieces of data in English.

The next OSCAR satellite was launched on August 12, 1986, at 20:45 UTC from Japan. This satellite was built and launched by the Japan Amateur Radio League, Japan Amsat, Nippon Electric Company, and the Japanese National Space Agency. It was first called JAS-1, for Japan Amateur Space missions manager 1, but its name was changed to Fuji OSCAR 12 (FO-12). It was placed in a relatively low orbit and permitted contacts at 4,000 kilometers. FO-12 had the capacity for digital communication.

On June 23, 1987, the Soviet Union again launched a spacecraft, Kosmos 1861, carrying two amateur radio translators and one special navigational transponder. The amateur radio translators were designated Radio 10 and Radio 11 and were identical except for operational frequencies. Like Radios 3 and 5 before them, each translator had a robot repeater.

OSCAR 13 was launched by an Ariane rocket on June 15, 1988, at 11:19 UTC from Kourou. This satellite had a RUDAK (regenerative transponder for digital amateur communications) built by German radio amateurs. This satellite was placed in a very elliptical orbit with a high apogee. Because the spacecraft would remain above one point on Earth for more than 11 hours per day, reception antennae on Earth would not need to be redirected often.

Between 1988 and 1996, seventeen additional OSCARs were launched by the United States, Russia, and the ESA. These satellites continue to link amateur radio users throughout the world and provide a means for conducting further experiments in the field of radio communications. Subsequent satellites maintained amateur access to worldwide communications. Many of these were financed at the grassroots level. One such example was AMSAT's OSCAR-E, or Echo, the first of an anticipated series of microsats (small satellites) destined for low-Earth-orbit radio links for amateurs. Set for a mid-2000's launch, this payload was financed in part by donations as little as $25, for which a donor would receive a pin with the satellite drawn in block diagram form and a slogan that read, "I gave Echo a lift."

### Contributions

The OSCAR satellites and Soviet amateur radio satellites have contributed to scientific knowledge, though not in the same way that a governmental program would have done. With the exception of "piggyback" rides on launch vehicles, the OSCAR series was entirely funded by nongovernmental sources; consequently, the satellites could be used for amusement as well as for research.

At a cost of $65, OSCAR 1 was the first nongovernmental satellite to be placed in orbit. OSCARs 6 and 7 were used experimentally to send medical data, such as electrocardiogram readings, from the West Coast to the East Coast. Though these satellites were not used in a medical emergency, this series of experiments proved that they could have been.

Another first occurred on January 6, 1975, when OSCAR 6 and OSCAR 7 were cross-linked: One radio operator sent a signal to OSCAR 7, OSCAR 7 relayed the signal to OSCAR 6, and OSCAR 6 transmitted the signal to a second operator. At the time, no one, including government or commercial interests, had ever achieved a cross-link of two satellites. Since then, many cross-links have been performed.

OSCAR 7 was employed in the "proof of concept" for the Search and Rescue Satellite-Aided Tracking (SARSAT) system in December of 1975. Signals from the Goddard Space Flight Center showed that a low-power uplink could provide tracking information with a potential accuracy of between 3 and 6 kilometers. The COSPAS/SARSAT program, a joint project of the United States, Russia, France, Canada, and several other nations, is functional. (COSPAS is a Russian acronym for Space System for Search of Vehicles in Distress.)

The Transpolar Skitrek/Project Nordskicomm, a joint Soviet and Canadian project begun on March 1, 1988, made use of the COSPAS/SARSAT system. A team of Soviet and Canadian skiers would ski from Cape Arktichesky, far north in the Soviet Union, to Cape Columbia in northernmost Canada, via the North Pole. The skiers would get their location information from COSPAS/SARSAT. They would obtain the information by broadcasting an emergency radio signal which would be picked up by a COSPAS/SARSAT satellite, relayed to the ground, and sent to the University of Surrey, where it would be relayed to OSCAR 11 (Uosat 2). Uosat 2 would then use its "talking computer" to relay the location information to the skiers on its next pass over them. The Uosat signal could be easily picked up by a small VHF radio.

### Context

In 1959, no one realized that a private group of citizens could put a satellite in orbit, but by 1961, the first OSCAR satellite was launched. Amateur radio satellites have made important contributions to education. Amateur radio operators who wished to talk through a translator on board a satellite had to learn a new vocabulary and new scientific concepts. In conjunction with OSCARs 6 and 7, especially, NASA developed programs designed to bring space into the classroom. Using readily available commercial equipment, students studied some aspects of space science and space communications, such as orbital mechanics and telemetry decoding.

Project OSCAR was honored in 1965 by the International Institute of Communications; it was called "the brightest success in the space age" and praised for sparking "new developments in the art of communication." Project OSCAR was also awarded the 1965 American Radio Relay League Technical Merit Award for pioneering work with artificial satellites.

Amateur radio satellites were important for several reasons. First, they engendered a pioneering spirit in many radio amateurs. Second, they brought space exploration into the classroom for many students. Third, they have shown potential for use in emergency relief systems; radio amateurs are known for their help in providing communications during hurricanes, tornadoes, floods, earthquakes, and other natural disasters.

**See also:** Get-Away Special Experiments; Landsat 1, 2, and 3; Seasat; Shuttle Amateur Radio Experiment.

### Further Reading

Butrica, Andrew J. *Beyond the Ionosphere: Fifty Years of Satellite Communications.* NASA SP-4217. Washington, D.C.: Government Printing Office, 1997. Part of the NASA History series, this book looks into the realm of satellite communications. It also delves into the technology that enabled the growth of satellite communications. The book includes many tables, charts, photographs, and illustrations, a detailed bibliography, and reference notes.

Collins, A. Frederick, and Robert Hertzberg. *The Radio Amateur's Handbook.* 15th ed. New York: Harper & Row, 1983. This handbook is revised every year. It covers all aspects of amateur radio, including voice communications, slow-scan television, fast-scan television, and the OSCARs. Includes a short history of the OSCAR program. Contains photographs, drawings, circuit diagrams, tables, and, when necessary, equations. Designed to be read by those who wish to acquire some technical background.

Davidoff, Martin. *The Satellite Experimenter's Handbook.* 2d ed. Newington, Conn.: American Radio Relay League, 1990. This book's purpose is to provide experiments for satellite users. In the process of describing experiments, it also informs the user about features of the satellite and, to some extent, about satellite mechanics; consequently, it helps the reader gain a general understanding of amateur radio satellites. Requires some technical knowledge.

Gavaghan, Helen. *Something New Under the Sun: Satellites and the Beginning of the Space Age.* New York: Copernicus Books, 1998. This book focuses on the history and development of artificial satellites. It centers on three major areas of development—navigational satellites, communications satellites, and weather observation and forecasting satellites.

Hambly, Richard M. "AMSAT OSCAR-E Project Status Update: A New LEO Satellite from AMSAT-NA." *AMSAT Journal* 25, no. 7 (November/December, 2002). This article focuses on the development of a new series of amateur radio satellite (referred to as microsats) and provides both technical and general information about the system's capabilities.

Ingram, Dave. *OSCAR Satellite Review: Anthology-Plus-Update Guide to Today's OSCAR Action.* Birmingham, Ala.: Author, 1988. This booklet is a collection of several articles, originally published in *CQ Magazine*, plus several additional pieces by the same author. Most of these articles provide information about a satellite and its orbital mechanics. The articles also describe the kind of equipment necessary to use these satellites. Suitable for those with some technical background.

Kalter, Joanmarie. "Look! Down in the Basement! It's Captain Video of the East Coast Tuning His Ham TV!" *TV Guide* 36 (August 27, 1988): 30. This article is for general audiences. It does not, however, discuss amateur radio satellites. Instead, it provides some general information about radio amateurs and describes the transmission and reception of amateur television.

Riportella, Vern. "Getting on the New OSCAR." *QST* 72 (September, 1988): 85. This magazine has a monthly column called "Amateur Satellite Communications." This issue's column describes how to communicate through OSCAR 13. Suitable for those with some technical background, the journal contains one or more articles per month on amateur satellites.

Shute, Nancy. "Homemade Satellites." *Air and Space,* December, 1986/January, 1987, 76. A well-written, short history of the OSCAR program, this article contains interviews with radio amateurs and features photographs, including one of a satellite being constructed on a dining room table. Suitable for general audiences.

Zimmerman, Robert. *The Chronological Encyclopedia of Discoveries in Space.* Westport, Conn.: Oryx Press, 2000. Provides a complete chronological history of piloted and robotic spacecraft and explains flight events and scientific results. Suitable for all levels of research.

*T. Parker Bishop*

# Ames Research Center

*Date:* Beginning October 1, 1958
*Type of facility:* Space research center

*The Ames Research Center, together with the separate Dryden Flight Research Facility, is NASA's West Coast research institute for the development of technology in the fields of aeronautics and the life, space, and Earth sciences. Achievements in these areas include the first supersonic and the first hypersonic flights, the world's largest wind tunnel, the first short takeoff and landing jet, the Pioneer space projects, and space shuttle and space station life science studies.*

## Key Figures

*Joseph S. Ames* (1864-1943), chair of the National Advisory Committee for Aeronautics (NACA), 1927-1939

*Charles F. Hall* (b. 1920), project leader and spacecraft principal investigator for numerous robotic probes

*Hans Mark,* Ames Center director, 1969-1977

*Hugh L. Dryden* (1898-1965), NASA chairperson, 1947-1958

*Scott Hubbard,* Ames Center director from 2002

## Summary of the Facility

The history of Ames Research Center began as long ago as 1915, only twelve years after the Wright brothers developed the first airplane. It was then that the United States established the National Advisory Committee for Aeronautics (NACA). By the end of the decade, NACA had achieved impressive results and was recognized at home and abroad. By 1936, NACA officials had become aware of two problems: European nations were rapidly building new research facilities, and the room for growth at the United States' established facility at Langley Research Center in Virginia was limited.

Because of political developments in Europe, the U.S. Congress authorized the establishment, in 1939, of a second aeronautical research center. The ground breaking took place on September 14 at Moffett Field, a U.S. Navy airfield 64 kilometers south of San Francisco. In 1940, the new aeronautical laboratory was named for Joseph S. Ames, president of The Johns Hopkins University, a man of

integrity and ability respected by colleagues and friends alike. He was one of the original members of the special committee on expanded facilities and served for twenty-five years without compensation as a member of NACA. On October 1, 1958, with the enactment of the National Aeronautics and Space Act, Ames became part of the National Aeronautics and Space Administration (NASA), along with other NACA facilities and certain Department of Defense facilities. In 1981, NASA merged the Dryden Flight Research Center with Ames. From time to time, Congress or NASA considers closing down or merging centers; presently Ames and Dryden exist as separate NASA facilities, although Ames continues to contribute significantly to aspects of aerospace research conducted at Dryden.

The choice of the Ames-Moffett site resulted from several considerations. Location near the growing aircraft companies was at that time of prime importance. By 1939, almost half of the aircraft industry was based on the West Coast. Joint meetings with engineers from Langley required time and energy for travel. The need for a flying field far from an area of high air-traffic density yet near a Navy base became evident. A moderate climate and good flying weather through most of the year were desirable, and adequate inexpensive electrical power on the site was a necessity. Proximity to research institutions and universities would also be helpful.

Mountain View, the geographical site of Ames-Moffett, is located in the Santa Clara Valley, more popularly called Silicon Valley, where the heart of the nation's electronics industry has developed. Under the capable leadership of Smith De France and his associate John Parsons, Ames's employees adjusted well to the move to the West Coast. At its beginning, the Ames staff included fifty-one persons; half had been Langley employees. In the late 1980's, the facility employed more than two thousand civil service personnel and about fifteen hundred contractor employees; approximately four hundred graduate students and university faculty members

worked at the center as well. As of 2005, Ames sported $3.5 billion in capital equipment, enjoyed a fairly steady $700 million annual operating budget, and employed just under three thousand people.

One of the major concerns of the center was to produce working wind tunnels. In the fall of 1940, three wind tunnels were under construction. By 1944, the then-largest wind tunnel in the world (measuring 12 by 24 meters) was completed at Ames. This wind tunnel, best known as the 80-by-120-foot wind tunnel, would remain in operation for decades. World War II placed demands on the Ames Research Center by adding defense-related tasks to its basic research. The Ames research staff was faced with the task of de-icing aircraft. Ice is a burden on propellers and power plants, changes the aerodynamic properties of wings, and hampers radio reception and visibility. Ames's de-icing research was among the most dramatic of its early successes.

During World War II and directly after it, the center's research efforts included investigation of the compressibility effects of transonic flight (flight at approximately the speed of sound) and examination of the characteristics of different wing shapes. As transonic and supersonic flight (flight at one to five times the speed of sound) became practicable, the aeronautical laboratory had to redesign wing and fuselage shapes entirely. Collaboration with the military and with industry produced aircraft such as the quiet short-haul research airplane, which reduced noise pollution, and the XV-15 tilt-rotor aircraft, which functioned as both a helicopter and an airplane.

The establishment of NASA ultimately transformed the character and goals of Ames. Notable were the differences between long-duration topical research and the new, mission-

*The Dirigible Hangar, Moffett Field, NASA Ames Research Center* (NASA) oriented projects required to assume leadership in the Space Race. NASA's missions were to assure American supremacy and to put a human on the Moon before the end of the decade. This meant that Ames received a larger budget and was able to expand.

The addition of the Ames Life Sciences Directorate was a major factor for future planning, as it was to focus on the basic biological effects of extraterrestrial environments and the possible molecular evidence of extraterrestrial life. The Life Sciences division was to direct its efforts toward the medical and behavioral aspects of medicine and biology, as they related to crewed spaceflight, and to the scientific problems pertaining to more fundamental investigation of metabolism, nutrition, and general life functions in space.

Aeronautical research is carried on at both Ames-Moffett and Ames-Dryden. The latter is located at Edwards, California, in the Mojave Desert approximately 128 kilometers northeast of Los Angeles. It occupies 520 acres adjacent to Edwards Air Force Base. Ames-Dryden's primary research tools include research aircraft such as the B-52 carrier, high-performance jet fighters, the HiMat (for Highly Maneuverable Aircraft Technology), the X-29 forward-swept-wing aircraft, the AFTI (for Advanced Fighter Technology Integration), the Oblique Wing, and the F-18.

Computer scientists at Ames use and help develop some of the world's leading computational systems. These researchers generate images for flight simulation and three-dimensional computed and experimental results. Researchers in the field of artificial intelligence have developed expert systems in support of physicists, pilots, astronauts, and flight and space experimenters. High-fidelity flight simulators are an example of the type of powerful tools being used for research and development of new aircraft and spacecraft. Test pilots, astronauts, and engineers "fly" and evaluate control systems at Ames Research Center's piloted flight simulation complex, probably the most sophisticated in the world.

Basic and applied research in space science at Ames includes the interaction between humans and machines in both aircraft and spacecraft. For example, planning the design of the Reagan-era station concepts—the Freedom space station in the early 1990's and the eventual International Space Station—required investigations for habitability, management, extravehicular activity, and the selection, training, and effective management of crew. The Life Sciences division conducts scientific and technical research with the following goals: to understand the origin and evolution of life in the universe, to investigate the role of biological processes in planetary evolution, and to study the effects of the space environment on humans and develop the foundation for living and working in space.

Space science has occupied an important position at Ames. The Pioneer series of projects, under the leadership of Charles F. Hall and Smith De France, transformed the concept of Ames as an aeronautical laboratory into that of a research center. Pioneers 6, 7, 8, and 9, identical spacecraft, were designed to explore interplanetary space, returning information about magnetic fields, particles, cosmic rays, and cosmic dust. Pioneers 10 and 11 became two of NASA's most interesting and visible projects. These low-budget vehicles conducted investigations of the interplanetary medium beyond Mars, crossed the asteroid belt, and provided information about the atmosphere and environment of Jupiter. Pioneer 11 performed a flyby of Saturn before continuing on beyond the solar system. Pioneer 10 passed the orbit of Pluto in the late 1980's and remains the most distant human-made object.

Management for the Pioneer Venus missions came under the aegis of Ames as well, and Charles F. Hall was placed in charge of the project. The program was designed to study Earth's neighboring planet with two vehicles: one, Pioneer 12, which would gather data while orbiting Venus, and the other, Pioneer 13, a multi-probe spacecraft. Pioneer 12, also known as Pioneer Venus 1, circled Venus after its launch in 1978, mapping the planet's surface by radar, studying its cloud systems and magnetic environment, and observing the interac-

*Aerial view of NASA Ames Research Center, Mountain View, California.* (NASA)

tions of the solar wind (ejections from the Sun's surface). Pioneer 13, also known as Pioneer Venus 2, consisted of four probes toted by a 2.7-meter bus. The probes and the carrier penetrated the Venusian atmosphere at widely separated locations in both day and night hemispheres and measured temperature, pressure, and density down to the planet's surface.

Astronomers and physicists at Ames make use of infrared (wavelengths of light beyond the visible spectrum) astronomy to study star formation and objects in the solar system. The scientists make observations from the Kuiper Airborne Observatory and from various ground-based telescopes. Remote sensing by way of satellites has been one of NASA's methods for studying Earth resources. In addition, high-altitude photography,

using NASA assets such as the Kuiper Airborne Observatory, the Learjet, the U-2, and the ER-U-2 research aircraft, remains a basic tool at Ames Research Center for obtaining astronomical and Earth data.

The development of the Galileo orbiter/probe mission was to be divided between the Jet Propulsion Laboratory, which later had the responsibility for the orbiter craft, and the Project Office at Ames, which had the responsibility for the probe. Thermal tiles for the space shuttle, investigative studies for the Space Station, overseeing the construction of the space telescope, and various collaborations with industry and other NASA centers are among the contributions of the Ames Research Center. In addition, Ames's scientists cooperate with investigators of other nations in biological research, astronomy, space science, and aeronautical studies in their continued scientific and technological research.

The accomplishments of the Ames Research Center may be listed under four headings: aeronautics, life, space, and Earth sciences. Engineers and scientists at Ames have developed four key interactive elements in the field of aeronautics. First, they help build large supercomputers for computational analysis. One of NASA's main research areas is in basic and applied computational fluid dynamics (CFD). Scientists use CFD to study the flow of air on computer model aircraft and spacecraft so that the data obtained may complement the data provided by wind-tunnel testing. Second, researchers at Ames helped build the world's largest and busiest complex of wind tunnels and arc-jet facilities for model-testing designs. Third, the world's most outstanding flight research simulators for determining aircraft and spacecraft capabilities and handling qualities prior to, during, and on return flights were built at Ames. Last, Ames is known for the construction of full-scale experimental vehicles for flight research.

By using and creating the state-of-the-art technological tools in conjunction with ground-based facilities, researchers at Ames have developed unusual aircraft concepts. One example is the Oblique Wing, a scissor-like asymmetrical aircraft. Another is the XV-15 tilt-rotor vehicle, a vertical takeoff and landing aircraft.

The Life Sciences Division's research in the biological sciences, space biomedicine, and life-support systems responded to the needs of the space shuttle Spacelab missions and development and operation of the International Space Station. Twenty-five space shuttle missions carried Spacelab research. The last one was the Neurolab mission, which was flown on *Columbia* in April, 1998. Extensive basic research to understand the effects of the low-gravity environment on biological systems has been under way for decades. The research also focuses on the development of technology to support a safe, productive, and extended-duration human presence in space.

The Ames Research Center has contributed significantly to the Mercury, Gemini, Apollo, and space shuttle missions. Among these contributions are improvements in atmospheric reentry systems and heating aerothermodynamics (heating processes of air motion). The center has contributed to the design of the shuttle orbiter and the materials for its heatshield. Dryden continues to handle the shuttle landing operations as well as the flight research on virtually every new military fighter and experimental aircraft built in the United States.

The Pioneer series of spacecraft, an Ames triumph, included the first probe to pass through the asteroid belt and go on to Jupiter and Saturn. In June, 1983, Pioneer 10 left the solar system, and twenty-eight years after launch, was 75 astronomical units away from Earth. The radar instruments aboard Pioneer Venus 2, which reached Earth's neighboring planet in 1978, produced the first topographical map of the surface of Venus. The Lunar Prospector mission, launched in January, 1998, was a one-year circular, polar mapping of key characteristics of the Moon, including elemental abundances, gravity fields, magnetic fields, and outgassing events. After a year and a half of exploration, Lunar Prospector was deliberately crashed into a permanently shadowed south polar crater to attempt the detection of water on the Moon. Although the data from the crash did not provide direct evidence of water, Lunar Prospector did tantalizingly suggest that there were regions on the Moon where ice could exist.

In Airborne Science and Earth Resources programs, the Ames Research Center's investigations range from collections of pollutant matter in the upper atmosphere to surveys of crops, watersheds, and forests, as well as observations of the galactic center and new star formations in the Orion nebula. Of special note is the C-141 Kuiper Airborne Observatory's discovery of rings around Uranus and an atmosphere around the planet Pluto. The achievement of the U-2's radiation measurements, the results of which supported the Big Bang theory of creation, is regarded as outstanding by the scientific community.

In collaboration with American industry, universities, the Department of Defense, and other NASA centers, Ames assisted in development efforts for the cancelled National AeroSpace Plane. At Ames, scientists studied the plane's basic shape and propulsion system with the use of wind tunnels and the center's supercomputers in both the numerical aerodynamic simulation system and the central computer facility. Researchers evaluated contractors' proposed airframe designs, studying thermal protection materials for the aircraft, and developing new techniques for cooling the leading edges of the nose and wings. Models were tested for responses to heat and weight loads, studies were conducted to understand the behavior of candidate materials in harsh hydrogen-rich environments, computational chemistry was used to determine the properties of shock-heated air and combustion gases, and computational fluid dynamics was used to study promising aircraft shapes.

**Context**

As a result of the accomplishments and the operations of the Ames Research Center, universities, industry, government agencies, and the entire populace of the United States have been affected in some way. One needs only to read NASA's annual volume "Spinoff" to realize the impact of NASA research on the political, social, and cultural aspects of life in the United States and throughout the world. Continuous strong relationships with the university community by way of interchange of scientists, engineers, and technologists with principal investigators capitalize on synergies between research disciplines.

Ames's established leadership in computational analysis, wind-tunnel research, flight simulation, and flight research has contributed new rotorcraft, powered-lift, high-performance vehicles to industry. It has advanced key elements of basic aeronautics, which determine component characteristics and contribute to total aircraft performance. The Ames facilities and capabilities have played a crucial role in supporting important national and industry needs.

Cooperative activities with the various branches of the military, the Federal Aviation Administration, and the United States Geological Survey on broad technical goals, with state and local governments on environmental and Earth resources problems, with the aerospace industry in virtually all areas demanding research and technology, and with scientists from the United Kingdom, Europe, Canada, and Japan further expand the scope of the Ames Research Center.

**See also:** Cape Canaveral and the Kennedy Space Center; Goddard Space Flight Center; Jet Propulsion Laboratory; Johnson Space Center; Langley Research Center; Marshall Space Flight Center; Mercury Project Development; Pioneer Missions 1-5; Pioneer Missions 6-E; Pioneer 10; Pioneer 11; Pioneer Venus 1; Pioneer Venus 2; Space Centers, Spaceports, and Launch Sites.

**Further Reading**

Anderson, Seth B. *Memoirs of an Aeronautical Engineer: Flight Testing at Ames Research Center, 1940-1970.* Washington, D.C.: NASA, 2002. This is number 26 in NASA's Monographs in Aerospace History. It covers the full range of NASA Ames aerospace research during the period indicated in the title.

Bilstein, Roger E. *Orders of Magnitude: A History of the NACA and NASA, 1915-1990.* 3d ed. NASA SP-4403. Washington, D.C.: Government Printing Office, 1989. The third edition of this small volume expands upon the first short history of NASA, written in 1965. It covers the country's air and space studies from World War I (1914-1918) to World War II (1939-1945) and beyond, to the threshold of spaceflight in the space shuttle era. Contains no reference notes and only a generalized bibliography.

Bugos, Glenn. *Atmosphere of Freedom: Sixty Years at the NASA Ames Research Center.* NASA SP-4314. Washington, D.C.: Government Printing Office, 2000. This NASA History series book provides a historical account of aeronautical and aviation research carried out at NASA Ames since 1940.

Hartman, Edwin P. *Adventures in Research: A History of Ames Research Center, 1940-1965.* NASA SP-4302. Washington, D.C.: Government Printing Office, 1970. The occasion of the twenty-fifth anniversary of Ames's founding triggered the writing of this history. The title of this weighty paperback volume (555 pages) is indicative of the excitement of the persons who lived and worked at Ames while history was being made. The text has appendices and references.

Muenger, Elizabeth A. *Searching the Horizon: A History of Ames Research Center, 1940-1976.* NASA SP-4304. Washington, D.C.: Government Printing Office, 1985. The foreword of this book was written by Hans Mark. He proclaims, "This is a book about a remarkable institution. It captures the soul of the place and the work of the people who made it what it is today." By reading this book, one can become familiar with the founders and developers of Ames. Included are an appendix covering the major events of Ames's history and a bibliographic essay.

National Aeronautics and Space Administration. *Ames Research Center.* Moffett Field, Calif.: Ames Research Center, 1984. A colorful twenty-four-page booklet which may be obtained from the Public Affairs Office at Ames. Illustrated with color photographs, this succinct brochure summarizes the eight branch divisions of the center and describes their achievements, their goals, and their vision of future development.

National Research Council. *Lunar Prospector End of Mission and Overview Press Kit.* Washington, D.C.: Government Printing Office, 1999. This document, available directly from NASA and on the Internet at http://lunar.arc.nasa.gov, contains details of the Lunar Prospector Mission and the role Ames took in the flight. It is highly illustrated and covers the flight from prelaunch through lunar impact.

_____. *A Review of the New Initiatives at the NASA Ames Research Center.* Washington, D.C.: National Academy Press, 2001. This is a collection of commissioned research articles written in response to a 2000 workshop that reviewed the Ames Research Center's efforts to develop a service and technology park in Silicon Valley. The park was designed to bring corporations and university groups together in ways that could capitalize upon work and facilities at Ames at Moffett Field.

*Clarice Lolich, updated by David G. Fisher*

# Ansari X Prize

*Date:* Fall, 1995, to October 4, 2004
*Type of program:* Piloted spaceflight

*The Ansari X Prize competition was an attempt to duplicate the challenge that the Orteig Prize, won by Charles A. Lindbergh in 1927, made to the aviation industry. This time the X Prize would challenge the private sector to take up where various national space programs had failed to go in developing less expensive reusable spacecraft. If this challenge could be met, it would open space to the private sector as well as the tourist trade.*

## Key Figures

*Peter H. Diamandis*, chairman, founder, and president of the X Prize Foundation
*Gregg E. Maryniak*, trustee and executive director of the X Prize Foundation
*Byron K. Lichtenberg*, trustee and cofounder of the X Prize Foundation
*Collette M. Bevis*, cofounder of the X Prize Foundation
*Robert K. Weiss*, vice chairman and executive producer of the X Prize Foundation
*Anousheh Ansari* and
*Amir Ansari*, contributors of several million dollars to the X Prize Foundation
*Burt Rutan* (b. 1943), head of the team that won the first Ansari X Prize, in 2004
*Paul G. Allen*, Microsoft cofounder who helped finance the building of SpaceShipOne

## Summary of the Program

The concept of the X Prize began in 1994 with the gift of a book titled *The Spirit of St. Louis* (1953), written by Charles A. Lindbergh. It was Lindbergh's personal account of how he had accomplished the first nonstop solo flight across the Atlantic Ocean. This gift from Gregg E. Maryniak to fellow space enthusiast and visionary Peter H. Diamandis was the spark that created the X Prize. In reading Lindbergh's vivid narrative, Diamandis saw history repeating itself. He then realized that if spaceflight was to evolve beyond the various national space programs, the private sector would have to begin developing its own space transportation systems.

The dawn of powered flight began on December 17, 1903, with the Wright brothers, Orville and Wilbur, at Kitty Hawk, North Carolina. The flight may have lasted only a few seconds but it proved that powered flight was possible, and many competitors would soon follow in their footsteps. The aviation industry quickly grew from this initial flight, and the excitement of taking to the air swept over the public. Between 1905 and 1935, hundreds of aviation prizes were offered to those willing to accept the challenges of flying higher, faster, and farther than anyone else. The most famous of these prizes was the $25,000 Orteig Prize, offered by hotel owner Raymond Orteig. His prize would be paid to the first person to fly solo across the Atlantic Ocean. In 1927, Lindbergh succeeded in winning the prize by using cheaper, "off the shelf" proven technology. His use of a smaller, faster approach showed that a small highly professional team could outperform both government and large industrial programs.

After reading about Lindbergh's accomplishments, Diamandis saw the potential for re-creating the same infectious excitement associated with the pioneer days of aviation and the first piloted spaceflights. If the challenge of winning a prize to advance aviation worked in 1927, why not offer the same challenge for spaceflight in the 1990's? Diamandis first needed to attract the attention of the aerospace industry. He decided that the best approach was to create a technological challenge, set an achievable goal, and then offer a prize that would challenge both aircraft designers and pilots to push the limits of technology and reach out into space.

In 1995, together with his friends and associates Byron K. Lichtenberg, Collette M. Bevis, and Maryniak, Diamandis created the X Prize Foundation as the first step toward achieving his goals. The X Prize Foundation, which was originally based in Rockville, Maryland, first needed to identify an appropriate community that could financially support the project. St. Louis, Missouri, soon became the prime candidate city. Local business leaders proposed the creation of a New Spirit of St. Louis organization to underwrite the startup costs of the project. This was also in keeping with the Lindbergh tradition: Nine St. Louis businessmen had provided the $25,000 needed to build his airplane.

The official announcement of the X Prize took place at a gala celebration under the famous St. Louis Arch on May 18, 1996. Dignitaries present included several former astronauts, National Aeronautics and Space Administration (NASA) Administrator Dan Golden, aviation designer Burt Rutan, and Morgan and Eric Lindbergh, the grandsons of Charles Lindbergh. It was at this inaugural dinner

*SpaceShipOne hangs on to its mothership,* White Knight, *during a flight test.* (NASA/Scaled Composites)

that Burt Rutan stepped forward and announced his intent to enter the competition, and that he expected to win it. Later that year, the X Prize Foundation moved into its new home in the McDonnell Planetarium of the St. Louis Science Center. Individual and corporate sponsorships soon followed, and the five-million-dollar mark was reached in May, 1998. Six years later, in May, 2004, the X Prize Foundation received a multimillion-dollar contribution from the Ansari family and recognized their generosity by changing the name to the Ansari X Prize. Like many other people, Anousheh and Amir Ansari had dreams of one day participating in a spaceflight, but the odds were still against it. For the moment, spaceflight would still be limited to the lucky few who had the opportunity to venture into space.

Burt Rutan's 1996 commitment to winning the Ansari X Prize took its first step to achieving that goal with the flight of SpaceShipOne, on June 21, 2004. Veteran pilot Mike Melvill took SpaceShipOne beyond the altitude of 80 kilometers (50 miles), which is the point arbitrarily used to define the start of outer space, and eventually reached a maximum altitude of just over 100 kilometers (62 miles). That qualified Melvill to become the first civilian pilot to obtain his astronaut's wings. The flight of SpaceShipOne also proved that the technology developed by Rutan and his team did work, and it positioned them as the front runners to win the Ansari X Prize.

The first flight to qualify to win the Ansari X Prize took place on September 29, 2004, as SpaceShipOne, with Mike Melvill once again at the controls, soared into the blue California sky over the Mojave Desert and reached an altitude of more than 100 kilometers. The flight did not lack tense moments. Melvill experienced several unexpected barrel rolls that challenged his ability to control the vehicle. Fortunately, his flying experience and his faith in the aircraft's design brought him safely home. He had successfully met the first prize requirements. Five days later, on October 4, SpaceShipOne was again heading toward space with pilot Brian Binnie at the controls. After an 80-second

rocket burn SpaceShipOne continued to climb until Earth's gravity overpowered the craft's upward momentum. This time the craft performed flawlessly and soon began its long spiral glide back to Earth. Upon its landing, it had won the Ansari X Prize.

**Contributions**

The knowledge gained through the competition for the Ansari X Prize resulted from the successful design and flight of a spacecraft that could launch three humans, or a pilot and the equivalent weight of two others, to a suborbital altitude of 100 kilometers (62 miles) on two consecutive flights with the same vehicle within a two-week period. This was a first in the history of space exploration. To answer this challenge, more than twenty teams from seven countries registered for the competition. Each team set its sights on a 2004 launch date. In order to win the Ansari X Prize, these teams would need to combine existing technology with new and innovative designs to build a vehicle capable of reaching space.

The team headed by Burt Rutan and financed by Microsoft cofounder Paul Allen developed a specially designed short-winged rocket plane named SpaceShipOne. In a manner similar to the historic X-15 flights, it is attached to the *White Knight*, a carrier plane. At an altitude of about 15.4 kilometers (50,000 feet) the *White Knight* releases it and then the spacecraft begins its rocket-powered flight into space. SpaceShipOne is powered by a hybrid fuel consisting of a mixture of nitrous oxide and tire rubber that creates a less volatile and safer propellant than that used in NASA's space shuttles.

Another innovation in design is in the technology used for reentry. In contrast to the space shuttle, which uses special heatshield tiles for reentry, SpaceShipOne employs a "feathered" wing system that prevents overheating by slowing down its reentry speed, allowing it to "float" gently back to Earth. Just like the space shuttle, SpaceShipOne glides back to Earth to an unpowered landing. Most space shuttle missions plan for a landing back at the Kennedy Space Center launch site in Florida, but quite often bad weather or some other consideration forces the shuttle to land at an alternate site in California or New Mexico. Unlike the space shuttle, SpaceShipOne does not venture far from its launch point, but basically goes straight up and then directly back to its launch site. Future commercial flights of craft like SpaceShipOne will be only suborbital missions, but plans are already being made to take the next step into orbit.

The flight of SpaceShipOne proved that spaceflight is possible with more economical and less sophisticated technology than currently used by the various national space programs. The aeronautical and astronautical knowledge gained through the successful flights of SpaceShipOne inspired space enthusiast and billionaire Richard Branson to form Virgin Galactic, a tourist company planning to send five passengers per flight on a modified version of SpaceShipOne. Once these private sector flights advance to orbital missions, they too will have to develop new strategies to meet the greater challenges of orbital flight.

### Context

The Ansari X Prize re-created the excitement of a competitive challenge to the aerospace industry to develop a new type of relatively inexpensive reusable spacecraft that would make space travel possible for the general public. This would create new opportunities for a much wider range of industrial users and even make spaceflight possible for the tourist industry.

From its conception to the successful flight of SpaceShipOne, the Ansari X Prize stimulated both the general public and the private sector with its vision of "spaceflight for everyone." Several designers of innovative experimental aircraft eagerly took up the challenge to be the "first in space." Their competitive spirit quickly took over, and the race was on to win the prize. They also realized that once this first venture into space had been accomplished, the way would be open for future commercial development.

On October 4, 2004, SpaceShipOne broke through the bonds of Earth's gravity for the second time in less than two weeks and became the winner of the Ansari X Prize. This historic achievement came on the forty-seventh anniversary of the launch of Sputnik, which in 1957 had ushered in the Space Age. In those forty-seven years, humanity has gone from orbiting small satellites in low-Earth orbit to landing humans on the Moon, and now finally to making space accessible for all those who are willing to accept the challenges of spaceflight.

**See also:** Private Industry and Space Exploration; SPACEHAB; SpaceShipOne.

### Further Reading

Adams, Eric. "X Prize Victory Redemption." *Aviation and Space.* October, 2004. http://www.popsci.com/popsi/aviation/article/0,20967,709317,00.html. Accessed November, 2004. A concise report on the successful flights of SpaceShipOne and the winning of the Ansari X Prize. The article also focuses on the personal side of the pilots who flew the missions.

Ansari X Prize home page. http://www.xprize.com. Accessed November, 2004. Originally the official home page for the Ansari X Prize and later for the X Prize Cup competition. Features of this site include a comprehensive presentation of the Ansari X Prize; the X Prize Foundation; its founders, sponsors, and advisory members; and recent press releases and up-to-date events.

David, Leonard. "SpaceShipOne Wins $10 Million Ansari Prize in Historic Second Trip to Space." http://www.space.com/missionlaunch/xprize2_success_041004.html. Accessed November, 2004. This excellent article reports on the October 4, 2004, flight of SpaceShipOne, which captured the Ansari X Prize. Also addresses the future of piloted spaceflight for the private sector.

Perry, Lacy. "Behind the X Prize." http://science.howstuffworks.com/x-prize.htm. Accessed November, 2004. A comprehensive overview of the X Prize and the people involved in it, serving as a good introduction to the entire X Prize concept. The article also looks at the different approaches the competitors took to designing their aircraft and preparing for their flights. The concept of space tourism is also evaluated in the light of SpaceShipOne's successful flights.

Taylor, Chris. "The Sky Is the Limit." *Time* 164, no. 22 (November 29, 2004): 64-68. A concise yet comprehensive look at SpaceShipOne and the sequence of events that led to the winning of the Ansari X Prize competition. The article also presents a detailed illustration of the SpaceShipOne vehicle.

*Paul P. Sipiera*

# Apollo Program

*Date:* May 28, 1964, to December 19, 1972
*Type of program:* Lunar exploration, piloted spaceflight

*The Apollo mission was the United States' bid for international leadership in space exploration, to be demonstrated by landing humans on the Moon and returning them safely to Earth during the 1960's. After eight years of development, the project achieved success with the return of Apollo 11 on July 24, 1969. The program also had political significance, heralding the superiority of the United States and effectively ending the Space Race with the Soviet Union.*

## Key Figures

*James E. Webb* (1906-1992), NASA administrator, 1961-1968

*D. Brainerd Holmes* (b. 1921), associate administrator for Manned Spaceflight, 1961-1963

*George E. Mueller* (1918-2001), associate administrator for Manned Spaceflight, 1963-1969

*George M. Low* (1926-1984), director of Manned Spaceflight programs, 1958-1964; deputy director of Manned Spacecraft Center, 1964-1967; manager of Apollo Spacecraft Project office, 1967-1969; NASA deputy administrator, 1969-1976

*Rocco A. Petrone* (b. 1926), director of Launch Operations, Kennedy Space Center, 1961-1969; director of Apollo Program Office, 1969-1973

*Samuel C. Phillips* (1921-1990), director of Apollo Program Office, 1964-1969

*Wernher von Braun* (1912-1977), director of Marshall Space Flight Center

*Robert R. Gilruth* (1913-2000), director of Manned Spacecraft Center

*Kurt H. Debus* (1908-1983), director of Kennedy Space Center

*Christopher C. Kraft, Jr.* (b. 1924), Apollo flight director

## Summary of the Program

Piloted flights to the Moon, either to land on it or merely to circumnavigate it, appeared in most of the early plans for the American space program. In July, 1960, the National Aeronautics and Space Administration (NASA) convened its first NASA-Industry Program Plans Conference, during which long-range goals for the Apollo Program (the advanced piloted spaceflight program) were outlined to executives of the aerospace industry. The plans included circumlunar flights and (sometime after 1970) a piloted lunar landing. In February, 1961, a Manned Lunar Landing Task Group, chaired by George M. Low, director of Manned Spaceflight programs, reported that a piloted lunar landing could be accomplished by the end of the 1960's. Almost simultaneously, NASA announced the award of a contract for a study of the navigation and guidance system required for a lunar mission; a few months later, it contracted studies on the spacecraft for such a mission.

The decisive stimulus, however, came in April, 1961, when the Soviet Union sent Yuri A. Gagarin into Earth's orbit. This development was widely perceived as a threat to American security and its technological preeminence in the world—a propaganda stroke to be countered by some dramatic demonstration of the United States' capability. On May 25, 1961, a scant ten days after America's first piloted, suborbital flight, President John F. Kennedy went before Congress to call for the United States to land humans on the Moon and return them safely to Earth "before this decade is out." The Space Race was on.

Inasmuch as NASA had yet to send a human being into Earth orbit, Kennedy's challenge was a formidable one. Neither the spacecraft nor the launch vehicles and facilities for a lunar landing existed, nor did the operational techniques for conducting it. Studies already under way were intensified following Kennedy's address, and by the end of 1961, the basic requirements for the launch vehicle and launch facilities had been outlined. Although the final design for a lunar spacecraft could not yet be specified, a contract for its design and development was awarded late in 1961.

The key decision of the project was made in July, 1962, when, after more than a year of studying alternatives, NASA Administrator James E. Webb announced that lunar orbit rendezvous had been chosen as the primary operational mode for the Apollo Program. Lunar orbit rendezvous required a large, three-stage rocket (the Saturn V) to launch two spacecraft: one, the Command and Service Module (CSM), to carry a three-person crew into orbit around the Moon and back; and another, the Lunar Module (LM), in which two astronauts would land on the Moon while the third tended the Moon-orbiting CSM. After a brief trip outside their lander to collect samples and emplace instruments, the lunar explorers would blast off in the upper stage of the LM to return to lunar orbit, rendezvous with the CSM, and then head back to Earth. This goal appeared operationally complex and difficult, but studies indicated that it was superior to other methods in terms of development time, cost, and simplicity of management.

Following this decision, NASA awarded contracts for the CSM, the LM, and the Saturn V rocket. In September, 1961, the Manned Spacecraft Center was established to manage development and preflight testing of the spacecraft, train astronauts, and conduct flight operations. Construction began on a 6.7-square-kilometer site

some 35 kilometers southeast of Houston, Texas, the following year; the center began operations in its new facilities in 1964. To launch the Saturn V, NASA built its own Launch Complex (later named Kennedy Space Center) on Merritt Island, adjacent to the Air Force Missile Test Center at Cape Canaveral, Florida. New facilities included two launch pads plus a Vehicle Assembly Building in which as many as four Saturn V's could be assembled simultaneously in a protected environment. Two diesel/electric-powered transporters were designed and built to move completed vehicles from the assembly building to the launch pad. The new site was activated in mid-1966.

While spacecraft and launch vehicles were being developed, other anticipated problems were addressed in parallel projects. Project Gemini investigated the operational problems of rendezvous in orbit, established that humans could function normally for up to fourteen days in weightlessness,

*President John F. Kennedy issued the challenge for the Apollo Program in his historic speech before a joint session of Congress on May 25, 1961.* (NASA)

qualified many spacecraft systems, and provided training for crews and flight controllers. To test Apollo's launch escape system, which could pull the spacecraft away from the Saturn rocket in case of a malfunction in the first few minutes of ascent, a solid-fueled rocket called Little Joe 2 was built and a test program was started at the White Sands Missile Range in southern New Mexico. The two-stage Saturn I and its more powerful version, the Saturn IB, were brought to operational status, to be used in Earth-orbital tests of the Apollo CSM. In late 1963, the Lunar Orbiter project was established to furnish high-resolution photographs of possible lunar-landing sites, and the Surveyor Program (a robotic lunar lander intended to gather scientific data) was called upon to provide information on the physical characteristics of the Moon's surface, data required in the design of the LM's landing gear.

Starting in 1962, NASA solicited advice from prominent American scientists in planning a science program for the Apollo Program. Summer conferences in 1962 and 1965 outlined the major objectives, defining three general types of scientific work: observations made by the astronauts, collection of samples for detailed study in the laboratory, and emplacement of instruments to return data to Earth over a long period of time. A third conference in 1967 defined an extensive program of piloted and robotic lunar exploration, from which NASA would later select experiments suited to the operational constraints of the missions. At this conference, NASA also set up an advisory body of scientists who would assist the Manned Spacecraft Center in planning the scientific work for each mission.

In spite of many delays in the development of two complex and sophisticated spacecraft, by late 1966, NASA was preparing to put the first piloted CSM through its paces in Earth orbit. On January 27, 1967, however, during a simulation two weeks before the scheduled launch, a flash fire swept through the spacecraft, killing all three crew members. Had it flown, the mission would have been called Apollo 1. Throughout the investigation of the accident, the mission was referred to only as Apollo-Saturn 204. As with past-piloted programs, Apollo flights were given names based upon their launch vehicle, using a three-digit number. Saturn IB flights were indicated by the number 2 followed by a two-digit sequence number. Saturn V flights used 5 instead of 2. The name Apollo 1 would have been used as the radio call sign.

Thorough investigation failed to pinpoint the exact cause of the fire, but it appeared to have been started by an electrical short-circuit, which produced a spark that ignited flammable material in the CSM. In the pure oxygen atmosphere used during flight, the fire had spread with a rapidity no one had anticipated. Apollo's first piloted mission was delayed some eighteen months, while every aspect of spacecraft manufacture and test procedures was reviewed to reduce the danger of fire.

Progress continued, however, in other phases of the Apollo Program. The first flight of a complete Saturn V rocket (Apollo 4) was successfully conducted on November 9, 1967; it carried a test version of the CSM to test the integrity of the heatshield under conditions approximating reentry from a lunar flight. Two months later, Apollo 5 carried a LM into Earth orbit for extensive robotic testing. On April 4, 1968, the second Saturn V flight test (Apollo 6) furnished more data on the CSM. Marred by significant failures in all three stages, Apollo 6 provoked months of corrective work on the Saturn systems.

Crewed flights of Apollo resumed in October, 1968, with an eleven-day Earth-orbit test of the CSM (Apollo 7). Successful in all respects, Apollo 7 restored confidence in the CSM and paved the way for the longest step taken so far: the Moon-orbiting mission of Apollo 8 (December 21-27, 1968). The excellent performance of both launch vehicle and spacecraft virtually assured achievement of the lunar landing within the decade. Only two more test flights remained: Apollo 9 tested both the CSM and LM (piloted) in Earth orbit (March 3-13, 1969), and Apollo 10 rehearsed all phases of a lunar-landing mission except the actual landing and return to orbit (May 18-26, 1969).

Apollo reached its primary goal on July 20, 1969, when the LM *Eagle* touched down gently in the Sea of Tranquility. During a 21.5-hour stay, Astronauts Neil A. Armstrong and Edwin E. "Buzz" Aldrin, Jr., spent 2.5 hours on the lunar surface collecting 21 kilograms of lunar rocks and soil and setting out two scientific instruments. The return of the Apollo 11 crew and their triumphant national and world tours marked the high point of public interest in the Apollo Program, although nine more flights remained on the schedule.

Following the initial success, NASA gave more attention to science in the Apollo Program. Modifications to the spacecraft to allow longer stays on the Moon were ordered, along with construction of a battery-powered vehicle (the Lunar Rover) to extend the astronauts' range of exploration. Flight operations planners began refining their techniques to allow precision landings (within 1,000 meters of a preselected spot) and to reach more difficult areas. Apollo 12 (November, 1969) landed less than 200 meters from its targeted spot, stayed 31.5 hours, and brought back 34 kilograms of samples.

Apollo 13, launched April 11, 1970, was to have been the first flight to a location of prime scientific interest, the Fra Mauro formation, 530 kilometers west and slightly south of the center of the lunar disk as viewed from Earth. Two days after the mission began, however, an explosion in the Service Module forced Mission Control to abort the flight. The crew returned safely to Earth, but the ensuing investigation and modification of the spacecraft delayed the next mission until January 31, 1971. Apollo 14, the last of the limited-duration early missions, landed at Fra Mauro, stayed 33.5 hours, and returned with 43 kilograms of lunar material.

By 1970, public interest in lunar exploration had waned, and a new administration committed to a space-based defense system and to reducing federal spending was in office. Budget cuts forced NASA to sacrifice current projects in order to support future ones. One Apollo Program mission was canceled in January, 1970; two more were dropped the following September. The remaining three,

numbered 15, 16, and 17, would be equipped to stay longer, travel farther, and conduct more experiments than had earlier missions.

Apollo 15 landed nearly 800 kilometers north of the lunar equator, at the foot of the Apennine Mountains near Hadley Rille, a sinuous valley of interest to scientists. In the 67 hours they stayed on the Moon, the Apollo 15 astronauts made three trips in their rover, driving a total of 20 kilometers, and changed their operational plans on the recommendation of scientists working in the Mission Control Center. While the others were on the Moon, the Command Module pilot operated remote-sensing instruments from the orbiting CSM, photographed numerous areas of interest, and released a subsatellite that would report scientific data to Earth after the crew left. Apollo 15 was the most productive mission of the series, not to be surpassed by either of the remaining two.

Apollo 16 visited a site near the crater Descartes, 550 kilometers to the east and south of the Moon's equator. Launched on April 16, 1972, Apollo 16 conducted much the same kind of explorations as had its predecessor. The final mission of the series, Apollo 17 (launched December 7, 1972), carried the only scientist to visit the Moon, geologist Harrison H. "Jack" Schmitt. Schmitt and Mission Commander Eugene A. Cernan landed more than 1,000 kilometers to the east and north of the Moon's equator in a valley called Taurus-Littrow, where they spent 75 hours on the surface and explored the valley for a total of 22 hours on three excursions. Their return with 110 kilograms of lunar samples marked the end of the most ambitious venture into space thus far attempted.

**Contributions**
Besides gathering 380 kilograms of lunar surface material, the astronauts of the Apollo Program left a variety of instruments, some of which operated for up to five years after the project ended. The data from samples and instruments have allowed scientists to piece together a coherent but still somewhat tentative picture of the Moon's structure, composition, and probable evolutionary his-

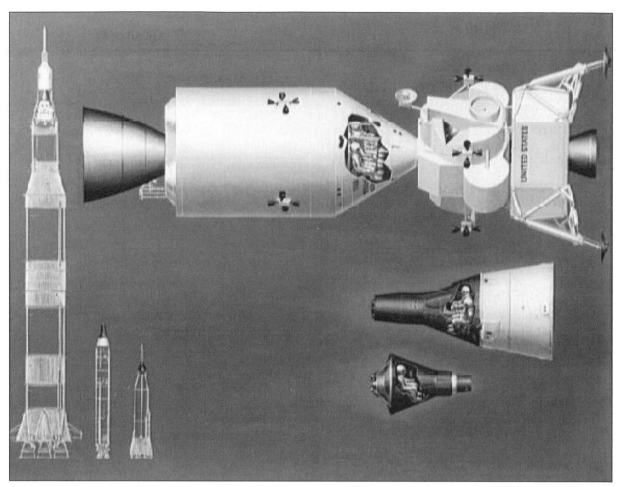

*A comparative view of the sizes and components of the one-person Mercury (bottom), two-person Gemini (center), and three-person Apollo spacecraft. Apollo's Command Module (top left) attaches to the Lunar Module. Diagrams of the booster vehicles appear on the far left.* (NASA)

tory. The Moon appears to be layered much like Earth, with a crust about 60 kilometers thick lying over a different layer extending to a depth of about 1,000 kilometers. A partially molten core may exist at the Moon's center. Periodic, weak tremors (moonquakes), which may be correlated with lunar tides, have been observed at depths of 800 to 1,000 kilometers, much deeper than earthquakes.

Chemically, the lunar samples contain the same elements found on Earth, and many of the minerals on the Moon are familiar to geologists. Yet the Moon is somewhat poorer in volatile elements

(those which are driven off at relatively low temperatures) than is Earth, and it has somewhat more radioactive elements than the cosmic average.

The oldest Moon rocks are about 4.5 billion years old, indicating that they solidified in the late stages of the formation of the solar system. A widely accepted view holds that the Moon never completely melted; only the outer layer, perhaps to a depth of 320 kilometers, melted in the beginning. As it lost heat and began to solidify, different minerals crystallized at different temperatures, and convection brought deeper, hotter material to the surface, partially remelting the crust. As cooling continued during a period of some 200 million years, a rigid crust of considerable thickness formed, composed mainly of light-colored minerals rich in aluminum and calcium. Beneath this crust, a mantle of dark material rich in iron and magnesium settled, and at the center, a core of dense, partially molten material rich in iron and sulfur may have formed. At some time, the Moon seems to have had a magnetic field generated by this liquid core, but the magnetism has now vanished.

For perhaps 300 million years after the crust formed, the Moon was bombarded by fragments of the primordial material of the solar system, some of them 50 to 100 kilometers in diameter. The impacts caused local melting of the crust, chemically altered the original material, and scattered debris over thousands of square kilometers of the surface. Below the crust, the mantle remained at least partially molten, in part because of radioactive heating. Toward the end of this bombardment, about 4.1 to 3.9 billion years ago, objects the size of large asteroids gouged out the large circular basins prominent on the Moon's visible face. Episodic flows of molten basalt from the interior filled these basins to roughly their present levels. Localized volcanic activity may have continued for another 2 billion years; the Apollo evidence is not conclusive, but the youngest rock that may have originated in volcanic action is 3.16 billion years old.

For roughly the last 3 billion years, the Moon has undergone no large-scale surface changes. Small meteorites have produced continuous slow erosion, adding to the layer of dust, and occasionally a larger body has produced another crater (an object weighing about 1,100 kilograms was detected in May, 1972, by Apollo's seismometers). Now and then, a boulder has been dislodged and rolled down a slope, leaving its track in the dust; many such tracks were photographed by the Lunar Orbiter, and one such boulder was sampled on Apollo 17. Otherwise, the Moon must look very much like it has for millennia.

Scientists were disappointed in their hope that Apollo would yield the secret of the Moon's origin.

Before 1969, it was thought that the Moon might have been knocked out of the primeval Earth by the impact of another body or that it might have formed about the same time as Earth out of similar materials. Neither theory was fully confirmed by the Apollo data, and the question is still unsettled. Newer theories hold that both collision and accretion may have been involved, but more evidence is needed before any theory can be confirmed.

### Context

Humans' first venture to another celestial body remains the supreme achievement of piloted space exploration, unlikely to be surpassed for some time. The Apollo Program provided the driving force for the U.S. space program through the 1960's, carrying the rest of the American effort with it. The sense of adventure inherent in the piloted lunar landing captured the imagination of the American public and the mass-communications media far more than did any scientific dividends that might result from lunar exploration. Unfortunately, it also led inevitably to a decline in public attention once the goal was reached on the first attempt. Try as they might, NASA's leaders could not sustain the momentum accumulated by mid-1969; after Apollo 11, new goals for piloted spaceflight seemed either anticlimactic or too expensive.

As early as 1959, NASA planners had argued for a lunar-landing mission on the grounds that it was an end in itself, requiring no justification in terms of its contribution to some greater goal. What they did not foresee was that the first lunar landing would come to be seen as an end to space exploration and that continuation of the Apollo Program would require more justification than the project had required at first. Much had changed between Kennedy's statement of the goal and NASA's achievement of it. As it became increasingly clear that the Soviet Union was not in the race for the Moon, as racial strife and other domestic problems intensified, and as the nation became more deeply entangled in southeast Asia, American priorities changed and the space program suffered. The administration of Richard M. Nixon, elected president in 1968, chose to concentrate on foreign policy and reduction of government spending, and although piloted spaceflight had many steadfast supporters in Congress, it faced an uncertain future as the Apollo Program entered its final stages. In addition, the administration wanted NASA to switch its priorities in the direction of a space-based defense system. To this end, a space station, serviced by a reusable shuttlecraft, would be placed into Earth orbit. There, astronauts could conduct experiments in orbit, while keeping an eye on covert activity on Earth. Additional funding for these programs would come from the Defense Department in exchange for the use of the shuttle to launch large military payloads.

The cutbacks suffered by the Apollo Program in 1970 were particularly unfortunate for lunar scientists, for only after the first two missions did mission planners begin to cater to their wants. The third flight landed at the site chosen by the scientists as the most interesting. The last three missions were marked by substantial increases in landed payload and returned samples, length of time spent on the Moon, and mobility for the astronauts. There was, as well, a noticeable increase in the willingness of flight planners to allow the scientists a greater say in mission operations, including changes to surface operations plans during the missions if that seemed warranted by scientific considerations.

Science had come late into the Apollo Program, however, and during its formative years, the ultimate capability of the Apollo system was defined without regard for its optimum scientific use. The launch vehicle and the two spacecraft were designed with adequate margins for operational success and safety but with little room for growth—a fact dictated by Kennedy's time limit and the choice of lunar orbit rendezvous for the mission mode (which in turn was largely determined by that limit). The result was a system suited to its narrowly conceived purpose rather than to the needs of scientific exploration. Given these limitations, scientists and engineers exploited the Apollo hardware almost to its limits. In so doing, they made possible a scientific understanding of the Moon that will be refined only, perhaps, by the establishment of a permanent base on the lunar surface.

For all of its magnificent achievements, the Apollo Program proved to be a technological dead end. The Saturn V was too costly to be economical for other applications, and the Apollo spacecraft were too small and too specialized. Their fatal limitation was that they could be used only once. In an era of shrinking budgets and demands for utility, piloted spaceflight had to go in a different direction.

**See also:** Apollo Program: Command and Service Module; Apollo Program: Developmental Flights; Apollo Program: Geological Results; Apollo Program: Lunar Lander Training Vehicles; Apollo Program: Lunar Module; Apollo Program: Orbital and Lunar Surface Experiments; Apollo 1; Apollo 7; Apollo 8; Apollo 9; Apollo 10; Apollo 11; Apollo 12; Apollo 13; Apollo 14; Apollo 15; Apollo 15's Lunar Rover; Apollo 16; Apollo 17; Apollo-Soyuz Test Project; Astronauts and the U.S. Astronaut Program; Escape from Piloted Spacecraft; Lunar Exploration.

### Further Reading

Allday, Jonathan. *Apollo in Perspective: Spaceflight Then and Now.* Bristol, England: Institute of Physics, 2000. This book takes a retrospective look at the Apollo space program and the technology that was used to land a man on the Moon. Allday aims to explain the basic physics and technology of spaceflight, and to convey the huge technological strides that were made and the dedication of the people working on the program.

Beattie, Donald A. *Taking Science to the Moon: Lunar Experiments and the Apollo Program.* Baltimore: Johns Hopkins University Press, 2001. This book describes, from the perspective of NASA Headquarters, the struggles to include science payloads and lunar exploration as part of the Apollo Program. Beattie discusses the experiments and details the decisions, meetings, and NASA infighting that got these important surface experiments on the flights to the Moon.

Benson, Charles D., and William Barnaby Faherty. *Moonport: A History of Apollo Launch Facilities and Operations*. NASA SP-4204. Washington, D.C.: Government Printing Office, 1978. A detailed history of the planning, construction, and operation of the Saturn launch facilities at Kennedy Space Center. A technical history, suitable for college level or professional readers.

Bilstein, Roger E. *Stages to Saturn: A Technological History of the Apollo/Saturn Launch Vehicles*. Gainesville: University Press of Florida, 2003. Starting with the earliest rockets, Bilstein traces the development of the family of massive Saturn launch vehicles that carried the Apollo astronauts to the Moon and boosted Skylab into orbit. *Stages to Saturn* not only tells the important story of the research and development of the Saturn rockets and the people who designed them, but also recounts the stirring exploits of their operations, from orbital missions around Earth testing Apollo equipment to their journeys to the Moon and back.

Brooks, Courtney G., James M. Grimwood, and Loyd S. Swenson, Jr. *Chariots for Apollo*. NASA SP-4205. Washington, D.C.: Government Printing Office, 1979. NASA's official history of the Apollo Program spacecraft projects, this volume also deals with the questions of flight planning, astronaut training, and the Apollo 11 mission. A companion to Bilstein's Saturn history, suitable for the serious general reader.

Cernan, Eugene A., and Don Davis. *The Last Man on the Moon: Astronaut Eugene Cernan and America's Race in Space*. New York: St. Martin's Press, 1999. The story of the last two men to walk on the Moon is told by Cernan, commander of Apollo 17, and Davis, an experienced journalist. This autobiography tells the story behind the story of the Gemini and Apollo Programs. Cernan, whose spaceflight career spanned both programs, narrates it. He was the first person to spacewalk in orbit around the Earth and the last person to leave footprints on the lunar surface.

Chaikin, Andrew. *A Man on the Moon: The Voyages of the Apollo Astronauts*. New York: Penguin Group, 1998. This is a comprehensive look at the Apollo Program, based upon hundreds of hours of in-depth interviews with each of the astronauts who went to the Moon. Chaikin also spoke with hundreds of people who contributed to the success of the piloted lunar landings. Every aspect of each mission is detailed, from preflight preparations to postflight activities. Appendices include astronaut biographical information, a list of the hundreds of persons interviewed, and pertinent data on each of the Apollo piloted flights. There is a bibliography and extensive author's notes. Tom Hanks, who starred in *Apollo 13*, the popular film dramatization of the ill-fated flight, wrote the introduction to this update of the 1994 book.

Collins, Michael. *Carrying the Fire: An Astronaut's Journeys*. New York: Farrar, Straus and Giroux, 1974. A sensitive, well-written autobiography by Apollo 11's Command Module pilot, giving a firsthand account of astronaut selection and training and mission planning. The best of the few books by former astronauts, suitable for any interested reader.

_____. *Liftoff: The Story of America's Adventure in Space*. New York: Grove Press, 1988. Many books have been written about the Apollo Program, most of them about Apollo 11. Few have given us an inside look at the delicate melding of man and machine. Contributing to this complete history of America's piloted space programs, Collins devotes a large portion to Apollo. He sets the record straight on some of the misconceptions of astronauts and space machines. The book is illustrated with eighty-eight line drawings by James Dean, former NASA art director, which add stark realism to an otherwise unreal world.

Cooper, Henry S. F., Jr. *Thirteen: The Apollo Flight That Failed.* Baltimore: Johns Hopkins University Press, 1995. An expanded version of Cooper's articles published in *The New Yorker* magazine, this is the best available description of the ill-fated Apollo 13 mission. Written by an experienced and capable space journalist.

Harland, David M. *Exploring the Moon: The Apollo Expeditions.* Chichester, England: Springer-Praxis, 1999. *Exploring the Moon* focuses on the exploration carried out by the Apollo astronauts while on the lunar surface, not on the technology of getting there. It is a story of the great adventure of exploring the Moon, and combines the words of the astronauts themselves with the photographs they took. *Exploring the Moon* is a lunar travelogue, a minute-by-minute account of what the astronauts did, said, and felt, enhanced by their subsequent reflections.

Kelly, Thomas J. *Moon Lander: How We Developed the Apollo Lunar Module.* Washington, D.C.: Smithsonian Books, 2001. Grumman Chief Engineer Kelly gives a firsthand account of designing, building, testing, and flying the Apollo Lunar Module. It was, he writes, "an aerospace engineer's dream job of the century." Kelly's account begins with the imaginative process of sketching solutions to a host of technical challenges with an emphasis on safety, reliability, and maintainability. He catalogs numerous test failures, including propulsion-system leaks, ascent-engine instability, stress corrosion of the aluminum alloy parts, and battery problems, as well as their fixes under the ever-present constraints of budget and schedule. He also recaptures the anticipation of the first unmanned Lunar Module flight with Apollo 5 in 1968, the exhilaration of hearing Apollo 11's Neil A. Armstrong report that "The *Eagle* has landed," and the pride of having inadvertently provided a vital "lifeboat" for the crew of the disabled Apollo 13.

Kraft, Christopher C., Jr. *Flight: My Life in Mission Control.* East Rutherford, N.J.: Penguin Putnam, 2002. Kraft gives an account of his life in Mission Control. The first NASA flight director would play an integral role in what would become the National Aeronautics and Space Administration.

Lambright, W. Henry. *Powering Apollo: James E. Webb of NASA.* Baltimore: Johns Hopkins University Press, 2000. Lambright explores James E. Webb's leadership role in NASA's spectacular success.

Liebergot, Sy, and David Harland. *Apollo EECOM: Journey of a Lifetime.* Burlington, Ont.: Apogee Books, 2003. This is the life story of Sy Liebergot, former NASA flight controller, with emphasis on his years working in Apollo Mission Control.

Light, Michael, and Andrew Chaikin. *Full Moon.* New York: Alfred A. Knopf, 1999. This coffee table book is filled with high-resolution images from the Apollo Program missions. The authors obtained the original master negatives of photographs taken by Apollo astronauts and scanned them electronically. The results are breathtaking. The pictures trace an entire flight from liftoff to splashdown, using the best images from each flight. The beauty of the color and black-and-white photographs is unmarred by captions. Instead, the authors placed thumbnail versions of the pictures with a brief description in the back of the book.

McDougall, Walter A. *The Heavens and the Earth: A Political History of the Space Age.* 2d ed. Baltimore: Johns Hopkins University Press, 1997. A wide-ranging treatment of the global politics that propelled the United States into the Space Race and the consequences of the technocratic bent of Kennedy, Johnson, Webb, and others. Winner of the Pulitzer Prize in history, this is one of the best books on the geopolitical aspects of space, thoroughly scholarly but extremely readable. College level.

MacKinnon, Douglas, and Joseph Baldanza. *Footprints: The Twelve Men Who Walked on the Moon Reflect on Their Flights, Their Lives, and the Future.* Washington, D.C.: Acropolis Books, 1989. Here is the Apollo Program lunar-landing story as told by the twelve men who walked on the Moon. Learn what it was like to be a part of the Apollo Program and take a stroll in the lunar sunlight. The authors sat down with each of the astronauts and discussed both the flights and their lives before the mission and since. Some of the tales of the behind-the-scenes activities are priceless. Only a few of the questions seem out of place, but it should be left to the reader to decide which, if any, are. The book is illustrated with color paintings by astronaut-artist Alan L. Bean. He has annotated each painting with the ideas he had when he created it.

Murray, Charles, and Catherine Bly Cox. *Apollo: The Race to the Moon.* Burkittsville, Md.: South Mountain Books, 2004. An intriguing look at the people behind the accomplishment, this book portrays those whose unenviable task it was to translate a president's dream into reality. The reader gets insight into the background of the Apollo Program team members, as well as their extracurricular activities. Reference notes, black-and-white photographs.

National Aeronautics and Space Administration. *Apollo Mission Press Kits.* http://www-lib.ksc.nasa.gov/lib/presskits.html. Provides detailed preflight information about each of the Apollo missions, Apollo 6 through Apollo 17. Accessed March, 2005.

Schefter, James L. *The Race: The Uncensored Story of How America Beat Russia to the Moon.* New York: Doubleday Books, 1999. In 1963, a young reporter for *Time-Life* named James Schefter was given a dream job: cover America's race to the Moon. Since the astronauts were under contract to *Life* for their stories, Schefter was given complete access to the biggest players at NASA. But at the time, his primary role was to excite the public about the new, expensive, experimental space program, and he could not write about everything he saw. In *The Race*, he does.

Scott, David, and Alexei Leonov. *Two Sides of the Moon: Our Story of the Cold War Space Race.* London: Thomas Dunne Books, 2004. Astronaut Scott and Cosmonaut Leonov recount their exceptional lives and careers in the context of the Cold War.

Shepard, Alan, and Donald K. "Deke" Slayton, with Jay Barbree and Howard Benedict. *Moon Shot: The Inside Story of America's Race to the Moon.* Atlanta: Turner, 1994. This is, indeed, the inside story of the Apollo Program as told by two men who actively participated in it. Some of their tales appear here for the first time. The book was adapted for a four-hour documentary in 1995.

Siddiqi, Asif A. *The Soviet Space Race with Apollo.* Gainesville: University Press of Florida, 2003.
_____. *Sputnik and the Soviet Space Challenge.* Gainesville: University Press of Florida, 2003. These two volumes represent the first comprehensive history of the Soviet space programs.

Stoff, Joshua, and Charles R. Pellegrino. *Chariots for Apollo: The Untold Story Behind the Race to the Moon.* New York: HarperCollins, 1999. This dramatic chronicle of the race to the

Moon takes us behind the scenes of this awesome quest, into the minds of the people whose lives were devoted to it and changed by it, and through the missions.

Sullivan, Scott P. *Virtual Apollo: A Pictorial Essay of the Engineering and Construction of the Apollo Command and Service Modules.* Burlington, Ont.: Apogee Books, 2002. This book allows the public to become acquainted with the Apollo spacecraft in detail and learn the story of its design and construction. Full-color drawings in exacting detail provide inside and out views of the Command and Service Modules complete with details of construction and fabrication.

_____. *Virtual LM.* Burlington, Ont.: Apogee Books, 2004. The intricacies of the LM design and the details of its manufacture, including some of the major problems that had to be overcome, are detailed in this book. *Virtual LM* shows the details of design and production using full-color renderings of the structures, components, subassemblies and the completed spacecraft, accompanied by supporting descriptions. It shows the Apollo Lunar Module as both an engineering masterpiece and a work of art.

Swanson, Glen E. *Before This Decade Is Out: Personal Reflections on the Apollo Program.* Gainesville: University Press of Florida, 2002. This significant collection of oral histories of the Saturn/Apollo Program recounts the unique adventure of the lunar landing program as witnessed by some of the political leaders, engineers, scientists, and astronauts who made it such a success. It includes recollections from James E. Webb, the NASA administrator whose political connections in Washington extended back to the New Deal of the 1930's; rocket pioneer and architect of the Saturn V rocket Wernher von Braun; the resolute Robert R. Gilruth, director of the Houston center; the engineering iconoclast Maxime A. Faget, whose designs of spacecraft made flights to the Moon possible; and astronauts such as Harrison H. "Jack" Schmitt and Charles M. Duke, Jr.

Turnill, Reginald. *The Moonlandings: An Eyewitness Account.* New York: Cambridge University Press, 2003. An eyewitness account of one of the most thrilling adventures of the twentieth century, by a verteran space journalist.

Ulivi, Paolo, and David M. Harland. *Lunar Exploration: Human Pioneers and Robotic Surveyors.* London: Springer-Verlag London Limited, 2004. A well-paced, rapidly moving, balanced, even-handed account of lunar exploration.

Wendt, Guenter, and Russell Still. *The Unbroken Chain.* Burlington, Ont.: Apogee Books, 2001. Autobiography of the only person who worked side by side with every astronaut bound for space.

*W. David Compton*

# Apollo Program: Command and Service Module

*Date:* 1959 to 1975

*Type of spacecraft:* Lunar exploration, piloted spacecraft

*The Apollo Command and Service Module was designed as the "mother ship" for the Apollo lunar landing program. The Command Module would ferry the three-man crew both to and from the Moon and remain in orbit while two of the crew members carried out their activities on the lunar surface.*

## Summary of the Program

In 1957, after Americans' initial shock at the Soviet success with the first artificial satellite, Sputnik, wore off, the newly created U.S. space agency, the National Aeronautics and Space Administration (NASA), started to plan for piloted spaceflight activities both near and long term. After getting the go-ahead for Project Mercury in October of 1958 as a simple one-person orbiter, NASA realized that a serious project was needed to exploit the knowledge gained from Mercury. In May, 1959, the Space Task Group (STG) recommended a number of advanced goals that would require a long-duration, multicrew vehicle. The most noteworthy recommendation was Project Horizon, which would land humans on the Moon by 1965. In May, 1960, this "advanced spaceflight program" was renamed "Apollo."

In November, 1961, the Space and Information System Division of North American Aviation, Inc., was selected to design and build the Apollo spacecraft. An order was placed for eleven flightworthy spacecraft, eleven mock-ups, and fifteen boilerplates. The boilerplates were low-fidelity models used primarily for testing when a full spacecraft was not needed. Much discussion would follow as to how to get humans to the Moon. The final candidate, called Lunar Orbit Rendezvous (LOR), was announced on July 11, 1962.

The Apollo would be divided into three sec-

tions: The Command Module (CM), which would act as the "lunar bus" taking the crew to and from the Moon; the Service Module (SM), which would house most of the equipment, rockets, and fuel needed for the journey; and the lightweight Lunar Module (LM), which would actually land on the Moon. The LM was originally known as the Lunar Excursion Module, until the "excursion" part of the name was deemed too frivolous and was therefore dropped. In its early life, it was referred to occasionally by NASA as the "bug," by its manufacturers as the LM ("el-em"), but nearly always by the astronauts as the "lem."

The CM very quickly took the form of a ballistic cone, 3.23 meters tall and about 3.9 meters in diameter. It weighed about 5.9 metric tons with the crew. The spacecraft was divided into three sections. The forward compartment contained the recovery equipment, such as the parachutes, mortars, and two small rocket engines. The crew compartment contained the couches, control panels, navigation equipment, and stowage lockers. The aft compartment was a ring-shaped structure that wrapped around the edge of the crew compartment, containing ten small rocket engines for reentry maneuvers, water, fuel, batteries, and other utility equipment.

The recovery equipment in the forward compartment consisted of three main parachutes 25 meters in diameter and three inflatable bags used to right the capsule if it tipped upside-down after landing. Additional utility equipment included dye markers, a beacon, mortars for launching the drogue parachutes, and line cutters to free the chutes after splashdown. The forward heatshield would be jettisoned at about 7.3 kilometers in altitude, followed immediately by deployment of the drogues. At 3 kilometers, the main chutes would be released, slowing the CM to less than 9 meters per second (m/sec) for all three chutes, or 11 m/sec should only two deploy.

Compared to the relatively snug size of the Mercury and Gemini spacecraft, the Apollo offered the crew a generous 5.95 cubic meters in interior volume. The crew of three would sit in collapsible couches, the commander (CDR) on the left, Command Module pilot (CMP) in the center, and Lunar Module pilot (LMP) on the right. Arcing over them was the main display panel, with circuit-breaker panels along the side. Underneath the couches were numerous stowage containers. At the foot of the couches was the lower equipment bay (LEB), which housed the navigational telescopes along with additional lockers, oxygen panels, a water dispenser, and access panels.

The CM heatshield was made out of brazed stainless steel honeycomb material with an outer layer of phenolic epoxy resin as an ablative material. It varied in thickness from 1.25 centimenters at the nose to about 5 centimenters at the rear.

The spacecraft had five windows: two on the side, two forward-looking rendezvous windows, and a large circular hatch window. Each window consisted of two panes, each 6.35 millimeters (mm) thick. The inner pane was made of tempered silica glass, and the outer window was amorphous-fused silicon capable of withstanding temperatures up to 1,500° Celsius.

The CM had two hatches: a large rectilinear side hatch used primarily for entry at liftoff and exit after landing, and a circular forward hatch used to gain access to the LM when the two spacecraft were docked. A docking tunnel up through the apex of the spacecraft provided an easy transfer path.

*The Apollo 1 Command and Service Module being prepared for mating to the Lunar Module Adapter in the Manned Spacecraft Operations Building.* (NASA)

At the top was the docking probe, used to link the CM with the LM a few hours after liftoff. The probe was a complicated and spindly-looking device that would be inserted into the shallow concave drogue of the LM during docking. A set of three small capture latches would lightly clip the two vehicles together. The probe could then be retracted, pulling the LM toward the CM until twelve latches mounted on the docking ring around the probe were able to close (only three were actually needed to ensure a pressure-tight seal). Upon pressurizing the 81-centimenter-diameter tunnel and removing the forward hatch, the probe, the drogue, and the LM hatch could then be opened.

Like the Mercury spacecraft, the Apollo had a Launch Escape System (LES) mounted on the very top. This was designed to pull the CM away from the launch vehicle in case of an emergency in the early stages of the launch. The LES measured 10 meters long and weighed about 3.6 metric tons. It carried three rockets: a pitch control motor, a tower jettison motor, and the main launch escape motor. In the case of an emergency within the first 100 seconds of flight, the launch escape motor would ignite, providing nearly 654 kilonewtons of thrust for about 14 seconds (the horsepower equivalent of 4,300 automobiles). During a normal launch, the LES would be jettisoned at an altitude of about 90 kilometers.

Guidance and navigation were accomplished with inertial, optical, and computer subsystems. The inertial system measured changes in the attitude and velocity of the spacecraft. The information was then sent to the Apollo Guidance Computer (AGC) for display and analysis. Periodic navigational fixes were made using the two telescopes in the lower equipment bay and by sightings of bright stars, the Earth, and the Moon. Midcourse corrections would then be made if needed.

The AGC was the first computer to use integrated circuits. (In fact, the first three computers accounted for the majority of the world's output of circuits, with each one consuming more than 4,000 units.) The AGC operated at a clock speed of about 1 megahertz and a memory of 72 kilobytes, of which 64 kilobytes consisted of rope-core memory, "programmed" by the placement of discrete premagnitzed ferrite cores, each representing a single bit storing software specific to the mission.

For most of the mission, the CM was attached to the cylindrical Service Module. The SM contained the main propulsion system, as well as fuel, oxygen, water supplies, and some experiments for later flights. It measured 7.34 meters long and 3.9 meters across, and weighed 25 metric tons fully fueled, 5.2 tons when empty. At the start of the mission, the SM would hold 7 tons of fuel and 11.4 tons of oxidizer.

Projecting out of the bottom of the SM was the large, bell-shaped service propulsion engine skirt. The engine, a part of the Service Propulsion System (SPS), provided the thrust needed to do the major trajectory changes: insertion into and escape out of lunar orbit, midcourse corrections, high-altitude launch aborts, and retrograde reentry burns for Earth-orbit flights. The SPS produced a thrust up to 97.86 kilonewtons and was designed to be as simple and reliable as possible. During the Apollo Program, the SPS performed flawlessly.

The Service Module was divided into six sectors of three different sizes. Sector 1 was reserved for later flights and used for high-resolution mapping cameras and a small subsatellite deployed into lunar orbit. Sectors 2, 3, 5, and 6 held the SPS fuel tanks. Sector 4 supported most of Apollo's electrical power subsystems, including the three fuel cells and supporting consumables. It was in this section that Apollo 13's oxygen tank was located when it exploded, blowing the entire side panel off the spacecraft.

Located around the sides of the SM are four "quads," clusters of four small rockets, used for the Reaction Control System (RCS). The RCS thrusters (each with a thrust of 441 newtons) were used to change the orientation of Apollo and for very small trajectory changes.

The CM came in two basic configurations: Block I and Block II. The Block I vehicle was the first design, made for Earth-orbit missions. The main difference was that it had none of the complicated docking equipment. In the end, only a couple of Block I spacecraft were scheduled to be flown in lieu of the more advanced Block II. However, it was a Block I unit behind the Apollo 1 pad fire that killed the crew of the first piloted mission. As a result, Block I spacecraft were deemed too dangerous, and none ever flew piloted.

The Soviets developed the durable Soyuz spacecraft in response to Apollo. It was to serve many of the same purposes, including possible lunar missions. Unlike Apollo, Soyuz used solar panels for power instead of fuel cells. It was made of two compartments instead of one, and it would land on land, instead of in the ocean.

The first Apollo flight occurred on February 26, 1966, with a 36-minute suborbital mission to test the overall performance of the complete rocket stack and how the Apollo CM would stand reentry. On January 27, 1967, the first planned Apollo (Apollo 1) crew perished in a fire, setting back the program by twenty months. The first piloted mission took place in October of 1968 with the Earth-orbiting flight of Apollo 7. Apollo 8 was the first piloted mission to the Moon; Apollo 9, the first to have the entire lunar vehicle in space; and in July, 1969, Apollo 11's CM, *Columbia*, would ferry Neil Armstrong and his crew to the first lunar landing. The only significant failure of a Block II spacecraft system was Apollo 13, when an oxygen tank exploded only two days into the mission in the Service Module. Apollo 14's SM would be outfitted with a backup oxygen tank and batteries in the previously empty section 1. Eleven piloted Command Modules flew in the Apollo lunar program from 1968 to 1972. However, that would not end the CM's usefulness. It would later be used as a crew transfer spacecraft for the three Skylab missions in 1974. The final CM flew in 1975 as the American entry into the Apollo-Soyuz Test Project, a joint mission between the United States and the Soviet Union.

### Contributions

The Apollo Command Module brought NASA valuable experience in operation in deep space. Important navigation techniques were pioneered, not to mention extensive in situ lunar mapping operations. After the tragedy of the Apollo 1 fire, many important safety requirements were implemented, leading to a dramatic lessening of risk. The Apollo Guidance Computer (AGC) pioneered many advanced techniques now in common use in personal computers and acted as a clear technology driver that greatly advanced digital computing in general.

### Context

The Command Module was the third and most advanced spacecraft built by NASA, tailored specifically for the demanding task of going to the Moon and returning safely back to Earth. Without it, a lunar landing would have been impossible as neither the Gemini nor the Mercury spacecraft had the capabilities in guidance or crew size to accomplish the mission. Nor, years later, would the space shuttle be able to climb higher than a few hundred miles. The versatile and elegant Apollo design is still referenced in literature promoting simple replacement spacecraft for the much more complicated and costly space shuttle.

It was not the first three-man spacecraft to fly. The Soviets' Voskhod 1 flew a three-man crew in 1964 on a risky one-day mission. (Voskhod was a converted single-pilot Vostok spacecraft rushed out the door due to delays in the Soyuz development.) However, when completed, the Soyuz would be the Soviet "Apollo," serving many of the same goals.

**See also:** Apollo Program; Apollo Program: Developmental Flights; Apollo Program: Lunar Lander Training Vehicles; Apollo Program: Lunar Module; Apollo 7; Apollo 8; Apollo 9; Apollo 10; Apollo 11; Apollo 12; Apollo 13; Apollo 14; Apollo 15; Apollo 16; Apollo 17; Apollo-Soyuz Test Project; Escape from Piloted Spacecraft.

### Further Reading

Murray, Charles, and Catherine B. Cox. *Apollo: The Race to the Moon*. 1989. Rev. ed. Burkittsville, Md.: South Mountain Books, 2004. An updated edition of an account of the Apollo Program from an engineer's perspective. Includes a CD-ROM of audio and images.

National Aeronautics and Space Administration. *The Apollo Spacecraft: A Chronology*. NASA SP-4009. 4 vols. Washington, D.C.: Scientific and Technical Information Division, Office of Technology Utilization, National Aeronautics and Space Administration, 1978. A detailed chronological listing of all important Apollo events, including management meetings, tests, decisions and missions. The four volumes are broken chronologically:

volume 1, through November 7, 1962 (edited by Ivan D. Ertel and Mary Louise Morse); volume 2, November 8, 1962-September 30, 1964 (edited by Morse and Jean Kernahan Bays); volume 3, October 1, 1964-January 20, 1966 (edited by Courtney G. Brooks and Ertel); and volume 4, January 21, 1966-July 13, 1974 (edited by Ertel and Roland W. Newkirk with Brooks). Includes bibliographical references, illustrations. Although the four-volume set has been out of print for a number of years, it is available online at http://www.hq.nasa.gov/office/pao/History/SP-4009/cover.htm.

National Aeronautics and Space Administration, Apollo Project Office. *Apollo Operations Handbook.* 2 vols. Washington, D.C.: Author, n.d. The bible of the Apollo spacecraft, used by astronauts and engineers. It details all of the systems, controls, and procedures. Original copies are nearly impossible to acquire, but many rare Apollo documents are now available on the Web. Go to the official NASA technical archives Web site at http://ntrs.nasa.gov/archive/nasa/casi.ntrs.nasa.gov/19730061045_1973061045.pdf for the Block I manual.

Sullivan, Scott P. *Virtual Apollo: A Pictorial Essay of the Engineering and Construction of the Apollo Command and Service Modules.* Burlington, Ont.: Apogee Books, 2002. Sullivan, a mechanical designer and artist, uses sophisticated design software to present every aspect of the spacecraft through a series of highly detailed color engineering renditions.

*Michael Smithwick*

# Apollo Program: Developmental Flights

*Date:* August 21, 1959, to April 5, 1968
*Type of program:* Lunar exploration, piloted spaceflight

*During the initial phases of the Apollo Program, space scientists developed increasingly powerful launch vehicles, including Little Joe II, Saturn I, Saturn IB, and Saturn V. They also tested escape mechanisms, should the boosters fail in flight, and the susceptibility of spacecraft to meteoroids. All these developments ultimately made possible several successful piloted missions to the Moon.*

## Key Figures

*James E. Webb* (1906-1992), NASA administrator during the early years of the Apollo Program

*Rocco A. Petrone* (b. 1926), director of launch operations, Kennedy Space Center, during the developmental phase of the Apollo Program

*Wernher von Braun* (1912-1977), director of George C. Marshall Space Flight Center and later of the John F. Kennedy Space Center

*Samuel C. Phillips* (1921-1990), director of the Apollo Program Office, 1964-1969

## Summary of the Program

On May 25, 1961, President John F. Kennedy, addressing a joint session of Congress, committed the United States to putting Americans on the Moon and returning them safely to Earth:

> I believe that this nation should commit itself to achieving the goal, before this decade is out, of landing a man on the Moon and returning him safely to Earth.

The U.S. space program seemed far from this goal, having only recently put the first American into space for just a few minutes. Even before Kennedy's speech, however, scientists and officials at the National Aeronautics and Space Administration (NASA) had begun planning for missions that included circumlunar flights and piloted lunar landings. However, much needed to be done, since neither powerful rockets nor sophisticated spacecraft yet existed for such ventures. Because President Kennedy insisted on the safety of the astro-

nauts, a primary concern of the Apollo Program was the integrity and reliability of space technologies. This concern was behind the robotic flights of the developmental phase of the Apollo Program.

Even before NASA formally elaborated plans for piloted lunar missions, progress toward this goal was already taking place. Having dramatic examples that American rockets could explode on the launch pad, scientists and engineers realized that an escape mechanism for the astronauts was essential. The purpose of the "beach abort" program was to develop a reliable escape system to rescue crew members in case of an accident before or during the launch. Beginning in 1959 as part of the Mercury program and using a "Little Joe" rocket as a booster, tests began on abort maneuvers and escape technologies. A cluster of small rockets would lift the vehicle to its desired test altitude. The first cross-section drawings of the vehicle showed four holes up and the project was dubbed "Little Joe," from the crap-game throw of a double deuce on the dice. In one of these tests a Little Joe booster from Wallops Island, Virginia, in a spacecraft topped by an escape tower, launched a monkey. The test was a success, and the monkey returned safely to Earth.

In order to loft the heavy Apollo spacecraft and its escape system, a very powerful booster was needed, and in the early 1960's NASA scientists developed Little Joe Senior, later known as Little Joe II. This booster would be capable of accelerating an Apollo spacecraft to speeds similar to those it would experience during the launch on a trip to the Moon. During 1963 and 1964, several tests were performed using a robotic Apollo Command Module (CM) and the Launch Escape System (LES) atop a Little Joe II booster. The first Little Joe II "qualification test vehicle" was launched from White Sands Missile Range in New Mexico on August 28, 1963, and it successfully accomplished its objective of gathering data on temperatures within and outside the spacecraft and pressures on it. The first Little Joe II to fly with an Apollo CM was launched on May 13, 1964. On December 8, 1964, a Little Joe II launch vehicle successfully demonstrated the Apollo's escape system.

While these robotic tests were going on, the Mercury and Gemini piloted space programs were also developing and testing technologies for the Apollo missions. All these programs needed detailed information on how certain materials behave in the very high frictional heat generated when spacecraft reenter the Earth's atmosphere at speeds of 40,250 kilometers per hour. Thus, in 1962, NASA officials announced the Project Fire program. Besides gathering data on heatshield performance, this project would also study radio signal loss during the high-heat phase of reentry. The first test of a Project Fire spacecraft occurred on April 14, 1964. An Atlas D booster topped by a subscale Apollo capsule was launched from Cape Kennedy to an altitude of 122,000 meters. Following separation from the booster, reaching apogee, and the start of reentry, the payload's solid-fueled Antares II rocket fired for 30 seconds, increasing the spacecraft's descent speed to 40,500 kilometers

per hour. The exterior of the spacecraft reached temperatures of 11,400 kelvins (K) before plunging into the Atlantic Ocean. A second Project Fire test on May 22, 1965, produced similar results.

Although satisfactory for these preliminary tests, the Little Joe II and Atlas D boosters were unsatisfactory for a Moon mission. By the late 1950's, Wernher von Braun and others had realized that an extremely powerful rocket would be necessary for the Apollo missions. Von Braun's group had developed the Jupiter rocket for military missiles and early NASA launches, and since Saturn is the next planet beyond Jupiter, this new rocket was called Saturn. Because of the intensified pace of the Apollo Program, several versions of Saturn vehicles were developed concurrently. The Saturn I and the more powerful Saturn IB both used von Braun's idea of the clustering of rockets rather than a single gigantic rocket. The first stage of the Saturn I also used clustered propellant tanks. They consisted of eight Redstone missile tanks: four painted white holding liquid oxygen (LOX) and four painted black holding RP-1 fuel. They were clustered around a central Jupiter rocket tank, containing LOX. The clustered-engine concept was tested on the Saturn I maiden flight, called Saturn-Apollo 1 (SA-1), from Cape Canaveral on October 27, 1961. The 416,250-kilogram launch vehicle, the largest in the world, had water-filled dummy upper stages. The Saturn I used eight modified Thor IRBM/Jupiter IRBM, S-3D, rocket engines. The modified engine was called the H-1 rocket engine. The Block I booster's eight clustered H-1 engines developed 5.8 million newtons of thrust in carrying its payload to an altitude of 137 kilometers. In later launches of the Saturn I, Block II, the thrust was improved to 6.7 million newtons.

One of the significant, but largely unheralded, accomplishments of the Saturn I was putting Pegasus satellites into orbit in 1965. Because its long solar panels were like wings, the satellite was named after the winged horse of ancient Greek mythology. The principal purpose of Pegasus was the detection of micrometeoroids in outer space. Because these tiny objects posed a potential hazard to equipment and crew, it was essential to discover their size, speed, and number. The final three Saturn I Block II vehicles—SA-9, SA-8, and SA-10—launched three Pegasus satellites in February, May, and July of 1965, providing the most information ever gathered on micrometeoroids. However, because the Saturn I was unable to launch the Apollo Command and Service Module (CSM) and the Lunar Module (LM) together, NASA officials canceled the Saturn I program after ten successful missions, and in its place, the Saturn IB program was accelerated.

The Saturn IB was basically an improved Saturn I, with an upgraded first stage. Its first launch was from Cape Kennedy on February 26, 1966, primarily to test the booster and the attached Apollo spacecraft. The flight was designated Apollo-Saturn 201 (AS-201). Apollo flights were given names based upon their launch vehicle, using a three-digit number. Saturn IB flights were indicated by the number "2" followed by a two-digit sequence number.

*The missions of the Apollo Program were designed to explore the features of the Moon and its relationship to Earth and the formation of the solar system. The program's official emblem expressed that quest. (NASA CORE/Lorain County JVS)*

Saturn V flights used "5" instead of "2." The spacecraft reentered Earth's atmosphere at a velocity of 29,000 kilometers per hour, and after the CM was recovered from the Atlantic Ocean, its interior structure and exterior heatshield were found to have performed as expected. Several other Saturn IB launches followed, with no failures. During these trials, the Saturn IB was improved, and extensive tests were performed on the Command Module, the Service Module, the spacecraft-Lunar Module adapter, and the Apollo escape system.

The Saturn V, the biggest of the Saturn family of boosters, had been developed by von Braun's team as NASA's response to the large Soviet boosters that had led to the Soviet Union's amazing series of "firsts" in the "Space Race." The three stacked stages of the Saturn V had a total length of about 111 meters, including spacecraft and launch escape system. In early plans, the first test of the Saturn V was supposed to use a live first stage and dummy upper stages, but time pressure and the successes of the Saturn IB led to an important test on November 9, 1967, in which all Saturn V stages were live. This robotic flight, also known as Apollo 4 (Apollo-Saturn 501), was a resounding success. The Saturn V rockets, which generated 33 meganewtons of thrust, all worked to perfection. After reaching an altitude of 16,100 kilometers, the spacecraft's engines propelled Apollo at an angle into the atmosphere at such high speeds that the heatshield reached a temperature of about 3,000 K.

On January 22, 1968, the Saturn IB launched Apollo 5 (AS-204). Its primary goal was to test the Lunar Module in orbit. The flight had several problems, including a premature rocket-engine shutdown, which ground controllers were able to correct. More worrisome was the second robotic Saturn V flight on Apollo 6 (AS-502), which lifted off the Kennedy launch pad on April 4, 1968. About two minutes into the flight, severe lengthwise oscillations began to shake the launch vehicle. Known as "pogo vibrations," after a pogo stick's motion, these had the potential to damage the rockets and the spacecraft. Also troubling was the failure of two rocket engines and the related shut-

*The unpiloted Apollo 4 Command Module now on display at the Kennedy Space Center was used to test the heat thermal protection system during reentry into Earth's atmosphere.* (NASA)

down of a third. Using the remaining engines, controllers were able to salvage the mission and return the spacecraft safely to Earth. After study, scientists were able to correct the Saturn V's problems. So certain were scientists and NASA officials that they understood the cause and solution of these failures that Apollo 6 became the final robotic Saturn V flight.

### Contributions

Though the Apollo Program was peppered with failures, even the tragic loss of human life, the developmental flights were able to overcome deficiencies and achieve goals at a rate much faster than initially planned. The early beach abort tests established the reliability of the launch boosters, the structural integrity of the spacecraft, and the workability of the escape system. Through knowledge gained in these tests, NASA scientists came to believe that the escape system would function properly even under the most severe conditions. Project Fire provided NASA scientists with knowledge of the Apollo capsule's thermal character-

istics. The knowledge gained about heatshield materials and communications blackouts during reentry proved helpful during the piloted Apollo flights. For example, improved understanding of aerobraking—the use of atmospheric friction to change a spacecraft's orbit—made reentry more of a science and less of a gamble.

Knowledge gained in these preliminary tests was refined during the later Little Joe II missions, which gathered information on the reliability of the launch escape system of the Apollo Command Module. The Little Joe II flights showed that, if things went wrong during liftoff or early in the flight, the Command Module could be safely jettisoned away from the launch rocket. The unpiloted Saturn tests proved that gigantic rockets could successfully launch huge payloads into orbit. The Pegasus satellites produced abundant data about high-energy radiation in space. The sensors in these satellites also collected quantitative data on the sizes, types, frequencies, directions, and penetrability of the many micrometeoroids that hit the spacecraft. These satellites also provided valuable data on the lifetimes of electronic components in the harsh space environment.

While there were some problems with the robotic Saturn flights, none of these failures seriously impaired the overall goals of the program. In sum, the developmental flights of various rockets and robotic spacecraft showed that the Saturn V and the Apollo CSM had an excellent chance to help fulfill the Kennedy-instigated goal of a piloted mission to the Moon.

### Context

Even though certain American politicians and NASA officials sometimes denied that the United States was in a "space race" with the Soviet Union, the competition between these superpowers was an important part of the context of the Apollo Program, especially in its developmental phase. The gauntlet that President Kennedy threw down in 1961 resulted in massive changes in the funding, research, and development of essential space technologies. The heatshields, escape systems, Saturn rockets, and other technologies that had been developed during Apollo's preliminary phases were the *sine qua non* of the program's ultimate success.

**See also:** Apollo Program; Apollo Program: Lunar Module; Apollo 1; Apollo 7; Apollo 8; Apollo 9; Apollo 10; Apollo 11; Apollo 12; Apollo 13; Apollo 14; Apollo 15; Apollo 16; Apollo 17; Apollo-Soyuz Test Project; Launch Vehicles; Mercury Project Development; Saturn Launch Vehicles; Skylab Program; Skylab Space Station; Soyuz Launch Vehicle; Space Shuttle; Space Stations: Origins and Development.

### Further Reading

Allday, Jonathan. *Apollo in Perspective: Spaceflight Then and Now.* Philadelphia: Institute of Physics, 1999. A retrospective look at the Apollo Program and the technology that was used to land humans on the Moon. The author explains the basic physics and technology of spaceflight and conveys the huge technological strides that were made and the dedication of the people working on the program. All major aspects of the Apollo Program are covered, including crews, vehicles, and space suits.

Baker, David. *The History of Piloted Spaceflight.* New York: Crown, 1982. This reprint of a book originally published in 1981 includes a popular account of the Apollo Program.

_____. *The Rocket: The History and Development of Rocket and Missile Technology.* New York: Crown, 1978. This illustrated history contains material on the development of space technologies for the Moon mission. Bibliography and index.

Bilstein, Roger E. *Stages to Saturn: A Technological History of the Apollo/Saturn Launch Vehicles.* Gainesville: University Press of Florida, 2003. Although this history of the development of the Saturn I, IB, and V rockets may be too technical for some general readers, it is an excellent source for those interested in the people, research, problems, and successes in constructing these (and other) launch vehicles.

Chaikin, Andrew. *A Man on the Moon: The Voyages of the Apollo Astronauts.* New York: Penguin Books, 1998. This reprint, which was published in connection with a successful television miniseries, contains a foreword by actor Tom Hanks. Although Chaikin emphasizes the human element in the Apollo story, he also treats the early developmental phases. Bibliography, extensive "author's notes," and an index.

Cortright, Edgar M., ed. *Apollo Missions to the Moon.* NASA SP-350. Washington, D.C.: Government Printing Office, 1975. The early chapters of this book, written by such experts as James E. Webb, Robert Gilruth, and Wernher von Braun, provide firsthand accounts of the development of the Apollo Program. Includes a chronology, "Key Events in Apollo," and an index.

Heppenheimer, T. A. *Countdown: A History of Space Flight.* New York: John Wiley & Sons, 1997. This account, which surveys significant developments from World War II rockets to the space shuttle, has the advantage of access to recently opened Soviet archives and declassified CIA documents. An extensive bibliography and an index.

Murray, Charles, and Catherine Bly Cox. *Apollo: The Race to the Moon.* New York: Simon & Schuster, 1989. Written to commemorate the twentieth anniversary of the Apollo 11 Moon landing, this book emphasizes the people behind the research and development of Apollo spacecraft and those who flew the missions. The material obtained in their many interviews of participants enlivens their account. Notes and an index.

National Aeronautics and Space Administration. *The Apollo Spacecraft: A Chronology.* NASA SP-4009. 4 vols. Washington, D.C.: Scientific and Technical Information Division, Office of Technology Utilization, National Aeronautics and Space Administration, 1978. This illustrated chronology of the early development of the Apollo spacecraft contains a wide variety of information "both directly and indirectly related to the program." The four volumes are broken chronologically: volume 1, through November 7, 1962 (edited by Ivan D. Ertel and Mary Louise Morse); volume 2, November 8, 1962-September 30, 1964 (edited by Morse and Jean Kernahan Bays); volume 3, October 1, 1964-January 20, 1966 (edited by Courtney G. Brooks and Ertel); and volume 4, January 21, 1966-July 13, 1974 (edited by Ertel and Roland W. Newkirk with Brooks). Includes bibliographical references. Although the four-volume set has been out of print for a number of years, it is available online at http://www.hq.nasa.gov/office/pao/History/SP-4009/cover.htm.

World Spaceflight News. *Saturn V: America's Apollo Moon Rocket.* Mount Laurel, N.J.: Author, 2000. This report features more than two hundred images, illustrations, drawings, schematics, tables, and charts. Although its layout leaves something to be desired, the book offers valuable information about Saturn V.

_____. *Saturn V Flight Manual: Astronaut's Guide to the Apollo Moon Rocket.* Mount Laurel, N.J.: Author, 2001. This flight manual (produced for the Apollo 8 mission) was designed for the astronauts and places special emphasis on flight systems, events, and crew interactions. There are no muddy photographs to waste space; every page is packed with details about the launch vehicle, clearly reproduced from an original NASA document.

*Robert J. Paradowski*

# Apollo Program: Geological Results

*Date:* July 20, 1969, to December 19, 1972
*Type of program:* Lunar exploration, piloted spaceflight

*The Apollo Program consisted of six piloted missions to the Moon. It demonstrated that human beings could land on other celestial bodies, function on them, explore them, and then return home safely. The missions collected and returned data and samples that yielded vast amounts of scientific information about the Moon's composition and structure.*

## Summary of the Program

Although the United States launched the Apollo Program to prove its superiority in space, it was also on a quest for knowledge. It was hoped that many questions would be answered during these piloted lunar landings. Geophysicists have long guessed at the internal structure of the Moon and its similarity to Earth. Experiments were designed with the expectation of gaining this specific knowledge. Particularly, scientists wanted to learn whether the Moon had a core and crust like the Earth, what its shape was, and the relationship between Earth-Moon tidal forces and the shape of the Moon. Other experiments were designed to learn if the Moon was seismically active, if it had a magnetic field, and how great the internal heat flow, if any, was.

Because geologists were interested in the lunar surficial features, most crew members were trained in geology. Scientists hoped to find the composition of the lunar rocks in the highlands and maria (seas). They also wanted to learn about volcanism, about the principal processes that made the topography (weathering, transport, and erosion), and whether there was any evidence of tectonic processes. Most laypersons wanted to know if there was any fossil evidence of life on the Moon.

Astronomers also had many questions. They hoped to find records of the history of the Moon and its interaction with the Earth, and how it related to the formation of the solar system. They wanted to learn the age of the Moon and the age of its exposed layers. They also wanted to obtain information concerning the interaction between the Moon and the Earth, the frequency of meteoric impacts, the thermal history, the flux of radiation, and what history the magnetic fields might reveal.

The main mission of Apollo 11 was to collect enough lunar rock to study—approximately 10 kilograms of matter was collected and returned to Earth. The astronauts set up a passive seismometer on the surface, an optical reflector, and a solar wind experiment to measure what type of solar energy reached the lunar surface.

Color television cameras mounted on the Apollo 12-17 descent stages provided live coverage of the explorations for those back on Earth. These missions provided scientists with opportunities to measure the pressure of the lunar atmosphere. Magnometers and seismometers were set up to study the lunar magnetic field and to collect seismic data from the impact of the Lunar Module. Solar wind and atmospheric pressure data were also collected.

The Apollo 12 landing site was near a ray of the crater Copernicus, where core samples were taken. Rock samples contained a variety of material ejected from local craters. The Apollo 14 mission landed on a relatively smooth area of the Moon, the Fra Mauro highlands, which is part of a larger deposit of ejecta (debris ejected from meteor impacts) from Mare Imbrium. It appeared older than the nearby maria (dark areas on the Moon that proved to be lava-filled basins).

Apollo 15 was quite different from the previous missions. Astronauts were able to "travel" for about four hours and could explore five times farther from the Landing Module by using the Lunar Rover—the first electrically powered surface vehicle. The time they stayed on the Moon's surface was twice as long as before. Apollo 15 was also the first mission to use diverse experiments to study the lunar surface from orbit. These experiments were designed primarily to find regional variations in the chemical composition and gravitational and magnetic fields on the Moon. The tests were carried out with x-ray and gamma-ray sensors, a subsatellite with magnetic sensors, and high-resolution cameras that photographed 12.5 percent of the surface.

The Apollo 15 surface studies measured the temperature gradient of the top few meters of the lunar surface. The results of this study (if it is representative of the whole Moon) suggest that the radioactive content of the Moon is substantially higher than that of Earth. The impact of the Saturn IVB stage had been recorded on both the Apollo 12 and 14 seismometers. These seismic data indicated a great difference in the sound-velocity contrasts in the lunar subsurface. The results were so anomalous it had to be concluded that the chemical composition of the Moon in the upper 70 kilometers must be quite varied. The Apollo 15 Lunar Ionosphere Detector was the first to detect water vapor in the lunar exosphere.

Apollo 16 continued the pursuit of the previous studies. The crew added soil mechanics studies that measured porosity, distribution of size, and angularity of the grains of the regolith (loose surface material). New measurements of the electromagnetic spectrum (ultraviolet, x-ray, cosmic-ray, and gamma-ray) were taken along with solar wind measurements to learn more about the Sun, solar flares, and galactic particles. Volcanics, ridges, rilles, arches, and craters were investigated in depth.

Apollo 17 continued to research earlier theories. New experiments tended to concentrate on the subsurface makeup of the Moon. Gravimeter studies were added to determine the thickness of the basalt valley plates, and electrical property experiments were used to help resolve the subsurface question. Heat flow experiments were set up to find the amount of subsurface radioisotopes. More stratigraphic studies of the surface, photographs, and samples were taken. Five mice were flown in a self-contained unit to study the effects of cosmic rays on living tissue, especially the brain. Four of the five mice survived.

These tests and experiments contributed to a huge database of

*The riverlike feature is Hadley Rille. It is 1.5 kilometers wide and 300 meters deep. Rilles are channels in which lava flowed during the eruption of mare basalts.* (NASA CORE/Lorain County JVS)

information on the celestial neighbor closest to Earth and helped to answer some of the questions that perplexed humankind from the beginning.

### Contributions

The Moon is made from rocky material similar to Earth's. The Moon was melted at high temperatures to great depths early in its history of concretion. Evidence shows that the lunar highlands are the remnants of low-density rocks that floated to the surface in a lava sea that covered the Moon to a depth of tens of kilometers. After the cooling of this dark basalt, the surface was bombarded by meteorites that created differentiation between the highlands and basins. The lunar basalt flows were then bombarded with more impacts, which created the huge craters called seas or maria. These were later filled by more dark-colored lava flows 3.2 to 3.9 billion years ago, giving them the look of seas from the Earth (hence the name *mare*, which is Latin for "sea"). Meteoric impacts, subsequently, made the topography of basin maria and mountain ranges that are currently visible. The large dark basins, such as Mare Imbrium, are, therefore, gigantic impact craters formed early in lunar history. Lunar volcanism occurred mostly as lava floods and lava fountains. Despite some similarity, the crystalline igneous rocks are differentiated from their Earth cousins and from meteorites.

Orbital studies proved that "mascon" basins (large "mass concentrations" that lie beneath the surface of many large lunar basins) were several kilometers lower than the surrounding topography. Gravitational anomalies in these basins demonstrated that they are filled to a depth of 10 kilometers with material that exceeds the average density of the crust. The astronauts brought back large samples from the highlands that are high in aluminum and silicon, which proved that the rocks beneath the lunar highlands are less dense than the rocks beneath the lunar marc, confirming the origin of the surface features.

All Moon rocks originated through high-temperature processes and can be subdivided into three types: anorthosites (light, low-density, feldspar-rich rocks) found in the highlands; basalts (volcanic or igneous, dark-colored lavas or magmas) that fill the maria or mare basins; and breccias (composite rocks made with cemented unweathered fragments). Evidence supports the premise that shock cementation or lithification due to meteoric impacts formed the breccias. Small glass-lined pits on the surfaces of some of the samples were an unexpected find. These glass vesicles were formed by common shock melting and impact metamorphism. An alternate theory suggested that these vesicles originated within the last hundred thousand years, when the Sun had a super-flare. This flare heated the lunar surface to a temperature that, due to a focusing effect, caused material inside the craters to melt, leaving the other surface material intact.

*This sample of rock and mineral fragments all mixed together is called a "breccia." It was collected in the lunar highlands, where the rocks have been modified by meteorite impact.* (NASA CORE/Lorain County JVS)

The surface of the Moon is covered by broken rock and rubble up to several meters thick that was formed by continuous bombardment by meteorites. This layer of debris, called the lunar regolith, makes it hard to find actual rock outcrops. The deeper regolith has been more highly impacted, which shows that the area is older. The regolith has inherent, unique records of the radiation history of the Sun that will help scientists learn more about climate changes on Earth. Craters, on the other hand, are the windows to the interior of the Moon and can be "used" like drill holes. Rock samples were taken with the knowledge that a single sample could not be representative of the entire region. It was also difficult to determine if a sample came from a specific region or if it was exotic (from far away). A hypothesis presented was that larger and lighter-colored rocks had a less complicated history and were closer to their origin. The ejecta patterns could then help to identify the source of the rock. Ejecta near the rim of the craters is probably from the deepest part of the crater, and material farthest away is probably from near the surface.

Erosion has occurred on the lunar surface. The older the crater, the more eroded it is due to impacts. Some lunar rocks found were rounded and some appeared sandblasted, also due to impact. It appears that no surface water has ever been present since none of the rocks were hydrated and there was no evidence of water erosion or sorting. No sedimentary rocks such as limestone or sandstone were found on the Moon. Less dust was found than expected. Surface material at the Apollo 11 landing site in Mare Tranquilitatis was unsorted fragmental debris ranging in size from 1 meter to dust. Older rock, which is bombarded more, is smaller in size than new fragments. Many rocks were rounded on top and angular where buried. Even though this layer of debris was about 5 meters thick, this lunar soil was not much different from its terrestrial cousin in mechanical behavior and composition. Extremely low amounts of organic material were found in the rock samples, and there was no evidence of biologic material. No fossil evidence has ever been found.

Four types of samples brought back were all chemically very similar, yet different in percentage composition from terrestrial rock. These samples contained from five to 10 percent titanium and also contained more zirconium (Zr), yttrium (Y), and chromium (Cr) than would be expected in comparable basalts. The measured samples were low in the alkali and volatile elements, such as sodium (Na), potassium (K), rubidium (Rb), lead (Pb), and bismuth (Bi). Surprisingly, the level of potassium found is similar to that found in chondritic meteorites. The rare-gas content of these samples was also measured by mass spectrometry and the amounts were surprisingly high. Scientists concluded that the measured rare gases were from the solar wind. Elements found in iron meteorites (nickel, cobalt, platinum) were highly depleted or not found at all. Quartz (silicon dioxide) was completely missing on the Moon.

The regolith was quite thin (a few centimeters) at the Apollo 12 landing site, which proved that it was much younger than the Apollo 11 site. Scientists were elated that the astronauts brought back material that was examined by Surveyor. In the thirty months that Surveyor was on the Moon, it did not show any accumulated dust, and photographs showed no dust on the rocks. Surface creep might be a long-term evolutionary process. The fine rocks collected at this site contained less rubidium (Rb), zirconium (Zr), uranium (U), and thorium (Th) and more iron (Fe), magnesium (Mg), and nickel (Ni) than those at the Apollo 11 landing site. Again, the chemistry of these two sites is similar. These rocks show the same characteristics of behavior as terrestrial rocks that cool and form fractional crystallization.

The Apollo 15 orbital tests were somewhat surprising. The landing site was bounded on three sides by the Apennine Mountains and on the fourth by the Hadley Rille, one of the freshest-appearing, largest, and most sinuous rilles on the Moon. Photographs of this region show a wide variety of materials and structures. The Apennine Mountains have parallel, continuous linear bands that show cross-bedding and some fracturing and three separate lines resembling "high lava" marks. The rille has exposed bedrock on the upper 15 percent that appears to be a result of fracturing in the mare crust. Rocks in these highlands contain more than 25 percent by weight aluminum oxides and plagioclase. The studies of the Hadley Rille region provided clear evidence that the area is underlain by a series of lava flows similar to those in the Pacific Northwest.

Rays from younger impact craters had concentrations of rocky angular pieces. Ejecta was scattered in irregular patterns, and rocks were anorthosite- or plagioclase-rich. A simple explanation of the enriched plagioclase is that it was intruded by a trace-element-rich liquid after its formation. The rock densities increased smoothly with depth at the Apollo 16 landing site and were erratic at the Apollo 15 site. The upper 60 kilometers of the lunar crust in the highlands proved to be rich in anorthosite.

There are two accepted theories of the lunar origin: a piece broke from the Earth or the Earth gravitationally captured the Moon. There is no doubt that the Moon and the Earth are related. The compositions of the two are very similar, except that the Moon has highly depleted iron and volatile elements. The Moon, it now seems, was not "captured" by the Earth but instead broke away from the Earth early in its formation with the "help" of an object the size of Mars.

The idea that the Moon is round was disproven early. Scientists found the lunar crust thicker on its far side while most of the volcanic basins and mascons are on the near side. Therefore, the center of mass is displaced toward the Earth by several kilometers, giving the Moon a slight bulge. The tidal pull of the Earth may also have an effect on the shape of the Moon.

Though no seismic signals similar to terrestrial earthquake waves were observed for the Moon, there appeared to be block faulting. This could mean that there is little present seismic activity or that the Moon's interior does not transmit waves readily. So-called moonquakes appear to be at the boundary of the crust and a homogeneous lithosphere at 1 kilometer. The signal amplitude of the Lunar Module-ascent stage increased slowly and then decreased slowly. The signal continued for approximately an hour. Subsequent studies suggested that the Moon "rings" like a bell.

Almost all data about the atmosphere were as expected, except for variable spectra at a lunar sunrise. This showed that the lunar atmosphere contains helium, neon, argon, and some contaminants. The lunar atmospheric pressure is less than $8 \times 10^{-8}$ torr. A permanent magnetic field was recorded and, though small, was higher than expected. Local magnetic anomalies seemed to show that the magnetic fields of the Moon were the result of the magnetized surface rock and not of an internal field. This may be attributable to natural magnetism in the rocks. Other magnetic samples showed that the Moon cooled in the presence of a strong magnetic field, which means that either the Moon once generated an internal magnetic field or it was once located near a strong external field. The highlands are much less radioactive than anticipated, suggesting that mare basalts have a higher radioactivity than highland materials.

Because lunar materials tend to darken with age, the albedo (ability to reflect light) can be used to determine the length of time that ejecta have been exposed at the surface and their comparative ages. Using the albedo model, the formation of the regolith at the Apollo 12 site appears to be younger and to have formed when the Moon was no longer growing appreciably by accretion. It is possible that the Apollo 12 landing site is one billion years younger than the Apollo 11 site possibly because of prolonged filling of the mare by basalts. The surfaces of maria proved older than expected; K-Ar dating set the age of the crystalline rocks at about 3.2 billion years to nearly 4.6 billion years. Because the oldest terrestrial rocks are only 3.5 billion years, the Moon has to be as old or older than the Earth.

While evidence of the first beginnings of the Earth is no longer visible, the Moon preserved its early history with its highly visible impact craters and ejecta. Moon rocks in the maria and in the highlands have been almost undisturbed since

*Astronauts found a pyroclastic deposit on the Moon at one of the landing sites. The soil is composed of numerous droplets of glass that formed by fire fountaining.* (NASA CORE/Lorain County JVS)

their formation. The Apollo Program has answered many questions and posed many more. Despite geologists' successes, definitive answers still elude them.

### Context

Even before Galileo first gazed at the Moon with a telescope in 1609, humankind questioned its origin. Folklore and science fiction flourished about the largest moon of the inner planets and the second brightest object in the night sky. In the three hundred years that preceded the Apollo lunar landings, humans were able to measure the size, distance, and gravity of the Moon. The Moon was mapped and theories proliferated about the maria and highlands.

In 1959 the first uncrewed lunar probes were launched. The Ranger spacecraft followed and were the first successful American probes. Ranger 7 sent back 4,316 excellent photographs of the Mare Cognitum before it crashed into the Moon on July 31,

1964. Rangers 8 and 9 sent back exceptional pictures of Maria Tranquilitatis and Alphonsus, respectively. By 1966-1968 the Lunar Orbiters had been placed in close orbit to map and choose crewed landing sites.

The Surveyor Program was designed for soft landings, and the first was launched on May 30, 1966. It sent back 11,000 pictures. Surveyor 7, the last of the series, sent back 21,000 photographs and some scientific data. This probe measured the temperature of the lunar surface. Surveyor also bounced radar waves off the surface so that the electrical properties and much of the chemical composition could be determined. The next logical step was a crewed spaceflight.

The Apollo Program subsequently established the new science of lunar geology. The astronauts brought back a total of 386 kilograms (850 pounds) of Moon rocks and two core samples to study. Many scientific experiments were accomplished on the lunar surface as well. Apollo 12 landed near the Surveyor site and returned with several pieces from the Surveyor 3 spacecraft. The Apollo series finally gave scientists a firsthand look at the Moon, giving rise to new theories of its origin and the origin of the solar system.

To continue the exploration, the United States launched Clementine and mapped the surface of the Moon. Clementine was then scheduled to rendezvous with a comet but failed in its attempt.

In March, 1998, mission scientists from the Lunar Prospector announced their first tentative findings of the presence of water ice in shadowed craters near the Moon's south and north poles. They estimated that up to six billion metric tons of water ice might be buried in these craters under about 46 centimeters of soil. The theory was based on the levels of hydrogen detected, which in water ice is conditional. While scientists watched, using sensitive spectrometers tuned to look for the ultraviolet emission lines expected from the hydroxyl ion, the Lunar Prospector crashed into the Moon. The dust and particulates kicked up with the impact of the 354-pound spacecraft did not reveal any evidence of water.

The Lunar Prospector did confirm that the Moon has a small core, giving credibility to the theory that the Moon formed when a Mars-sized body struck the Earth early in its formation and split off a chunk that formed the Moon. The size of the lunar core was debated during the Apollo missions. The Earth's core is about 30 percent of its mass, while the lunar core is only about 3 percent of its mass. Analysis of the Prospector data will help refine estimates of the amounts of gold, iridium, and platinum that are concentrated with metallic iron in the lunar rocks. These data may continue to support the theory that the Earth and Moon share a common origin. A dedicated seismic study with two or more seismic penetraters is needed to refine this theory, and a new project is anticipated during the first decade of the twenty-first century. Lunar geology will surely receive an added boost if the proposed Bush Moon-Mars Initiative is carried out. That expansion of human exploration beyond low-Earth orbit calls for a return to the Moon between 2015 and 2020. Robotic missions would have to precede the return of astronauts, and in the process return valuable geological data; the first proposed one being the Lunar Reconnaissance Orbiter, presently scheduled for launch in 2008.

**See also:** Apollo Program; Apollo Program: Orbital and Lunar Surface Experiments; Apollo 8; Apollo 11; Apollo 12; Apollo 14; Apollo 15; Apollo 16; Apollo 17; Clementine Mission to the Moon; Lunar Exploration; Mercury Project.

## Further Reading

Beattie, Donald A. *Taking Science to the Moon: Lunar Experiments and the Apollo Program.* Baltimore: Johns Hopkins University Press, 2001. Discusses the struggles that led to the inclusion of science payloads and goals in the Apollo Program. Provides a detailed insider's story of the Apollo Lunar Surface Experiments Package, and the scientific results.

Brush, Stephen G. *Fruitful Encounters: The Origin of the Solar System and of the Moon from the Chamberlin to Apollo.* New York: Cambridge University Press, 1996. This book traces the theories of the solar system from the early twentieth century to the present. The lunar rock samples brought back by Apollo suggested a theory that the Moon was created by a collision between Earth and a Mars-sized planet and contains material from both. The three books in this series present a survey of different theories of the origin of the solar system, the Earth, and the Moon. Using the data from the Apollo Program, the author discusses the theories of the origin of the Moon and the one most accepted currently, the impact theory.

Harland, David M. *Exploring the Moon: The Apollo Expeditions.* Chichester, England: Springer-Praxis, 1999. Focuses on the exploration carried out by the Apollo astronauts while on the lunar surface and not on the technology of getting there. It is a story of the great adventure of exploring the Moon, and combines the words of the astronauts themselves with the photographs they took. *Exploring the Moon* is a lunar travelogue, a minute-by-minute account of what the astronauts did, said, and felt, enhanced by their subsequent reflections.

Hartmann, William K. *Moons and Planets.* 5th ed. Belmont, Calif.: Thomson Brooks/Cole, 2005. Provides detailed information about all objects in the solar system. Suitable on three separate levels: high school student, general reader, and the college undergraduate studying planetary geology.

Moore, Patrick. *The Moon*. New York: Rand McNally, 1981. The history of lunar observations and lunar probes from 1958 to 1980. The author discusses the lunar orbit, eclipses, tides, and features of the Moon in detail. Good background on the effects of the Moon on the Earth. Lunar features such as the maria, the highlands, and craters are discussed in an easy-to-follow manner. Excellent color plates, maps, and photographs of both the near and far sides of the Moon.

Morrison, David, and Tobias Owen. *The Planetary System*. 3d ed. San Francisco: Addison Wesley, 2003. Organized by planetary object, this work provides contemporary data on all planetary bodies visited by spacecraft since the early days of the Space Age. Suitable for high school and college students and for the general reader.

Spudis, Paul D. *The Geology of Multi-Ring Impact Basins: The Moon and Other Planets*. London: Cambridge University Press, 2005. Discusses the formation of huge basins by large impacting bodies and their evolution. Comparisons are made between lunar features and similar structures on other bodies such as Mercury and moons in the outer solar system.

Wilhelms, Don E. *To a Rocky Moon: A Geologist's History of Lunar Exploration*. Tucson: University of Arizona Press, 1993. A detailed scientific history of lunar exploration from the early Pioneer probes through the final crewed Apollo lunar landing.

*Judith Belsky Farrin*

# Apollo Program: Lunar Lander Training Vehicles

*Date:* 1960 to 1971

*Type of program:* Lunar exploration, piloted spaceflight

*When NASA undertook the challenge set forth by President John F. Kennedy to put an American on the Moon before 1970, the need for an inflight lunar landing trainer became apparent. New aircraft had to be designed to simulate, in the Earth's atmosphere and gravity, the landing characteristics of a not-yet-designed lunar landing craft.*

## Key Figures

*Warren J. North*, chairman of the LLRV Coordination Panel

*John Ryken*, Bell Aerosystems LLRV project manager

*Hubert Drake*, credited with originating the idea of an LLRV

*Donald Bellman*, LLRV project manager

*Gene Matranga*, senior engineer on the LLRV project at the Flight Research Center

## Summary of the Program

How does one design an aircraft to simulate the handling characteristics of a spacecraft? Engineers at the National Aeronautics and Space Administration (NASA) Flight Research Center (FRC, now the Dryden Flight Research Center) in California were confronted with this very question in 1960. What later became known as the Apollo Program was in its infancy and plans to land a piloted spacecraft on the Moon were still in the conceptual phase. In order to study and analyze piloting techniques needed to fly and land the tiny Apollo Lunar Module (LM) in the Moon's airless environment, a vehicle was designed to simulate the handling characteristics of the lunar craft.

Having only a wisp of an atmosphere, the Moon cannot support a winged aircraft. For a lunar flying craft to function at all, movement—pitch, yaw, roll, plus descent and ascent—would have to come from a complex propulsion system. In addition, the pull of lunar gravity is only one-sixth that of the Earth. For training purposes, an ordinary

airplane would never be able to simulate a lunar landing.

Apollo Program planners developed three concepts to address the need for a device that would simulate the anticipated handling characteristics of a lunar vehicle: an electronic simulator, a tethered device, and the ambitious FRC contribution, a free-flying vehicle. All three became serious projects, but eventually the FRC's Lunar Landing Research Vehicle (LLRV) became the most significant one. Hubert Drake is credited with originating the idea, while Donald Bellman and Gene Matranga were senior engineers on the project, with Bellman the project manager.

The LLRV, as well as the production-version Lunar Landing Test Vehicle (LLTV), was built of tubular steel like a giant four-legged bedstead. In order to simulate a lunar landing profile, the LLRV (nicknamed the "flying bedstead") had a jet engine with 19,000 newtons of thrust. The engine got the vehicle up to the test altitude of about 1,220 meters and was then throttled back to support five-sixths of the vehicle's weight, simulating the reduced gravity of the Moon. It could hover, fly horizontally, and land at the velocity of a Lunar Module under the influence of the Moon's gravity. Two lift rockets with thrust that could be varied from 450 to 2,200 newtons handled the vehicle's rate of descent. Sixteen smaller rockets, mounted in pairs, gave the pilot control in pitch, yaw, and roll. As safety backups, six 2,220-newton rockets could take over lift function and stabilize the craft for a moment if the main jet engine failed. The pilot had a zero-zero (zero velocity, zero altitude) ejection seat that would then pull him away to safety.

NASA held conceptual planning meetings with engineers from Bell Aerosystems, Buffalo, New York, and issued them a $50,000 study contract in December of 1961. Bell, a company with experience in vertical takeoff and landing (VTOL) aircraft, had independently conceived a similar, free-flying simulator. Out of this study came the NASA Headquarters' endorsement of the LLRV concept, resulting in a $3.6 million production contract awarded to Bell for delivery of the first of two vehicles for flight studies at the FRC within fourteen months. The contract required the vehicles to take off and land on their own power, reach an altitude of 1,200 meters, hover, move horizontally, and remain off the ground for fourteen minutes.

Bell Aerosystems was a subsidiary of the Bell Aircraft Corporation, a U.S. aircraft manufacturer that had built several types of fighter aircraft for World War II. Perhaps the most famous was the X-1, the first supersonic aircraft. Bell also developed the Reaction Control System (RCS) for the Pro-

ject Mercury spacecraft and the Lunar Module Ascent Engine. One of Bell's more daring ventures was the highly successful and unique Rocket Belt. The initial Small Rocket Lift Device (SRLD), or Rocket Belt, was a minimum strap-on backpack system intended to lift a human in controlled flight over a short distance.

The two LLRVs were delivered to FRC in April, 1964. Immediately, handling evaluation tests on a tilt table constructed at the FRC began. These tests permitted the assessment of the engines without actually flying. After several tests, LLRV number 1 was shipped to Edwards Air Force Base in the California desert to begin flight tests. On the day of the first flight, October 30, 1964, research pilot Joe Walker flew it three times for a total of just under 60

*A Lunar Landing Research Vehicle at the Lunar Landing Research Facility, Langley Research Center, in 1963.* (NASA)

seconds to a peak altitude of 3 meters. Later flights were shared between Walker, Don Mallick (another Dryden pilot), Jack Kleuver of the U.S. Army, and two pilots from NASA's Manned Spacecraft Center in Houston, Texas: Joseph Algranti and H. E. "Bud" Ream.

By April, 1966, the LLRV had performed more than one hundred successful flights and NASA had accumulated enough data from the LLRV flight program at the FRC to give Bell a contract to deliver three LLTVs at a cost of $2.5 million each. The LLTV was similar to the LLRV, but the pilot had a three-axis side control stick and a more restrictive cockpit view, both features of the real Lunar Module. The precise fly-by-wire system simulated the motions and control system response Apollo astronauts would later face while nearing the Moon's surface in the LMs.

After testing at the FRC, the LLRVs were sent to Houston, where research pilots learned to become LLTV instructor pilots. In December of 1966, vehicle No. 1 was shipped to Houston, followed by No. 2 in January, 1967, within weeks of its first flight. In December, 1967, the first of the LLTVs joined the LLRVs, eventually making up the five-vehicle training and simulator fleet.

In all, NASA built five LM trainers of this type. During training flights at Ellington Air Force Base near Houston, Texas, three of the five vehicles were destroyed in crashes. Neil Armstrong was flying LLRV-1 on May 6, 1968, when it went out of control. The helium pressurization system for the steering jets on the LLRV failed, leaving Armstrong no means of controlling the vehicle. He safely ejected before the LLRV became a pile of crumpled metal. Two of the LLTVs were lost in

*Test-firing of the first Lunar Landing Research Vehicle at Edwards Air Force Base in 1964. Note the location of the test pilot on the right.* (NASA)

crashes on December 8, 1968 (piloted by Joe Algranti) and January 29, 1971 (piloted by Stuart Present). The pilots also ejected safely from the crashing LLTVs. The two accidents in 1968, before the first lunar landing, did not deter Apollo program managers, who enthusiastically relied on the vehicles for simulation and training.

**Contributions**

The flights of the Lunar Landing Research Vehicle represented the first flight tests of a vehicle with control characteristics similar to the LM. Successfully flown by experienced test pilots, the LLRV gave NASA the confidence to permit astronauts to practice lunar landings in the Lunar Landing Training Vehicles. It also gave these test pilots the skills to train astronauts to fly the LLTV. In turn, the LLTV gave the astronaut a sense of how a lunar landing would feel under actual flight conditions. By canceling out five-sixths of the Earth's gravity, the LLTV allowed the astronaut to experience the effect of their input to the jet-propelled control system. This allowed them to develop the skills to land a Lunar Module safely during their one opportunity to do so 360,000 kilometers from home.

All Apollo mission commanders and their back-ups flew many hours in the LLTVs before their Apollo flights. Nearly all of the Apollo astronauts offered high praise for the experience—and confidence—they gained from their LLTV flight time. The worth of the LLRV-LLTV program was realized during the final moments before Apollo 11 astronauts Neil Armstrong and Edwin "Buzz" Aldrin completed the first Moon landing in the LM *Eagle*. As the two men were getting close to the Moon's surface, Armstrong saw that they were nearing a rocky area. He disregarded the LM's automatic landing system and switched to manual control during the last moments of descent. Armstrong landed the LM on a safer, more suitable spot. He later said his practice flights in the LLTVs had given him the confidence to override the automatic flight control system and control *Eagle* manually during that epic Apollo 11 mission. Donald "Deke" Slayton, then NASA's astronaut chief, said there was no other way to simulate a Moon landing except by flying the LLTV.

### Context

One of the greatest technological feats in history was accomplished on July 20, 1969, when two Americans in a spiderlike craft swooped down over the Moon and gently touched down on the Sea of Tranquility. Electronic simulators, using movies and animation, gave the astronauts in training an idea of how the landing would look. However, it took an equally ungainly looking vehicle, the LLTV, to give the "feel" for the approach and landing.

The Soviet Union had a plan to send humans to the Moon. They had the cosmonauts and they had the launch vehicles and spacecraft to take them there. There is no evidence, however, they had a vehicle even remotely similar to the LLRV and the LLTV. It would be impossible to tell whether cosmonauts would have had the skill to land their vehicle on the Moon.

When the space shuttle was being developed, it was determined that a flying simulator would be of vital importance to the success of the program. The shuttle orbiter was designed to land on a runway following a long glide through the atmosphere. The Shuttle Training Aircraft, a modified Grumman Gulfstream II, was configured to resemble the cockpit of the orbiter. In addition, modifications were made to the external features of the craft in order to simulate the feel of the orbiter during descent.

LLTV A1, one of the two original LLRVs, was returned to Dryden, where visitors can see it. LLTV B3, the last of the three training vehicles built, is on public display at the Johnson Space Center in Houston, Texas.

**See also:** Apollo Program; Apollo Program: Lunar Module; Apollo 15's Lunar Rover; Apollo 17; Lunar Exploration; Mariner 8 and 9; Surveyor Program.

### Further Reading

Brooks, Courtney G., James M. Grimwood, and Loyd S. Swenson, Jr. NASA SP-4205. *Chariots for Apollo: A History of Manned Lunar Spacecraft*. Washington, D.C.: Scientific and Technical Information Division, Office of Technology Utilization, National Aeronautics and Space Administration, 1979. Still one of the best accounts of the Apollo Program. Although out of print for a number of years, the text is available online at http://www.hq.nasa.gov/office/pao/History/SP-4205/cover.html. Both the LLRV and the LLTV are discussed, including their contributions to the lunar landing missions.

Darling, David. *The Complete Book of Spaceflight: From Apollo 1 to Zero Gravity.* 4th ed. Hoboken, N.J.: John Wiley & Sons, 2002. The author provides in-depth coverage of all key piloted and robotic missions and space vehicles—past, present, and projected—along with clear explanations of the technologies involved. Darling briefly describes the Apollo hardware, including the LLRV and LLTV. A good desktop reference.

Ertel, Ivan D., and Roland W. Newkirk, with Courtney G. Brooks, eds. *The Apollo Spacecraft: A Chronology.* Vol. 4. Washington, D.C.: Scientific and Technical Information Division, Office of Technology Utilization, National Aeronautics and Space Administration, 1978. The fourth of four volumes detailing the creation and development of the Apollo Command and Service Modules and the Lunar Module. This part covers the period from January 21, 1966, through July 13, 1974. The LLRV and LLTV are chronicled in this day-by-day accounting of the Apollo Program. Although the four-volume set has been out of print for a number of years, it is available online at http://www.hq.nasa.gov/office/pao/History/SP-4009/cover.htm.

Stoff, Joshua, and Charles R. Pellegrino. *Chariots for Apollo: The Untold Story Behind the Race to the Moon.* New York: HarperCollins, 1999. One of the more detailed books on the development of the Lunar Module from Grumman's point of view, including some coverage of the LLRV and LLTV.

*Russell R. Tobias*

# Apollo Program: Lunar Module

*Date:* 1959 to 1975

*Type of spacecraft:* Lunar exploration, piloted spacecraft

*The Apollo Lunar Module would take two astronauts down to the lunar surface while leaving one crew member in lunar orbit with the Command Module return vehicle.*

## Summary of the Program

In 1957, the United States was reeling from the Soviets' Space Race victory in placing the first artificial satellite, Sputnik 1, in orbit around Earth. The newly created U.S. space agency, the National Aeronautics and Space Administration (NASA), started to plan for piloted spaceflight and began Project Mercury in October of 1958. After succeeding in the first step, to develop a simple one-person orbiter, NASA realized that it would need to plan a project to exploit the knowledge gained from Mercury. In May, 1959, the Space Task Group (STG) recommended a number of advanced goals that would require a long-duration, multiperson vehicle. The most noteworthy recommendation was Project Horizon, which would land humans on the Moon by 1965. In May of 1960, this "advanced spaceflight program" would be renamed Apollo. Much discussion would follow as to how to accomplish this goal. The final candidate, called Lunar Orbit Rendezvous (LOR), was announced on July 11, 1962.

Apollo would be divided into three sections: The Command Module (CM) would take the crew to and from the Moon, the Service Module (SM) would house most of the equipment and fuel for the journey, and the lightweight Lunar Module or LM (pronounced "lem" from NASA's early name for it, the Lunar Excursion Module) would actually land on the Moon.

Unlike the massive and hulking vehicles frequently depicted in science-fiction films, the LM would be a small and fragile craft. It would be the first "true" spacecraft, never able to fly in an atmosphere, as it never needed to operate there. In November, 1962, Grumman Aircraft was selected as the surprise prime contractor of the LM above eight other much larger candidates. Grumman had already submitted proposals for other parts of the Apollo Program and conducted their own studies on the LOR techniques.

As specified by NASA, the LM would consist of two main sections: the descent stage and the ascent stage. The descent stage would have the larger engine needed to take the entire stack safely down to the lunar surface. It would also contain the landing gear and experiment payload. The ascent stage would comprise the crew cabin, life support, a smaller engine for lunar liftoff, and equipment to handle docking with the Command Module in orbit around the Moon.

The descent stage was 3.2 meters high and 4.2 meters wide. Its fueled mass ranged from 10,246 kilograms for Apollo 11 to 11,463 kilograms for Apollo 17.

The ascent stage was 3.76 meters high and 4.0 by 4.3 meters wide. Its fueled mass ranged from 4,819 kilograms (Apollo 11) to 4,985 kilograms (Apollo 17). The ascent stage's structure was made of a fusion-welded and mechanically fastened assembly of aluminum alloy. It was then covered by hand-crinkled thermal blankets consisting of multiple layers (at least 25 layers) of aluminized sheet (H-film or mylar). Each polymide composite layer was only 0.00381 milimeter (mm) thick and coated on one side with a micromillimeter thickness of aluminum.

On top of the LM were antennae for radar and communications. On each corner was an "RCS quad," consisting of four Reaction Control System (RCS) engines used for attitude and small translational maneuvers. Each engine could produce 445 newtons of thrust. The top had a circular hatch and docking tunnel used for crew transfer from the Command Module. The larger square front hatch was the exit door out to the porch, mounted at the top of one of four sets of landing gear on the descent stage. Each leg ended in an aluminum-honeycomb footpad about 1 meter in diameter. Two triangular windows provided the pilots with large, sweeping views forward and downward.

The descent stage had three compartments for experiments. On the left side of the ladder was the Modularized Equipment Stowage Assembly (MESA), which housed the lunar rock boxes and collection tools, television camera, and other assorted items. On the right side would be the Lunar Rover, used on the final three landing missions. Opposite the MESA in the back, the Apollo Lunar Surface Experiments Package (ALSEP) would be kept.

The Descent Propulsion System (DPS, pronounced "dips") was a gimbaled, throttleable engine with a thrust ranging from 4,500 newtons to 45,000 newtons. It could easily be restarted if needed for emergency use (as was the case during Apollo 13's return to Earth after an inflight accident). The smaller Ascent Propulsion System (APS, pronounced "apps") had a thrust of about 15,000 newtons. Both engines needed to be as reliable as possible with a minimum of failure modes, yet very lightweight. A hypergolic design was selected because it required the fewest moving parts and used a fuel mixture called Aerozine-50 and an

*Apollo 9's Lunar Module* Spider *in a photograph taken by astronaut David Scott from the Command and Service Module* Gumdrop. *The picture shows the ascent stage before docking with the CSM.* (NASA David Scott)

oxidizer called nitrogen tetroxide. Hypergolic propellants ignite upon contact and require no ignition source. Aerozine-50 is a mixture of 50 percent hydrazine and 50 percent unsymmetrical dimethyl hydrazine. The Aerozine-50/nitrogen tetroxide combination was the same used by the Gemini Titan II launch vehicle.

The crew cabin had a volume of nearly 7 cubic meters. The front portion was 2.33 meters in diameter and just over 1 meter deep. It contained the main control panels, computer display, attitude and altitude indicators, emergency controls, windows, cameras, and a docking-assist telescope. The commander stood on the left and the pilot on the right. The midsection in the aft part of the cabin contained the main storage areas, where the lunar extravehicular activity suits, water dispensers, food, film, reference documentation, and other assorted items could be found.

At liftoff atop the Saturn V, the LM would be stowed under the Command and Service Module (CSM) in the Spacecraft Lunar Module Adapter (SLA). Shortly after translunar injection, the CSM pulled away from the adapter, after which the four SLA upper panels were explosively separated from the base, exposing the LM. The CSM rotated 180°, docked with the LM, and pulled it out of its cocoon. Since it served no inflight purposes, the LM was virtually ignored, short of a quick checkout, until lunar orbit was reached.

On landing day, the two astronauts climbed on board about 2 hours before undocking for a final checkout about 4.5 hours before touchdown. The actual powered descent took about twelve minutes. The commander flew the LM during the final part of the landing phase, taking over from the computer if needed, while the Lunar Module pilot (LMP) monitored systems, read out data to Mission Control, and prepared for any emergency situations. On touchdown, the crew would keep the LM in a launch configuration for an emergency liftoff during the course of the lunar stay.

The crew exited through the front hatch, the commander going first, with the LMP leaving some forty-five minutes later. The exit order was dictated

by the inward-swinging hatch and not by any protocol. During a typical mission, the two men would gather a few quick samples around the landing site in case they had to leave early, erect the United States flag, and deploy the experiments package and the optional Lunar Rover. Any further EVAs would explore the region beyond the immediate landing site. At liftoff, four explosive bolts would release the two stages, with the descent stage now serving as the launch platform. Orbit was achieved some seven minutes later. After docking and the transfer of crew and samples, the now useless ascent stage was released and sent either into solar orbit or, in later flights, back to crash on the lunar surface as a seismological experiment.

Grumman's initial contract called for the construction of nine ground test vehicles and eleven flight units. The LM finally flew on January 22, 1968, in the Apollo 5 mission. The goal was to test the propulsion systems, restart operations, LM staging, and structural tests without a crew on board. LM-2 was a backup for Apollo 5 but was never needed and is now on display in the Smithsonian Institution's Air and Space Museum. The first piloted flight was originally scheduled to be on Apollo 8, but delays forced the LM-3 mission to wait until Apollo 9, in March of 1969. That mission demonstrated nearly the entire flight profile except for the actual landing, while in Earth orbit, including abort scenarios and a test of the lunar space suit out on the front porch.

Apollo 10 in May of 1969 repeated the previous mission, but this time in lunar orbit. The only serious problem occurred during staging, when an incorrectly set switch caused the ascent stage to gyrate wildly for a few frightening seconds. The next mission, Apollo 11, took LM-5, *Eagle*, down to the lunar surface for the first Moon landing. Neil Armstrong and Buzz Aldrin spent a total of 21 hours and 36 minutes on the surface with a single, short extravehicular activity (EVA), or "moonwalk."

Four months later, Pete Conrad and Alan Bean would take LM-6, *Intrepid*, for a pinpoint landing in the Ocean of Storms. The touchdown was within walking distance, 183 meters, of the Surveyor 3 spacecraft, which had touched down there two and one-half years earlier. They completed two EVAs with a total lunar stay time of 31 hours, 31 minutes.

Apollo 13's crew was saved thanks to LM-7. With the Service Module incapacitated after an oxygen tank exploded, the Lunar Module *Aquarius* had to serve as a lifeboat. Grumman engineers had to extend the vehicle's capabilities well beyond their designed limits on power, heating, and navigation. Some of the abort scenarios tested on Apollo 9 now came into use, as the crew needed the LM's descent engine to readjust their trajectory to get home before running out of supplies and oxygen. *Aquarius* was released shortly before reentry and any parts that survived crashed south of New Zealand.

Apollo 14's LM-8 landed in the Fra Mauro region of the Moon on February 5, 1972, following the original Apollo 13 time line. Two EVAs were completed and total lunar stay was just over 33 hours. This ended the "H" missions, making way for the long-term "J" missions starting with Apollo 15. However, LM-9 was configured only to be used for the shorter H flights. Therefore, it was retired to be a museum display at the Saturn V Center at the Kennedy Space Center, making way for the more advanced LM-10 and her descendants.

For longer-duration missions and heavier payloads, LM-10 needed refinements in the propulsion and guidance systems. In addition to this, the new Lunar Roving Vehicle was now included, attached to the descent stage immediately under the commander's position. Two additional tanks were added for oxygen and water, along with a new battery to double lunar stay time. Apollo 15's *Falcon* landed on July 30, 1971. Jim Irwin and Dave Scott completed three EVAs and a total lunar stay time of 66 hours.

LM-11, *Orion*, landed in the Descartes region of the Moon on April 20, 1972. *Challenger*, LM-12, became the last Lunar Module to fly when it completed the Apollo 17 mission. It landed at Taurus-Littrow on December 11, 1972. The crew spent a total of 75 hours on the Moon and 22 hours, 3 minutes outside the LM.

Budget cuts led to the cancellation of the final two Apollo missions, Apollo 18 and 19. This left two additional vehicles, LM-13 and LM-14, which now serve as museum pieces.

During the time of America's lunar exploration, the Soviets were likewise working on a piloted lunar lander—a closely held secret until the early 1990's. Due to the lower payload capacity of the Soviet N1 Moon rocket, about 70 percent of the Saturn V, the lunar hardware was much smaller. Only two cosmonauts would go on the mission instead of three, and only one would make the trip to the surface aboard the "LK" lander. The LK was just a third the weight of the LM, 5.5 metric tons, and was 5.2 meters tall. Unlike the LM, it had no formal "descent stage." The small spherical cabin sat on top an equally small landing platform with four legs. The LK would deorbit with a larger engine, nicknamed "the crasher," that would be jettisoned about 4 kilometers above the surface. Its own tiny engine would take it all the way down and then serve as the ascent engine as well. There was no docking tunnel, so the crew transfer required a spacewalk. The limited payload capacity would limit the first planned lunar stay to only a few hours. Three actual LKs flew robotic Earth-orbit missions in 1970 and 1971. However, when the piloted lunar program was canceled in 1974 in the wake of difficulties with the launch vehicle, remaining hardware was relegated to museums.

### Contributions

The main goal of the Lunar Module was to provide a way for two astronauts to land on the Moon and get back to lunar orbit. The difficulty of the task led to many new engineering techniques needed to save weight yet produce a vehicle strong enough to withstand the mission. Weight increases threatened the entire program as the LM grew to more than 1.3 kilograms beyond its limit. An extensive review of redundant systems, changing materials used, manufacturing processes, and guidance techniques reduced the weight and made the LM flyable.

The software algorithms needed for the landing phase would find use in other missions, such as the Mars Exploration Rovers in 2004. New techniques were pioneered in lightweight restartable engine design, wiring techniques, and leak prevention in the fluid systems.

## Context

The LM was the key component to the lunar landing. Its success paved the way for NASA to meet the challenge President John F. Kennedy had made years earlier. The Soviet counterpart lander was much smaller and less capable, and while it might have been able to do the job had the N1 launch vehicle succeeded, its victory would have been symbolic and a far less significant scientific triumph.

**See also:** Apollo Program; Apollo Program: Command and Service Module; Apollo Program: Developmental Flights; Apollo Program: Geological Results; Apollo Program: Lunar Lander Training Vehicles; Apollo Program: Orbital and Lunar Surface Experiments; Apollo-Soyuz Test Project; Lunar Exploration; Lunar Orbiters.

## Further Reading

Kelly, Thomas J. *Moon Lander: How We Developed the Apollo Lunar Module.* Washington, D.C.: Smithsonian Institution Press, 2001. The Lunar Module chief engineer tells the "inside story" of the lander's creation.

Murray, Charles, and Catherine B. Cox. *Apollo: The Race to the Moon.* 1989. Rev. ed. Burkittsville, Md.: South Mountain Books, 2004. An updated rerelease of the account of the Apollo Program from an engineer's perspective. Includes a CD-ROM of audio and images.

National Aeronautics and Space Administration. *The Apollo Spacecraft: A Chronology.* NASA SP-4009. 4 vols. Washington, D.C.: Scientific and Technical Information Division, Office of Technology Utilization, National Aeronautics and Space Administration, 1978. A detailed chronological listing of all important Apollo events, including management meetings, tests, decisions and missions. The four volumes are broken chronologically: volume 1, through November 7, 1962 (edited by Ivan D. Ertel and Mary Louise Morse); volume 2, November 8, 1962-September 30, 1964 (edited by Morse and Jean Kernahan Bays); volume 3, October 1, 1964-January 20, 1966 (edited by Courtney G. Brooks and Ertel); and volume 4, January 21, 1966-July 13, 1974 (edited by Ertel and Roland W. Newkirk with Brooks). Includes bibliographical references, illustrations. Although the four-volume set has been out of print for a number of years, it is available online at http://www.hq.nasa.gov/office/pao/History/SP-4009/cover.htm.

Stoff, Joshua. *Building Moonships: The Grumman Lunar Module.* Mount Pleasant, S.C.: Arcadia, 2004. The story of the engineers who built and tested the Lunar Modules. Provides a visual history of the design, construction, and launch of the Lunar Module. Includes many rare photos of Lunar Modules in production at Grumman.

Sullivan, Scott P. *Virtual Apollo: A Pictorial Essay of the Engineering and Construction of the Apollo Command and Service Modules.* Burlington, Ont.: Apogee Books, 2002. Sullivan, a mechanical designer and artist, uses sophisticated design software to present every aspect of the spacecraft though a series of highly detailed color engineering renditions.

_____. *Virtual LM.* Burlington, Ont.: Apogee Books, 2004. Like *Virtual Apollo*, this book shows the details of design and production using full-color renderings of the structures, components, subassemblies, and the completed spacecraft, accompanied by supporting descriptions.

*Michael Smithwick*

# Apollo Program: Orbital and Lunar Surface Experiments

*Date:* May 28, 1964, to December 19, 1972
*Type of program:* Lunar exploration, piloted spaceflight

*The principal goal of the Apollo Program was to put an American on the Moon before the Soviets. Before President John F. Kennedy made his famous "before this decade is out" speech on May 25, 1961, he asked NASA officials if this goal was achievable. He did not inquire into the scientific experiments that could be performed during such a mission, nor did he inquire about the knowledge that could be obtained from the rocks brought back by the astronauts. The scientific community, however, eagerly looked forward to the opportunity to study the Moon at close range.*

## Key Figures

*James E. Webb* (1906-1992), NASA administrator, 1961-1968

*George M. Low* (1926-1984), director of Manned Spaceflight Programs, 1958-1964; deputy director of the Manned Spacecraft Center, 1964-1967; manager of the Apollo Spacecraft Project Office, 1967-1969; and NASA deputy administrator, 1969-1976

*Samuel C. Phillips* (1921-1990), director of Apollo Program Office, 1964-1969

*Christopher C. Kraft, Jr.* (b. 1924), Apollo flight director

## Summary of the Program

In early 1962, the National Aeronautics and Space Administration (NASA) began soliciting scientific experiments for the Apollo flights. Chemists, physicists, geologists, geophysicists, and astrobiologists were brought together for conferences during the summers of 1962 and 1965 to identify three general types of scientific work: astronaut observations, sample collection, and instrument placement to return data to Earth. At a conference during the summer of 1967, NASA established an advisory body of scientists who would assist the Manned Spacecraft Center (Houston, Texas) in planning the scientific work for each mission.

The first lunar scientific package was the Early Apollo Scientific Experiments Package (EASEP) deployed by the Apollo 11 crew. EASEP consisted of a Passive Seismograph Experiment (PSE) and a Laser Ranging Retroreflector (LRR). These experiments, along with the Lunar Dust Detector (LDD) and the Solar Wind Composition (SWC) by Foil Entrapment experiment, were the only ones available at the time of Apollo 11.

EASEP had a central station that relayed data to and from the experiments and routed power to the experiments. The PSE monitored lunar seismic activity and detected meteoroid impacts and free oscillations of the Moon. The LRR is a retroreflector—a device that sends light or other radiation back where it came from regardless of the angle of incidence—built of cubes of fused silica. The LDD experiment measured high-energy radiation damage to solar cells, reduced solar cell output due to dust accumulation, and reflected infrared energy and temperatures for use in computing lunar surface temperatures. The SWC determined the elemental and isotopic composition of the solar wind by measurement of particle entrapment on an exposed aluminum foil sheet. The astronauts deployed the experiment, and the foil was brought back with them to Earth for analysis.

Subsequent missions carried Apollo Lunar Surface Experiments Packages (ALSEPs). The exact makeup of the ALSEP for each mission varied with the location of the landing and the amount of time available for deployment. Each ALSEP included a central processing station and a SNAP-27 radioisotope thermal generator (RTG). The RTG provided power for the experiments using heat generated by the radioactive decay of an artificially produced isotope. Their fuel capsules were kept in a separate cask (mounted outside the Lundar Module's descent stage) for safety.

ALSEP included components of the EASEP, as well as Active Seismic Experiment, Lunar Seismic Profiling Experiment, Charged Particle Lunar Environment Experiment, Cold Cathode Gauge, Cosmic Ray Detector, Far UV Camera/Spectrograph, Lunar Surface Gravimeter, Traverse Gravimeter Experiment, Heat Flow Experiment, Lunar Ejecta and Meteorites Experiment, Lunar Neutron Probe Experiment, Lunar Surface Magnetometer, Lunar Portable Magnetometer, Solar Wind Spectrometer, and Suprathermal Ion Detector Experiment.

The Active Seismic Experiment consisted of a string of three geophones, instruments that profiled the internal structure of the Moon to a depth of approximately 460 meters. Two seismic sources were included: an astronaut-activated thumper device containing twenty-one small explosive initiators, and a rocket grenade launcher that was capable of launching four grenades at known times and distances from the seismometer. High-frequency natural seismic activity was also monitored with the geophones. The Lunar Seismic Profiling Experiment was an extension of the Active Seismic Experiment carried on Apollo 14 and 16. The data were used to determine the internal characteristics of the lunar crust to a depth of several kilometers.

The Charged Particle Lunar Environment Experiment measured the ambient fluxes of charged particles, both electrons and ions. It consisted of a box supported by legs, containing two similar physical charged-particle analyzers oriented in different directions for minimum exposure to the ecliptic path of the Sun. The Cold Cathode Gauge measured the tenuous lunar atmosphere. Only the amount of gas was measured with this unit, not its composition. The Cosmic Ray Detector consisted of a four-panel array of passive particle track detectors to observe cosmic-ray and solar wind nuclei and thermal neutrons; it included metal foils to trap light solar wind gases.

The Far Ultraviolet Camera/Spectrograph was a miniature observatory that acquired imagery and spectra in the far-ultraviolet range (below 160 nanometers). The experiment determined composition and structure of the upper atmosphere of Earth from its spectra and the structure of the geocorona. It obtained direct evidence of intergalactic hydrogen in distant galaxy clusters and spectra and imagery of the solar wind and other gas clouds in the solar system. It detected gases in the lunar atmosphere, including volcanic gases; obtained spectra and colors of external galaxies in the far UV; obtained spectra and colors of stars and nebulae in the Milky Way; and evaluated the lunar surface as a site for future astronomical observatories.

The Lunar Surface Gravimeter made very accurate measurements of the lunar gravity and of its variation with time. It was essentially a sensitive spring balance, which also functioned as a one-axis seismometer. The Traverse Gravimeter Experiment made relative gravity measurements at a number of locations in the Apollo 17 landing area and used these to obtain information about the geological substructure.

The Heat Flow Experiment made temperature and thermal-property measurements in the lunar subsurface in order to determine the rate at which heat flows out of the interior of the Moon. Two slender temperature-sensing probes, placed in predrilled holes in the subsurface, made the measurements. The Lunar Ejecta and Meteorites (LEAM) Experiment detected secondary particles ejected by meteorite impacts on the lunar surface and pri-

*Buzz Aldrin after deploying the Early Apollo Scientific Experiments Package (EASEP). The Passive Seismic Experiment Package (PSEP) is in the foreground, the U.S. flag is in the center background, the lunar surface television camera is in the left background, and the Lunar Module is in the far right background.* (NASA Neil A. Armstrong) Armstrong)

mary micrometeorites themselves. The experiment measured particle speed, radiant direction, particle momentum, and particle kinetic energy.

The Lunar Neutron Probe Experiment measured time-integrated fluxes of thermal neutrons as a function of depth in the regolith. This information has geological relevance to the speed of regolith turnover and for the understanding of radiation protection required for longer human occupancy of the Moon.

The Lunar Surface Magnetometer measured the magnetic field on the lunar surface and determined from these measurements some of the deep-interior electrical properties of the Moon. The Lunar Portable Magnetometer measured the steady magnetic field at different locations at the Apollo 14 and 16 sites.

The Solar Wind Spectrometer measured the charged-particle flux entering a Faraday cup. It compared the solar wind properties at the lunar surface with those measured in space near the Moon. It determined whether there were any subtle effects of the Moon on the solar wind properties and related these to properties of the Moon. It studied the motion of waves or discontinuities in the solar wind and made inferences as to the length, breadth, and structure of the magnetospheric tail of the Earth from continuous measurements made for four or five days around the time of the full Moon.

The Suprathermal Ion Detector Experiment (SIDE) consisted of two positive ion detectors, which provided information on the energy and mass spectra of the positive ions close to the lunar surface that result from solar-ultraviolet or solar wind ionization of gases. It measured the flux and energy spectrum of positive ions in the Earth's magnetotail and magnetosheath during those periods when the Moon passes through the magnetic tail of the Earth, provided data on the plasma interaction between the solar wind and the Moon, and determined a preliminary value for the electric potential of the lunar surface.

Sector 1 of the Apollo Service Module (SM) was usually filled with ballast to maintain the SM's center of gravity. On Apollo 15-17, it housed a Scientific Instrument Module (SIM) for lunar study. The equipment included camera systems, a lunar subsatellite (Apollo 15 and 16), a laser altimeter, an S-band transponder, a lunar sounder (Apollo 17), an ultraviolet spectrometer, and an infrared radiometer.

The main photographic tasks during lunar orbit were performed with cameras in the SIM. There were two photographic packages: the Mapping Camera Subsystem and the Panoramic Camera, which obtained high-quality metric/mapping photographs and high-resolution panoramic photographs in both stereoscopic and monoscopic modes. The subsystem included the metric camera, the stellar camera, and the laser altimeter. The photography collected established a unified lunar reference system, provided photo mapping at scales as large as 1:250,000, and synoptic interpretation of geological relationships and surface material distribution. The 610-meter ITEK Panoramic Camera obtained high-resolution panoramic photographs, in both stereoscopic and monoscopic modes, of the lunar surface. The camera provided 1- to 2-meter-resolution photography from an orbital altitude of 111 kilometers.

Prior to their return to Earth, the Apollo 15 and 16 crews deployed small, 38-kilogram subsatellites into lunar orbit. These subsatellites were identical and performed three experiments that continued the study of the lunar environment. Accurate tracking of the spacecraft from Earth using the S-band transponder allowed details of the Moon's gravity field to be mapped, which provided information about the distribution of mass in the Moon's interior. The strength and orientation of the magnetic field near the Moon were measured with a magnetometer. The density and energy of electrons and protons near the Moon were measured with the Charged Particle Lunar Environment Experiment.

In the Laser Altimeter Experiment (Apollo 15-17), a pulse from a laser was aimed at the lunar surface. The reflection of the pulse from the surface was then observed with a small telescope. The length of time the pulse took to travel from the spacecraft to the Moon and back is related to the height of the spacecraft above the surface of the Moon. Measurements were made roughly every 30 kilometers across the Moon's surface. These measurements are sufficiently accurate to distinguish height variations of 10 meters between adjacent measurement points.

In the S-band Transponder Experiment (Apollo 14-17), the frequency of radio waves transmitted by the spacecraft was accurately measured at Earth and compared with the frequency of the waves as transmitted by the spacecraft. Changes in the frequency of the radio waves occur because the spacecraft is moving, a phenomenon known as the Doppler effect. Measurements of the frequency of the radio waves at the spacecraft and at Earth allowed the spacecraft's velocity to be determined with very high accuracy.

The Apollo Lunar Sounder Experiment on Apollo 17 replaced the subsatellite and used radar to study the Moon's surface and interior. Radar waves with wavelengths between 2 and 60 meters were transmitted through a series of antennae near the back of the Service Module. After the waves were reflected by the Moon, they were received using the same antennae and the data were recorded on film for analysis on Earth. The experiment "saw" into the upper 2 kilometers of the Moon's crust in a manner somewhat analogous to using seismic waves to study the internal structure of the

Moon. This experiment also provided very precise information about the Moon's topography. In addition to studying the Moon, the experiment measured radio emissions from the Milky Way galaxy.

The Ultraviolet Spectrometer Experiment (Apollo 17) measured the density and composition of the tenuous lunar atmosphere by observing how the atmosphere scattered solar ultraviolet radiation. A transient atmosphere was detected after the Lunar Module (LM) completed its descent, but this disappeared within a few hours as the rocket exhaust gases dissipated. During the return back to Earth, the ultraviolet spectrometer studied a number of astronomical targets, including Earth, the Milky Way galaxy, and selected stars.

The Infrared Radiometer Experiment (Apollo 17) measured surface temperatures and cooling rates at night on the Moon by measuring the emission of infrared radiation from the lunar surface. This experiment measured temperatures between 85 and 400 kelvins on different parts of the Moon. Results from this experiment also determined the size of rocky blocks in the ejecta blankets around impact craters to determine how such material was distributed as a function of distance from the crater.

### Contributions

The Apollo lunar experiments program acquired scientific data to aid in determining the internal structure and composition of the Moon and the composition of the lunar atmosphere. New insights into the geology and geophysics of Earth were gained, and the state of the interior of the Moon and genesis of lunar surface features were ascertained.

Before Apollo, the state of the Moon was a subject of almost unlimited speculation. It is now known that the Moon is made of rocky material that has been variously melted, erupted through volcanoes, and crushed by meteorite impacts. The Moon possesses a thick crust (approximately 60 kilometers thick), a fairly uniform lithosphere (60-1,000 kilometers), and a partly liquid asthenosphere (1,000-1,740 kilometers); a small iron core

at the bottom of the asthenosphere is possible but unconfirmed. Some rocks give hints of ancient magnetic fields, although no planetary field exists today.

The extensive record of meteorite craters on the Moon provides a key for unraveling time scales for the geologic evolution of Mercury, Venus, and Mars based on their individual crater records. Before Apollo, the origin of lunar impact craters was not fully understood. Ages of Moon rocks range from about 3.2 to 4.6 billion years. The distinctively similar oxygen isotopic compositions of Moon rocks and Earth rocks show common ancestry. Relative to Earth, the Moon was highly depleted in iron and in volatile elements needed to form atmospheric gases and water. Extensive testing revealed no evidence for life, past or present, among the lunar samples. All Moon rocks originated through high-temperature processes with little or no involvement with water.

Early in its history, the Moon was melted to great depths to form a "magma ocean." The Moon is slightly asymmetrical in bulk form, possibly as a consequence of its evolution under Earth's gravitational influence. Its crust is thicker on the far side, while most volcanic basins—and unusual mass concentrations—occur on the near side. Large mass concentrations ("mascons") lie beneath the surface of many large lunar basins and probably represent thick accumulations of dense lava. Relative to its geometric center, the Moon's center of mass is displaced toward Earth by several kilometers. The surface of the Moon is covered by a rubble pile of rock fragments and dust, called the lunar regolith, that contains a unique radiation history of the Sun, which is of importance to understanding climate changes on Earth.

Seismic data indicate that there are three primary types of activity: deep moonquakes, shallow moonquakes, and meteoroid impacts. Deep moonquakes are repetitive, occurring at fixed locations and at monthly intervals with remarkable regularity. They are clearly correlated with tidal deformation of the Moon. Shallow moonquakes are located on or near the surface of the Moon, leaving a large gap in seismic activity between the zone of the shallow moonquakes and the deep moonquakes. There are no marked regularities in their occurrence. Meteoroid impacts have a distinctive seismic characteristic in contrast to the two types of moonquake characteristics. The meteoroid impacts generate the largest observed signals.

Laser-ranging beams from Earth are aimed at the Laser Ranging Retroreflectors on the Moon and are reflected back to their point of origin. Precise measurements of Earth-Moon distance, motion of the Moon's center of mass, lunar radius, and Earth geophysical information are available. The round-trip travel time of the beam pinpoints the Moon's distance with staggering precision: better than a few centimeters out of 385,000 kilometers, typically.

Photographs taken from lunar orbit provide synoptic views for the study of regional lunar geology. The photographs were used for lunar mapping and geodetic studies and were valuable in training the astronauts for future missions.

### Context

As of 2005, the Laser Ranging Retroreflectors from Apollo 11, 14, and 15 were still being used to increase the accuracy of the Earth-Moon measurements to near-millimeter accuracy. More than thirty years later, they were the only Apollo science experiments still functioning.

In 1994, the Clementine spacecraft systematically mapped the 38 million square kilometers of lunar surface, in eleven colors in the visible and near-infrared parts of the spectrum, over the course of seventy-one days. In addition, the spacecraft took high-resolution and mid-infrared thermal images, and mapped the topography of the Moon with a laser-ranging experiment. It improved our knowledge of the surface gravity field of the Moon through radio tracking and carried a charged-particle telescope to characterize the solar and magnetospheric energetic particle environment.

Lunar Prospector, one of the NASA Discovery Program missions, performed a low-polar-orbit investigation of the Moon, including mapping the surface composition and locating lunar resources, measuring magnetic and gravity fields, and studying outgassing events. In 1998, Lunar Prospector discovered water ice at both of the lunar poles and provided the first operational gravity map of the Moon.

As a first step in implementing the lunar aspect of NASA's Vision for Space Exploration, development of the Lunar Reconnaissance Orbiter (LRO) began. Launch is set as early as 2008 to map the resources of the Moon.

Although we continue to send robotic probes to the Moon, piloted missions have yielded the most information about our celestial companion. Plans for piloted U.S. missions to the Moon have been announced. Unless the U.S. government provides money to develop these missions, the six lunar experimental stations left by Apollo astronauts will forever remind us of the missed opportunities to explore the Moon.

**See also:** Apollo Program; Apollo Program: Command and Service Module; Apollo Program: Geological Results; Apollo Program: Lunar Lander Training Vehicles; Apollo Program: Lunar Module; Apollo 9; Apollo 11; Clementine Mission to the Moon; Lunar Exploration; Lunar Orbiters; Lunar Prospector; Manned Orbiting Laboratory; Materials Processing in Space.

### Further Reading

Beattie, Donald A. *Taking Science to the Moon: Lunar Experiments and the Apollo Program.* Baltimore: Johns Hopkins University Press, 2001. Describes the struggles that took place to include science payloads and lunar exploration as part of the Apollo Program. Beattie—who served at NASA from 1963 to 1973 in several management positions and finally as program manager, Apollo Lunar Surface Experiments—supplies a detailed, insider's view of the events leading up to the acceptance of science activities on all the Apollo missions.

Harland, David M. *Exploring the Moon: The Apollo Expeditions*. Chichester, England: Springer-Praxis, 1999. Focuses on the exploration carried out by the Apollo astronauts while on the lunar surface. Includes a "before and after" discussion of the results of the Apollo Program.

National Aeronautics and Space Administration. *ALSEP Termination Report*. NASA RP-1036. Washington, D.C.: Author, 1979. The complete documentation of all lunar surface experiments deployed during the Apollo Program. Experiments are detailed and results are discussed. Available online at ftp://nssdcftp. gsfc. nasa. gov/miscellaneous/documents/b32116. pdf. Accessed May, 2005.

_____. *Apollo 11 Preliminary Science Report*. NASA SP-214. A detailed report of the Apollo 11 scientific experiments and results. Available online at http://www.hq.nasa .gov/office/pao/History/alsj/a11/a11psr.html. Accessed May, 2005.

_____. *Apollo 12 Preliminary Science Report*. NASA SP-235. A detailed report of the Apollo 12 scientific experiments and results. Available online at http://www.hq.nasa .gov/office/pao/History/alsj/a12/a12psr.html. Accessed May, 2005.

_____. *Apollo 14 Preliminary Science Report*. NASA SP-272. A detailed report of the Apollo 14 scientific experiments and results. Available online at http://www.hq.nasa .gov/office/pao/History/alsj/a14/a14psr.html. Accessed May, 2005.

_____. *Apollo 15 Preliminary Science Report*. NASA SP-289. A detailed report of the Apollo 15 scientific experiments and results. Available online at http://www.hq.nasa .gov/office/pao/History/alsj/a15/a15psr.html. Accessed May, 2005.

_____. *Apollo 16 Preliminary Science Report*. NASA SP-315. A detailed report of the Apollo 16 scientific experiments and results. Available online at http://www.hq.nasa .gov/office/pao/History/alsj/a16/a16psr.html. Accessed May, 2005.

_____. *Apollo 17 Preliminary Science Report*. NASA SP-330. A detailed report of the Apollo 17 scientific experiments and results. Available online at http://www.hq.nasa .gov/office/pao/History/alsj/a17/a17psr.html. Accessed May, 2005.

Ulivi, Paolo, and David M. Harland. *Lunar Exploration: Human Pioneers and Robotic Surveyors*. London: Springer-Verlag London, 2004. Ulivi covers the robotic programs of the 1950's and 1960's, as well as recent lunar exploration. Although it does not cover the piloted exploration, it does provide comparative studies of Ranger, Lunar Orbiter, Surveyor, Clementine, and Lunar Prospector.

*Russell R. Tobias*

# Apollo 1

*Date:* January 27, 1967
*Type of mission:* Piloted spaceflight

*The AS-204 mission (later renamed Apollo 1) was scheduled to be the first piloted test flight of the Apollo-Saturn space vehicle that would eventually take American astronauts to the Moon. On January 27, 1967, during a routine launch pad test sequence, a fire broke out inside the Apollo Command Module, killing three astronauts. The tragedy set the Apollo Program's schedule back more than twenty-one months, but it led to the development of a safer and improved spacecraft.*

## Key Figures

*Virgil I. "Gus" Grissom* (1926-1967), Apollo 1 mission commander
*Edward H. White II* (1930-1967), Apollo 1 senior pilot
*Roger B. Chaffee* (1935-1967), Apollo 1 pilot
*Walter M. Schirra, Jr.* (b. 1923), backup Apollo 1 mission commander
*Donn F. Eisele* (1930-1987), backup Apollo 1 senior pilot
*Walter Cunningham* (b. 1932), backup Apollo 1 pilot
*Maxime A. Faget* (1921-2004), chief, NASA flight systems division
*Charles W. Frick*, Apollo project manager
*Wernher von Braun* (1912-1977), director of Marshall Space Flight Center

## Summary of the Mission

The AS-204 (Apollo 1) mission was to be the first piloted flight of the Apollo-Saturn spacecraft that would eventually take humans to the Moon. The thirteenth flight of the Saturn I rocket had been completed on August 25, 1966, and it had fulfilled all its major mission objectives. Everything appeared to be ready for the ultimate test of placing the Apollo Command Module into Earth orbit. The flight crew consisted of veteran astronauts Command Pilot Virgil I. "Gus" Grissom, Senior Pilot Edward White II, and Pilot Roger Chaffee, a first-time astronaut. Their fourteen-day flight would test the launch operations, ground tracking and control facilities, and finally the performance of the Apollo-Saturn spacecraft. Grissom was determined to get the most out of this mission and hoped to earn a place on the first flight to the Moon.

The Apollo spacecraft measured 11 meters (about 37 feet) long, and when fully fueled it weighed about 27 metric tons. Apollo was considerably larger and its design much more complex than either of the Mercury or Gemini spacecraft. North American Aviation in Downey, California, was the principal manufacturer of the capsules for all three programs. Perhaps the most significant change in the design of the Apollo capsule was its inward-opening hatch and the time required to open it. Even during routine operations in the launch pad area, it required technicians a minimum of ninety seconds to open the Apollo hatch from the outside. Time did not seem to be a concern, because no one really foresaw a need for a quick egress from the spacecraft while it was on the launch pad. In comparison, the Mercury and Gemini capsules opened from the inside out and, if necessary, provided a faster exit for the astronauts. There was a concern, however, about a premature blowing of the hatch, which had happened during Gus Grissom's Mercury flight and which had almost cost him his life.

The Apollo astronauts were all volunteers and knew the risks of flying experimental aircraft. Many astronauts may have had their personal misgivings about the program and a few actually believed there would be some serious accidents, but they were all dedicated to the goal of landing a human being on the Moon. Safety was always a primary concern, but staying on schedule was often the priority.

Prior to the fire there were several technical reports that raised concerns about the design of the Apollo spacecraft and in particular the pure oxy-

gen atmosphere in the cabin. In 1964, Dr. Emmanuel Roth of the Lovelace Foundation for Medical Education and Research prepared a research report for National Aeronautics and Space Administration (NASA), "The Selection of Space-Cabin Atmospheres." In this report, he concluded that combustible materials will burn violently and uncontrollably in a pure oxygen atmosphere. A second 1964 report by Dr. Frank J. Hendel, a staff scientist at North American Aviation, stated:

Pure oxygen at five pounds per square inch of pressure . . . presents a fire hazard which is especially great on the launching pad. . . . Even a small fire creates toxic products of combustion: no fire-fighting methods have yet been developed that can cope with a fire in pure oxygen.

*The crew members of Apollo 1 in their space suits in January, 1967 (from left): Virgil I. "Gus" Grissom, Edward H. White II, and Roger B. Chaffee.* (NASA)

The United States Air Force and Navy had previous experience with oxygen fires and had developed some preventive procedures that could have been used, but NASA simply missed or ignored the warning signs of a potential disaster.

On January 27, 1967, a launch pad simulated countdown was set to take place for the AS-204 crew. The simulation would take the crew from just before liftoff through several hours of flight. The launch vehicle itself was not fueled, but the crew would be in their space suits and breathing pure oxygen to duplicate the conditions they would experience in space. The day was filled with minor problems, but the simulation proceeded as planned. At one point Grissom reported a "foul" odor in the oxygen that was being supplied to his suit. A second problem with the oxygen supply that was never resolved produced a high oxygen flow that triggered the master alarm. The third serious problem, which persisted for more than two hours, was faulty communications between the capsule and various support facilities. This failure in communications forced a hold of the countdown at 5:40 P.M. At 6:31 P.M. they were ready to resume the countdown when the call of "Fire. I smell fire," followed two seconds later by "Fire in the cockpit," came from the capsule. It now became apparent that all the earlier warning signs that had been ignored were proved correct and the astronauts were in serious danger.

The technicians in the blockhouse could not believe their eyes as they saw flames and smoke fill the capsule. As the orders were given to begin emergency procedures, the Command Module ruptured and intense heat and dense smoke drove the rescuers back from the capsule. By the time the rescuers reached the astronauts, they were dead. A medical board later determined that the astronauts had died of carbon monoxide asphyxia, with thermal burns as a contributing cause.

*Apollo 1's Command Module the day after the fire in which astronauts Grissom, White, and Chaffee lost their lives.* (NASA)

### Contributions

The deaths of Grissom, White, and Chaffee caused immediate grief and raised concerns about the future of the space program. In spite of all the mistakes that led to the Apollo 1 fire, NASA was determined to start over and do it right this time. The lives of Grissom, White, and Chaffee were not lost in vain. The tragedy forced a comprehensive investigation into the circumstances surrounding the cabin fire that led to the deaths of the three astronauts.

In the past very little attention was given to the possibility of an on-the-ground emergency, especially during flight preparations and prelaunch simulations. NASA's approach to fire prevention also proved to be wrong. It did not consider the potential hazard of having an overpressurized oxygen-filled cabin in the spacecraft. They felt that minimizing the chances of ignition was the primary concern, but the Apollo 1 fire proved that containing the spread of the fire should be an even more serious concern. The Apollo 1 fire also demonstrated that the existing emergency procedures had to be revised to include the possibility of a spacecraft cabin fire. New safeguards would have to be developed, and the Apollo Command Module would have to be redesigned to meet much tougher standards.

The Apollo 1 investigative board identified six major problems that were directly responsible for the fire in the Command Module:

(1) A sealed cabin, pressurized with pure oxygen, was extremely dangerous.
(2) The distribution of combustible materials easily spread the fire through the cabin.
(3) Wiring that carried the spacecraft's power could easily short out.
(4) Plumbing that carried a combustible and corrosive coolant was prone to leakage.
(5) Inadequate equipment (improper hatch opening) and improper escape procedures.
(6) Ground test procedures were inadequate for rescue or medical assistance.

As a result of the accident investigations. the various committees made recommendations to initiate major design and engineering modifications to the Apollo Command Module immediately. The Apollo 1 fire also proved that it was safer to have a mixture of breathable gases in the cabin than a pure oxygen atmosphere. In addition, the board advised revisions in test planning, test discipline, manufacturing processes and procedures, and quality control. It was recommended that combustible materials be replaced with nonflammable materials wherever possible, and the new designs needed to incorporate fire breaks in order to reduce the possibility of spreading a fire. Those systems used for oxygen or liquid combustibles had to be fire-resistant and tested before installation.

Perhaps the most important change was made to the capsule's hatch. As originally designed, the hatch opened inward and required assistance from the technicians on the outside. This design never anticipated the need for a rapid exit by the astronauts inside. The hatch had to be redesigned so that it could provide a more rapid and easier egress in the case of an emergency. If the hatch had been configured with an outward motion, perhaps one or more of the astronauts could have escaped the burning capsule.

**Context**

Where does the Apollo 1 tragedy fit into the decade-long "Race to the Moon"? In his 1961 speech, President John F. Kennedy challenged the American people to land a man on the Moon by the end of that decade. In many respects it seemed almost impossible. New technologies would have to be developed that required materials and components that had not yet been invented. To accomplish this goal, NASA chose to follow an "evolving technology" approach through a series of flights that comprised the Project Mercury and the Gemini Program. The Apollo Program was the third step on the road to the Moon.

In the six years leading up to the Apollo 1 fire, NASA had many close calls with disaster in the Mercury and Gemini missions, but a man had never been lost. This success rate fed the idea that the new technologies and procedures they were developing were adequate to get the job done in the allotted time. A sense of invincibility seemed to prevail, and sometimes caution was put aside in favor of staying on schedule.

The Apollo 1 fire brought NASA back to reality. NASA would have to develop a new infrastructure and redesign a safer spacecraft. It is unfortunate that the lessons learned by the Apollo 1 fire did not last long after the end of the Apollo Program. Once again, problems arising from scheduling and cost-cutting would infect the space shuttle program, and the inevitable tragedies of *Challenger* and *Columbia* were just waiting to happen.

**See also:** Apollo Program; Apollo Program: Command and Service Module; Apollo Program: Developmental Flights; Apollo 7; Astronauts and the U.S. Astronaut Program; Escape from Piloted Spacecraft; Saturn Launch Vehicles.

**Further Reading**

*Aviation Week and Space Technology.* "Apollo Accident Report." February 6, 1967, 29-35; February 13, 1967, 33-36; February 20, 1967, 22-23. This series of articles discusses the various approaches NASA took to maintain its schedule of landing a man on the Moon by the end of 1969 in spite of the setback from the Apollo 1 fire.

Bergaust, Erik. *Murder on Pad 34.* New York: G. P. Putnam's Sons, 1968. A highly critical account of the investigation into the Apollo 1 accident, based primarily on published reports from Congress and the media. It offers little new insight into the tragedy, yet it does offer a very different viewpoint regarding the circumstances surrounding the Apollo 1 fire and the Apollo Program in general.

Biddle, Wayne. "Two Faces of Catastrophe." *Air and Space/Smithsonian* 5 (August/September, 1990): 46-49. An interesting comparison of the ways NASA handled its investigation into the 1967 Apollo 1 fire and the loss of space shuttle *Challenger* in 1986. His conclusion is that NASA's infrastructure became more fragile and lost its direction after the successful Apollo Program that concluded in 1972.

Brooks, Courtney G., James M. Grimwald, and Loyd S. Swenson, Jr. *Chariots for Apollo: A History of Manned Lunar Spacecraft.* NASA SP-4205. Washington, D.C.: Government Printing Office, 1979. This publication presents, in the section "Tragedy and Recovery 1967," a concise yet comprehensive report of the events leading up to and following the Apollo 1 fire.

Chaikan, Andrew. *A Man on the Moon: The Voyages of the Apollo Astronauts.* New York: Viking Penguin, 1994. This book presents a comprehensive and accurate description of the various Apollo missions, but its particular brilliance comes from Chaikan's humanistic depiction of the people involved in the Apollo Program. This is evident in chapter 1, "Fire in the Cockpit," which specifically deals with the Apollo 1 fire. Chaikan states, "The greatest irony was that Gus Grissom, who had almost drowned after his Mercury mission because of a hatch that opened prematurely, was claimed by a hatch that could not be opened at all."

Kennan, Erlend A., and Edmund H. Harvey, Jr. *Mission to the Moon: A Critical Examination of NASA and the Space Program.* New York: William Morrow, 1969. An overall criticism of NASA management in its apparent lack of perspective and flexibility at all key levels. The book does not balance the criticism with appropriate references to the success of the Apollo Program.

*Paul P. Sipiera*

# Apollo 7

*Date:* October 11 to October 22, 1968
*Type of mission:* Piloted spaceflight

*Apollo 7 was the first piloted flight of the Apollo spacecraft, the vehicle that would be used to carry humans to the Moon. During its eleven-day, 163-orbit mission, the three-person crew tested the spacecraft systems, ground tracking, and support network that would be essential for the success of later lunar flights.*

## Key Figures

*Walter M. Schirra, Jr.* (b. 1923), Apollo 7 mission commander
*Donn F. Eisele* (1930-1987), Apollo 7 senior pilot
*Walter Cunningham* (b. 1932), Apollo 7 pilot
*Thomas P. Stafford* (b. 1930), Apollo 7 backup mission commander
*John W. Young* (b. 1930), Apollo 7 backup senior pilot
*Eugene A. Cernan* (b. 1934), Apollo 7 backup pilot

## Summary of the Mission

Apollo 7 was the first test flight of the spacecraft that would fulfill President John F. Kennedy's pledge to have the United States land a man on the Moon and return him safely to Earth before 1970. Coming as it did after the Apollo 1 fire, the Apollo 7 mission was also an important test of the American ability to overcome adversity and continue with the conquest of space. Prior to Apollo 7, the National Aeronautics and Space Administration (NASA) had used the highly successful one-person Mercury and two-person Gemini spaceflights to develop and master many of the skills necessary for the lunar flights. Among many other diverse procedures, NASA had learned how to put a spacecraft into orbit, how to maneuver it once in orbit, and how to rendezvous and dock piloted spacecraft. The Apollo Program, with its three-person crew, was designed to put those skills to practical use in flying two separate spacecraft, a Command and Service Module (CSM) and a Lunar Module (LM), to the Moon.

In order to meet their 1970 deadline, NASA had to move very rapidly in developing the launch vehi-

cle and spacecraft to be used on the first Apollo flight. Apollo 1 was scheduled for launch on February 27, 1967. Virgil I. "Gus" Grissom, who had flown previously on the second Mercury flight and on Gemini 3, the first Gemini mission, was chosen to command the maiden Apollo mission. Edward H. White II, a veteran of Gemini IV and the first American to walk in space, and Roger B. Chaffee, a rookie, were the senior pilot and pilot, respectively.

Unfortunately, NASA moved too quickly. On January 27, 1967, a flash fire in the spacecraft during a training session at the launch pad took all three astronauts' lives. The board of inquiry into the tragedy found numerous human errors and design defects in the spacecraft. A twenty-one-month delay in the Apollo launch schedule and an extensive redesign of the Apollo spacecraft and ground support systems resulted.

The Apollo 1 backup crew, which subsequently became the Apollo 7 prime crew, consisted of Walter M. Schirra, Jr., commander, Donn F. Eisele, senior pilot, and Walter Cunningham, pilot. The new

Apollo 7 backup crew, Thomas P. Stafford, John W. Young, and Eugene A. Cernan, later flew to the Moon as the prime crew for Apollo 10. Apollo 7's mission essentially duplicated that of Apollo 1. Schirra, Cunningham, and Eisele were to stay in Earth orbit for eleven days, longer than would be necessary for a lunar-landing mission, and perform a variety of tests and experiments to check the hardware and software systems in the Apollo spacecraft.

The spacecraft in which the Apollo 7 crew would fly was unlike any that had been flown before. Whereas the Mercury Capsule was so small that its lone astronaut could do nothing but remain stationary on his couch and the Gemini had much the

*A Saturn IB rocket propelled Apollo 7 into orbit.* (NASA CORE/ Lorain County JVS)

same arrangement for its two-person crew, the Apollo Command Module (CM) had more than 18.6 square meters of usable cabin space and two separate work areas. The Apollo spacecraft was 3.7 meters high and nearly 4 meters in diameter at its base and weighed 5,556 kilograms. The Service Module (SM), which would provide propulsion and power to the CM in space, was nearly 4 meters in diameter and 7 meters long.

While later lunar flights would use the powerful Saturn V rocket, the Earth-orbital Apollo 7 flight, which did not carry a Lunar Module (LM), was launched by a smaller Saturn IB launch vehicle. The Saturn IB was a combination of an S-IB first stage that used eight engines to produce more than 7.3 million newtons of thrust at launch and an S-IVB second stage with a single engine that generated more than 1 million newtons of thrust to place the Apollo Command and Service Module (CSM) into orbit. On lunar missions, the S-IVB stage, the third stage of the Saturn V rocket, would be restarted and used to place the CSM and the LM on a lunar trajectory.

The Apollo 7 spacecraft and Saturn IB rocket were launched from Launch Pad 34 at Cape Canaveral at 15:02:45 Coordinated Universal Time (UTC, or 11:02:45 A.M. eastern daylight time) on October 11, 1968. Once the craft was in a 222-by-280-kilometer (138-by-174-mile) orbit, the crew separated the CSM from the S-IVB and practiced rendezvous and docking maneuvers with the now-spent rocket booster. These were the same procedures that future flights would use to remove the LM from its berth atop the S-IVB. Schirra also moved the CSM close to the booster and performed station-keeping maneuvers, procedures used to keep the two objects together in orbit. Later in the flight, after changing orbits and moving many kilometers from the booster, the astronauts performed orbital maneuvers that again brought the CSM close to the S-IVB to practice the methods to be used by Command Module pilots on lunar-landing missions to overtake and dock with a returning Lunar Module ascent stage.

Much of the Apollo 7 flight plan involved simulating tasks that would be necessary on future lunar missions, including several firings of the large Service Propulsion System (SPS) engine at the rear of the CSM. This engine was used to change orbits around Earth and to practice the smaller mid-course correction burns future crews would use to fly to the Moon. One SPS firing—the longest of the flight, lasting for 66 seconds—boosted the CSM to an orbital height of 433 kilometers above Earth.

Apollo 7 also gave NASA an important opportunity to fine-tune the ground tracking and data-relay systems it had been building in preparation for the coming lunar flights. Because the Apollo Program missions were far more complex than those of Projects Mercury and Gemini, the agency needed to redefine the nature and scope of its contact with piloted spaceflight missions.

While the Apollo 7 mission was as successful as any that had come before it, it was not without its troubles. Shortly after reaching orbit, first Schirra and then the other crew members developed severe head colds, which caused considerable discomfort. Combined with an agenda packed with equipment tests and scientific experiments, this caused a number of emotional exchanges with Mission Control in Houston. During the first day of the mission, when Mission Control asked that Schirra turn on a small black-and-white television camera to broadcast images to Earth, he refused, stating that the demands of the crowded flight plan and his own physical discomfort made such public-

*From left, astronauts Walter Cunningham, Walter Schirra, and Donn Eisele, the crew of Apollo 7, which was launched on October 11, 1968.* (NASA CORE/Lorain County JVS)

relations efforts unnecessary intrusions into the mission. Later, however, the television camera was activated and millions watched as the Apollo 7 crew displayed their sophisticated new spacecraft and the spectacular views of Earth from the Command Module windows. These first live television images from an Apollo spacecraft in space, particularly after the Apollo 1 fire, helped reassure the American public.

On another occasion during the flight, Schirra declined to take orders from Mission Control, stating that from that point on he was going to be the "onboard flight director." Near the end of the mission, Schirra, Eisele, and Cunningham all objected to wearing their pressure-suit helmets during reentry. The new safety procedures after Apollo 1 called for the astronauts to wear their space suits in case the spacecraft should be damaged during reentry and experience an emergency depressurization, but the astronauts were afraid that wearing the helmets would prevent them from tending to their still-severe colds and relieving the pressure in their ears caused by congestion. Thus, the crew reentered without their helmets.

On October 22, 1967, Apollo 7 splashed down in the Atlantic Ocean, less than ten miles from the *Essex*. Upon splashdown, the spacecraft tipped over so that its nose was under water, leaving the astronauts hanging upside down inside. After a brief delay, balloons were inflated at the spacecraft's nose and the craft was returned to an upright position. Recovery crews met the spacecraft within thirty minutes after splashdown.

During the nearly eleven days of its flight, Apollo 7 made 163 orbits of Earth and traveled nearly 7.24 million kilometers. The success of Apollo 7 was one of the major reasons that NASA advanced its mission timetable and decided to launch the next mission, Apollo 8, directly to the Moon.

### Contributions

Apollo 7 was a mission designed to prove the "working science" that was at the heart of the most crucial phase of the American piloted space effort. While later Apollo missions would try to find ways to put space travel to practical use on Earth, Apollo 7 was the first flight to prove that the miles of wires and millions of nuts and bolts that made up the Apollo spacecraft could actually carry humans into space and return them safely to Earth.

This first Apollo mission also showed that the key aspects of a lunar-landing mission were possible with the fledgling Apollo hardware. These milestones included the successful launching of a multistage, multiengine piloted space vehicle;

short- and long-distance rendezvous and station-keeping maneuvers with the S-IVB booster in Earth orbit; the firing of the Service Propulsion System (SPS) several times during the mission to change orbital planes; and the conducting of several vital guidance, navigational, and communications tests of the new spacecraft.

It is also believed by some observers that Schirra helped to redefine the role of the flight crew and the flight crew commander for future Apollo missions. Before the Apollo Program, almost all control for both American and Soviet spaceflights was exercised by the respective Mission Control centers; thus, the crews in space were, in many respects and with a few notable exceptions, merely glorified passengers. Schirra's, Eisele's, and Cunningham's assertions for self-determination gave future flight crews valuable models for playing larger roles in the in-flight decision-making processes. This new relationship with Mission Control, particularly on flights such as Apollo 13 and Skylabs 2, 3, and 4, would prove to be one of the vital differences between success and failure on those missions.

The Apollo 7 crew tested literally every switch and system on the new Apollo spacecraft. This examination gave NASA the best possible insight into the reliability and quality of its new space equipment and the Program that would, in less than a year from its first flight, land humans on the Moon. In short, even though Apollo 7 never left Earth orbit and did not carry a Lunar Module, it was a key dress rehearsal for later Moon landings.

### Context

Apollo 7 was the seventeenth U.S. piloted spaceflight, following the six Mercury and the ten Gemini piloted missions. It was also the first spaceflight by either the Soviet Union or the United States to take place after the twin tragedies of the Apollo 1 fire on January 27, 1967, and the crash of Soyuz 1 on April 23, 1967, which took the life of Soviet cosmonaut Vladimir M. Komarov, the first man to die during a spaceflight. Because of its unique role after these tragedies, Apollo 7 helped show that NASA was capable of overcoming failure and learning from its mistakes.

Aside from building upon the lessons NASA had learned from earlier programs, the Apollo 7 mission played an essential role in demonstrating progress in the American space program. To the public, spaceflight had become almost commonplace after several nearly identical Gemini missions. Apollo 7 was the first milestone flight since Gemini IV's first American spacewalk and Gemini VIII's first docking in space.

Apollo 7 also helped the piloted spaceflight program to demonstrate to the public the value of placing humans in space as opposed to smaller, less complicated robotic satellites. After 1965, the year the space agency's budget began to shrink under increasing pressure from Congress, there was considerable debate within NASA about the benefits of placing humans in orbit when, some believed, machines could do the same jobs without the risk or the expense. Many of the complex procedures practiced by the Apollo 7 and later Apollo crews were beyond the capabilities of the robotic program.

Occurring when it did, on the heels of the Apollo 1 fire and immediately before the first pi-

loted flight around the Moon, Apollo 8, Apollo 7 has received somewhat less attention from historians. It is this very position in the march to the Moon, however, that made the Apollo 7 mission pivotal to later Apollo successes. Had Apollo 7 failed or even been seen as a failure by the American public, the entire piloted space program might have been canceled or seriously impaired. Without the successful test flight of the spacecraft and support systems on Apollo 7, there might not have been an Apollo 8 or any of the lunar-landing missions. Sadly, some say that Apollo 7 was also the beginning of the end of NASA's glory days, its time of highest public support and funding. Once the lunar-landing phase of Apollo became a reality, the space agency's budgets were cut dramatically by Congress and many missions were canceled or scaled back to save money.

**See also:** Apollo Program; Apollo Program: Command and Service Module; Apollo Program: Developmental Flights; Apollo Program: Lunar Module; Apollo Program: Orbital and Lunar Surface Experiments; Apollo 8; Apollo 9; Apollo 10; Cape Canaveral and the Kennedy Space Center; Mercury Project; Mercury-Atlas 8; Saturn Launch Vehicles.

**Further Reading**

Allday, Jonathan. *Apollo in Perspective: Spaceflight Then and Now.* Philadelphia: Institute of Physics, 1999. This book takes a retrospective look at the Apollo Program and the technology that was used to land humans on the Moon. The author explains the basic physics and technology of spaceflight, and conveys the huge technological strides that were made and the dedication of the people working on the program. All of the major aspects of the Apollo Program are covered, including crews, vehicles, and space suits.

Braun, Wernher von, and Frederick I. Ordway III. *Space Travel: A History.* Rev. ed. New York: Harper & Row, 1985. This book, written by one of the pioneers of the American space program, is an enjoyable portrait of the history of rocketry as a science and space travel as a modern technological phenomenon. In easy-to-understand language, this volume reveals the human drama and the sense of history in the making that helped the Space Race capture the imagination of the post-World War II generation.

Godwin, Robert, ed. *Apollo 7: The NASA Mission Reports.* Burlington, Ont.: Apogee Books, 2000. This book includes the preflight Apollo 7 press kit that describes the mission and its aims. The crew technical debriefing includes details on the Apollo spacecraft and launch vehicle systems, as well as the performance of the equipment and crew. In-flight photographs and film are included on the accompanying CD-ROM.

Kraft, Christopher C., Jr. *Flight: My Life in Mission Control.* East Rutherford, N.J.: Penguin Putnam, 2002. Kraft gives an account of his life in Mission Control. The first NASA flight director would play an integral role in what would become the National Aeronautics and Space Administration.

Lambright, W. Henry. *Powering Apollo: James E. Webb of NASA.* Baltimore: Johns Hopkins University Press, 2000. Lambright explores James E. Webb's leadership role in NASA's spectacular success.

Liebergot, Sy, and David Harland. *Apollo EECOM: Journey of a Lifetime.* Burlington, Ont.: Apogee Books, 2003. This is the life story of Sy Liebergot, former NASA flight controller, with emphasis on his years working in Apollo Mission Control.

National Aeronautics and Space Administration. *Apollo Mission Press Kits.* http://www-lib.ksc.nasa.gov/lib/presskits.html. Provides detailed preflight information about each of the Apollo missions, Apollo 6 through Apollo 17. Accessed March, 2005.

————. *Apollo 7 Mission Report.* Washington, D.C.: National Technical Information Services, 1968. This is the official report of the results of the Apollo 7 flight. The original is out of print, but copies can be obtained through the National Technical Information Services. An Internet search will turn up a copy in Adobe Acrobat format for easy download. Nowhere else will you find as in-depth a document on the technical, scientific, and human aspects of the flight.

Osman, Tony. *Space History.* New York: St. Martin's Press, 1983. This book, written for British space enthusiasts, is a layperson's view of the American and Soviet space programs. Filled with more subjective observations on the Space Race as a worldwide cultural phenomenon, it provides an interesting, if somewhat complex, perspective.

Schirra, Walter M., Jr., with Richard N. Billings. *Schirra's Space.* Annapolis, Md.: Naval Institute Press, 1995. Walter Schirra knows spacecraft. As the only person to pilot the Mercury, Gemini, and Apollo spacecraft, Schirra knew what he wanted in his vehicle and in his mission. Interference from ground control was definitely not what his Apollo 7 crew wanted. In fact, they nearly staged the first mutiny in space when mission controllers wanted to cram more work into their already busy schedule. Schirra tells about his contributions to Apollo and his feelings toward the entire program. Black-and-white photographs add to the text, which is a well-written look at the life of a "space jockey."

Shepard, Alan B., Jr., and Donald K. "Deke" Slayton, with Jay Barbree and Howard Benedict. *Moon Shot: The Inside Story of America's Race to the Moon.* Atlanta: Turner, 1994. This is, indeed, the inside story of the Apollo Program as told by two men who actively participated in it. Some of their tales appear here for the first time. The book was adapted for a four-hour documentary in 1995.

Slayton, Donald K., with Michael Cassutt. *Deke! U.S. Manned Space: From Mercury to the Shuttle.* New York: Forge, 1995. This is the autobiography of the last of the Mercury astronauts to fly in space. After being grounded from flying in Project Mercury for what turned out to be a minor heart murmur, Slayton was appointed head of the Astronaut Office. During his reign he assigned all of the Apollo crew members to their flights. Later, he commanded the Apollo-Soyuz Test Project flight in 1975. This is a behind-the-scenes look at America's attempt to land humans on the Moon.

Sullivan, Scott P. *Virtual Apollo: A Pictorial Essay of the Engineering and Construction of the Apollo Command and Service Modules.* Burlington, Ont.: Apogee Books, 2002. This book allows the public to become acquainted with the Apollo spacecraft in detail and learn the story of its design and construction. Full-color drawings in exacting detail provide inside and out views of the Command and Service Modules complete with details of construction and fabrication.

Wendt, Guenter, and Russell Still. *The Unbroken Chain.* Burlington, Ont.: Apogee Books, 2001. Autobiography of the only person who worked side by side with every astronaut bound for space.

*Eric Christensen*

# Apollo 8

*Date.* December 21 to December 27, 1968
*Type of mission:* Lunar exploration, piloted spaceflight

*Apollo 8, the world's first piloted mission to the Moon, collected in-flight data on prolonged human and engineering performance, topographic features, and the appearance of the lunar surface from close lunar orbit. This mission also tested the entire technological support system for human flight to the vicinity of the Moon and back to Earth.*

## Key Figures

*Frank Borman* (b. 1928), Apollo 8 mission commander

*James A. Lovell, Jr.* (b. 1928), Apollo 8 Command Module pilot

*William A. Anders* (b. 1933), Apollo 8 Lunar Module pilot

*Neil A. Armstrong* (b. 1930), Apollo 8 backup mission commander

*Edwin E. "Buzz" Aldrin, Jr.* (b. 1930), Apollo 8 backup Command Module pilot

*Fred W. Haise, Jr.* (b. 1933), Apollo 8 backup Lunar Module pilot

*Robert R. Gilruth* (1913-2000), leader of the Space Task Group

*George M. Low* (1926-1984), Apollo Program manager

*Rocco A. Petrone* (b. 1926), Apollo Program director

*Kurt H. Debus* (1908-1983), director of the Kennedy Space Center

*Wernher von Braun* (1912-1977), director of the Marshall Space Flight Center

*Christopher C. Kraft, Jr.* (b. 1924), Apollo flight director

*Donald K. "Deke" Slayton* (1924-1993), chief astronaut

## Summary of the Mission

The decision to fly Apollo 8 to the Moon and have it make several orbits was made in a conference at the Marshall Space Flight Center at Huntsville, Alabama, in the summer of 1968—barely six months before the mission actually flew. All the primary directors of the American space program were at the meeting: Rocco A. Petrone, Christopher C. Kraft, Jr., Robert R. Gilruth, Director Kurt H. Debus of the Kennedy Space Center, Director Wernher von Braun of the Marshall Space Flight Center, Chief Astronaut Donald K. "Deke" Slayton, and the man who developed the idea, George M. Low, manager of the Apollo Program in Houston, Texas.

After three hours of deliberation, it was determined that the mission was indeed possible and, even though the change was a critical departure from Apollo 8's original intended mission, the feat could in fact be done. Final top-level National Aeronautics and Space Administration (NASA) authorization came in October, but the switch of plans had already been implemented, and personnel and machines were being readied for lunar orbit.

At 12:51 Coordinated Universal Time (UTC, or 7:51 A.M. eastern standard time) on the morning of December 21, 1968, the Apollo 8 mission left Launch Complex Pad 39A at the Kennedy Space Center. It was the first time that the mighty Saturn V rocket had been used to loft human beings into space. The Saturn first-stage fuel, 2,009,835 liters of a mixture of kerosene and liquid oxygen, was consumed in 2 minutes and 34 seconds. The second-stage Saturn II rocket carried 1,358,815 liters of liquid hydrogen and liquid oxygen; it pushed the third-stage Saturn IVB and the Command and Service Module (CSM) ahead of it for 6 minutes and 10 seconds to reach an altitude of about 175 kilometers.

After the second stage was spent, the Saturn IVB took over for 2 minutes and 40 seconds in order to reach Earth orbit. During the second Earth orbit, the Apollo men and machines achieved translunar injection, the point at which the spacecraft leaves Earth orbit and begins the long journey to the Moon. At that vital point, the Saturn IVB engines burned for 5 minutes and 19 seconds, pushing the craft's speed to 39,110 kilometers per hour—the speed necessary for a payload to escape Earth's gravity. After the burn, the Saturn IVB separated

*Apollo 8, launched December 21, 1968, was the first piloted flight to use the Saturn V rocket to enter into outer space.* (NASA CORE/ Lorain County JVS)

from the CSM and was placed on a path toward the Sun. Meanwhile, Apollo 8 was on its way to the Moon.

Apollo 8 was the second in a series of four Apollo missions designed to test the men, equipment, and ground and recovery support systems in preparation for the Apollo 11 Moon landing. The crew for the Apollo 8 mission consisted of two experienced astronauts—Mission Commander Frank Borman, who had flown on the Gemini VII mission; Command Module Pilot James A. Lovell, a veteran of Gemini missions VII and XII—and Lunar Module Pilot William A. Anders, making his first flight into space. The crew had trained for many months in simulation exercises, neutral buoyancy water pools, and centrifuge trials, studying a large number of procedures and alternative procedures in case of emergencies. On December 21, they were more than prepared for the flight. Meanwhile, their flight backup team included Neil A. Armstrong, Edwin E. "Buzz" Aldrin, Jr., and Fred W. Haise, Jr., all of whom played a vital role in the mission by working in Mission Control at the Johnson Manned Spacecraft Center in Houston, Texas.

The Kennedy Space Center Launch Complex at the Cape consisted of twin launch pads (A and B), each covering approximately 648,000 square meters. The complex was built specifically to launch piloted Apollo spacecraft for the lunar missions and was constructed during the period from 1961 through 1966. Located 3.5 miles (roughly 5.6 kilometers) from the Vehicle Assembly Building (VAB), where the Apollo spacecraft was mounted atop the Saturn rocket, the complex was connected to the VAB by a track called a "crawlerway" over which the assembled Apollo-Saturn configuration, sitting upright on a unique transporter platform, slowly rode on its way to the launch site. The 3.5-mile ride from the VAB to Complex 39, Launch Pad A, required several hours to accomplish.

*The spacecraft splashed down in the Pacific Ocean on December 27, 1968. The crew and craft were recovered by the USS* Yorktown. (NASA CORE/Lorain County JVS)

The Apollo spacecraft, built by Rockwell International in Downey, California, was actually two compartments tightly fixed together: the Command Module (CM) and the Service Module (SM). The flight crew of three rode in a sitting position on their backs in the CM, which was the in-flight living quarters and cockpit; it was 3.35 meters long, measured 4 meters in diameter, and weighed 6,000 kilograms. The SM contained the equipment and instruments needed to fly the CSM: life-support, in-flight fuel, communications, electronics, and power generation. The Service Module was a cylinder 4 meters in diameter and 7.3 meters long, weighing 26,000 kilograms fully loaded. The CSM, topped by a separate emergency rocket for launch abort, sat atop the three stages of the Saturn rocket.

The outward voyage to the Moon took two days. Each day at about 20:00 UTC (3:00 P.M. eastern standard time) the astronauts appeared live on national television on Earth. At 09:59 UTC (4:59 A.M. eastern standard time) on December 24, the Service Propulsion System engines inserted Apollo 8 into lunar orbit, and the Apollo craft was captured

by the Moon's gravitational pull, traveling at a speed of 10,593 kilometers per hour. At that time, the Apollo 8 craft rounded the back side of the Moon and remained out of communication with the Earth for about forty-five minutes. The orbit Apollo 8 circumscribed around the Moon at entry was between 111 and 313 kilometers; the orbit was later circularized at 112 kilometers.

For about twenty hours, the Apollo craft continued to orbit the Moon, finally completing ten full orbits, each orbit taking two hours. The crew initiated live telecasts during the second and ninth orbits, describing the lunar surface 112 kilometers below and discussing their own reactions to flight conditions and to being the first humans to see the Moon up close. During the lunar orbits, the astronauts took sextant readings, photographed the Moon's topography, and made notes about specific features with a view to determining the condition of several alternate landing sites.

At 02:31 UTC (9:31 P.M. eastern standard time) on Christmas Eve, the astronauts performed a special live Christmas telecast during which they read from the Book of Genesis. Anders started: "For all the people on Earth, the crew of Apollo 8 has a message we would like to send you." After a moment's pause, he began reading again: "In the beginning God created the Heaven and the Earth." After four verses, Lovell took over; at the end of the eighth verse, Borman continued, finally ending with the words, "And from the crew of Apollo 8, we close with good night, good luck, a Merry Christmas, and God bless all of you—all of you on the good Earth."

What did the astronauts see on the lunar surface? Lovell said, "The vast loneliness up here is awe-inspiring," and commented about the appearance of Earth from 386,160 kilometers (240,000 miles). Borman described the lunar surface as looking like clouds of pumice stone, while Anders said it was whitish gray, like dirty beach sand and fine haze dust; he mentioned especially the innumerable pockmarks from possible asteroid and meteorite bombardments over the Moon's long history. The astronauts also described the Moon's major geological features, such as long trenches called "rilles," cliffs, and the sizes of various craters, some with high peaks in their centers.

In order to communicate with the Earth, the Apollo 8 mission required fourteen land stations, four instrumented ships, and eight instrumented aircraft, at times using orbiting satellite relays, cable, telephone, teletype, and radio. Over this unbelievable network came not only the live telecasts enjoyed by millions of viewers around the world but also the vital communications between Mission Control and the astronauts and all the telemetered monitoring of each astronaut's life signs (heartbeat, temperature, respiration, pulse). The astronauts' life functions were also closely monitored while they slept, to be compared with readings taken during waking periods. Telemetry and radio signals between Earth and the Moon take three seconds round-trip.

In the tenth orbit, at 06:10 UTC (1:10 A.M. eastern standard time) on the morning of December 25, the Service Propulsion System engines again were ignited, this time for a 303-second burn, which put the astronauts on a trajectory headed for Earth. This maneuver is referred to as trans-Earth injection. At this time, Apollo 8 was moving at a speed of 8,850 kilometers per hour. By the time the craft reached Earth's atmosphere, it was again traveling at a speed of about 40,225 kilometers per hour under the strong pull of Earth's gravity.

Reentry is a very precise and difficult task. The craft must enter a "window" in the atmosphere measuring 643 kilometers by 42 kilometers. If the angle of reentry is too steep, the craft will be incinerated by the enormous friction with the atmosphere; if the angle is too shallow, the craft will skip off it like a flat stone skipping over water and shoot out into space, never to be recovered. The window was correctly entered by the Apollo 8 spacecraft at 128 kilometers above Earth. The Apollo 8 reentry was actually photographed by a Pan-American Air-

ways commercial airliner pilot on his way from Honolulu to Fiji. The photograph, taken in the dead of night, disclosed a fiery tail estimated by the pilot to be about 200 kilometers in length issuing from the rear of the Command Module. A returning Apollo spacecraft would heat up to about 2,760° Celsius. Most materials cannot withstand a temperature this high and quickly incinerate. The Apollo craft, however, was covered on the outside with ablative material—a special heat-resistant covering with an extremely high kindling temperature.

The unneeded Service Module was separated from the Command Module and left to burn up on reentering Earth's atmosphere. During the actual reentry maneuver, the decelerating astronauts withstood six times the gravity of Earth, finding themselves pushed heavily down into their seats. At 7,315 meters during descent, three special braking parachutes called "drogue chutes" were activated. The purpose of the chutes was twofold: They acted as an additional braking mechanism for the swiftly falling craft, and they stabilized the craft to keep it from spinning and maintain its upright position for the splashdown in the ocean below. The chutes slowed the craft to 483 kilometers per hour. At 3,048 meters in altitude, with the craft slowed to about 225 kilometers per hour, the drogue chutes were cut loose and the three Apollo main parachutes opened to land the craft. Each of these main parachutes was 25.5 meters in diameter, colored alternately orange and white to aid searchers in quickly locating the craft as it floated lazily downward.

Apollo 8 landed in its intended target zone in the Pacific Ocean at 15:51 UTC (10:51 A.M. eastern standard time) on Monday, December 27. The Pacific Ocean was still dark, however, because local time was 4:50 A.M. The total recovery time was about 1 hour, 20 minutes, while special balloons in the top of the cone-shaped Apollo Command Module automatically inflated to keep the craft from rolling over on its side if the seas were rough. At the same time, an automatic radio signal and a flashing beacon went into operation to aid in the craft's location, and a green dye was released which quickly colored the water for some distance around the craft.

The astronauts were helped out of the craft by a team of U.S. Navy frogmen trained in underwater maneuvers and technology. The team fitted the flotation collar and helped each astronaut out of the Command Module and into a life raft, then into a platform seat dangling from a helicopter. Each astronaut was flown to a nearby recovery ship. For Apollo 8, the recovery ship was the USS *Yorktown*, an aircraft carrier and part of a vast fleet of recovery ships and airplanes waiting for the returning astronauts and their craft. After the astronauts were safe, the vacated Command Module then was emptied of precious cargo (film packs, tape recordings, and the like) and hauled aboard the recovery ship by helicopter. Afterward, the craft was carefully inspected by a team of scientists who studied the craft's condition after its long journey and fiery reentry.

### Contributions

The flight of Apollo 8 allowed the first visual inspection of the Moon's surface at close range. An idea of the lunar surface—rough, cratered, dusty, possibly dangerous—was gained, although the astronauts' descriptions did not necessarily answer the host of questions that plagued NASA scientists. The crew also described conditions aboard the spacecraft throughout the entire flight—information that was invaluable for future spaceflights to the Moon. Apollo 7 was the only previous mission to provide such data, and it had remained in Earth orbit. The first journey to the Moon and back revealed that a crew on a lunar flight not only would survive, but also could function well in their spacecraft. Biological and engineering data gathered during the Apollo 8 mission served as a base of information for planning future missions.

### Context

Probably the greatest value of the Apollo 8 mission was political. The flight put the United States undeniably ahead of the Soviet Union in the race to the Moon and let the world know that Americans were back in the leadership position. The flight was also an achievement in its own right; it had taken humans to the Moon for the first time in recorded history and, in that context, scored a lasting spiritual triumph for the people of the United States. It proved that travel to the Moon was possible in the space hardware then designed for the job. In a larger sense, however, Apollo 8 also proved that humankind was capable of "leaving"—leaving the restricting and limiting confines of planet Earth to reach another body in space.

**See also:** Apollo Program; Apollo Program: Command and Service Module; Apollo Program: Developmental Flights; Apollo Program: Geological Results; Apollo Program: Lunar Module; Apollo 1; Apollo 7; Apollo 9; Apollo 10; Apollo 11; Apollo 13; Apollo 17; Clementine Mission to the Moon; Saturn Launch Vehicles; Space Task Group.

### Further Reading

Borman, Frank, with Robert J. Serling. *Countdown: An Autobiography*. New York: William Morrow, 1988. Astronaut Borman writes of the Apollo 204 accident and the investigation that followed from an insider's point of view. He was one of the sleuths appointed by NASA to find the cause of the fire and a cure. He compares this inquiry with the one that followed the *Challenger* accident nineteen years later. As part of the Apollo spacecraft redesign team, Borman reveals some of the problems and infighting that occurred during the critical rebuilding stage. Of course, the portion of the book that is the most riveting is the chapter devoted to his Apollo 8 mission. Borman gives insight into the human aspect of spaceflight.

Godwin, Robert, ed. *Apollo 8: The NASA Mission Reports*. Burlington, Ont.: Apogee Books, 1999. This collection contains reprints of the Apollo 8 press kit, "Pre-mission Report and Objectives," "Supplemental Technical Report," and the "Post Flight Summary." It also includes a CD-ROM featuring 850 70-millimeter images taken during the mission.

Kraft, Christopher C., Jr. *Flight: My Life in Mission Control.* East Rutherford, N.J.: Penguin Putnam, 2002. The first NASA flight director gives an account of his life in Mission Control.

Lambright, W. Henry. *Powering Apollo: James E. Webb of NASA.* Baltimore: Johns Hopkins University Press, 2000. Explores James E. Webb's leadership role in NASA's spectacular success.

Liebergot, Sy, and David Harland. *Apollo EECOM: Journey of a Lifetime.* Burlington, Ont.: Apogee Books, 2003. This is the life story of Sy Liebergot, former NASA flight controller, with emphasis on his years working in Apollo Mission Control.

Lindsay, Hamish. *Tracking Apollo to the Moon.* London: Springer-Verlag London Limited, 2001. This lively chronicle features interviews, quotes, and extensive photographs, including some that appear here for the first time.

National Aeronautics and Space Administration. *Apollo 8 Mission Report.* Washington, D.C.: National Technical Information Services, 1969. This is the official report of the results of the Apollo 8 flight. The original is out of print, but copies can be obtained through the National Technical Information Services. An Internet search will turn up a copy in Adobe Acrobat format for easy download. Nowhere else will you find as in-depth a document on the technical, scientific, and human aspects of the mission.

——————. *Apollo Mission Press Kits.* http://www.lib.ksc.nasa.gov/lib/presskits.html. Provides detailed preflight information about each of the Apollo missions, Apollo 6 through Apollo 17. Accessed March, 2005.

Shepard, Alan B., Jr., and Donald K. "Deke" Slayton, with Jay Barbree and Howard Benedict. *Moon Shot: The Inside Story of America's Race to the Moon.* Atlanta: Turner, 1994. This is, indeed, the inside story of the Apollo Program as told by two men who actively participated in it. Some of their tales appear here for the first time. The book was adapted for a four-hour documentary in 1995.

Slayton, Donald K., with Michael Cassutt. *Deke! U.S. Manned Space: From Mercury to the Shuttle.* New York: Forge, 1995. This is the autobiography of the last of the Mercury astronauts to fly in space. After being grounded from flying in Project Mercury for what turned out to be a minor heart murmur, Slayton was appointed head of the Astronaut Office. During his reign he assigned all of the Apollo crew members to their flights. Later, he commanded the Apollo-Soyuz Test Project flight in 1975. This is a behind-the-scenes look at America's attempt to land humans on the Moon.

Sullivan, Scott P. *Virtual Apollo: A Pictorial Essay of the Engineering and Construction of the Apollo Command and Service Modules.* Burlington, Ont.: Apogee Books, 2002. Full-color drawings provide inside and outside views of the Command and Service Modules complete with details of construction and fabrication.

Wendt, Guenter, and Russell Still. *The Unbroken Chain.* Burlington, Ont.: Apogee Books, 2001. The autobiography of the only person who worked with every astronaut bound for space.

Zimmerman, Robert. *Genesis: The Story of Apollo 8: The First Manned Flight to Another World.* New York: Dell, 1999. Zimmerman is a science journalist who delivers an account not only of the flight around the Moon but also of the political climate in America at the time.

*Thomas W. Becker*

# Apollo 9

*Date:* March 3 to March 13, 1969
*Type of mission:* Piloted spaceflight

*During the Earth-orbital flight of Apollo 9, the complete Apollo spacecraft, Command and Service Module, and Lunar Module were tested in both docked and independent rendezvous flight phases. Apollo 9 paved the way for Apollo 10's dress rehearsal of the piloted lunar-landing mission.*

## Key Figures

*James A. McDivitt* (b. 1929), Apollo 9 mission commander
*David R. Scott* (b. 1932), Apollo 9 Command Module pilot
*Russell L. Schweickart* (b. 1935), Apollo 9 Lunar Module pilot
*Charles "Pete" Conrad, Jr.* (1930-1999), backup Apollo 9 mission commander
*Richard F. Gordon, Jr.* (b. 1929), backup Apollo 9 Command Module pilot
*Alan L. Bean* (b. 1932), backup Apollo 9 Lunar Module pilot

## Summary of the Mission

Apollo 9's flight plan called for the Lunar Module, *Spider*, to separate from the Command and Service Module (CSM), *Gumdrop*, and perform independently up to a distance of 160 kilometers. *Spider* was to perform an active rendezvous, ultimately redocking with *Gumdrop*. For the first time, two piloted spacecraft on the same mission would be in flight. During their separate maneuvers, the spacecraft would use radio call signs to help ground controllers differentiate between the CSM and the LM. Apollo 9 was to integrate fully all Apollo facilities and hardware with a crew. It was the first piloted mission (American or Soviet) in which a successful rendezvous absolutely had to be accomplished, because the paper-thin Lunar Module was incapable of withstanding reentry and contained no parachutes. Apollo 9 was also to test the moonsuit and its portable life-support system during extravehicular activity. The National Aeronautics and Space Administration (NASA) selected James A. McDivitt, David R. Scott, and Russell L. Schweickart to fly this ambitious mission.

James Alton McDivitt was born on June 10, 1929, in Chicago, Illinois. In 1959, he received a bachelor of science degree in aeronautical engineering from the University of Michigan. Three years later, NASA selected McDivitt as part of the second astronaut group, and he flew as command pilot of Gemini IV in 1965. After Apollo 9, McDivitt served as Apollo Spacecraft Program manager at the Manned Spacecraft Center in Houston. Following that assignment, in 1972 McDivitt resigned from both NASA and the U.S. Air Force, where he had achieved the rank of brigadier general.

David Randolph Scott was born June 6, 1932, in San Antonio, Texas. In 1954, Scott received a bachelor of science degree from the United States Military Academy. Eight years later, he received a master of science degree in aeronautics and astronautics and an engineer of aeronautics and astronautics degree from the Massachusetts Institute of Technology. Scott was chosen in 1963 by NASA during the third astronaut selection. His first flight assignment was as the pilot of Gemini VIII. After Apollo 9, Scott served as backup commander of

Apollo 12 and flew as commander of Apollo 15 (the fourth lunar-landing mission). Scott became special assistant for Mission Operations for the Apollo-Soyuz Test Project in 1972. In April, 1975, he was appointed director of NASA's Dryden Flight Research Center at Edwards Air Force Base in California. In 1977, he resigned from that position. Before leaving the Air Force, Scott achieved the rank of colonel.

Russell Louis Schweickart was born October 25, 1935, in Neptune, New Jersey. In 1956, he received a bachelor of science degree in aeronautical engineering, and in 1963, a master of science degree in aeronautics and astronautics, both from the Massachusetts Institute of Technology. Before joining NASA, Schweickart served as an aviator in the U.S. Air Force. Selected by NASA as part of the third astronaut group, he was assigned as backup commander for Skylab 2 and served as NASA's director of User Affairs in the Office of Applications and as assistant for Payload Operations in the Office of Planning and Program Integration after Apollo 9. Following a leave of absence from NASA to serve as the California governor's science adviser, Schweickart resigned from the space agency in 1979.

Apollo 9 was rolled out to Launch Complex Pad 39A on January 3, 1969. Originally, launch was scheduled for February 28, but cold symptoms experienced by the prime crew forced a three-day delay. Apollo 9 lifted off at 16:00 Coordinated Universal Time (UTC, or 11:00 A.M. eastern standard time) on March 3 and entered a 168-by-189-kilometer orbit. Following a simulation of the parking orbit spacecraft checkout, McDivitt separated *Gumdrop* from the spent Saturn V third stage and performed the transposition and docking maneuver to extract *Spider.* With the Apollo spacecraft assembled, the single J-2 engine was commanded to ignite, sending the spent stage off into a solar orbit that would carry it far from the paths of *Gumdrop* and *Spider.* The Service Propulsion System engine was fired to raise the spacecraft into a higher orbit as a test of the autopilot system and the docked Command and Service and Lunar Modules' dy-

namics. During the next several days, many such burns were performed.

Schweickart awoke on March 5 with a queasy stomach, ultimately vomiting. His condition improved, and he was able to proceed, along with McDivitt, with the transfer from *Gumdrop* into *Spider* via the docking tunnel connecting the two spacecraft. (This was not the first crew transfer attempted in space. A pair of spacewalking cosmonauts transferred from Soyuz 5 to Soyuz 4 in January, 1969.) *Spider's* systems were powered up and McDivitt and Schweickart prepared for independent flight maneuvers that they would execute later. Live television pictures provided views inside the Lunar Module cabin. The Descent Propulsion System engine was burned for six minutes to enlarge the docked vehicles' orbit. After six hours inside *Spider,* McDivitt returned to *Gumdrop;* after eight hours, Schweickart did the same, but not before becoming nauseated again.

Schweickart's condition forced a one-day delay in his scheduled spacewalk to test the moonsuit and a suited astronaut's ability to exit the Lunar Module and climb down its ladder. An extravehicular activity transfer from *Spider's* porch to *Gumdrop's* open hatch was canceled altogether because of Schweickart's weakened state. All three astronauts donned pressure suits on March 6. Both *Gumdrop* and *Spider* were depressurized, and both the Lunar and Command and Service Modules' hatches were opened. Schweickart backed out of *Spider's* hatch wearing the moonsuit and portable life-support system and stood on the Lunar Module porch. Scott stood up in *Gumdrop's* open hatch, retrieving an experiment package designed to measure contamination near the spacecraft. Schweickart returned to the Lunar Module cabin, closing the hatch thirty-eight minutes after opening it, after a test of the life-support capability and flexibility of the moonsuit. It performed superbly.

After a six-hour rest period, the crew prepared to perform the separation and rendezvous maneuvers on March 7. After Lunar Module activation, *Gumdrop* and *Spider* undocked. *Spider* executed an orbital ballet as Scott visually inspected the Lunar

*One of the main objectives of the Apollo 9 mission was to test the Lunar Landing Module in space. Here the spacecraft is in Earth orbit.* (NASA CORE/Lorain County JVS)

Module's exterior. McDivitt ignited the Descent Propulsion System engine, simulating the initial portion of a powered descent to the lunar surface. *Spider* entered an orbit 21 kilometers above *Gumdrop*, thus slowing down and falling behind *Gumdrop*. Two hours later, the two spacecraft had separated 88 kilometers from each other. At this point, the Descent Propulsion System engine was fired once more to increase the separation distance to 136 kilometers, whereupon the ascent and descent stages separated. An Ascent Propulsion system (APS) engine burn simulated a lunar liftoff. Based on data from spacecraft radar, the Lunar Module's guidance system calculated the various maneuvers necessary to return *Spider*'s ascent stage to *Gumdrop* from a distance of 160 kilometers.

*Spider* and *Gumdrop* had separated at 12:45 UTC (7:45 A.M.). About six hours later, the Lunar Module ascent stage completed the active rendezvous, and for nearly twenty-five minutes, the

two spacecraft flew in formation, examining each other's exterior. *Spider*'s ascent stage appeared to have suffered no major damage as a result of its separation from the descent stage. McDivitt maneuvered *Spider* into docking attitude; sunlight reflecting off the Command Module produced glare, however, making the docking (accomplished at 19:15 UTC (2:15 P.M.) quite difficult. McDivitt remarked that the maneuver was not a docking but an eye test.

The astronauts had trouble opening the hatch to return to the Command Module. After they had left *Spider*, the hatch was closed, and the ascent stage was jettisoned following some difficulties. *Spider*'s Ascent Propulsion System (APS) engine was fired until fuel exhaustion, propelling the Lunar Module into a highly elliptical orbit (6,965 by 235 kilometers) that would keep it safely away from Earth-orbiting satellites. At 02:00 UTC (9:00 P.M.), the crew concluded their long but eventful day. Beginning a well-deserved rest period, they requested that they not be disturbed the next morning.

By the conclusion of flight day five, 97 percent of all Apollo 9 mission objectives had been fulfilled. The remaining five days of the flight had a lighter crew schedule. Experiments and photography consumed most of the astronauts' time.

A Service Propulsion System engine burn was aborted on March 8 when the autopilot malfunctioned, causing propellant-tank ullage motors to fail. One revolution later, Mission Control was ready to attempt a potential solution to the problem. Ninety minutes behind schedule, the Service Propulsion System engine was ignited to lower Apollo 9's orbit to 192 by 168 kilometers so that a safe reentry would be possible even if the engine failed.

Weather conditions in the prime recovery zone (southwest of Bermuda) turned foul as the flight drew to a close. McDivitt, Scott, and Schweickart became orbiting weathermen, monitoring the meteorological developments. Flight controllers considered returning Apollo 9 one day early to avoid the heavy seas and gale-force winds brewing in the recovery zone. Eventually the recovery zone was changed, and one extra orbit was added to the flight. While Apollo 9 prepared for reentry, the USS *Guadalcanal* steamed toward Grand Turk Island, where acceptable weather conditions prevailed.

Apollo 9's Service Propulsion System engine burned for 11.8 seconds at 16:32 UTC (11:32 A.M.) on March 13 to begin the computer-guided reentry. The Command Module was sighted shortly after main parachute deployment, remaining visible until splashdown in a "stable-one" position (with apex upright) at 17:00:53 UTC (12:00:53 P.M.), less than 2 kilometers from target and only 6 kilometers from the aircraft carrier.

### Contributions

Apollo 9's primary goal was to verify the operational capability of Apollo spacecraft to undertake a piloted lunar landing. While doing so, a few procedures requiring alteration for future flights were discovered. One such item involved the Lunar Module's ascent-stage redocking after lunar activities. The Command and Service Module's surface was so shiny that it could produce glare serious enough to hinder astronaut vision. On future flights, the Command Module pilot would actively perform the docking maneuver.

During Apollo 9's flight, the first clear episode of space sickness occurred. Mercury and Gemini astronauts were confined in their spacecraft, but Apollo had sufficient interior space to permit intravehicular activity. Schweickart made his first spaceflight on Apollo 9, and he had fewer flying hours than other astronauts. He had a history of motion sickness; thus, he ingested one fifty-milligram Marezine tablet prior to launch as a precaution. Nevertheless, he experienced dizziness (without nausea) when he left his couch and turned his head rapidly. Later in the mission, he experienced

*Astronauts James McDivitt, David Scott, and Russell Schweickart spent ten days in Earth orbit on the Apollo 9 mission.* (NASA CORE/Lorain County JVS)

several bouts of vomiting. Fortunately, Schweickart was able to use a disposal bag. Vomiting and sensations of nausea were not directly connected; vomiting episodes came without advance warning but appeared to occur during or after periods of extensive motion within the spacecraft. After a second episode, Schweickart felt relief, but a significant loss of appetite and aversion to food odors were residual symptoms. He consumed only liquids and freeze-dehydrated fruits through flight day 6.

Schweickart was subjected to extensive postflight vestibular tests to ascertain whether his inflight difficulties were unique or a more general problem that could be experienced by future astronauts. It was determined that Schweickart's vestibular apparatus functioned normally and that he had no greater susceptibility to motion sickness than other astronauts. Thus, medical attention was focused on prediction and prevention. It was discovered that vestibular adaptation to weightlessness might proceed less effectively if astronauts moved extensively early in the flight. Symptoms could be minimized by a combination of planned head movements and medication, such as Scopolamine and Dexedrine. Motion-sickness symptoms were experienced by numerous Apollo (and shuttle) astronauts, but the lessons of Apollo 9 had helped NASA medical personnel to understand the problem.

After completion of the spacewalk, Schweickart doffed the moonsuit and portable life-support system but recharged the system's consumables. This was done as a precaution for contingency reuse and as a demonstration of normal operation under lunar-landing flight conditions. Apollo 9 gave a thorough checkout of the Apollo pressure-suit systems. Each astronaut wore pressure garments for nearly fifty-two hours, much of the time minus gloves and helmet. Suit design was adequate for all aspects of a lunar mission.

Apollo 9 tested early infrared photography systems that would prove useful in later Earth resources imaging systems flown on Skylab and Landsat. The astronauts used a system of four cameras complete with battery power and several filters to locate new sites of natural resources, demonstrating the potential of such techniques.

**Context**

All Apollo hardware necessary for a piloted lunar landing had been human-rated by the end of 1968 except for one essential part. On Apollo 9, the Lunar Module made its piloted Earth-orbital debut. The Lunar Module had been flight-tested without a crew in Earth orbit during Apollo 5's mission in January, 1968. Its Descent Propulsion System engine was test-fired but shut down prematurely as a result of computer control errors. A second robotic test flight was canceled as needless duplication. Thus, the Lunar Module was piloted on its next test. Apollo 8 originally was scheduled to fly the Module. Design and ground-test delays, however, forced NASA to postpone its flight to Apollo 9.

Apollo 9 astronauts orbited Earth 151 times, traveling 6.96 million kilometers in 10 days, 1 hour, and 53 seconds. President Richard M. Nixon described the flight as ten days that thrilled the world. During its mission, Apollo 9 accumulated a number of important firsts. These included the first American docking of two piloted spacecraft, the first astronaut transfer from one piloted spacecraft to another through an internal passage connecting the two, the first spacewalk of the Apollo Program, the first demonstration of the television system to be used for recording the initial steps down to the lunar surface on Apollo 11, and the first piloted Command Module to land in an apex-upright, or stable-one, position.

This flight was the first American piloted spaceflight of 1969, the year in which a piloted lunar landing had to become reality if President John F. Kennedy's bold goal, to land a person on the Moon before decade's end, was to be met. Only a few weeks before, the Soviets had attempted to beat many of Apollo 9's spectacular achievements. Soyuz 4 and 5, independently launched piloted spacecraft, had rendezvoused and docked in orbit. Unlike Apollo 9, the Soyuz 4/Soyuz 5 combination had no internal passageway for astronaut crew transfers. Two cosmonauts from Soyuz 5 transferred to Soyuz 4, launched with a single cosmonaut, using handrails to crawl along the spacecraft's exteriors from one open hatch to the other. A similar procedure would be used in Apollo only in the event that the Lunar Module's ascent stage was prevented from docking by a malfunction.

Apollo 9, on the other hand, was part of a logical progression toward the goal of a piloted lunar landing. Apollo 7 and Apollo 8 had human-rated the Command and Service Module, proving that the spacecraft could navigate the 384,000 kilometers of space between Earth and the Moon and precisely return to a safe, water landing. Apollo 9 proved that the Lunar Module could sustain life long enough for two astronauts to perform a landing, that the Descent Propulsion System engine could be flown manually and automatically to reduce spacecraft velocity as desired in a powered descent, that the Ascent Propulsion System (APS) engine could lift the Lunar Module's ascent stage off the lunar surface using the descent stage as a launch pad, and that the Lunar Module could navigate independently of the Command and Service Module using positional data from its onboard radar systems. A lunar landing was closer to reality as a result of lessons learned on Apollo 9.

**See also:** Apollo Program; Apollo Program: Command and Service Module; Apollo Program: Developmental Flights; Apollo Program: Lunar Module; Apollo Program: Orbital and Lunar Surface Experiments; Apollo 1; Apollo 7; Apollo 8; Apollo 10; Gemini V; Space Stations: Origins and Development; Space Task Group.

### Further Reading

Allday, Jonathan. *Apollo in Perspective: Spaceflight Then and Now.* Philadelphia: Institute of Physics, 1999. A retrospective look at the Apollo Program and the technology that was used to land humans on the Moon.

Brooks, Courtney G., James M. Grimwood, and Loyd S. Swenson, Jr. *Chariots for Apollo: A History of Manned Lunar Spacecraft.* NASA SP-4205. Washington, D.C.: Government Printing Office, 1979. An excellent history of the design, development, production, testing, and flight operations of Apollo spacecraft. Contains charts, tables, and photographs. Extensive footnotes list original papers, memos, and communications.

Ertel, Ivan D., and Roland W. Newkirk. *The Apollo Spacecraft: A Chronology.* Vol. 4. NASA SP-4009. Washington, D.C.: Government Printing Office, 1978. A chronological history of important decisions, development, tests, assemblies, planning, and flights of Apollo hardware. Contains photographs, extensive appendices, and references.

Godwin, Robert, ed. *Apollo 9: The NASA Mission Reports.* Burlington, Ont.: Apogee Books, 1999. This collection contains reprints of the Apollo 9 press kit, "Preflight Mission Operation Report," "Post Launch Mission Operation Report" and the "Mission Operation Report Supplement." It also includes a CD-ROM featuring images taken during the mission and detailed drawings of the spacecraft.

Johnston, Richard S., Lawrence F. Deitlein, and Charles A. Berry. *Biomedical Results of Apollo.* NASA SP-368. Washington, D.C.: Government Printing Office, 1975. Prepared by doctors intimately associated with the Apollo Program, this official report details medical data collected on the adaptation of the human body to weightlessness and spaceflight stresses. Designed for those with background in physiology. Numerous charts, graphs, and tables.

Kelly, Thomas J. *Moon Lander: How We Developed the Apollo Lunar Module.* Washington, D.C.: Smithsonian Books, 2001. Grumman Chief Engineer Kelly gives a firsthand account of designing, building, testing, and flying the Apollo Lunar Module.

Kraft, Christopher C., Jr. *Flight: My Life in Mission Control.* East Rutherford, N.J.: Penguin Putnam, 2002. Kraft gives an account of his life in Mission Control as the first NASA flight director.

Lambright, W. Henry. *Powering Apollo: James E. Webb of NASA.* Baltimore: Johns Hopkins University Press, 2000. Lambright explores James E. Webb's leadership role in NASA's spectacular success.

Levine, Arnold S. *Managing NASA in the Apollo Era.* NASA SP-4102. Washington, D.C.: Government Printing Office, 1982. This NASA History series text outlines management of the Apollo Program. Includes organization, contractual research and development, determination of NASA policy, budgetary decisions, and planning. Easy reading. Charts, tables, and references are included.

Liebergot, Sy, and David Harland. *Apollo EECOM: Journey of a Lifetime*. Burlington, Ont.: Apogee Books, 2003. This is the life story of Sy Liebergot, former NASA flight controller, with emphasis on his years working in Apollo Mission Control.

National Aeronautics and Space Administration. *Apollo Mission Press Kits*. http://www-lib.ksc.nasa.gov/lib/presskits.html. Provides detailed preflight information about each of the Apollo missions, Apollo 6 through Apollo 17. Accessed March, 2005.

_____. *Apollo 9 Mission Report*. Washington, D.C.: National Technical Information Services, 1969. This is the official report of the results of the Apollo 9 flight. The original is out of print, but copies can be obtained through the National Technical Information Services. An Internet search will turn up a copy in Adobe Acrobat format for easy download. Nowhere else will you find as in-depth a document on the technical, scientific, and human aspects of the mission.

Shepard, Alan B., Jr., and Donald K. "Deke" Slayton, with Jay Barbree and Howard Benedict. *Moon Shot: The Inside Story of America's Race to the Moon*. Atlanta: Turner, 1994. This is, indeed, the inside story of the Apollo Program as told by two men who actively participated in it. Some of their tales appear here for the first time. The book was adapted for a four-hour documentary in 1995.

Slayton, Donald K., with Michael Cassutt. *Deke! U.S. Manned Space: From Mercury to the Shuttle*. New York: Forge, 1995. This is the autobiography of the last of the Mercury astronauts to fly in space. After being grounded from flying in Project Mercury for what turned out to be a minor heart murmur, Slayton was appointed head of the Astronaut Office. During his reign he assigned all of the Apollo crew members to their flights. Later, he commanded the Apollo-Soyuz Test Project flight in 1975. This is a behind-the-scenes look at America's attempt to land humans on the Moon.

Sullivan, Scott P. *Virtual Apollo: A Pictorial Essay of the Engineering and Construction of the Apollo Command and Service Modules*. Burlington, Ont.: Apogee Books, 2002. Full-color drawings provide inside and outside views of the Command and Service Modules complete with details of construction and fabrication.

_____. *Virtual LM*. Burlington, Ont.: Apogee Books, 2004. Shows the details of design and production using full-color renderings of the structures, components, subassemblies and the completed spacecraft, accompanied by supporting descriptions.

Wendt, Guenter, and Russell Still. *The Unbroken Chain*. Burlington, Ont.: Apogee Books, 2001. The autobiography of the only person who worked with every astronaut bound for space.

*David G. Fisher*

# Apollo 10

*Date:* May 18 to May 26, 1969
*Type of mission:* Lunar exploration, piloted spaceflight

*Apollo 10 successfully orbited the Moon and provided the final testing of all systems needed for an actual landing. The Lunar Module descended to within 15 kilometers of the lunar surface before returning to and docking with the Command and Service Module.*

### Key Figures

*Thomas P. Stafford* (b. 1930), Apollo 10 mission commander
*John W. Young* (b. 1930), Apollo 10 Command Module pilot
*Eugene A. Cernan* (b. 1934), Apollo 10 Lunar Module pilot
*L. Gordon Cooper, Jr.* (1927-2004), Apollo 10 backup mission commander
*Donn F. Eisele* (1930-1987), Apollo 10 backup Command Module pilot
*Edgar D. Mitchell* (b. 1930), Apollo 10 backup Lunar Module pilot

### Summary of the Mission

In May of 1969, the thousands of people who went to Florida to witness the liftoff of Apollo 10 from Launch Complex Pad 39B were treated to an impressive sight. The Saturn V rocket stack rose thirty-six stories from the flatlands of Cape Kennedy. At the bottom of the stack stood the huge first stage, whose five engines would generate more than 346,698 newtons of thrust. Above that stood the second stage. Its five smaller engines would burn an efficient hydrogen/oxygen fuel as opposed to the kerosene/oxygen-burning lower stage. Atop the second stage rested the upper stage, known as the S-IVB. This stage also burned hydrogen and would provide the final bit of lift necessary to put the ship into Earth orbit. Nestled above the S-IVB, secured inside a protective conical shroud, lay the Lunar Module (LM). Fitting snugly onto the upper end of the shroud, the Command and Service Module (CSM) carried the three astronauts: Thomas P. Stafford, the commander; John W. Young, the Command Module pilot; and Eugene A. Cernan, the Lunar Module pilot. For this mission, Stafford had chosen the names *Charlie Brown*

and *Snoopy* for the CSM and LM, respectively. Finally, atop the CSM and completing Apollo 10, the needlelike solid-fueled rocket of the launch escape system was poised to fire and pull the Command Module (CM) free of the Service Module (SM) and the rest of the Saturn V stack in the event of an emergency abort. After being pulled free of danger, the astronaut-carrying Command Module could then parachute to safety. As mission commander, Stafford could initiate the escape system if it became necessary to do so.

All three astronauts were space veterans, making the Apollo 10 crew the most experienced ever to fly. Young had flown in the first piloted Gemini mission, and he later commanded the successful Gemini X. Stafford had been aboard Gemini VI-A and helped his commander, Walter M. Schirra, conduct the first space rendezvous with another piloted spacecraft, Gemini VII. Stafford had also had his turn to command a Gemini, flying Gemini IX with Cernan. Unlike his fellow crew members, Cernan had been up in space "only" once—but what a trip it had been. Cernan performed a heroic spacewalk, struggling valiantly to wrestle a huge maneuvering unit into place without a foothold or an armhold or gravity to help. The crew's combined experience would be welcome on this mission, which, if completely successful, would allow the next Apollo mission to attempt a Moon landing.

The countdown proceeded normally for this translunar mission, and at 16:49 Coordinated Universal Time (UTC, or 12:49:00 eastern standard time), Apollo 10 lifted off into the noonday sky. The first stage treated the astronauts to some bone-jarring, up-and-down oscillations called the pogo effect (so named because the sensation of such movements resembles that of a ride on a huge pogo stick). After two and a half minutes, the first stage completed its burn and was automatically jettisoned. Immediately, the second stage was fired. The pogo effect came back with this stage, harder than before, but the experienced crew ignored the turbulence. It was only when the second stage quit and the third stage was fired and gave them not pogo, but a jolting, side-to-side shaking, that the astronauts wondered if they would reach orbit intact. They did, and no damage had been done to any of the systems. Apollo 10 orbited Earth while the crew prepared for the next, crucial maneuver. Called the translunar injection, the move involved firing the S-IVB engine again to boost the ship out of Earth orbit and on its way to the Moon. The S-IVB ignited as planned but gave the crew such a turbulent and vibrating ride for its entire six-minute burn that they briefly considered aborting. They stuck with it, however, and when the S-IVB shut off, they discovered that it had put them on such an accurate trajectory that only one of the planned four course-correction maneuvers would be needed. For all of its rumblings, the Saturn V had performed well.

The spacecraft was now on course for the Moon. The trajectory was called "free return"—if nothing were done to stop it, the ship would swing once around the Moon, whose gravity would slingshot it back to Earth. This type of trajectory was considered the safest, because the ship would automatically return to Earth even if the Service Propulsion System (SPS) on the CSM failed to operate. Young, the pilot of CSM *Charlie Brown*, detached the ship from the shroud's adapter and pulled away from it. Then the shroud's panels were swung away, exposing the LM, *Snoopy*, which was attached to the end of the S-IVB. Young carefully turned the CSM around so that its nose was facing *Snoopy*. Then he brought the craft in for a nearly perfect docking. *Snoopy*, now attached to *Charlie Brown* like a faithful dog, was freed from the S-IVB. The "boy-and-his-dog" combination separated from the third stage, which was ordered to use its remaining propellant to send itself into solar orbit.

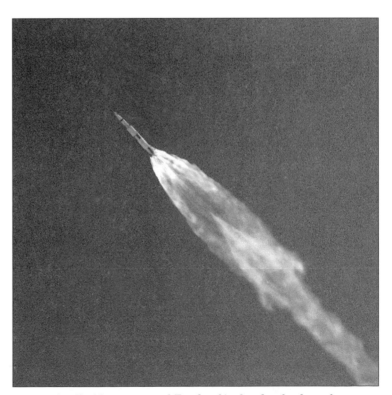

*Apollo 10 soars toward Earth orbit shortly after launch.*

Now began the long, coasting voyage to the Moon. Pilot Young put Apollo 10 into a slow roll so that the heat of solar radiation would evenly distribute itself over the entire surface of the ship. This "barbecue" maneuver, so named because the ship turned as if on a spit, prevented overheating of any one side of the ship.

The four-day trip to the Moon was punctuated by many television broadcasts from the crew. They seemed to delight in their marvelous color camera and eagerly sent back footage. Not only did they film the removal of *Snoopy* from the third stage, but they also allowed home viewers the opportunity to see the gradually shrinking globe of Earth from their spaceborne perspective. They telecast their own antics inside the Command Module as well, demonstrating life in space for everyone at home.

When not on camera, they kept themselves busy with housekeeping and navigational duties. Their only complaint concerned stomach cramps from the highly chlorinated water they drank. The Moon began to loom larger and larger in their windows as its gravity pulled them closer.

Finally, the time came for the lunar orbit-insertion maneuver. In order to prevent the spacecraft from simply following its free-return trajectory, the crew had to fire *Charlie Brown*'s engine as it circled behind the Moon. This would slow their speed enough to allow the Moon's gravity to bring them into a stable orbit. To use the minimum amount of fuel for this maneuver, the insertion had to take place at their closest approach to the Moon. That point would come on the Moon's far side, the side not seen from Earth. Because the Moon would be between Apollo 10 and Mission Control on Earth, they would be on their own, without radio contact until the maneuver had been completed.

Mission Control lost their signal as the craft swung behind the Moon. Young oriented the ship properly and, when the time came, fired the CSM's engine. The nearly six-minute thrusting went smoothly and precisely. When Apollo 10 regained radio contact with Earth, it was in an elliptical orbit with its apoapsis, the farthest point from the surface, at 315.6 kilometers, and its periapsis, or near-est point, at 110.4 kilometers—less than a kilometer from their ideal orbit. At the appropriate time, the engines were fired again, this time in a maneuver designed to change the orbit by lowering apoapsis to the level of periapsis. The burn successfully circularized the orbit.

Now secure in lunar orbit, it was time to prepare for the next stage of the mission. The hatch connecting *Charlie Brown* to *Snoopy* was opened for the first time, and Cernan crawled through to give the LM a thorough testing. He checked the control systems and then activated *Snoopy*'s power and communications circuits. Everything worked, and Cernan crawled back into the Command Module. The three astronauts took a much-needed rest period to fortify themselves for the busy day ahead.

Mission Control expressed concern over an observed misalignment of *Snoopy* and *Charlie Brown*. Instead of being docked together in a perfect line, they were off by an angle of 3.5°. Mission Control feared that when the two separated, they might damage the docking adapter and so prevent *Snoopy* from relinking with the CSM after its mission was completed. After the situation had been considered, it was decided to go ahead as planned, but extra care was to be taken during separation. Cernan and Stafford crawled through the tunnel into the LM, leaving Young with a spacecraft all to himself. *Charlie Brown* and *Snoopy* were then sealed and, while Apollo 10 was orbiting on the far side of the Moon, separated. When Mission Control regained contact, they had two ships with which to communicate; the reasons for giving the craft separate names became clear. Far from being a formality, the names kept communications straight.

While Young watched from above in his now-spacious CSM, Cernan and Stafford fired *Snoopy*'s descent engine in a maneuver called descent orbit insertion. This insertion put *Snoopy* into an orbit with an apoapsis of 113 kilometers and a periapsis of only 15.5 kilometers. The low gravity and the absence of air on the Moon allow orbits so low that lunar satellites may simply skim the tops of the mountains. Periapsis was planned to occur at the point over the lunar surface favored for an actual landing

*After splashdown, Apollo 10's Command and Service Module* Charlie Brown *floats on the South Pacific as divers assist the astronauts into a lifeboat.* (NASA)

mission, thus allowing Apollo 10 the opportunity to photograph the approach the landing mission would take. Both Stafford and Cernan were delighted by the impressive view their low orbit offered. They took special note of the landing approach, which they dubbed "U.S. 1"—as if it were a four-lane highway.

After closest approach, the descent engine fired again to make the orbit even more elliptical. It would still return for a 15.5-kilometer swoop over the projected landing site, but at the moment it would rise to 352 kilometers. The new apoapsis gave the slowly orbiting CSM time to catch up and pass the LM from below, putting the two craft into better positions for docking. Though there was some difficulty in putting *Snoopy* into the proper attitude for the burn, Cernan and Stafford accomplished the maneuver and coasted in their longer

orbit toward a second pass over the site. While the astronauts busied themselves with preparations for returning to the CSM, cameras recorded the future landing path in minute detail. *Snoopy's* crew planned to disengage the descent stage of the LM, and, at the moment of closest approach, fire the ascent engine in a series of maneuvers which would place them in a low, circular orbit beneath and behind the CSM.

Immediately after releasing the descent stage, however, the LM began to gyrate wildly, prompting a mild expletive from Cernan as they fought to regain control. Fortunately, the attitude control thrusters worked admirably and responsively without the added mass of the descent stage. They were able to whip the ship back around and recover stability in short order. The burn was nominal, with the ascent engine performing optimally. They coasted in their new orbit until they were in the proper position relative to the CSM and then began a series of burns to put them in range for a visual docking. Young did not have to touch the controls on the CSM; this docking was the responsibility of the LM crew. Despite the earlier worries about the docking adapter, *Snoopy* and *Charlie Brown* reunited with precision. Soon thereafter, Stafford and Cernan crawled back aboard the CSM to rejoin Young. They carried with them some equipment that had proved troublesome, so that an inspection could be performed on Earth.

After making some final checks, the crew waved good-bye to *Snoopy*, which they released into a solar orbit. The astronauts were given an extra day of lunar orbiting to rest after their busy day. Upon awakening, they prepared for the most crucial burn, the trans-Earth injection. The CSM's engines would have to fire, at the precise time and for a carefully controlled duration, so as to break Apollo 10 free of the Moon's pull and send the astronauts earthward. The SPS on the CSM had been designed for maximum simplicity and reliability with this particular maneuver in mind. After Apollo 10 disappeared behind the Moon on its thirty-first orbit, Mission Control anxiously waited for the craft's reemergence. If the burn was successful, the ship would swing around the Moon quickly. As hoped, the CSM resumed radio contact far in advance of an orbital prediction, confirming that the engines had ignited and Apollo 10 was heading home. The crew seemed particularly pleased with this burn, and for good reason. Their trajectory was accurate, and several course-correction maneuvers were abandoned. Indeed, the small thrust of a routine wastewater dump was sufficient to make the only minor course control adjustment that was needed.

The trip back to Earth took only two days. The astronauts occupied the time with several television transmissions for the benefit of people on Earth. As a joke on Mission Control personnel, the astronauts played a tape of Dean Martin singing "Going Back to Houston" that had been smuggled aboard. They also continued to perform several checks on the operation of the CSM; thus, every aspect of its functioning was known in detail.

Earth grew larger, until it filled the crew's windows. The crew of Apollo 10 made one last course correction and then jettisoned the now unnecessary SM part of the CSM. The conical CM had the only heatshield, so only it could survive reentry. The Module entered the atmosphere with its blunt end first. The astronauts prepared for a hard reentry. Unlike the Mercury and Gemini craft, Apollo craft returned to Earth from an escape velocity of 11 kilometers per second. Apollo 10 blazed through the atmosphere. The astronauts inside felt a force seven times the pull of normal gravity.

The engineers who had designed the CM had distributed its weight in such a way that the ship's center of mass was displaced from the central axis. This had the effect of tilting the ship slightly as it flew through the atmosphere, thereby generating an aerodynamic lift force, just like that on an airplane wing. Though small, the Module's lift could thus be used to steer the vessel during descent through the atmosphere.

Because the reentry speed was putting too great a load on the heatshield, the crew rotated the Module so that the line of the lift force would point straight up. The lift raised the Module out of the dense lower atmosphere long enough for it to cool. Then, it was allowed to sink back and finish reentry at a more sedate gravitational force of 4. Parachute deployment occurred in two stages: First the drogues came out, and then the main chutes unfurled and slowed Apollo 10 to a graceful, upright splashdown. The crew and the CM were recovered in record time by the recovery vessel *Princeton*. As it had been for Apollo 8 and would be for all successive lunar-launch Apollos, the recovery area was in the Pacific Ocean. The quick recovery time was the result of a precise reentry, a fitting end to a successful mission.

**Contributions**

After splashdown, the mission was over for the crew of Apollo 10, although there would be endless debriefing sessions. The real work that remained was for the scientists and engineers, who were to examine in detail every bit of information the vessel and its crew had gathered during the voyage. Apollo 10 had proved the reliability of the LM in lunar orbit and had made several passes at a proposed landing site. The critically important propulsion systems, two on the LM and one on the CSM, all had done their jobs. The SPS on the CSM had demonstrated not only its crucial restart ability but also a remarkable accuracy in its steady, lambent burns.

The descent engine on the LM had performed its several restarts successfully as it simulated an actual landing. For the real thing, however, it would also have to deliver thrust evenly enough to allow the craft to hover briefly over the surface while the pilot scanned for a place to land. The engineers were now confident that it could do this.

Finally, the ascent engine on the LM had to function perfectly or there would be no going home for the two lunar explorers. As it turned out, the engine itself had done what it had to do, but the wild gyrations that had occurred after separation from the descent stage had to be examined. Eventually, it was discovered that the cause of the problem was as simple as a switch that had been left in the wrong position.

The LM contained, as part of its emergency equipment, a backup navigation system. This piece of equipment was called the Abort Guidance System (AGS). It was designed to activate should a landing abort be called. It would detect the abort by sensing the release of the descent stage prior to a landing. Because Apollo 10 did not land, when the descent stage was released the Abort Guidance System went into abort mode. A single switch told it what to do in the event of an abort. If the switch had been set properly for this mission, it would have instructed the system to respond to the abort by doing nothing more than maintaining the spacecraft's present attitude and allowing the crew time to determine what needed to be done.

In this case, however, ground crews had left the switch in the automatic position. In this position, the system was effectively told to find the way home. Thus, it would activate, firing the thrusters to search the sky for a beacon on the orbiting CSM. Once found, it would lock on with its radar to give data for the rendezvous. This is what happened to Stafford and Cernan. At the moment of separation, the system caused *Snoopy* to pivot around, looking for *Charlie Brown*. The crew interpreted it as an uncontrolled gyration and quickly went to manual control.

The pogo oscillations of the first two stages and the shaking of the third stage of the Saturn V were never adequately explained. These problems had developed on Apollo 6, a robotic mission, and had been presumed fixed. Nevertheless, the magnitude of the pogo effect had been more severe on Apollo 6—it had actually done damage to the ship.

One final result of the mission was greater confidence in the practical performance of the rocket: It had delivered Apollo 10 to a near-perfect tra-jectory, although the price of lunar exploration seemed to include a bumpy ride.

**Context**

The Apollo Program traced its origins to a speech made in 1961 by President John F. Kennedy. He had made a piloted lunar landing a national goal, to be completed by the end of the decade. For a long time, the effort was seen as mainly a "Space Race" with the Soviet Union, whose efficient space program had been the first to launch a satellite and the first to put a human in space. It was clear that the Soviets were also working on a lunar mission, probably a circumlunar orbiting vessel. Their achievements spurred American efforts considerably. The tragedy of Apollo 1, which had killed three U.S. astronauts, had set the American program back a year. It also provoked an acrimonious response from the Soviet Union, which maintained that the Americans were carelessly wasting human life in their blind struggle for a dubious space supremacy. Ironically, the Soviets experienced their own space disaster shortly thereafter, losing one of their own cosmonauts.

By the summer of 1968, the United States had yet to launch a single piloted Apollo mission. To meet Kennedy's challenge by the end of 1969, everything would have to work perfectly. Apollo 7, the first piloted mission, went up in October. Then came Apollo 8: Launched in late December, it put an end to the Space Race by successfully orbiting humans around the Moon. The Soviets seemed to lose heart and never bothered to follow up on their initial plans. Nevertheless, Apollo 8 had no Lunar Module; thus, it could not be a final test of all the systems necessary for a landing. Apollo 9 was launched in March of 1969. It did have an LM, but it remained in Earth orbit and conducted crucial docking tests. The responsibility for a final dress rehearsal fell to Apollo 10. With a planned circumlunar trajectory and a fully functioning LM, this mission would have to prove, once and for all, the complete reliability of all Apollo systems. This it did. The crew's views of the landing sites, combined with the carefully produced lunar maps made by robotic probes, gave the National Aeronautics and Space Administration planners the final bit of confidence they needed. It was decided that Apollo 11 would be the world's first attempt at a piloted lunar mission. Already, the Moon had been visited by orbiting and soft-landing probes from both the United States and the Soviet Union. Now, however, with a scheduled launch in the summer of 1969, thanks to the success of Apollo 10, the United States was ready to fulfill the nation's dream.

**See also:** Apollo Program; Apollo Program: Command and Service Module; Apollo Program: Developmental Flights; Apollo Program: Lunar Module; Apollo 1; Apollo 7; Apollo 9; Apollo 11; Apollo 12; Apollo 13; Apollo 17; Apollo-Soyuz Test Project; Gemini V; Mercury-Atlas 9; Space Shuttle Flights, 1983; Space Stations: Origins and Development.

### Further Reading

Allday, Jonathan. *Apollo in Perspective: Spaceflight Then and Now.* Philadelphia: Institute of Physics, 1999. This book takes a retrospective look at the Apollo Program and the technology that was used to land humans on the Moon.

Baker, David, ed. *Jane's Space Directory, 2005-2006.* Alexandria, Va.: Jane's Information Group, 2005. This reference book, updated annually, is invaluable for obtaining general information on virtually any piloted or robotic mission, both those launched by the United States and those launched by other countries. Includes illustrations and a useful index.

Collins, Michael. *Liftoff: The Story of America's Adventure in Space.* New York: Grove Press, 1988. Written by an Apollo 11 astronaut, this is an excellent account of the American space program. It provides, in narrative fashion, a thorough description of the human side of space exploration.

Cortwright, Edgar, ed. *Apollo Expeditions to the Moon.* NASA SP-350. Washington, D.C.: Government Printing Office, 1975. A collection of essays on the Apollo Program covering all phases, from the planning stages through the conclusion of each mission; ends with a useful article on the scientific legacy of Apollo. Illustrated.

Ertel, Ivan D., and Roland W. Newkirk. *The Apollo Spacecraft: A Chronology.* Vol. 4. NASA SP-4009. Washington, D.C.: Government Printing Office, 1978. This document contains all the information on the actual Apollo Program missions and is a detailed chronology of memos, dispatches, and conferences that provides information on the specific procedures used for operating the Apollo Program. The text includes an appendix on the Apollo organization and the contractors responsible for building the various Apollo subsystems. Illustrated.

Godwin, Robert, ed. *Apollo 10: The NASA Mission Reports.* Burlington, Ont.: Apogee Books, 1999. This collection contains reprints of the Apollo 10 press kit, "Pre-Launch Mission Operation Report," "Post Launch Mission Operation Report," and the "Post Launch Mission Operation Report Supplement." It also includes a CD-ROM featuring images taken during the mission and detailed drawings of the spacecraft.

Kelly, Thomas J. *Moon Lander: How We Developed the Apollo Lunar Module.* Washington, D.C.: Smithsonian Books, 2001. Grumman Chief Engineer Kelly gives a firsthand account of designing, building, testing, and flying the Apollo Lunar Module.

Lambright, W. Henry. *Powering Apollo: James E. Webb of NASA.* Baltimore: Johns Hopkins University Press, 2000. Explores James E. Webb's leadership role in NASA's spectacular success.

Lindsay, Hamish. *Tracking Apollo to the Moon.* London: Springer-Verlag London Limited, 2001. Features interviews, quotes, and extensive photographs, including some that appear here for the first time.

National Aeronautics and Space Administration. *Apollo Mission Press Kits.* http://www-lib.ksc.nasa.gov/lib/presskits.html. Provides detailed preflight information about each of the Apollo missions, Apollo 6 through Apollo 17. Accessed March, 2005.

_____. *Apollo 10 Mission Report.* Washington, D.C.: National Technical Information Services, 1969. This is the official report of the results of the Apollo 10 flight. The original is out of print, but copies can be obtained through the National Technical Information Services. An Internet search will turn up a copy in Adobe Acrobat format for easy download. Nowhere else will you find as in-depth a document on the technical, scientific, and human aspects of the mission.

Shepard, Alan B., Jr., and Donald K. "Deke" Slayton, with Jay Barbree and Howard Benedict. *Moon Shot: The Inside Story of America's Race to the Moon.* Atlanta: Turner, 1994. This is, indeed, the inside story of the Apollo Program as told by two men who actively participated in it. Some of their tales appear here for the first time. The book was adapted for a four-hour documentary in 1995.

Slayton, Donald K., with Michael Cassutt. *Deke! U.S. Manned Space: From Mercury to the Shuttle.* New York: Forge, 1995. This is the autobiography of the last of the Mercury astronauts to fly in space. After being grounded from flying in Project Mercury for what turned out to be a minor heart murmur, Slayton was appointed head of the Astronaut Office. During his reign he assigned all of the Apollo crew members to their flights. Later, he commanded the Apollo-Soyuz Test Project flight in 1975. This is a behind-the-scenes look at America's attempt to land humans on the Moon.

*Greg Tomko-Pavia*

# Apollo 11

*Date:* July 16 to July 24, 1969
*Type of mission:* Lunar exploration, piloted spaceflight

*Apollo 11 was the first mission to land humans on the Moon and return them to Earth, meeting a goal articulated by President John F. Kennedy at the height of the Cold War. Observations by the astronauts, experiments using equipment they deployed, and studies of samples they brought back greatly increased our knowledge of the Moon and prepared for further discoveries by the Apollo Program.*

## Key Figures

*Neil A. Armstrong* (b. 1930), Apollo 11 mission commander
*Edwin E. "Buzz" Aldrin, Jr.* (b. 1930), Apollo 11 Command Module pilot
*Michael Collins* (b. 1930), Apollo 11 Lunar Module pilot
*James A. Lovell, Jr.* (b. 1928), Apollo 11 backup mission commander
*William A. Anders* (b. 1933), Apollo 11 backup Command Module pilot
*Fred W. Haise, Jr.* (b. 1933), Apollo 11 backup Lunar Module pilot
*Thomas O. Paine* (1921-1992), NASA administrator (1969-1970)
*George E. Mueller* (1918-2001), associate administrator for Manned Spaceflight
*Wernher von Braun* (1912-1977), director of the Marshall Space Flight Center
*Robert R. Gilruth* (1913-2000), director of the Manned Spacecraft Center, Houston
*Christopher C. Kraft, Jr.* (b. 1924), director of Flight Operations, Mission Control, Houston
*Kurt H. Debus* (1908-1983), director of the Kennedy Space Center
*Rocco A. Petrone* (b. 1926), director of Launch Operations, Kennedy Space Center

## Summary of the Mission

The Apollo 11 mission was the first to land people on the Moon and bring them back, meeting President John F. Kennedy's challenge made in an address to Congress on May 25, 1961:

> I believe that this nation should commit itself to achieving the goal, before this decade is out, of landing a man on the Moon and returning him safely to Earth.

Astronauts Neil A. Armstrong, Edwin E. "Buzz" Aldrin, Jr., and Michael Collins took with them to the Moon a collection of instruments known as the Early Apollo Scientific Experiments Package (EASEP) and brought back to Earth photographs of the lunar landscape and 22 kilograms of rock and soil samples for later study.

The vehicle, known as the Lunar Module (LM), which carried Armstrong and Aldrin to the Moon's surface, was not streamlined; it had no need to be, because it operated only in airless space. With an overall height of nearly 7 meters and a diagonally measured width of 9.5 meters, the LM had a launch weight slightly greater than 15 metric tons on Earth. The cylindrical crew compartment, 2.35 meters in diameter and 1.07 meters deep, provided two stand-up flight stations with harness restraints, front windows, display panels, and landing controls. The LM included a throttleable descent engine, providing up to 458 newtons of thrust, and an ascent engine, providing up to 163 newtons of thrust, each operating on a supply of Aerozine-50 rocket fuel. Other major equipment included liquid propellant (monomethylhydrazine and nitrogen tetroxide), reaction thrusters for fine control of attitude (the orientation of the LM and Command and Service Module, or CSM), VHF and S-band antennae, a platform and ladder leading from the crew compartment to the lunar surface, and a docking hatch and radar for linking with the Command and Service Module.

The CSM consisted of a pressurized Command Module (CM), in which all three astronauts traveled, and a Service Module (SM), which contained electrical power and environmental control equipment, a movable S-band antenna array, and the Service Propulsion System (SPS), which could develop 948 newtons of thrust. The conical CM had a length of 3.2 meters, a maximum diameter of 3.9 meters, and a launch weight (with astronauts) of 5.9 metric tons. The cylindrical SM had a length of 7.4 meters, a diameter of 3.9 meters, and a launch weight of 24.5 metric tons.

On June 12, 1969, two weeks after the successful conclusion of the Apollo 10 mission, NASA announced that the next Apollo flight would include the first landing on the Moon. Nevertheless, it was stated that this crew would not be launched on the expected date unless everything was ready; in addition, the public was informed that NASA would not hesitate to bring the crew home if problems were encountered. All was ready at the expected time, and Apollo 11 lifted off from Launch Complex Pad 39A at the Kennedy Space Center on a Saturn V launch vehicle at exactly 13:32 Coordinated Universal Time (UTC, or 9:32 A.M. eastern daylight time) on July 16, 1969.

*The crew of Apollo 11 (from left): Mission Commander Neil A. Armstrong, Command Pilot Michael Collins, and Lunar Module Pilot Buzz Aldrin. (NASA)*

The launch and the first phases of the mission so closely followed the nominal plan that only one of four scheduled midcourse corrections was required during the trip to the Moon. On July 19, the SPS was fired to put Apollo 11 into an elliptical lunar orbit, which was made more nearly circular with a later rocket burn. The following day, after all three astronauts had slept, Armstrong and Aldrin crept through the 0.8-meter-diameter tunnel from the CM, named *Columbia*, to the LM, named *Eagle*. At the start of the thirteenth orbit, 15 kilometers above the lunar surface, *Eagle* separated from *Columbia* and began its descent toward a target in the Sea of Tranquility, one of five smooth landing sites near the lunar equator that had been selected from photographs supplied by robotic Lunar Orbiters. At 2.2 kilometers, with the landing site about 8 kilometers ahead, the guidance computer controlling the descent engine ordered ma-

*July 16, 1969: Liftoff of Apollo 11 aboard a Saturn V launch vehicle. Armstrong, Collins, and Aldrin are on their way to the Moon.* (NASA)

get as a result of errors in the data fed to the guidance computer, but the landing was well within the planned ellipse, 12 kilometers long and 5 kilometers wide. Armstrong reported, "Houston, Tranquility Base here. The *Eagle* has landed."

At 02:56 UTC on July 21 (10:56 P.M. on July 20 eastern time), Neil A. Armstrong became the first person from Earth to set foot on another celestial body, saying, "That's one small step for man, one giant leap for mankind." The article "a," missing in the voice transmission and the original transcript, was later inserted in the record; thus, the message includes the phrase "one small step for *a* man," referring to the short gap between the LM's landing pad and lunar surface, over which Armstrong was required to jump. While coming down the ladder, he had pulled a lanyard that opened an equipment compartment and deployed a monochrome television camera. Armstrong's first priority was to collect a "contingency sample" of soil and small rocks, sealed in a plastic bag and stored in a pocket above his left

neuvers putting *Eagle* into a vertical orientation and giving Armstrong and Aldrin their first close view of the plain below. At an altitude of 150 meters, dropping toward the surface at about 6 meters per second, Armstrong suddenly took over manual control when he saw that the automatic system was taking the Module onto a surface littered with boulders and rocks. For about 90 seconds, he adjusted the craft's position, searching for a clear spot, and he retained semimanual control until touchdown, allowing the computer to control the firing of the descent engine. The Module's legs touched the surface at 20:17 UTC (4:17 P.M. eastern daylight time) on July 20. It landed at a point about 5 kilometers to the west of the nominal tar-

knee, as an assurance that some lunar material would be returned to Earth even if the mission had to be cut short. Edwin E. "Buzz" Aldrin, Jr., joined him on the lunar surface 19 minutes later. They set a black-and-white television camera away from *Eagle* to give people on Earth a broad view of the Sea of Tranquility, and then read the statement engraved on a plaque attached to the front landing gear: "Here men from the planet Earth first set foot upon the Moon, July, 1969 A. D. We came in peace for all mankind." They planted a 0.9-meter-by-1.5-meter American flag, stiffened with wires so as to appear to be flying, and had a brief radio-telephone conversation with President Richard M. Nixon.

Armstrong and Aldrin's primary scientific tasks were to gather samples of rocks and soil for return to Earth and to deploy the EASEP. The first experiment was to unroll a sheet of aluminum foil to trap particles of the solar wind (the high-temperature ions and electrons that flow out from the Sun). After 77 minutes, it was rolled up and placed in a vacuum box, which would be returned to Earth for analysis. The second experiment, left on the Moon, was an array of four solar-powered seismometers; these instruments were to record and broadcast information on any tremors, whether caused by internal seismic or volcanic activity or by external meteorite or other bombardment. The first signals were of the astronaut's footsteps. The final apparatus left on the Moon was a square reflector, made up of one hundred fused-silica prisms, designed to return pulses of laser light sent from Earth, thus allowing precise determinations of the distance from Earth to the Moon. Because of the difficulty of bending to the ground in their pressurized space suits, Aldrin and Armstrong used core tubes, a scoop on an extension handle, and long tongs to gather lunar soil and a representative sample of lunar rocks. These were placed into two boxes, each formed from a single piece of aluminum with a lining of the soft metal indium around its lip. When a box was closed and had straps drawn tightly around it, a sharp strip around the edge bit deeply into the indium, thus sealing the samples in a vacuum to protect them from contamination. After 2.5 hours on the Moon, Armstrong returned to the *Eagle* (Aldrin had returned before him) to rest and prepare for the return trip.

At 17:55 UTC (1:55 P.M.) on July 21, the ascent engine of *Eagle* was ignited, lifting off and leaving behind the Module's lower part, while Michael Collins piloted *Columbia* through its twenty-fifth lu-

*A rare photograph, taken by Lunar Module Pilot Buzz Aldrin, of Neil Armstrong on the Moon's surface; most Apollo 11 images are of Aldrin. Armstrong is standing near the Lunar Module's Modularized Equipment Stowage Assembly.* (NASA)

nar orbit. After a series of four maneuvers with the thruster jets, the two craft docked at 21:35 UTC (5:35 P.M.) 110 kilometers above the lunar surface. The trip back to Earth was uneventful. At 16:22 UTC (12:22 P.M.) on July 24, the astronauts jettisoned the SM to start the final descent into Earth's atmosphere. The CM landed upside down, but it was soon righted, only 18 kilometers from the recovery aircraft carrier, the USS *Hornet*. The three astronauts spent the next three days in a trailer aboard the carrier, which returned to Hawaii; then they were transported by plane and truck to the Lunar Receiving Laboratory in Houston, where they remained in quarantine until August 11, undergoing extensive postflight debriefings.

**Contributions**

Analysis of the soil and rock samples showed no fossil life, living organisms, or any organic material except for minute traces that scientists believed were the result of contamination from the collection boxes or the Lunar Receiving Laboratory. To test for pathogens (disease-producing agents), biologists inoculated two hundred mice that had been bred in a completely sterile environment and thus had essentially no immunity to disease, with finely ground particles of lunar material. The mice experienced no ill effects.

The material in the Sea of Tranquility is igneous (fire-formed) rock, once molten at a temperature in excess of 1,200° Celsius. Though resembling terrestrial basalt in terms of its elements (mostly oxygen, silicon, iron, aluminum, titanium, calcium, and magnesium), the rocks are not like Earth rocks in their proportions. Lunar basalts are rich in titanium and other refractories (high-melting-point materials) and relatively poor in volatiles (low-melting-point materials), such as sodium and potassium. Dating by the rate at which radioactive potassium has been converted into argon indicates that the Sea of Tranquility has been solid for at least three billion years.

Glass makes up fully half of the soil samples returned; most of it is in irregular fragments, though about 5 percent of the glass is in the form of drops and globules in brown, green, deep red, and yellow. Erosion processes analogous to sandblasting have smoothed and rounded the surfaces of the rocks. Most specimens have small glass-rimmed pits or glassy areas, suspected to be the result of micrometeorite bombardment, because the rocks and soil show ample evidence of such impacts.

There was no evidence for the existence of water on the Sea of Tranquility at any time since the rocks first came to the surface; they show no evidence of rounding by water erosion and are now extremely dry, and their mineral composition indicates that the liquid from which they crystallized had negligible amounts of water chemically bound within it.

The particles of solar wind trapped in the aluminum foil were liberated as the foil was vaporized in high vacuum. Mass spectrographic analysis of the isotopic composition of helium, neon, and argon led to subtle refinements in models of the origin of the solar system.

The 3-meter telescope at the Lick Observatory on Mount Hamilton in California and the 2.7-meter telescope at the McDonald Observatory on Mount Locke in Texas were used to bounce laser light from the retroreflector; the distance to the Moon was determined with an uncertainty of not more than 15 centimeters.

Information returned by the lunar seismometer, one hundred times as sensitive as terrestrial seismometers at the time of this mission, showed that the Moon is much quieter internally than is Earth. Several trains of high-frequency waves were attributed to landslides, which were the result of stresses associated with the change from extreme heat to extreme cold, in a young, nearby crater. Several fairly strong shocks with lower frequencies, at first thought to be moonquakes, were later attributed to the Module's release of gases or to abnormalities within the instruments. The instruments themselves survived one lunar "day" (14 Earth days of continuous sunlight) and one "night," during which they did not broadcast to Earth because of the lack of solar-generated electrical power. The electronics failed in the next light cycle.

*Buzz Aldrin walking away from the Lunar Module,* Eagle, *and facing a solar wind experiment.* (NASA)

### Context

Apollo 11 was the third piloted Apollo mission to carry American astronauts into orbit around the Moon, having been preceded by Apollo 8 and Apollo 10. It was followed by six more Apollo missions; each of these later missions took three astronauts into orbit around the Moon. On all subsequent missions except Apollo 13, two of the astronauts made a lunar landing. Thus, twelve Americans walked on the surface of the Moon during the Apollo Program; a total of twenty-four saw the surface from lunar orbit.

Since the invention of telescopes around 1600, the Moon has been known to be covered with crat-

ers, mountains, valleys, and the vast plains called maria, or seas—a term dating to the time when astronomers believed that there were oceans on the Moon. Although only one side of the Moon ever faces Earth, it is nevertheless possible to see 59 percent of the Moon's surface from Earth. Because the Moon's axis of rotation is not exactly perpendicular to the plane of its orbit, a small portion of the far side beyond the lunar north or south pole can sometimes be seen. Because the Moon's orbital speed in its elliptical path varies according to Kepler's third law, while its rotational speed on its axis is essentially constant, astronomers can occasionally see slightly beyond the eastern and western limbs (the visible edges) of the Moon. Nevertheless, features near the limbs or poles are viewed so obliquely that their shapes are difficult to determine; objects less than about 0.5 kilometer in diameter cannot be resolved in any earthbound telescope.

Exploration of the Moon by spacecraft began in the early 1960's with three successful hard landings of the Ranger series; the two wide-angle and four narrow-angle television cameras aboard each craft transmitted close-up views during the descents. The resulting 13,453 images included some from altitudes of less than 1 kilometer; they showed details as small as 1 meter in diameter. The obvious next step in preparing for the piloted landings, soft-landing instrument packages, was accomplished during the Surveyor series. Shortly after the first of these missions, the Lunar Orbiter missions began; they used photographic film that was processed and electronically scanned. Five successful orbiters transmitted data to Earth, where 1,950 high-resolution photographs were reconstructed, covering more than 99 percent of the lunar surface; these photographs formed the basis for the selection of the Apollo Program's landing sites.

The great age of the rocks returned by Apollo 11 was initially surprising to many geologists, and whether the source of the heat that formed them was from a volcanic lunar interior or impacts by large meteorites remains undetermined. While many questions about the nature and history of the Moon and the solar system were answered by data from the Apollo 11 mission, others were raised.

Materials brought back by later Apollo missions proved helpful in understanding new puzzles. Data were also broadcast back to Earth by the five Apollo Lunar Surface Experiments Packages (ALSEPs) that they deployed, which were redesigned on the basis of experience gained from this mission. For example, ALSEP electronics were protected from the heat of the lunar "day" by heat-radiating thermal blankets.

More than ten years of precise laser testing of the distance between Earth and the Moon using the retroreflectors—which were part of EASEP and all ALSEPs—provided stringent tests of the general theory of relativity, Albert Einstein's theory of gravity, and the first direct experimental measurement of the rate of continental drift. These are important in understanding the geologic activity of the crust and have wide-ranging implications for the environment on Earth.

**See also:** Apollo Program; Apollo Program: Command and Service Module; Apollo Program: Geological Results; Apollo Program: Lunar Lander Training Vehicles; Apollo Program: Lunar Module; Apollo Program: Orbital and Lunar Surface Experiments; Apollo 10; Apollo 12; Apollo 13; Apollo-Soyuz Test Project; Astronauts and the U.S. Astronaut Program; Extravehicular Activity; Gemini Program; Lunar Exploration; National Commission on Space.

### Further Reading

Aldrin, Edwin E. "Buzz," Jr., and Malcolm McConnell. *Men from Earth*. New York: Bantam Books, 1991. Everyone knows Neil A. Armstrong's first words as he stepped onto the Moon, but few people know that during the walk he touched Buzz Aldrin on the shoulder and said, "Isn't it fun?" As one of the first men on the Moon, Aldrin tells the story of America's journey to the lunar surface from a perspective unavailable to most. His description of the Apollo 11 landing reads like a James Michener novel. One of the most difficult questions for an astronaut to answer is, "What's it like?" Aldrin has found a way to answer that question so that the reader can easily envision the desolation of the lunar surface. A briefly annotated but extensive bibliography and photographs are included.

Allday, Jonathan. *Apollo 11: The NASA Mission Reports*. Burlington, Ont.: Apogee Books, 1999. Includes the postflight debriefing of the astronauts, as well as the entire uncut television broadcast of the moonwalk on CD and an exclusive interview with Buzz Aldrin conducted in June, 1999.

_____. *Apollo in Perspective: Spaceflight Then and Now.* Philadelphia: Institute of Physics, 1999. This book takes a retrospective look at the Apollo Program and the technology that was used to land humans on the Moon.

Armstrong, Neil, Michael Collins, and Edwin E. Aldrin. *First on the Moon.* New York: Williams Konecky Associates, 2002. Armstrong, Collins and Aldrin give us the exclusive story of Apollo 11, from the earliest preparations to the final touchdown back on planet Earth. Photographs accompany the text.

Baker, David. *The History of Manned Space Flight.* New York: Crown, 1982. A comprehensive, chronological account of human flight beyond Earth's atmosphere from pre-World War II theorizing and rocket development through the beginnings of the first successes in the space shuttle program. Illustrated with monochrome and color photographs. Contains tables about hardware, missions, and participants.

Bedini, Silvio A., Wernher von Braun, and Fred L. Whipple. *Moon: Man's Greatest Adventure.* New York: Harry N. Abrams, 1969. This profusely illustrated volume includes essays on the human fascination with the Moon throughout history and transcripts of conversations from the lunar surface. Concludes with a chronology of significant flights up to the lunar landing; includes charts, maps, and a fold-out diagram of the Apollo 11 flight.

Chaikin, Andrew. *A Man on the Moon: The Voyages of the Apollo Astronauts.* New York: Penguin Group, 1998. The Apollo Moon landings are retold through the eyes and ears of the people who were there. Based on interviews with twenty-three Moon voyagers, as well as those who struggled to get the program moving, this book conveys every aspect of the missions with breathtaking immediacy: from the rush of liftoff to the heart-stopping lunar touchdown to the final hurdle of reentry.

Collins, Michael. *Carrying the Fire: An Astronaut's Journeys.* New York: Farrar, Straus and Giroux, 1974. Everyone knows Neil A. Armstrong was the first human to walk on the Moon. Most remember that Buzz Aldrin went with him to the lunar surface. Few can say who waited in orbit while the other two "made history." Michael Collins was the only person who could not have watched the lunar landing on television. He was too busy in *Columbia* trying to locate the LM *Eagle* through his low-powered telescope and making sure Armstrong and Aldrin had a place to which to return. Collins relays his feelings about the flight and what he would have done if *Eagle* and its crew had been stranded on the Moon. This personal glimpse at the historic flight emphasizes the human element of the story—perhaps the most neglected aspect.

_____. *Liftoff: The Story of America's Adventure in Space.* New York: Grove Press, 1988. Many books have been written about the Apollo Program, most of them about Apollo 11. Few have given us an inside look at the delicate melding of man and machine. Contributing to this complete history of America's piloted space programs, Collins devotes a large portion to Apollo. He sets the record straight about some of the misconceptions of astronauts and space machines. The book is illustrated with eighty-eight line drawings by James Dean, former NASA art director, which add stark realism to an otherwise unreal world.

Godwin, Robert, ed. *Apollo 11: The NASA Mission Reports.* Vol. 1. Burlington, Ont.: Apogee Books, 1999. This collection contains reprints of the Apollo 11 press kit, preflight mission operation report, postflight mission operation report, and the Apollo 11 postflight press conference. The CD-ROM includes two thirty-minute movies of the flight as well as more than thirteen hundred still pictures taken during the mission.

_____. *Apollo 11: The NASA Mission Reports.* Vol. 2. Burlington, Ont.: Apogee Books, 1999. This collection contains a reprint of the Apollo 11 technical crew debriefing. The CD-ROM includes an exclusive interview with Dr. Buzz Aldrin, a unique interactive panoramic image of Tranquility Base, and the entire unedited television broadcast from Tranquility Base.

_____. *Apollo 11: The NASA Mission Reports.* Vol. 3. Burlington, Ont.: Apogee Books, 2002. Volume 3 of this set includes the Apollo 11 mission report, detailing the results of the first piloted lunar-landing mission. The book includes a DVD of an exclusive movie, *Apollo 11: Moon Walk,* a 140-minute composition from two camera angles with unique panoramas and still images.

Kaufmann, William J. *Exploration of the Solar System.* New York: Macmillan, 1978. A comprehensive survey of the solar system sciences, including a chapter that synthesizes knowledge about the Moon gained from space exploration with that available previously. Includes a glossary, tables of data about planets and their satellites, a log of missions launched from 1962 through 1977, and an index.

Kelly, Thomas J. *Moon Lander: How We Developed the Apollo Lunar Module.* Washington, D.C.: Smithsonian Books, 2001. Grumman Chief Engineer Kelly gives a firsthand account of designing, building, testing, and flying the Apollo Lunar Module.

Kraft, Christopher C., Jr. *Flight: My Life in Mission Control.* East Rutherford, N.J.: Penguin Putnam, 2002. The first NASA flight director gives an account of his life in Mission Control.

Lambright, W. Henry. *Powering Apollo: James E. Webb of NASA.* Baltimore: Johns Hopkins University Press, 2000. Lambright explores James E. Webb's leadership role in NASA's spectacular success.

Levinson, Alfred Abraham, et al., eds. *Proceedings of the Apollo 11 Lunar Science Conference.* 3 vols. Elmsford, N.Y.: Pergamon Press, 1970. A complete record of a four-day discussion of the results of three months of intensive investigation of the lunar samples returned by Apollo 11. Mostly accessible to a college-level, nontechnical student audience. Each of the 180 articles ends with references for further study.

Liebergot, Sy, and David Harland. *Apollo EECOM: Journey of a Lifetime.* Burlington, Ont.: Apogee Books, 2003. This is the life story of Sy Liebergot, former NASA flight controller, with emphasis on his years working in Apollo Mission Control.

Lindsay, Hamish. *Tracking Apollo to the Moon.* London: Springer-Verlag London Limited, 2001. Features interviews, quotes, and extensive photographs, including some that appear here for the first time. There are numerous pictures and illustrations.

Mason, Brian, and William G. Melson. *The Lunar Rocks.* New York: Wiley-Interscience, 1970. A concise and coherent review of the scientific effort expended in analyzing the lunar samples returned by Apollo 11, this work gives an interpretation of the results, including implications for understanding lunar history. Illustrated with monochrome photographs, graphs, and tables. Concludes with an extensive list of references and index. College-level material.

National Aeronautics and Space Administration. *Apollo 11 Mission Report.* Washington, D.C.: National Technical Information Services, 1969. This is the official report of the results of the Apollo 11 flight. The original is out of print, but copies can be obtained through the National Technical Information Services. An Internet search will turn up a copy in Adobe Acrobat format for easy download. Nowhere else will you find as in-depth a document on the technical, scientific, and human aspects of the mission.

——————————. *Apollo Mission Press Kits.* http://www-lib.ksc.nasa.gov/lib/presskits.html. Provides detailed preflight information about each of the Apollo missions, Apollo 6 through Apollo 17. Accessed March, 2005.

Shepard, Alan B., Jr., and Donald K. "Deke" Slayton, with Jay Barbree and Howard Benedict. *Moon Shot: The Inside Story of America's Race to the Moon.* Atlanta: Turner, 1994. This is, indeed, the inside story of the Apollo Program as told by two men who actively participated in it. Some of their tales appear here for the first time. The book was adapted for a four-hour documentary in 1995.

Siddiqi, Asif A. *The Soviet Space Race with Apollo.* Gainesville: University Press of Florida, 2003.

——————————. *Sputnik and the Soviet Space Challenge.* Gainesville: University Press of Florida, 2003. This is the first comprehensive history of the Soviet piloted space programs, covering a period of thirty years.

Slayton, Donald K., with Michael Cassutt. *Deke! U.S. Manned Space: From Mercury to the Shuttle.* New York: Forge, 1995. This is the autobiography of the last of the Mercury astronauts to fly in space. After being grounded from flying in Project Mercury for what turned out to be a minor heart murmur, Slayton was appointed head of the Astronaut Office. During his reign he assigned all of the Apollo crew members to their flights. Later, he commanded the Apollo-Soyuz Test Project flight in 1975. This is a behind-the-scenes look at America's attempt to land humans on the Moon.

Swanson, Glen E. *Before This Decade Is Out: Personal Reflections on the Apollo Program.* Gainesville: University Press of Florida, 2002. This collection of oral histories of the Saturn/Apollo Program recounts the unique adventure from the perspective of political leaders, engineers, scientists, and astronauts who made it such a success.

Turnill, Reginald. *The Moonlandings: An Eyewitness Account.* New York: Cambridge University Press, 2003. Turnill spent his career covering all the piloted space missions as well as planetary missions like Mariner, Pioneer, Viking, and Voyager.

Ulivi, Paolo, and David M. Harland. *Lunar Exploration: Human Pioneers and Robotic Surveyors.* London: Springer-Verlag London Limited, 2004. The authors provide a well-paced, rapidly moving, balanced, even-handed account of lunar exploration in this popular history.

Wagener, Leon. *One Giant Leap: Neil Armstrong's Stellar American Journey.* New York: Forge Books, 2004. This first biography of Neil A. Armstrong relies on hundreds of interviews with family and friends and on NASA files.

Wendt, Guenter, and Russell Still. *The Unbroken Chain.* Burlington, Ont.: Apogee Books, 2001. The autobiography of the only person who worked with every astronaut bound for space.

*John J. Dykla*

# Apollo 12

*Date:* November 14 to November 24, 1969
*Type of mission:* Lunar exploration, piloted spaceflight

*Apollo 12 was the second successful lunar landing. The astronauts completed two separate moonwalks, set up an automated scientific analysis package, and recovered equipment from the Surveyor 3 probe.*

## Key Figures

*Charles "Pete" Conrad, Jr.* (1930-1999), Apollo 12 mission commander
*Richard F. Gordon, Jr.* (b. 1929), Apollo 12 Command Module pilot
*Alan L. Bean* (b. 1932), Apollo 12 Lunar Module pilot
*David R. Scott* (b. 1932), Apollo 12 backup mission commander
*Alfred M. Worden* (b. 1932), Apollo 12 backup Command Module pilot
*James B. Irwin* (1930-1991), Apollo 12 backup Lunar Module pilot

## Summary of the Mission

The day before the planned liftoff of Apollo 12, its monstrous Saturn V rocket stood on Launch Pad 39A under dark and ominous clouds. Heavy rain drenched the Florida launch site, while threatening thunderstorms raised concerns at Kennedy Space Center (KSC). The projected space mission had a launch window of only a few hours. If weather were to delay liftoff longer than that, Apollo 12 would be unable to achieve the desired trajectory for the planned lunar-landing location. The weather was not the only problem facing the launch crew; a fuel cell leakage had been detected in the Command and Service Module (CSM). The cells combined hydrogen and oxygen to form water, generating electricity in the process; they were the vital power supply for the CSM and the source of potable water for the crew. There was no way of repairing the damage in time for launch. Fortunately, the Module to be used in Apollo 13 was at the Cape. The crews quickly pirated parts from that craft to repair Apollo 12.

Meanwhile, the astronauts went through their final medical examinations. Charles "Pete" Conrad, Jr., the commander, was a veteran of the Gemini V and Gemini XI missions. Gemini IX had been Richard F. Gordon, Jr.'s only previous space mission, but he had had the privilege of conducting a spacewalk to the docked Agena rocket. Alan L. Bean was the rookie of the Apollo 12 crew. He was to be responsible for operating the Lunar Module (LM), which was stowed in its launch configuration, secured inside a special conical adapter that also held the CSM, locked on top of the towering Saturn V stack.

Despite the rain and thundering clouds, it was decided to launch on schedule. At 16:22 Coordinated Universal Time (UTC, or 11:22 A.M. eastern standard time), on November 14, 1969, the Saturn V blasted off, shooting the drenched spacecraft through the heavy clouds. The astronauts' ride was a comfortable one—for the first thirty seconds. Then they reported a flash of blinding light that filled the cabin. On their instrument panels, warning lights lit up. The Saturn V had been hit by lightning. In a fraction of a second, most of the CSM's electrical systems had failed. Most critical were the guidance control gyroscopes, which were now spinning aimlessly instead of keeping track of the spacecraft's orientation. Also, all three fuel cells had completely shut down. It was fortunate that the Saturn V's own guidance control platform had not been affected by the strike. Had that occurred, the rocket would have immediately gone out of control, and the crew's only hope for survival would have been the solid rocket motor on the launch escape system, which was built to pull the Control Module free of the Saturn V.

For some twenty seconds, lightning continued to strike Apollo 12, fouling the CSM's electrical systems even more. With fingers ready to trigger the abort sequence, the crew and Mission Control put their faith in the five Saturn V first-stage engines. Producing their own thunder with 346,700 newtons of thrust, the engines sped the rocket up

*Apollo 12 ready for liftoff atop the Saturn V booster.* (NASA)

above the storm clouds. When the first stage separated and the second stage fired, Apollo 12 was out of the storm. While the Saturn V's second and third stages propelled the ship into a precise orbit, the astronauts examined the damage. The fuel cells restarted quickly; the electrical panel's circuit breakers had shut them down before any problems could occur. Within minutes, they had everything but the guidance system operating. The gyroscopes needed to be completely restarted. With the help of Mission Control the crew managed to square them away in time for the translunar injection maneuver.

To break out of Earth orbit and head for the Moon, they needed to fire up the Saturn V's third-stage engine once more. The burn would use up most of the rocket's remaining fuel, but it would impart escape velocity to Apollo 12. The maneuver was accomplished safely and so accurately that the first planned course correction maneuver was not needed. Now Apollo 12 would coast the rest of the way to the Moon. The Apollo trajectory was of a kind called free return. This meant that if the crew did nothing, the spacecraft would simply loop around the Moon, whose gravity would sling the vessel back to Earth. The flight plan called for a maneuver while looping behind the Moon, the lunar orbit insertion, which would brake the ship into a stable lunar orbit. Free return was a safety feature, for if the insertion failed because the CSM's engine refused to fire, the mission would automatically abort back to Earth rather than strand the astronauts in space. Unfortunately, the specific landing site desired for Apollo 12 would not allow a free return trajectory. Thus, at some point in their coasting voyage, Gordon would have to fire the CSM engine once to break them free of the return trajectory and onto the new "nonreturn" one. As might be expected, neither the crew nor Mission Control was eager to commit Apollo 12 to the nonreturn trajectory any sooner than was necessary.

Gordon separated the CSM from the adapter and turned it around to perform a nose-first docking with the LM. As on previous missions, names serving the purpose of radio call signs had been given to the two ships. The CSM was given the name *Yankee Clipper*, while the LM was dubbed *Intrepid*. Once the LM was extracted from its position on the Saturn V's third stage (called the S-IVB), the Apollo 12 craft was in its final translunar configuration. The S-IVB used its remaining fuel to send itself out of the way, while Bean entered the *Intrepid* through the docking tunnel and began checking it. A fully functional LM was the extra bit of insurance that the crew needed to have the confidence

to break from the free return trajectory. If the CSM's engine failed while on the nonreturn trajectory, the astronauts could use the LM's engine to abort the mission and return the spacecraft to Earth.

Such contingency plans were not needed on this mission. After the days-long translunar voyage was completed, Apollo 12 disappeared behind the Moon on its nonreturn trajectory and came back around the other side in a perfect lunar orbit. The astronauts had performed all the necessary maneuvers and system checks on both the *Yankee Clipper* and the *Intrepid*. They had also found the time to send several television transmissions back to Earth. The crew was given plenty of rest for the demanding days to follow. The mission plan was for the LM to land in the Ocean of Storms, a spot considerably farther east on the Moon's face than Apollo 11's Sea of Tranquility landing site. The particular landing target was the Surveyor 3 space probe, which had soft-landed on the Moon previously. One of the mission objectives was to land close enough to Surveyor for the two astronauts to retrieve some parts.

*Alan Bean descends from the Lunar Module* Intrepid *on* November 19, 1969. (NASA Charles Pete Conrad)

When the time came for the two halves of the Apollo 12 spacecraft to separate, Conrad and Bean crawled through the docking tunnel and entered the LM. Gordon, now with the CSM to himself, would watch and wait from his lunar-orbiting CSM while his partners voyaged to the lunar surface. Separation occurred normally, and shortly thereafter, the *Intrepid* fired its descent engine for a burn which performed a maneuver called the descent orbit insertion. Its purpose was to set up the LM for the landing. From the 100-kilometer, near-circular orbit in which the *Yankee Clipper* remained, the insertion placed *Intrepid* in an elliptical orbit which dipped down to 15 kilometers above the lunar surface before arcing back up to 100 kilometers. The point of closest approach, periapsis (or sometimes "pericynthion" for specifically lunar orbits), was made to occur exactly over the landing area. As the *Intrepid* coasted toward periapsis, Mission Control acquired its precise position and velocity and corrected the readings of the LM's own internal guidance. In order to land *Intrepid* next to Surveyor 3, pinpoint accuracy would be required; the LM made a few correcting burns to line itself up precisely for the landing.

*Alan Bean holds a container of lunar soil for transport back to Earth.* (NASA Charles Pete Conrad)

As periapsis approached, Conrad and Bean examined the lunar terrain, reassured by the match between the actual lunar surface and the camera shots with which they had trained. A steady, continuous burn from the *Intrepid*'s descent engine began breaking the ship out of orbit. The ship's landing radar locked on to the surface at about 12 kilometers altitude and began giving them the precise descent speed and altitude readings they would need. Matters were progressing perfectly for the landing near the Surveyor 3, which had soft-landed at the rim of a crater. The automated landing sequence directed *Intrepid* to a landing in the center of the shallow crater. Conrad was amazed at the instrument's navigational ability but decided that the crater floor was not an appropriate landing site. He took over manual control for the final landing sequence and piloted the vessel across the crater to the side opposite the Surveyor, bringing the LM down to a soft, if very dusty, landing. Unlike that of Apollo 11, this landing had been highly fuel efficient, and the lander crew had to vent out all the remaining propellant from the descent engine's tanks.

Conrad and Bean then prepared for their first lunar excursion. Conrad was first through the

hatch, and, after deploying the modularized equipment stowage assembly, he dropped down to the lunar surface. Bean followed, and the two began setting up the equipment they would need to accomplish the goals set for their first planned excursion. In addition to remaining longer on the Moon than did Apollo 11, they were going to make a second excursion the next day. The extra time would allow the completion of more information gathering activities. First, however, came public relations. Conrad moved to set up the color television camera so that people on Earth could watch them bustle about on the Moon. The camera was much better than the simple black-and-white camera that Apollo 11 had used. In fact, this camera was the highly reliable one that Apollo 10 had used for its many broadcasts. Unfortunately, while setting it up, Bean inadvertently allowed the camera lens to catch the sunlight directly. This instantly and permanently put an end to all video transmissions for the remainder of their stay.

The Surveyor 3 lunar probe stood in plain sight, only 200 meters from their touchdown point, but for now, they would have to ignore it and concen-

*Charles "Pete" Conrad stands before the flag during the November 19, 1969, moonwalk on the Ocean of Storms.* (NASA Alan Bean)

trate on some rather tricky activities. Apollo 12 carried a full complement of scientific instruments, called the Apollo Lunar Surface Experiments Package (ALSEP). Among the equipment in the ALSEP was a seismometer, for detecting any rumblings on the lunar surface; a solar wind collector, to measure the strength and composition of the ion flux striking the Moon's surface from the Sun; a magnetometer to detect any magnetic field disturbances; and a laser reflector, which would allow Earth-based researchers to measure the distance to the Moon with unprecedented accuracy.

The ALSEP was expected to remain in operation for about a year. In order to provide it with power for this long period, an efficient and reliable electrical supply was needed. Thus, Apollo 12 became the first ship to carry a radioisotope thermal generator. Essentially a small nuclear plant, the

generator provided energy by using the heat derived from the radioactive decay of plutonium 238. Because plutonium is an extremely dangerous substance, the thermal fuel unit itself was carried inside a special graphite cask. The cask was capable of withstanding reentry heat, so that, should the LM be forced to reenter Earth's atmosphere from orbit, because of either a disaster or a mission abort, the thermal unit's integrity would be preserved, and it could be recovered without releasing its deadly contents over the world. Now the *Intrepid*'s crew had to remove the fuel unit from the cask and install it in the generator. The plutonium produced so much heat that they could feel it through their thick pressure-suit gloves.

Next, the ALSEP had to be assembled. The astronauts walked out about 150 meters from the LM—away from the expected range of the dust from their liftoff. It took them more than an hour to get the package properly installed. After they finished, they barely had time for a quick trip out to the rim of a large nearby crater before returning to the *Intrepid*. Their lunar excursion had lasted 4 hours, considerably longer than the 2.5-hour jaunt of Apollo 11. Now the crew could rest before starting another excursion the next day.

After about eight hours of sleep, the crew quickly prepared for the second trip. Eager explorers, they could not wait to be back outside, managing to be more than an hour ahead of schedule. This excursion was to be a real lunar hike. The astronauts would first walk back to the ALSEP to remove one of the instruments to be returned to Earth. On their way to the ALSEP, the sensitive seismometer recorded their every footstep. After that task was completed, Conrad and Bean next proceeded to inspect several neighboring craters, gathering samples and photographing the interiors. Finally, they walked over to the Surveyor, a probe which had been on the Moon for thirty months. They photographed the probe, examined its landing footprint, and removed several pieces of it for study back home. That completed, they gathered up some final samples and boarded the *Intrepid*. Altogether, they had spent three times as much time on the lunar surface as had the astronauts of Apollo 11. Next, the crew prepared to lift off and dock with the orbiting *Yankee Clipper*.

Meanwhile, Gordon, circling quietly overhead, had performed some investigations of his own. He carefully photographed the region of lunar surface called Fra Mauro, near the crater Descartes in the Sea of Rains. This location was geologically interesting and was the proposed landing site for the next Apollo mission. During one orbit, ground control contacted Gordon to report Earth observation of a strange occurrence in the crater Alphonsus. Since the early days of reticular lunar observations, such events, called transient lunar phenomena, had been observed. They consisted of a noticeable brightening of the floor of certain craters. Alphonsus was frequently the site of these events, which were believed to result from em-

anations of gas from below the surface—minor lunar eruptions. *Yankee Clipper*'s path passed over Alphonsus, but Gordon detected no visual difference in the crater.

*Intrepid*'s crew fired the LM's ascent engine, lifting them from the lunar surface and sending them on their way toward a rendezvous with the CSM. Their picture-perfect docking with *Yankee Clipper* was sent to Earth by the CSM's television camera. Once the crew loaded all of their samples aboard the CSM, they jettisoned the LM, which had served so well. It was programmed to fire its engines one last time to slam it into the Moon. The impact was precisely timed and the seismometer on the ALSEP gained valuable information about the inner composition of the Moon.

Finally, the *Yankee Clipper* ignited its Service Propulsion System (SPS) for the crucial burn that would take it out of lunar orbit and send it back to Earth. After this was successfully accomplished, the astronauts retired to a much-needed rest period.

The return trip was highlighted by several television transmissions. During one question-and-answer segment, the crew responded to reporters' questions relayed through Mission Control. A course correction maneuver put them on a neat reentry trajectory. The Command Module, carrying the astronauts, separated from the bulky Service Module and began its fiery descent. Clever designers had managed to supply the reentering Apollo Control Module with some aerodynamic lift, which was used to raise the craft out of the denser, lower atmosphere briefly, to allow the reentry heat to dissipate. The lift was also used to steer the vessel to its splashdown point. The parachutes deployed properly, and Apollo 12 completed its mission, splashing down in the Pacific Ocean less than 7 kilometers from the recovery vessel, the USS *Hornet*. The three astronauts were given isolation garments and, as had been the practice on Apollo 11, quarantined. Meanwhile, back on the Moon, the ALSEP was faithfully returning data. It would continue to do so for eight years, eight times longer than its projected life—a mark of the success of Apollo 12.

## Contributions

As the second piloted expedition to the Moon, the Apollo 12 mission was free from some of the tension that had pervaded Apollo 11. The crew concentrated on gathering scientific data. The Apollo 12 list of mission objectives contained several technical and experimental entries, all of which were achieved. Among the technical objectives achieved, perhaps the most significant was Conrad's successful pinpoint landing. He had managed to land so close to the Surveyor probe that moondust kicked up by the landing rocket settled in a thin layer on the probe's surface. Another technical feat was accomplished by Gordon, who precisely photographed the future landing site in the Fra Mauro region.

Nevertheless, it is the scientific accomplishments of Apollo 12 that deserve mention. As had the Apollo 11 astronauts before them, Conrad and Gordon gathered soil and rock samples from the lunar surface. Unlike its predecessor, however, Apollo 12 also deployed the ALSEP Modules. The central Module contained the antenna, which could transmit as many as nine million instrument readings to Earth daily. The radioisotope thermal generator generated about 75 watts; it used more than four hundred thermocouples to convert the heat from the slowly decaying plutonium into electricity.

The ALSEP also contained two instruments to investigate the lunar "atmosphere," a wispy and transient collection of gases seeping out of the surface and drifting off into space. The cold cathode gauge experiment measured the density and changes in this layer, while the suprathermal ion detector experiment examined positive ions in it. These ions were formed from interactions between the gases and the energy reaching the Moon from the Sun.

Scientists obtained valuable data on the solar wind (particle flux emanating from the Sun, as opposed to sunlight, or radiative flux). Another instrument, the solar wind spectrometer, measured the range of particle energies in the solar wind directly. Earth-based observations on the solar wind are impossible, because Earth's magnetic field deflects the charged particles. The Moon has no such field, however, and is therefore the ideal location for such experiments.

Even without its own magnetic field, the Moon does interact with both terrestrial and solar fields. For that reason, Apollo 12 placed a lunar surface magnetometer at the ALSEP site. This enabled scientists to deduce critical information about the composition of the Moon's interior as well as interesting facts about its history.

Another instrument, the passive seismic experiment module, revealed key data on the lunar interior. The four seismographs in the package could measure seismic disturbances with periods as small as 0.05 second or as long as 250 seconds. The amplitude of the subtle moonquakes was magnified by a factor of ten million before transmission to Earth. Apollo 12's seismographs accurately registered the impact of the abandoned LM. For true scientific work, however, several seismograph stations are needed at various locations on the surface. Thus, the later Apollo missions also deposited similar instruments at their landing sites, allowing Earth researchers to chart the epicenters and propagation of moonquakes.

A final instrument, the laser ranging retroreflector, completed the ALSEP. A basic principle of optics requires that a beam of light fired from any angle into a set of three mirrors forming the inner corner of a cube will be reflected back whence it came. The lunar ranging retroreflector consisted of an array of such tiny "corner reflectors." The beam from an Earth-based laser could be bounced back from the reflector and, by measuring the time it takes the beam to complete its circuit, the distance between the laser and reflector could be determined to within about 15 centimeters. This allowed the study of minute variations of the Moon's orbit.

## Context

Before the Apollo Program, some believed that the Moon was considerably younger than the Earth. In fact, one theory held that the Moon formed from material wrenched from the primordial, partially cooled, but still molten Earth. Analysis of Apollo rock samples established the Moon as a terrestrial contemporary. Both bodies formed some 4.6 billion years ago. The interesting thing about the rock samples from sites such as that of Apollo 12 is their resemblance to Earth's lava rock. The dark lunar "seas," or maria, were formed by huge lava flows. There are two methods which could produce these oceans of cooled lava. One is the terrestrial process of volcanism. If the Moon had a similar molten interior, it, too, could sponsor major volcanic eruptions. Lunar lava could flow for great distances because of the Moon's low surface gravity (one-sixth that of Earth). The other process involves bombardment by meteors. The impact of a fair-sized meteor would generate huge amounts of heat and could produce a substantial lava flow. The key might be to examine the Moon for current activity.

Apollo 12's ALSEP instruments provided the first opportunity to do just that. Among other things a series of moonquakes was observed at every lunar perigee (point in the Moon's elliptical orbit when it is closest to Earth). The moonquakes originated deep within the surface, some 800 to 1,000 kilometers below it. These quakes were probably caused by tidal forces from the Earth, but they revealed interesting data on lunar composition. Certain seismic waves, called shear waves, were severely damped at depths below the 1,000-kilometer layer. This indicates that there is some partial melting below that level. The ALSEP magnetometer and solar wind interaction detectors have shown that the outer surface, down to that 1,000-kilometer mark, is solid. Thus, there is the potential for some volcanism, depending on how hot those inner layers are. The lack of a global magnetic field seems to indicate that the Moon's core is not like Earth's. The terrestrial field is caused by the planet's molten iron and nickel core, and the interior heat drives the volcanoes. Thus, current lunar volcanic activity is ruled out both by observation and by inference. Interestingly, though, residual magnetization detected in lunar rock samples indicates that the Moon may once have had a substantial field, hence a completely molten core; thus, a history of volcanic activity earlier in its life seems likely.

The cold cathode experiment did detect a minute lunar atmosphere, about a billion times less than Earth's sea-level pressure. These gases are believed to be by-products of radioactive decay in the lunar surface. The gases seep to the surface and gradually bleed off into space, leaving the Moon an almost totally airless world.

**See also:** Apollo Program; Apollo Program: Command and Service Module; Apollo Program: Developmental Flights; Apollo Program: Geological Results; Apollo Program: Lunar Module; Apollo Program: Orbital and Lunar Surface Experiments; Apollo 11; Apollo 13; Space Task Group; Surveyor Program.

## Further Reading

Allday, Jonathan. *Apollo in Perspective: Spaceflight Then and Now.* Philadelphia: Institute of Physics, 1999. This book takes a retrospective look at the Apollo Program and the technology that was used to land humans on the Moon.

Bean, Alan, with Andrew L. Chaikin. *Apollo: An Eyewitness Account by Astronaut/Explorer Artist/Moonwalker.* Shelton, Conn.: The Greenwich Workshop Press, 1988. Alan Bean, the artist, painted what he saw from his unique perspective as an astronaut. Many of the paintings included are fantasy moonscapes. To pay tribute to his Apollo 12 Command Module pilot, Richard F. Gordon, Jr., Bean included him in several scenes on the lunar surface. Gordon, circling high above in the Command Module, never actually saw the surface up close. With Chaikin, Bean tells his story of the second lunar landing.

Beattie, Donald A. *Taking Science to the Moon: Lunar Experiments and the Apollo Program.* Baltimore: Johns Hopkins University Press, 2001. This book describes, from the perspective of NASA Headquarters, the struggles to include science payloads as part of the Apollo Program.

Chaikin, Andrew. *A Man on the Moon: The Voyages of the Apollo Astronauts.* New York: Penguin Group, 1998. The Apollo Moon landings are retold through the eyes and ears of the people who were there. Based on interviews with twenty-three Moon voyagers, as well as those who struggled to get the program moving, this book conveys every aspect of the missions with breathtaking immediacy: from the rush of liftoff to the heart-stopping lunar touchdown to the final hurdle of reentry.

————————. *Space.* London: Carlton Books, 2002. The book contains more than 300 photographs chosen from thousands of images in the space archives of NASA, Russia, and Europe.

Collins, Michael. *Liftoff: The Story of America's Adventure in Space.* New York: Grove Press, 1988. This work provides a comprehensive description of the human side of space exploration as seen by an intimate participant, one of the Apollo 11 astronauts.

Ertel, Ivan D., and Roland W. Newkirk. *The Apollo Spacecraft: A Chronology.* Vol. 4. NASA SP-4009. Washington, D.C.: Government Printing Office, 1978. This account of the Apollo missions opens up the project's operations for the reader. Illustrated. Contains an appendix on both the Apollo organization and its contractors.

Godwin, Robert, ed. *Apollo 12: The NASA Mission Reports.* Vol. 1. Burlington, Ont.: Apogee Books, 1999. This collection contains reprints of the Apollo 12 press kit, prelaunch mission operation report, postlaunch mission operation report and the postlaunch mission operation report supplement. It also includes a CD-ROM featuring over 2,000 70-millimeter images taken during the mission, several unique interactive panoramic images of the Ocean of Storms, a twenty-five-minute movie of the voyage of Apollo 12, plus the complete TV footage from the lunar surface and the in-flight press conference.

————————. *Apollo 12: The NASA Mission Reports.* Vol. 2. Burlington, Ont.: Apogee Books, 2004. This collection contains a DVD with footage from the mission. The book highlights even more details from the Moon encounters of November, 1969, from the pinpoint landing to leaky space suits and more.

Kelly, Thomas J. *Moon Lander: How We Developed the Apollo Lunar Module.* Washington, D.C.: Smithsonian Books, 2001. Grumman Chief Engineer Kelly gives a firsthand account of designing, building, testing, and flying the Apollo Lunar Module.

Kraft, Christopher C., Jr. *Flight: My Life in Mission Control.* East Rutherford, N.J.: Penguin Putnam, 2002. The first NASA flight director gives an account of his life in Mission Control.

Lambright, W. Henry. *Powering Apollo: James E. Webb of NASA.* Baltimore: Johns Hopkins University Press, 2000. Lambright explores James E. Webb's leadership role in NASA's spectacular success.

Liebergot, Sy, and David Harland. *Apollo EECOM: Journey of a Lifetime.* Burlington, Ont.: Apogee Books, 2003. This is the life story of Sy Liebergot, former NASA flight controller, with emphasis on his years working in Apollo Mission Control.

Lindsay, Hamish. *Tracking Apollo to the Moon.* London: Springer-Verlag London Limited, 2001. Features interviews, quotes, and extensive photographs, including some that appear here for the first time.

National Aeronautics and Space Administration. *Apollo Mission Press Kits.* http://www-lib.ksc.nasa.gov/lib/presskits.html. Provides detailed preflight information about each of the Apollo missions, Apollo 6 through Apollo 17. Accessed March, 2005.

_____. *Apollo 12 Mission Report.* Washington, D.C.: National Technical Information Services, 1970. This is the official report of the results of the Apollo 12 flight. The original is out of print, but copies can be obtained through the National Technical Information Services. An Internet search will turn up a copy in Adobe Acrobat format for easy download. Nowhere else will you find as in-depth a document on the technical, scientific, and human aspects of the mission.

Shepard, Alan B., Jr., and Donald K. "Deke" Slayton, with Jay Barbree and Howard Benedict. *Moon Shot: The Inside Story of America's Race to the Moon.* Atlanta: Turner, 1994. This is, indeed, the inside story of the Apollo Program as told by two men who actively participated in it. Some of their tales appear here for the first time. The book was adapted for a four-hour documentary in 1995.

Slayton, Donald K., with Michael Cassutt. *Deke! U.S. Manned Space: From Mercury to the Shuttle.* New York: Forge, 1995. This is the autobiography of the last of the Mercury astronauts to fly in space. After being grounded from flying in Project Mercury for what turned out to be a minor heart murmur, Slayton was appointed head of the Astronaut Office. During his reign he assigned all of the Apollo crew members to their flights. Later, he commanded the Apollo-Soyuz Test Project flight in 1975. This is a behind-the-scenes look at America's attempt to land humans on the Moon.

Swanson, Glen E. *Before This Decade Is Out: Personal Reflections on the Apollo Program.* Gainesville: University Press of Florida, 2002. The Saturn/Apollo Program as witnessed by some of the political leaders, engineers, scientists, and astronauts who made it such a success.

Turnill, Reginald. *The Moonlandings: An Eyewitness Account.* New York: Cambridge University Press, 2003. By a veteran space journalist who spent his career covering all the piloted space missions as well as planetary missions like Mariner, Pioneer, Viking, and Voyager.

Ulivi, Paolo, and David M. Harland. *Lunar Exploration: Human Pioneers and Robotic Surveyors.* London: Springer-Verlag London Limited, 2004. A well-paced, rapidly moving, balanced, even-handed account of lunar exploration.

Wendt, Guenter, and Russell Still. *The Unbroken Chain.* Burlington, Ont.: Apogee Books, 2001. Wendt is the only person who worked with every astronaut bound for space.

*Greg Tomko-Pavia*

# Apollo 13

*Date:* April 11 to April 17, 1970
*Type of mission:* Lunar exploration, piloted spaceflight

*The third scheduled lunar landing of the Apollo Program, Apollo 13 was destined never to reach its goal. Instead, an explosion on board the spacecraft caused the landing to be aborted, and the crew used its Lunar Module as a lifeboat.*

## Key Figures

*James A. Lovell, Jr.* (b. 1928), Apollo 13 mission commander

*John L. "Jack" Swigert, Jr.* (1931-1982), Apollo 13 Command Module pilot

*Fred W. Haise, Jr.* (b. 1933), Apollo 13 Lunar Module pilot

*John W. Young* (b. 1930), Apollo 13 backup mission commander

*Thomas K. "Ken" Mattingly II* (b. 1936), original Apollo 13 Command Module pilot

*Charles M. Duke, Jr.* (b. 1935), Apollo 13 backup Lunar Module pilot

*Eugene F. Kranz* (b. 1933), Apollo 13 flight director

## Summary of the Mission

While much of the world paid little attention to the start of Apollo 13, events transpired during the flight to make it one of the most dramatic and difficult missions of the entire American piloted space effort. The National Aeronautics and Space Administration (NASA) itself later termed the flight the most "successful failure" it had ever had.

Apollo 13 followed the Projects Mercury and Gemini, four successful Earth- and lunar-orbital Apollo missions, and the lunar-landing flights of Apollos 11 and 12. Its mission, the longest and most complex to date, was to use the standard Apollo Command and Service Module (CSM) and Lunar Module (LM) to explore the Fra Mauro crater and its environs, a sprawling, mountainous region below the equator on the near side of the Moon, the side that always faces the Earth. Once on the Moon, the astronauts were to conduct two extravehicular activities (EVAs), or "moonwalks," to gather rock and dirt samples from an area of the surface that was suspected of having been formed by volcanic activity billions of years before. Scien-

tists believed that studying the formation of this older area of the Moon would provide a clearer picture of Earth's own origins.

The crew for Apollo 13 was commanded by James A. Lovell, Jr., a veteran of the Gemini VII and XII missions and of the first piloted lunar-orbital mission, Apollo 8. At the time of the Apollo 13 mission, Lovell had spent 572 hours in space, more than any other living human being. His crew on the Apollo 13 flight was to include rookie astronauts Fred W. Haise, Jr., as Lunar Module pilot, and Thomas K. "Ken" Mattingly II, as Command Module pilot. The backup crew consisted of John W. Young, veteran of Gemini 3, Gemini X, and Apollo 10, along with John L. "Jack" Swigert, Jr., and Charles M. Duke, Jr.

Less than a week before the April 11, 1970, launch date, however, backup crew member Duke was unknowingly exposed to German measles at the home of a friend. At the time, primary and backup crews received identical training, often together. Thus, when Duke returned to his training schedule, he exposed the other members of both crews. Within a short time, Duke himself came down with the virus, and NASA doctors tested the other astronauts for immunities. Mattingly, the primary Command Module pilot, who did not have such an immunity, was removed from the primary crew two days before launch. Swigert took his place.

Apollo 13, like all the Apollo lunar-landing missions, was launched atop the most powerful American rocket booster, the three-stage Saturn V. The first stage of the Saturn V, the S-IC, used five F-1 rocket engines to develop a thrust of more than 33.4 million newtons at launch. The second stage, the S-II, employed five smaller J-2 engines to provide more than 4.4 million newtons of thrust. The third stage, the S-IVB, had a single J-2 engine that would first boost the piloted payload into Earth orbit and then power the Apollo spacecraft through translunar injection and on its way to the Moon.

Once safely out of Earth's orbit, the Apollo 13 CSM, *Odyssey*, would separate from the S-IVB stage, turn around, and remove the LM, *Aquarius*, from its berth atop the booster. The S-IVB would then be sent on a trajectory to crash on the Moon's surface near Apollo 12's landing site to test the seismic experiments placed there by astronauts Charles "Pete" Conrad, Jr., and Alan L. Bean. According to Lovell, this was the only part of the Apollo 13 mission that took place as planned.

Apollo 13 was launched on schedule from Launch Pad 39A at the Kennedy Space Center (KSC). Shortly after the second stage ignited, one of the booster rocket's five engines shut down, requiring the other four to compensate by firing 34 seconds longer than planned. Apart from this, the mission's initial phases were nearly flawless.

Things changed dramatically, however, approximately 56 hours into the flight, when the spacecraft was more than 320,000 kilometers from Earth. Shortly after completing a live television broadcast in which Lovell and Haise gave earthbound viewers a tour of *Aquarius*, the astronauts heard a loud bang from the Service Module (SM), quickly followed by a drop in pressure in the SM's oxygen tanks and a loss of electrical power to the Command Module (CM). Within an hour, the SM was dead in space, without oxygen or power for the CM, which had only a few hours of each for use during reentry.

Later investigation showed that a short circuit in a heater in the SM's no. 2 oxygen tank had exploded. This damaged the no. 1 tank and parts of the interior of the Service Module and blew off the bay no. 4 cover, spreading a cloud of debris and oxygen vapor outward for more than 48 kilometers from *Odyssey*.

Fortunately for the crew, the accident happened on the way to the Moon and not on the return phase of the mission. The LM, *Aquarius*, was still attached to the CM and was fully stocked for its part of the mission. Never-before-used emergency plans called for using the LM as a lifeboat to ferry the CSM and the crew back to Earth in an emergency. Less than an hour after the explosion, Apollo 13's lunar landing was canceled and the flight became a lifeboat mission.

With the power and oxygen supply to *Odyssey* all but gone, Lovell, Haise, and Swigert moved into *Aquarius*'s smaller crew compartment. Because the LM was designed to sustain two men for approximately two days on the Moon's surface, the onboard oxygen, water, and power supplies had to be strictly rationed over the four-day return to Earth. Thus, the temperature in the spacecraft was allowed to drop to an almost-unbearable 3.3° Celsius (38° Fahrenheit), and frost formed on the inside of the windows. It also meant that each astronaut had only 177 milliliters (6 ounces) of water a day to drink and use in the preparation of dehydrated foods, far below normal amounts.

Once they reached the far side of the Moon, the astronauts needed to fire *Aquarius*'s descent-stage main engine to slingshot the spacecraft free of the Moon's gravitational pull and send it back toward Earth. This gravity-assist maneuver, never before tried with an LM in that configuration, was the most delicate part of the revised mission. While the spacecraft flew around the Moon, millions of peo-

*Astronauts Fred Haise, James Lovell, and Thomas Mattingly (the original Apollo 13 crew complement) at the Apollo 13 launch site.* (NASA CORE/Lorain County JVS)

crew prepared for the final leg of their flight. Because *Aquarius* did not have a heatshield and was not equipped to reenter Earth's atmosphere, the crew would have to return to *Odyssey*, the CM, and abandon the LM at the last possible moment. This was accomplished less than an hour before the CM reentered the upper atmosphere. After such a mission, *Odyssey* had a picture-perfect splashdown in the Pacific Ocean, less than four miles from the recovery ship, the USS *Iwo Jima* on April 17, 1970.

**Contributions**

The lessons learned from the flight of Apollo 13 were not those intended prior to its launch. While the scientific discoveries that were expected from its anticipated landing on the Moon were postponed until Apollo 14, the near-disaster of Apollo 13 showed the men and women of the American space program that, even after the many successes they had enjoyed on previous flights, piloted spaceflight was still an extremely dangerous venture each and every time it was attempted. Tragically, the Soviet Union learned the same lesson fourteen months later with the deaths of the Soyuz 11 crew members.

Apollo 13 brought together the thousands of people who worked for NASA and its various contractors in a single effort to save three men in space. It showed them that they could use ingenuity and drive to overcome a crisis unlike any they had ever experienced and that through their efforts, they could directly affect the success of a mission 400,000 kilometers away from Earth. The mission also gave NASA a renewed appreciation for the communications link between Mission Control and the astronauts in space. Technicians on Earth worked around the clock to test options the crew could use in space. Staying in constant contact with their colleagues on Earth helped boost the morale

ple on Earth waited nervously for news that the burn had successfully taken place. Three other burns speeded the spacecraft's return to Earth and corrected its course.

During the approximately four days it took for Apollo 13 to return to Earth, the three astronauts overcame several minor but potentially serious problems. The most notable of these challenges was the rigging of the makeshift system to remove carbon dioxide from the LM cabin when that craft's own system had proved inadequate. Haise developed a bladder infection and Swigert also became ill because of the primitive conditions in which they had to live during the return flight.

At 138 hours into the flight, as Apollo 13 approached Earth, the SM was jettisoned and the of the Apollo 13 crew, easing somewhat the despair felt after the cancellation of their lunar landing and the discomfort they had to endure during the return flight.

From a technical perspective, Apollo 13 also showed NASA the importance of redundant systems on board the spacecraft. After that flight, the SMs were redesigned to include additional fuel cells and oxygen tanks, which would serve as reserves in case something happened to the primary systems. The explosion in the SM also caused NASA to revise and correct quality-control problems in the construction and prelaunch testing of equipment. Minor design changes were also made to the LM, improving its ability to handle the demands of possible future rescue missions. This is believed to have saved the lives of other flight crews and to have contributed substantially to the success of the missions that followed Apollo 13.

### Context

Apollo 13 marked the midpoint in the Apollo Program: Four missions had preceded it to the Moon, and four followed it. The mission also signified the beginning of the end of the Apollo Program's lunar-landing portion: Shortly after Apollo 13 flew, the last three planned missions, Apollos 18, 19, and 20, were canceled because of budget cuts. Moreover, many of the early astronauts would leave the program as the American piloted program began a dormant period that would last until the advent of the Space Transportation System, otherwise known as the space shuttle.

The launch of Apollo 13 represented for millions of Americans the start of "routine" trips to the Moon. Many, in fact, questioned the value of repeated lunar missions, believing that the money used for such ventures could be better spent in other areas. When Apollo 13 became a life-and-

death struggle to return the crew to Earth, these opponents of the space program argued that it was an example of placing humans at unnecessary risk to pursue questionable goals. The postflight investigation of the Apollo 13 mishap caused a nine and one-half month delay in the Apollo launch schedule and a significant redesign of the Apollo SM. Because the Apollo 13 landing in the Fra Mauro region was canceled, the final four mission profiles had to be revised to allow Apollo 14 to carry out Apollo 13's assignment.

One result of the flight of Apollo 13 was more humorous than substantive. After the successful splashdown and recovery of the crew, engineers at Grumman Aerospace, the Bethpage, New York, based contractor that built the LM, sent Rockwell International, the builder of the CSM, a $400,000 bill for "towing and road service" for the use of *Aquarius* as a rescue craft. Word of the joke eventually reached the news media; for months after the flight, Grumman received hundreds of dollars in contributions from private citizens who wanted to help Rockwell pay its debt to the builder of the LM.

In 1995, *Apollo 13*, a feature film directed by Ron Howard and starring Tom Hanks, Kevin Bacon, Bill Paxton, Gary Sinise, and Ed Harris, retold the story of the ingenuity and team effort that succeeded in returning the spacecraft and astronauts to Earth. The film was nominated for nine Academy Awards, winning two. It captured the imagination of a new generation and helped remind the public of the true heroism that characterized not only these early astronauts but also their families and colleagues on the ground.

**See also:** Apollo Program; Apollo Program: Command and Service Module; Apollo Program: Lunar Module; Apollo 11; Apollo 12; Apollo 14; Apollo 15; Apollo 16; Apollo 17; Escape from Piloted Spacecraft; Lunar Exploration.

### Further Reading

Allday, Jonathan. *Apollo in Perspective: Spaceflight Then and Now.* Philadelphia: Institute of Physics, 1999. This book takes a retrospective look at the Apollo space program and the technology that was used to land humans on the Moon.

Beattie, Donald A. *Taking Science to the Moon: Lunar Experiments and the Apollo Program.* Baltimore: Johns Hopkins University Press, 2001. This book describes, from the perspective of NASA Headquarters, the struggles that took place to include science payloads as part of the Apollo Program.

Chaikin, Andrew. *A Man on the Moon: The Voyages of the Apollo Astronauts.* New York: Penguin Group, 1998. The Apollo Moon landings are retold through the eyes and ears of the people who were there. Based on interviews with twenty-three Moon voyagers as well as those who struggled to get the program moving, this book conveys every aspect of the missions with breathtaking immediacy: from the rush of liftoff to the heart-stopping lunar touchdown to the final hurdle of reentry.

Cooper, Henry S. F., Jr. *Thirteen: The Apollo Flight That Failed.* 1973. Reprint. Baltimore: Johns Hopkins University Press, 1995. A fascinating look at the efforts to rescue three astronauts "lost in space" aboard the Apollo 13 Command Module. Although it reads like a novel, it is based upon news reports and interviews with the participants.

Godwin, Robert, ed. *Apollo 13: The NASA Mission Reports.* Burlington, Ont.: Apogee Books, 2000. In this book, some of the rare official documentation of the voyage of Apollo 13 is collected and made readily available for the first time. The Apollo 13 press kit details the plans for the mission and the lunar exploration that would never be. Also included is the Apollo 13 review board report, detailing the accident, its causes, and the recommendations to prevent a reoccurrence.

Kelly, Thomas J. *Moon Lander: How We Developed the Apollo Lunar Module.* Washington, D.C.: Smithsonian Books, 2001. Chronicles the development and building of the world's first true piloted spaceship, the Apollo Lunar Module. Grumman's pride in having inadvertently provided a vital "lifeboat" for the crew of the disabled Apollo 13 comes through clearly.

Kraft, Christopher C., Jr. *Flight: My Life in Mission Control.* East Rutherford, N.J.: Penguin Putnam, 2002. The first NASA flight director gives an account of his life in Mission Control.

Kranz, Eugene F.. *Failure Is Not an Option: Mission Control from Mercury to Apollo 13 and Beyond.* New York: Simon & Schuster, 2000. The autobiography of the flight director during Apollo 13.

Lambright, W. Henry. *Powering Apollo: James E. Webb of NASA.* Baltimore: Johns Hopkins University Press, 2000. Explores James E. Webb's leadership role in NASA's spectacular success.

Liebergot, Sy, and David Harland. *Apollo EECOM: Journey of a Lifetime.* Burlington, Ont.: Apogee Books, 2003. Aboard Apollo 13, Lovell, Haise and Swigert performed wonders battling for their lives, but without the expertise, quick thinking and technical support of Mission Control, they never could have come home. Sy Liebergot was there and relates the details.

Lindsay, Hamish. *Tracking Apollo to the Moon.* London: Springer-Verlag London Limited, 2001. Features interviews, quotes, and extensive photographs, including some that appear here for the first time. There are numerous pictures and illustrations to help foster this approach. The chronological order allows the reader to see how the various decisions and events shaped the direction of the U.S. space program.

Lovell, Jim, and Jeffrey Kluger. *Lost Moon: The Perilous Voyage of Apollo 13.* New York: Houghton Mifflin Company, 2000. A blow-by-blow account of the harrowing Apollo 13 mission by

Mission Commander James A. Lovell, Jr. In addition, activities within the Mission Control Center are chronicled in detail. Appendices include a mission time line, a list of the personnel, and mission information about each piloted Apollo flight.

National Aeronautics and Space Administration. *Apollo Mission Press Kits.* http://www-lib.ksc.nasa.gov/lib/presskits.html. Provides detailed preflight information about each of the Apollo missions, Apollo 6 through Apollo 17. Accessed March, 2005.

_____. *Apollo 13 Mission Report.* Washington, D.C.: National Technical Information Services, 1970. This is the official report of the results of the Apollo 13 flight. The original is out of print, but copies can be obtained through the National Technical Information Services. An Internet search will turn up a copy in Adobe Acrobat format for easy download. Nowhere else can one find as in-depth a document on the technical, scientific, and human aspects of the mission.

Pellegrino, Charles R., and Joshua Stoff. *Chariots for Apollo: The Making of the Lunar Module.* New York: Atheneum, 1985. This book is unquestionably one of the most informative and enjoyable discussions of the early days of the American piloted space effort. Written with all the drama and excitement of a popular novel, it brings the reader into the minds and hearts of the men and women who built the spiderlike ships that carried twelve astronauts to the surface of Earth's nearest celestial neighbor.

Shepard, Alan B., Jr., and Donald K. "Deke" Slayton, with Jay Barbree and Howard Benedict. *Moon Shot: The Inside Story of America's Race to the Moon.* Atlanta: Turner, 1994. This is, indeed, the inside story of the Apollo Program as told by two men who actively participated in it. Some of their tales appear here for the first time. The book was adapted for a four-hour documentary in 1995.

Slayton, Donald K., with Michael Cassutt. *Deke! U.S. Manned Space: From Mercury to the Shuttle.* New York: Forge, 1995. This is the autobiography of the last of the Mercury astronauts to fly in space. After being grounded from flying in Project Mercury for what turned out to be a minor heart murmur, Slayton was appointed head of the Astronaut Office. During his reign he assigned all of the Apollo crew members to their flights. Later, he commanded the Apollo-Soyuz Test Project flight in 1975. This is a behind-the-scenes look at America's attempt to land humans on the Moon.

Ulivi, Paolo, and David M. Harland. *Lunar Exploration: Human Pioneers and Robotic Surveyors.* London: Springer-Verlag London Limited, 2004. Covers lunar exploration, including programs like Ranger and other American probes in the late 1950's.

Wendt, Guenter, and Russell Still. *The Unbroken Chain.* Burlington, Ont.: Apogee Books, 2001. Autobiography of the only person who worked with every astronaut bound for space.

*Eric Christensen*

# Apollo 14

*Date:* January 31 to February 9, 1971
*Type of mission:* Lunar exploration, piloted spaceflight

*Early in 1971, the Apollo 14 mission succeeded in landing the third American scientific team on the lunar surface. Providing important new data on the structure and geologic history of the Moon, the Apollo 14 mission paved the way for advanced, long-duration scientific missions that together produced a vast compendium of knowledge on the Earth-Moon system.*

**Key Figures**

*Alan B. Shepard, Jr.* (1923-1998), Apollo 14 mission commander
*Stuart A. Roosa* (1933-1994), Apollo 14 Command Module pilot
*Edgar D. Mitchell* (b. 1930), Apollo 14 Lunar Module pilot
*Eugene A. Cernan* (b. 1934), Apollo 14 backup mission commander
*Ronald E. Evans* (1933-1990), Apollo 14 backup Command Module pilot
*Joseph H. Engle* (b. 1932), Apollo 14 backup Lunar Module pilot
*Paul W. Gast*, chief scientist, Apollo Lunar Science staff

**Summary of the Mission**

After the explosion in Apollo 13's Service Module (SM) and the resulting cancellation of the mission's planned lunar exploration, Apollo 14 was designed to restore confidence in the lunar exploration program and bring back data and samples from the Fra Mauro area. Careful examination of photographs from orbital scanning had led scientists to believe that this area contained some of the Moon's oldest rocks, thrust up from the ancient lunar crust during the cataclysmic formation of the Sea of Rains.

Apollo 14 was launched from the Kennedy Space Center (KSC) in Florida at 21:03 Coordinated Universal Time (UTC, or 4:03 P.M. eastern standard time) on January 31, 1971. The astronauts were sitting atop a 111-meter-high Saturn V rocket that could deliver more than 40 million newtons of thrust during its three-stage burn. Saturn V utilized liquid oxygen and kerosene in the first stage and liquid oxygen and liquid hydrogen in the second and third stages. Remembering the

lightning that had caused problems for Apollo 12's takeoff, National Aeronautics and Space Administration (NASA) launch specialists, in collaboration with mission specialists operating from the Manned Spacecraft Center in Houston, Texas, held the launch for 40 minutes, waiting for clouds to dissipate. Eleven minutes and 49 seconds after launch, the "live" upper stage of the Saturn V rocket, the instrumentation unit harboring the Lunar Module (LM) and the Command and Service Module (CSM), was placed into Earth "parking" orbit for systems check, docking maneuvers, and ultimately a precision burn toward the Moon.

Operating at an orbit of 183 by 189 kilometers, with all systems checked, the Apollo 14 crew completed the translunar injection burn of the SPS, propelling them from Earth's orbit.

While coasting toward their lunar target, the astronauts had to dock the CSM and LM, a necessary preamble to lunar orbiting. A hard dock was completed after five unsuccessful attempts. After a check of the docking mechanism for possible malfunction, a critical burn of the rocket's upper stage guided the CSM and LM toward a rendezvous with the Moon. The SM's main engine placed the joined spacecraft into an elliptical 109-by-18-kilometer lunar orbit at 07:00 UTC on February 4, 1971. After an orbital correcting burn, the two spacecraft were placed in a nearly circular orbit, roughly 110 kilometers above the lunar surface.

Alan Shepard and Edgar D. Mitchell piloted the LM, dubbed *Antares*, and separated from Stuart Roosa in the CSM, dubbed *Kitty Hawk*, to effect lunar landing in the Fra Mauro region. Because of problems with a faulty switch in the Abort Guidance System (AGS), the *Antares* was forced into a manual landing at 08:37 UTC on February 5, at 3°40′24″ south latitude and longitude 17°27′55″ west. Throughout the history of piloted American space travel, astronauts had pressed NASA to relinquish control of spacecraft to them as pilots. The lunar landings of Apollo 11 and 14 would not have been possible had the pilots not been prepared to land manually. Apollo 14's landing was very successful, deviating only 50 meters from the proposed landing site in the Fra Mauro highlands.

Two extravehicular activity (EVA) periods were planned for the Apollo 14 expedition. The first EVA, which lasted 4 hours and 49 minutes, concentrated on deployment of equipment and scientific packages and collection of local geological samples. The second EVA lasted 4 hours and 20 minutes and involved an extended geological traverse to Cone Crater, northeast of the landing site.

The first EVA began forty minutes late on February 5 at 14:00 UTC because of problems in the LM's communications system. The EVA was extended thirty minutes beyond what had been planned, to allow for completion of the experiments and sampling. Initially, the astronauts collected grab samples of rock and soil near the LM. Next, Shepard erected a television camera to broadcast their activities back to Earth, and together they deployed the S-band antenna that transmitted their television signals. The astronauts unpacked and set up the Apollo Lunar Surface Experiments Package

(ALSEP), which was stowed in *Antares*'s modularized equipment stowage assembly. The scientific experiment package contained a solar wind detector and a spectrometer to measure the frequency and energy of charged particles in the solar wind, ionosphere and atmosphere measuring devices, and a laser mirror to measure changes in the Earth-Moon distance and fluctuations in rotation rates.

Shortly thereafter, the astronauts performed an active seismic experiment that sought to define the depth of the lunar regolith (unconsolidated soil and lunar rubble) at the landing site. The experiment involved use of a string of three geophones, an instrument package, and charges that produced seismic waves. The velocities of seismic waves are directly proportional to the density of the crustal material through which they travel. Small explosive charges were therefore fired against the ground in various places; the geophones then recorded the different seismic velocities as a function of depth. These data allowed a determination of the crustal structure at the *Antares* landing site.

The television cameras permitted mission scientists on Earth to direct the collection of geologic samples. Shepard and Mitchell, using the two-wheeled cartlike Modularized Equipment Transporter, collected roughly 3.5 kilograms of rocks and soil under the watchful eye of the chief scientist of the Apollo Lunar Science staff, Paul Gast, in Houston. After the active seismic experiment and geologic specimen collecting, a passive seismograph was deployed from the equipment stowage assembly and activated to record moonquakes and impacts. Instantly, it began recording weak moonquakes (and the footfalls of astronauts) and transmitted seismic velocity data back to Earth. The first EVA ended as a success, and the astronauts entered *Antares* for a well-earned rest period.

The second EVA began more than two hours early (at the request of the crew) on February 6, with the goal of making a trek of roughly 1.5 kilometers northeastward toward Cone Crater. Pulling the Modularized Equipment Transporter uphill, the astronauts were to sample, near Cone Crater, rocks spewed from the Sea of Rains; these rocks were believed to have originated deep in the lunar crust. In addition, the astronauts were to perform magnetic surveys along their route and collect soil and rock samples.

Their progress was slowed by the necessity of pulling the nonmotorized equipment transporter, by their bulky space suits, and by their numerous stops to perform magnetometer surveys, characterize the landscape, and photograph and sample the lunar surface. Soon, it became evident that they would be unable to reach Cone Crater and perform meaningful investigations there in the allotted time. Therefore, under direction from Mission Control, the astronauts collected peripheral samples, dug a trench to observe the layering and mechanical strength of the regolith, and collected more rock and soil samples from a boulder field immediately south of Cone Crater. Returning to *Antares*, they checked the scientific instruments and the alignment of the Earth-transmitting antenna system before entering the LM for ascent to the CSM, *Kitty Hawk*, which was orbiting overhead.

During his lunar orbit in the CSM, Roosa had performed important scientific duties, including photographic surveying of large continuous tracts of the Moon to produce detailed views of future Apollo landing sites. Radar experiments, S-band transponder experiments, and visual tracking of lunar surface features, including the LM landing site, were also completed.

With the liftoff systems checked, at 18:48 UTC on February 6 the ascent stage of the LM performed as planned by launching members of the Apollo 14 exploration team and their samples into lunar orbit. Rendezvous was accomplished in one orbit, and docking with the CSM took place at 20:35 UTC. Shepard and Mitchell joined Roosa in the CSM, and they jettisoned the LM ascent stage. It impacted the lunar surface between the Apollo 12 and 14 seismic experiment stations. The energy of its impact produced a signal that lasted for 1.5 hours; this signal provided important data on the structure of the Moon's crust.

While in lunar orbit, the Apollo 14 crew continued to perform scientific experiments and photographic surveys of the Moon, Earth, and galactic space. The crew was catapulted toward Earth at 01:39 UTC on February 7 with a critical burn of the CSM propulsion system and continued to perform experiments in zero-gravity conditions en route to Earth. After a minor midcourse correction, the Apollo 14 crew splashed down in the Pacific Ocean at 20:24 UTC on February 9, 1971, at 27°2′24″ south latitude and longitude 172°41′24″ west, roughly 1 kilometer from the target point.

In a hugely successful mission, following on the heels of a near-disastrous Apollo 13 attempt, the Apollo 14 crew set new records for human presence on the Moon; brought back 43 kilograms of lunar rock, soil, and core samples; and procured abundant scientific and photographic data for years of research for scientists back on Earth.

### Contributions

Data gathered during lunar fieldwork and scientific experiments carried out on the Moon's surface, in orbit, and during flights to and from the

*The crew of Apollo 14 (from left): Stuart A. Roosa, Alan B. Shepard, Jr., and Edgar D. Mitchell.* (NASA)

*Apollo 14 Lunar Module Pilot Ed Mitchell gets moondust on his boots as he travels across the lunar surface, map in hand.* (NASA)

Moon contributed significantly to knowledge of lunar geology. The prime objective of the Apollo 14 mission was to examine and collect rocks of the Fra Mauro formation (debris thrust up from the ancient lunar crust and spewed radially outward during formation of the Sea of Rains). Because the Fra Mauro formation is a blanketing of ejecta distributed in lines around the Sea of Rains, sampling and dating of the Fra Mauro terrain would help define the age of the Sea of Rains. Cone Crater—a younger feature, roughly 340 meters in diameter, punched through the Fra Mauro formation—offered access to the underlying strata.

Not surprisingly, of the Apollo 14 samples returned to Earth, almost all are breccias (highly fractured rocks composed of angular clasts of pre-existing rocks). On the Moon, as on Earth, brecciation of crystalline rocks is a product of impact processes. In contrast to the collection of Apollo 12, which was dominated by volcanic rock,

and that of Apollo 11, which was about 50 percent volcanic, the Apollo 14 collection was 90 percent fragmental in origin. The fragmental rocks of the Fra Mauro formation, which locally exhibit brecciated fragments as clasts within breccias, record a complex crustal evolution of initial cratering followed by formation of the Sea of Rains itself. Dating of these rocks has placed creation of the Sea of Rains at about 3.9 billion years before the present.

The active seismic experiment, described earlier, defined an unconsolidated layer of surface regolith 8.5 meters thick at the scientific instrumentation site and confirmed a nearly identical result that had been attained at the Apollo 12 landing site. A 50-meter-thick layer that possesses the seismic properties of a fragmental unit, like the Fra Mauro formation, underlies the regolith. The passive seismic experiment formed a network with the Apollo 12 station, roughly 180 kilometers away. Impacts of meteoroids and lunar quakes were distinguished. The moonquakes increased in frequency when the Moon was in perigee (closest to Earth), probably because of a release of internal tidal strain. The impacts of the Saturn's upper stage and the jettisoned LM ascent stage provided strong signals for the Apollo seismic network and prepared the way for testing of early hypotheses regarding the lunar interior.

The Apollo 14 spectrometer recorded an increase in charged particles as the Moon passed through Earth's magnetosphere (an envelope of charged particles produced by the interaction of Earth's magnetic field and the solar wind). A laser retroreflector added to the Apollo 11 mirror baseline helped provide data on the lunar interior and rotation of both Earth and the Moon; it helped scientists discover that the Moon was receding from Earth at about 3 centimeters per year. The portable magnetometer used during the EVAs verified Apollo 12 measurements of magnetic field strength and showed that the field varied considerably from place to place. Other experiments conducted during the Apollo 14 mission are described in detail in the books listed below.

## Context

The voyage of Apollo 14 restored faith in the Apollo Program after the nearly disastrous Apollo 13 mission and added an important new set of data on the geologic evolution of the Moon. The Apollo 11 and 12 missions to the Sea of Tranquility and the Ocean of Storms had identified the dark rocks of the lunar maria, or basins, as volcanic basalts; the Apollo 14 mission provided samples of different rock types (ejecta units) and data that helped geologists estimate the age of the Sea of Rains, a multiringed impact basin. These data have led to significant advances in studies of planetary formation during the period of intense cratering that persisted throughout the early history of the solar system. Of lasting significance, scientific instrumentation left at the Apollo 14 site provided an important link for future Apollo science stations, allowing for precision triangularization and corroboration of experimental results.

**See also:** Apollo Program; Apollo Program: Command and Service Module; Apollo Program: Developmental Flights; Apollo Program: Geological Results; Apollo Program: Lunar Module; Apollo Program: Orbital and Lunar Surface Experiments; Apollo 1; Apollo 11; Apollo 12; Apollo 13; Apollo 15; Apollo 16; Apollo 17; Materials Processing in Space.

## Further Reading

Allday, Jonathan. *Apollo in Perspective: Spaceflight Then and Now.* Philadelphia: Institute of Physics, 1999. This book takes a retrospective look at the Apollo Program and the technology that was used to land humans on the Moon.

Beattie, Donald A. *Taking Science to the Moon: Lunar Experiments and the Apollo Program.* Baltimore: Johns Hopkins University Press, 2001. This book describes, from the perspective of NASA Headquarters, the struggles that took place to include science payloads and lunar exploration as part of the Apollo Program. Beattie discusses the experiments and details the decisions, meetings, and NASA infighting that got these important surface experiments on the flights to the Moon.

Benson, Charles D., and William B. Faherty. *Moonport: A History of Apollo Launch Facilities and Operations.* NASA SP-4204. Washington, D.C.: Government Printing Office, 1978. A well-illustrated, highly detailed account of the development of the Apollo launch complex at the Kennedy Space Center in Florida. The genesis and development of the Saturn program and the technical achievements of mission engineers are the highlights of this book, which is accessible to high school and college-level readers. The Apollo missions are briefly described, but the strength of this volume lies in its account of the Apollo support systems.

Chaikin, Andrew. *A Man on the Moon: The Voyages of the Apollo Astronauts.* New York: Penguin Group, 1998. The Apollo Moon landings are retold through the eyes and ears of the people who were there. Based on interviews with twenty-three Moon voyagers as well as those who struggled to get the program moving, this book conveys every aspect of the missions with breathtaking immediacy: from the rush of liftoff to the heart-stopping lunar touchdown to the final hurdle of reentry.

Godwin, Robert, ed. *Apollo 14: The NASA Mission Reports.* Burlington, Ont.: Apogee Books, 2001. In this book, some of the rare official documentation of the voyage of Apollo 14 is collected and made commercially available for the first time. Included are the Apollo 14 press kit, detailing the plans for the mission and the postmission operation report, and a brief record of the mission's accomplishments.

Kelly, Thomas J. *Moon Lander: How We Developed the Apollo Lunar Module.* Washington, D.C.: Smithsonian Books, 2001. Grumman Chief Engineer Kelly gives a firsthand account of designing, building, testing, and flying the Apollo Lunar Module.

Kraft, Christopher C., Jr. *Flight: My Life in Mission Control.* East Rutherford, N.J.: Penguin Putnam, 2002. The first NASA flight director gives an account of his life in Mission Control.

Lambright, W. Henry. *Powering Apollo: James E. Webb of NASA.* Baltimore: Johns Hopkins University Press, 2000. Explores James E. Webb's leadership role in NASA's spectacular success.

Liebergot, Sy, and David Harland. *Apollo EECOM: Journey of a Lifetime.* Burlington, Ont.: Apogee Books, 2003. This is the life story of Sy Liebergot, former NASA flight controller, with emphasis on his years working in Apollo Mission Control.

Lindsay, Hamish. *Tracking Apollo to the Moon.* London: Springer-Verlag London Limited, 2001. Features interviews, quotes, and extensive photographs, including some that appear here for the first time.

National Aeronautics and Space Administration. *Apollo 14 Mission Report.* Washington, D.C.: National Technical Information Services, 1971. This is the official report of the results of the Apollo 14 flight. The original is out of print, but copies can be obtained through the National Technical Information Services. An Internet search will turn up a copy in Adobe Acrobat format for easy download. Nowhere else will you find as in-depth a document on the technical, scientific, and human aspects of the mission.

_____. *Apollo 14 Preliminary Science Report.* NASA SP-272. Washington, D.C.: Government Printing Office, 1971. A detailed technical description of the experimental goals and results of the Apollo 14 mission. This collection of reports is directed toward persons with a good background in astronomy and some knowledge of engineering and geology.

_____. *Apollo Mission Press Kits.* http://www-lib.ksc.nasa.gov/lib/presskits.html. Provides detailed preflight information about each of the Apollo missions, Apollo 6 through Apollo 17. Accessed March, 2005.

Shepard, Alan B., Jr., and Donald K. "Deke" Slayton, with Jay Barbree and Howard Benedict. *Moon Shot: The Inside Story of America's Race to the Moon.* Atlanta: Turner, 1994. This is, indeed, the inside story of the Apollo Program as told by two men who actively participated in it. Some of their tales appear here for the first time. The book was adapted for a four-hour documentary in 1995.

Slayton, Donald K., with Michael Cassutt. *Deke! U.S. Manned Space: From Mercury to the Shuttle.* New York: Forge, 1995. This is the autobiography of the last of the Mercury astronauts to fly in space. After being grounded from flying in Project Mercury for what turned out to be a minor heart murmur, Slayton was appointed head of the Astronaut Office. During his reign he assigned all of the Apollo crew members to their flights. Later, he commanded the Apollo-Soyuz Test Project flight in 1975. This is a behind-the-scenes look at America's attempt to land humans on the Moon.

Thompson, Neal. *Light This Candle: The Life and Times of Alan Shepard—America's First Spaceman.* New York: Crown, 2004. Until his death in 1998, Shepard guarded the story of his life zealously. Thompson's exclusive access to private papers and interviews with Shepard's family and closest friends offers a riveting account of Shepard's life.

Turnill, Reginald. *The Moonlandings: An Eyewitness Account.* New York: Cambridge University Press, 2003. This work is a unique eyewitness account of one of the most thrilling adventures of the twentieth century by a veteran space journalist.

Ulivi, Paolo, and David M. Harland. *Lunar Exploration: Human Pioneers and Robotic Surveyors.* London: Springer-Verlag London Limited, 2004. The authors cover the robotic programs like Ranger and other American probes in the late 1950's.

Wendt, Guenter, and Russell Still. *The Unbroken Chain.* Burlington, Ont.: Apogee Books, 2001. Wendt is the only person who worked with every astronaut bound for space.

*Charles Merguerian*

# Apollo 15

*Date:* July 26 to August 7, 1971
*Type of mission:* Lunar exploration, piloted spaceflight

*The Apollo 15 mission placed the United States' fourth scientific team on the lunar surface. The primary objectives of the mission were fulfilled: It returned to Earth with an abundance of new data on the geologic evolution of the Moon and set the stage, as an exploration prototype, for the Apollo 16 and 17 missions.*

## Key Figures

*David R. Scott* (b. 1932), Apollo 15 mission commander
*Alfred M. Worden* (b. 1932), Apollo 15 Command Module pilot
*James B. Irwin* (1930-1991), Apollo 15 Lunar Module pilot
*Richard F. Gordon, Jr.* (b. 1929), Apollo 15 backup mission commander
*Vance DeVoe Brand* (b. 1931), Apollo 15 backup Command Module pilot
*Harrison H. "Jack" Schmitt* (b. 1935), Apollo 15 backup Lunar Module pilot

## Summary of the Mission

Following the success of Apollo 14 six months earlier in the Fra Mauro area of the Moon, the Apollo 15 mission was designed to continue the United States' lunar investigations in the Hadley-Appenine Mountains, located along the southeastern rim of the Sea of Rains. National Aeronautics and Space Administration (NASA) scientists were convinced that ancient rocks found in the mountainous regions peripheral to the multiringed Sea of Rains held the clue to the age and genesis of the primordial lunar crust.

Apollo 15 was launched on schedule from Launch Pad 39A of the Kennedy Space Center in Florida on July 26, 1971, at 13:34 Coordinated Universal Time (UTC). The launch vehicle was an 111-meter-high Saturn V rocket, capable of delivering more than 40 million newtons of thrust with its liquid oxygen and kerosene first stage and liquid oxygen and liquid hydrogen second and third stages. Eleven minutes and 44 seconds after liftoff, the Command and Service Module (CSM), the Lunar Module (LM), and the Saturn IVB booster were placed into a nearly circular Earth orbit at an altitude of roughly 170 kilometers. Immediately following the launch, the Manned Spacecraft Center in Houston, Texas, assumed control of the mission. After a routine systems check, a translunar injection burn of the booster was accomplished at 16:24 UTC. The CSM and LM were separated from the Saturn IVB booster after the injection burn, and the booster was directed to impact the Moon and provide a seismic signal for the experiment packages in place on the surface. The booster made lunar impact at 20:58 UTC on July 29; Apollo 12 and 14 seismograph installations recorded its strong signal. Because the booster's kinetic energy was known, calibration of these seismograph stations was made possible.

After two minor midcourse corrections to prepare for lunar orbit and a series of complicated maneuvers during lunar orbit injection, Apollo 15 was firmly placed into a nearly circular lunar orbit of 121 by 101 kilometers at 19:12 UTC on July 30. After separation from the CSM, dubbed *Endeavor,* the LM, dubbed *Falcon,* began its descent and landed at approximately 22:16 UTC at 26°26′00″ latitude and longitude 3°39′20″ east in a rugged valley between the Apennine highlands region and Hadley Rille. (A rille is a linear surface feature of unknown origin—perhaps a collapsed lava tube, a linearly fractured crust, or an ancient river valley.)

A stand-up extravehicular activity (EVA) was accomplished when Commander David R. Scott depressurized the LM, opened a hatch, and examined and photographed the lunar landscape for 33 minutes. Three subsequent periods of EVA employed the new Lunar Roving Vehicle (LRV), a mechanized vehicle built by Boeing Aircraft Company, which enabled Scott and James B. Irwin, pilot of the Lunar Module, to travel greater distances and carry more complex scientific instrumentation than had been possible in previous missions. The LRV weighed 209 kilograms on Earth but only 34 kilograms in the lunar gravity field. It was capable of carrying Scott, Irwin, their life-support

*The Apollo Command and Service Module* Endeavor *in lunar orbit, as seen from the approaching Lunar Module* Falcon. *Note the scientific instruments on the exposed section of the Service Module.* (NASA CORE/ Lorain County JVS)

systems, communications equipment, scientific instruments, and 27 kilograms of lunar samples. Use of the LRV permitted total EVA time to be more than double the nine-hour EVA performed during Apollo 14.

The first EVA began at midday UTC on July 31 and lasted 6 hours and 34 minutes; this was 26 minutes less than planned because of excessive oxygen consumption. Collecting samples of lunar basalt (an iron- and magnesium-rich volcanic rock forming the dark areas, or maria, of the Moon and the oceanic crust of Earth), Scott and Irwin skirted the LRV along the rim of Hadley Rille and collected specimens along the rille's edge. They traveled southwestward toward Elbow and St. George Craters and collected rocks ejected from the craters during their formation. After careful observation and collection of rock, soil, and core samples, the Apollo exploration crew climbed aboard the

LRV for the 1.5-kilometer return trip to the *Falcon.* Following their LRV tracks, the crew returned to the landing site and set up a seismograph station, a solar wind experimental package, and a laser mirror to add to the retroreflector mirror network begun by the Apollo 11 and 14 missions.

The second EVA occurred at midday UTC on the following day and lasted 7 hours and 12 minutes. The crew drove the LRV southeastward toward Mount Hadley, a 3.47-kilometer-high mountain uplifted during the impact that created the Sea of Rains to the northwest. Vigorous collecting of geologic specimens and recording of field data ensued, as the crew visited many craters and then turned their attention to Mount Hadley. Here, 12-meter-thick layers were exposed on the face of the mountain, suggesting deposition of volcanic lavas or ash flows. After completing the 12.7-kilometer journey, the exploration team attempted to drill holes into the lunar crust for heat probes. Because of design problems and unexpected hardness of the lunar regolith (unconsolidated "soil" resting above the bedrock), the drill would penetrate only to about 1 meter of depth, rather than the 3 meters desired by project scientists. Heat probes were placed in the shallow holes and connected to the instrumentation. Finally, another attempt to drill (this time a drilled core) resulted in a 2.4-meter continuous boring of the lunar regolith. Irwin dug a trench, and soil mechanics experiments were carried out.

Apollo 15's third EVA began on the morning of August 2, with the express purpose of studying Hadley Rille. Layers of varying thickness in the rille's walls were interpreted to represent basaltic lavas. The astronauts climbed down the rille's sloping walls, sampling the rocks and photographing the structure. After 4 hours and 50 minutes, the EVA was over; the LM crew abandoned the LRV, set

up a television camera to transmit the takeoff, and entered the LM's ascent stage for liftoff.

While the *Falcon* crew was extending knowledge of lunar geology, Alfred M. Worden had accomplished thirty-four orbits in *Endeavor* and had carried out an impressive array of scientific tasks. Utilizing panoramic and mapping cameras, he managed to photograph roughly 10 percent of the lunar surface. A laser altimeter, which failed late in the mission, procured altitude data to allow the scaling of photographs and the measuring of topographic variations of the lunar surface. In addition, lunar surface features were photographed by hand, and detailed photographs of the planned Apollo 17 landing site near Littrow Crater were taken.

Use of the camera attached to the LRV allowed images of the LM's ascent stage liftoff to be transmitted to Houston and witnessed in real time by people on Earth. Liftoff occurred on August 2 at

*Commander David R. Scott salutes the flag with lunar mountains about 5 kilometers in the distance and the Lunar Module* Falcon *on the right.* (NASA James B. Irwin)

17:11 UTC, with rendezvous and docking successfully performed two hours later. After men, samples, and equipment were transferred from the LM's ascent stage, it was jettisoned, and made impact with the Moon within 22 kilometers of the target site. The seismographs of Apollo 12, 14, and 15 recorded the event.

Following a correction of orbit in preparation for a critical burn that would return Apollo 15 to Earth, a 38-kilogram, 78-centimeter unnamed satellite built by the TRW Company was ejected from the CSM. It would orbit the Moon for roughly one year, and for several months it would transmit data gathered from onboard experiments on lunar geologic structure, density variations, electric and magnetic fields, and charged particles. During Apollo 15's return journey, photographs of the Moon, Earth, and space were taken, and numerous scientific and medical experiments were conducted.

*Endeavor* returned to Earth's atmosphere on August 7 and splashed down 2 kilometers from the target point in the Pacific Ocean at 20:45 UTC, within 10 kilometers of the prime recovery ship, USS *Okinawa*. The crew of Apollo 15 returned roughly 78 kilograms of lunar rocks, soils, and core, as well as a vast collection of photographic data. They had spent 18 hours and 30 minutes in three EVAs and 66 hours and 54 minutes on the Moon. The Apollo 15 mission provided a treasure trove of new data on the genesis, structure, and geologic development of the Moon.

### Contributions

One of the prime objectives of the Apollo 15 mission was to analyze the Hadley mountain area, as it was thought to contain exposures of ancient lunar crust that predated the formation of the Sea of Rains. Prominent layers exposed on the fresh face of Mount Hadley defied precise analysis, because of poor lighting condi-

tions, but they were thought to represent gently dipping regolithic compositional strata, or perhaps vertical fault structures. Extensive sampling and photography of Hadley Rille, on the other hand, allowed scientists to conclude that this topographic feature represents lava flows up to 60 meters thick, of variable composition. Rocks from the Hadley Delta show evidence of intense fracturing or shock-metamorphism (brecciation) and partial melting resulting from impact.

Sea of Rains lavas sampled from the Marsh of Decay were dated at 3.1 to 3.4 billion years old. They show high iron content and low sodium compared with terrestrial basalts (which are much younger). They vary in texture; some being dense and fine-grained while others are rich in bubbles (scoria), like lavas found in volcanic provinces on Earth. The soils sampled are similar to those collected on previous missions: shock-melted spherical glass droplets, angular shards of glass, basalt and mineral fragments, and fine-grained breccias. The spheri-

*Apollo 15's Command Module* Kitty Hawk *descends to a safe splashdown despite the failure of one of the three main parachutes.* (NASA)

cal glass represents shock-melted airborne debris that cooled in flight, while the angular shards suggest impact-related fracturing of preexisting glassy volcanic crusts.

Apollo 15's seismometer, joined to the Apollo 12 and 14 network, returned data showing that there is a boundary between the lunar crust and mantle and that the crust's thickness varies from 45 to 70 kilometers. The number of moonquakes was minor, compared with the number of quakes on Earth, indicating a relatively inactive, highly evolved geologic body. The Apollo 15 magnetometer experiment showed a slight field, compared to measurements made at the Apollo 12 and 14 sites, and allowed calculations of interior mantle and core temperatures. The heat-flow experiment showed that lunar heat flow was greater than expected—roughly half that of Earth—which suggested that the lunar interior has a greater radioactive component than had been thought. X-ray fluorescence measurements taken in orbit confirmed the chemical contrasts of the lunar maria and highland areas and helped define differences between these chemical-physiographic provinces.

The lunar-orbital mass spectrometer experiment proved inconclusive as gas molecules measured in the Moon's airless atmosphere were apparently related to the Command Module (CM). The subsatellite measured lunar magnetic variations associated with specific craters and found interplanetary fields interacting with the Moon's field. In addition, the solar wind was found to be disturbed as it passed around the Moon, and anomalous gravity increases (mascons) were mapped; these findings confirmed earlier results from robotic Lunar Orbiter missions. Other experiments performed during the Apollo 15 mission are described in detail in the volumes cited below.

**Context**

The voyage of Apollo 15 marked the first of three extended lunar exploration missions using the motorized Lunar Rover, a vehicle that extended the range, quantity, and quality of scientific investigations of the Moon. Apollo research had shown that the Moon was similar to Earth in total age, although its geologic evolution had halted much earlier. Data from Apollo 14 had yielded an approximate date of 4 billion years before the present for the cataclysmic creation of the Sea of Rains; its lavas, dated at roughly 3.2 billion years old, mark the cessation of major volcanism on the Moon.

Mare basalts of the Apollo 15 site were strikingly similar to those collected by Apollos 11 and 12 and the Soviet Union's Luna 16 mission. Apparently, the Moon experienced a number of internal heating episodes that resulted in an ancient layered crust, much thicker than Earth's. The Moon's primordial highland crust proved to be similar to the ancient terrestrial igneous rocks known as anorthosites (rocks composed primarily of calcium, aluminum, silicon, and oxygen in combination). Radioactive dating demonstrated that the crust formed early in the solar system's history, roughly 4.5 billion years ago. Earth-based studies of collected geologic specimens, photographs, and raw scientific data have painted a new view of lunar geologic history, shed new light on the early history of Earth and the solar system, and produced dramatic new visuals of Earth in space.

**See also:** Apollo Program; Apollo Program: Command and Service Module; Apollo Program: Developmental Flights; Apollo Program: Geological Results; Apollo Program: Lunar Lander Training Vehicles; Apollo Program: Lunar Module; Apollo Program: Orbital and Lunar Surface Experiments; Apollo 11; Apollo 14; Apollo 15's Lunar Rover; Apollo 16; Apollo 17.

**Further Reading**

Allday, Jonathan. *Apollo in Perspective: Spaceflight Then and Now.* Philadelphia: Institute of Physics, 1999. This book takes a retrospective look at the Apollo Program and the technology that was used to land humans on the Moon.

Beattie, Donald A. *Taking Science to the Moon: Lunar Experiments and the Apollo Program*. Baltimore: Johns Hopkins University Press, 2001. This book describes, from the perspective of NASA Headquarters, the struggles that took place to include science payloads as part of the Apollo Program.

Benson, Charles D., and William B. Faherty. *Moonport: A History of Apollo Launch Facilities and Operations*. NASA SP-4204. Washington, D.C.: Government Printing Office, 1978. A well-illustrated account of the development of the Kennedy Space Center's Apollo launch complex. The origins and development of the Saturn program and the technical achievements of mission engineers are discussed in great detail. The Apollo missions are briefly described, but the focus of this volume is the Apollo support systems. Accessible to high school and college-level readers.

Chaikin, Andrew. *A Man on the Moon: The Voyages of the Apollo Astronauts*. New York: Penguin Group, 1998. The Apollo Moon landings are retold through the eyes and ears of the people who were there. Based on interviews with twenty-three Moon voyagers as well as those who struggled to get the program moving, this book conveys every aspect of the missions with breathtaking immediacy: from the rush of liftoff to the heart-stopping lunar touchdown to the final hurdle of reentry.

Criswell, David R., ed. *Proceedings of the Third Lunar Science Conference*. 3 vols. Cambridge, Mass.: MIT Press, 1972. Presenting the research of hundreds of scientists, this massive three-volume set reports on the findings of Apollos 11, 12, 14, and 15 and the Soviets' Luna 16 mission. Meticulously detailed accounts of the exploration of the lunar surface and of Earth-based analysis of returned samples and data.

Godwin, Robert, ed. *Apollo 15: The NASA Mission Reports*. Vol. 1. Burlington, Ont.: Apogee Books, 2001. In this book, the amazing fourth lunar landing is explained in depth with documents, such as the mission's press kit, which detailed the plans for the flight, and the postmission operation report and technical crew debriefing, which give an in-depth look at the results of the mission.

Harland, David M. *Exploring the Moon: The Apollo Expeditions*. Chichester, England: Springer-Praxis, 1999. Focuses on the exploration carried out by the Apollo astronauts while on the lunar surface, and not on the technology of getting there. Harland concentrates on the final three Moon landings: Apollo 15, 16, and 17. The three missions accounted for three-quarters of all lunar surface activity to date. There is a before-and-after discussion of the results of the Apollo Program. The author's unique insights into these three Apollo missions answer most questions. Using the actual transcripts of what the astronauts said to each other while carrying out their duties and numerous photographs taken at each step of the exploration, this book provides a graphic illustration of Apollo's contribution to the exploration of the Moon.

Kelly, Thomas J. *Moon Lander: How We Developed the Apollo Lunar Module*. Washington, D.C.: Smithsonian Books, 2001. Kelly gives a firsthand account of designing, building, testing, and flying the Apollo Lunar Module.

Kraft, Christopher C., Jr. *Flight: My Life in Mission Control*. East Rutherford, N.J.: Penguin Putnam, 2002. Kraft gives an account of his life in Mission Control.

Liebergot, Sy, and David Harland. *Apollo EECOM: Journey of a Lifetime*. Burlington, Ont.: Apogee Books, 2003. This is the life story of Sy Liebergot, former NASA flight controller, with emphasis on his years working in Apollo Mission Control.

Lindsay, Hamish. *Tracking Apollo to the Moon*. London: Springer-Verlag London Limited, 2001. Features interviews, quotes, and extensive photographs, including some that appear here for the first time.

National Aeronautics and Space Administration. *Apollo 15 Mission Report*. Washington, D.C.: National Technical Information Services, 1971. This is the official report of the results of the Apollo 15 flight. The original is out of print, but copies can be obtained through the National Technical Information Services. An Internet search will turn up a copy in Adobe Acrobat format for easy download. Nowhere else will you find as in-depth a document on the technical, scientific, and human aspects of the mission.

_____. *Apollo 15 Preliminary Science Report*. NASA SP-289. Washington, D.C.: Government Printing Office, 1972. A detailed technical description of the scientific experiments and findings of the Apollo 15 mission. Requires of its readers some knowledge of geology and engineering as well as familiarity with astronomy and the history of the space program.

_____. *Apollo Mission Press Kits*. http://www-lib.ksc.nasa.gov/lib/presskits.html. Provides detailed preflight information about each of the Apollo missions, Apollo 6 through Apollo 17. Accessed March, 2005.

Scott, David, and Alexei Leonov. *Two Sides of the Moon: Our Story of the Cold War Space Race*. London: Thomas Dunne Books, 2004. In this unique dual autobiography, Astronaut Scott and Cosmonaut Leonov recount the drama of the Space Race set against the conflict that once held the world in suspense.

Shepard, Alan B., and Donald K. "Deke" Slayton, with Jay Barbree and Howard Benedict. *Moon Shot: The Inside Story of America's Race to the Moon*. Atlanta: Turner, 1994. This is, indeed, the inside story of the Apollo Program as told by two men who actively participated in it. Some of their tales appear here for the first time. The book was adapted for a four-hour documentary in 1995.

Slayton, Donald K., with Michael Cassutt. *Deke! U.S. Manned Space: From Mercury to the Shuttle*. New York: Forge, 1995. This is the autobiography of the last of the Mercury astronauts to fly in space. After being grounded from flying in Project Mercury for what turned out to be a minor heart murmur, Slayton was appointed head of the Astronaut Office. During his reign he assigned all of the Apollo crew members to their flights. Later, he commanded the Apollo-Soyuz Test Project flight in 1975. This is a behind-the-scenes look at America's attempt to land humans on the Moon.

Turnill, Reginald. *The Moonlandings: An Eyewitness Account*. New York: Cambridge University Press, 2003. This work is a unique eyewitness account of one of the most thrilling adventures of the twentieth century, as told by a veteran space journalist.

Ulivi, Paolo, and David M. Harland. *Lunar Exploration: Human Pioneers and Robotic Surveyors*. London: Springer-Verlag London Limited, 2004. The authors provide a well-paced, yet balanced account of lunar exploration in this popular history.

Wendt, Guenter, and Russell Still. *The Unbroken Chain*. Burlington, Ont.: Apogee Books, 2001. Because of his unique perspective from the launch pad, Wendt's story is filled with important accounts and rich anecdotes, many of which are published here for the first time.

*Charles Merguerian*

# Apollo 15's Lunar Rover

*Date:* July 31 to August 2, 1971
*Type of spacecraft:* Lunar exploration, piloted spacecraft

*The Lunar Rover greatly expanded the area available for exploration by astronauts. It also demonstrated the use of a new technology: four-wheeled, piloted vehicles to explore celestial bodies.*

### Key Figures

*David R. Scott* (b. 1932), Apollo 15 mission commander
*James B. Irwin* (1930-1991), Apollo 15 Lunar Module pilot
*John W. Young* (b. 1930), Apollo 16 mission commander
*Charles M. Duke, Jr.* (b. 1935), Apollo 16 Lunar Module pilot
*Eugene A. Cernan* (b. 1934), Apollo 17 mission commander
*Harrison H. "Jack" Schmitt* (b. 1935), Apollo 17 Lunar Module pilot

### Summary of the Technology

The Apollo 15 Lunar Roving Vehicle (LRV) was the first piloted vehicle to travel on the Moon's surface. It permitted the astronauts to collect soil samples from a range of sites in the Moon's Hadley-Apennine region. Previously limited to a small radius around the Lunar Module (LM), astronauts now could explore an area about ten times as large. By allowing the astronauts to use less energy, the LRV doubled the length of stay possible on the lunar surface. The vehicle also tested wheeled-vehicle operations in the hostile space environment.

LRV planners had decided that light weight and simplicity of design and operation were overriding needs. Engineers had wanted to minimize travel time without lessening stability or control of the LRV or in any way jeopardizing the astronauts' safety. To do this, the designers needed to predict how the LRV and soil would interact. Thus, soil experiments were conducted in all earlier Apollo piloted missions, and data from the Soviet Luna lunar lander and Lunokhod lunar rover were analyzed.

The LRV was built by the Boeing Aircraft Company's Aerospace Group, under the direction of the Marshall Space Flight Center. Boeing built special trainers for Earth tests, because the lightweight LRV could not rest on its wheels in Earth gravity. The trainers were tested in Rio Grande Gorge, a terrestrial model of Hadley Rille, outside Taos, New Mexico. Colonel David R. Scott, commander of the Apollo 15 mission, and Lieutenant Colonel James B. Irwin, pilot of the LM for Apollo 15, tested the trainer there. The astronauts nicknamed the LRV *Rover* and the trainer, *Grover.* Three rovers and trainers were built at a cost of about $21 million.

The completed rover looked like a cross between a dune buggy and a stripped-down Jeep. It was 3.1 meters long, slightly more than 1.83 meters wide, 1.14 meters high, and had a 2.29-meter wheelbase. It massed 207 kilograms but could carry 490 kilograms—181.5 kilograms for each astronaut plus life-support equipment and 127 kilograms for other equipment, tools, communication gear, and lunar samples. In comparison, an average family car carries only one-half of its own weight. When fully loaded, the rover could negotiate obstacles 30 centimeters high, cross 70-centimeter-wide crevasses, and climb slopes of 20 to 23°.

The rover was powered by two 36-volt silver-zinc batteries that drove independent, one-quarter-horsepower direct current motors in each wheel. Each wheel had a harmonic-drive gear, which allowed continuous application of power to the wheels without gear shifting. Mechanical brakes were used, with equalizer cables ensuring the same force on either side of the rover. There were separate brakes for front and rear wheels. The front and rear wheels could be steered independently, using rack-and-pinion steering.

With help from General Motors, Boeing developed tires of woven piano wire. They were faced with titanium strips in a chevron pattern that covered 50 percent of the tread area. This combination gave sufficient flotation on the lunar soil without impeding traction. Fiberglass fenders topped off the wheels. The chassis was made of aluminum-alloy tubing welded at structural joints and suspended from each wheel by a pair of parallel triangular suspension arms. Loads were transmitted

from these arms through the torsion bars to the chassis.

The crew area was equipped with tubular aluminum seats spanned with nylon and fitted with nylon seat belts. There were footrests, side restraints, and handholds for help in boarding. An armrest supported the driver's arm while he manipulated the hand controller. This T-shaped controller, located in the center display console, could be used by either astronaut. The console had navigation displays on the top and monitoring controls below. Stowage areas were located both under and behind the seats, for equipment and samples.

The instruments on the rover included a navigational system, radio and television communication system, and a 16-millimeter film camera. The navigation system had a directional gyroscope aligned with the Sun for inertial guidance. The system automatically calculated the direction and distance to the LM, and odometers recorded the distance traveled. Finally, its small, solid-state computer determined the heading, bearing, range, and speed of the rover. The radio and television communicated from the rover directly to Earth, using both high- and low-gain antennae mounted on the rover. These enabled scientists at Mission Control to interact directly with the two astronauts during their sample-gathering missions. The television camera could even be operated by radio control from Earth. The film camera had some problems; its film feed intermittently malfunctioned.

The LM carried the rover to the surface folded into a pie-shaped quadrant of the LM descent stage. The rover came out of the LM almost like a pull-down Murphy bed. One astronaut could deploy the rover by paying out a nylon tape. First, the rover swung out from the LM's bay. Next, the rear chassis unfolded and locked, and the rear wheels unfolded. Then

*Apollo 15's Lunar Rover at the Hadley-Apennine landing site.* (NASA)

the front chassis and wheels snapped out. Finally, it was lowered to the surface using a second nylon tape, and the astronaut unfolded the seats and footrests. The rover took about seven minutes to deploy.

To get into the rover in the Moon's low gravity, the astronauts stood facing forward next to it and jumped about half a meter, pushing off sideways and kicking their feet out ahead, then settled slowly into their seats. The hand controller acted as an accelerator and a steering and brake control. The best method for acceleration was to push the controller forward to apply full throttle, then back off to the desired speed. Pushing the controller sideways turned the rover, and pulling it backward applied the brakes. Pulling it back about 8 centimeters set the parking brake.

Scott and Irwin began their first extravehicular activity (EVA) at 13:13 Coordinated Universal Time (UTC) on July 31, 1971. Mission rules restricted travel to no more than about 9 kilometers from the LM—the distance that the astronauts could safely walk back, with sufficient oxygen, to the LM. During the first EVA, they could not get the front-wheel steering to operate; instead, they used only rear-wheel steering. They went southwest to the Hadley Rille, journeyed along it southward to Elbow Crater and an area near St. George Crater, and returned north across the mare to the LM. The EVA lasted 6 hours and 33 minutes and covered 10.3 kilometers. They lost part of the front left fender during the EVA.

Between EVAs, Scott and Irwin parked in a north-south orientation to reduce overheating of the vehicle. Before the second EVA, Mission Control discovered the cause of the steering problem. Scott jokingly accused them of sending up Marshall technicians to fix the rover while the astronauts slept.

The second EVA began at 11:48 UTC on August 1. Scott and Irwin traveled south-southwest across the mare, near Index, Arbeit, Crescent, Dune, and Spur Craters, returning along their outward route. The EVA lasted 7 hours, 12 minutes, and covered 12.5 kilometers.

The third EVA began at 08:52 UTC on August 2. It covered 5.1 kilometers west to Scarp Crater, northwest along the rille, and back east across the mare. This EVA lasted 4 hours and 50 minutes.

For the liftoff of the LM, the rover was parked about 91 meters east, and its television was set to monitor liftoff. Unfortunately, the television's elevation clutch had begun to slip during the second EVA, and it deteriorated steadily. Thus, the camera could not be turned to follow the swift liftoff. The rover was left parked on the lunar surface at 26°5′ north latitude, longitude 3°40′ east.

**Contributions**

Apollo 15 astronauts were ordered to conduct a "lunar grand prix" during their third EVA to evaluate the rover's performance under controlled conditions. This test, with Scott driving and Irwin observing, only confirmed their previous experience. In general, the rover was a well-built vehicle and could stand the stress of exploring lunar terrain well.

The average speed of the rover was 9.6 kilometers per hour, although 13 kilometers per hour was achieved on level ground. The rover could accelerate smoothly with little wheel slippage. Braking was different from that of the trainers on Earth—it took twice as far to stop.

Handling was responsive and the rover maneuverable at speeds up to 5 kilometers per hour. Above that speed, lateral skidding could occur in maximum turns. Once the crew took a downhill slope too quickly, and front wheels dug in while the rear did a 180 degree skid. Front-wheel steering with rear wheels locked was recommended for these higher speeds. During the "lunar grand prix," Scott got all four wheels off the ground at once, although he failed to notice. The astronauts also encountered slopes up to 15°, which the rover had no problem climbing, although there was a problem in parking crosswise on a slope. The lightweight vehicle, without passengers, started to slide down the slope; the astronauts had to take turns holding on to it.

The ride, although stable, was bouncy. The wheels sank only 1.25 centimeters into the soil. Irwin called the ride "a combination of a bucking bronco

*James B. Irwin loads the Lunar Rover next to the Lunar Module* Falcon, *with Hadley Delta and the Apennine Front in the left background.* (NASA)

and [a] rowboat in a rough sea." Unfortunately for the crew, the seatbelt designers had failed to take into account the astronauts' increased girth with their space suits and the fact that they rode lighter in the seats than on Earth. They also had difficulty locating the belts, which tended to slide out of stowage. Better restraints might have even increased the road feel through the hand controller.

The major visibility problem occurred near small craters 1 meter wide or less. Because of the poor light, these craters were not visible until the rover was within 2 or 3 meters of them. They tried to avoid them, but still bottomed out the chassis three times during their EVAs. Lunar dust was not a problem, and no soil accumulated in the wheels to slow their progress.

### Context

The Apollo 15 rover was the first piloted vehicle to travel the lunar surface but not the last. After its success, rovers were sent with both the Apollo 16 and Apollo 17 missions. The seatbelts had been redesigned, but all three rovers suffered from trouble with fenders (each of the other missions lost rear fenders). Apollo 17 landed in a very dusty region. Because the dust slowed progress severely, astronauts Eugene A. Cernan and Harrison H. "Jack" Schmitt had to make an emergency repair. Following instructions from Mission Control, they constructed a new fender out of four pages from a lunar surface map book by taping them together and attaching them to the fender guide with emergency lighting clips. All three vehicles could be reclaimed if future lunar missions required their use.

Piloted vehicles were not the only method used to investigate the lunar surface. On November 10, 1970, the Soviets' Luna 17 discharged a small automated roving vehicle, Lunokhod 1, down a ramp. It massed 757 kilograms and was similar in size to a small car. It had a wheelbase of 2.2 meters and a track width of 1.6 meters. Its instrument container was a squat cylinder that sat on top of eight wire-mesh wheels, four per side. The "lid" of the cylinder flipped open, exposing solar cells to power the vehicle. It had automatic brakes to prevent harm from obstacles and overly steep inclines. Its instruments conducted soil and cosmic radiation tests. It also had a laser reflector, used to obtain accurate distance measurements between the Moon and Earth. The Lunokhod's television and camera sent more than twenty thousand photographs back to Earth. Controlled by a five-man team, Lunokhod 1 worked beyond its three-month program, until October, 1971. It explored more than 80,000 square meters. Luna 21 carried Lunokhod 2 to the Moon in January, 1973. It traveled more than 37 kilometers and conducted more than eight hundred soil tests within four months.

Robotic vehicles are still less versatile than piloted ones. It is difficult to plan in advance which measurements will be taken and thus which instruments will be needed. The human presence provides more latitude in missions as circumstances change and discoveries are made. Nevertheless, robotic vehicles are cheap, costing a fraction of what piloted vehicles cost. Oxygen, food, and water are not considerations, and robotic vehicles can be sent to locations on planets where it may be too hazardous for humans to venture. Future planetary exploration will no doubt rely on both types of vehicles to meet specific mission needs.

**See also:** Apollo Program; Apollo Program: Geological Results; Apollo Program: Lunar Lander Training Vehicles; Apollo Program: Lunar Module; Apollo Program: Orbital and Lunar Surface Experiments; Apollo 1; Apollo 15; Apollo 16; Apollo 17.

### Further Reading

Allday, Jonathan. *Apollo in Perspective: Spaceflight Then and Now.* Philadelphia: Institute of Physics, 1999. This book takes a retrospective look at the Apollo Program and the technology that was used to land humans on the Moon. The author explains the basic physics and technology of spaceflight, and conveys the huge technological strides that were made and the dedication of the people working on the program. All of the major aspects of the Apollo Program are covered, including crews, vehicles, and space suits.

Boeing Aircraft Company, The. *Lunar Roving Vehicle Operations Handbook.* Huntsville, Ala.: The Boeing Company, 1971. The Lunar Rover (LRV) was the first vehicle driven on an extraterrestrial landscape. The handbook describes the LRV this way: "The LRV is a four-wheeled, self-propelled, manually controlled vehicle to be used for transporting crewmen and equipment on the lunar surface. The vehicle has accommodations for two crew members and the stowed auxiliary equipment designed for the particular mission. The LRV system comprises Mobility Subsystem, Electrical Power Subsystem, Control and Display console, Navigation Subsystem, Crew Station, Thermal Control Subsystem, and Space Support Equipment." This handbook, prepared by the LRV contractor Boeing, provides complete information on this fantastic little "moon buggy" that made the last Apollo landings unique. It is available online at http://www.hq.nasa.gov/alsj/lrvhand.html. Accessed March, 2005.

Godwin, Robert, ed. *Apollo 15: The NASA Mission Reports.* Vol. 1. Burlington, Ont.: Apogee Books, 2001. In this book, the amazing fourth lunar landing is explained in depth with documents, such as the mission's press kit that detailed the plans for the flight and the postmission operation report and technical crew debriefing that gives an in-depth look at the results of the mission.

Harland, David M. *Exploring the Moon: The Apollo Expeditions.* Chichester, England: Springer-Praxis, 1999. Focuses on the exploration carried out by the Apollo astronauts while on the lunar surface and not on the technology of getting there. Harland concentrates on the final three Moon landings: Apollo 15, 16, and 17. The three missions accounted for three-quarters of all lunar surface activity to date. There is a before-and-after discussion of the results of the Apollo Program. The author's unique insights into these three Apollo missions answer most questions. Using the actual transcripts of what the astronauts said to each other while carrying out their duties and numerous photographs taken at each step of the exploration, this book provides a graphic illustration of Apollo's contribution to the exploration of the Moon.

National Aeronautics and Space Administration. *Apollo Expeditions to the Moon*. NASA SP-350. Washington, D.C.: Government Printing Office, 1975. A fine overview of the entire Apollo Program, from its conception to the general knowledge gained from the program. Covers the boosters, the spacecraft, and early robotic lunar probes. Discusses both astronaut selection and training and Mission Control's role in the program's success.

_____. *Apollo 15 at Hadley Base*. NASA EP-94. Washington, D.C.: Government Printing Office, 1971. This short, educational publication contains plentiful color photographs of the astronauts at work during the Apollo 15 mission. A chronological listing of mission highlights and a description of the Rover's three traverses are also provided.

Shepard, Alan B., Jr., and Donald K. "Deke" Slayton, with Jay Barbree and Howard Benedict. *Moon Shot: The Inside Story of America's Race to the Moon*. Atlanta: Turner, 1994. This is, indeed, the inside story of the Apollo Program as told by two men who actively participated in it. Some of their tales appear here for the first time. The book was adapted for a four-hour documentary in 1995.

Slayton, Donald K., with Michael Cassutt. *Deke! U.S. Manned Space: From Mercury to the Shuttle*. New York: Forge, 1995. This is the autobiography of the last of the Mercury astronauts to fly in space. After being grounded from flying in Project Mercury for what turned out to be a minor heart murmur, Slayton was appointed head of the Astronaut Office. During his reign he assigned all of the Apollo crew members to their flights. Later, he commanded the Apollo-Soyuz Test Project flight in 1975. This is a behind-the-scenes look at America's attempt to land humans on the Moon.

Sullivan, Scott P. *Virtual LM*. Burlington, Ont.: Apogee Books, 2004. The intricacies of the LM design and the details of its manufacture, including some of the major problems that had to be overcome are detailed in this book. *Virtual LM* shows the details of design and production using full-color renderings of the structures, components, subassemblies, and the completed spacecraft, accompanied by supporting descriptions. It shows the Apollo Lunar Module as both an engineering masterpiece and a work of art.

Turnill, Reginald. *The Moonlandings: An Eyewitness Account*. New York: Cambridge University Press, 2003. This work is a unique eyewitness account of one of the most thrilling adventures of the twentieth century, by a veteran space journalist.

Ulivi, Paolo, and David M. Harland. *Lunar Exploration: Human Pioneers and Robotic Surveyors*. London: Springer-Verlag London Limited, 2004. The authors provide a well-paced, balanced account of lunar exploration.

*Mary Walsh*

# Apollo 16

*Date:* April 16 to April 27, 1972
*Type of mission:* Lunar exploration, piloted spaceflight

*Apollo 16 was the second to last of the Apollo lunar flights. With no new primary objectives, Apollo 16 explored more difficult terrain than had previous Apollo missions and provided extensive far-ultraviolet photography of Earth and space.*

### Key Figures

*John W. Young* (b. 1930), Apollo 16 mission commander
*Thomas K. "Ken" Mattingly II* (b. 1936), Apollo 16 Command Module pilot
*Charles M. Duke, Jr.* (b. 1935), Apollo 16 Lunar Module pilot
*Fred W. Haise, Jr.* (b. 1933), Apollo 16 backup mission commander
*Stuart Roosa* (1933-1994), Apollo 16 backup Command Module pilot
*Edgar D. Mitchell* (b. 1930), Apollo 16 backup Lunar Module pilot
*R. L. Fleischer,* principal investigator, cosmic ray detector experiment
*George R. Carruthers,* principal investigator, far ultraviolet camera/spectrograph experiment

### Summary of the Mission

Apollo 16 was the thirty-first of 106 launches during 1972, two of which were piloted. It was the thirteenth and next-to-last Apollo mission, the tenth piloted Apollo mission, and the fifth successful lunar landing. In the excitement of the first piloted lunar landing, the National Aeronautics and Space Administration (NASA) had announced, on July 29, 1969, nine more flights and their proposed dates and landing sites. This mission, then identified as Apollo 17, had been set for September, 1971, but by September 2, 1970, a rearrangement of the original schedule had shifted number 17 to number 16 in the Apollo mission sequence. On October 1, 1970, the launch date was reset to January, 1972; when crew selection was announced on March 2, 1971, the launch date was again reset, to March, 1972.

A site in the Descartes region in the central lunar volcanic highlands was selected on June 17, 1971. The Lunar Roving Vehicle (LRV) was delivered to the Kennedy Space Center (KSC) on September 1. The time line for March 17 to March 29, 1972, was defined on October 21, 1971, with major events—lunar landing, three extravehicular activities (EVAs), and the departure of the Lunar Module (LM) from the Moon—set on four successive evenings from March 21 to 24 for prime-time television.

The erection of the launch vehicle, a Saturn V three-stage booster, began on October 6, 1971. The spacecraft consisted of the LM, given the code designation *Orion* for the constellation emblematic of the Apollo Program, and the Command and Service Module (CSM), code-named *Casper* for the friendly cartoon ghost, because its pilot, Thomas K. "Ken" Mattingly II, had noted that television images of men on the Moon had a ghostly character. The LRV was given its prelaunch check and installed in the LM on November 17, 1971. The total assemblage was rolled out on December 13, 1971, from the KSC's Vehicle Assembly Building, to Launch Complex Pad 39A.

On January 4, 1972, the LM's pilot, Charles M. Duke, Jr., was admitted to the hospital for bacterial pneumonia but was released one week later. On January 7, launch was rescheduled to the next window of April 16 to 23. There were problems with the fitting of the extravehicular mobility units (space suits), with the LM batteries, and with the docking-ring jettison device that would separate the Command Module (CM) from the Service Module (SM). On January 25, a fuel tank ruptured in the Reaction Control System that is used to steer the SM. The spacecraft was removed from the launch pad and returned to the assembly building at a cost of $200,000; it was the only Apollo to require such a return. Working overtime, Apollo mechanics replaced the fuel tanks, and the craft was re-erected February 5, returned to the launch pad February 8, and reassembled by February 11.

*Apollo 16 astronauts John Young, Thomas Mattingly, and Charles Duke emerge from the recovery helicopter after splashdown.* (NASA CORE/ Lorain County JVS)

The crew began their quarantine March 27 and passed their final physical examinations April 11. Because of heart irregularities detected on the crew of Apollo 15, more stringent preflight medical examinations of the Apollo 16 astronauts were conducted, and limitations were placed on their work schedule. In addition, potassium supplements were prescribed and changes were made in their diets.

The six-day countdown began April 10 at 13:30 Coordinated Universal Time (UTC, or 8:30 A.M. eastern standard time). Launch occurred on schedule on April 16 at 17:54 UTC (12:54 P.M.); it was watched by half a million viewers who had gathered near the KSC and by an estimated 38 million via television. Following the common Apollo flight plan the S-IVB (Saturn V's third stage) and spacecraft entered an Earth parking orbit, whose elliptical configuration had an apogee (the point farthest from Earth) of 175.9 kilometers and a perigee

(the point nearest to Earth) of 166.7 kilometers. Despite minor leakage of nitrogen from the Instrument Unit's temperature control system and continuous venting of helium regulating the S-IVB's Auxiliary Propulsion System, Apollo 16 was inserted into lunar trajectory 2 hours and 34 minutes into the flight, at 20:28 UTC (3:28 P.M.). Thirty-one minutes later, color television showed the CSM separating from the S-IVB and docking with the LM that had been carried to space inside the S-IVB. A separating burn sent the S-IVB on its own route to the Moon, but because of the helium depletion there was no second correction. Tracking was lost, but impact, detected by seismometers left by Apollos 12, 14, and 15, occurred on April 19 at 21:02 UTC (4:02 P.M.) at a location 1.8° north, 23.3° west.

Translunar coasting of the docked CSM and LM did not require a first midcourse correction. The crew identified light-colored particles seen streaming from the LM at docking as shredded thermal paint. An electrophoresis demonstration (the motion of charged particles through a liquid under the influence of an electrical field), a technique hampered on Earth by heat convection and sedimentation, was initiated at 25:05 GET (ground-elapsed time, or the time since liftoff, of 25 hours, 5 minutes). At 107,300 kilometers, 216,600 kilometers, and 327,800 kilometers, ultraviolet photographs of Earth were taken. A second midcourse correction was made at 30:39 GET.

John W. Young, the mission commander, entered the LM with Duke to begin a two-hour check of its systems at 53:50 GET. When the scientific instrument module door was jettisoned at 69:59 GET, all instruments were ready to function, and the astronauts returned to the CSM. Lunar orbit insertion on April 19 at 20:22 UTC (3:22 P.M.) placed Apollo 16 into an ellipse whose apolune (the farthest point from the Moon) was 314 kilometers and perilune (the point closest to the Moon), 107.7 kilometers. Young and Duke reentered the LM, powered its systems, and undocked and separated it from the CSM at 96:14 GET on the far side of the Moon. As the LM and then the CSM came from behind the Moon, each reported malfunctions to the Manned Spacecraft Center. In the LM, Duke's hel-

met contained excess potassium-enriched orange juice because of microphone entanglement with the supply tube. A helium-control tank had ruptured in the Reaction Control System. Mattingly, in the CSM, was unable to conduct orbit circularization. A delay in separating the vehicles beyond quick redocking distance lasted some five and a half hours before it was determined that sufficient backup systems permitted mission continuation.

The LM began descent on the sixteenth orbit. *Orion* touched down April 21 at 01:23 UTC (April 20 at 9:23 P.M. eastern standard time) within Cayley Plains in the Descartes region at 9°00′1″ south, 15°30′50″ east, some 230 meters northwest of target. To conserve electricity after the delay in orbit, only a minimum of systems on the LM remained activated while Young and Duke slept.

The mission's first EVA began when Young climbed down the ladder onto the lunar surface on April 21 at 16:59 UTC (11:59 A.M.); Duke followed. Their first steps were not viewed on Earth. The inoperable LM antenna was replaced by a portable system activated from the LRV. The Apollo Lunar Surface Experiments Package (ALSEP) was deployed at 8°59′34″ south, 15°30′41″ east. Young accidentally disconnected the electronic cable and rendered useless the heat-flow experiment. Mission Control decided repair of the $1.2-million experiment would be too complex and time-consuming to risk other activities. All other ALSEP components functioned properly. The U.S. flag was positioned. The LRV was deployed, and within 40 minutes all systems came into operation. The round trip westward to Flag Crater and the return by Spook, Plum, and Buster Craters covered 4.2 kilometers and were transmitted by the LRV television camera. Some 19 kilograms of surface samples were collected. The activity terminated at

*The Lunar Rover during a moonwalk.* (NASA CORE/Lorain County JVS)

126:16 GET, and Young and Duke rested a second time.

The next extravehicular activity left the LM by a south-southeast traverse on April 22 at 16:33 UTC (11:33 A.M.) to explore Survey Ridge and take samples from the Cinco Craters on Stone Mountain. The astronauts returned in a northwesterly direction with stops at Stubby and Wreck Craters. They collected 32.3 kilograms of material over 11.1 kilometers. The EVA, which ended at 150:14 GET, was televised in color with the LRV camera.

The final EVA began thirty minutes early, at 165:45 GET, to allow more time at North Ray Crater where an immense basalt chunk, House Rock, was sampled. They visited Shadow Rock to the southeast, but four other scheduled stops were canceled. Altogether, 44.3 kilograms of material were gathered. Television coverage continued while the LRV was speed-tested, reaching 17 kilometers per hour on a 15° downward slope. After covering 11.4 kilometers, Young and Duke returned to the LM at 171:25 GET.

Mattingly, meanwhile, orbited in *Casper*. His main tasks included lunar and stellar photography in the far ultraviolet range. Through Mission Control, he also kept Young and Duke advised as to the best land approaches to Stone Mountain and North Ray Crater. Previous missions had detected a highly radioactive area near the Ocean of Storms south of the Apollo 14 landing site; this was photographed using gamma-ray spectroscopy. Mattingly noted that the large basins on the Moon's far side appeared to have surfaces similar to the Apollo 16 landing site.

Young and Duke depressurized *Orion*, discarded excess equipment, and repressurized the ascent stage for liftoff on April 24 at 01:26 UTC (April 23 at 8:26 P.M. eastern standard time). The LRV's camera televised the liftoff for live viewing on Earth. Docking with the CSM was not seen, because the LM antenna remained inoperable. Samples, film, and equipment were transferred, and the crew rested in the CSM while in lunar orbit. The LM was jettisoned at 195:12 GET but could not be placed in proper orbit because of the astronauts' failure to position a switch correctly in its guidance system.

A subsatellite was launched from the Service Module at 196:14 GET; it was intended to measure interplanetary and Earth magnetic fields near the Moon. The shaping burn by the spacecraft to obtain orbit for the device, however, was not performed before its ejection. It crashed on the far side of the Moon near 10.2° north, 111.9° east on May 29, after 425 revolutions.

Trans-Earth injection at 200:33 GET sent the CSM back toward Earth after 65 revolutions. Images of the receding Moon from inside the CSM and of the lunar surface from the LRV were transmitted for 2 hours. From 243:35 until 244:59 GET, Mattingly made two in-flight EVAs to recover film cassettes from the scientific instrument module. He observed the external condition of the spacecraft and deployed for ten minutes a microbial ecological evaluation device, which exposed 60 million microbes to the direct ultraviolet rays of the Sun. The astronauts described the Moon's far side for an eighteen-minute television press conference on the mission.

The CM separated from the SM fifteen minutes before reaching Earth interface at an altitude of 121.9 kilometers. Splashdown in the mid-Pacific near American Samoa, approximately 5 kilometers away from the recovery carrier USS *Ticonderoga*, occurred on April 27 at 19:44 UTC (2:44 P.M.). The Command Module flipped upside down but was righted by inflating air bags. The astronauts emerged in fresh flight suits, were taken by helicopter to the biomedical area of the recovery ship, and were flown the next day via Hickam Air Force Base, Hawaii, to Ellington Air Force Base, Texas. The lunar samples, data, and equipment went to Ellington; the retrieved CM remained to be unloaded at San Diego. It went on public display at Alabama Space and Rocket Center in Huntsville on February 5, 1974.

**Contributions**

Apollo 16 reinforced the discoveries of the previous lunar-landing missions and extended exploration into a fifth lunar locale. Of the three primary objectives—to inspect and sample the lunar surface; to activate emplaced experiments; and to conduct orbital experiments, including photography—none was novel. Despite the accident with the heat-flow component, nine other experiments within ALSEP functioned well. In conjunction with the science stations set up by Apollo 12 at 3°11′ south, 23°23′ west, and by Apollo 15 at 26°06′ north, 3°39′ east, Apollo 16's positioning created an equilateral triangle with legs about 1,200 kilometers apart for passive seismic and surface magnetometer measurements.

The two science experiments unique to Apollo 16 were the cosmic-ray detector and the far ultraviolet camera/spectrometer. The latter provided a detailed view of Earth's corona and three atmospheric rings. It supported the theory that solar evaporation of water into the high atmosphere provides a more important source of Earth's oxygen than photosynthesis by green plants. The same instrument took two hundred views of Earth, the Milky Way, and other galaxies in wavelengths of ultraviolet light—1,600 to 500 angstroms—not observable from Earth. Spatial distribution and intensity of emission of atomic hydrogen, atomic oxygen, and molecular nitrogen were imaged, and spectra never before observed were obtained—significant for theories related to the universe and its dynamics over time.

Ninety percent of soil and rock samples brought from the Descartes region had high concentrations of aluminum (26.5 percent) and calcium (15 percent), lessening any probability of volcanic formation. X rays of core tubes showed a greater abundance of metallic iron than samples from other Apollo sites, a factor in the highly localized magnetic field. The age of rock fragments brought back by Apollo 16 ranged from 3.98 to 4.25 billion years. The Apollo 16 samples, like those from the Apollo 14 site, were rich in potassium, uranium, and thorium. The older, premare crust had a very low carbon content, even less than the later mare regions. Moonquakes occur most frequently at perigee; both Apollo 15 and 16 data indicated that some dark-floored or rimmed craters emit gas puffs at the same time.

The gamma-ray spectrometers included in the orbital experiments of Apollo 15 and 16 mapped radioactivity over 20 percent of the surface, with highest levels on the near side in the Sea of Rains and the Ocean of Storms, the site of the Apollo 12 landing. Overall radioactivity was higher on the far side of the Moon. Experiments conducted with data from the Apollo 14 and 16 astronauts' helmets provided measurement of the higher doses of heavy-particle cosmic radiation, an indication that the level of Apollo spacecraft shielding would be inadequate for piloted, deep-space missions.

By March, 1974, lunar soil samples from the six Apollo lunar landings were used to produce a scale of relative age. Scientists determined that the Moon's crust was more than 60 kilometers thick and that there was a negligible magnetic field. On December 16, 1974, a map of the gravitational forces of the far side of the Moon was completed with data supplied by Lunar Orbiter 5 (1967) and the two subsatellites from Apollos 15 and 16. The basins on the far side exerted less pull on orbiting spacecraft than the lava-filled basins on the near side.

By November 7, 1976, though information was still being received from some thirteen experiments of Apollos 12, 15, 16, and 17, it was possible to affirm that the Moon has a millimeter-thick atmosphere and that basaltic rock and silicon are the chief constituents of the first two layers under the surface. All surviving ALSEP stations had their telemetries deactivated on September 30, 1977.

**Context**

Plans for the achievement of the national goal of placing a human being on the Moon by the end of the 1960's had taken into account neither the effect of total cost on the national budget nor the national mood during the unpopular Vietnam conflict. With the primary mission accomplished, funds diminished. By October, 1969, Apollo 20 was rescheduled to May, 1973, and then canceled on January 4, 1970; production of its LM and Saturn V launch vehicles was suspended. The first in-flight failure, the explosion of an oxygen tank in the SM

*Apollo 16 astronaut Charles M. Duke, Jr., collects soil samples at the edge of Plum Crater near the Descartes landing site. The Lunar Rover is parked in the background.* (NASA John W. Young)

of Apollo 13 on April 13, 1970, precipitated not merely the limping back to Earth under LM power of a badly damaged spacecraft, but also the reevaluation of the Apollo Program.

Apollo flights 15 and 19 were canceled September 2, 1970. Those originally numbered 16, 17, and 18 were renumbered 15, 16, and 17, and experiments intended for six flights were placed within this revised series of three. Time between flights was stretched to make December, 1972, the program's end.

Meanwhile, the Soviet robotic Luna craft were functioning regularly and with novel results. Luna 16 landed on the Moon and returned samples to Earth (September 12 to 24, 1970); Luna 17 (November 10 to 27, 1970) landed a remote-controlled

self-propelled roving craft, Lunokhod 1, which operated for ten months. The next year saw Apollo 14 (January 31 to February 9) and Apollo 15 (July 26 to August 7), with its subsatellite spring-ejected from its SM to remain in lunar orbit. The robotic Luna 18 and 19 probes entered orbit in September, but Luna 18 crashed on attempted landing. In 1972, Luna 20 landed and returned samples to Earth (February 14 to 24). Apollo 16 flew in April. The year ended with the Apollo 17 mission (December 7 to 19).

The year 1973 began with Luna 21 (January 8) and its placing on the Moon's surface of Lunokhod 2, which operated for six months. The last scheduled U.S. mission occurred in June when NASA launched Explorer 49, designated Radio Astronomy Explorer 2, as a Lunar Orbiter; its purpose was to use the Moon as a shield from Earth's radio noise in viewing outer space. Soviet lunar probes continued with Luna 22 (May 29, 1974), Luna 23 (October 28, 1974), and Luna 24 (August 9 to 25, 1976); the latter returned large core sections to Earth. Space exploration then turned from the Moon to piloted working stations in Earth orbit, with reusable shuttles and endurance records for human presence and scientific experiments; to planetary and solar system probes, including robotic Mars landers; and to Pioneers that escaped into deep space, surveying the giant outer planets en route.

The overall cost of the Apollo Program reached $25 billion for the eleven flights, not including launching and ancillary facilities, some of which were museum pieces by mission's end. The program's value, however, cannot be measured against expenditure or technology demonstrated, even if only six pairs of men lived on the Moon for a total of 298 hours and 47 minutes.

**See also:** Apollo Program; Apollo Program: Command and Service Module; Apollo Program: Developmental Flights; Apollo Program: Geological Results; Apollo Program: Lunar Module; Apollo Program: Orbital and Lunar Surface Experiments; Apollo 15; Apollo 15's Lunar Rover.

### Further Reading

Allday, Jonathan. *Apollo in Perspective: Spaceflight Then and Now.* Philadelphia: Institute of Physics, 1999. This book takes a retrospective look at the Apollo Program and the technology that was used to land humans on the Moon. The author explains the basic physics and technology of spaceflight, and conveys the huge technological strides that were made and the dedication of the people working on the program. All of the major aspects of the Apollo Program are covered, including crews, vehicles, and space suits.

Beattie, Donald A. *Taking Science to the Moon: Lunar Experiments and the Apollo Program.* Baltimore: Johns Hopkins University Press, 2001. This book describes, from the perspective of NASA Headquarters, the struggles that took place to include science payloads and lunar exploration as part of the Apollo Program. Beattie discusses the experiments and details the decisions, meetings, and NASA infighting that got these important surface experiments on the flights to the Moon.

Benson, Charles D., and William Barnaby Faherty. *Moonport: A History of Apollo Launch Facilities and Operations.* NASA SP-4204. Washington, D.C.: Government Printing Office, 1978. A history of the development of the complex Apollo launching facility, against the backdrop of the socioeconomic problems experienced by the U.S. space program. Spans the years 1957 to 1977 when it seemed that the immense and costly enterprise of piloted spaceflight was becoming a mere tourist attraction. Detailed but readable by the advanced student.

Brooks, Courtney G., James M. Grimwood, and Loyd S. Swenson, Jr. *Chariots for Apollo: A History of Manned Lunar Spacecraft.* NASA SP-4205. Washington, D.C.: Government Printing Office, 1979. The detailed and complicated story of various components that ultimately fit together to form the Apollo spacecraft. The lunar landing of Apollo 11 is the climax; all the other flights are treated as mere epilogue. Necessary for understanding the spacecraft capable of supporting piloted lunar landing.

Chaikin, Andrew. *A Man on the Moon: The Voyages of the Apollo Astronauts.* New York: Penguin Group, 1998. The Apollo Moon landings are retold through the eyes and ears of the people who were there. Based on interviews with twenty-three Moon voyagers as well as those who struggled to get the program moving, this book conveys every aspect of the missions with breathtaking immediacy: from the rush of liftoff to the heart-stopping lunar touchdown to the final hurdle of reentry.

Godwin, Robert, ed. *Apollo 16: The NASA Mission Reports.* Vol. 1. Burlington, Ont.: Apogee Books, 2002. Compiled in this work are several documents about the mission including the complete debriefing in the crew's own words. The Apollo 16 press kit details the plans for the mission.

Harland, David M. *Exploring the Moon: The Apollo Expeditions.* Chichester, England: Springer-Praxis, 1999. Focuses on the exploration carried out by the Apollo astronauts while on the lunar surface, and not on the technology of getting there. Harland concentrates on the final three Moon landings: Apollo 15, 16, and 17. The three missions accounted for three-quarters of all lunar surface activity to date. There is a before-and-after discussion of the results of the Apollo Program. The author's unique insights into these three Apollo missions answer most questions. Using the actual transcripts of what the astronauts said to each other while carrying out their duties and numerous photographs taken at each step of the exploration, this book provides a graphic illustration of Apollo's contribution to the exploration of the Moon.

Kelly, Thomas J. *Moon Lander: How We Developed the Apollo Lunar Module.* Washington, D.C.: Smithsonian Books, 2001. Grumman Chief Engineer Kelly gives a firsthand account of designing, building, testing, and flying the Apollo Lunar Module.

Kraft, Christopher C., Jr. *Flight: My Life in Mission Control.* East Rutherford, N.J.: Penguin Putnam, 2002. The first NASA flight director gives an account of his life in Mission Control.

Liebergot, Sy, and David Harland. *Apollo EECOM: Journey of a Lifetime.* Burlington, Ont.: Apogee Books, 2003. This is the life story of Sy Liebergot, former NASA flight controller, with emphasis on his years working in Apollo Mission Control.

Lindsay, Hamish. *Tracking Apollo to the Moon.* London: Springer-Verlag London Limited, 2001. Features interviews, quotes, and extensive photographs, including some that appear here for the first time.

National Aeronautics and Space Administration. *Apollo Mission Press Kits.* http://www-lib.ksc.nasa.gov/lib/presskits.html. Provides detailed preflight information about each of the Apollo missions: Apollo 6 through Apollo 17. Accessed March, 2005.

_____. *Apollo 16 at Descartes.* NASA EP-97. Washington, D.C.: Government Printing Office, 1973. Illustrated, condensed statement describing the mission for the casual reader, including the schoolchild.

_____. *Apollo 16 Mission Report.* Washington, D.C.: National Technical Information Services, 1972. This is the official report of the results of the Apollo 16 flight. The original is out of print, but copies can be obtained through the National Technical Information Services. An Internet search will turn up a copy in Adobe Acrobat format for easy download. Nowhere else will you find as in-depth a document on the technical, scientific, and human aspects of the mission.

_____. *The Apollo Spacecraft: A Chronology.* 4 vols. NASA SP-4009. Washington, D.C.: Government Printing Office, 1969-1978. This chronicle of events in the Apollo Program, arranged by date, includes a bibliographical annotation for each entry. Volumes 1 through 3 cover the concept of Apollo, designs, and contracts, with definitions of mission requirements in terms of orbital procedure, hardware, software, and the fabrication and testing of the various components. Volume 4 covers the actual preparations for the flights, the setback incurred by the Apollo 1 fire, and the achievements through Apollo 17. Each volume lists key events. Appendices provide flight summaries, funding data, organizational charts, personnel lists, and other tabularized data. Valuable to anyone interested in the details of the program.

Shepard, Alan B., Jr., and Donald K. "Deke" Slayton, with Jay Barbree and Howard Benedict. *Moon Shot: The Inside Story of America's Race to the Moon.* Atlanta: Turner, 1994. This is, indeed, the inside story of the Apollo Program as told by two men who actively participated in it. Some of their tales appear here for the first time. The book was adapted for a four-hour documentary in 1995.

Slayton, Donald K., with Michael Cassutt. *Deke! U.S. Manned Space: From Mercury to the Shuttle.* New York: Forge, 1995. This is the autobiography of the last of the Mercury astronauts to fly in space. After being grounded from flying in Project Mercury for what turned out to be a minor heart murmur, Slayton was appointed head of the Astronaut Office. During his reign he assigned all of the Apollo crew members to their flights. Later, he commanded the Apollo-Soyuz Test Project flight in 1975. This is a behind-the-scenes look at America's attempt to land humans on the Moon.

Swanson, Glen E. *Before This Decade Is Out: Personal Reflections on the Apollo Program.* Gainesville: University Press of Florida, 2002. The lunar-landing program as witnessed by some of the political leaders, engineers, scientists, and astronauts who made it such a success.

Turnill, Reginald. *The Moonlandings: An Eyewitness Account.* New York: Cambridge University Press, 2003. As the BBC aerospace correspondent, Turnill spent his career covering all the piloted space missions as well as planetary missions like Mariner, Pioneer, Viking, and Voyager.

Ulivi, Paolo, and David M. Harland. *Lunar Exploration: Human Pioneers and Robotic Surveyors.* London: Springer-Verlag London Limited, 2004. Even-handed account of lunar exploration as a popular history. Covers the robotic programs like Ranger and other American probes in the late 1950's.

Wendt, Guenter, and Russell Still. *The Unbroken Chain.* Burlington, Ont.: Apogee Books, 2001. Wendt is the only person who worked with every astronaut bound for space.

*Clyde Curry Smith*

# Apollo 17

*Date:* December 7 to December 19, 1972
*Type of mission:* Lunar exploration, piloted spaceflight

*Apollo 17 was the last and perhaps the most ambitious flight of the Apollo Program. It landed in the rugged terrain of the Taurus-Littrow Valley to explore the lunar highlands and the multiringed basin nearby. Apollo 17 provided data that confirmed the earlier results of Apollo 15's heat-flow measurements, which had suggested that the Moon's interior was hotter than previously believed. Apollo 17 also returned the largest number of lunar rocks for study.*

**Key Figures**

*Eugene A. Cernan* (b. 1934), Apollo 17 mission commander
*Ronald E. Evans* (1933-1990), Apollo 17 Command Module pilot
*Harrison H. "Jack" Schmitt* (b. 1935), Apollo 17 Lunar Module pilot
*John W. Young* (b. 1930), Apollo 17 backup mission commander
*Stuart Roosa* (1933-1994), Apollo 17 backup Command Module pilot
*Charles M. Duke, Jr.* (b. 1935), Apollo 17 backup Lunar Module pilot

**Summary of the Mission**

The Apollo 17 mission was the last of the piloted missions to the Moon. Its predecessors had included the first piloted lunar-orbital mission (Apollo 8) and a later test descent in Apollo 10. Apollos 11 through 16 included five successful Moon landings. Unfortunately, financial and political constraints eliminated the last three proposed Apollo missions. Thus, Apollo 17 would offer the last chance for decades to collect observational data directly from the Moon's surface.

In retrospect, the Apollo missions were a logical step in humankind's attempt to learn about the universe. With the creation of the National Aeronautics and Space Administration (NASA) in 1958, a systematic approach was detailed for the piloted exploration of space; its emphasis lay on one day reaching the Moon—the fulfillment of the romantic and scientific dreams of centuries. In 1961, that dream seemed nearer to reality than ever before as President John F. Kennedy challenged the world: "I believe that this nation should commit itself to achieving the goal, before this decade is out, of landing a man on the Moon and returning him safely to Earth." Over the next several years, at the cost of billions of dollars and several lives, that dream was realized; finally, the landing of Apollo 11 was achieved on July 20, 1969. A seemingly impossible feat had been accomplished.

Apollo 17 may have been the last mission of the Apollo Program, but it was in no way a routine one. It began as the first night launch of any American piloted spacecraft and the first for the Saturn V launch vehicle as well. Originally scheduled for a late-evening launch on December 6, the mission was delayed for nearly two and one-half hours because of a faulty computer, which would not accept a manual correction order. The flight was halted at T minus 30 seconds, apparently ending hopes of a flawless mission. Finally, at 05:33 Coordinated Universal Time (UTC, or 12:33 A.M.) on December 7, 1972, the hopeful crowd of spectators witnessed a spectacular sight as the ignition

*A view seen from Apollo 17 in orbit around the Moon. The large crater is 20 kilometers across.* (NASA CORE/Lorain County JVS)

of the Saturn V booster rocket turned night into day.

The Apollo 17 crew consisted of Eugene A. Cernan as commander, Ronald E. Evans as Command Module pilot, and Harrison H. "Jack" Schmitt as Lunar Module pilot. As commander, Cernan brought with him the experience of having flown aboard Gemini IX and Apollo 10. Although new to space, Evans had served as a fighter pilot in Vietnam, and he had been backup Command Module pilot for Apollo 14 as well as part of the support crew for Apollos 7 and 11. The final crew member, Schmitt, was a professional scientist and not a career pilot, as all previous astronauts had been. Schmitt's background included a doctorate in geology from Harvard University. It was hoped that his expertise would improve the chances of making an exciting discovery on the lunar surface.

As the last in the series, Apollo 17 benefited from the experiences of the earlier missions—as seen in the design of the spacecraft itself. The predecessors to the Apollo spacecraft were the

Mercury (one-person) and Gemini (two-person) vehicles, which lived up to their designation as spacecrafts. Merely enclosures to protect the astronauts and their equipment, they were not designed for comfort. These missions helped develop the techniques of maneuvering, docking, and working outside the spacecraft. These procedures would later become commonplace during the Apollo flights.

Compared to the earlier Mercury and Gemini spacecraft and boosters, the Apollo spacecraft and launch vehicle were huge. Fully assembled, the spacecraft and launch vehicle stood 111 meters (364 feet) high and weighed about 3 million kilograms (7 million pounds). Within the three-person Command Module (CM), named *America*, thousands of items of equipment, support systems, and personal gear were stored and arranged around the astronauts. Nevertheless, the crew's quarters provided 6 cubic meters (212 cubic feet) of space, or 2 cubic meters (71 cubic feet) of habitable volume per person—the average space contained in a telephone booth. Fortunately for the astronauts, weightlessness made moving around the spacecraft relatively easy.

In addition to the CM, there was the Service Module (SM), which held the fuel for the main propulsion engine, numerous scientific packages, oxygen and helium tanks, and various other support systems. Below this stage sat the Lunar Module (LM), a 7-meter-tall vehicle that would take the astronauts and their Lunar Roving Vehicle (LRV) to the Moon's surface. This LM, named *Challenger*, was a two-stage vehicle, with the upper portion serving as home while on the surface and as a return vehicle for the astronauts. The lower portion contained the descent engine for getting to the surface and provided storage space for the Lunar Rover. It also served as a launch platform for the return flight to the CM. The ascent stage, after returning the two astronauts to the CM, was deliberately crashed into the Moon to provide a controlled impact event for seismic experiments to record. The lower stage still remains on the Moon, with a plaque bearing an inscription recording the human presence there. It will remain intact for millions of years if left undisturbed and will serve as a reminder to the future of an earlier era's hopes and dreams.

Following its dramatic nighttime liftoff, the Apollo 17 spacecraft established Earth orbit. It would orbit Earth once before a burn from the S-IVB stage would propel it toward the Moon. That burn occurred about three hours later and lasted for almost six minutes, giving Apollo 17 a 10,440-meter-per-second (34,250-foot-per-second) velocity. No further thrust would be required, beyond minor adjustments to assure an accurate rendezvous with the Moon. During the three days spent coasting toward the Moon, the astronauts watched Earth gradually shrink in size while the Moon loomed ever larger. The CM entered lunar orbit at

19:48 UTC (2:48 P.M.) on December 10, with the landing scheduled for the next day.

In keeping with the tradition established in previous missions, Cernan named his craft: After separation from *America, Challenger* descended toward the lunar surface. Contact was made at 18:55 UTC (1:55 P.M.) on December 11 with the words: "Okay Houston, the *Challenger* has landed. Tell *America* that *Challenger* is at Taurus-Littrow."

Shortly after landing, astronauts Cernan and Schmitt left the protection of the LM and stepped onto the surface of the Moon. Inspection of the lander and deployment of the LRV were the first order of business, with a test drive of the vehicle following shortly thereafter. Both astronauts initially experienced difficulty with working in one-sixth Earth gravity, but with time they were able to adapt to their new environment.

Time is of the essence during lunar exploration, and the astronauts had to work fast with little time for rest or sightseeing. Their first extravehicular uactivity (EVA) lasted for 7 hours, 12 minutes; it covered 4.4 kilometers (approximately 2 nautical miles) and collected 13 kilograms (29 pounds) of rock specimens. Numerous scientific experiment packages were also deployed during this first day on the Moon. The job that proved to be most difficult was the drilling by Astronaut Cernan for the heat probe experiment and the neutron flux experiment. During these activities, his pulse rate increased considerably, and he suffered from the strain of working inside a pressure suit. Even with the lower gravity of the Moon, working on the surface proved to be no easy task.

The second EVA began almost twenty-four hours after the astronauts had first set foot on the Moon. The first task after making the usual inspections of their equipment was to

*Accidents do happen, even in space. Here is how the astronauts repaired a broken fender on their Lunar Rover using a map and duct tape.* (NASA CORE/Lorain County JVS)

repair a fender that had been damaged on the Lunar Rover during the previous day's activities. While the astronauts had rested after their first day, NASA engineers had devised a procedure to repair the damage. The astronauts were instructed to tape four plastic sheets from a map book together, fold them once, and fasten them with clamps cannibalized from another experiment. It worked perfectly, thus preventing dust from being thrown up and covering the astronauts during their excursions.

This second EVA was to last 7 hours and 37 minutes, and it covered 20 kilometers (11 nautical miles), during which the astronauts collected 36 kilograms (80 pounds) of rock. After deploying additional experimental equipment, photography and sample collecting were the priorities. At a crater called Shorty, Astronaut Schmitt discovered what he called "orange soil." The significance of this reddish-orange material around a crater is that gases may have been extruded as the result of a volcanic event. This suggests that there was volcanic activity both before and after the formation of the lighter-colored mantle material.

The third EVA began as before with the usual check-out procedures. Scheduled activities would cover 11.6 kilometers (6 nautical miles) and collect 66 kilograms (145 pounds) of rock. Extensive studies of a number of craters and the collection of specimens at the base of a mountain called the North Massif were typical of the day's activities. Toward the end of the EVA, Astronaut Schmitt discovered an intriguing layer of rock that he believed to be representative of the latest mantling in the area and the light-colored material representative of impact debris. These were significant finds, because the darkest rocks (basalt) are the youngest, and the lighter rocks (anorthosites) are the oldest. Astronaut Schmitt believed that he had found a possible transition rock between the two extremes; his training as a professional geologist proved valuable in this instance, because the untrained eye might have missed this feature. Schmitt's expertise gave the Apollo 17 mission an added dimension that no other mission could claim.

After all the experiments had been completed and preparations for their return to Earth had been made, Cernan and Schmitt were able to pause and reflect on the three days they had just spent on the Moon. Cernan read from the inscription on the plaque mounted to one of the lunar lander's legs: "Here man completed his first exploration of the Moon December, 1972 A. D. May the spirit of peace in which we came be reflected in the lives of all humankind." Liftoff came at 22:55 UTC (5:55 P.M.) on December 14. In less than three days, the astronauts would be back on Earth, having completed the most successful lunar exploration to date.

While Cernan and Schmitt were exploring the lunar surface, Evans was monitoring scientific experiments from on board the CM. Rendezvous with *America* was accomplished without difficulty, and the astronauts remained in lunar orbit for one last day of collecting data. The Lunar Module's ascent stage was placed into a trajectory that would crash it into the surface within 16 kilometers (9 nautical miles) of the landing site. The force of the crash would provide a seismic shock that would be used to determine the nature of the Moon's interior.

Because of the stronger gravitational pull of Earth on the spacecraft, the flight home was shorter than the outbound leg had been. The highlight of the return trip was an EVA by Evans to retrieve some film cassettes from an instrument bay. With all the data now aboard the CM, only the safe reentry through the atmosphere and a splashdown in the Pacific remained. Splashdown occurred with pinpoint accuracy at 19:24 UTC (2:24 P.M.) on December 19. The Apollo Moon missions had safely concluded.

**Contributions**

The principal reason that the Taurus-Littrow region was selected as the landing site for Apollo 17 was that it comprises three distinct highland sites that form a complex area where ejecta from four basins (the Seas of Crises, Rains, Serenity, and Tranquility) could have accumulated. Emphasis was placed on the chemistry and physical characteristics of the rocks present in this area. Questions

*An Apollo 17 astronaut examines a large boulder at the landing site. Rocks were taken from different layers in order to learn about lunar geology, which includes knowing how different rock types relate to each other.* (NASA CORE/Lorain County JVS)

to be answered concerned the nature of the Moon's primitive crust and the effects of meteorite bombardment on these ancient rocks. Rocks that formed as a result of those large impacts were also interesting. The field observations conducted by the astronauts established that the highland massifs (mountainous regions) and hills are composed primarily of a type of rock called breccia (a rock formed of rock fragments cemented together by a cooling lava). Three distinct types, based principally on color differences, were recognized: light gray, blue gray, and greenish gray. These breccias, as well as a few less abundant types, are associated with the major impact events that occurred in an older parent rock that was either a coarse-grained gabbro or an anorthosite.

Basalt samples representative of the floor of the Sea of Serenity are almost identical to those in the Apollo 11 Sea of Tranquility samples. The Apollo 17 basalts are 3.7 billion years old, are very dark in color because of a relatively high titanium content, and are extremely low in silica. Chemically, these samples match those taken from the other landing sites, but they differ in age. As expected, age is rela-

tive to the impact event that created the basin.

Chemical analysis of the "orange soil" described by Schmitt proved that the sample was nonvolcanic in origin—definitely not a product of fumarolic activity as initially thought. The soil in fact proved to be an accumulation of yellow, orange, and brown glass spherules. Their ages were about the same as those of the maria basalts (3.7 billion years). Observations made from orbit also showed that other orange-tinted areas exist; the color is not unique to the Apollo 17 landing site.

Other observations made from orbit provided data to develop a lunar altitude profile, produced a thermal map of the lunar surface, and searched (unsuccessfully) for evidence of a lunar atmosphere by means of an ultraviolet spectrometer. In addition, hundreds of photographs were taken to add to those acquired from previous missions.

Heat-flow data acquired from the drill holes confirmed the higher temperature values recorded by Apollo 15 and gave additional support to the theory of a much hotter lunar interior than had been previously believed. Gravimetric measurements made during the three traverses across the lunar surface showed that high-density material fills the valley between the North and South massifs. This indicates that the valley depth at one time was much greater than it is now.

The data returned from the impacts of the LM's ascent stage and the S-IVB into the lunar surface suggest that the lunar crust is thinner than previously believed. These data, combined with the lunar sounder experiment on board the CM, provided evidence to construct a better model for an explanation of the nature and history of the lunar crust.

### Context

Humankind's view of the Moon prior to the Apollo Program was based on high-quality telescopic observations and a few reflectivity studies of the lunar surface material. Indirect measurements of the Moon's mass, size, and density were easily calculated from Earth, but questions relating to the depth of the dust layer that covers the Moon's surface or the origin of its craters remained for piloted exploration to answer. Before the Apollo Program, some of these questions were answered by the Ranger, Surveyor, and Lunar Orbiter probes, but it would take the keen human eye and the ability to adapt to changing situations to produce the best scientific results.

Was the Apollo Program worth all the expense, risk, and commitment? What were the benefits? Even the casual observer of the space program realized the importance of landing people on the Moon and returning them safely to Earth. It was a technological achievement without parallel. The more subtle rewards realized from the Apollo Program were not as well publicized. It must be remembered that the Apollo Program had to develop a completely new technology governed by the need for small and lightweight equipment to do the job ordinarily handled by larger devices. Miniaturization of computer circuits and components, plus the development of a computer system that could perform a wide variety of tasks almost instantaneously, was necessary if the mission was to succeed. Even before the conclusion of the Apollo Program, these technological achievements sifted down into the public sector as transistor radios, pocket calculators, and home computers. No doubt these advances would eventually have happened, but the needs of the space program speeded their development.

The wealth of data returned by Apollo 17, combined with that of the previous five missions, enabled scientists to construct a detailed picture of the Moon's history. Apparently, the Moon formed with the rest of the solar system about 4.5 billion years ago. Its upper portion probably melted to a depth of 100 to 300 kilometers during the first 500 million years of its existence. This melting produced a two-component crust, with one element being a lighter anorthositic-gabbroic member, the second a denser pyroxene-rich lower member. Both parts would have been subjected to intense meteoroid bombardment, which produced the large basin structures such as the Sea of Serenity. Extensive volcanism followed between 3.8 and 3.2 billion years ago; eventually, the large basins were filled in with lava. Little else has happened since, with the exception of a few large impacts.

The Apollo missions did not answer all the questions concerning the Moon's origin, chemical makeup, and potential as an economic resource. What the Apollo Program did accomplish was to give the peoples of Earth a sense of unity. Photographs of Earth taken by the astronauts showed a relatively tiny blue world against the blackness of space. No political boundaries were visible, and the effects of human habitation on the planet could be seen only at night, when the lights of cities appeared. Nevertheless, the planet is teeming with life. Despite political conflicts, famines, and pollution, Earth stands out as a hospitable place in the solar system when compared with the bleak Moon. As many of the astronauts stated upon their return to Earth, their appreciation for Earth as a whole was enhanced by the journey to the Moon. Apollo 17 should be remembered not as the last piloted mission to the Moon but as a tribute to what humankind can accomplish.

**See also:** Apollo Program; Apollo Program: Command and Service Module; Apollo Program: Developmental Flights; Apollo Program: Geological Results; Apollo Program: Lunar Lander Training Vehicles; Apollo Program: Lunar Module; Apollo Program: Orbital and Lunar Surface Experiments; Apollo 16; Clementine Mission to the Moon; Gemini Program.

### Further Reading

Allday, Jonathan. *Apollo in Perspective: Spaceflight Then and Now.* Philadelphia: Institute of Physics, 1999. This book takes a retrospective look at the Apollo Program and the technology that was used to land humans on the Moon. The author explains the basic physics and technology of spaceflight, and conveys the huge technological strides that were made and the dedication of the people working on the program. All of the major aspects of the Apollo Program are covered, including crews, vehicles, and space suits.

Beattie, Donald A. *Taking Science to the Moon: Lunar Experiments and the Apollo Program.* Baltimore: Johns Hopkins University Press, 2001. This book describes, from the perspective of NASA Headquarters, the struggles that took place to include science payloads and lunar exploration as part of the Apollo Program. Beattie discusses the experiments and details the decisions, meetings, and NASA infighting that got these important surface experiments on the flights to the Moon.

Cernan, Eugene A., and Don Davis. *The Last Man on the Moon: Astronaut Eugene Cernan and America's Race in Space.* New York: St. Martin's Press, 1999. The story of the last two men to walk on the Moon is told by Cernan, commander of Apollo 17, and Davis, an experienced journalist. This autobiography tells the story behind the story of the Gemini and Apollo Programs. Cernan, whose spaceflight career spanned both programs, fittingly narrates it. He was the first person to spacewalk in orbit around the Earth and the last person to leave footprints on the lunar surface.

Chaikin, Andrew. *A Man on the Moon: The Voyages of the Apollo Astronauts.* New York: Penguin Group, 1998. The Apollo Moon landings are retold through the eyes and ears of the people who were there. Based on interviews with twenty-three Moon voyagers as well as those who struggled to get the program moving, this book conveys every aspect of the missions with breathtaking immediacy: from the rush of liftoff to the heart-stopping lunar touchdown to the final hurdle of reentry.

Godwin, Robert, ed. *Apollo 17: The NASA Mission Reports.* Vol. 1. Burlington, Ont.: Apogee Books, 2002. This book contains the preflight press kit detailing the plans for the mission. The postmission operation report and technical crew debriefing describe the results of the flight and the observations of the crew. Bonus CD-ROM includes the complete television downlink from the lunar surface, more than eleven hours of video, and a series of unique interactive panoramas of the lunar surface.

Harland, David M. *Exploring the Moon: The Apollo Expeditions.* Chichester, England: Springer-Praxis, 1999. Focuses on the exploration carried out by the Apollo astronauts while on the lunar surface and not on the technology of getting there. Harland concentrates on the final three Moon landings: Apollo 15, 16, and 17. The three missions accounted for three-quarters of all lunar surface activity to date. There is a before-and-after discussion of the results of the Apollo Program. The author's unique insights into these three Apollo missions answer most questions. Using the actual transcripts of what the astronauts said to each other while carrying out their duties and numerous photographs taken at each step of the exploration, this book provides a graphic illustration of Apollo's contribution to the exploration of the Moon.

Kelly, Thomas J. *Moon Lander: How We Developed the Apollo Lunar Module.* Washington, D.C.: Smithsonian Books, 2001. Grumman Chief Engineer Kelly gives a firsthand account of designing, building, testing, and flying the Apollo Lunar Module.

Kraft, Christopher C., Jr. *Flight: My Life in Mission Control.* East Rutherford, N.J.: Penguin Putnam, 2002. The first NASA flight director gives an account of his life in Mission Control.

Liebergot, Sy, and David Harland. *Apollo EECOM: Journey of a Lifetime*. Burlington, Ont.: Apogee Books, 2003. This is the life story of Sy Liebergot, former NASA flight controller, with emphasis on his years working in Apollo Mission Control.

Lindsay, Hamish. *Tracking Apollo to the Moon*. London: Springer-Verlag London Limited, 2001. A complete, detailed narrative of the Apollo Moon Landing Program, telling the story of each mission.

National Aeronautics and Space Administration. *Apollo Mission Press Kits*. http://www-lib.ksc.nasa.gov/lib/presskits.html. Provides detailed preflight information about each of the Apollo missions: Apollo 6 through Apollo 17. Accessed March, 2005.

_____. *Apollo 17 Mission Report*. Washington, D.C.: National Technical Information Services, 1973. This is the official report of the results of the Apollo 17 flight. The original is out of print, but copies can be obtained through the National Technical Information Services. An Internet search will turn up a copy in Adobe Acrobat format for easy download. Nowhere else will you find as in-depth a document on the technical, scientific, and human aspects of the mission.

Shepard, Alan B., Jr., and Donald K. "Deke" Slayton, with Jay Barbree and Howard Benedict. *Moon Shot: The Inside Story of America's Race to the Moon*. Atlanta: Turner, 1994. This is, indeed, the inside story of the Apollo Program as told by two men who actively participated in it. Some of their tales appear here for the first time. The book was adapted for a four-hour documentary in 1995.

Slayton, Donald K., with Michael Cassutt. *Deke! U.S. Manned Space: From Mercury to the Shuttle*. New York: Forge, 1995. This is the autobiography of the last of the Mercury astronauts to fly in space. After being grounded from flying in Project Mercury for what turned out to be a minor heart murmur, Slayton was appointed head of the Astronaut Office. During his reign he assigned all of the Apollo crew members to their flights. Later, he commanded the Apollo-Soyuz Test Project flight in 1975. This is a behind-the-scenes look at America's attempt to land humans on the Moon.

Swanson, Glen E. *Before This Decade Is Out: Personal Reflections on the Apollo Program*. Gainesville: University Press of Florida, 2002. A collection of oral histories of the Saturn/Apollo Program from astronauts such as Apollo 17's Lunar Module Pilot Harrison H. "Jack" Schmitt and Charles M. Duke, Jr.

Turnill, Reginald. *The Moonlandings: An Eyewitness Account*. New York: Cambridge University Press, 2003. An eyewitness account of one of the most thrilling adventures of the twentieth century. Turnill started work in Fleet Street at the age of fifteen, and he covered the launch of Sputnik 1 and all major missions thereafter.

Ulivi, Paolo, and David M. Harland. *Lunar Exploration: Human Pioneers and Robotic Surveyors*. London: Springer-Verlag London Limited, 2004. The authors provide a well-paced, rapidly moving, balanced, even-handed account of lunar exploration as a popular history. They cover the robotic programs like Ranger and other American probes in the late 1950's. They also look at recent and future planned lunar exploration.

Wendt, Guenter, and Russell Still. *The Unbroken Chain*. Burlington, Ont.: Apogee Books, 2001. Wendt's autobiography relates the glory days of spaceflight.

*Paul P. Sipiera*

# Apollo-Soyuz Test Project

*Date:* July 15 to July 24, 1975
*Type of program:* Piloted spacecraft

*Cooperation in piloted spaceflight is an important aspect of the future of space exploration for several reasons, not the least of which is the sharing of knowledge and the cost of space exploration. The Apollo-Soyuz Test Project (ASTP) was the first such joint effort.*

## Key Figures

*Thomas P. Stafford* (b. 1930), Apollo 18 mission commander

*Vance DeVoe Brand* (b. 1931), Apollo 18 Command Module pilot

*Donald K. "Deke" Slayton* (1924-1993), Apollo 18 Docking Module pilot

*Alexei A. Leonov* (b. 1934), Soyuz 19 commander

*Valery Kubasov,* Soyuz 19 flight engineer

*Alan L. Bean* (b. 1932), Apollo 18 backup mission commander

*Ronald E. Evans* (1933-1990), Apollo 18 backup Command Module pilot

*Jack R. Lousma* (b. 1936), Apollo 18 backup Docking Module pilot

*Anatoli Filipchenko,* Soyuz 19 backup commander

*Nikolai Rukavishnikov,* Soyuz 19 backup flight engineer

*Alexei Eliseyev,* ASTP Soviet flight director

*Peter Frank,* ASTP American flight director

*Konstantin Bushuyev,* ASTP Soviet technical director

*Glynn Lunney,* ASTP American technical director

## Summary of the Mission

The opening of the Space Age brought about a perceived competition between the Soviet Union and the United States. Whether this "Space Race" was real or not, its participants looked for many firsts—the first satellite, the first piloted flight, the first two-person flight, the first human on the Moon, and the like. Early in the Space Race, however, groundwork for joint space cooperation was laid.

The first agreement for such cooperation was prepared by the National Aeronautics and Space Administration (NASA) and the Soviet Academy of Sciences in June, 1962—less than five years after the beginning of the Space Race with the Soviet launch of Sputnik 1 on October 4, 1957. Many ad-

ditional agreements were devised thereafter, including meteorological and communications links, a sharing of data on Earth's magnetic field, and joint work in space biology and medicine. Despite these efforts, work on a possible joint piloted mission, along with additional cooperative ventures, began to slow toward the end of the 1960's.

With a renewed effort that began in 1969, Soviet and American space technicians and officials began discussing flight safety and space rescue. This discussion included a look at possible problems, especially the vast differences in the two nations' rendezvous and docking techniques. This series of talks led to a second series in October, 1970, in which working groups began to examine docking procedures and equipment that would resolve the disparities. Their work included an examination of the techniques for rendezvous and the associated radio, radar, and visual procedures. They examined the difficulties of docking two spacecraft with incompatible docking systems and the design of proposed docking hardware. They also discussed other problems: differences in the environmental and communications systems of the two spacecraft and the overall costs of such a joint effort.

A proposal was completed at a Soviet-American summit meeting in May, 1972, in a five-year plan entitled *Agreement Concerning Cooperation in the Exploration and Use of Outer Space for Peaceful Purposes.* This agreement called for planning systems that would ensure compatibility between a Soviet and an American spacecraft, culminating in a joint mission with an American Apollo and a Soviet Soyuz spacecraft in 1975. Early meetings between the Soviets and the Americans revealed five major areas of difficulties, including the rendezvous and docking of the two spacecraft. If the Apollo-Soyuz Test Project (ASTP) was going to work, the problems had to be resolved.

The rendezvous problem was cleared when the American procedures were chosen. The American version had the advantage of the Apollo's larger fuel supply (a result of its lunar mission capabilities) and its type of rendezvous system (a strictly manual system versus the manual/automated one in the Soyuz). The American system required the addition of a special transponder (a passive receiver-transmitter that responds when the correct signal is received) aboard the Soyuz to make it compatible. Additionally, the use of the Apollo optical tracking system required that a portion of the Soyuz be painted white instead of its usual green. (The Apollo optical system required the white color for proper system use, but the Soviet Soyuz required the green for correct spacecraft temperature.) Additional flashing beacons, orientation lights, and docking targets were also added to the Soyuz.

The docking of the two spacecraft presented major difficulties, because the systems were not compatible. After much discussion, it was decided that a newly designed docking link, called the Docking Module, would be used. The Apollo-Saturn IB combination would carry the module into orbit. One end of the module would be compatible with the Apollo spacecraft, which would dock with it first, and the other end would mate with the Soyuz docking mechanism. During launch it would be stored below the Command and Service Module (CSM) in the adapter where the Lunar Module was kept in earlier missions. Later, the Apollo CSM/Docking Module configuration would dock with the waiting Soyuz. After a number of ground tests and verifications, the Soviets tested the Soyuz end of the Docking Module in space aboard Soyuz 16. This test confirmed that the system would work.

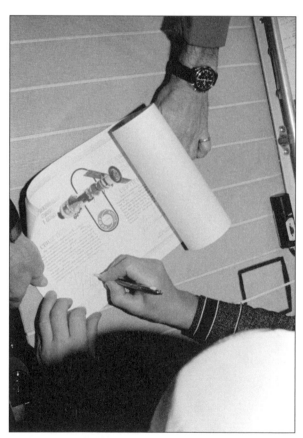

*Signing of the Russian-American certificate commemorating the docking of Apollo and Soyuz; the hands are those of Soviet cosmonaut Valery Kubasov and American astronaut Deke Slayton.* (NASA)

The third major problem was that of the atmospheres used in each of the two craft. The Apollo's atmosphere was pure oxygen at a pressure of 5 pounds per square inch. The Soyuz's atmosphere used a mixture of 19 to 32 percent oxygen and 66 to 78 percent nitrogen at a pressure of 14.7 pounds per square inch. (The accepted average pressure at sea level on Earth is also 14.7 pounds per square inch.)

For a crew transfer to occur—one of the major goals of ASTP—the Docking Module had to have an air lock and hatches. This air lock would allow astronauts to pass into the Docking Module with an atmosphere similar to that of the Apollo spacecraft, seal the Apollo Docking Module hatches, and allow the atmosphere of the module to adjust slowly to that of the Soyuz. Then the Soyuz Docking Module hatches could be opened, allowing the American astronauts to join the Soviet cosmonauts. For this operation to work in a reasonable amount of time, the Soviets agreed to reduce the pressure in the Soyuz to 10 pounds per square inch. This major change required alterations of the Soyuz and reverification of the integrity of the entire system.

Additional problems facing the ASTP dealt with communications between the two spacecraft and their respective mission controllers and the methods employed by both countries in controlling spaceflight. The communications problems were solved by the addition of radio equipment aboard both spacecraft. Operational problems were solved by various compromises and modifications.

The cosmonauts and astronauts chosen for this mission had extensive training in a variety of fields, including instruction in speaking the non-native English or Russian. The Soviets chose Alexei Leonov as the Soyuz commander. Leonov had flown on one previous flight and had performed the first extravehicular activity (EVA), or spacewalk, in 1965. The flight engineer for Soyuz was Valery Kubasov. He flew as commander aboard Soyuz 6 in 1969, a mission that had attempted a three-way space rendezvous with Soyuz 7 and 8. The American crew consisted of one astronaut who had flown previ-

*Installation of the Soyuz launch vehicle and spacecraft at Baikonur Cosmodrome in Kazakhstan in July, 1975. (NASA)*

ously and two others who had yet to experience spaceflight. Thomas P. Stafford, Apollo-Soyuz commander, had flown in both the Gemini and Apollo Programs. He had commanded Gemini IX-A and Apollo 10. Vance DeVoe Brand was the Apollo Command Module pilot. Donald K. "Deke" Slayton served as the Docking Module pilot. Slayton had been selected as one of the seven original Mercury astronauts. He had been grounded before his first scheduled flight in 1962 because of a heart murmur. After his condition disappeared, however, he was assigned to the ASTP mission.

The Soviets launched Soyuz 19 atop a Soyuz-U booster at 12:20 Coordinated Universal Time (UTC, or 8:20 A.M. eastern daylight time and 5:20 P.M. Moscow time) on July 15, 1975, from the

Baikonur Cosmodrome in Kazakhstan (also known as Tyuratam). The Soyuz-U is a standardized, modernized version of the R-7 launch vehicle (used to launch Sputnik and piloted spacecraft since Vostok 1) with higher performance first and second-stage engines. As planned, Leonov maneuvered twice to circularize the orbit of the Soyuz at a height of 225 kilometers. The Americans in Apollo 18 followed the Soviets into orbit at 19:50 UTC (3:50 P.M. eastern daylight time)—7 hours and 30 minutes after the Soyuz launch. The orbit of the Apollo was established at a perigee (the point in its orbit when an object is closest to Earth) of 149 kilometers and an apogee (the point in its orbit when an object is farthest from Earth) of 167 kilometers. Both spacecraft had the same orbital inclination (the angle of orbit relative to Earth's equator).

After launch, the Apollo crew began procedures to dock with and remove the Docking Module from its launch position. After checking the Docking Module and other Apollo spacecraft systems, the Apollo began to approach the Soyuz craft. Docking of the two spacecraft was achieved on July 17, 1975, at 16:09 UTC (12:09 P.M. eastern daylight time). This operation was conducted without problems and was in fact accomplished a few minutes earlier than planned. This part of the mission demonstrated that a compatible docking system could be used in spaceflight conditions.

The first crew transfer saw Stafford and Slayton joining Leonov and Kubasov in their Soyuz. A total of four meetings between the two crews occurred. These meetings were not only ceremonial (documents were signed, presentations were made, and messages were sent from the leaders of both countries, Leonid Ilyich Brezhnev and Gerald Ford) but also experimental. After the first undocking, the Soyuz crew directed a docking with the Apollo and the Docking Module. Many additional experiments were carried out by both crews. These experiments represented the fields of astronomy, biology, chemistry, physics, Earth science, and medicine.

The Soyuz landed about 87 kilometers northeast of Arkalyk, Kazakhstan, in the Soviet Union on July 21, 1975, after a six-day mission. The Ameri-

cans landed in the Pacific Ocean three days later, on July 24, 1975. They experienced some difficulty, however, when a malfunction led to nitrogen tetroxide gas leaking into the spacecraft, causing discomfort.

**Contributions**

The ASTP demonstrated that two nations could cooperate, plan, and implement a piloted mission. Many obstacles had to be overcome to make ASTP a success scientifically, technically, and politically. The primary goal of ASTP—to rendezvous and dock an American Apollo and a Soviet Soyuz spacecraft—was accomplished successfully. This part of the mission added knowledge to both countries' space programs. It gave both countries an opportunity to study each other's methodologies and systems. Careful planning and extensive training led to this goal's achievement. It demonstrated that space rescue by other countries was possible in that two differing systems, such as the Soyuz and the Apollo, could be made compatible for such a mission. Problems such as different atmospheres, communications, and control of the spacecraft had to be resolved for the missions to go beyond the planning stages.

The many experiments carried out by the crews of the Soyuz and Apollo extended knowledge about the universe, space, and Earth in general. A total of thirty-two scientific experiments were performed during the missions, five of which were joint experiments. The Apollo spacecraft was used as an artificial eclipsing disk, creating an artificial eclipse of the Sun for the Soyuz crew, who photographed and studied the phenomenon. This allowed collection of data on the corona (the outer atmosphere of the Sun) and other such solar features usually visible only during a total solar eclipse or with specialized equipment. An experiment to measure concentrations of oxygen and nitrogen in the upper atmosphere of Earth was done by flashing light beams from Apollo to a reflector on the Soyuz, which in turn reflected the light back to the Apollo, where it was analyzed. Another experiment studied the effects of the space environment on the

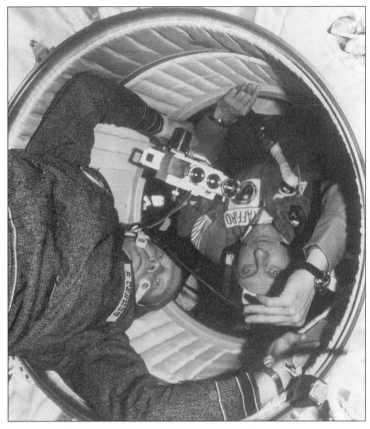

*Astronaut Thomas P. Stafford and cosmonaut Aleksei A. Leonov met in the hatchway connecting the Apollo Docking Module (DM) to the Soyuz Orbital Module (OM). Leonov was holding a camera.* (NASA)

provided an extremely stable platform above the atmosphere to conduct such a study. Astronomers were interested in using the telescope on a number of potential objects, including nearby stars, white dwarf stars (stars that have exhausted most or all of their fuel and have collapsed to a very small size), and the planet Jupiter.

### Context

The ASTP was a joint American-Soviet attempt to fly piloted missions, rendezvous and dock the two spacecraft, and have the astronauts and cosmonauts actually meet in space. This mission was highly successful because of the dedication of the two countries and the careful planning and training that went along with such a mission. Most of the test actually occurred in the original groundwork, planning, and implementation.

For Apollo-Soyuz to work, both the United States and the Soviet Union had to be ready to compromise on several points. This was done in a number of important systems, such as the communications systems, the atmospheres aboard the Soyuz spacecraft and the Docking Module, and the actual operations of the missions themselves.

ASTP demonstrated that a Docking Module could be designed and constructed to meet the requirements of both spacecraft. The Docking Module worked superbly on the mission and demonstrated that two apparently incompatible spacecraft could be made compatible for such a mission.

The experiments carried out, jointly and separately, extended the knowledge of both nations. The mission provided important information about the Sun and its outer atmosphere. Other astronomy experiments performed aboard Apollo provided new data and information about the extreme-ultraviolet spectrum. The production of crystals

cellular activity of a particular fungus selected for the experiment.

The crews also studied microbes in both spacecraft and skin swabs taken from the astronauts and cosmonauts before, during, and after the missions. Experiments conducted in orbit using an electric furnace studied crystallization in the microgravity space environment. The Soviets had performed a number of crystallization experiments in previous flights aboard their Salyut space stations. Individual projects and experiments conducted by the two crews included astrophysical, biological, biomedical, and technological research.

The Apollo spacecraft was equipped with a special telescope to measure radiation in the extreme ultraviolet portion of the spectrum. The spacecraft and the science of crystallization have much potential in the microgravity environment of near-Earth orbit. Additional information about the formation and processing of such materials was made available through this mission. The size, purity, and quality of crystals formed aboard spacecraft have led scientists to believe that crystal manufacture in space may have major applications in industry as well as science. In addition, a study of the extent of Earth's atmosphere was made possible, information on the concentration of the two major elements in Earth's atmosphere—oxygen and nitrogen—was collected, and additional information on the biology of spaceflight was collected, both independently and jointly.

The Apollo-Soyuz Test Project was successful beyond the original goals outlined by both the Soviet Union and the United States. It led to major international cooperation in space. The ASTP opened the door for future cooperation on a number of missions and projects—both piloted and robotic. It also provided an opportunity for cooperation, cooperation that could benefit not only the two nations involved but also the entire planet.

**See also:** Apollo Program; Apollo Program: Command and Service Module; Apollo Program: Lunar Module; Applications Technology Satellites; Cooperation in Space: U.S. and Russian; Launch Vehicles; Mercury Project; National Aeronautics and Space Administration.

## Further Reading

Ezell, Edward C., and Linda N. Ezell. *The Partnership: A History of the Apollo-Soyuz Test Project.* Washington, D.C.: Government Printing Office, 1978. One title in the NASA History series, this book chronicles the Apollo-Soyuz Test Project, the first joint American-Soviet crewed spaceflight. The origins of the ASTP, immediately following the successful completion of the Apollo 11 Moon-landing mission, are traced. Also discussed are the political aspects of the venture. There are dozens of black-and-white photographs of the flight. Color photographs, taken from orbit, are in the back of the book. Line drawings show the inner workings of much of the equipment related to the mission. There is an impressive 70-page appendix listing source notes and bibliographic references, and an index. Other appendices include: NASA organization charts, development of American and Soviet crewed spaceflight, a summary of the American/Soviet meetings, and descriptions of the ASTP launch vehicles.

Hall, Rex, and David J. Shayler. *Soyuz: A Universal Spacecraft.* Chichester, England: Springer-Praxis, 2003. The authors review the development and operations of the reliable Soyuz family of spacecraft, including lesser-known military and unmanned versions. While most works on Soviet/Russian space operations focus on space station activities, the story of the Soyuz spacecraft has been largely neglected.

Lee, Chester M. *Apollo Soyuz Mission Report.* San Diego, Calif.: Univelt, 1977. This is the official report of the results of the Apollo-Soyuz flight. Nowhere else will you find as in-depth a document on the technical, scientific, and human results of the U.S. portion of the mission.

Margon, Bruce, and Stuart Bowyer. "Extreme-Ultraviolet Astronomy from Apollo-Soyuz." *Sky and Telescope,* July, 1975, 4-9. A detailed article about the ultraviolet astronomy telescope aboard the Command Module of the Apollo as part of the Apollo-Soyuz Test Project. Includes illustrations and photographs of the equipment. College-level material.

National Aeronautics and Space Administration. *Apollo Mission Press Kits.* http://www-lib.ksc.nasa.gov/lib/presskits.html. Provides detailed preflight information about the Apollo-Soyuz Test Project mission. Accessed March, 2005.

Slayton, Donald K., with Michael Cassutt. *Deke! U.S. Manned Space: From Mercury to the Shuttle.* New York: Forge, 1995. This is the autobiography of the last of the Mercury astronauts to fly in space. After being grounded from flying in Project Mercury for what turned out to be a minor heart murmur, Slayton was appointed head of the Astronaut Office. During his reign he assigned all of the Gemini and Apollo crew members to their flights. Later, he commanded the Apollo-Soyuz flight in 1975. Slayton recounts his long-awaited trip into space and how it affected his view of the world.

Stafford, Thomas P., with Michael Cassutt. *We Have Capture: Tom Stafford and the Space Race.* New York: Smithsonian Institution Press, 2002. Stafford's Apollo-Soyuz team was the first group of Americans to work at the cosmonaut training center, and also the first to visit Baikonur, the top-secret Soviet launch center, in 1974.

Sullivan, Scott P. *Virtual Apollo: A Pictorial Essay of the Engineering and Construction of the Apollo Command and Service Modules.* Burlington, Ont.: Apogee Books, 2002. This book allows the public to become acquainted with the Apollo spacecraft in detail and learn the story of its design and construction. Full-color drawings in exacting detail provide inside and outside views of the Command and Service Modules complete with details of construction and fabrication.

Wendt, Guenter, and Russell Still. *The Unbroken Chain.* Burlington, Ont.: Apogee Books, 2001. Wendt is the only person who worked with every astronaut bound for space.

*Mike D. Reynolds*

# Applications Technology Satellites

*Date:* December 6, 1966, to June 30, 1979
*Type of spacecraft:* Communications satellites

*The Applications Technology Satellites (ATS's) were part of a multifaceted satellite program designed to demonstrate the promise of artificial satellites for direct broadcast communications and meteorological monitoring of Earth's surface.*

## Summary of the Satellites

Applications Technology Satellites (ATS's) were developed by the National Aeronautics and Space Administration (NASA) to demonstrate that artificial satellites could be employed in the private sector in a number of productive ways. They were also developed to test new technology, particularly onboard television equipment. These cylindrical craft, 1.5 meters in diameter, weighed between 1,430 and 1,738 kilograms. Thousands of solar cells ensured that the ATS's would enjoy long lives in space. They were launched into geostationary orbits between 1966 and 1974, the first satellites to serve a truly global population. The ATS spacecraft were fitted with communications relay equipment and color cameras for photographing Earth's surface.

ATS 1 was the first in the series. Launched on December 6, 1966, it was placed over Christmas Island in the Pacific Ocean to act as a relay point for intercontinental telephone and television transmissions in the Pacific basin region. The first artificial satellite to be equipped with television cameras, it provided the first continuous photographs of Earth's surface in that region, helping meteorologists study weather patterns on a global scale. It relayed a picture of the entire Pacific basin region once every twenty-four minutes. ATS 1 also relayed communications between aircraft and ground controllers on an occasional basis.

The next successfully launched ATS satellite was ATS 3, which was placed in orbit on November 5,

1967. ATS 2, launched April 5, 1967, and later ATS 4, launched August 10, 1968, were never operational, both failing to achieve the required orbital altitude. ATS 3 was placed near the Galápagos Islands at longitude 95 west: Its primary mission was to deliver weather imagery. Equipped with a multicolor spin-scan camera, it was later moved to longitude 47 west to provide the first color photographic reconnaissance of the Caribbean region during hurricane season. ATS 3 was instrumental in providing important information about the massive Hurricane Camille, which slammed into the southern United States in the fall of 1969. Later, the satellite was returned to its original position along the equatorial plane, where it resumed its weather-monitoring duties. ATS 3 also had served as a communications link for video and audio transmissions from the Olympic Games in Mexico City in 1968. Soon after that task was completed, it was placed at longitude 85 west, where it was employed in a study designed to measure the dimensions of Earth. The ATS 3 mission lasted for eight years. NASA controllers deactivated the satellite in January, 1976, but it remained in orbit, transmitting data to other users well beyond its designed lifetime.

ATS 5, perfectly launched on August 12, 1969, was only partially successful; during the phase between separation and parking orbit, the satellite began to wobble on its spin axis, making it impossible to perform the primary gravity-gradient experiment. Nevertheless, nine of the thirteen experiments returned useful data to Earth until 1976.

The last in this spacecraft series was ATS 6, launched aboard a Titan III-C carrier rocket on May 30, 1974. Once deployed, it resembled an umbrella, with a 9.1-meter antenna and two large solar panels located at opposite ends of a long, slender boom that was anchored to the top of the antenna. A direct-broadcast television satellite, often called the "educational satellite," ATS 6 was used to transmit a variety of educational programming to Third World countries. Among the educational projects it supported was the Satellite Instructional Television Experiment, or SITE, which provided programming for some twenty-four hundred remote

*The "Space Flower," an Applications Technology Satellite in an open position. Prior to being launched into space, the mesh and ribbed "blossom" would be wrapped around the antenna's hub, later to open in space when a signal would cut the cable holding the antenna in place.* (NASA)

villages in India. A control center was established in Ahmadabad and operated by the Indian Space Research Organization (ISRO). During the year-long experiment, from August, 1975, to the end of July, 1976, the satellite was stationed above Lake Victoria on the African continent.

ATS 6 was also used for long-distance medical diagnosis in Alaska and to relay medical and hygiene information to teachers in Appalachia as part of a health telecommunications experiment. The project, named Telemedicine, was designed to provide visual and voice contact between doctors at large medical centers and health aides in remote locations around the world.

During three months following completion of the SITE program, NASA and the Agency for International Development (AID) combined forces to provide similar programming for twenty-seven countries in Latin America, Africa, and Asia. The three-million-dollar project included the installation of uplink/downlink equipment in the capitals of each of the participating nations, and as many as five downlink receiving sites in remote areas of each of those countries. Named AIDSAT, for AID Space-Age Technology, the program called for the showing of one of three technology films in each of the host countries, followed by a two-way discussion between officials in the host country and space professionals in the United States. One film discussed communications technology for national development. Another discussed the advantages of using satellites for natural resource monitoring, and a third illustrated the value of satellites in predicting natural disasters and supporting relief efforts. For many of the participating countries, AIDSAT broadcast the first color television programs to be seen by their citizens.

In July of 1975, ATS 6 was moved to a position over Africa at 35 east longitude, where it was used to relay television and audio communications from the Apollo-Soyuz Test Project (ASTP) to the United States. The ASTP mission, which was conducted between July 15 and July 24, 1975, involved the docking in space of two spacecraft, one from the United States and the other from the Soviet

Union. The first "handshake" between American and Soviet astronauts following the maneuver occurred over the Soviet Union, requiring satellite-to-satellite relay of communications from the Soviet Union to the northern latitudes of the Western Hemisphere in order to provide real-time television coverage of the event in the United States. This was the first time a satellite had been used to provide a communications link between a crewed spacecraft and ground controllers. Further, the new position of ATS 6 effectively increased communications coverage from less than 20 percent of Earth's surface during each orbit to more than 50 percent.

During that same month, NASA publicly invited international corporate entities, educational institutions, and foreign governments to propose uses for ATS 6 following the Indian SITE program. It invited proposals that dealt with societal problems, technology, and communications.

Also in 1975, ATS 6 was used to track GEOS 3, one of the Geodetic Earth-Orbiting Satellite series, as part of a satellite-to-satellite tracking experiment designed to gather data on the physical nature of Earth and its motion. Perhaps the most active of the ATS series, ATS 6 was fully functional until mid-1979, when it began to fail and was removed from orbit.

### Contributions

The ATS satellites demonstrated the flexibility of space technology designed for scientific as well as communications applications. The first to carry color television cameras, these satellites helped scientists understand the nature of ever-changing weather patterns, and gave many developing nations their first look at color television programming. The achievements in both areas represent major events in the evolution of space technology.

One of the most significant meteorological discoveries that resulted directly from ATS photographic imagery dispelled the long-held belief that the equator acts as a barrier to weather systems, separating those in the Northern Hemisphere from those in the Southern Hemisphere. ATS im-

agery of storm systems originating in the Caribbean region of the Atlantic Ocean showed that such systems often do cross the equator and that weather systems in each hemisphere often affect those in the other. ATS also demonstrated the value of satellites in gathering data from hundreds, even thousands, of unstaffed remote weather stations in inaccessible regions of Earth and on its vast oceans. Such information is an important tool for timely forecasting of fast-changing weather patterns and the prompt dissemination of those forecasts. ATS technology was also able to refine the technique of snow mapping, which helps meteorologists predict levels of snow melt. With such data, they can anticipate spring flooding and forecast moisture levels in watershed regions below the snow-melt areas, thereby providing important information for use by farmers and municipalities that rely on the watershed. ATS data also advanced the technique of remote sensing of Earth's surface with infrared and microwave sensors able to function at night as well as during the day, thus providing continuous coverage of Earth's weather. This instrumentation can also detect agricultural crop yields, which has led to the forecasting of world food supplies using satellite data.

Yet another application of remote-sensing instrumentation aboard the ATS's is to the detection and location of forest fires in remote regions, even before smoke and flame are seen by land-based forestry personnel. ATS technology has made it possible to spot changes in ocean currents, even circulatory patterns among strata of ocean water—of great value to the maritime industry, the fishing industry, and environmental scientists, among others.

The ATS's were among the first with weather applications to be placed in high orbit. They were preceded by other weather satellites, whose lower orbits resulted in limited coverage. The ATS orbits, 35,800 kilometers above the equator, gave clear, crisp color images of 40 percent of Earth's surface, providing meteorologists with a broader understanding of how weather systems interact with one another. When these images were overlaid in chronological order, dramatic weather systems and cloud sequences could be detected, adding significantly to the ability of forecasters both to predict the weather and to explain it visually. It was meteorological imagery from the ATS series that resulted in the first use of satellite photographs in televised weather newscasts.

The ATS series also demonstrated that satellites can be moved from one location to another along the equatorial plane as required to accomplish different tasks. This mobility added significantly to the utility of artificial satellites and provided backup capabilities for an increasing number of communications satellites being deployed in the late 1960's and during the decade of the 1970's. The ATS's were the first to be employed in the tracking of land-based and airborne craft that contained small radio transmitters. This tracking technique demonstrated promise for a system of air traffic control that would provide safer, more efficient air transportation and aid in air and sea rescue operations.

The ATS's were also among the first to serve well beyond their planned life span. The last in the series, ATS 6, was retired for lack of programming support rather than technical deterioration. Its huge array of solar panels functioned well, providing enough power to keep performance levels high for virtually all the tasks, planned and unplanned, it was asked to do over a busy five-year period.

### Context

At the time the ATS satellites were placed in orbit, the international space community had already performed remarkable feats with artificial satellites. Still, the high cost of placing these satellites in orbit and operating them over long periods of time created political pressure to find ways to apply satellite technology in the marketplace. The breakneck pace of piloted and robotic space exploration during the early 1960's had resulted in a cornucopia of technological innovation, much of it with a potential for productive use that needed to be demonstrated in some tangible way. Already, the international consortium known as Intelsat had begun to build and deploy communications satellites, and large telecommunications companies in the United States had begun to deploy the same for domestic communication. The ATS series was launched as a catalyst for further development, to give users the opportunity to experiment with a variety of applications that might lead to a vibrant, productive marketplace for space technology. Such a development would serve NASA and the entire space industry, and would relieve the federal government of some of the burdensome financial responsibility for the space program.

*An Applications Technology Satellite during a test at the Space Environment Simulation Laboratory, Johnson Space Center. This satellite was launched into orbit in 1974.* (NASA)

There was also pressure to make space technology available to all nations in addition to those engaged in space exploration and exploitation. While satellite weather forecasting and oceanography served the entire population of the planet without regard for national boundaries, satellite communications were limited to those nations with well-developed telecommunications infrastructures. The ATS satellites provided the opportunity to transmit educational programming to regions where formal education was unavailable. The SITE program in India gave millions of people their first exposure to television and, more important, to critical information about health, hygiene, and other social issues. The ATS's also provided information about agriculture and animal husbandry. Later studies concluded that these programs were effective teaching tools, immensely helpful to the populations they served.

Following the success of the SITE program, NASA and AID expanded the scope of the educational satellite by creating another program to serve twenty-seven additional developing nations. Each was provided with several receiving antennae and one uplink, or transmitting, antenna, to engage in its own educational programs. The result was an expansion of the potential space technology marketplace, as well as growth in the availability of satellite transmission of educational programming around the globe.

There was also a need for communications satellites to support the growing U.S. space program. As crewed spacecraft increasingly circled the globe, and as probes were launched toward the Moon and beyond, there was a need for satellites that could be temporarily located in the proper position along the geostationary path to support individual missions. ATS could perform that function in addition to its other duties. Indeed, its versatility was its most significant asset. That versatility was demonstrated twice: once when ATS 1 was called upon to provide coverage of the 1968 Olympics in Mexico City, and again during the summer of 1975, when ATS 6 served as a communications relay for television coverage of the first joint statement from space by U.S. astronauts and Soviet cosmonauts during the Apollo-Soyuz linkup over the Soviet Union.

Perhaps equally important, the ATS's were the first to be made available for projects created and conducted outside the traditional space community. This led to a greater global awareness of satellites as appropriate technology for social and scientific research and communication.

**See also:** Air Traffic Control Satellites; Amateur Radio Satellites; Apollo-Soyuz Test Project; Intelsat Communications Satellites; Meteorological Satellites; Mobile Satellite System; SMS and GOES Meteorological Satellites; Telecommunications Satellites: Private and Commercial; Tracking and Data-Relay Communications Satellites.

### Further Reading

Bilstein, Roger E. *Orders of Magnitude: A History of the NACA and NASA, 1915-1990.* 3d ed. NASA SP-4403. Washington, D.C.: Government Printing Office, 1989. This history of the U.S. space program is written for the layperson and is well illustrated with black-and-white photographs. A small volume, it takes a broad approach with a focus on the political climate that prevailed during the time span covered. Particular emphasis is placed on the crewed programs, but the proliferation of satellites during the mid-1960's is described in the context of the exploitation of space science that proved beneficial to large segments of Earth's population. Included is a discussion of the events that led to what is described as "the new space program." Contains a bibliography and index.

Elbert, Bruce R. *Introduction to Satellite Communication.* Cambridge, Mass.: Artech House, 1999. This is a comprehensive overview of the satellite communication industry. It discusses the satellites and the ground equipment necessary to both the originating source and the end user.

Hirsch, Richard, and Joseph John Trento. *The National Aeronautics and Space Administration.* New York: Praeger, 1973. Describes the agency, its organization, its programs, and its relationship to other government agencies. Appendices include information (perhaps dated) on obtaining employment with NASA, and a particularly useful list of U.S. and Soviet crewed spaceflights from 1961 to 1971.

Ley, Willy. *Events in Space.* New York: David McKay, 1969. Contains summaries of all satellite programs, national and international, that were carried out during the 1960's. Particularly useful is a series of tables and glossaries that define space jargon and describe many of the satellite and rocket series of the decade. Also included are lists of satellite launches complete with launch dates and other information relative to the satellites and the programs with which they were and are associated.

Martin, Donald H. *Communication Satellites.* 4th ed. New York: American Institute of Aeronautics and Astronautics, 2000. This work chronicles the development of communications satellites and worldwide networks over the last four decades of the twentieth century, from Project Score to modern satellite communication systems.

National Aeronautics and Space Administration. *NASA, 1958-1983: Remembered Images.* NASA EP-200. Washington, D.C.: Government Printing Office, 1983. This oversize paperback book, heavily illustrated with color photographs, traces various programs of the space agency from its beginnings. Included is a chapter on space sciences and another on Earth orbit applications, including communications.

Paul, Günter. *The Satellite Spin-Off: The Achievements of Space Flight.* Translated by Alan Lacy and Barbara Lacy. Washington, D.C.: Robert B. Luce, 1975. A survey of the commercial, scientific, and communications applications that developed from the space research of the 1960's and early 1970's. Contains a comprehensive account of the ATS program. This book is written from the perspective of the European community, which makes it necessary reading for those desiring an international perspective on the U.S. space program. In addition to communications applications, space medicine, meteorology, cartography, agriculture, and oceanography are discussed.

Shelton, William Roy. *American Space Exploration: The First Decade.* Boston: Little, Brown, 1967. A historical account of the space program, with comprehensive listings of every American spaceflight launched between 1957 and 1967.

Zimmerman, Robert. *The Chronological Encyclopedia of Discoveries in Space.* Westport, Conn.: Oryx Press, 2000. Provides a complete chronological history of all piloted and robotic spacecraft and explains flight events and scientific results. Suitable for all levels of research.

*Michael S. Ameigh*

# Asteroid and Comet Exploration

*Date:* Beginning February 17, 1996
*Type of program:* Planetary exploration

*Near-Earth objects (NEOs) are asteroids and comets maintaining orbits that bring them close to Earth. NEOs represent unchanged remnant debris from the earliest phase of solar system formation. Near-Earth rendezvous missions are designed to gather data on the chemical constituents of this debris, giving clues to the primordial mixture from which planets form.*

**Key Figures**

*Edward Weiler,* NASA's associate administrator, Office of Space Science

*Carl Pilcher,* NASA science director for Solar System Exploration

*Thomas B. Coughlin,* project manager at the Applied Physics Laboratory (APL), Near Earth
    Asteroid Rendezvous (NEAR) mission

*Andrew Francis Cheng* (b. 1951), project scientist at APL for NEAR

*Thomas Morgan,* project scientist at NASA for NEAR, Stardust, and Deep Impact

*Mark Dahl,* Stardust program executive

*Donald E. Brownlee,* principal investigator for Stardust

*Peter Tsou,* deputy principal investigator for Stardust

*Kenneth L. Atkins,* Stardust mission manager at the Jet Propulsion Laboratory (JPL)

*Lindley Johnson,* Deep Impact program executive

*Michael F. A'Hearn,* Deep Impact mission principal investigator

*Rick Grammier,* JPL project manager for Deep Impact

*David Spencer,* JPL mission manager for Deep Impact

## Summary of the Missions

Active exploration of comets and asteroids began in the the mid-1980's. In 1985, the robotic spacecraft the International Cometary Explorer (ICE), overseen by the National Aeronautics and Space Administration (NASA), passed through the tail of Comet Giacobini-Zinner, where it collected data on the interaction between the solar wind and cometary atmospherics. In March of 1986, it made a flyby of Comet Halley. In 1986, five robotic spacecraft were launched from Earth to encounter Comet Halley: Giotto, from the European Space Agency; Vega 1 and Vega 2, from the former Soviet Union; and Sakigake and Suisei from Japan. These missions collected data and images of the comet's nucleus. Mission Giotto also went on to encounter and collect data on Comet P/Grigg-Skjellerup.

While crossing the asteroid belt on its way to enter orbit around Jupiter, the Galileo spacecraft became the very first robotic probe to encounter an asteroid at relatively close range and return images of the pockmarked surface of the body. On October 29, 1991, Galileo flew within 1,600 kilometers of Gaspra while flying by at a relative speed of 8 kilometers per hour. This asteroid was very irregular in shape (20 by 12 by 11 kilometers), heavily cratered, and covered with a thin layer of dust and rubble. Then, on August 28, 1993, Galileo encountered a larger asteroid named Ida (its longest dimension was 55 kilometers), but at a greater distance from the asteroid than the Gaspra encounter. Ida showed an older surface and a sported a 1.5-kilometer-diameter small moon which was called Dactyl. Galileo detected magnetic fields originating from both of these asteroids. A wealth of information came from these encounters, but as a fortunate byproduct of the Galileo mission rather than a primary objective.

The first of four major NASA *Discovery* missions to near-Earth objects was the Near Earth Asteroid Rendezvous (NEAR) mission, launched February 17, 1996. NEAR's goal was to help answer many fundamental questions about the nature and origin of asteroids and comets. NEAR was the first in a series of small-scale uncrewed spacecraft designed to be built and launched on a three-year schedule for a cost of less than $150 million. The NEAR spacecraft was shaped like an octagonal prism, just under 2 meters long per side, with four solar panels and a fixed X-band high-gain radio antenna. The NEAR spacecraft was built by The Johns Hopkins University's Applied Physics Laboratory and equipped with an x-ray/gamma-ray spectrometer, a near-infrared imaging spectrograph, a multispectral camera, a charge-coupled device (CCD) imaging detector, a laser altimeter, and a magnetometer.

In 2000, the NEAR project acquired a more personal surname: Shoemaker. Eugene M. Shoemaker was a pioneer in the science of astrogeology. He had trained astronauts who landed on the Moon in lunar geology and field work to maximize the scientific returns from the Apollo lunar landings. Later he and his wife, Carolyn, championed the search for near-Earth objects in order to alert the world to hazards from potential impacts by these bodies. In 1997, he died in a road accident while on a research trip in Australia. The new name for the project, NEAR Shoemaker, is in his honor.

The Stardust mission was launched February 7, 1999, to encounter the Comet Wild 2 five years later at a distance of about 640 million kilometers from Earth. Stardust is the first American space mission dedicated solely to studying a comet.

While traveling toward its cometary encounter, Stardust would make three loops around the Sun and collect interstellar dust particles. Stardust was aimed to come within 100 kilometers of the comet's nucleus and take detailed photographs of its surface. When passing through the comet's coma—the gas and dust envelope surrounding the nucleus—Stardust was designed to capture materials spewed from the Sun-activated comet. The capture mechanism for the samples is a unique substance called aerogel attached to the spacecraft's panels to soft-catch and preserve the material. Aerogel is a silica-based solid with a porous, spongelike structure in which 99 percent of the volume is empty. In February, 2000, Stardust deployed its aerogel collector and began collecting interstellar dust from a stream of particles from outside the solar system.

Startdust entered Comet Wild 2's coma on December 31, 2003. The spacecraft's instruments and systems were protected from bombardment by a forward shield. Indeed, the spacecraft encountered what a principal investigator termed swarms of particles. Instead of a uniform increase in particle impacts, Stardust hit regions of intense particle bombardment followed by a decrease in particle concentration before hitting yet another swarm of particle bombardment. Moving at 6.1 kilometers per second, images of Comet Wild 2 were snapped and particles captured in the aerogel. Stardust imaged the 5-kilometer-wide nucleus of the comet and survived the closest approach on January 2, 2004. Just minutes after that closest approach, Stardust began sending a stream of data back to Earth. For the next thirty hours unprecedented information about a comet was received by the Science Team. Six hours after the encounter concluded, the collector grid was retracted into the sample return spacecraft to keep the cometary particles pristine until opened in a sterile and inert environment in a laboratory on Earth.

Stardust was expected to deliver its sample return container to Earth after a plunge through the atmosphere over the northwestern Utah desert salt flats within the Air Force Utah Test and Training

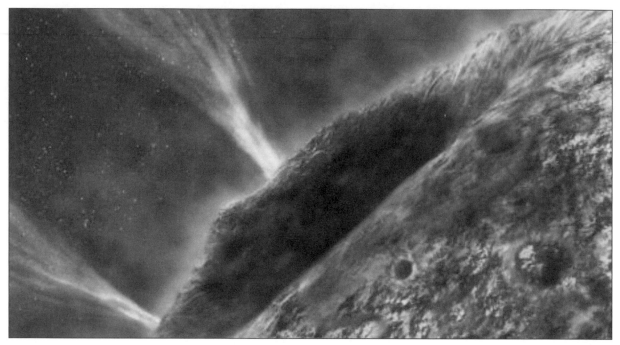

*An artist's conception of Comet Wild 2 from the vantage point of the Stardust spacecraft in January, 2004.* (NASA-JPL)

Range on January 15, 2006. The sample return container would separate from the main spacecraft, while the main portion of Stardust would continue to travel in a long-lived orbit through space. After atmospheric heating slowed the sample return container, which was equipped with a carbon-based thermal protection system, a parachute system would further diminish its speed so that a special helicopter mid-air recovery could be made before the container hit the ground. Stardust's return system and engineering software were virtually identical to those on the Genesis spacecraft, which collected solar wind particles. After atmospheric entry on September 8, 2004, the Genesis return spacecraft's parachute system did not properly slow the vehicle for the special helicopter mid-air recovery and instead Genesis slammed into the ground and cracked open. Fortunately, the samples were not completely destroyed. Given the outcome of Genesis, some concern remained that Stardust might suffer a similar fate upon return unless a software change could be made to prevent it.

The Comet Nucleus Tour (CONTOUR) was a NASA Discovery mission that launched on July 3, 2002, at 06:47:41 Coordinated Universal Time (UTC). Its primary objectives were close flybys (within approximately 100 kilometers) of the Comets Encke, Schwassmann-Wachmann-3, and d'Arrest to gather data concerning the makeup of comet nuclei. Coming that close to the nuclei of these comets, the spacecraft would encounter the period of maximum activity at each comet. CONTOUR was equipped with a high-gain radio transmitter, an imager, a dust analyzer, a neutral gas ion mass spectrometer, and a remote imager/ spectrograph. This equipment would enable CONTOUR to image data down to a resolution of 4 meters, perform spectral mapping down to resolutions of 100 meters, and obtain detailed compositional data on cometary gases and dust. If, during CONTOUR's flight, a fourth comet was discovered to be within range and suit the research needs of the mission, it would be retargeted to encounter that additional comet. Unfortunately, the CONTOUR mission was lost on August 15, 2002, when it failed to contact Earth after attempting a firing of its main rocket motor.

The NASA Deep Impact mission launched on January 12, 2005, from Cape Canaveral's Complex 17. The purpose of this mission was to impact the sunlit side cometary surface of Comet P/Tempel 1 approximately six months after launch, observe how the crater formed as a result of the impact, measure the crater's depth and diameter, measure the composition of the crater's interior and ejecta blanket, and determine changes in the comet's outgassing produced by the collision.

Just four days after launch, the Deep Impact spacecraft was pointed at the Moon in order to test its imaging systems and spectrometer. An image was taken from a distance of 1.65 million kilometers from the Moon and 1.27 million kilometers from the Earth. Then, on February 11, 2005, the flight control team successfully conducted the initial Deep Impact spacecraft trajectory-correction maneuver. As a result of this thruster firing, Deep Impact was more precisely targeted toward Comet Tempel 1.

The impact sequence was initiated on July 3 as the main Deep Impact spacecraft released its 500-kilogram copper probe/impactor toward the comet. The impactor was made of copper in order to minimize corruption of observed spectral emissions created by its high-speed striking of the comet. On the Fourth of July, the impactor hit the comet almost directly on target, moving at roughly 37,000 kilometers per hour. Prior to the flight, scientists had taken bets as to the size of the crater that would be formed. Some estimated that a hole approximately 120 meters wide and 25 meters deep might be excavated by Deep Impact's celestial fireworks.

To record the actual event, the flyby portion of the spacecraft had to execute an avoidance maneuver to get out of the Tempel 1's flight path. Then the spacecraft turned back to observe the impactor as it hit the comet and to record both the creation of the crater and the plume produced by the collision. Scientists were ecstatic that all went so well. The event and the availability of data on the Internet captured the public's attention; up to four times as many people used Web-based resources to learn about Deep Impact's collision with Comet Tempel 1 as had when the Mars Exploration Rovers landed on the Red Planet eighteen months earlier.

Initial images caught the expansion of the plume and the ejecta coming out of the new crater. The surface of the comet appeared to be covered by dust and not composed of dirty water ice. The data gathered would require months of analysis before a full picture of Comet Tempel 1's composition and structure would be available. After the encounter,

*In this artist's rendition, Deep Space 1 flies within 10 miles of asteroid 1992 KD. At the time it was the closest encounter with an asteroid ever attempted.* (NASA)

flight controllers sought and received funding to continue the flyby spacecraft by retargeting it toward another comet rendezvous.

Another spacecraft on its way toward a comet rendezvous was Rosetta, launched on March 2, 2004, on an Ariane 5 rocket from the Kourou, French Guiana, launch site used by the European Space Agency (ESA). Although Rosetta is an ESA mission primarily, American scientists were involved in the project. The Rosetta spacecraft was targeted to intercept Comet 67P/Churyumov-Gerasimenko in May, 2014.

### Contributions

NEAR Shoemaker's mission goal was to help answer many fundamental questions about the nature and origin of asteroids and comets. In March, 1996, NEAR Shoemaker took photographs of Comet Hyakutake. On June 27, 1997, NEAR Shoemaker made a 1,200-kilometer flyby of asteroid 243 Mathilde and of asteroid 433 Eros on December 23, 1998. During its flyby of 243 Mathilde, NEAR Shoemaker discovered the density of 243 Mathilde to be 1.3 grams per cubic

*Asteroid 243 Ida is about 56 kilometers long, is irregularly shaped, and has numerous craters on its surface. This image was taken during the Galileo mission to Jupiter as the spacecraft flew through the asteroid belt.* (NASA CORE/Lorain County JVS)

centimeter—if it were a bit lighter, it could float in water, which has a density of 1 gram per cubic centimeter.

NEAR Shoemaker was scheduled to return and orbit 433 Eros in January of 1999, but engine trouble resulted in the spacecraft's flying by the asteroid. After the engine trouble was diagnosed and rectified, NEAR Shoemaker was able to return to 433 Eros. The NEAR Shoemaker spacecraft's rendezvous with 433 Eros provided the first close-up look at one of thousands of near-Earth asteroids (NEAs) that orbit the Sun and inner planets of the solar system.

On February 14, 2000, NEAR Shoemaker successfully rendezvoused with 433 Eros and began a yearlong orbit to map the asteroid's surface and determine the asteroid's mass, structure, geology, chemical composition, rotational characteristics, gravity, and magnetic field. Data indicated that 433 Eros was smaller than predicted, has many craters on its surface, and has a density comparable to the average density of Earth, 2.6 grams per cubic centimeter. Images indicate that 433 Eros is an asteroid with layered geology, most likely of volcanic origin, and probably is a fragment of a larger asteroid that was broken apart by a collison. One striking feature of 433 Eros's surface is the abundance of large boulders ejected from the impact craters during their explosive formation. The large concentration of craters could indicate that 433 Eros is an older asteroid.

NEAR Shoemaker made four important firsts in space exploration: It was the first spacecraft to orbit an asteroid; it was the first to encounter a C-type asteroid (a carbonaceous asteroid rich in carbon compounds); and on March 2, 2000, NEAR Shoemaker provided the first data used to diagnose the composition of an asteroid by x-ray signature. Then, on February 12,

2001, NEAR was gently maneuvered to softly impact the surface of Eros. Images were returned up through a successful touchdown in an area bordering a saddle-shaped depression named Himeros. Consideration was given to attempting to relaunch NEAR in order to look at the impact site, but the spacecraft did not respond and remains the first asteroid lander.

Stardust will be the first space mission ever to bring back extraterrestrial material from outside the orbit of the Moon, and the first to bring back cometary material for analysis by scientists worldwide. Plans call for the spacecraft to capture up to one thousand particles bigger than 15 microns in diameter. These materials are believed to consist of ancient presolar interstellar grains and nebular condensates left over from the formation of the solar system. The analysis of this material is expected to yield important insights into the evolution of the Sun and planets and possibly the origins of life on Earth.

The primary objectives of the lost CONTOUR mission were close flybys of comets to gather data concerning the makeup of comet nuclei. CONTOUR's data would have allowed scientists to improve their knowledge of key cometary characteristics and obtain detailed analyses of comet nuclei and to assess their diversity among comets.

The purpose of the Deep Impact mission was to impact a cometary surface and generate debris ejecta for analysis. Comets are thought to have rocky nuclei dating from the early formation of the solar system. By impacting the nucleus and discharging debris, scientists hope to glean relatively unspoiled primordial material, existing since the early history of the solar system, for analysis. The impact ejecta provided data showing the difference between the comet's core and its surface.

## Context

The latest estimates indicate there are between 500 and 1,000 near-Earth objects larger than 1 kilometer in diameter and more than 100,000 objects larger than 100 meters in diameter. Near-Earth objects, both asteroids and comets, represent possible remnants of unchanged debris from the earliest phase of solar system formation. Near-Earth rendezvous missions have been designed to gather data on the chemical constituents of this debris, giving geologists and astronomers clues to the primordial mixture from which the planets may have formed.

The Near-Earth rendezvous missions have been planned as a series of small-scale uncrewed probes designed to be built and launched on a relatively brief schedule for a cost of less than $150 million, conforming to a self-imposed NASA-mandated policy of constructing and launching exploration missions "faster, cheaper, and better" than in previous decades.

These missions were designed to increase understanding of primitive bodies that have the potential to impact planetary bodies. Scientists generally agree that impacts of large bodies such as asteroids and comets have been responsible for mass extinctions on Earth. (The dinosaurs may have been forced into extinction 65 million years ago when an object struck the Yucatan Peninsula and sent a fireball and ejecta far from the impact site.) Understanding bodies such as asteroids and comets may assist in the development of means to deflect or destroy any object that threatens life on Earth.

**See also:** Cooperation in Space: U.S. and Russian; Dawn Mission; Deep Impact; Hubble Space Telescope: Science; Orbiting Astronomical Observatories; Planetary Exploration; Search for Extraterrestrial Life; Solar and Heliospheric Observatory; Stardust Project; Telescopes: Air and Space; Voyager Program.

## Further Reading

Anderson, C. M. "NEAR Shoemaker Goes to Work." *The Planetary Report* 20, no. 3 (May/June, 2000): 12-17.

Brandt, John C., and Robert D. Chapman. *Introduction to Comets.* New York: Cambridge University Press, 2004. Provides a detailed exposé about virtually every cometary phenomenon.

Crovisier, Jacques, and Thérèse Encrénaz. *Comet Science: The Study of Remnants from the Birth of the Solar System.* New York: Cambridge University Press, 2000. Provides a comprehensive overview of comets, including discoveries regarding Comets Hale-Bopp and Hyakutake. Describes the links and differences between comets, asteroids, and Kuiper Belt objects. Beautifully illustrated.

Gehrels, Tom, ed. *Hazards Due to Comets and Asteroids.* Tucson: University of Arizona Press, 1995. Contains information gathered and compiled from a number of workshops conducted on the hazards to Earth from comets and asteroids. Identifies hazardous comets and asteroids and outlines possibilities for intercepting and altering their orbits. Also considers defenses against approaching bodies and applying existing technologies as defenses against Earth impacts.

Lewis, John S. *Mining the Sky: Untold Riches from the Asteroids, Comets, and Planets.* New York: Perseus Books Group, 1997. Provides a scenario in which the resources of the solar system could be used in order to relieve stresses on Earth's diminishing resources and energy reserves.

Remo, John L., ed. *Near-Earth Objects: The United Nations Conference on Near-Earth Objects.* Annals of the New York Academy of Sciences 822. New York: New York Academy of Sciences, 1997. Includes forty-seven papers on evaluating current astronomical observations on Near-Earth objects and outlines future exploration missions to NEOs. A good starting place for anyone interested in the study of NEOs and their possible interaction with the Earth.

Russell, C. T., ed. *The Near Earth Asteroid Rendezvous Mission.* Boston, Mass.: Kluwer Academic, 1998. Describes the NEAR mission, its instruments, investigation plans, and data center, as well as the first launch of NASA's *Discovery* missions. Of interest to all potential users of the NEAR observations and to those interested in planning future low-cost missions.

Sagan, Carl. *Comet.* Rev. ed. New York: Ballantine Books, 1997. Written by the noted popularizer of science, this work describes the pristine remnants left over from the origin of the solar system called comets. Liberally illustrated with photographs, this novel humanizes the study of science and teaches a history of astronomy along the way.

Verschur, Gerrift L. *Impact: The Threat of Comets and Asteroids.* London: Oxford University Press, 1997. Describes the nature of Near Earth asteroids and comets, and the threat they pose for devastating impacts on Earth.

*Randall L. Milstein and David G. Fisher*

# Astronauts and the U.S. Astronaut Program

*Date:* Beginning April, 1959
*Type of program:* Piloted spaceflight

*NASA's astronaut program has supplied flight crew members and mission design assistance for all the United States' piloted spaceflights, from Freedom 7, piloted by Alan B. Shepard, Jr., to the Space Transportation System, popularly known as the space shuttle.*

## Key Figures

*Robert R. Gilruth* (1913-2000), first director of the Manned Spacecraft Center
*Donald K. "Deke" Slayton* (1924-1993), astronaut, first assistant director of Flight Crew Operations, Manned Spacecraft Center, and flight director astronaut
*Alan B. Shepard, Jr.* (1923-1998), astronaut and first head of the Astronaut Office, Manned Spacecraft Center

## Summary of the Program

The U.S. astronaut program has been one of the most vital and visible aspects of the American space effort since its inception in the late 1950's. The United States and the Soviet Union, the first two nations to engage in space exploration, largely defined the progress of their space programs by how successful they were in using humans in space.

Some experts believed, both in the early years of the U.S. space program and later, that space exploration could best be conducted through the use of robots and computerized satellites controlled by radio from the ground. Others believed that trained, knowledgeable astronauts were the key to gaining a thorough understanding of space. Although both opinions were reasonable, piloted space travel more readily captured the public's imagination.

After the launch of Sputnik 1 by the Soviet Union in October of 1957, there was much public and governmental interest in the United States in matching the Soviets' achievement. This interest caused President Dwight D. Eisenhower to form the National Aeronautics and Space Administra-

tion (NASA) in 1958 and, a year later, to announce the advent of Project Mercury, a program intended to put a human into space and return that astronaut safely to Earth.

It was during the early development of Project Mercury that NASA scientists and engineers first outlined the challenges and duties that the pilot of a spacecraft would face on a spaceflight. The astronaut would have to withstand the force of a launch by a rocket booster that could exert pressures equal to fifteen times the gravitational pull of Earth, and would have to adapt to, function in, and on more complicated missions, control that spacecraft, all while weightless. Astronauts would have to be trained observers who could accurately relate what they saw and experienced on the flight. In an emergency, they would need to react with split-second timing and complete confidence. Finally, astronauts would have to withstand the large gravitational forces of reentry through Earth's atmosphere and the impact of a splashdown in the ocean.

This profile of capabilities led NASA managers to narrow the field of prospective astronauts to military test pilots, aviators who are trained and experienced in flying unconventional and experimental aircraft. Test pilots fly the prototypes of new aircraft to determine their safety, performance capabilities, and most appropriate control procedures before the aircraft are mass-produced and put to general use. The test pilot's suggestions add a practical perspective to the theories upon which the original design was based, and these suggestions are incorporated into the final version of the plane. These pilots are among the most highly skilled professional fliers, and their work often calls upon the abilities defined by NASA engineers and physicians as desirable in astronauts. Many test pilots also have advanced degrees in subjects such as engineering and thermodynamics, so they are qualified to participate in mission design and control.

Consequently, NASA's first call for astronaut candidates in 1959 went out to the military services. Later astronaut candidate calls would also go out to civilian test pilots and, eventually, to scientists. The space agency stipulated that qualified test pilots have a minimum of fifteen hundred hours of specialized flying time; be less than 5 feet, 11 inches tall, so they could fit into the small Mercury Capsule; be forty (later, thirty-five) years old or younger; weigh less than 180 pounds; and possess a college degree in engineering, a biological science, or another area related to spaceflight.

Initially, 508 military test pilots met the basic requirements. After an involved screening process, this group was narrowed to 110 and later to 69, 32, and, finally, 18 candidates. These men were put through a battery of physical and psychological tests and in-depth interviews by physicians, psychiatrists, and other experts. In April of 1959, the seven astronauts who would fly the Mercury missions were announced and the NASA astronaut program was under way. Future astronauts would go through a similar winnowing process.

Navy officers Alan B. Shepard, Jr., M. Scott Carpenter, and Walter M. Schirra, Jr.; Air Force pilots Virgil I. "Gus" Grissom, L. Gordon Cooper, and Donald K. "Deke" Slayton; and Marine Corps aviator John H. Glenn, Jr. became the "Original Seven" and gained instant fame. They, like the dozens of astronauts who have followed them, brought to the program diverse experiences and talents and quickly became immersed in the design process for the spacecraft they would one day fly.

Because the American piloted space program as a whole is far too complex for any individual to master, once in the program each new astronaut is responsible for an area of specialization. Communications and navigation, ground tracking and recovery, the design of the instrument panel, the pressure suit and the life-support system, the spacecraft control systems, and rocket booster monitoring each become the responsibility of an individual

*Astronaut Guion S. Bluford working and moving about inside the shuttle during STS-8.* (NASA CORE/Lorain County JVS)

*The astronauts of the Apollo-Soyuz Test Project (from left, Tom Stafford, Deke Slayton, and Vance Bland) speak with President Ford.* (NASA CORE/Lorain County JVS)

astronaut. Each astronaut works with NASA engineers, contractors, and subcontractors on the design and development of specific pieces of equipment or systems, offering practical feedback just as test pilots do, and keeps the other astronauts informed of progress in the area.

During the Gemini, Apollo, Skylab, and space shuttle programs, groups of astronauts were responsible for flight crew participation and communication in areas such as mission planning, environmental control systems, training and simulators, in-flight experiment packages, range operations and crew safety, biomedical flight requirements, extravehicular activities (EVAs), special payloads, planning for future missions, and the training of new groups of astronauts.

Once a candidate has been accepted into the astronaut program, he or she must undergo a rigorous training schedule to become familiar with the intricacies of travel in space. In the U.S. space program's early days, after President John F. Kennedy committed the nation to an effort to land a human on the Moon by the end of the 1960's, this training program involved extensive study in subjects such

as astronomy, aerodynamics, rocket propulsion, space medicine, flight mechanics, computer technology, meteorology, physics, guidance and navigation, and communications. Apollo astronauts also underwent many hours of training in geology to prepare them for collecting and handling samples of lunar soil and rocks.

New astronauts also receive several days of survival training in desert and jungle terrains to prepare them for the unlikely event of their spacecraft's returning to Earth in a remote area. Nonpilot astronauts, such as the scientists who went on later Apollo and Skylab flights and were included as mission specialists on the space shuttles, have been required to undergo aircraft flight instruction as part of their basic astronaut training.

Once an astronaut is scheduled to go on a specific flight, he or she is given a new training schedule geared to the requirements of the mission. For pilot astronauts, this training includes many hours in spacecraft simulators practicing the maneuvers they will have to perform in space. For mission specialists on the space shuttle, the training varies depending on the nature of the mission; in some cases, mission specialists must spend hours practicing EVA techniques in specially adapted swimming pools that simulate weightlessness.

Early in the space program, astronaut flight assignments were made principally by Donald K. "Deke" Slayton, assistant director of Flight Crew Operations at the Manned Spacecraft Center (MSC, later Johnson Space Center) in Houston, and by Alan Shepard, the head of MSC's Astronaut Office. Astronauts were assigned first to support teams for specific missions and then to a mission's backup crew. The backup crew received training identical to that of the mission's primary crew and was prepared—as in the case of Gemini IX, when the first crew was killed before the mission in an airplane crash—to fly the mission in the primary

*Commander Eileen Collins consults a checklist while seated at the flight deck Commander's station in the Shuttle Columbia during STS-93.* (NASA)

crew's place. After an astronaut was assigned to a backup crew, he could expect to skip two missions and be assigned to the primary crew of the third one. This "leapfrogging" system allowed for flight crews to receive a maximum amount of training as efficiently and effectively as possible.

It has always been a major goal of the American space effort to make space travel accessible to the average person. That is why, on later space shuttle flights, many of the mission specialists were not astronauts but rather civilians who had been involved in the development of a specific satellite or experiment. These individuals received basic astronaut training and usually participated only in the segment of the mission that directly concerned them. A U.S. senator, a U.S. congressman, a Saudi Arabian prince, and scientists and engineers associated with private contractors in West Germany, Canada, France, the Netherlands, and the United States have all flown on space shuttles as "special" mission specialists, called payload specialists.

This practice, which would one day include journalists and persons from many other professions, was temporarily discontinued when S. Christa McAuliffe, an American schoolteacher, was killed, along with six other astronauts in the space shuttle *Challenger* accident in January, 1986.

### Contributions

The U.S. astronaut program has changed human beings' perceptions of themselves. It has demonstrated that Earth is only the first stepping stone in what many believe will be an interplanetary evolutionary process. Moreover, U.S. piloted spaceflights have been responsible for much of what scientists know about Earth and the solar system.

Piloted spaceflights have shown that humans can safely and reliably be placed in Earth orbit, can perform a variety of tasks while in orbit for long periods, and can survive the rigors of reentry through Earth's atmosphere. They can also survive through the 370,000 kilometers that separate Earth and its nearest neighbor, the Moon. They can operate complex and delicate equipment in order to land on the Moon and explore its surface, and they can use that same equipment to return safely to Earth.

Less directly, American piloted space travel has given rise to new medical monitoring equipment and treatment techniques that are credited with having saved countless lives. New synthetic materials and unconventional uses for known organic materials have resulted from the piloted space effort. The revolutionary computer technology of the 1980's originated at least partly in technology designed to serve the needs of NASA's piloted missions.

### Context

The development of the U.S. astronaut program took place simultaneously with two other critical processes in the American and Soviet space efforts, each of which, in turn, affected the evolution of the astronaut's role in space exploration.

The first process was the development of the technology needed to fly robotic satellites and sophisticated robotic probes into Earth orbit and deep space. While the American piloted space program grew and absorbed the lion's share of the dwindling NASA budget in the 1960's and 1970's, the proponents of robotic programs argued that robotic spacecraft could do as much as piloted missions for a tenth of the cost.

Whether or not it was true, the argument had an impact on the piloted program's long-term planning, particularly as it gave NASA's opponents a reason to cut spending for piloted flights after the first Apollo lunar landings. Some of this money was channeled to robotic satellite projects, but much of it was lost to the space program forever.

It was not, in fact, until the advent of the Space Transportation System (the space shuttle) that piloted and robotic activities reached an accommodation of sorts. When the space shuttle became NASA's primary launch system, supporters of robotic missions found that they had to rely on the piloted shuttle to carry satellites and robotic probes into space. The shuttle has had some difficulties in launching payloads, but the success some shuttle crews have had in repairing and restoring impaired satellites could be said to have more than offset such problems.

The other process that affected the U.S. astronaut program was the evolution of the program's counterpart in the Soviet Union. With the exception of major milestones such as the Apollo Moon landings, the Soviet cosmonaut program had a significant share of "firsts" in space. With respect to the quality and reliability of its spacecraft and support systems, however, the Soviet Union was very secretive, and that helped to spur American efforts to surpass Soviet space achievements.

**See also:** Apollo Program; Apollo Program: Orbital and Lunar Surface Experiments; Apollo 1; Apollo 11; Apollo 15; Apollo 17; Asteroid and Comet Exploration; Ethnic and Gender Diversity in the Space Program.

### Further Reading

Ackmann, Martha. *The Mercury 13: The Untold Story of Thirteen American Women and the Dream of Space Flight.* New York: Random House, 2003. The story of the thirteen remarkable women who underwent secret testing and personal sacrifice in the hope of becoming America's first female astronauts. Ackmann interviewed not only these women but also Charles E. "Chuck" Yeager, John H. Glenn, Jr., M. Scott Carpenter, and others.

Aldrin, Edwin E. "Buzz," Jr., and Malcolm McConnell. *Men From Earth.* New York: Bantam Books, 1991. In his book, Aldrin interweaves the story of U.S. and Soviet efforts to reach the Moon with his firsthand experience flying both the Gemini and Apollo missions during the height of the Space Race. His recounting of his two spaceflights is compelling, especially the account of the nearly aborted Apollo 11 lunar landing.

Allen, Joseph P., and Russell Martin. *Entering Space: An Astronaut's Odyssey.* Rev. ed. New York: Stewart, Tabori and Chang, 1984. Astronaut Allen's dramatic account of his flights on board the space shuttle will interest laypersons and experts alike. Allen, one of the two astronauts on the shuttle mission that repaired two disabled satellites in orbit, provides a unique view of life above Earth.

Borman, Frank, with Robert J. Serling. *Countdown: An Autobiography.* New York: William Morrow, 1988. This is an account of Borman's NASA years and with Eastern Airlines. Also memorable is the tribute to Susan Borman's poignant struggle to be "the Perfect Wife married to the Perfect Husband who was the Perfect Astronaut in a Perfect American Family raising Perfect Children."

Braun, Wernher von, and Frederick I. Ordway III. *Space Travel: A History.* Rev. ed. New York: Harper & Row, 1985. This book, written by one of the pioneers of the American space program, is an enjoyable portrait of the history of rocketry as a science and space travel as a modern technological phenomenon. In clear language, the authors bring out the sense of history in the making that helped the Space Race capture the imagination of the post-World War II generation.

Burgess, Colin, Kate Doolan, and Bert Vis. *Fallen Astronauts: Heroes Who Died Reaching for the Moon.* London: Bison Books, 2003. Near the end of the Apollo 15 mission, David R. Scott and fellow moonwalker James B. Irwin placed on the lunar soil a small tin figurine called "The Fallen Astronaut," along with a plaque bearing a list of names. This book tells the stories of those sixteen astronauts and cosmonauts who died reaching for the Moon.

Burrows, William E. *This New Ocean: The Story of the First Space Age.* New York: Random House, 1998. This is a comprehensive history of the human conquest of space, covering everything from the earliest attempts at spaceflight through the voyages near the end of the twentieth century. Burrows is an experienced journalist, who has reported for *The New York Times, The Washington Post*, and *The Wall Street Journal.* There are many photographs and an extensive source list. Interviewees in the book include Isaac Asimov, Alexei Leonov, Sally K. Ride, and James A. Van Allen.

Carpenter, M. Scott, and Kris Stoever. *For Spacious Skies: The Uncommon Journey of a Mercury Astronaut.* New York: Harcourt, 2003. Coming from a family of early Colorado pioneers, Astronaut M. Scott Carpenter grew up with a vibrant frontier tradition of exploration. He went on to become one of seven Project Mercury astronauts to take part in America's burgeoning space program in the 1960's. Here he writes of the pioneering science, training, and biomedicine of early spaceflight and tells the heart-stopping tale of his famous spaceflight aboard *Aurora 7.* Carpenter also shares a family story of tenderness and fortitude.

Cernan, Eugene A., and Don Davis. *The Last Man on the Moon: Astronaut Eugene Cernan and America's Race in Space.* New York: St. Martin's Press, 1999. The story of the last two men to walk on the Moon is told by Cernan, commander of Apollo 17, and Davis, an experienced journalist. This autobiography tells the story behind the story of Gemini and Apollo. Cernan, whose spaceflight career spanned both programs, fittingly narrates it. He was the first person to spacewalk in orbit around the Earth and the last person to leave footprints on the lunar surface.

Collins, Michael. *Carrying the Fire: An Astronaut's Journeys.* New York: Farrar, Straus and Giroux, 1974. This easy-to-understand autobiography provides a fascinating look at the world within the American space program. Collins, the Command Module pilot for Apollo 11, offers unique insights into astronauts' lives.

_____. *Liftoff: The Story of America's Adventure in Space.* New York: Grove Press, 1988. In this book, Collins begins with the origins of America's space program, Mercury, proceeds through Gemini and Apollo, and assesses post-Apollo programs through the space shuttle and plans for the Space Station.

Cooper, L. Gordon, with Bruce Henderson. *Leap of Faith: An Astronaut's Journey into the Unknown.* New York: Harper Torch, 2000. In this compelling, down-to-earth memoir, L. Gordon Cooper, Jr., recalls his adventurous life pushing the envelope in the cockpits of planes and spacecraft alike. He also looks toward the next millennium of space travel, offering deeply held views on the existence of extraterrestrial intelligence—including the distinct possibility that we have already made contact.

Glenn, John, with Nick Taylor. *John Glenn: A Memoir.* New York: Bantam Books, 1999. Glenn tells his life story in this well-written and very interesting autobiography. The book covers his career from his days as a Marine pilot, through his Mercury-Atlas 6 orbital mission, and onward to his flight aboard STS-95 in 1998. In between, he was a husband, a father, and a senator from Ohio.

Kevles, Bettyann. *Almost Heaven: The Story of Women in Space.* New York: Basic Books, 2003. The story of spacefaring women, from Valentina Tereshkova to Kalpana Chawla. Kevles illuminates what makes these women tick, including whether they were truly accepted into the Astronaut Corps or were merely tokens?

Leonov, Alexei, Boris Belitsky, and Vladimir Lebedev. *Space and Time Perception by the Cosmonaut.* Honolulu: University Press of the Pacific, 2001. This is a book about the dynamics of spaceflight and the role of the cosmonaut. The authors discuss the changes that take place in environmental conditions and, accordingly, in the psycho-physiological mechanisms of space and time perception when humans emerge into outer space. They consider the effects of weightlessness, prolonged confinement, emotional stress, and other spaceflight factors on the human perception of time.

Pogue, William R. *How Do You Go to the Bathroom in Space?* New York: Tom Doherty Associates, 1985. This and many other interesting questions about living in space are answered by former Astronaut Pogue. He was one of three crew members who spent eighty-four days in orbit aboard Skylab. Actually, there are 156 questions that the author has gleaned from thousands that he and his fellow astronauts have received in a quarter-century of piloted spaceflight. Many of the questions lead to humorous answers, but most provide a great deal of firsthand information of interest to general readers. The questions touch on each of the American piloted space programs. Several appendices provide additional information, including a summary of the physiological effects of spaceflight, a list of Earth features recognizable from space, a guide to information and resources, and a bibliography.

Santy, Patricia A. *Choosing the Right Stuff: The Psychological Selection of Astronauts and Cosmonauts.* Westport, Conn.: Praeger, 1994. The history of the psychological and psychiatric evaluation of astronaut and cosmonaut candidates is detailed. This book documents how NASA underutilized, downplayed, then ultimately ignored psychiatric and psychological characteristics in selecting astronauts until very recently.

Schefter, James L. *The Race: The Uncensored Story of How America Beat Russia to the Moon*. New York: Doubleday Books, 1999. In 1963, Schefter was assigned to cover America's race to the moon. His primary role was to excite the public about the space program. Here, he writes about everything he saw.

Schirra, Walter M., Jr., with Richard N. Billings. *Schirra's Space*. Annapolis, Md.: Naval Institute Press, 1995. Irreverent, provocative, and filled with fascinating anecdotes, this autobiography by one of America's first astronauts offers a revealing inside look at the early days of spaceflight and the men who captured the heart of the nation.

Shepard, Alan B., Jr., and Donald K. "Deke" Slayton, with Jay Barbree and Howard Benedict. *Moon Shot: The Inside Story of America's Race to the Moon*. Atlanta: Turner, 1994. This is, indeed, the inside story of the Apollo Program as told by two men who actively participated in it. Some of their tales appear here for the first time. It is almost as good as being a fly on the wall during the heyday of piloted spaceflight. The book was turned into a four-hour documentary in 1995.

Slayton, Donald K., with Michael Cassutt. *Deke! U.S. Manned Space: From Mercury to the Shuttle*. New York: Forge, 1995. This is the autobiography of the last of the Mercury astronauts to fly in space. After being grounded from flying in Project Mercury for what turned out to be a minor heart murmur, Slayton was appointed head of the Astronaut Office. During his reign he assigned all of the Apollo crew members to their flights. Later, he commanded the Apollo-Soyuz Test Project flight in 1975. This is a behind-the-scenes look at America's attempt to land humans on the Moon.

Thompson, Milton O. *At the Edge of Space: The X-15 Flight Program*. Washington, D.C.: Smithsonian Books, 2003. Thompson was one of the twelve X-15 test pilots.

Thompson, Neal. *Light This Candle: The Life and Times of Alan Shepard—America's First Spaceman*. New York: Crown Publishing Group, 2004. This book is based on Thompson's exclusive access to private papers and interviews with Shepard's family and closest friends, including John H. Glenn, Jr., Walter M. Schirra, Jr., and L. Gordon Cooper. It offers a riveting, action-packed account of Shepard's life.

Wendt, Guenter, and Russell Still. *The Unbroken Chain*. Burlington, Ont.: Apogee Books, 2001. Wendt is the only person who worked side by side with every astronaut that left the Cape bound for space.

Woodmansee, Laura S. *Women Astronauts*. Burlington, Ont.: Apogee Books, 2002. Includes interviews with many past and current women astronauts.

_____. *Women of Space: Cool Careers on the Final Frontier*. Burlington, Ont.: Apogee Books, 2003. Covers careers in space behind the scenes, including the stories of Mars Pathfinder engineer Donna Shirley, director of the Center for SETI Research Jill Tarter, astrophysicist and celestial musician Fiorella Terenzi, astronomer Sandra Faber, and space artist Lynette Cook.

*Eric Christensen*

# Atlas Launch Vehicles

*Date:* Beginning June 11, 1957
*Type of technology:* Expendable launch vehicles

*The Atlas launch vehicle was initially developed by the U.S. Air Force to launch and carry thermonuclear warheads a distance of 8,200 kilometers. Its reliability and power made it possible to boost humans into space. Thereafter, the Atlas would serve, in various configurations with other hardware, as a booster for many other space missions.*

### Key Figures

*Karel J. Bossart* (1904-1975), director of Project MX-774
*Krafft Ehricke*, rocket engineer who advanced the concept of Atlas
*Hans Friedrich*, who participated in the development of Atlas
*Dwight D. Eisenhower* (1890-1961), thirty-fourth president of the United States, 1953-1961

### Summary of the Technology

Following the conclusion of World War II, the United States was complacent with its atomic power. Experimentation with self-propelled bombs had earlier proved disappointing, and military authorities believed that the aircraft-carried weapons would be the decisive factor in a future confrontation. Fission-type weapons used in World War II were extremely heavy and would require an unfeasibly large launch vehicle to carry them a sufficient distance. The attitude of government was that balancing the budget was the top priority; plans for space exploration were left entirely to the future.

Three factors came together in 1953 that prepared the way for the development of the Atlas launch vehicle. First, the United States became aware of the Soviet Union's success in developing a powerful, long-range rocket weapon. Second, advances in nuclear engineering resulted in lightweight fusion reaction weaponry, requiring less launch power. Third, a new generation of scientists and military leaders came into authority. They were not content to rely on past weaponry success and were concerned with developing measures to protect the nation's future.

The history of Atlas actually began in October of 1945, when scientists were sorting through the components of the German V-2s shipped to the United States. A letter was sent to Convair (then Consolidated-Vultee Aircraft Corporation), requesting development of a rocket with a range of 8,200 kilometers. The Air Force agreed to the construction of ten vehicles, with three different developmental stages: The A-stage, Teetotaler, was a subsonic, self-navigational jet plane; the B-stage, Old Fashioned, was a test missile to try out the design work for the final stage; and the C-stage, Manhattan, was to be the end result, a rocket with a range of 8,200 kilometers. Under the direction of Karel J. Bossart, Project MX-774 came into being. In 1947, however, the Air Force felt the budget restrictions imposed by the administration of Harry S. Truman, and it canceled the project. Convair requested permission to continue with unexpended funds and managed to launch three MX-774's, but with less than satisfactory results. Premature burnout plagued all three launches, and the Air Force funds were depleted, leaving the Manhattan 8,200 kilometers C-stage incomplete.

At the time Convair was starting its developmental research, three of the V-2 engines were shipped to North American Aviation, and within a year its researchers were working on a liquid propellant engine for their Navajo, which was an air-breathing cruise missile. After the RocketDyne Division was created in 1949, the Atlas propulsion system was developed, based upon the Navajo design.

In September of 1949, the Soviet Union detonated its first atomic bomb. The realization that the United States no longer had a monopoly on nuclear weapons caused a resurgent interest in the MX-774 project, which was recoded MX-1593 by the Air Force. Because of its familiarity with prior developments, Convair was again awarded the contract. The same team, headed by Bossart, changed the Air Force name from "Hiroc" to "Atlas," recalling the mythological figure who bore the weight of the world on his shoulders. Convair subcontracted the engine studies to North American Aviation, in Los Angeles, and Aerojet-General, in Azusa, California.

By December of 1954, the three-engine configuration had been approved, and in 1956 the first static tests took place. There were two static test sites: Sycamore Canyon, northeast of San Diego, and the Air Force Missile Static Test Site, which overlooked the runway at Edwards Air Force Base, California. The static tests checked the propulsion system; the longest firing time was eighteen minutes. Component testing for reliability was performed at Point Loma in San Diego.

The first full Atlas test occurred on June 11, 1957, at Cape Canaveral. The Atlas-4A flight lasted only one minute before losing control and was destroyed in flight by the range safety officer. Even so, by October, a vigorous space program was in effect.

The Soviets' early lead in powerful rocketry had given them a head start in the space program. Although the American public was dismayed, the Soviets' launch of Sputnik 1 had the positive effect of spurring the expenditures and expansion of the American program. It also encouraged the American engineers, for they now knew the heavy launch vehicles could work. Nevertheless, they continued

to gear payloads to fit the already developed vehicles. Only the lunar program was considered important enough to develop its own vehicle, Saturn. Even the Mercury capsule was designed to fit the Atlas orbital capacity.

The Atlas launch vehicle was a "one-and-a-half-stage rocket." Developers did not know whether an engine could be ignited in space, so they built the booster and main stages into one. All three of the engines were fueled from the same source, but the two MA-3 booster engines burned out and dropped off after the initial thrust and acceleration, leaving the main stage, or sustainer engine, to continue. The MA-3 engines produced 67,415.73 newtons of thrust, and the sustainer engine provided 13,483.15 newtons of thrust. The Atlas height was 21.95 meters, and its diameter was 3.05 meters. Payload weight at low orbit was 1,225.05 kilograms; no escape missions were possible without upper stages.

The four basic components of the Atlas are the propulsion system, or power source; the nose cone, or package carried; the guidance system, which monitors speed and position (trajectory) of the nose cone; and the flight control system, which is responsible for corrections in flight. Because Atlas has no stabilizing fins, it is aerodynamically unstable. The flight control system must monitor the direction of the engine thrust to stabilize the flight. Two vernier engines provide the fine control required upon sustainer shutoff, to steady the nose cone. Both the guidance and the flight control systems are dependent on propulsion; no corrections are possible without motion.

The Atlas uses a liquid propellant, rocket propellant 1 (RP-1) with liquid oxygen as the oxidizer. RP-1 is a hydrocarbon resembling kerosene, which, when mixed with liquid oxygen, becomes an explosive gel. When a spark is introduced to the mixture in the combustion chamber, it provides the Atlas rocket with a total thrust of 80,898.88 newtons and can accelerate it up to a speed of 26,240 kilometers per hour.

The Atlas has been called a "paper bag with engines on it," in view of its dime-thin shell. It has the characteristics of a stainless steel balloon, except

*The Atlas-Agena, a two-stage launch vehicle used to propel unpiloted spacecraft into orbit.* (NASA CORE/Lorain County JVS)

for the exceptional strength of the shell when pressurized. Its strength and rigidity come from internal pressure when it is filled with nitrogen gas. This thin-skinned structure is also the fuel tank, containing only a bulkhead to keep the oxygen separate from the RP-1. This concept was developed to keep the vehicle's weight as low as possible. The weight has to be countered by thrust; with a low vehicle weight, less thrust is required for a launch. The Atlas tank shell is extremely lightweight: only 2 percent of the fuel weight. This remarkable tank was constructed by welding 91-centimeter-wide strips of thin stainless steel together into 3.05-meter rings, then welding the rings together until the length was achieved. The pressurizing process took weeks; as the tank was filled, tiny wrinkles and indentations would fill out, creating the sleek appearance of a launch vehicle. During the process, surgical cleanliness had to be maintained, be-

cause even a trace of contaminants could react with fuel or gases and cause corrosion, which could destroy the spacecraft. When Atlas arrived in space, the internal pressure was maintained by helium, released from bottles in the propulsion sections.

Tracking was accomplished with the Azusa system, designed by Jim Crooks in 1946. This system was capable of predicting the impact location, after flight information was fed into the ground computer. The transponder in the nose cone weighed only 17.24 kilograms and was placed in an aluminum case that measured only 30 by 25 centimeters.

The second Atlas test flight (6A) was, like 4A, unstable; still, 95 percent of the desired test data were obtained. Finally, on December 17, 1957, Atlas-12A was launched and was pronounced a complete success.

In 1958, Krafft Ehricke calculated the possibility of putting Atlas into orbit. It was decided to try putting Atlas 10B into orbit, with only the top personnel aware of the plan. Workers at the launch site became curious and concerned when the nose cone underwent unusual preparations. The flight was dedicated to Dr. Hans Friedrich, who had suffered a fatal heart attack just prior to the planned flight. The first real hint to the uninformed ground crew that orbit was being attempted came when the computer indicated that no impact was predicted. The launch took place on December 18, 1958, and Project Score placed Score, the first communications satellite, in space. President Dwight D. Eisenhower relayed his famous Christmas message to the American public via satellite relay.

Project Mercury was in full development, with many new types of tests required to ensure the safety of humans in space. On September 9, 1959,

Atlas-10D, adapted and called Big Joe, was launched to a height of 164 kilometers to test the capsule and escape systems. Unfortunately, the capsule failed to separate completely from the launch vehicle before firing thrusters. The result was an inaccurate trajectory, because the thrusters attempted to turn the entire launch vehicle and ran out of power. On the same day, Atlas-12D launched the first full-range ballistic missile from the Pacific Missile Range.

The Mercury-Atlas was 29.06 meters (m) in overall length, with 20.52 meters taken up by motors and tank. The diameter was 1.78 m. The Atlas-D was chosen for the Mercury flight because it alone had enough power to boost the capsule into orbit. The only alteration needed was an adapter to carry the capsule. Atlas-93D carried Enos the chimpanzee into his famous orbital flight, and Atlas launch vehicles were used in each succeeding crewed orbital Mercury flight. Atlas has been paired with the Agena-B to launch heavy satellites and lunar and interplanetary spacecraft. Both Ranger and Mariner were launched by the Atlas-Agena configuration. Other successful combinations were the Atlas-Able and the Atlas-Centaur. Five decades after program inception, the Atlas family of boosters continued to reliably place payloads in low-Earth orbit, geosynchronous positions, and on escape trajectories.

### Contributions

Technologically, the Atlas launch vehicle was a giant leap forward from the German V-2 rocket in every design aspect, from the propulsion unit to the guidance system. Probably the most important overall knowledge gained from the Atlas system was the fact that a guided rocket system was indeed feasible and would actually work.

Many technological advances were made by the Atlas launch vehicle program. The "metal balloon" structure was developed, as well as the bulkhead separation of fuel from oxidizer. The structural innovation was vital in lowering the weight of the vehicle to allow for larger payload launch capabilities.

Engine development, with the three-engine cluster, was an Atlas innovation. The guidance system, including ground control, which was developed for the Atlas, was vital. (There was no guidance system in the V-2.) Flight control provisions, which included the vernier engine additions, were also an important technical advance. Flight control in the V-2 was a matter of computing the direction of launch with the flight duration and hoping for the best.

Management took an innovative step, impelled by the needs of developing the intercontinental ballistic missile (ICBM) weapons system. "Concurrency" was a new term, coined to explain the simultaneous creation of the various launch facilities, research, and construction necessary for the total project. Both authority and responsibility had to reside in one organization, and technical competence was essential. Outside authority for funding and program decisions had to be accomplished with a minimum of delay. The entire new management process stressed performance, availability, and strong financial control.

The Western Development Division (WDD), later called the Air Force Ballistic Missile Division of the Air Research and Development Command, satisfied the research and development functions of the weapons system. The procurement and production end was handled by the Special Aircraft Project Office of the Air Materiel Command. There were actually four vehicles under development at the same time. The Strategic Air Command (SAC) of the Air Force worked on the ICBM, but following the ICBM came the intermediate range ballistic missile (IRBM), which was developed separately by the Air Force, Army, and Navy, resulting in the Thor, Jupiter, and Polaris vehicles. This new management attitude and process made possible the supervision of the massive space program.

### Context

Probably the most significant aspect of the Atlas story is the comparison between the original design purpose and the actual use of the launch vehicle. Atlas was meant to be a weapon delivery system;

in fact, in some research sources, the word "Atlas" refers to the first American ICBM. Its purpose was simply stated: to accelerate a nose-cone package to an exact velocity and direct it to an exact location. The weapons race following World War II and the Soviets' early long-range rocket advancements were responsible for the final development of the Atlas. Because the Soviets were successful in launching the world's first long-range, powerful rocket, they were ahead in every early space endeavor, placing the first satellite in orbit, the first man in space, and the first woman in space, and launching the first probe to the Moon. Once the need was identified by the U.S. government, the Atlas was quickly finalized for launch. The early testing and research by Convair had provided the developmental springboard.

The Atlas vehicle, after demonstrating its versatility and reliability, was used for many types of space-related launches. Of 134 launches attempted between June, 1961, and July, 1963—during the months of the Mercury flight program—28 were made by the Atlas. Moreover, the Atlas continued to prove useful while other vehicles became outmoded or overpowered. As part of the Atlas-Centaur configuration, the Atlas was flown more than five hundred times across more than four decades. Its promise for commercial launch ventures is impressive: Private industry is sure enough of the Atlas-Centaur combination to risk construction costs of between $75 million and $80 million per vehicle.

The Atlas launch vehicle was created in the era of mistrust and anxiety that accompanied the creation of the atom bomb. Conventional warfare would no longer be adequate; it became apparent to the leaders of the United States that future wars would be fought by hurling nuclear warheads from a great distance. People have so far been able to avoid a confrontation of such globally destructive force and have altered the giant rockets to accept more peaceful tasks. The Atlas launch vehicle, along with the others of its kind, has been instrumental in redirecting U.S. efforts toward an ancient dream: space travel.

In 1989, General Dynamics Commercial Launch Services introduced four versions of the Atlas-Centaur launch vehicle. The first, designated Atlas-1, is a basic satellite launcher for geostationary payloads. To increase the thrust of the main engines, the Atlas-2 and subsequent vehicles eliminated the vernier engines. The body of the Atlas-2 was lengthened 3 meters and the Centaur 1 meter to provide additional engine burn-time for both stages. The first Atlas 2-Centaur carried a Department of Defense satellite into orbit on February 10, 1992. The RL-10 engines on the Centaur stage were later upgraded to provide additional thrust and the launcher was designated the Atlas-2A, which carried a telecommunications satellite into orbit on June 9, 1992. The workhorse of the Atlas fleet is the Atlas-2AS. On December 15, 1993, it became the first of its breed to use strap-on solid rocket boosters, giving it the capability to loft a 3,800-kilogram payload into orbit.

The Atlas family of vehicles continues to provide reliable commercial launch services to the world, consisting of three basic families: the Atlas-2 (2A and 2AS), the Atlas-3 (3A and 3B) and the Atlas-5 (300, 400, 500 and Heavy Series). The Atlas-2 family is capable of lifting payloads ranging in mass from 2,812 kilograms to 3,719 kilograms to geosynchronous transfer orbit (GTO). The final Atlas-2 boosters were launched in 2004. The Atlas-3 family is capable of lifting payloads up to 4,500 kilograms to GTO. The Atlas-3 family builds upon the design of Atlas-2 with the use of a new single-stage Atlas main engine, the Russian RD-180. The Atlas-3A uses a twin RD-180 configuration and has a single-engine Centaur atop it. The changes to Centaur for Atlas-3B are a stretched tank (1.68 meters) and the addition of second engine. The first Atlas-3 launched from Cape Canaveral's Pad 36 on May 24, 2000. This design was meant only as a transitional form from the Atlas-2 version to the more powerful Atlas-5. The final Atlas-3 booster launched from Cape Canaveral on February 3, 2005. The Atlas-5 family is capable of lifting payloads up to 8,200 kilograms to GTO and over 5,940 kilograms directly to geosynchronous orbit (Atlas 5-Heavy).

The Atlas-5 family uses a single-stage Atlas main engine, the Russian RD-180, and the newly developed Common Core Booster® (CCB) with up to five strap-on solid rocket boosters. The CCB is 3.8 meters in diameter by 32.5 meters long and uses 284,453 kilograms of liquid oxygen and RP-1 propellants. The Atlas 5-Heavy configuration will use three CCB stages strapped together to provide capability necessary to lift the heaviest payloads. Both Atlas-5 400 and 500 configurations incorporate a stretched version of the Centaur upper stage (CIII), which can be configured with a single-engine or a dual engine. The first launches of the Atlas-5 400 and 500 took place in August, 2002, and July, 2003, while the first Atlas-5-Heavy flights had not yet been built. As the Bush Moon-Mars Initia-

tive develops, the Atlas-5 is under consideration as the launch vehicle for the Crew Exploration Vehicle, the first new American piloted spacecraft since the space shuttle. However, in the second half of 2005, NASA studies investigated the advantages of using shuttle-based hardware for a separate heavy-lift booster for cargo and a solid rocket booster-based launch vehicle for the Crew Exploration Vehicle. A decision was expected to lead to contracts for new boosters to support astronauts' continued exploration beyond low-Earth orbit.

**See also:** Delta Launch Vehicles; Gemini Program; Launch Vehicles; Mariner 1 and 2; Mariner 6 and 7; Mercury Project; Mercury Project Development; Saturn Launch Vehicles; Soyuz Launch Vehicle; Titan Launch Vehicles; Vanguard Program.

**Further Reading**

Chapman, John L. *Atlas: The Story of a Missile.* New York: Harper and Brothers, 1960. An excellent narrative of the history of the Atlas program, this book provides descriptive detail and brings people, places, and events to life. For general audiences.

Colucci, Frank. "Atlas for Sale." *Space* 4 (July/August, 1988): 4-6. This article discusses the reasons for the rising commercial uses of the Atlas rocket in its function as booster for the Centaur 1 and 2A. For general audiences.

Emme, Eugene M., ed. *The History of Rocket Technology: Essays on Research, Development, and Utility.* Detroit: Wayne State University Press, 1964. A technical history that also discusses the political and social background of the Atlas program. This is a compilation of essays on the development, research, and use of the various vehicles. College level.

Goodwin, Harold L. *All About Rockets and Space Flight.* New York: Random House, 1964. This is an elementary-level book, and it explains basic rocketry and spaceflight theories in easily understood terms. Good for beginning studies at any age.

Isakowitz, Steven J., Joseph P. Hopkins, Jr., and Joshua B. Hopkins. *International Reference Guide to Space Launch Systems.* 3d ed. Reston, Va.: American Institute of Aeronautics and Astronautics, 1999. This best-selling reference has been updated to include late twentieth century launch vehicles and engines. It is packed with illustrations and figures and offers a quick and easy data retrieval source for policymakers, planners, engineers, launch buyers, and students. New systems included are Angara, Beal's BA-2, Delta 3 and 4, H-2A, VLS, LeoLink, Minotaur, Soyuz 2, Strela, Proton M, Atlas-3 and -5, Dnepr, Kistler's K-1, Shtil, with details on Sea Launch using the Zenit vehicle.

"Launch Vehicles." *Aviation Week and Space Technology,* January 17, 2000, 144-145. This table details the specifications for each of the 2000 launch vehicles and spacecraft, as well as their status.

Ley, Willy. *Rockets, Missiles, and Men in Space.* Rev. ed. New York: Viking Press, 1967. This is the fourth revision of a book first written in 1944. The very earliest thoughts on space are presented and followed up to the time of revision. A good general reference; college-level reading.

Zaehringer, Alfred J., and Steve Whitfield. *Rocket Science: Rocket Science in the Second Millennium.* Burlington, Ont.: Apogee Books, 2004. Written by a soldier who fought in World War II under fire from German V-2 rockets, this book includes a history of the development of rockets as weapons and research tools, and projects where rocket technology may go in the near future.

Zimmerman, Robert. *The Chronological Encyclopedia of Discoveries in Space.* Westport, Conn.: Oryx Press, 2000. Provides a complete chronological history of all crewed and robotic spacecraft and explains flight events and scientific results. Suitable for all levels of research.

*Ellen F. Mitchum, updated by David G. Fisher*

# Atmospheric Laboratory for Applications and Science

*Date:* March 24, 1992, to November 14, 1994
*Type of program:* Scientific platforms

*The Atmospheric Laboratory for Applications and Science (ATLAS) consisted of a package of instruments to record data relating to the upper atmosphere, especially ozone depletion, and irradiance from the Sun and outer space. The three planned ATLAS missions were completed by late 1994.*

## Key Figures

*Charles F. Bolden, Jr.* (b. 1946), mission commander of STS-45, which carried ATLAS-1 into space

*Brian Duffy* (b. 1953), STS-45 pilot

*Kenneth D. Cameron* (b. 1949), mission commander of STS-56, which carried ATLAS-2 into space

*Stephen S. Oswald* (b. 1951), STS-56 pilot

*Donald R. "Don" McMonagle* (b. 1952), mission commander of STS-66, which carried ATLAS-3 into space

*Curtis L. Brown, Jr.* (b. 1956), STS-66 pilot

## Summary of the Missions

The Atmospheric Laboratory for Applications and Science (ATLAS) program identifies a series of experiments carried out by the National Aeronautics and Space Administration (NASA) in cooperation with other countries to study the Earth's atmosphere, natural and human-induced changes in the Earth's atmosphere, the effects of solar radiation upon the Earth's atmosphere, and information from distant galaxies. This far-reaching program was developed as part of the larger study, Mission to Planet Earth (MTPE), an international science program to monitor and predict changes in the Earth's environment. The ATLAS component of this study consisted of a complement of various scientific instruments to monitor conditions in the Earth's atmosphere at various altitudes during an eleven-year solar cycle. The first ATLAS package, ATLAS-1, was launched aboard the space shuttle in 1992. This

was followed by the ATLAS-2 launch in 1993 and ATLAS-3 in 1994. Each ATLAS unit carried several monitoring devices, each of which concentrated on a specific aspect of the environmental study. ATLAS-1 was the most comprehensive of the missions launched, containing twelve experimental units. Seven of these units flew again as part of the ATLAS-2 and ATLAS-3 missions. With the exception of one of the ATLAS-1 spectrometers directed toward distant galaxies, all units were designed for environmental studies. ATLAS-1 was directed toward measurements in the Southern Hemisphere. ATLAS-2 measurements were concentrated toward the Northern Hemisphere. ATLAS-3, launched in November rather than March or April, recorded observations during an alternate part of the year.

The first of the ATLAS missions (STS-45) was launched aboard the space shuttle *Atlantis* on

March 24, 1992. The mission commander was Colonel Charles F. Bolden, Jr., U.S. Marine Corps, and the spacecraft pilot was Lieutenant Colonel Brian Duffy, U.S. Air Force. The mission lasted approximately nine days, landing on April 2, 1992. *Atlantis* circled the Earth at an altitude of 340 kilometers. In addition to overseeing ATLAS package of instruments, its crew carried out other scientific measurements. The second ATLAS mission (STS-56) was launched with the space shuttle *Discovery* on April 8, 1993, and after approximately nine days in space returned to the Kennedy Space Center on April 17, 1993. *Discovery* circled the Earth at an altitude of 300 kilometers. This mission was commanded by Colonel Kenneth D. Cameron, U.S. Marine Corps. The pilot was Stephen S. Oswald. The night launch of this probe and its emphasis on the Earth's Northern Hemisphere rather than the Southern Hemispheric data gathered from ATLAS-1 complemented the information gained from ATLAS-1 rather than duplicating measurements already recorded. Other scientific projects in addition to ATLAS were also on board *Discovery*. ATLAS-3 (STS-66) was launched with the space shuttle *Atlantis* on November 3, 1994, and was commanded by Lieutenant Colonel Donald R. "Don" McMonagle, U.S. Air Force. The pilot was Lieutenant Colonel Curtis L. Brown, Jr., U.S. Air Force. *Atlantis* circled the Earth at an altitude of 300 kilometers. Carrying instrumentation similar to that aboard ATLAS-1 and ATLAS-2, this November mission was designed to record observations relating to seasonal ozone concentrations that would differ from those of the earlier flights, which took place in March and April. Other scientific studies aboard this mission included an experiment relating to the Sun, which complemented the measurements of ATLAS-1 and ATLAS-2 measurements and which was designed

by college students as part of the Colorado Space Grant College program at the University of Colorado, Boulder.

Additional ATLAS launches were planned to acquire data at various times and under various conditions of seasonal change. A total of eleven ATLAS missions were anticipated, corresponding to the eleven-year solar cycle of the Sun. The program was halted, however, after the third flight.

Of the seven ATLAS units common to all the flights, three were directed toward study of components within the Earth's atmosphere. The Atmospheric Trace Molecules Spectroscopy experiment (ATMOS) utilized infrared radiation measurements to monitor the concentrations and concentration changes of chemicals over time and altitude above the Earth's surface. Its goal was to understand and predict future changes related to ozone concentrations in the atmosphere. This was accomplished through use of an infrared spectrometer that recorded the emissions of specific substances in the troposphere and stratosphere. More than

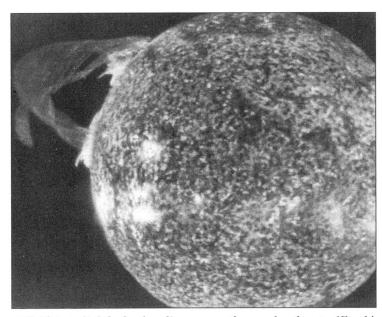

*ATLAS 1 studied the Sun's radiant energy that reaches the top of Earth's atmosphere. The total amount of energy may vary when parts of the Sun explode violently, forming solar flares like the one in this photograph.* (NASA CORE/Lorain County JVS)

thirty individual substances were monitored, including ozone, nitrous oxide, nitric acid, methane, carbon monoxide, water vapor, hydrochloric acid, hydrofluoric acid, and chlorofluorocarbons. Also measured were particulate concentrations and the effects they had on the reactions between the various chemicals present. ATLAS-1 was the second flight of the ATMOS spectrometer. It had first flown into space on the STS 51-B/Spacelab 3 mission in 1985. An ATMOS unit will be placed on the completed International Space Station (ISS), where it will record data over extended periods of time.

The Millimeter-Wave Atmospheric Sounder (MAS) is a microwave spectrometer detecting radiation in the 1- to 10-millimeter range. It was designed to monitor daily concentrations of ozone, chlorine oxide, and water vapor within the region from 20 to 100 kilometers above the surface of the Earth. Temperature and pressure measurements over this region were also recorded. The MAS unit aboard ATLAS-3 was an improved version with twice the sensitivity for monitoring chlorine oxide, thus providing more accurate data from which chlorine oxide and ozone interactions could be followed. Regrettably, it ceased to transmit data after only a few hours' operation, thus preventing the gathering of further useful information on chlorine oxide. It was the only instrument in the ATLAS-3 package designed to monitor this ozone antagonist.

The Shuttle Solar Backscatter Ultraviolet (SSBUV) spectrometers record emissions for selected ultraviolet energies from ozone, nitrous oxide, and sulfur dioxide. The data from ozone were particularly important, as measurements were made over a more restricted vertical distance, from 40 to 60 kilometers, allowing for more accurate evaluation of ozone formation and degradation in this region. One of the spectrometer measurements taken

*An onboard photograph of the ATLAS 1 module in the open cargo bay of the space shuttle* Atlantis *(STS-45).* (NASA)

was that of the Earth's surface brightness: The spectrometer recorded the 370-nanometer energy reflected from Earth's surface. In addition to measurements pertaining to Earth, the SSBUV spectrometers were also directed toward the Sun, where they accurately monitored ultraviolet rays from 160 to 405 nanometers, thus gathering data to be used as part of the eleven-year solar cycle study. The SSBUV spectrometer had flown in space as early as 1989. SSBUV-4 was aboard as part of the first ATLAS experiment. The fifth trip (SSBUV-5) was aboard ATLAS-2, and SSBUV-7 was carried on the ATLAS-3 mission. An SSBUV-8 study was airborne in January, 1996.

The remaining four ATLAS units functioning aboard all three ATLAS flights were directed toward the Sun with the purpose of understanding how solar fluctuations affected atmospheric conditions on Earth. The Active Cavity Radiometer

Irradiance Monitor (ACRIM) unit combined ultraviolet, visible, and infrared spectrometers to measure the total solar irradiance emitted by the Sun to an overall accuracy of 0.1 percent. Built-in tracking devices allowed the unit to point directly toward the Sun during measurements, adjusting for the motion of the spacecraft.

The Measurement of Solar Constant (SOLCON) spectrometer is similar in content to the ACRIM unit and was designed to also measure solar irradiance. Data from the two units (ACRIM and SOLCON) are correlated to produce a more exact description of solar irradiation than could be achieved by either unit alone. The ACRIM unit was developed at the Jet Propulsion Laboratory, while the SOLCON spectrometers are of Belgian origin.

The Solar Spectrum Measurement (SOLSPEC) instrument also consisted of three spectrometer units designed to measure solar irradiance from ultraviolet, visible, and infrared radiations, 180 to 3,200 nanometers. Data received from SOLSPEC aboard all three ATLAS flights were consistent and provided reliable measurements of emissions from the Sun that could be compared with those from other instruments. The unit was overseen by French scientists.

The Solar Ultraviolet Spectral Irradiance Monitor (SUSIM) is an ultraviolet spectrometer that, like SOLSPEC, was designed to measure solar irradiance. Here, however, only ultraviolet radiation was recorded over the energy region from 120 to 400 nanometers. The unit complemented and, to some extent, overlapped the SOLSPEC spectrometer. The SUSIM data were applied to the effects of ultraviolet radiation on ozone depletion. It also served in the study of instrument degradation from the continuous exposure to ultraviolet radiation.

ATLAS-1, in addition to the seven instruments described, consisted of other units for further exploration of our solar system and beyond. These included the Grille spectrometer, an infrared measuring unit that uses a plate (grille) that has a large area and a set of alternating reflective and transparent zones. It first flew aboard Spacelab 1 in 1983. Its purpose was to measure trace amounts of various middle atmospheric components, including carbon dioxide, ozone, and methane. Data from the Grille spectrometer complemented those obtained from the ATMOS unit. The Atmospheric Lyman-Alpha Emission (ALAE) ultraviolet spectrometer monitored the hydrogen-heavy hydrogen ratio within the lower thermosphere by observing the Lyman-alpha wavelengths of hydrogen and its heavier isotope, deuterium. The Imaging Spectrometric Observatory (ISO) experiment, which previously flew aboard Spacelab 1 in 1983, is designed to measure trace amounts of molecules observed as faint nighttime emissions above the surface of the Earth. The Atmospheric Emissions Photometric Imaging (AEPI) spectrometer is designed to measure auroras; luminous flashes of light that sometimes appear in the night sky. Both naturally occurring auroras and those artificially produced by the Space Experiments with Particle Accelerators (SEPAC) unit, also aboard the *Atlantis* flight of ATLAS-1, were studied. The Far Ultraviolet Space Telescope (FAUST) was designed to monitor ultraviolet radiation from distant galaxies rather than study the Earth's atmosphere and the effects imposed upon it by humankind and solar radiation. Pointing toward the distant heavens, it recorded ultraviolet signals originating from young stars, signals that could not be monitored directly from the Earth's surface because of the absorption of these rays in the atmosphere. The FAUST spectrometer first flew aboard Spacelab 1 in 1983.

### Contributions

Atmospheric changes occur slowly over a considerable time span, often years. The ATLAS program, designed as a series of flights during an eleven-year solar cycle, had as one of its goals to monitor both the atmospheric conditions above Earth and fluctuations in the irradiance of the Sun. As part of the global environmental study Mission to Planet Earth (MTPE), the ATLAS program was one of international cooperation employing scientific expertise from Belgium, France, Germany, Japan, the Netherlands, Switzerland, the United Kingdom, and the United States. Three of the

ATLAS instruments that flew on all three missions to date were directed toward the Earth. These units, ATMOS, MAS, and SSBUV, each monitored specific chemicals over wide regions above the Earth's surface from pole to pole and at varying altitudes. Flying at different times of the year and recording at different times during a twenty-four-hour period, they were able to gather data pertaining to seasonal ozone depletion, especially as it proceeded over Antarctica. These data supported other evidence for the buildup of human-made chemicals in the atmosphere, especially chlorofluorocarbons, and the interactions that these chemicals undergo, contributing to ozone depletion.

The solar study instruments (ACRIM, SOLCON, SOLSPEC, SUSIM) were directed toward the Sun. Solar radiation (x-ray, ultraviolet, visible, infrared) plays a significant role in atmospheric changes, changes that affect not only ozone depletion but also temperature conditions and changes in weather. It is estimated that the total irradiance from the Sun varied by about 0.1 percent over a recent eleven-year solar cycle. Whether this change is sufficient to cause any notable change in the Earth's climate is under study. To monitor irradiance changes as small as 0.1 percent over an eleven-year span requires highly accurate instrumentation. Because of continual exposure, especially to ultraviolet radiation, and through natural deterioration, spectrometers in continuous use can drift from their calibrated, accurate settings. For this reason the data from several similar instruments with similar capabilities are compared. After only a few days in space, the returned instruments are checked by scientists for accurate calibrations.

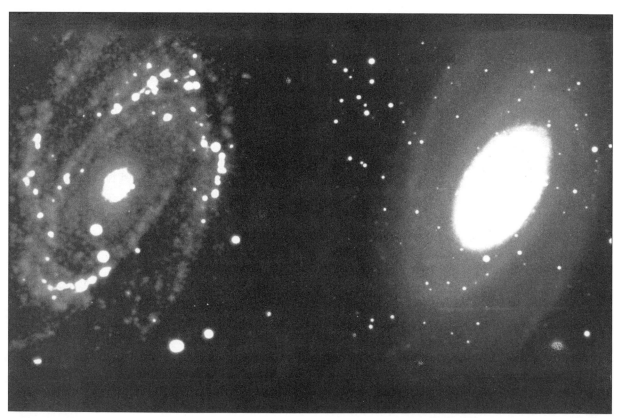

*ATLAS 1 studied the ultraviolet light in space that is usually absorbed by the Earth's ozone layer. The onboard telescope imaged galaxies, groups of galaxies, supernova remnants, and individual stars.* (NASA CORE/Lorain County JVS)

ATLAS units are sent into space again after adjustments are made. Data from the various ATLAS missions were compared with similar data from the Upper Atmosphere Research Satellite (UARS), which went into orbit in 1992 and was designed to transmit data regarding ozone depletion and related atmospheric changes for a two-year period. Comparison of data from the two programs allowed for adjustment of the UARS instruments for possible drift and provided for an alternate set of values from which conclusions regarding the composition and chemistry of the upper atmosphere could be drawn.

### Context

Ozone depletion in the upper atmosphere is a recognized fact. Various data-gathering studies support this, among them the ATLAS projects. Measurements made by the various ATLAS instruments strongly support the contention that human-made chemicals, most notably chlorofluorocarbons, released into the atmosphere, decompose, providing reactive species that attack ozone. Data from the ATMOS spectrometer aboard ATLAS-1 in 1992 measured hydrochloric acid and hydrofluoric acid concentrations, showing an increase of 37 percent and 62 percent, respectively, over previous concentration data gathered in 1985. These increases were directly attributed to the buildup of atmospheric chlorofluorocarbons. Although it might be argued that some of the chlorine increase was due to natural causes, for example, as the result of volcanic eruptions, the fluorine increase could result only from human-made sources. Concentration measurements of other species—for example, nitrogen oxides, sulfur dioxide, and chlorine oxide—further supported data gathered elsewhere and the reaction pathways proposed for the breakdown of ozone.

The presence of ultraviolet radiation also contributes to the breakdown of ozone. The various ATLAS instruments directed toward the Sun gathered data necessary to find the contribution of ultraviolet radiation in ozone depletion. Total solar irradiance data from the various solar measurements provided information helpful in understanding the extent to which the Sun influences other atmospheric conditions and chemical reactions. It also gave insight into the interpretation of data from the extended UARS experiment, whereby continuous erosion from the ultraviolet rays of the Sun was shown to cause a fourfold error in irradiance readings at one of the monitored wavelengths. ATLAS-1 had, as part of its instrumental complement, FAUST. Directed toward outer space, away from our solar system, FAUST was designed to measure ultraviolet radiations from distant galaxies and from various star clusters within our galaxy. These data were intended to assist in locating white dwarf stars throughout the Milky Way, to aid in examining how new stars are formed, and ultimately to assist in understanding details relating to the structure of the universe.

**See also:** Apollo Program: Orbital and Lunar Surface Experiments; Atlas Launch Vehicles; Mission to Planet Earth; Space Shuttle; Space Shuttle Flights, 1992; Space Shuttle Flights, 1993; Space Shuttle Flights, 1994.

### Further Reading

"ATLAS: Mission to Earth, Sun, and Stars." *Astronomy* 20 (July, 1992): 26-27. This brief description of the ATLAS-1 mission emphasizes the FAUST experimental measurements.

Cohen, M., T. P. Sasseen, and Stuart Bowyer. "The Galactic Far-Ultraviolet Sky as Seen by FAUST: Modeling and Observations." *The Astrophysical Journal* 427 (1994): 848-856. This article provides the names and numerical data obtained from FAUST measurements of the ultraviolet emissions of various stars.

"FAUST Glimpses the Far-Ultraviolet Sky." *Sky and Telescope* 84 (September, 1992): 247. The extent of the FAUST experiment program from its inception in 1970 to the flight of ATLAS-1 is briefly presented.

Gunson, M. R., M. C. Abrams, L. L. Lowes, E. Mahieu, R. Zander, C. P. Rinsland, M. K. W. Ko, N. D. Sze, and D. K. Weisenstein. "Increase in Levels of Stratospheric Chlorine and Fluorine Loading Between 1985 and 1992." *Geophysical Research Letters* 21 (1994): 2223-2226. Details of the ATMOS measurements from the ATLAS-1 flight provide support for considering chlorofluorocarbon emissions responsible for increased chlorine/fluorine concentrations in the upper atmosphere.

Schwartz, A. T., D. M. Bunce, R. G. Silberman, C. L. Stanitski, W. J. Stratton, and A. P. Zipp. *Chemistry in Context: Applying Chemistry to Society.* 2d ed. Dubuque, Iowa.: Wm. C. Brown, 1997. Chapter 2 of this chemistry textbook contains an extensive discussion of ozone and ozone depletion as a result of chemical interactions and irradiation.

Seinfeld, John H., and Spyros Pandis. *Atmospheric Chemistry and Physics: Air Pollution to Climate.* New York: John Wiley and Sons, 1997. This is an extensive reference on atmospheric chemistry, aerosols, and atmospheric models. While this book may be too complex for the average reader, it is extremely useful as a research tool on the science of atmospheric phenomena.

Tilford, S. G., G. Asrar, and P. W. Backlund. "Mission to Planet Earth." *Advances in Space Research* 14 (1994): 5-9. Details of the various proposed flights that would constitute the MTPE program are summarized, including the ATLAS flights.

Torr, M. R. "ATLAS-1 and Middle Atmosphere Global Change." *Advances in Space Research* 14 (1994): 189-199. This paper summarizes the objectives of the first ATLAS mission, especially as they relate to solar measurements and the correlation of data with those of other spaceflights.

*Gordon A. Parker*

# Attack Satellites

*Date:* Beginning September 9, 1966
*Type of spacecraft:* Military satellites

*Beginning in the 1950's, the Soviet Union and the United States experimented with satellite weapons, developing satellites capable of delivering nuclear weapons from outer space and other satellites capable of destroying foreign satellites in orbit.*

## Summary of the Satellites

In the early 1960's, the United States was numerically and technologically superior to the Soviet Union in both intercontinental ballistic missiles (ICBMs) and photo reconnaissance satellites. Having military bases around the world, the United States dispersed its missile forces, placing its short-range missiles close to the Soviet border and its long-range missiles in the United States. Thus, American missiles threatened the Soviet Union from many different directions, and the Soviets had to disperse their radar defenses to watch for possible attacks. Another disadvantage faced by the Soviets was the short range of their ICBMs. Because they had no bases near the United States, the Soviets had to place their ICBMs within their own borders; thus, Soviet missiles could reach the United States only by crossing the North Pole. As a result, the United States, threatened from only one direction, concentrated its radar defenses. Finally, an inferiority in reconnaissance satellites put the Soviets at a disadvantage because the United States had better information. To remedy the situation, the Soviets experimented with new space weapons.

Since the end of World War II, the idea of placing nuclear weapons into permanent orbit around the Earth and bringing them down on an enemy in wartime has been considered. Fearing the development of an orbital bombardment system, the General Assembly of the United Nations unanimously adopted a resolution on December 31, 1963, calling upon all nations to refrain from placing nuclear weapons into orbit. Nevertheless, it was only a resolution and not binding. In the late 1960's, the Soviet Union tested an alternative method of delivering nuclear bombs. Known as the Fractional Orbit Bombardment System (FOBS), an ICBM would place a satellite armed with a nuclear bomb into a low orbit around Earth. After making only a fraction of an orbit, retrorockets would slow the satellite and it would reenter the atmosphere over its target. By sending the FOBS over the South Pole, as Nikita Khrushchev had suggested in 1961, the Soviets could avoid American radar defenses around the North Pole. On September 17 and November 2, 1966, the Soviets launched satellites but destroyed them before they completed a full orbit. Neither launch was announced, but the United States monitored them closely, concluding that they were FOBS tests because of the flight paths.

In January, 1967, the Soviet Union, the United States, Great Britain, and fifty-seven other countries signed the Outer Space Treaty. Among other things, this treaty established general rules for the peaceful exploration of outer space and prohibited the placement of weapons of mass destruction (nuclear weapons) in space. Nevertheless, in 1967, the Soviets conducted more FOBS tests. Each test flight was announced, but the Soviets gave each satellite a Kosmos number, normally reserved for scientific satellites. A two-stage variant of the SS-9 ICBM, known as the F-1r, carried the Kosmos into space, although to put these satellites into orbit, a smaller payload had to be used. If used as an ICBM, the SS-9 could carry a 20- to 25-megaton warhead weighing about 4,500 kilograms. When used for FOBS's, however, the SS-9 could carry only a 10-megaton bomb weighing only 3,200 kilograms.

Nine unarmed Kosmos satellites, following the flight path of a FOBS, were launched from the Baikonur Cosmodrome test center in the Soviet Union: Kosmos 139 on January 25, Kosmos 160 on May 17, Kosmos 169 on July 17, Kosmos 170 on July 31, Kosmos 171 on August 8, Kosmos 178 on September 19, Kosmos 179 on September 22, Kosmos 183 on October 18, and Kosmos 187 on October 28. Each of these test flights followed a basic pattern. The second stage of the SS-9, the satellite's covering, and the unarmed Kosmos satellite went into orbit. The flight path took the Kosmos over Siberia, Japan, the Central Pacific, South America, the South Atlantic, Africa, and the Mediterranean at heights between 135 and 215 kilometers. Before a full orbit was completed, retrorockets brought the satellite back to Earth over the Soviet Union, where its reentry could be monitored. Because of the low orbit, gravity quickly pulled the second stage and the satellite's covering back into Earth's atmosphere, and they burned up. These FOBS tests were successful; only Kosmos 160 failed. After the nine launches in 1967, the Soviets made only six more test flights from 1968 to 1971. The Soviet FOBS was operational during those years, and silos were built in special facilities west of the Baikonur Cosmodrome. Of more than three hundred SS-9's built by the Soviets, only eighteen were used as FOBSs—an apparently token force.

On November 3, 1967, the U.S. Secretary of Defense Robert McNamara, announced that the Soviets were testing a FOBS. According to McNamara, however, the Soviets had not technically broken the Outer Space Treaty, because the Kosmos satellites had neither carried a nuclear weapon into space nor completed a full orbit. In 1978, the United States and the Soviet Union signed the second Strategic Arms Limitation Treaty. Among other things, this treaty banned the development, testing, and deployment of Earth-orbital nuclear weapons. It specifically banned FOBS's. Although the Senate never ratified the treaty, both nations lived up to its terms. The Soviets retired their SS-9's, and they did not test FOBS's with a newer rocket.

Fearing that the Soviets would place armed satellites into orbit, the United States began studying antisatellite (ASAT) systems in the 1950's. Although a co-orbital antisatellite (the Satellite Interceptor, or SAINT) was considered in the early 1960's, American tests have centered on the use of the direct ascent method of attacking satellites. A co-orbital A-sat is placed into the same orbit as its target and catches it from behind, but a direct-ascent A-sat, fired from a ground station or an airplane, intercepts its target just as a surface-to-air missile intercepts an airplane. The first direct ascent A-sats were not very accurate and used nuclear weapons with a wide kill range to destroy their targets. Early proposals included launching a Skybolt missile from an airborne B-52, a Minuteman missile from the ground, or a Polaris missile, but each was rejected.

Eventually, the United States tested two different programs: Program 505 and Program 437. The U.S. Army tested the first operational American A-sat, Program 505, at the White Sands Missile Range on December 16, 1959. This A-sat used a three-stage, solid-fueled Nike-Zeus antiballistic missile. In 1962, the Nike-Zeus A-sat was based at Kwajalein Island in the Pacific Ocean. The U.S. Air Force conducted Program 437 from Johnson Island in the Pacific, using a liquid-fueled Thor intermediate range missile to carry its A-sat into space. Four tests of Program 437 took place between February and May, 1964. On June 10, 1964, the Johnson Island facilities became operational.

The superior altitude of the Thor missile made Program 505 obsolete, and it was canceled after May, 1966. Program 437 was also soon canceled, however, because money needed for it was spent on the Vietnam War. Delayed deliveries and test failures added to the problems, and a hurricane destroyed the facilities on Johnson Island in 1972. Furthermore, the small number of Thor missiles could not have matched the growing number of Soviet satellites. On April 1, 1975, the program was terminated.

Still interested in an aircraft-launched A-sat, the Air Force studied a proposal for an inexpensive A-sat. Named Project SPIKE, it called for an F-106 fighter to launch a missile at low-orbiting satellites. The idea was dropped, but in the 1980's, an updated version reappeared. Called Project 1005, the new program used an F-15 fighter to launch a two-stage, solid-fueled Miniature Homing Vehicle (MHV) from 15,000 meters. The MHV was successfully tested on September 13, 1985, off the coast of California. Using an internal guidance system to track its target, the MHV can distinguish its target from the stars and other nearby objects. Unlike the previous American A-sats, the MHV does not destroy its target with an explosion. Instead, it collides with the target at 4,181 meters (13,716 feet) per second. A great advantage of the MHV is its freedom from extensive ground facilities. The only necessity is a place for an F-15 to take off, which includes aircraft carriers.

American photo reconnaissance satellites particularly worried the Soviets. Before 1963, the Soviet Union publicly denounced the American use of photo reconnaissance satellites. The Soviets tried to have these satellites banned by the United Nations, but these demands were dropped after the Soviets put their own reconnaissance satellites into space. Nevertheless, the possibility that an enemy might put orbital weapons into space (the United States chose not to build a FOBS for a number of reasons, but the Soviets were worried about a Chinese FOBS) and the superiority of American reconnaissance satellites prompted the Soviets to test a system for destroying them in orbit.

The Soviets tested A-sat systems twice, preferring a co-orbital method of attack. In the first test program, from 1968 to 1971, they would put a target satellite into space. Then, they would launch a second satellite with a variant of the SS-9 (the F-1m) into the same orbit to intercept the target satellite. This interceptor satellite used radar to track down its target after a few orbits. When it came within

1 kilometer of the target, a nonnuclear bomb on the interceptor satellite would explode, making a "hot-metal kill." That is, the exploding A-sat created a cloud of metal fragments that destroyed the target satellite. Given Kosmos numbers, the first Soviet A-sats attacked target satellites at various altitudes to simulate different types of American reconnaissance satellites.

The second Soviet A-sat test program employed a new method to attack a satellite. On February 12, 1976, Kosmos 803, a target satellite, was placed into orbit. Two months later, on April 13, an F-1m rocket carried Kosmos 814, the A-sat, into space. By following a lower orbit, Kosmos 814 quickly caught up to the target satellite. Then, an onboard rocket propelled Kosmos 814 up to the target's orbit for a fast flyby. The interception, from launch to flyby, took only forty-two minutes. Afterward, Kosmos 814 fell back to Earth and burned up on reentry. The interception method used by Kosmos 814 is called "pop up" because of the way it moved up from its original orbit to attack the target in a higher orbit. By using the pop-up method, the Soviets no longer had to announce their intention by chasing a satellite for several orbits. They could destroy an American satellite in less than one orbit and while it was out of range of American tracking stations. The Soviets tested variations of A-sat intercept orbits through the end of 1977.

American and Soviet negotiators tried to ban A-sats in the late 1970's, but they failed. In the 1980's, the Soviets resumed A-sat testing. On April 18, 1980, Kosmos 1174 attacked a target satellite, but it missed by at least 8 kilometers. Kosmos 1174, however, demonstrated an ability to intercept several satellites by changing its orbit two more times. After its third interception of the target satellite, Kosmos 1174 blew itself up on April 20. In January, 1981, Kosmos 1243 passed within lethal range of a target satellite after making two and one-half orbits and then burned up in the atmosphere. Six weeks later, Kosmos 1258 successfully repeated the accomplishments of Kosmos 1243, again coming close enough to the target satellite to have destroyed it.

### Contributions

After intensive testing in 1966 and 1967, the Soviet Union deployed an operational weapons system that overcame the limitations of its ICBMs, by using an ICBM to place a nuclear weapon into a low orbit. The flight path of the Kosmos satellites used to test the FOBS proved that the Soviet Union had the capability to strike a target anywhere on Earth's surface from bases inside the Soviet Union. Furthermore, the Soviets demonstrated their ability to attack the United States from any direction. Nevertheless, the Soviet FOBS's did not require any advances in technology, other than the SS-9 missile, being merely an application of reentry techniques already practiced by the Soviet-crewed space program.

The American A-sat programs changed from complex operations requiring large, land-based facilities to a more mobile system. Using sophisticated internal tracking systems, American A-sats are accurate enough to eliminate the need for nuclear weapons. The American A-sat program relies upon advancing technology to be effective.

The Soviet Union also had the ability to destroy an enemy's satellites. Their first A-sat test program proved that an interceptor satellite could be placed into the same orbit as another satellite. Equipped with radar, the interceptor could make minor course corrections, track its target, and get close enough to make a hot-metal kill. Soviet A-sats showed technological progress, as was demonstrated in the second testing program. By using better homing devices, thrusters, and interception techniques, they cut down the time needed to intercept a satellite in orbit. With Kosmos 1174, Soviet A-sats progressed technologically to the point that one ASAT could attack a number of different satellites.

### Context

On October 4, 1957, the Soviet Union placed the first human-made satellite, Sputnik 1, into orbit around Earth. Being able to place satellites into Earth orbit meant that the Soviet Union could place weapons there also. With its FOBS and A-sat satellites, the Soviet Union lived up to its potential.

The same technological advances that have made possible the peaceful exploration of space have aided its military use. The technology needed to place a crewed spacecraft into orbit and bring it safely back to Earth at a precise location is used to place a nuclear weapon into orbit and bring it disastrously back to Earth. The skills and equipment needed to perform the intricate docking procedures used to put men on an orbiting space station are the same ones needed to intercept a satellite.

It is now known that the Soviet FOBS and A-sat programs resulted from a technological inferiority to the United States. With bases close to the Soviet Union and a superiority in land- and submarine-based ICBMs—which are more accurate than FOBS's—the United States had no reason to develop an orbital bombing system. Aware of the disparity, Soviet premier Nikita Khrushchev tried to close the gap by placing missiles in Cuba; after the Cuban Missile Crisis in 1962, however, he was forced to remove them. Thus, the FOBS's offered a logical way to get around the American ICBMs and early-warning radar. Once the Soviet Union built more and better land- and submarine-based ICBMs of its own, the less accurate FOBS's were no longer needed, as indicated by the Soviets' willingness to do away with them under the Strategic Arms Limitation Talks. As a result, the movement to ban orbital nuclear weapons, which had started with the United Nations resolution in 1963, found success in the late 1970's.

Both the United States and the Soviet Union experimented with A-sats. American superiority in photo reconnaissance satellites and the threat of Chinese FOBS's stimulated the Soviet A-sat program, and the Soviet FOBS's forced the United States to experiment as well. The elimination of Soviet FOBS's removed part of the reason for the U.S. A-sat program, but the growing number of Soviet reconnaissance satellites kept alive the United States' interest in A-sats. Similarly, the increasing reliance of the United States on satellites for reconnaissance and communications kept the Soviets interested in improving its A-sats.

**See also:** Ames Research Center; Atmospheric Laboratory for Applications and Science; Early-Warning Satellites; Electronic Intelligence Satellites; Nuclear Detection Satellites; Ocean Surveillance Satellites; Spy Satellites; Strategic Defense Initiative; United States Space Command.

### Further Reading

Bulkeley, Rip, and Graham Spinardi. *Space Weapons: Deterrence or Delusion?* Totowa, N.J.: Barnes and Noble, 1986. Provides a basic discussion of the various types of space weapons and their roles in international relations. Includes illustrations. Suitable for general audiences.

Burrows, William E. *Deep Black: Space Espionage and National Security.* New York: Random House, 1986. Primarily about reconnaissance satellites, this study contains information about antisatellite systems and other space weapons. Illustrated.

Butrica, Andrew J. *Beyond the Ionosphere: Fifty Years of Satellite Communications.* NASA SP-4217. Washington, D.C.: Government Printing Office, 1997. Part of the NASA History series, this book looks into the realm of satellite communications. It also delves into the technology that enabled the growth of satellite communications. The book includes many tables, charts, photographs, and illustrations, a detailed bibliography, and reference notes.

Cox, Christopher. *The Cox Report: U.S. National Security and Military/Commercial Concerns with the People's Republic of China.* Washington, D.C.: Regnery Publishing, 1999. Investigates U.S.-Chinese security interaction and reports that China successfully engaged in harmful espionage and obtained sensitive military technology from the United States. Some of the technology obtained includes information on American reconnaissance and attack satellites.

Manno, Jack. *Arming the Heavens: The Hidden Military Agenda for Space, 1945-1995.* New York: Dodd, Mead, 1984. Describes the political as well as technological aspects of the militarization of space.

Meyer, Stephen M. "Space and Soviet Military Planning." In *National Interests and the Military Use of Space*, edited by William J. Durch. Cambridge, Mass.: Ballinger, 1984. Part of a collection considering the political reasons for the militarization of space, this article is a discussion of Soviet plans to use its space weapons. Illustrated with graphs and charts. College-level material.

Peebles, Curtis, with Kenneth Gatland. *Battle for Space.* New York: Beaufort Books, 1983. A discussion of the origins, developments, and future of space weapons. Includes chronological breakdowns of treaties and satellite launches. Illustrated with drawings, charts, and color and black-and-white photographs.

Sheldon, Charles S. *Review of the Soviet Space Program, with Comparative United States Data.* New York: McGraw-Hill, 1968. Designed to be a ten-year review of the Soviet space program with comparisons to the space program of the United States, this work provides a contemporary view of the Soviet space program. Suitable for college-level readers.

Shelton, William Roy. *Soviet Space Exploration: The First Decade.* New York: Washington Square Press, 1968. An older, general survey of the Soviet space program. It contains a small amount of information on military applications.

Stockholm International Peace Research Institute. *Outer Space: Battlefield of the Future?* New York: Crane, Russak, 1978. Provides a factual description of Soviet and American space weapons. It emphasizes the efforts to control space weapons through international treaties. One chapter is devoted to Fractional Orbit Bombardment Systems and antisatellite systems. Includes tables listing Soviet launches by Kosmos number. Suitable for all audiences.

Yenne, Bill. *Secret Weapons of the Cold War: From the H-Bomb to SDI.* New York: Berkley Books, 2005. A contemporary examination of Cold War superweapons and their influence on American-Soviet geopolitics.

Zaehringer, Alfred J., and Steve Whitfield. *Rocket Science: Rocket Science in the Second Millennium.* Burlington, Ont.: Apogee Books, 2004. Written by a soldier who fought in World War II under fire from German V-2 rockets, this book includes a history of the development of rockets as weapons and research tools, and projects where rocket technology may go in the near future.

*Jeffrey S. Underwood*

# Biosatellites

*Date:* December 14, 1966, to November 15, 1970
*Type of program:* Scientific platforms

*The U.S. biosatellite program began with a series of three spacecraft that investigated the effects of spaceflight on basic life processes. A later program, the Orbiting Frog Otolith (OFO), investigated the effects of weightlessness on vestibular, or balance-sensing, organs.*

### Key Figures

*Thomas P. Dallow,* program manager
*Joseph F. Saunders,* program scientist
*Charles A. Wilson,* project manager
*Bonne C. Cook,* Spacecraft Systems manager
*H. M. Wittner,* program manager from General Electric Company

### Summary of the Satellites

The United States' biosatellite program began in 1962, when the Space Sciences Board of the National Academy of Sciences recommended that the National Aeronautics and Space Administration (NASA) investigate the effects of the space environment on the organization and function of living organisms. In June, 1962, the director of NASA's bioscience programs asked the agency's Project Designation Committee to recommend a name for the type of satellite such a project would require. It was decided that the word "biosatellite" would be used, as a contraction of "biological satellite." In December, 1962, NASA Administrator James E. Webb formally accepted the biosatellite program and selected the Ames Research Center in California to manage the program.

Spacecraft feasibility studies began in March, 1963, and mission planners established three different mission profiles. The first was a three-day mission to study the effects of the microgravity environment (so-called weightlessness) and microgravity combined with gamma radiation on plants, insects, and simple animals. The second was a thirty-day flight to study the effects of extended weightlessness on the cardiovascular and central nervous systems of small primates. The third proposed mission was a twenty-one-day flight to investigate the effects of spaceflight on biological rhythms and the effects of prolonged weightlessness on cellular processes in plants and small animals.

The General Electric Company received the contract to build six biosatellites, a prime vehicle and a backup for each mission type, in July, 1963. Designers at General Electric based the biosatellite on their already proven Discoverer series of recoverable satellites. The biosatellites comprised a 102-centimeter-diameter reentry body attached to a 145-centimeter-diameter cylindrical adapter section. The reentry body consisted of an exterior forebody, a recovery spacecraft (which contained the payload), an aft thermal cover, and a thrust cone (which contained the 4,450-newton-thrust retrorocket). An ablative heatshield made of phenolic nylon with a fiberglass liner, to protect the payload from the searing heat of reentry, covered the forebody. The spacecraft life-support system provided an oxygen-nitrogen atmosphere at sea-level composition and pressure. Except for the oxygen and nitrogen tanks, which were in the adapter section, the entire life-support subsystem was in the reentry spacecraft. The biosatellite also had a thermal control system that maintained spacecraft temperatures from 18.3° to 23.9° Celsius. Electrical power for the three-day missions came from silver-zinc batteries. For longer missions, the biosatellites carried a fuel cell, which combined oxygen and hydrogen to generate electricity.

To prevent accelerations caused by tumbling or other unwanted spacecraft motions, six nitrogen-gas thrusters coupled with gyroscopes maintained an in-flight rotation rate of less than one revolution every twenty minutes. At the end of the mission, two infrared Earth sensors and a magnetometer provided data for spacecraft alignment for reentry.

The biosatellite carried two parachutes. The first was a small drogue parachute used to stabilize the spacecraft; it was deployed once the satellite had descended to 24,400 meters. At 3,048 meters, the main parachute was deployed. After the main parachute was deployed, a U.S. Air Force cargo aircraft retrieved the spacecraft in mid-air by "air-snatch" recovery. For this technique, the aircraft trailed a pair of poles with a cable between them. As the spacecraft descended beneath its parachute, the pilot snared the parachute with the cable. Once the parachute was snared, the crew hauled the poles, cable, and spacecraft into the airplane. The Air Force developed this technique in the late 1950's to recover the Discoverer reconnaissance satellites. If the air-snatch retrieval were to fail, the biosatellite spacecraft's radio beacon and dye marker allowed for a recovery from the ocean.

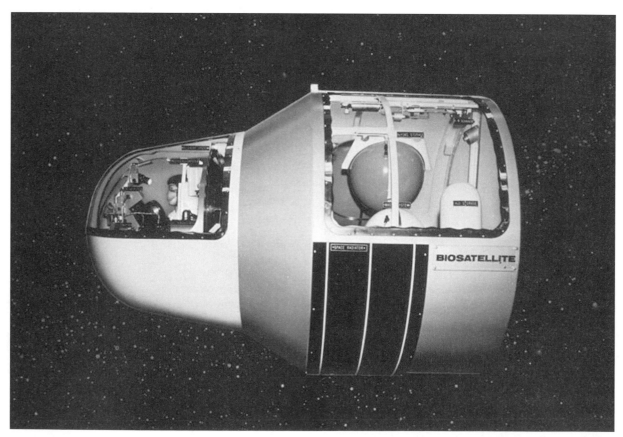

*A model of a biosatellite with a monkey in the front and the life-support package at the rear (right).* (NASA-ARC)

Speedy recovery was essential, because project scientists needed the spacecraft for a postflight examination within six hours of reentry. The first two spacecraft were 206 centimeters long and weighed about 430 kilograms. Biosatellite 3, which was used for a thirty-day primate mission, weighed 680 kilograms.

During 1962 and 1963, NASA solicited proposed experiments for the three missions. Out of 185 proposals, they selected twenty. Of these, fifteen eventually flew. The experiments examined the effects of spaceflight on life processes from the cellular level up to that of a primate. They included studies of the orientation of wheat seedling roots and shoots, development of frog eggs, and the central nervous and cardiovascular systems of primates.

Thrust-Augmented Delta (TAD) launch vehicles powered the first two biosatellites into orbit. The TAD stood 28 meters tall and had a liftoff thrust of 1.46 million newtons. The TAD second stage produced a thrust of 34,700 newtons. Biosatellite 3, the heaviest satellite of the series, used a 32.3-meter Long Tank Thrust-Augmented Delta newtons two-stage booster. Even though both stages of this launch vehicle had the same thrust as the TAD, Biosatellite 3 had a greater payload lifting capacity because it carried more propellants. NASA's Goddard Space Flight Center in Greenbelt, Maryland, managed the Delta launch vehicle program.

Biosatellite 1 lifted off from Cape Kennedy, Florida, on December 14, 1966. It carried thirteen experiments to study the effects of weightlessness and weightlessness with gamma radiation on living organisms. The TAD performed flawlessly and placed the spacecraft in a nearly circular orbit with a perigee (orbital low point) of 307 kilometers and an apogee (orbital high point) of 317 kilometers. Biosatellite 1 carried seven radiation experiments in the forward section of the spacecraft. Specimens included bacteria, bread mold, flour beetles, plants, wasp larvae, and fruit flies. The spacecraft's aft section contained control versions of these experiments, which were shielded from the radiation, along with six other experiments. The latter group included pepper plants, wheat seedlings, frog eggs, and amoebas. Throughout the mission, which lasted three days, all spacecraft systems performed well, and during the forty-seventh orbit, the reentry spacecraft separated from the adapter section according to schedule. Flight controllers waited for the retrorocket to fire and slow the craft for its fiery plunge through the atmosphere. Nothing happened, and Biosatellite 1 remained in orbit. As a result, no useful scientific data were obtained from the first biosatellite. Biosatellite 1's spacecraft finally reentered on February 14, 1967, after its orbit decayed. In an operation code-named Lost Ball, Air Force aircraft carried out a search for Biosatellite 1. One of the search aircraft detected radio signals over western Australia, and an intensive search was made of that area. Yet efforts to locate the spacecraft were unsuccessful.

The second biosatellite flight began on September 7, 1967. A TAD launch vehicle placed it in a 302-by-327-kilometer orbit inclined 33.5° to the equator. It carried virtually the same set of experiments as Biosatellite 1. At first, the mission went well. Then, after nearly two days in orbit, the spacecraft began experiencing communications problems. Because a tropical storm was moving into the recovery area, mission managers terminated the flight 45 hours after launch, shortening the mission by 17 orbits. An Air Force pilot successfully retrieved Biosatellite 2 using the air-snatch technique, and the recovery spacecraft arrived at Hickam Air Force Base in Honolulu, Hawaii, in good condition. Shortening the mission reduced the scientific yield to only 71 percent of what was expected, but the mission was still called a success.

The third biosatellite flight was to last thirty days and include a primate. The passenger in the recovery spacecraft was a male pigtail monkey named Bonny, 6.4 kilograms in weight and 79 centimeters in height. Researchers selected the pigtail monkey, which is indigenous to Malaysia and Thailand, because it is similar to people in major organ functions, brain organization, and cardiovascular functions. Ten electrodes implanted in Bonny's brain showed the effects of weightlessness on his memory, learning ability, alertness, emotional state, and sleep patterns. These were implanted six months prior to flight so that Bonny's behavior could be correlated to instrument data. An eleventh sensor measured brain temperature. Other instrumentation included sensors to measure muscle tension and activity, cardiovascular function, and blood pressure. A catheter in the monkey's bladder permitted three types of urine analyses to be conducted four times each day.

Bonny wore a nylon-mesh vest that was zipped and laced to a hammock-like couch. The vest restrained the monkey so that he could not tamper with any of the medical sensors but permitted enough freedom of movement so Bonny could operate food and water dispensers. Food was in the form of square pellets. Bonny could receive up to sixty pellets per day: up to twenty per session as a reward for behavioral task testing twice each day, and twenty during a daily free period. Biosatellite 3 used a hydrogen-oxygen fuel cell to generate electricity. One of the by-products of the fuel cell was water, which was used for drinking.

Biosatellite 3 lifted off on June 29, 1969. The booster placed the spacecraft in a 356-by-386-kilometer orbit. After only a few days in orbit, however, Bonny's condition began to deteriorate. On the launch pad, the monkey's success rate on the psychomotor tests averaged 73 percent. On the third day after launch, his performance fell to 27 percent, then 23 percent. On July 6, he refused to drink any water. His body temperature and heart rate began to drop, and his breathing became shallow. He began losing bodily fluids through urine and perspiration and became dehydrated. On July 7, after only nine days in space, ground controllers ended the flight. The recovery aircraft missed the air-snatch recovery, and Bonny splashed down in the Pacific Ocean about 40 kilometers north of Kauai, Hawaii. The onboard radio beacon led recovery crews to the spacecraft, which was quickly plucked from the ocean and flown to Hickam Air Force Base. About twelve hours later, in a state of serious dehydration, Bonny suffered a massive fatal heart attack. A necropsy re-

vealed that Bonny had become ill because the spacecraft temperature of 20.0° to 21.1° Celsius had been too low.

Despite the death of Bonny, Biosatellite 3 had supplied scientists with nine days of in-flight physiological measurements. Some 80 percent of the mission's data were obtained from space telemetry, with the remainder expected from postflight examination of the monkey. By the time Biosatellite 3 flew, NASA managers had already decided to terminate the program. In April, 1969, they asked Congress to redirect biosatellite funds to the Apollo Applications Program, which later became known as Skylab. It was believed that Skylab's long-duration crewed spaceflights would provide opportunities for broader research than would be possible with an uncrewed satellite.

Yet the end of the biosatellite program was not the end of biological satellites. A later project, the Orbiting Frog Otolith (OFO), sought to measure the effects of weightlessness and acceleration on the vestibular, or balance-sensing, organs in the inner ears of frogs. Frogs were selected because their inner-ear structure is similar to that of human beings. The otoliths are the sensor cells in the inner ear. Two male bullfrogs, each weighing about 320 grams, had microelectrodes attached to their vestibular nerves. They were housed in a water-filled, self-contained centrifuge inside a pressure-tight spacecraft housed in a larger spacecraft. OFO was 76 centimeters in diameter and 119 centimeters in length and weighed 133 kilograms. NASA's Ames Research Center, which had managed the biosatellite program, also managed OFO.

A four-stage Scout B launch vehicle propelled the OFO into a 290-by-507-kilometer orbit on November 9, 1970. The vehicle was launched from the NASA facility at Wallops Island, Virginia. At first, the frogs' otoliths signaled to the nervous system that gravity, a prime reference for balance, was missing. Within three days, however, the frogs' otoliths adjusted to the new environment, and ground monitors received normal signals. The mission lasted six days, and no attempt was made to recover the frogs.

## Contributions

Despite the problems encountered throughout the Biosatellite series of flights, the program yielded a considerable quantity of scientific data. Because of the retrorocket failure, there was no return of scientific data from Biosatellite 1. Yet some engineering data were obtained during the flight on the overall spacecraft performance.

The second mission, which orbited for 45 hours, returned so much information that scientists spent two years analyzing the Biosatellite-2 experiments. Among their findings, the scientists observed that the seventy-five wheat seedlings flown on Biosatellite 2 grew as much in two days of weightlessness as they would normally grow in three days on Earth. They also found that the greatest effects of weightlessness were noted in young and actively growing cells and tissues. Rapidly dividing cells with a high metabolic rate showed more effects than mature cells, which divide more slowly. In general, plants could not maintain proper orientation as they grew in weightlessness. The root systems of plants grown in weightlessness spread in an erratic, tangled manner. Radiation effects produced greater damage in weightlessness than on Earth, again with the greatest damage occurring to young, rapidly dividing cells or reproductive cells. The least damage was noted in mature structural cells. Some of the experiments also indicated that the absence of gravity slowed the growth and metabolism of rapidly reproducing cells. Overall, animal cells seemed less affected than plant cells.

The OFO mission provided data that helped researchers understand a specific problem affecting astronauts and cosmonauts: space adaptation syndrome. The signals telemetered to the ground indicated that gravity was the dominant environmental factor affecting humans' sense of balance. At first, the frogs' otoliths frantically signaled their nervous systems that something was wrong. As the mission progressed, however, their otoliths acclimated to the new environment, and within three days began sending normal signals.

## Context

The Biosatellite and OFO flights represented efforts to quantify the effects of the space environment on living things. The first biological spaceflights had occurred in the late 1940's, when researchers placed monkeys, plant seeds, insects, and mice aboard V-2 missiles fired from White Sands Proving Grounds in New Mexico. The flights with the plant seeds and fruit flies were more closely related to the later biosatellite program than the monkey flights were, for they sought to determine if there were any long-term genetic effects from spaceflight. The flights featuring primates were essentially engineering demonstrations of life-support systems and opportunities to observe the gross effects of weightlessness on humanlike forms. During the same period as these flights, U.S. Air Force scientists were flying small animals, insects, and tissue samples aboard high-altitude balloons in an effort to understand the biological effects of cosmic radiation on living things.

In November, 1957, the Soviets launched a dog named Laika aboard their second satellite, Sputnik 2. Laika was the first living creature to orbit Earth. Laika was not recovered from orbit, but ground controllers monitored her condition for a week via telemetry.

The first successful recovery of live animals from orbit came in 1960, with the flight of the Soviet Sputnik 5. Two dogs, named Strelka and Belka, orbited Earth for one day aboard this prototype of the crewed Vostok spacecraft and then returned. Although subsequent canine space travelers were not as fortunate, two successful Sputniks flew during March, 1961, each carrying a single dog. These orbited Earth only once, rehearsing the flight of the first man in space, Yuri A. Gagarin, on April 12, 1961.

The Americans also used animals to test their first crewed spacecraft. On January 31, 1961, a chimpanzee named Ham flew a suborbital flight lasting 16 minutes during the Mercury-Redstone 2 mission. Ham's name was an acronym for Holloman Aero Med, which was itself a shortened form for the Holloman Aero-Medical Field Laboratory in southern New Mexico, where the Mercury chimpanzees trained. Later that same year, another chimpanzee, Enos, orbited Earth twice during the Mercury-Atlas 5 flight. The chimpanzee flights were important steps in the process of qualifying the Mercury spacecraft for crewed flight. While the animal passengers were examined, these flights did not seek to unravel the mysteries of how spaceflight affected basic life processes. Such investigations did not occur until biosatellite flights.

In the mid-1970's, American researchers participated in another biological satellite, the Soviet Kosmos 782. The Kosmos 782 experiments were housed in a spherical return spacecraft similar in construction to the Vostok spacecraft that had carried the first Soviet cosmonauts into space in the early 1960's. The spacecraft, which was launched on November 25, 1975, orbited for nineteen days. Two of the American experiments involved the effects of weightlessness on plant tissue and cell development. Another studied the development of fish embryos in weightlessness. Kosmos 782 also carried five rats and a colony of fruit flies. The Soviets also provided tissue samples from the rats and fruit flies for American scientists to study after the flight.

Bioscience investigations in space continued aboard the Spacelab missions. Because Spacelab could accommodate animal cages, scientists were able to conduct their research in space at first hand. Later, this firsthand experimentation would continue on other space shuttle life-science laboratories. Bioscience research also was performed aboard the International Space Station beginning as early as its initial construction phase. Much of that research involved plant growth experiment programs or biomedical research using the Human Research Facility inside the Destiny Laboratory. After the *Columbia* accident, the Bush administration directed the mission of the International Space Station to focus upon research that would directly support expansion of human exploration of space, namely bioscience and biomedical research.

**See also:** Ames Research Center; Attack Satellites; Cooperation in Space: U.S. and Russian.

## Further Reading

Clement, Gilles. *Fundamentals of Space Medicine (Space Technology Library)*. New York: Springer, 2003. Covers the effects of spaceflight on bacteria, animals, plants, and the human beings who have lived and worked in orbit for prolonged periods of time.

Muenger, Elizabeth A. *Searching the Horizon: A History of Ames Research Center, 1940-1976*. NASA SP-4304. Washington, D.C.: Government Printing Office, 1985. Included in this text is a description of the formation of the Life Sciences Directorate and the origins of the biosatellite program.

National Aeronautics and Space Administration. *Biosatellite II*. NASA Facts NF-3. Washington, D.C.: Government Printing Office, 1969. One of the NASA Facts Series of educational publications. Provides a nontechnical summary of the major results of Biosatellite 2. Contains photographs of the test specimens flown aboard the spacecraft. Suitable for general audiences.

Pitts, John A. *The Human Factor: Biomedicine in the Manned Space Program to 1980*. NASA SP-4213. Washington, D.C.: Government Printing Office, 1985. A volume in the NASA History series, this work was written for a general audience. It provides an overview of NASA's life science programs, including biosatellites. It discusses some of the behind-the-scenes considerations that have affected NASA's biomedical research.

Planel, Hubert. *Space and Life: An Introduction to Space Biology and Medicine*. Cleveland: CRC Press, 2004. Covers the full range of space research on how weightlessness affects the human body.

Saunders, Joseph F., ed. *The Experiments of Biosatellite II.* NASA SP-204. Washington, D.C.: Government Printing Office, 1972. A compilation of the scientific results of Biosatellite 2. Starts with a description of the overall biosatellite program and provides background on the methods used to select missions and experiments. This source features papers presented by principal investigators. Rather technical in nature, but includes an introductory chapter.

*Gregory P. Kennedy*

# Cape Canaveral and the Kennedy Space Center

*Date:* Beginning July 24, 1950
*Type of facility:* Spaceport

*Cape Canaveral Air Force Station (CCAFS) and the Kennedy Space Center (KSC) are two of the United States' three principal facilities for launching satellites and piloted spacecraft. CCAFS is the older of the two and is used for all U.S. robotic launches going into easterly orbits. KSC is the prime launch center of NASA, with space shuttle facilities located on Merritt Island and robotic facilities located on Cape Canaveral.*

## Key Figures

*John F. Kennedy* (1917-1963), thirty-fifth president of the United States, 1961-1963

*Lyndon B. Johnson* (1908-1973), thirty-sixth president of the United States, 1963-1969

*Rocco A. Petrone* (b. 1926), Saturn Project officer and director of Launch Operations

*Kurt H. Debus* (1908-1983), director of the Kennedy Space Center (KSC), 1960-1974

*Lee R. Scherer* (b. 1919), KSC director, 1975-1979

*Richard G. Smith*, KSC director, 1979-1986

*Forrest McCartney*, KSC director, 1986-1992

*Robert L. Crippen, Jr.* (b. 1937), KSC director, 1992-1995

*Jay F. Honeycutt*, KSC director, 1992-1997

*Roy D. Bridges, Jr.* (b. 1943), KSC director, 1997-2003

*James W. Kennedy*, KSC director from 2003

## Summary of the Facilities

What is popularly known as Cape Canaveral is actually two separate space launch facilities on adjacent bodies of land. These are Cape Canaveral Air Force Station, located on Cape Canaveral, about midway down the Florida peninsula and separated from the mainland by the Banana and Indian Rivers, and NASA's Kennedy Space Center (KSC), located on Merritt Island to the northwest. Although the two are separate, the name Cape Canaveral, or simply "The Cape," is often used to refer to both.

The Cape's involvement with the space program originated with the U.S. government's need to test long-range military missiles. In July, 1946, the joint chiefs of staff recommended that an overwater test site be established because of the anticipated long flights of missiles then being tested or planned. This need was made apparent when an Army V-2 fired from the White Sands Missile Range in New Mexico accidentally headed south and landed outside Ciudad Juárez, Mexico, on May 29, 1947. Two overwater sites were recommended a month later. The El Centro Naval Air Station in California was preferred, but the Mexican government balked at allowing U.S. sovereignty over the tracking stations that would have been needed along the Mexican coast. The Banana River Naval Air Station was selected when the British government agreed to allow the United States to use the Bahama Islands. On May 11, 1949, President Harry S. Truman signed a bill authorizing the Joint Long Range Proving Ground Base. It was established on June 10, 1949. On May 16, 1950, sole control of the base was passed to the Air Force. The air station became known as Patrick Air Force Base, the launch facilities eventually became Cape Canaveral Air Force Station, and the complete range today is known as the Eastern Test Range.

The first launches from the Cape were made on July 24 and 29, 1950, when two Bumper rockets (German V-2 rockets with small solid-fueled-rocket upper stages) made low-angle firings programmed to pitch over and fly horizontally at an altitude of 16 kilometers. One of these rockets set an atmospheric speed record before hitting the ocean 322 kilometers from the shore. Longer flights soon followed, and by October, 1957, the "range" extended more than 11,000 kilometers, past Ascension Island in the Central Atlantic Ocean.

Early missile tests at Cape Canaveral were as likely to fail as to succeed, as engineers learned the subtleties of systems, structures, propellants, and guidance. The Snark, an early cruise missile, crashed so many times that residents spoke of "Snark-infested waters." One Polaris missile, destroyed by range safety upon heading inland, was dubbed the "InterBanana River Ballistic Missile."

Most of the early launches were affected by the United States' early ballistic missiles (Redstone, Jupiter, Thor, and Atlas) and early cruise missiles (Matador and Snark). Many of these evolved into today's small and medium space launchers: The Thor intermediate range ballistic missile (IRBM) became the Delta family, the Atlas intercontinental

*Aerial view of Missile Row, Cape Canaveral Air Force Station, Florida, in 1964. At this time, the Vehicle Assembly Building (upper left-hand corner) was under construction.* (NASA)

ballistic missile (ICBM) became the Atlas-Agena and Atlas-Centaur, and the Titan ICBM became the Titans III and IV. As these missiles evolved into space carriers, their launch pads were already the best U.S. location for reaching orbit. It had long been recognized that launches into orbit would have a "running start" if fired eastward and at a low latitude to take advantage of Earth's rotation. Cape Canaveral, at 28.5° north latitude, and with its launch pads and tracking facilities, was an ideal launch site.

Although the first satellite would be launched by the Soviet Union, the capability was demonstrated in September, 1956, by the U.S. Army when a Jupiter-C missile was fired down range. It was clear that the missile needed only an upper stage to insert a small payload into orbit. Yet the U.S. Navy's Vanguard had been assigned the task of launching the first U.S. satellite, and the Army was given strict orders not to repeat the attempt. On June 12, 1957, the first Vanguard test vehicle lost thrust at liftoff and collapsed in a ball of fire. The Army team was given permission to proceed, and on January 31, 1958, Explorer 1 was orbited by a Jupiter-C.

Notable launches since that time include the Ranger, Surveyor, and Lunar Orbiter series of lunar probes; the Pioneer, Mariner, Viking, and Voyager series of deep-space missions; the Orbiting Geophysical Observatories, Orbiting Astronomical Observatories, and High-Energy Astronomical Observatories; and dozens of Explorers of various types (as well as weather and communications satellites).

Construction of the first launch pad for the United States' first civilian super booster, the Saturn I, began at Cape Canaveral on June 3, 1959.

*The Vehicle Assembly Building in 1966, shortly after construction. A Saturn V launch vehicle is slowly rolling toward Launch Complex 39A.* (NASA)

Launch Complex 34 was dedicated on June 5, 1961, and saw its first launch on October 27, 1961, a suborbital test of the Saturn I first stage. Tragedy would visit the same site in 1967, when fire swept through the Apollo 1 spacecraft, killing its three-man crew and delaying the Moon landing program by nearly twenty-one months.

There were numerous other important launch pads at Cape Canaveral, but many are no longer functional. Their location by the Atlantic Ocean exposes them to corrosive salt spray, and intense effort is required to keep the pads in working order. Some of the oldest have become a part of an Air Force museum. The Titan III and IV facilities at Cape Canaveral, however, represent more modern developments. Built in the 1960's, the facilities include a Vertical Integration Building, where the heavy-lift Titan core vehicle is assembled and tested, and a Solid Motor Assembly Building, where the twin large solid boosters are attached to the core vehicle. The complete booster is hauled, by diesel train, to Launch Complex 40 or 41. Titan launch operations came to a close at Cape Canaveral when a Titan IV booster departed Complex 40 at 00:50 Coordinated Universal Time (UTC) on April 30, 2005. This booster delivered a National Reconnaissance Office-classified payload to orbit and put on a spectacular show in the skies along the East Coast of the United States and Canada as it ended more than five decades of heavy-lifting launch services from Cape Canaveral. This was the 168th Titan to launch. In total, there were forty-seven Titan I ICBMs, twenty-three Titan II ICBMs, twelve Titan II Gemini Program support boosters, four Titan III-As, thirty-six Titan III-Cs, seven Titan III-Es, eight Titan 34Ds, four Commercial Titans, and twenty-seven Titan IVs. After one final launch attempt from the Vandenberg site, the venerable Titan family would be retired, leaving a rich history that has served the national interest in wartime and peacetime, having dispatched both astronauts and highly ambitious robotic probes into space.

In addition to launching satellites and deep-space probes, the facilities at Cape Canaveral launched every American piloted mission into space until the Skylab program and served as the first Mission Control center during Project Mercury and the first piloted Gemini flight. Cape Canaveral also continued its original program of supporting development flights on ballistic missiles. The Polaris, Poseidon, Trident, Minuteman, and Pershing missiles were all tested in their early stages at Cape Canaveral. Submarine-launched missiles, although fired from offshore, use the Cape's tracking and support facilities.

The next major step in the development of the Cape as a space mission launch site was President John F. Kennedy's 1961 decision to send people to the Moon. A joint NASA-Department of Defense study assessed a number of possible launch sites for the program, including White Sands, Christmas Island, and Hawaii. White Sands and Cape Canaveral were favored because they offered existing facilities that could be expanded at less cost than building anew. Cape Canaveral was finally selected because launches over the ocean pose less threat to the civilian population.

To accommodate the massive launch facilities, NASA acquired some 300 square kilometers of citrus groves and swampland on Merritt Island, to the immediate northwest of Cape Canaveral. What became known as the Merritt Island Launch Area (MILA) covered much acreage in order to protect Florida residents from possible accidents with the large launch vehicles contemplated for the Moon program. Many of the citrus groves were leased back to their former owners, and most of the swampland was designated as a national wildlife refuge. The greatest concentration of endangered species in the United States can be found in this refuge.

KSC's history is almost as old as the Cape's. The Experimental Missiles Firing Branch was established at the Cape in 1951 by the Army Ordnance Guided Missile Center in Huntsville, Alabama. On July 1, 1960, this division was transferred to the newly established Marshall Space Flight Center, and on March 7, 1962, it became a separate Launch Operations Directorate. The Directorate was formally activated as the Launch Operations Center on July 1, 1962, and was designated as the executive agent for MILA on January 17, 1963. The center was renamed in honor of President Kennedy on November 28, 1963, and on July 26, 1965, its headquarters moved into new offices on Merritt Island and absorbed MILA. The name MILA is still used as a designator for a tracking station that supports launches and piloted space missions.

Kurt H. Debus, a member of the original team of World War II German rocket scientists, was the KSC's original director, from 1961 to 1974. Lee R. Scherer, a former Langley Research Center engineer, then Richard G. Smith, a former deputy director of the Marshall Space Flight Center, succeeded him. Following the *Challenger* accident in 1986, Smith resigned and was succeeded by Forrest McCartney, a retired Air Force general. In 1992, Robert L. Crippen, Jr.—the pilot of the first space shuttle mission—was appointed as director and served until January, 1995. His successor, Jay F. Honeycutt, ran the space center for the next two years. Roy D. Bridges, Jr., became the director of the KSC on March 2, 1997. Bridges, a retired Air Force major general, held many key space-related roles during his career. Prior to his last USAF assignment at Wright-Patterson Air Force Base, he was the commander of the Air Force Flight Test Center, Edwards Air Force Base, California. He also was commander of the Eastern Space and Missile Center at Patrick Air Force Base, Florida, and commander of the 6510th Test Wing at Edwards Air Force Base, California. As a NASA astronaut, he piloted the space shuttle *Challenger* on STS 51-F in July and August of 1985. Bridges remained the KSC director through the *Columbia* accident and the initial phases of the accident investigation period. On June 13, 2003, he was transferred to the center director position at the Langley Research Center. James W. Kennedy became acting KSC director on that date and assumed the Center director position officially as of August 10, 2003.

Although KSC is NASA's principal launch site, several years passed before it could host all major launch operations, as each NASA center had maintained its own launch operations office that inter-

acted separately with the Air Force at the Cape. In December, 1964, KSC absorbed the Florida office of the Manned Spacecraft Center, thus giving KSC control of piloted spacecraft from arrival until launch. In October, 1965, over the protests of the Goddard Space Flight Center, KSC assumed responsibilities for all robotic launches on both coasts. Also in October, KSC absorbed the independent Pacific Launch Operations Center, NASA's office at Vandenberg Air Force Base. At about the same time, the Air Force transferred to NASA five launch complexes because NASA was using them more than the Air Force. Indeed, NASA's activities were so intense that the Navy often had to delay test launches of its Polaris missile to accommodate spaceflights and NASA tests.

Most of the activity had now shifted to KSC's Merritt Island property, where the U.S. Army Corps of Engineers undertook a massive construction effort. Simply to prepare the site for construction, the Corps and its contractors had to dredge up 1.15 million cubic meters of soil to make new channels so rocket stages could be brought in by barge and raise the land from 46 centimeters to 2.1 meters above sea level to safeguard against flooding of the facilities.

KSC is divided into two major areas, the industrial area on the central part of the island and Launch Complex 39 on the northern part (the southern part of the island was not acquired). The industrial area includes administrative offices, the Operations and Checkout Building, and the Consolidated Instrumentation Facility (CIF). The operations and checkout building encloses a long "clean room," where complete payloads are assembled and tested under pristine conditions before flight. This facility served the Apollo spacecraft in the 1960's and 1970's and now handles space shuttle payloads. Astronaut crew quarters are also located in the Operations and Checkout Building. Launch Complex 39 includes the Vehicle Assembly Building (VAB), the Launch Control Center (LCC), and two launch pads (Pad 39A and Pad 39B). To support the space shuttle program, Orbiter Processing Facility Bays and Shuttle Landing

*Aerial view of Launch Complex 39 in 1998 showing the Vehicle Assembly Building, the Orbiter Processing Facilities, the Launch Control Center, the press site, and crawlerways to Launch Pads 39A (top left) and 39B. (NASA)*

Facility were added to Launch Complex 39, and a Vertical Processing Facility and Hypergolic Propellant Handling Facility were added to the industrial area.

For the Saturn V vehicle, the Huntsville rocket team organized the preparations such that the launcher would be assembled and tested in an enclosed facility, the VAB, and then carried to the launch pad. This setup would consolidate most assembly and checkout work and leave the launch pad free for other vehicles.

At the end of the lunar program, the facilities at KSC were modified to support the Skylab program. The "milk stool," a platform added atop the mobile launch platform from which the smaller Saturn IB could be launched, was one of the most visible changes. Following the Apollo 7 launch in 1968, the Saturn I complex at Cape Canaveral was moth-

balled for use in later years. As the Skylab program approached, however, it became apparent that it would be less expensive to modify Saturn V facilities for Skylab. This renovated setup saw the launch of three Skylab crews in 1973 and the Apollo-Soyuz Test Project in 1975.

Following the Apollo era, extensive modifications were made to KSC to support the space shuttle program. Although NASA analyzed potential alternate sites, there was little doubt that KSC would be the shuttle port because the cost of modification would be less than building a new facility at a separate location. By the same token, some limits were placed on the shuttle design to keep the modifications to KSC at a minimum.

Many facilities were added to KSC to support the space shuttle. The Orbiter Processing Facility (OPF), one addition, is a twin hangar in which the shuttle orbiter is prepared for subsequent missions. Large servicing platforms can be extended around the shuttle for access to any location. Most payloads are installed and tested in concert with launch and Mission Control centers. After the work is completed, the orbiter is rolled to the Vehicle Assembly Building (VAB). An Orbiter Modification and Refurbishment Facility, a single-bay OPF, was built in the mid-1980's.

The Shuttle Landing Facility (SLF), another component of the shuttle system, is a concrete runway 4.6 kilometers long and 91 meters wide aligned on a northwest-southeast axis. Missions land at Edwards Air Force Base in California when weather conditions at KSC are not favorable and the vehicle cannot remain aloft an additional day to await better weather in Florida. Regardless of whether a shuttle lands at KSC, Edwards Air Force Base, or elsewhere, KSC assumes responsibility for the shuttle after wheelstop.

Solid-Fueled Rocket Booster Processing Facilities were built to handle the many solid motor segments that must be stockpiled in advance of a mission; two buildings were added near the VAB to house booster segments. A nearby Rotation and Processing Building rotates the segments from a horizontal position to a vertical position. After a launch, the boosters are towed to Port Canaveral and then up the Banana River to a hangar at Cape Canaveral for removal of range safety ordinance and disassembly of the boosters.

The Hypergolic Maintenance and Checkout Facility consists of three buildings in an isolated section of the industrial area for cleaning and refurbishing storable propellant tanks and thrusters. The Vertical Processing Facility (VPF) handles the upper stages and their satellites, as well as special payloads such as the Hubble Space Telescope. Hazardous payloads are processed in the VPF and carried directly to the launch pad for installation in the orbiter. The Logistics Facility, a semiautomated warehouse 30,176 square meters in area, houses more than 190,000 shuttle parts. Other processing facilities at KSC are designated for payload preparation. Hangar L, for example, is equipped

to handle life science payloads. Another facility sterilizes satellites destined for landings on other planets.

**Context**

Cape Canaveral Air Force Station and Kennedy Space Center continue to fulfill the roles for which they were created, although the vehicles and technologies employed have evolved significantly over the years. Cape Canaveral is the primary launch site for nonshuttle payloads placed in easterly orbits. Yet this role was greatly diminished in the mid-1980's, when the space shuttle program was ascendant; the U.S. Air Force even considered closing the Eastern Test Range in the late 1980's. In 1984, however, the Air Force argued successfully that there should be a complementary expendable launch vehicle that could provide much of the robotic launch capability of the shuttle in case an accident should befall the piloted program, and the *Challenger* accident in 1986 also renewed interest in expendable launchers. The Titan IV was chosen for this role. During 1987, however, many launch vehicles proved faulty. Eventually, the Delta family of launchers was upgraded to the Delta II, III, IV, and IV-Heavy versions. Delta and Atlas launch pads were eventually given over to control by the booster manufacturer, a part of the Reagan administration's efforts to commercialize access to space.

A secondary reason for revamping the expendable launcher capability was the desire to maintain NASA's stature as a research and development agency rather than solely an operational agency. Besides many functional changes in shuttle activities, a major revision at KSC was the installation of a deputy manager for the Space Transportation System from NASA's headquarters. This deputy manager has the final say in authorizing a shuttle launch and may cancel a launch for safety or other reasons.

KSC remains the sole piloted launch facility for the Western world. Although many payloads have been transferred to expendables to relieve the burden on the shuttle, it is expected that the demand for space travel will exceed KSC's capability. The center may also host new expendable launchers. To accommodate more flights, additional ground facilities, including launch platforms and outfitted VAB high bays, will be required.

Besides functioning as a launch site, KSC has branched into the area of research and development. Engineers there have developed a closed-environment life-support system simulator for testing systems that would recycle air and water, thus reducing the need to resupply space stations with these two consumables. Lightning research is conducted as an extension of launch safety and weather monitoring. KSC engineers are also engaged in development work on automation and robotics to reduce costs associated with recycling booster hardware for each mission.

Both the Cape Canaveral Air Force Station and KSC have information centers that recount the history of rocketry and spaceflight for the visitor. Cape Canaveral's center is in and around the blockhouse that controlled the launches of the Mercury-Redstone flights and features a rocket park. Kennedy's Visitor Center is located on the west side of the main installation and includes a museum, IMAX movie theaters, and a rocket park. Bus tours for both facilities originate from the KSC Visitor Center.

Half a century after its beginning, Cape Canaveral continues to serve as an active launch facility. The same range hardware that helped launch John H. Glenn, Jr., in 1962 guided his space shuttle mission thirty-six years later. A significant human-power upgrade promises to permit the launching of different vehicles on successive days after a mere twenty-four-hour turnaround. At the start of the new century, planners opened Delta IV launch facilities at a remodeled Complex 37 and launched the Atlas-5 from Complex 41. Both boosters held the potential to support piloted exploration beyond low-Earth orbit as early as 2014.

**See also:** Apollo Program; Apollo Program: Developmental Flights; Atlas Launch Vehicles; Deep Impact; Delta Launch Vehicles; Johnson Space Center; Langley Research Center; Launch Vehicles; Lewis Field and Glenn Research Center; Marshall Space Flight Center; Space Centers, Spaceports, and Launch Sites; Space Shuttle; Space Shuttle: Approach and Landing Test Flights; Titan Launch Vehicles.

**Further Reading**

Baker, David. *The Rocket: The History and Development of Rocket and Missile Technology.* New York: Crown, 1978. A thorough, well-researched history of rocketry and specific vehicles. Includes extensive descriptions of launch preparations and the discussion of early failures.

Benson, Charles D., and William B. Faherty. *Gateway to the Moon.* Gainesville: University Press of Florida, 2001. This text was originally part of a NASA history series. It provides a detailed management and engineering history of the construction of the Kennedy Space Center in support of the Apollo Program.

Benson, Charles D., and William Barnaby Faherty. *Moonport: A History of Apollo Launch Facilities and Operations.* NASA SP-4204. Washington, D.C.: Government Printing Office, 1978. The official NASA history of the development of the Saturn V facilities. Although Saturn V no longer exists, the procedures involved with that vehicle serve to illustrate preparation of launch facilities in general.

Bilstein, Roger E. *Stages to Saturn: A Technological History of Apollo/Saturn Launch Vehicles.* Gainesville: University Press of Florida, 2003. Starting with the earliest rockets, Bilstein traces the development of the family of massive Saturn launch vehicles that carried the Apollo astronauts to the Moon and boosted Skylab into orbit.

Chaikin, Andrew. *Space.* London: Carlton Books, 2002. A large picture book spanning piloted and robotic exploration of space. Provides pictures of Earth and special resources as well.

Covault, Craig. "Cape Gears for New Vehicles, Launch Surge." *Aviation Week and Space Technology,* March 13, 2000, 54. This article looks at Cape Canaveral and the Kennedy Space Center as the fiftieth anniversary of the Cape approached. It discusses some of the Cape's history and looks at its ongoing operations.

Gatland, Kenneth. *The Illustrated Encyclopedia of Space Technology: A Comprehensive History of Space Exploration.* New York: Salamander, 1989. A broad overview of space exploration, with a chapter devoted to the history of launch facilities—including Cape Canaveral and other U.S. launch sites. Well illustrated, with maps and a time line.

Grimwood, James M. *Project Mercury: A Chronology.* NASA SP-4001. Washington, D.C.: Government Printing Office, 1963. A detailed chronology of the United States' first piloted space missions, from postwar beginnings through the start of the Gemini Program. The book provides insights into launch preparation activities.

Lay, Beirne. *Earthbound Astronauts: The Builders of Apollo-Saturn.* Englewood Cliffs, N.J.: Prentice-Hall, 1971. Describes the development of the Saturn V launch vehicles from the standpoint of several engineers and managers who worked on the program. Includes many details about launch preparations.

Levine, Arnold S. *Managing NASA in the Apollo Era.* NASA SP-4102. Washington, D.C.: Government Printing Office, 1982. A highly detailed history of the difficulties encountered and lessons learned from managing a growing agency organized to send Americans to the Moon. The activities of Kennedy Space Center are discussed in the context of the agency as a whole.

Pellegrino, Charles R., and Joshua Stoff. *Chariots for Apollo: The Making of the Lunar Module.* New York: Atheneum, 1985. A well-written history of the development of the Lunar Module. Although it focuses on the problems of building the spacecraft, much detail is given about the integration of the payload with the vehicle and its actual launch.

Swanson, Glen E. *Before This Decade Is Out: Personal Reflections on the Apollo Program.* Gainesville: University Press of Florida, 2002. This oral history of the Apollo Program provides insights into the thoughts of the people who accepted President Kennedy's lunar challenge, sent astronauts to walk upon the Moon, and returned them safely to Earth.

*Dave Dooling, updated by David G. Fisher*

# Cassini: Saturn

*Date:* Beginning October 15, 1997
*Type of program:* Planetary exploration

*Data from the Cassini mission was designed to investigate the atmosphere, rings, moons, and gravitational and magnetic fields of Saturn and to deposit a probe on the large moon Titan. The data from this mission would rewrite the textbooks on Saturn and assist in comparative planetology of the other gas giants: Jupiter, Uranus, and Neptune.*

## Key Figures

*Robert Mitchell*, Cassini program manager at the Jet Propulsion Laboratory (JPL)
*Earl H. Maize*, deputy program manager at JPL
*Dennis L. Matson*, Cassini project scientist at JPL
*Linda J. Spilker*, deputy project scientist at JPL
*Mark Dahl*, Cassini program executive NASA Headquarters
*Denis Bogan*, Cassini program scientist at NASA Headquarters
*Jean-Pierre Lebreton*, Huygens mission manager and project scientist at the European Space Agency

## Summary of the Mission

The Cassini orbiter and its piggybacked entry and hard landing probe Huygens together constitute the largest interplanetary spacecraft yet designed. Its mission is the most complex investigation of a single planet and its moons ever planned. A joint project of the National Aeronautics and Space Administration (NASA), the European Space Agency (ESA), and the Italian Space Agency (ASI), the Cassini-Huygens project has employed more than forty-three hundred people in seventeen countries in devising new technology and eighteen major scientific experiments. Program planners expect the probes to shed light on some of the deepest mysteries about giant gas planets, moons, and the solar system in general.

Under the overall direction of program Director Earle K. Huckins, the mission began on October 15, 1997, with a Titan IV launch from Cape Canaveral during a monthlong launch window with the most favorable arrangement of planets for the trip to Saturn. After a twenty-minute ascent, the Centaur upper stage inserted Cassini-Huygens into a 445-kilometer-high parking orbit. After separation from the Centaur, the spacecraft established communications with the Deep Space Network, received thorough systems checks, and deployed a number of spacecraft structures needed for the power generation system and for scientific measurements.

Because the spacecraft weighs 5,650 kilograms, a record for an interplanetary probe, the launch vehicle, Centaur upper stage, and Cassini's engines were not powerful enough for a direct course to Saturn. Instead, the cruise segment of the mission involved a complex trajectory that swung the spacecraft by Venus twice, Earth once, and Jupiter once. Each encounter, called a gravity-assist maneuver, increased its velocity relative to the Sun, saving fuel and shortening flight time. The first Venus flyby, on April 26, 1988, boosted the spacecraft

*The fully assembled Cassini spacecraft at the Jet Propulsion Laboratory in Pasadena, California. The Huygens probe (mounted to the left side of the craft) was provided by the European Space Agency and the radio antenna by the Italian Space Agency.* (NASA)

teroid was quite different from the asteroids Gaspra and Ida that the Galileo spacecraft had imaged during closer flyby encounters. By April 14, 2000, Cassini had completed its passage through the asteroid belt, on its way to a Jupiter flyby at the end of the year.

Cassini's main engine was fired for six seconds on June 14, 2000, in order to refine the spacecraft's approach to Jupiter. This was done not for any Jupiter science objectives but to ensure that Cassini would pass close by Saturn's moon Phoebe on its final approach to the ringed planet. On October 1 Cassini's imaging systems were activated to take initial pictures of Jupiter. The spacecraft came within 10 million kilometers of Jupiter on December 30. The encounter boosted Cassini's speed by 2.2 kilometers per second and sent it off toward the Saturn system. During the interplanetary cruise, a number of serious problems with systems and instruments on both Cassini and Huygens developed and were overcome.

Cassini is a hardy, versatile craft. The size of a motor home, 4 meters wide and 6.7 meters long, it holds 3,130 kilograms of liquid fuel for its two main engines and sixteen thrusters. One high-gain and two low-gain antennae can transmit as much as 249 kilobits of information per second to Earth at a distance that can take the signal from sixty-eight to eighty-four minutes to cross. Three generators, producing power from the radioactive decay of plutonium, will supply 650 watts of direct current power to the spacecraft's command-and-control, communications, and instrument systems. These systems require seven computers, 1,630 interconnect circuits, and more than 14 kilometers of cable. The main engines will burn nitrogen tetroxide and monomethylhydrazine to provide propulsion, whereas the thrusters, which burn hydrazine, will make small course corrections and partly control the spacecraft's attitude (orientation relative to its motion). Reaction wheels will provide further attitude control. These heavy wheels inside the spacecraft can be spun; in reaction the orbiter will turn in the opposite direction, so that controllers can adjust spin rates and point instruments at targets. Among the

speed by 7 kilometers per second; the second Venus flyby came on June 29, 1999, and boosted the spacecraft speed by 6.7 kilometers per second. The Earth flyby on August 18, 1999, provided a 5.5-kilometer-per-second increase in Cassini's speed. Instruments were operated during each of these encounters with Venus and Earth to test resolution and perform calibrations. During the cruise, the probe performed periodic equipment checks and made minor course adjustments. It entered the asteroid belt between Mars and Jupiter in mid-November of 1999. On January 23, 2000, Cassini took the first images of asteroid 2685 Masursky from a distance of 1.6 million kilometers. This as-new technology for the mission are solid-state data recorders, which have no moving parts, twenty types of Application-Specific Integrated Circuits (ASICs), Very High-Speed Integrated Circuits (VHSICs), a micro-miniature radio receiver, solid-state power switches, and an advanced gyroscope. Radiation hardening of the electronics and improved stability and control will give Cassini-Huygens more protection from the dangers of space than previous deep-space probes have had.

Cassini passed within 2,068 kilometers of the unusual outer moon Phoebe on June 11, 2004. Phoebe is the largest of Saturn's outermost moons and shows a heavily cratered surface that displays considerable variation in brightness. All instruments aboard the spacecraft provided data, determining Phoebe's composition, mass, and density. Scientists believed that Phoebe might be an object left over from early in the formation of the solar system. Five days after the Phoebe flyby, Cassini performed a modest trajectory correction maneuver to position itself for the Saturn orbit encounter.

All preparations for Saturn insertion were accomplished without incident, and on July 1, 2004, Cassini initiated its longest planned main-propulsion system burn. The spacecraft produced retrograde thrust for 95 minutes, and at 04:12 Coordinated Universal Time (UTC) entered Saturn orbit in such a way that as many as 76 orbits about the giant ringed planet might be possible. The initial orbital period was 116.3 days. Cassini came within 20,000 kilometers of Saturn's cloudtops and passed through the ring plane using its high-gain antenna as a dust shield against the high-speed impacts of small particles on the spacecraft's experiments and subsystems. This would be the closest Cassini would intentionally come to the cloudtops or the outer edge of the ring plane.

Just 36 hours after orbital insertion, Cassini encountered Titan for the first of its many planned close flybys. The distance of closest approach would vary from one flyby to the next. Eventually the spacecraft's radar would be expected to provide a nearly complete surface map of this curious moon. This first encounter served to provide confidence that Cassini's instruments had survived the seven-year journey from Earth to Saturn.

Cassini was on a collision course with Titan on December 24, 2004, when it released the Huygens probe. The probe then continued along that trajectory toward entry into the atmosphere of Saturn's largest moon. On December 27, Cassini performed an avoidance maneuver so that it would be safely in position to relay data from the Huygens probe to Earth and avoid hitting Titan. Early on the morning of January 14, 2005, the Green Bank Telescope picked up radio

*Cassini's view of Saturn, with its largest moons, Titan (lower left) and Rhea (top center).* (NASA/JPL/Space Science Institute)

waves indicating that the timer onboard the Huygens probe had turned on critical systems and experiments. The probe safely proceeded through atmospheric entry and touchdown on an unusual cryogenic (frozen) mud. The probe relayed data to the Cassini orbiter for storage until Cassini could turn to face Earth and transmit the data. Huygens lasted far longer than had been anticipated and provided a considerable album of images during descent and after landing. All onboard experiments produced data except one, but examination of Doppler shifts of the probe's signal was able to supply the wind-speed data that the probe had not been able to collect due to a software error.

Apart from its status as the target of the Huygens probe, Titan is of crucial importance to the Cassini mission for another reason: With a diameter of 5,150 kilometers, it has enough mass to generate a powerful gravitational field, which mission controllers will use to alter the direction and shape of Cassini's orbit while expending a minimum of fuel. In the Jupiter system, the Galileo spacecraft was able to use four large moons to effect orbit changes by gravity assists, but Cassini can use only Titan for that purpose, because Titan is the only large Moon in the Saturn system. As many as forty-four times the orbiter will fly by Titan at speeds ranging from 5 to 6 kilometers per second and at distances of 950 kilometers to 16,200 kilometers, although the closest approach may be lowered to 800 kilometers, atmospheric conditions permitting. By maneuvering slightly just before passing the moon, the orbiter will exit the flybys on drastically altered courses. The flybys will also give scientists a chance to map Titan's surface, monitor atmospheric changes, chart its gravitational field, and search for magnetic and particle fields.

The remainder of the four-year primary mission will repeat extensive investigations with other moons, the rings, and Saturn itself, compiling the most detailed scientific portrait of an outer planet by any probe. Cassini will also assemble a visual record of 300,000 color images. Accomplishing these goals will require about sixty orbits of the planet. Cassini will swing in as close as 180,000 kilometers from the cloudtops in orbits from ten to one hundred days long. The inclinations of the orbits (degrees of latitude above Saturn's equator) will vary from zero to more than 60°, about the same latitude above Saturn's equator as Anchorage, Alaska, is above Earth's. The high-inclination orbits will bring Saturn's poles into view, allowing scientists to study the peculiarities of its atmosphere and magnetic fields there. Also, during these inclined orbits several important occultations will occur; that is, the Earth, Sun, or stars will, on these occasions, be hidden from Cassini by Saturn, a moon, or the rings. The occultations will offer scientists special opportuni-

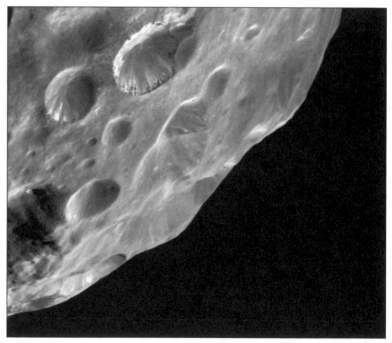

*Phoebe, one of Saturn's moons, in this image obtained from the Cassini orbiter on June 11, 2004, shows evidence of ice and a dark overlying material as well as heavy cratering.* (NASA/JPL/Space Science Institute)

ties to analyze the structure and composition of these bodies with radio waves or visual light.

Of particular interest during the tour will be the moons controlling Saturn's rings. Like a cowboy riding among cattle to herd them, small moons influence the positions of the myriad particles that compose the planetary rings. The interaction between a ring and its moons is gravitational in nature, but not fully understood, so mission scientists will scrutinize them as thoroughly as possible; however, the ice, dust, and small rocks nearby could damage Cassini, and so controllers will not risk extremely close flybys. On the other hand, flybys of the icy moons beyond the rings, a key objective of the tour, are possible, although requiring much fuel consumption by the thrusters. Of the icy satellites, Mimas, Enceladus, Dione, Rhea, and Iapetus will be targeted. Their surfaces show signs of collisions—a huge crater on Mimas, for example—and are expected to provide spectacular images. Iapetus displays a dichotomy in that one side is highly reflected and the other is one of the darkest surfaces seen in the solar system.

After about four years and hundreds of course adjustments, Cassini may run out of fuel for its main engines. Its planned mission would end if the fuel budget or financial budget ran out. Mission managers hope, however, that there will be enough electrical power and attitude control fuel left to continue collecting data and transmitting it to Earth a little while longer, although budget constraints may limit the amount of time that the Deep Space Network can devote to the probe. In any case, Cassini will continue to orbit. Eventually, perhaps as long as decades later, its attitude control will fail and it will tumble and break apart or impact destructively on a moon.

### Contributions

During its tour, the Cassini-Huygens spacecraft is expected to help answer a host of questions about Saturn, its rings, its icy moons, and Titan. Even before arriving at the planet, however, it engaged in gathering scientific data. Scientists used the spacecraft to search for gravitational waves in the solar system, and as it approaches Saturn its science instruments were gradually activated for testing and long-range sensing and imaging.

The scientific payload is diverse. Aboard Cassini are twelve sensor subsystems. The imaging science subsystem produces photographs in visible, ultraviolet, and near-infrared light. The radar is to locate and measure surface features through cloud layers on Titan. The radio science subsystem will study gravitational fields and the atmospheres and rings by detecting subtle variations in radio waves sent out from the orbiter. The ion and neutral mass spectrometer will sense neutral and charged particles near Titan, Saturn, and the icy moons that are part of their tenuous outer atmospheres and ionospheres; the visual and infrared mapping spectrometer will assay the chemical composition of the bodies' surfaces and atmospheres; the composite infrared spectrometer will measure their heat; and the ultraviolet imaging spectrometer will gather data about the atmospheres and rings from the ultraviolet energy they emit. The cosmic dust analyzer will be concerned with ice and dust particles in the system, and the radio and plasma wave science instrument will examine natural radio emissions and ionized gases flowing from the Sun or trapped in Saturn's magnetic field, aided by the plasma spectrometer. The magnetospheric imaging instrument and dual technique magnetometer will determine the nature of Saturn's magnetic field and its interaction with the solar wind (protons and electrons ejected from the Sun).

The Huygens probe carried six instrument packages. The Doppler Wind Experiment attempted to measure Titan's winds from their effects on the probe as it fell. The Surface Science Package investigated the physical properties of the moon's after the probe survived impact. The descent imager and spectral radiometer took photographs and measured temperature. The atmospheric structure instrument analyzed the relative abundance of the atmosphere's components, while the gas chromatograph and mass spectrometer determined the gases in it and the Aerosol Collector and Pyrolyser (ACP) examined clouds and suspended particles.

Cassini scientific objectives are ambitious. On Saturn, scientists expect to collect data on the abundance of gases, the types of elements and their isotopes, Saturn's wind patterns and cloud formations, its internal structure, the rotation of the deep atmosphere, the ionosphere and its interaction with the magnetic field, lightning in the atmosphere and its relation to radio emissions, the magnetosphere, and the formation and evolution of the planet. On Titan, instruments examined gases, looking especially for organic molecules, in order to understand how the moon can maintain such a thick, hazy atmosphere that is 60 percent thicker at the surface than Earth's. Huygens data will also help scientists determine how the atmosphere formed, its present dynamics, whether the surface is liquid or solid, its composition, and the composition of the thin upper

*The "photonegative" side of Saturn's rings as the Sun shines through: The B ring (center) blocks much of the Sun's light, while the less dense rings are brighter in the image because they scatter and transmit the light.* (NASA/JPL/Space Science Institute)

atmosphere and ionosphere. For the other moons, Cassini will continue to provide data on geological development, internal structure, and interaction with the rings. Cassini will study the rings themselves to assess their chemical makeup, distribution of material, and interaction with Saturn's magnetosphere, ionosphere, and atmosphere in order to discover how they change and what gravitational and magnetic mechanisms control their intricate structure. Within its first six months in the Saturn system, Cassini produced more high-resolution images of Saturn, its moons, and rings than had been achieved throughout preceding history. Cassini data has rewritten the textbooks on Saturn and assist in comparative planetology of the other gas giants: Jupiter, Uranus, and Neptune.

Some proposed that Titan would reveal a planetwide ocean of liquid hydrocarbons; other proposed that Titan had only separated lakes of these liquids. Huygens landed without a splash or a thud; one scientist described the probe's impact as being a splat, as though the probe had encountered cryogenic mud laced with hydrocarbons. The probe's penetrometer deployed 15 centimeters into this unusual soil. Images from the probe's descent imager/spectral radiometer indicated ice blocks strewn about and features indicative of fluid flow. As the temperature at the surface was only 93 kelvins, that fluid was cryogenic hydrocarbons. The data provided evidence of physical processes shaping Titan's surface: precipitation, erosion, mechanical abrasion, and fluvial activity.

Although the greatest focus apart from the planet itself was placed upon Titan, Cassini also investigated a number of the other larger moons at close range. Cassini passed within 123,400 kilometers of Iapetus on January 1, 2005. From a range ten times closer than Voyager 2's approach, images revealed one side of the moon to be as bright as snow and the the other to be as dark as tar. This dark side might well be rich in carbon-based compounds and is the side of the moon that leads in the direction of its orbital motion about Saturn. Thus the moon might have been dusted with this orbital material; however, it was possible that it originated from within the moon and spewed out on the surface. Further investigation would be needed to determine which scenario was the case. The spacecraft detected a 400-kilometer circular crater in the southern hemisphere and a line of mountains around the moon's equator. The latter gave the moon the appearance characteristic of a walnut.

Cassini made its first close encounter with the icy moon Enceladus on February 17, 2005, coming within 1,167 kilometers of the surface. The spacecraft's magnetometer picked up a bending of the planet's magnetic field by the icy moon caused by molecules interacting with the field by spiraling along field lines. This was evidence of gases arising from the surface or from the interior of Enceladus, which suggested a tenuous atmosphere around this moon. The icy moon has regions that are old and retain a large number of craters, but younger areas do display tectonic troughs and ridges.

Late in 2005, Cassini's suite of instruments determined that Enceladus had hot spots near its south pole, suggesting that this moon may still be geologically active. In a way, Enceladus's relationship to the Saturn system might be considered a more benign version of Io's relationship to the Jupiter system. High-resolution images of Mimas provided an accurate crater-density map for this unusual moon. Cassini scientists declared that Mimas is probably the most heavily bombarded moon in the solar system.

**Context**

Cassini is named after Italian-French astronomer Gian Domenico Cassini (1625-1712), who discovered four of Saturn's moons and a gap in the rings, now known as the Cassini Division. Huygens honors the Dutch natural philosopher and astronomer Christiaan Huygens (1629-1695), who first recognized Saturn's rings for what they are. Both men relied upon the recently invented astronomical telescope. The probe and orbiter, by furthering human investigation of Saturn and the solar system in general, belong in the tradition of modern astronomy to pursue knowledge aggressively with innovative technology. The mission also advances the joint goal of American and European space agencies to cooperate on complex exploration projects.

Cassini-Huygens capitalizes on and extends knowledge gained during the reconnaissances of Saturn by Pioneer 11 (1979), Voyager 1 (1980), and Voyager 2 (1981).

Although these probes merely hurtled past the planet, they detected new moons, and divisions and structure within the rings. Cassini's tour of the Saturn system and Huygens's descent to Titan's surface will bring in vastly more data from far more advanced instruments. Furthermore, the four years of observations will allow scientists to track changing atmospheric conditions on Saturn and learn more about the dynamics of a large gaseous planet, just as the Galileo probe did for scientists studying Jupiter. Specifically, scientists hope to learn how atmospheric storms start and end, the nature of the atmosphere below the outer cloud layer, and why Saturn emits 79 percent more heat than it gets from the Sun.

Although there is probably no life on Titan, as some scientists have speculated, Huygens data will probably settle the issue finally and make it possible to determine whether life could evolve there in the future. To do so it searched for organic chemicals and oceans. Because Titan's current atmospheric conditions resemble those of the Earth shortly after its formation, the moon may help scientists better understand terrestrial evolution after detailed data analysis was completed.

Discoveries about the origin and evolution of the Saturn subsystem will provide evidence about the origin and evolution of the entire solar system and the processes responsible for planetary formation. Theories developed from the information will aid astronomers in analyzing planets around other stars and, perhaps, in estimating the chance that conditions suitable for life exist outside the solar system.

**See also:** Asteroid and Comet Exploration; Hubble Space Telescope: Science; Huygens Lander; Jet Propulsion Laboratory; Lewis Field and Glenn Research Center; Mars Climate Orbiter and Mars Polar Lander; Nuclear Energy in Space; Pioneer 11; Planetary Exploration; Voyager 1: Jupiter; Voyager 1: Saturn; Voyager 2: Jupiter; Voyager 2: Saturn; Voyager Program.

## Further Reading

Burrows, William E. *Exploring Space: Voyages in the Solar System and Beyond.* New York: Random House, 1990. Burrows chronicles the history of the "far travelers," deep-space probes. He discusses the technology and politics of exploration programs and describes missions at length, all for general readers. The Cassini probe is mentioned only briefly; nonetheless, the book provides the background information on technology and scientific goals that place the probe in the context of planetary exploration. With photographs and an extensive bibliography.

Harland, David M. *Mission to Saturn: Cassini and the Huygens Probe.* London: Springer-Praxis, 2002. A detailed scientific and engineering history of the Cassini program, but also includes extensive discussion of Voyager program events and results.

Hartmann, William K. *Moons and Planets.* 5th ed. Belmont, Calif.: Thomson Brooks/Cole Publishing, 2005. Provides detailed information about all objects in the solar system. Suitable on three separate levels: high school student, general reader, and the college undergraduate studying planetary geology.

Irwin, Patrick G. J. *Giant Planets of Our Solar System: Atmospheres, Composition, and Structure.* London: Springer-Praxis, 2003. Provides an in-depth comparison of Jupiter, Saturn, Uranus, and Neptune, incorporating data obtained from astronomical observations and planetary spacecraft encounters.

Kohlhase, Charles. "Meeting with a Majestic Giant: The Cassini Mission to Saturn." *The Planetary Report,* July/August, 1993, 5-11. A lucidly written, nontechnical prospectus of the mission by its Science and Engineering manager. Kohlhase describes the hardware, gravity-assisted course, Saturn orbit, Titan probe, and prospects for scientific discovery. The entertainingly written article is accompanied by dramatic photographs, drawings, and diagrams of the probe, its course, and the Saturn subsystem.

Kohlhase, Charles, and Craig Peterson, eds. *The Cassini/Huygens Mission to Saturn and Titan Pocket Reference: Postlaunch Update.* NASA JPL 400-711. Pasadena, Calif.: Jet Propulsion Laboratory, 1998. A forty-page illustrated update on the missions to Saturn and Titan. Some illustrations are in color.

Lorenz, Ralph, and Jacqueline Mitton. *Lifting Titan's Veil: Exploring the Giant Moon of Saturn.* London: Cambridge University Press, 2002. Written prior to the Huygens landing, this text reviews the state of understanding about Titan prior to the initiation of Cassini scientific returns. Anticipates the excitement of discovery that Huygens would bring.

Morrison, David. *Voyages to Saturn*. NASA SP-451. Washington, D.C.: National Aeronautics and Space Administration, 1982. A lovely book with many color photographs from space probes, providing richly detailed background information. Morrison summarizes the history of Saturn observations and recounts the Pioneer and Voyager missions to Saturn. He explains theories about the planet's development and that of its rings and moon clearly for general readers.

Morrison, David, and Tobias Owen. *The Planetary System*. 3d ed. San Francisco: Addison Wesley, 2003. Organized by planetary object, this work provides contemporary data on all planetary bodies visited by spacecraft since the early days of the Space Age. Suitable for high school and college students and for the general reader.

National Aeronautics and Space Administration, Jet Propulsion Laboratory. *Outward to the Beginning: The CRAF and Cassini Missions*. 2d ed. JPL 400-341. Pasadena, Calif.: Author, 1991. This pamphlet outlines the mission, hardware, and science investigations for Cassini and a related project, the Comet Rendezvous Flyby Mission (CRAF). It describes the probes, Saturn and its magnetosphere, Titan, the icy satellites, mission operations, and management and development. With color illustrations.

*Planetary and Space Science* 46, nos. 9/10 (1998). A special issue edited by P. Cerrini et al., "The Cassini/Huygens Mission to Titan and Saturnian System." Contains a selection of papers presented at two international symposia held in Bologna, Italy, in November, 1996, and Vienna in April, 1997. Includes illustrations and bibliographical references.

Spilker, Linda J., ed. *Passage to a Ringed World: The Cassini-Huygens Mission to Saturn and Titan*. Washington, D.C.: National Aeronautics and Space Administration, 1997. Edited by a Cassini deputy project scientist, this collection provides a preview of the Cassini-Huygens mission. Chapters include details about the spacecraft, as well as Saturn and its moon Titan. There are numerous photographs and illustrations, a glossary of terms, acronyms and abbreviations lists, and a bibliography.

*Roger Smith and David G. Fisher*

# Chandra X-Ray Observatory

*Date:* Beginning July 23, 1999
*Type of spacecraft:* Space telescope

*The Chandra X-Ray Observatory (CXO) was designed to image and measure the temperatures of extremely hot objects such as supernova remnants, neutron star accretion disks, and cosmic gas clouds. It is far more sensitive and capable of revealing much finer detail than previous x-ray telescopes.*

### Key Figures

*Edward Weiler,* NASA's associate administrator for Space Science
*Martin C. Weisskopf,* MSFC project scientist
*Robert Kirschner,* associate director, Harvard-Smithsonian Center for Astrophysics
*Harvey Tananbaum,* director of the Chandra X-Ray Center

### Summary of the Satellite

The Chandra X-Ray Observatory is named for Subrahmanyan Chandrasekhar (1919-1995), a Nobel Prize-winning astrophysicist who became famous for his studies of dense matter and general relativity. Martin Rees, Great Britain's Astronomer Royal, said that Chandra (as he was generally called) "probably thought longer and deeper about our universe than anyone since Einstein." Because the observatory's mission is to probe more deeply into cosmic x-ray sources than has ever been done before, it seems very appropriate to name it Chandra.

The moving-van-sized (13.4-meter-long) observatory was lifted into low-Earth orbit by the space shuttle *Columbia* on July 23, 1999. The observatory consists of three major parts: the x-ray telescope, the science instruments, and the spacecraft.

Solar arrays on the spacecraft provide approximately two kilowatts of power. The spacecraft has systems to control the temperatures of critical components such as the x-ray mirrors, which can be affected by temperature changes as small as 0.1 degree Celsius. Communication equipment in the spacecraft stores data from the science instruments, sends them to the ground, and receives commands from the ground. The spacecraft also

points the telescope and tracks the target object during observations.

After being deployed by the space shuttle, the spacecraft was boosted by its hydrazine-fueled rockets into an elliptical orbit with a low point of about 10,000 kilometers and a high point of nearly 140,000 kilometers, more than one-third the distance to the Moon. This orbit was chosen so that Chandra would be outside the Earth's radiation belts most of the time and so that the Earth would block less of the sky. Chandra's orbit allows it to observe an object continuously for up to fifty-five hours without the Earth blocking the view. In comparison, the Earth blocks the view of the Hubble Space Telescope for thirty minutes out of every ninety minutes. The disadvantage of such a high orbit is that shuttle crews cannot reach Chandra for repair operations.

The x-ray telescope is the heart of Chandra. Ordinary lenses or mirrors would simply absorb or transmit x-ray photons without focusing them. However, just as bullets ricochet when they strike stone at a grazing angle, x-ray photons ricochet when they strike metal at a grazing angle. Chandra uses two sets of grazing-incidence mirrors to bring x rays to a focus. Both sets consist of four cylinders nested one inside another like a set of nested measuring cups (without bottoms). The inside surfaces of the cylinders are subtly shaped so that x rays ricochet from the first set to the second set and then come to a focus. To maintain focus, the 2.7-meter-long mirror assembly must remain aligned to within 1.3 microns, about one-fiftieth the width of a human hair. The mirrors were polished to within a handful of atomic diameters of the ideal shape and are the smoothest and cleanest mirrors ever made. They are coated with iridium since it is very corrosion resistant and also a good x-ray reflector.

The telescope collects x rays and focuses them onto one of two science instruments: the High Resolution Camera (HRC) or the Advanced CCD Imaging Spectrometer (ACIS). The HRC has two microchannel plates. Each plate is a 10-centimeter-square cluster of 69 million lead-oxide glass tubes. These tiny tubes are only ten microns in diameter and 1.2 millimeters long, and it is their small size that allows Chandra to produce high-resolution images. When an x ray strikes a thin metal layer (the photocathode) at a tube's front surface, it knocks electrons free from the metal. Next, high voltage accelerates these electrons and gives them enough energy to free more electrons within the tube. Soon a cascade of electrons from a tube in the first plate strikes the second plate, which is placed behind the first plate. What began as a single electron from the photocathode of the first tube is now an avalanche of thirty million electrons at the rear of the second tube. Accurately recording the position of the tube that had the electron avalanche allows an image of the x-ray source to be constructed.

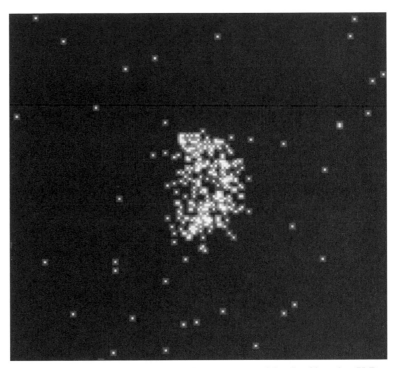

*The first x-ray image of the planet Venus, acquired by the Chandra X-Ray Observatory in January, 2001. The x radiation forms a half-crescent on the side oriented toward the Sun and is caused by the fluorescence from oxygen and other atoms in the Venusian atmosphere. (NASA/MPE/K.Dennerl et al.)*

The Advanced CCD Imaging Spectrometer uses an array of charge-coupled devices (CCDs) to record images. These CCDs are sophisticated versions of those used in camcorders. CCDs in camcorders convert visible light images into electronic signals that can then be stored in computer memory. The spectrometer part of the ACIS consists of two diffraction gratings that can be swung into position in front of the CCDs. The Low Energy Transmission Grating (LETG) is constructed with a series of fine gold wires placed parallel to each other and spaced 1.0 micrometer apart. The LETG is for x ray with energies from 0.08 to 2,000 electron volts (one thousand electron volts is a kiloelectron volt). The gold bars of the High Energy Transmission Grating (HETG) are only 0.2 micrometer apart and can analyze x rays with energies from 0.4 to 10 kiloelectron volts. Both gratings bend beams of x rays in a fashion similar to the action of a prism that sorts white light into the colors of the rainbow.

Energies for x rays are like colors for visible light. For example, suppose that a doughnut-shaped nebula were emitting x rays. ACIS would form a series of doughnut-shaped images—one image for each energy, or "color," emitted by the nebula. Because chemical elements emit x rays with energies characteristic of that element, ACIS can identify which elements are present in an x-ray source. ACIS can also measure temperatures since gases at higher temperatures emit higher-energy x rays.

### Contributions

Only objects that release copious amounts of energy are likely to emit x rays. A list of such objects includes hot young stars, supernova remnants, pulsars, black hole candidates, colliding galaxies, and quasars. All of these objects are of great interest to astronomers, and Chandra has produced several important discoveries concerning these objects. Chandra found nearly a thousand x-ray-emitting stars in the Orion nebula, a prolific stellar nursery only 1,500 light-years away. These x rays are thought to originate in violent flares on the surfaces of young, unstable stars. The Sun probably went through such a phase before it settled into its more stable, middle-aged state. Surprisingly, three of the faint x-ray sources are stars with only 5 percent the mass of the Sun. Such small stars lack enough mass to evolve into normal stars and are destined to become brown dwarfs.

Massive stars end their lives in the fiery cataclysm of a supernova explosion. So much energy is released in these explosions that a supernova may shine as brightly as an entire galaxy for days or even for weeks. When a star goes supernova, the outer parts of the star are flung into space as roiling clouds (nebulae) of hot gas and dust. The star's core becomes a pulsar, a rapidly spinning neutron star, or if it is massive enough, it becomes a black hole candidate. Chandra has examined several supernova remnants, including the Crab nebula and pulsar. Chandra produced an absolutely stunning image of the Crab nebula, an image that brought cheers from scientists when they first saw it because it clearly showed features that had been predicted for years but were only hinted at in previous images. The Chandra image shows the pulsar with powerful jets shooting outward from the poles and surrounded by rings of high-energy particles thrown outward at near light speed. The largest ring is two light-years in diameter.

Imaging Cassiopeia A, another nebula that is the remnant of a supernova, Chandra identified silicon, sulfur, and iron, elements that play key roles in the evolution of massive stars. Chandra's sharp resolution shows a small bright spot at the center of the nebula that is thought to be the neutron star. If so, this is the first time it has been detected. Chandra found oxygen and neon in the supernova remnant E0102-72 in the Small Magellanic Cloud, 190,000 light-years from Earth. Although it is a thousand years old, the temperature of the remnant nebula is still several million kelvins. Another target clearly seen by Chandra, PSR 0540-69, is a pulsar that rotates twenty times per second. A highly collapsed object, it probably has about twice the mass of the Sun but is no larger than a small city.

Peering at the center of our own galaxy, the Milky Way, Chandra is the first x-ray telescope to unambiguously image Sagittarius A* (pronounced "A star"), a 2.6-million-solar-mass black hole candidate that marks the galactic center. Curiously, a much fainter x ray discovered that the cloud surrounding the thirty-million-solar-mass black hole candidate in the nearby Andromeda galaxy has a temperature of only a few million kelvins. This is cooler than expected.

Hydra A is a cluster of galaxies 840 million light-years away in the direction of the constellation Hydra. Chandra shows that this cluster is permeated by a huge cloud of very hot gas, 35 to 40 million kelvins. Outer parts of this cloud have been shaped by magnetic fields into loops and tendrils. Similarly, Chandra shows that the galaxy cluster associated with radio galaxy 3C295 is filled with a vast cloud of fifty-million-kelvin gas. Because this cluster is five

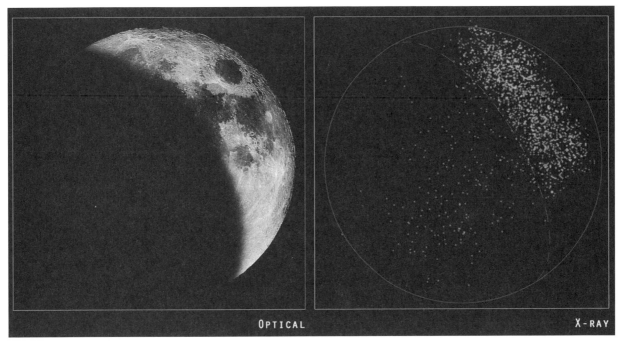

OPTICAL

X-RAY

*A comparative view of Earth's moon from an optical telescope (left) and an x-ray image from the Chandra instrument. The Chandra data have revealed that x radiation once thought to originate from the Moon actually comes from Earth's geocorona, the extreme outer atmosphere through which orbiting spacecraft move.* (optical: Robert Gendler; X-ray: NASA/CXC/SAO/J.Drake et al.)

billion light-years away, scientists see it as it was five billion years ago. Even more distant, at six billion light-years away, the luminous quasar PKS 0637-72 displays a powerful x-ray jet extending several hundred thousand light-years outward from its core. The jet is probably a beam of extremely high-energy particles.

### Context

Chandra is one in a series of x-ray satellites, but it can see objects ten times fainter and resolve details twenty-five times smaller than its predecessors. Uhuru, meaning freedom in Swahili, was launched on December 7, 1970, to mark the seventh anniversary of Kenyan independence. Uhuru was the first satellite devoted to the study of nonsolar x rays. The Einstein Observatory, launched in 1979, was the first satellite with focusing x-ray mirrors enabling it to see fainter sources. Launched in 1990 (by Germany, the United Kingdom, and the

United States), ROSAT became the first satellite to make an all-sky survey with an imaging telescope, while ASCA, launched in 1993 by Japan and the United States, was the first to use the new-generation x-ray detectors, CCDs.

Three greatly improved x-ray observatories were planned to build on the legacy of these previous satellites: Chandra, X-Ray Multi Mirror (XMM-Newton, launched by the European Space Agency and the United States), and Astro E (launched by Japan and the United States). Astro E would have been able to detect very high-energy x rays up to 700 kiloelectron volts, but it failed to reach a usable orbit; a replacement facility, the Astro E-2, also called Suzaku, launched on July 10, 2005. XMM-Newton was successfully launched December 10, 1999, and is performing beautifully. XMM-Newton can detect fainter sources than Chandra, but Chandra has better resolution, enabling it to record finer details. Chandra is the third of NASA's

Great Observatories series. The other Great Observatories are the Hubble Space Telescope, launched in 1990; the Compton Gamma Ray Observatory, launched in 1991; and the Spitzer Space Telescope, launched in 2003.

Chandra's results often confirm what previously had only been suspected, but occasionally completely unanticipated discoveries are made. The facts that the Milky Way's black hole candidate emits fewer x rays than expected and that the environment of Andromeda's black hole candidate is cooler than expected strongly suggest that the processes involved are more complex and less well understood than previously supposed.

To see what faint sources might be present, scientists pointed Chandra at a small patch of sky in the direction of the constellation Canis Venatici and collected data for 27.7 hours. Since the early 1960's scientists have known that space is filled with a faint x-ray glow, but they did not know if that glow came from very hot diffuse gas spread throughout the universe or if it came from a large number of discrete sources. ROSAT had previously shown that much of the lower-energy x-ray background comes from distant objects such as quasars or active galactic nuclei (AGNs). AGNs are thought to be supermassive (billion-solar-mass) black hole candidates that are rapidly accreting more mass. Mass spiraling inward forms an accretion disk about the black hole candidate, and the gravitational energy released heats the disk so that it emits gamma rays, x rays, and visible light. With better resolution and sensitivity, Chandra confirmed the ROSAT result and extended it to higher-energy x rays. Most of the x-ray background does come from discrete sources. Chandra found that about one-third of the sources are AGNs with brightly shining cores, but that another third of the x-ray sources are galactic nuclei that emit little or no visible light from their cores. Perhaps dust or gas surrounding their cores blocks visible light. If so, there may be tens of millions of similar objects over the whole sky, and the optical surveys of AGNs are very incomplete. Chandra found that the final third of the x-ray sources are in ultrafaint galaxies, galaxies that are barely detectable, if at all, in visible light. If they are so faint because they are far away, they would be among the most distant objects ever discovered.

**See also:** Cassini: Saturn; Compton Gamma Ray Observatory; Explorers: Astronomy; Galaxy Evolution Explorer; Galileo: Jupiter; Gamma-ray Large Area Space Telescope; High-Energy Astronomical Observatories; Hubble Space Telescope; Infrared Astronomical Satellite; Orbiting Astronomical Observatories; Space Shuttle Flights, 1999; Space Shuttle Mission STS-93; Spitzer Space Telescope; Swift Gamma Ray Burst Mission; Telescopes: Air and Space.

### Further Reading

Cowen, R. "X-Ray Data Reveal Black Holes Galore." *Science News*, January 15, 2000, 36. A brief article about Chandra's stunning discovery that the x-ray cosmic background glow comes from numerous point sources. Most of these sources are thought to be associated with massive black hole candidates at the centers of galaxies.

Fabian, A. C., K. A. Pounds, and R. D. Blandford. *Frontiers of X-Ray Astronomy.* London: Cambridge University Press, 2004. Discusses the revolution in research provided by the Chandra and Newton X-Ray Observatories, and those spacecraft that preceded them. Puts the data in cosmological context.

Haisch, Bernhard, and Jurgen Schmitt. "The Solar Stellar Connection: How Other Stars Are Shedding Light on the Mysteries of Solar Activity." *Sky and Telescope* 9 (October, 1999): 46. Tells how studying the magnetic behavior of stars, including through the use of x-ray telescopes, teaches us about the Sun.

Hawley, Steven A. "How We'll Deliver Chandra to Orbit." *Sky and Telescope* 2 (August, 1999): 54. A brief article on how the space shuttle and crew planned to launch Chandra into space. The follow-up article on Chandra's successful launch is "Collins and *Columbia* Launch Chandra," *Sky and Telescope*, October, 1999, 16.

Schlegel, Eric M. *The Restless Universe: Understanding X-Ray Astronomy in the Age of Chandra and Newton*. London: Oxford University Press, 2002. An accessible introduction to the history, methods of observation, and data of x-ray astronomy. Provides data from the Chandra, Newton, and Astro-E missions as well as from spacecraft that preceded them.

Still, Martin. "X-Ray Astronomy's Golden Age." *Sky and Telescope* 2 (August, 1999): 56-57. A comparison of the capabilities of Chandra, XMM-Newton, and Astro E.

Tucker, Wallace H., and Karen Tucker. *Revealing the Universe: The Making of the Chandra X-Ray Observatory*. Cambridge, Mass.: Harvard University Press, 2001. In addition to providing a detailed history of the Chandra program, this work explains the basic physics involved in x-ray astronomy.

Wali, Kameshwar C. *Chandra: A Biography of S. Chandrasekhar*. Chicago: University of Chicago Press, 1991. A lively, warm, sensitive, and revealing biography of the rather private but remarkable person who gave his name to the observatory. Readers will see why he deserves the honor of having the satellite named for him.

*Charles W. Rogers, updated by David G. Fisher*

# Clementine Mission to the Moon

*Date:* January 25 to May 10, 1994
*Type of program:* Lunar exploration

*The Clementine mission to the Moon was designed to employ evolving state-of-the-art technologies within a basic low-cost budget. It demonstrated the feasibility of cooperation between the Department of Defense and civilian space agencies. Scientifically, it accomplished the first complete mapping of the Moon, and renewed lunar exploration after a dearth of nearly two decades of new data.*

## Summary of the Mission

The Clementine mission to the Moon was the result of cooperation in space exploration between the Department of Defense and civilian space agencies. In the seventy days that Clementine operated in lunar orbit, it recorded more than 1.5 million images of the Moon's surface. Clementine also provided data to help determine diversity in lunar rock compositions and to search for the presence of water-ice in the polar regions.

Clementine was the latest, at the time, in a long line of spacecraft to visit the Moon. The United States launched three probes toward the Moon between October 11, and December 6, 1958. These first three Pioneer spacecraft either failed to reach the Moon or suffered from a launch systems failure. It would be the Soviet Luna 1 that first flew by the Moon at a distance of 6,000 kilometers in early January, 1959. Later that year in September, Luna 2 hit the Moon, and in October, Luna 3 returned the first picture of the lunar far side. The quality of the photograph was poor, but it hinted at the discoveries that were to come.

The "Race to the Moon," which began in early 1961, was a systematic approach to exploration that challenged the scientists of both the American and Soviet space programs. Large booster rockets had to be built and advanced computer technology had to be developed in order to place a human on the Moon. It would be a series of robot spacecraft that would pave the way. In total there were forty-nine launches to the Moon before the crew of Apollo 8 entered lunar orbit in December, 1968. Twenty-seven others would follow, which included the six crewed landings, uncrewed Soviet rovers and sample return vehicles, and the Jupiter-bound Galileo spacecraft.

The United States' Apollo Program to the Moon ended when astronauts Eugene A. Cernan and Jack Schmitt left the lunar surface in December, 1972, near the end of the Apollo 17 mission. As they lifted off, neither man was certain when the next person would return. Certainly, six crewed missions to the lunar surface and a couple of automated rovers and sample return missions did not answer all questions about the Moon. Many important questions remain unanswered. Was the Moon formed from the debris created from a large impact event with the Earth? Does the Moon have natural resources that could be mined for industrial use? Or simply, is there water-ice on the Moon? The immense cost of crewed missions to seek answers to questions like these seems prohibitive. National debts and the high cost of technology have hurt the space programs of many nations. International cooperation and more economical spacecraft and science packages seem to be the only sensible approach to future lunar and planetary exploration.

Clementine had an unlikely beginning for a lunar science mission since it was originally designed to be part of the Strategic Defense Initiative Organization (now called the Ballistic Missile Defense Organization). Designed to test various electronic components and tracking sensors, the spacecraft would be placed into lunar orbit, where it would have to operate under the harsher conditions of deep space. Its primary mission would be to monitor and track stars as a simulation of acquiring enemy missiles in flight. The ever-changing world political climate put an end to the "Star Wars Initiative," and a new approach had to be found in order to develop new spacecraft technology. Basic science would prove to be the logical partner for the military. A lunar mapping mission and a rendezvous with an asteroid would provide the technological challenge the military was looking for—Clementine was the result.

Originally named the Deep Space Program Science Experiment, the name Clementine was chosen for simplicity and to give the mission a personality. As spacecraft costs go, Clementine was a very inexpensive mission. The actual cost of the spacecraft was $55 million, and the Titan II-G launch vehicle added $20 million more. Even Mission Control was a model of low-cost efficiency. It was housed in a remodeled warehouse in Alexandria, Virginia, and was staffed by a small select group of experts. Both Mission Control and the spacecraft were deemed "a lean, mean team." That phrase carried a great deal of truth in it.

Specific details of the Clementine spacecraft indicate that it weighed approximately 1,365 kilograms at launch. This included the solid rocket engine that would deliver it to the Moon. The scientific package included six cameras and a laser transmitter. Primary to the mission were the two wide-field cameras for the star tracking experiment. In addition, an ultraviolet/visible camera along with a near infrared and a long wave infrared camera would feature prominently in the mission. Completing this array would be a high-resolution camera for detailed examination of features as small as 10 meters wide. The laser transmitter was used for measuring distance from the spacecraft to the lunar surface. This would provide precise data to develop accurate topographic maps of the Moon. All data transmitted from the spacecraft would be collected at the National Aeronautics and Space Administration's (NASA) Deep Space Network and sent along to Mission Control in Virginia for analysis.

The launch date for Clementine took place on January 25, 1994, and it was placed into low-Earth orbit. For the next several days Clementine would undergo rigorous testing in preparation for its departure to the Moon. During these first few days two problems arose that were potentially threatening to the success of the mission. The most dangerous was the excessive use of the onboard batteries and the resulting power drain. Fortunately these problems were quickly resolved, and Clementine left Earth orbit for the Moon on February 3, 1994.

Clementine entered lunar orbit on February 19, 1994, and began its mapping mission one week later. Five times each day it would transmit more than five thousand images from its 1.9 gigabyte computer memory. This activity continued daily until the lunar survey ended on April 21, 1994. At that time Clementine had completely mapped the lunar surface from its south to north polar orbit. Clementine's orbital path had been highly elliptical, with its perilune (closest point to the Moon) being only 400 kilometers and its apolune (farthest point from the Moon) at 2,900 kilometers. From such an orbit, Clementine provided detailed images of the Moon's polar regions. Earlier missions had concentrated on the Moon's equatorial region in preparation for the Apollo landings. The Jupiter-bound Galileo spacecraft also photographed the lunar poles, but not with the detail of Clementine.

A very important part of the Clementine mission was the planned rendezvous with the earth-crossing asteroid 1620 Geographos. Previously the Galileo spacecraft encountered the asteroids 951 Gaspra and 243 Ida. In both cases these brief flybys provided scientists with many surprises, notably the discovery of Ida's moon Dactyl. It was hoped that Clementine would not only test its navigational and intercept capabilities but also return extensive data that would answer many basic questions about asteroids. One persisting question is, how do the main

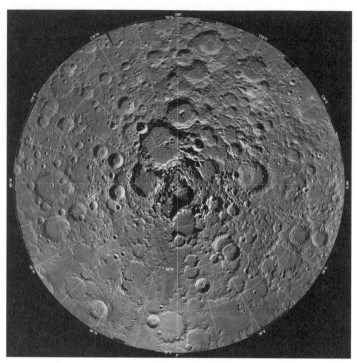

*This image of the north polar region of Earth's Moon was assembled from approximately 1,500 images taken by Clementine.* (NASA-JPL)

belt asteroids differ from the earth-crossing asteroids? One theory suggests that the earth-crossing asteroids may be the nuclei of old spent comets. Clementine may have answered that question.

On May 5, 1994, Clementine's main engine fired and it left lunar orbit for parking orbit around Earth. There it would be prepared for its voyage to Geographos. There was even hope that a second asteroid rendezvous could be made. However, mission controllers' hopes were shattered when, on May 7, a serious accident occurred. An onboard computer malfunction caused four of Clementine's attitude control thrusters to fire continuously for eleven minutes until they ran out of gas. This put the spacecraft into an eighty-revolution-per-minute spin that rendered the asteroid rendezvous mission impossible because of lack of control for positioning the spacecraft.

Although Clementine's mission to an asteroid was over, the spinning spacecraft's life was far from finished. The main engine was fired once again to

slow down the spin rate and to regain some control. This was done by using its reaction wheels. These are flywheels that are used to maintain or change attitude control of the spacecraft in conjunction with its thrusters. Help was also to come from the Earth's magnetosphere as it exerted a slowing effect on the spacecraft. Once stabilized, it was hoped that Clementine would resume its primary objective of testing its electronic components, but fate took a turn. Clementine's highly elliptical orbit placed it under the continuous pull of the Moon's gravity. This unexpected force altered the angle of sunlight striking Clementine's solar panels, and that led to a drastic reduction in available power. Power levels eventually reached zero, and all communications and computer functions ceased. As time passed, the Moon's gravitational influence pulled Clementine into a solar orbit that would never allow it to return home. Yet all may not have been lost. Hopes were high that once Clementine's batteries recharged, contact could be restored, and its systems monitoring would resume as planned.

### Contributions

Thousands of photographs of the lunar surface had been returned by the Ranger, Surveyor, Lunar Orbiter, and Apollo missions long before the Clementine mission. Continuous study of the almost 400 kilograms of lunar soil and rock have provided scientists with a wealth of information about our nearest neighbor in space, but many gaps exist in the database. Clementine was designed to close many of those gaps.

One of the many unanswered questions about the Moon is why the near side is dominated by the large dark maria basins, while the far side is almost devoid of them. The far side is basically the lighter-colored highland terrain. A relatively simple explanation can be found in comparative crustal thickness. A thinner crust is more susceptible to large impacts generating massive outpourings of lava than a thicker crust. Previous data indicated such a variation in the thickness between the near and far sides, with the near side being the thinner. The thicker far side crust would logically inhibit large eruptions of lava following an impact event. Clementine data confirmed the overall near side-thin, far side-thick crustal relationship for the Moon, but it also indicated that there are some exceptions. Some of the thinnest crust can be found in isolated areas on the far side, especially near the South Pole-Aitken Basin. These determinations were made from a combination of topographic and gravity measurements. Additional analysis of Clementine data is needed to further define the near side-far side controversy.

Geologically, Clementine achieved a first by successfully completing a multispectral mapping of the entire Moon, using eleven different wavelengths. It digitally imaged the entire lunar surface under constant geometry and lighting conditions.

This data, combined with the high-resolution altimetry, should provide detailed information concerning the mineral and rock variations across the lunar surface. The multispectral images provided by Clementine have made it possible to develop a global map of lunar mineral variations based on low-resolution color ratios. This has been demonstrated well with the analysis of the large-impact crater Aristarchus. Preliminary findings show significant variation in rock compositions when associated with depth within the crater. The near-surface compositions tend to be more suggestive of an anorthositic (feldspar-rich) rock when compared with the deeper level's more gabbroic (higher iron and magnesium content) rock. Information such as this provides geologists with a greater understanding of the early crustal development of the Moon.

In consideration of the technological aspects of the mission, Clementine has provided new information about spacecraft systems, design, and their

*An enhanced Clementine image of sunrise on the Moon with Venus in the background.* (Photo/image provided courtesy of the Naval Research Laboratory)

potential capabilities. Among the many innovative components tested are the superfast computer for rapid processing of images, new and more efficient solar cells, and the use of an entirely new nickel-hydride battery. It was this battery that initially saved the mission while still in Earth orbit. When the power level dropped dangerously low, its recharge capability permitted the mission to continue. The older nickel-cadmium version would not have survived the incident.

### Context

Clementine was the first of a new breed of spacecraft to visit the Moon. By its very design it was a multipurpose spacecraft that would test the reliability of its electronic components in deep space and then use those same components to collect valuable scientific data about the Moon and an asteroid. Its success has proven the value of low-cost/high-technology spacecraft that are adaptable for a variety of missions. It is also important to note that even when malfunctions threatened the spacecraft, it was able to circumvent the problems with alternative procedures. After two major technical setbacks, Clementine was still capable of fulfilling its primary mission. Only the ubiquitous force of gravity finally defeated the spacecraft and sent it into solar orbit far from home.

The history of space exploration will show that the Clementine mission represented the beginning of cooperative research between the military and civilian space programs. When government budgets were less restrictive and money was plentiful, it was common practice for each sector to develop its own spacecraft. Quite often they would develop similar spacecraft with only the slightest difference in design to accommodate their specific needs. This redundancy was and still is very costly and inefficient. Presently, with very restrictive budgets, it has become apparent that some form of

standardization is the only answer to continued space exploration. The design of low-cost basic spacecraft that can be easily modified to suit a particular need is the logical approach. This is evident in NASA's *Discovery* series spacecraft.

The Clementine mission was the first United States probe to the Moon since Explorer 49 was launched on June 10, 1973. After landing twelve astronauts on the Moon, the United States virtually gave up direct lunar exploration. Experiments left on the Moon by the astronauts returned valuable data about moonquakes for several years, but they were eventually turned off as a cost-cutting measure. Clementine seems to have reversed those years of neglect by returning so much data for such a relatively small cost. It has proven that there is still much to be learned from lunar exploration, and that information will be vital if humankind is ever to establish a permanent presence on the Moon.

As the data from Clementine are analyzed and the results become public, a greater awareness of the value of lunar research will become evident. What Clementine has done for the Moon has not yet been done for the Earth. Complete global maps such as those produced by Clementine do not exist for the Earth. It is extremely important that we apply such technology for a global inventory of our own natural resources and then implement a sensible program for their management. The value of space exploration and the knowledge we gain of other worlds continues to be of benefit to us in our own world.

**See also:** Apollo Program; Apollo Program: Geological Results; Apollo Program: Orbital and Lunar Surface Experiments; Apollo 1; Apollo 8; Apollo 17; Jet Propulsion Laboratory; Lunar Exploration; Lunar Orbiters; Lunar Prospector; Planetary Exploration; Space Centers, Spaceports, and Launch Sites; Strategic Defense Initiative.

### Further Reading

Goldman, Stuart J. "Clementine Maps the Moon." *Sky and Telescope* 87 (August, 1994): 20-24. This is a very well written article that describes the Clementine mission from its inception to its eventual demise in lunar orbit. Included in the article are brief explanations of the science experiments and the operation of the spacecraft. It is written for a general audience.

Graham, David. "Fuel for Fusion Power Abundant on Lunar Maria." *Astronomy* 27 (November, 1999). Images from the Clementine mission reveal rich caches of helium-3 on the Moon's surface. The rare gas is an efficient fuel for nuclear fusion.

Guest, John. *Geology on the Moon.* New York: Crane, 1977. Somewhat dated, this book describes the state of knowledge about the Moon as advanced by Apollo Program exploration and investigations.

Hartmann, William K. *Moons and Planets.* 5th ed. Belmont, Calif.: Thomson Brooks/Cole Publishing, 2005. Provides detailed information about all objects in the solar system. Suitable on three separate levels: high school student, general reader, and the college undergraduate studying planetary geology.

Lewis, John S. *Physics and Chemistry of the Solar System.* San Diego: Academic Press, 1995. This is a comprehensive survey of planetary physics and the physical chemistry of the solar system. Chapter 9 provides very good basic information about the Moon. The book contains a wealth of good references for further study. It is best suited for the college level or professional scientist.

McBride, Neil, and Iain Gilmour. *An Introduction to the Solar System.* New York: Cambridge University Press, 2004. This work provides a comprehensive tour of the solar system. Suitable for a high school or college course on planetary astronomy.

Pater, Imke de, and Jack J. Lissauer. *Planetary Science.* New York: Cambridge University Press, 2001. This is an advanced text about the physical, chemical, and geological processes at work in the solar system.

Spudis, Paul D. *The Geology of Multi-Ring Impact Basins: The Moon and Other Planets.* London: Cambridge University Press, 2005. Discusses the formation of huge basins by large impacting bodies and their evolution. Comparisons are made between lunar features and similar structures on other bodies such as Mercury and moons in the outer solar system.

United States Geological Survey. *Mission to the Moon: Clementine.* Washington, D.C.: National Aeronautics and Space Administration, 1997. This is a series of fifteen computer compact discs containing a mosaic of the Moon (more than 43,000 images partitioned into fourteen geographic zones). The images were taken by the Clementine spacecraft at a resolution of 100 meters per pixel.

Wagner, Jeffrey K. *Introduction to the Solar System.* Philadelphia, Pa.: Saunders College Publishing, 1991. A very good basic introduction to the varied topics that deal with the solar system. Chapter 15 is devoted to the Moon and offers an excellent overview of lunar science. Recommended for college level and the more advanced students of planetary science.

Wilhelms, Don E. *To a Rocky Moon: A Geologist's History of Lunar Exploration.* Tucson: University of Arizona Press, 1993. This book gives the reader an excellent perspective on lunar exploration from its earliest beginnings to the drama of the Apollo landings. Although it is written from a scientist's viewpoint, it is very readable for the average person interested in astronomy. Definitely suitable for a general audience.

*Paul P. Sipiera*

# Compton Gamma Ray Observatory

*Date:* April 5, 1991, to June 4, 2000
*Type of spacecraft:* Space telescope

*The Compton Gamma Ray Observatory (CGRO or simply GRO), the most massive robotic civilian spacecraft the United States has ever built, has provided dramatic new insights into some of the highest-energy phenomena in the universe.*

## Summary of the Satellite

Signals that emanate from some of the most energetic and exotic phenomena in the universe may traverse great distances, even billions of light-years (a light-year is the distance light travels in a year, about 10 trillion kilometers), only to be absorbed by the thin atmosphere that veils Earth. This deprives scientists on the ground of the opportunity to unravel the information these signals carry. To counter this, systems for detecting them can be flown in stratospheric balloons or in space.

When space shuttle mission STS-37 lifted off at 9:22 A.M. eastern standard time or 14:22 Coordinated Universal Time (UTC) on April 5, 1991, its principal payload was a spacecraft designed to provide its detectors a view of the universe unobscured by Earth's atmosphere. Stowed in *Atlantis*'s payload bay was the 15,620-kilogram Gamma Ray Observatory (renamed the Arthur Holly Compton Gamma Ray Observatory the following September), the most massive robotic civilian spacecraft the United States had ever built.

Two days after reaching orbit, the astronauts used the orbiter's mechanical arm to raise the spacecraft above the payload bay. When its solar arrays unfolded, they extended more than 21 meters from tip to tip. Later, after astronauts conducted an unscheduled spacewalk to release a balky antenna boom, ground controllers determined that Compton was ready to begin its mission. The astronauts released the arm's grasp on the satellite at 10:37 UTC on April 7, and *Atlantis* slowly moved away.

During the next few weeks, controllers continued to check out the spacecraft and to activate its sophisticated instruments designed to detect gamma rays, the highest energy form of electromagnetic radiation. Visible light, extending from a wavelength of about 400 nanometers (one nanometer is a billionth of a meter) for violet light to 700 nanometers for deep red, is the small region of the electromagnetic spectrum that human eyes detect. This light exhibits some properties of waves and some of particles. Longer wavelengths extend from infrared to microwaves through radio waves, where the wavelike properties dominate. At shorter wavelengths, the resemblance to particles is more pronounced, and it becomes more convenient for physicists to refer to individual packets of electromagnetic energy, known as photons, rather than waves. The energy of a photon of violet light at a wavelength of 400 nanometers is about 3 electron volts. At higher energy (shorter wavelength), the spectrum turns to ultraviolet and then to x rays. Photons of the highest energies, greater than a few hundred thousand electron volts, are known as gamma rays.

Gamma rays travel from their origins largely unaffected by intervening gravitational, electric, or magnetic fields, and they can penetrate interstellar dust and other obstacles before they reach Earth. Thus, they can bring undistorted information on the high-energy conditions that created them. Yet this useful property is responsible for an equally significant difficulty. Because gamma rays do not interact readily with matter, they cannot be focused or reflected like visible light, and it is difficult to detect and measure them. Lower-energy photons can be detected with small sensors (as small as, or even much smaller than, human eyes), but scientists have to construct large sensors to capture gamma rays. Compounding the problem, the number of gamma-ray photons from celestial sources is minute compared with visible photons, so it becomes even more important to build large detectors in order to make sensitive measurements. To perform its mission, Compton carried a scientific payload of about 6,000 kilograms.

Compton's four complementary instruments were built to make accurate measurements of the energy, intensity, and source location of gamma rays. The instruments made use of the findings of Arthur Holly Compton, who shared the Nobel Prize in Physics in 1927 for his research on how high-energy photons can interact with matter.

Some of the observatory's instruments incorporated special materials that would scintillate (produce a small flash of light) when a gamma ray interacted with them. Sensors trained on these scintillators then could convert the flash to an electronic signal. Another detector, known as a spark chamber, allowed the gamma rays to pass through layers of metallic foil. In doing so, the photons would produce pairs of electrons and their antimatter counterparts, positrons. In traveling through the gas-filled chamber, these particles would cause sparks, leaving trails that detectors could record.

The Imaging Compton Telescope (COMPTEL) was designed to detect gamma rays in the range

*The Compton Gamma Ray Observatory grappled by the Remote Manipulator System (robotic arm) of the shuttle during mission STS-37 in April, 1991.* (NASA)

from 1 to 30 megaelectron volts, forming images of the sources and measuring their spectra, or the number of gamma rays at different energies. COMPTEL stood over 2.6 meters high on Compton and was more than 1.7 meters in diameter. The instrument could view a patch of sky 64° wide, but because it was so large and had a mass of 1,324 kilograms, it could not be pointed separately from the spacecraft. Compton had to point COMPTEL in order for the instrument to make its measurements.

Starting its measurements where COMPTEL began to taper off in sensitivity, the Energetic Gamma Ray Experiment Telescope (EGRET) measured photons from 20 megaelectron volts up to 30 billion electron volts (or 30 gigaelectron volts). Combining a spark chamber with a scintillator enabled scientists to reconstruct the intensity, spectrum, and direction of the gamma rays within the 45° patch of sky EGRET could monitor. The 1,830-kilogram instrument was aligned with COMPTEL so that simultaneous measurements could be made across the gamma-ray spectrum.

In contrast to COMPTEL and EGRET, the Oriented Scintillation Spectroscopy Experiment (OSSE) could point without the entire spacecraft moving. It contained four assemblies of scintillators that could be separately oriented. Making its principal measurements from 50,000 electron volts (one thousand electron volts is a kiloelectron volt) to 10 megaelectron volts, it could be pointed rapidly to respond to unexpected transient events without disrupting the observing program of COMPTEL and EGRET. Although the 1,810-kilogram OSSE performed comprehensive observations in this lower energy range, it did have some capability to make measurements up to 250 megaelectron volts. Each OSSE spectrometer had a field of view of 4° by 11°.

The Burst and Transient Source Experiment (BATSE) consisted of eight identical 95-kilogram modules on the corners of the spacecraft, allowing coverage of the entire sky with Compton in any orientation. Principally spanning the range from 30 kiloelectron volts to 1.9 megaelectron volts, BATSE could measure the variation of gamma rays from sources that changed in brightness in as little as one ten-thousandth of a second. This capability was needed for searches for enigmatic phenomena known as gamma-ray bursts. Producing flashes of gamma rays that persist for a few thousandths of a second to a few minutes, these cosmic sources were among the key mysteries Compton was designed to unlock. When BATSE detected a gamma-ray burst, in addition to making its own measurements, it could signal the other instruments so they could interrupt their observations and switch to modes optimized for studying brief bursts of gamma rays.

By May 16, 1991, the spacecraft was ready to begin its scientific observations. Through November 17, 1992, COMPTEL and EGRET performed the first complete survey of gamma-ray sources from 1 megaelectron volts to 30 gigaelectron volts. With Compton pointing the instruments in one direction for about fourteen days at a time, they would measure the sources within their fields of view as OSSE observed those sources and others. BATSE kept watch on the entire sky. Following this initial reconnaissance, Compton was devoted through September 7, 1993, to making follow-up observations of the interesting sources discovered in its first phase and to viewing some new targets. For about a year after that, COMPTEL and EGRET made very thorough, very sensitive observations of a few selected targets. During its continuing mission, Compton undertook observations of a wide variety of gamma-ray-emitting objects.

Compton stored most of its observations on tape recorders before relaying the data to scientists through the National Aeronautics and Space Administration's (NASA) orbiting Tracking and Data-Relay Satellite System (TDRSS). The observatory carried two recorders for increased reliability, but both developed problems. By March, 1992, they were adding significant noise to the precious data. Rather than conduct an expensive space shuttle repair mission, NASA decided to have Compton transmit its data through TDRSS as soon as they were collected. This led to a reduction in its scien-

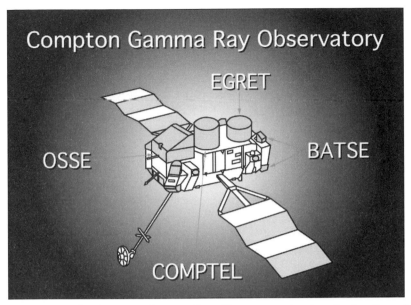

*Diagram of the Compton Gamma Ray Observatory showing the location of the telescopes COMPTEL and EGRET and the experiments BATSE and OSSE.* (P. J. T. Leonard/NASA/GSFC)

tific productivity, because TDRSS was not always available or within view of Compton. In March, 1994, NASA opened a new TDRSS ground station in Australia and repositioned one of the TDRSS satellites, dedicating it exclusively to Compton's data relay. With these modifications, Compton could return a substantial volume of data as soon as they were collected. This change turned out to provide an unexpected benefit, as it allowed scientists to respond rapidly to unanticipated discoveries that Compton made. When the spacecraft reported that it was detecting a gamma-ray burst, scientists could immediately begin turning other spacecraft and ground-based instruments with sensitivity in visible light and other wavelengths to the location Compton identified.

Compton began its mission in orbit 455 kilometers above Earth. If it had been much higher, Earth's radiation belts would have caused interference in the detectors. Small but continuous friction with Earth's tenuous outer atmosphere caused Compton to fall gradually to 346 kilometers by October, 1993. Without further action by controllers, by the following April the spacecraft

would have fallen to 290 kilometers, where there would have been too much friction for it to remain stable. Despite several problems with Compton's propulsion system, the orbit was raised in stages so that by December, 1993, it was at an altitude of 452 kilometers, ready to continue adding clues to the mysteries of the gamma-ray sky.

Early in 2000, NASA decided to end the mission of the Observatory and have it reenter the atmosphere on June 4, splashing down in the Pacific Ocean on the equator approximately 4,000 kilometers southeast of Hawaii. One of its gyroscopes had failed, and its control fuel was nearly depleted. Rather than risk having it fall intact as a 17,000-kilogram missile, the agency brought it down on a trajectory that caused it to break up during its fiery return. In this first deliberate and controlled crash of a satellite, NASA engineers directed the Compton through a series of retrograde rocket firings that dropped it from a high orbit and sent it plunging destructively through the upper atmosphere. A large portion of the spacecraft vaporized, and most of the pieces that survived to impact at sea were about the size of a pea or a grain of sand. Although large chunks were expected, none were ever found. More important, no one at sea was harmed. During its nine-year flight, the telescope had detected more than four hundred gamma-ray sources, ten times more than were previously known. Compton recorded more than twenty-five hundred gamma-ray bursts; before Compton, only about three hundred had been detected.

### Contributions

Compton provided data for scientists studying a great variety of high-energy phenomena. Its instruments located and measured atomic nuclei in outbursts from the Sun, radioactive elements expelled into space in supernova explosions, the mutual annihilation of matter and antimatter, and exotic objects such as pulsars, quasars, and possibly black holes.

Gaining an understanding of gamma-ray bursts was one of the primary goals of the mission, with BATSE always on the alert. Compton recorded an average of nearly six bursts per week throughout its mission. As more and more were detected, scientists were surprised to discover that the sources were evenly distributed across the sky. It was believed that the sources of these bursts were in the Milky Way galaxy. Yet the bursts detected by Compton came from all directions in the sky, not only those regions scientists would expect for objects in the galaxy.

Occasionally a gamma-ray burst occurred within the field of view of COMPTEL and EGRET, the instruments that cannot move quickly. The largest such burst occurred on Sunday, January 31, 1993, and became known as the Super Bowl Burst. It was ten times brighter than any other yet detected and one hundred times brighter at its peak than the brightest steady source in the Milky Way. Despite searches for counterparts to gamma-ray bursts at other energies, no objects were definitively identified using optical and radio telescopes. As Compton has added to the catalog of gamma-ray bursts, scientists continue to test, and in some cases discard, theories for their origins.

An unexpected discovery showed that even sources closer than the celestial targets Compton was designed to study could reveal new aspects of themselves when viewed in this end of the electromagnetic spectrum. Scientists were surprised and puzzled to detect bursts of gamma rays lasting a few thousandths of a second from high in the Earth's atmosphere above large thunderstorms.

The flexibility in Compton's design allowed the regular observing program to be interrupted when needed. In June, 1991, for example, the Sun gave evidence of increasing activity. On June 8, Compton's measurements of targets in the constellation Cygnus were suspended so it could be reoriented

for a week to focus its attention on the Sun. It was rewarded with several large solar flares (temporary outbursts of high-energy ions, subatomic particles, and photons). In addition to seeing the resulting gamma rays, COMPTEL was able to measure neutrons that emanated from the Sun.

One of the surprises from EGRET's survey of the sky was an extraordinarily strong gamma-ray source at a distance of about four billion light-years, a significant fraction of the size of the universe. Eventually scientists found more than forty of these objects, which emit extremely high levels of energetic gamma rays. Forming a new class of galaxies, now known as gamma-ray blazars, their discovery has spurred theorists to try to devise explanations for them.

**Context**

The first satellite to detect gamma rays from space was the first dedicated to astronomical measurements. During its five months of operation in 1961, Explorer 11 detected about thirty gamma-ray photons. The entire spacecraft was only 37 kilograms, a small fraction of even one of Compton's instruments, but its success led to the inclusion of gamma-ray detectors on many space science missions from the United States and other countries. Still, the next spacecraft to be dedicated to gamma-ray astronomy was not launched until 1972, when NASA launched its Explorer 48 (also known as Small Astronomy Satellite-2), and 1975, when the European Space Agency's Cosmic Observation Satellite-B was launched.

NASA equipped a number of spacecraft with gamma-ray detectors, including some of the High-Energy Astronomical Observatories and Orbiting Solar Observatories, the Solar Maximum Mission, the X-Ray Timing Explorer, and others. India and Japan have incorporated gamma-ray instruments in some of their spacecraft, and other countries have built or participated in international collaborations on gamma-ray missions. The first Russian-built instrument launched on a United States robotic mission was a gamma-ray detector on *Wind* in 1994. The former Soviet Union installed gamma-ray instruments on some of its space stations as well as many robotic missions.

Some deep-space missions, including the Pioneer Venus orbiter, Ulysses, Mars Observer, the former Soviet Union's Phobos missions to Mars, and International Sun-Earth Explorer-3, carried gamma-ray detectors, providing very long baselines for triangulating on sources of gamma-ray emission. Active at the same time as Compton, Ulysses observed the same Super Bowl Burst; Mars Observer's gamma-ray instrument was not recording data at the time.

Despite the many scientific spacecraft carrying gamma-ray detectors, one of the most important discoveries in gamma-ray astronomy was made with the United States Department of Defense Vela satellites in the late 1960's and early 1970's. These spacecraft searched for gamma rays from high-energy phenomena, but the sources they were built to detect were Soviet nuclear detonations above the atmosphere in violation of a treaty banning such tests. It came as a surprise when they detected strong bursts that were found to come not from nuclear weapons but rather from celestial sources. This led to more gamma-ray missions, including Compton, to try to understand these mysterious gamma-ray bursts.

Compton has been the premier gamma-ray astronomy mission, achieving at least a factor of ten improvement in sensitivity at all energies over earlier missions. Spanning the gamma-ray spectrum from 30 kiloelectron volts to 30 gigaelectron volts, Compton surpassed the range of coverage of any other astrophysics mission. Yet using only gamma rays to study astrophysical objects would be just as limiting as using only red light, visible light, or any other single portion of the electromagnetic spectrum. Compton was the second in NASA's Great Observatories Program. It was preceded by the Hubble Space Telescope, designed for observations in the ultraviolet and visible, with some capability extending into the infrared range. The Chandra X-Ray Observatory (formerly the Advanced X-Ray Astrophysics Facility) was launched aboard the space shuttle *Columbia* on July 23, 1999. The Space Infrared Telescope Facility (renamed the Spitzer Space Telescope once in orbit) was launched on August 25, 2003, atop a Delta rocket from Cape Canaveral. Astrophysicists hope that with all of the remaining observatories operating at the same time, targets could be studied across the electromagnetic spectrum, thus providing greater insight into their nature.

**See also:** Chandra X-Ray Observatory; Gamma-ray Large Area Space Telescope; High-Energy Astronomical Observatories; Hubble Space Telescope; Infrared Astronomical Satellite; International Ultraviolet Explorer; Orbiting Astronomical Observatories; Skylab Program; Space Shuttle; Space Shuttle Flights, 1991; Spitzer Space Telescope; Swift Gamma Ray Burst Mission; Telescopes: Air and Space.

### Further Reading

Clark, Stuart. *Stars and Atoms: From the Big Bang to the Solar System.* New York: Oxford University Press, 1995. Beautifully illustrated, with many helpful diagrams, this book provides the scientific background on many of the objects and phenomena Compton and other space-based observatories have been designed to study. Readers at all levels will find this clear and interestingly written.

Gehrels, Neil, Carl E. Fichtel, Gerald J. Fishman, James D. Kurfess, and Volker Schönfelder. "The Compton Gamma Ray Observatory." *Scientific American* 269 (December, 1993): 68-77. Written by scientists intimately involved with Compton's mission, this article in a popular science magazine provides an overview of the observatory and its mission and presents fascinating descriptions of the scientific discoveries it has enabled. Includes paintings of the spacecraft and its instruments and illustrations helping to explain its scientific results.

Goldsmith, Donald. *Supernova! The Exploding Star of 1987.* New York: St. Martin's Press, 1989. The astrophysicist author uses the discovery of a relatively nearby and easily observed exploding star in 1987 as the core of his descriptions of many topics in astrophysics. He discusses the production of gamma rays and other electromagnetic signals from celestial sources and how measurements reveal their nature. The book also contains some interesting perspectives on science from inside the astronomical community.

Leverington, David. *New Cosmic Horizons: Space Astronomy from the V2 to the Hubble Space Telescope.* New York: Cambridge University Press, 2001. This is a broad treatise exploring the development of space-based astronomical observations from the end of World War II to the Hubble Space Telescope and other major NASA space-based observatories.

Ramana Murthy, Poolla V., and Arnold W. Wolfendale. *Gamma-Ray Astronomy.* 2d ed. New York: Cambridge University Press, 1993. This book, which is a fully updated new edition of the authors' earlier volume published in 1986, will prove invaluable in providing the background science to this important field. In assessing the current state of the art, the book also indicates the exciting basis from which new discoveries will be made. The concentration on phenomenology makes this book a fine introduction to gamma-ray astronomy. It will be of use to all students and professional astronomers who are working in this developing field.

Signore, M., P. Salati, and G. Vedrenne, eds. *The Gamma Ray Sky with Compton GRO and Sigma: Proceedings of the NATO Advanced Study Institute, Les Houches, France, January 25-February 4, 1994.* Norwell, Mass.: Kluwer Academic, 1995. This report covers the early results from the Gamma Ray Observatory, as well as the French gamma-ray telescope SIGMA, the first imaging experiment on a satellite to observe gamma rays. Observations from ground-based gamma-ray telescopes are also discussed.

Time-Life Books. *Voyage Through the Universe.* Richmond, Va.: Author, 1988-1991. This beautiful series of books is excellently written and has an impressive list of technical consultants. Several of the volumes cover the exotic phenomena probed by Compton and other astrophysics missions. Also included are descriptions of the spacecraft and complementary ground-based techniques as well as clear presentations of the nature of the electromagnetic spectrum and what can be revealed by studying it. The books are superbly illustrated and can be read by all audiences.

Trefil, James F. *Space, Time, Infinity: The Smithsonian Views the Universe.* New York: Pantheon Books, 1985. Entertainingly written by a physicist, this book contains many beautiful and interesting photographs and paintings depicting astrophysical objects and the instruments and people that study them. The author provides a very readable insight into the challenges and rewards of the study of the universe.

Wheeler, Craig J. *Cosmic Catastrophes: Supernovae, Gamma-Ray Bursts, and Adventures in Hyperspace.* New York: Cambridge University Press, 2000. A complete treatise of high-energy processes at work in the universe. Covers the contributions of the Compton Gamma Ray Observatory contributions to that understanding.

*Marc D. Rayman*

# Cooperation in Space: U.S. and Russian

*Date:* Beginning January 12, 1958
*Type of issue:* Sociopolitical

*U.S.-Soviet/Russian space cooperation sprang from the diplomatic and political aftershocks of Sputnik 1 and remained driven by the diplomacy and politics of the moment. Significant accomplishments, such as the Apollo-Soyuz Test Project in 1975 and the space shuttle-Mir missions, have emerged during periods of favorable East-West relations.*

## Key Figures

*Hugh Odishaw,* a scientist and pioneer in space cooperation

*Msitslav Vsevolodovich Keldysh,* president of the Soviet Academy of Sciences

*Boris Petrov,* chairperson of the Academy Council for International Cooperation

*Arnold Frutkin,* director of International Programs, NASA

*Cary F. Milliner, Jr.,* U.S. manager, Joint American-Soviet Particle Intercalibration project

*G. S. Ivanov-Kholodni,* Soviet manager, Joint American-Soviet Particle Intercalibration project

*John A. Simpson,* a professor of physics at the University of Chicago and a Vega Comet-Dust Experiment scientist

*Roald Z. Sagdeyev* (b. 1932), director of the Space Research Institute, Soviet Academy of Sciences

*Lew Allen, Jr.* (b. 1925), director of the Jet Propulsion Laboratory

*Kenneth A. Souza,* an Ames Research Center scientist and Kosmos 1129 project manager for the United States

*Carl Sagan* (1934-1996), a Cornell University professor and president of the Planetary Society

*Anatoli A. Blagonravov* (1895-1975), with the Soviet Academy of Sciences

*Hugh L. Dryden* (1898-1965), NASA deputy administrator

## Summary of the Issue

In 1958, three months after the Soviet launch of the world's first satellite, Sputnik 1, U.S. President Dwight D. Eisenhower proposed a U.S.-Soviet agreement that would limit the two countries' space programs to peaceful purposes. Earlier informal visits by Soviet scientists to U.S. space facilities and informal suggestions for cooperative ventures had been largely unproductive. Eisenhower's tentative move was rebuffed by Soviet premier Nikolai Bulganin a month later, when it was tied to elimination of nuclear-weapons testing and foreign military bases.

President John F. Kennedy continued American efforts to involve the Soviets in cooperative ventures in a United Nations General Assembly speech late in 1961, identifying meteorology, communications, and interplanetary probes as three areas of possible cooperation. A brief thaw in Soviet intransigence came with the Soviets' support late that year for United Nations Resolution 1721, proposed by the United States, which established guidelines for cooperation for the four years that followed. Nevertheless, the use of space achievements to showcase alleged national superiority and as a lever to achieve military goals impeded progress until 1962.

In a congratulatory letter sent to Kennedy after the first American piloted orbital flight, Soviet premier Nikita Khrushchev suggested pooling the scientific, technical, and material efforts of the two nations to advance science instead of the Cold War and arms race. An American proposal followed two weeks later, calling for five areas of cooperation: creation of a worldwide weather satellite system, an exchange of spacecraft tracking services, mapping of Earth's magnetic field, feasibility studies for an intercontinental communications satellite system, and exchange of space-medicine information. Khrushchev's reply extended consideration to the optical tracking of lunar and planetary spacecraft, search-and-rescue support for spacecraft landing on foreign soil, and space law. The exchange of letters resulted in the Dryden-Blagonravov Agreement of June 8, 1962, which led to the first U.S.-Soviet cooperation in the three areas of meteorology, geomagnetic studies, and telecommunications by satellite.

Kennedy followed this initial success with a United Nations General Assembly speech on September 20, 1963, in which he formally invited discussions with the Soviets on a joint flight to the Moon, a suggestion that had been informally voiced to Khrushchev in person two years earlier. Although the invitation was not overwhelmingly supported even within the United States, it was reiterated by President Lyndon B. Johnson shortly after Kennedy's assassination in November of that year. The Soviet Union never made a substantive response to the suggestion.

The first significant results of the Dryden-Blagonravov Agreement came early in 1964, when the 41-meter U.S. passive communications satellite Echo 2 was used to bounce a twelve-minute radio signal from Jodrell Bank in England to the Zimenski Observatory in the Soviet Union. Limited

*American astronaut and Russian cosmonaut working together on the Apollo-Soyuz Test Project. This was the first international docking of two spacecraft and the last Apollo mission.* (NASA CORE/Lorain County JVS)

progress with the meteorological and geophysical aspects of the agreement was attributed largely to technical problems experienced by both nations rather than any reluctance to cooperate. In the area of space medicine, however, the Soviets invoked an escape clause in the agreement and declined to exchange data.

In 1966 and 1967, U.S. proposals for early exchange of lunar-soil data and requests for details on Soviet Venus probe experiments, to avoid duplication of efforts on U.S. missions, were largely declined. Also, a lack of concrete assurances of Soviet adherence to United Nations sterilization guidelines for all spacecraft components likely to have contact with other celestial bodies was a frequent concern of the international community.

After the U.S. Apollo 11 Moon landing, a new spirit of cooperation surfaced. Soviet cosmonauts visited with President Richard M. Nixon and toured research centers of the National Aeronautics and Space Administration (NASA); their

American counterparts visited Soviet space installations. Meetings began to address cooperation in the expanding field of piloted spaceflight. In 1970 an agreement was reached to design compatible systems for the rendezvous and docking of U.S. Apollo and Soviet Soyuz piloted spacecraft; this work culminated in the 1975 joint Apollo-Soyuz Test Project (ASTP) mission. The two countries also exchanged lunar material from areas of the Moon explored by Apollo missions and automated Soviet Luna spacecraft. American spaceflights received coverage in the Soviet news media, Soviet scientists attended U.S. launches, and space-medicine data exchanges were initiated. The launch of Soyuz 19 on July 15, 1975, beginning the ASTP mission, marked the first time a Soviet launch had been televised live.

The U.S.-Soviet Agreement on Space Cooperation, signed by Nixon in 1972 and renewed by President Jimmy Carter in 1977, called for cooperation in four areas, each with a joint working group: space biology and medicine; near-Earth space, the Moon, and planets; study of the natural environment; and space meteorology. A search-and-rescue satellite system, added at the 1977 renewal, was later expanded to multinational involvement under separate agreements that included Canada and France.

Under the auspices of the 1972/1977 agreement, joint sounding-rocket launches were conducted in the summer of 1978 as part of the Joint American-Soviet Particle Intercalibration (JASPIC) project to investigate ionization sources in the upper atmosphere during geomagnetic activity.

American university and research-organization scientists participated in the Soviet biological satellite missions of Kosmos 782 in 1975, Kosmos 936 in 1977, and the nineteen-day Kosmos 1129 flight in 1979, managed in the United States by NASA's Ames Research Center. The biological satellites used modified versions of the Vostok spacecraft that had carried Yuri A. Gagarin on the world's first piloted spaceflight in 1961.

Worsening international relations prompted President Ronald W. Reagan not to extend the landmark space cooperation agreement, and it expired on May 24, 1982, ending most but not all joint exchanges. Cooperation in the Soviet Kosmos biosatellite program, planetary exploration, and the international Search and Rescue Satellite project continued under separate agreements. In December, 1983, the Kosmos 1514 biological satellite again included American experiments, and one U.S. experiment was included on the July, 1985, Kosmos 1667 mission.

In December, 1984, twin Soviet Vega spacecraft were launched toward Halley's comet by way of Venus, carrying identical comet-dust detectors designed by University of Chicago physicist John A. Simpson. Other American scientists contributed a design for a neutral mass spectrometer to the mission and served on the imaging and plasma physics experiment teams for the Halley encounter; NASA cooperated with French participants in the Venus phase of the Vega mission by providing tracking support for twin instrumented balloons dropped into the atmosphere of the planet. "Vega" is the Russian acronym for Venus-Halley (Venera-Gallei).

By 1985, U.S.-Soviet relations were improving once again. A general agreement was negotiated late that year between Lew Allen, Jr., director of NASA's Jet Propulsion Laboratory, and Roald Z. Sagdeyev, director of the Space Research Institute of the Soviet Academy of Sciences, and on April 15, 1987, a new U.S.-Soviet space science agreement was signed. Joint working groups were established in five areas: space biology and medicine, solar system exploration, space astronomy and astrophysics, solar-terrestrial physics, and Earth sciences. Sixteen specific projects contained in the agreement called for coordination, tracking, and scientist participation in the Soviet missions to Phobos, a Martian moon, and in the U.S. Vesta asteroid and Mars Observer missions; exchange of data on cosmic gamma rays, gamma-ray bursts, x rays and other rays, cosmic dust, meteorites, lunar material, radio astronomy, solar terrestrial physics, the surface of Venus, global and natural resources changes, the Soviet Kosmos biosatellite program, piloted

*Workers encapsulate the WIND spacecraft and attach the payload assist module-D booster stage. The mission marked the first flight of a Russian scientific instrument on a U.S. spacecraft, part of an international effort to learn more about the interaction between the Earth and the Sun.* (NASA)

space missions, and spaceflight-induced changes in human physiology; studies of the feasibility of joint experiments on space biology; and publishing a second edition of a book on space biology and medicine that had originated from earlier cooperative agreements.

After the signing of the new agreement, NASA signed a protocol to coordinate position tracking of the small Phobos landers during their April, 1989, landing attempts on that Martian moon. The Soviets, in a goodwill move, agreed to include on one of the Phobos craft a plaque commemorating the discovery of Phobos in 1877 by the American astronomer Asaph Hall. Discussions were begun on using NASA's 1992 Mars Observer mission as a communications relay for a joint Soviet-French Mars balloon mission in 1994.

NASA participation in the Kosmos biosatellite program continued with twenty-seven experiments

aboard the thirteen-day Kosmos 1887 flight in late September of 1987, and discussions were begun on further participation in similar 1989 and 1991 flights.

Talks by Reagan and Soviet General Secretary Mikhail Sergeyevich Gorbachev at a Moscow summit in 1988 were followed by discussions of a future Mars rover and sample return flight, joint exploration of the planet Mercury, a solar probe mission, and the resumption of lunar flights. The Soviets also sought access to NASA's large vacuum chamber at the Johnson Space Center for extended testing of the Phobos laser experiment to obtain spectra of soil on the surface of Phobos.

Indications of high-level Soviet willingness to open discussions on possible joint U.S.-Soviet piloted flights to Mars further illustrated this new Soviet spirit of cooperation, but they were largely dismissed as premature by the Reagan administration and by NASA as 1988 drew to a close.

After the breakup of the Soviet Union in 1991, Russia was left with very little in the way of a space program. Its struggling economy needed a boost from the West to ensure future space activities. At the same time, the U.S. space station program was suffering from its own economic anemia. A joint U.S./Russian space station program was developed to provide a continuing scientific platform in space. The new International Space Station is a cooperative effort of fifteen nations. In the first of a three-phase program, Americans and Russians worked together aboard the space shuttle and Mir Complex. Several missions were carried out involving Americans living aboard Mir and Russians riding the space shuttle. Between March, 1995, and May, 1998, the Russian space station Mir hosted a series of NASA astronauts as crew members.

The Phase I program operated under a complicated logistical scheme. In the history of human spaceflight, no previous program has required so many transport vehicles, so much interdependent operation between organizations, and so much good timing. Shuttle-Mir experience gave participants an opportunity to gear up for the formidable cooperative effort the International Space Station requires.

More than forty spaceflights and at least three space vehicles—the space shuttle, the Russian Soyuz rocket, and the Russian Proton rocket—were planned to deliver the various components of the International Space Station (ISS) to Earth orbit. The first flight for ISS assembly used a Russian Proton rocket that lifted off in November, 1998, and placed the Zarya Control Module in orbit. In early December of that same year, the STS-88 mission saw the shuttle *Endeavour* attach the Unity Module to Zarya, initiating the first ISS assembly sequence.

### Contributions

Biomedical results from extended Soviet piloted missions to space stations were exchanged, and coordinated studies of bone mineral measurements were undertaken, which provided invaluable information on long-term human exposure to microgravity. Human physiological changes became more pronounced with increased exposure to the microgravity of space. American experiments carried aboard the series of Soviet Kosmos biological satellites found changes—similar to those experienced by astronauts and cosmonauts—in enzymes, animal bone strength, growth rate, and mineral content in specimens that spent more than eighteen days in space aboard Kosmos 1129 in the fall of 1979. On that same flight, U.S. studies of changes in animal muscle fibers showed that muscle mass was lower in the flight animals than in the ground controls; a 20 percent deficiency in total bone mineral content compared with that of control subjects was found also. The livers of flight rats contained less glycogen and less of some complex fats than their controls; another group of flight rats showed a shift of body water from skin and muscles to internal organs. Studies of carrot embryos and plantlets showed that plants suffered no ill effects from spaceflight, which provided en-

couragement for prolonged space missions. The growth rate of plant tumors on carrot slices was the same in space as on the ground, suggesting that metabolism in carrot tissue is unaffected by spaceflight.

In planetary exploration, samples of lunar materials were exchanged, providing both the United States and the Soviet Union with a much broader sample of soils and rocks from different parts of the Moon than would have been possible without cooperation. American Pioneer Venus spacecraft radar maps of Venus were given to the Soviets in preparation for landing-site selection for their Venera missions; in return, data from those landings were given to the United States. The Soviet Union continued to be the only source of surface photographs of Venus. Information from the Vega encounters with Halley's comet was quickly disseminated, including striking photographs of the comet's nucleus and gaseous envelope. Soviet scientists routinely complied with requests from NASA to reduce radio frequency interference from their military early-warning and scientific satellites during critical deep-space planetary encounters by U.S. spacecraft.

Records at NASA's Goddard Space Flight Center indicated that in the first five years of the cooperative Soviet Space System for Search of Vessels in Distress/U.S. Search and Rescue Satellite-Aided Tracking (COSPAS/SARSAT) search-and-rescue program, begun in September, 1982, two U.S. and two Soviet satellites rescued more than nine hundred persons from ships and downed aircraft by pinpointing their locations with onboard transponders sensitive to the frequencies used by ship and aircraft emergency pointing/indicating radio beacon equipment.

In 1987, a cooperative study of the effect of the ozone layer on Earth's climate was initiated using U.S. polar-orbiting meteorological satellites and Soviet ground-based measurements in Antarctica. A hole in the ozone layer that opens over the South Pole every September and October has shown signs of deepening with each successive year, indicating possible global ozone depletion.

### Context

The period immediately following Sputnik 1 was one of intense international competition, not only between the United States and the Soviet Union but between social, economic, and political systems as well. The Soviets justifiably took intense pride in the accomplishments of their space program and the new image it projected of a technological power. Being first with each new type of mission was almost an obsession.

The United States assumed a defensive posture in the aftermath of Sputnik and a number of spectacular U.S. space failures. American leaders' suggestions for cooperation were frequently interpreted as efforts to gather intelligence data on Soviet rocket systems; Soviet reluctance, in turn, raised questions about just how technologically sophisticated and capable their systems really were.

Soviet leaders did not express genuine willingness to cooperate until it became evident that Soviet setbacks and American accomplishments were likely to result in the United States' winning the race to land humans on the Moon. After the United States achieved that goal, in 1969, the Soviet posture shifted from competition to cooperation, and a new openness prevailed. Once it was seen that the United States had a clear technological edge, concerns about technology transfer led to U.S. doubts about cooperation in some areas and insistence that exchanges of information be balanced in some way.

The fortunes of space cooperation were frequently at the mercy of international events unrelated to matters of science, as when Soviet support for the declaration of martial law in Poland led to the United States' allowing the long-running U.S.-Soviet Agreement on Space Cooperation to expire. Even in the worst international climates, however, some cooperation continued among scientists and government agencies and through limited agreements.

The argument for cooperation is strong. It would take years for the United States to accumulate the biomedical data that the former Soviet Union has obtained from its completed and ongoing long-duration piloted space missions. Given the expense of sophisticated missions to the planets of the solar system, it is imperative to plan missions that complement, rather than duplicate, one another. Global environmental concerns necessitate the sacrifice of some sovereignty concerns for the sake of studies involving the land masses and territorial waters of many nations.

At a time when the United States was determined to fly people to the Moon and hoped to be the first nation to do so, President Kennedy proposed a joint U.S.-Soviet flight. Nearly twenty-five years later, with the former Soviets equally determined to fly one day to the planet Mars, former General Secretary Gorbachev expressed interest in a joint Soviet-U.S. flight to the Red Planet. As the new millennium began, the first components of the International Space Station were heading into orbit from both nations.

**See also:** Apollo-Soyuz Test Project; Ethnic and Gender Diversity in the Space Program; Funding Procedures of Space Programs; International Space Station: Development; National Aeronautics and Space Administration; Space Shuttle-Mir: Joint Missions.

### Further Reading

Belitsky, Boris, and A. S. Piradov. *International Space Law.* Honolulu: University Press of the Pacific, 2000. The writers of this book have given special attention to the most important principles and norms of international space law, and to analyzing the bilateral and multilateral agreements on individual aspects of space activities. Many specific examples are given concerning the solution of the international space problems that arise on a day-to-day basis.

Bizony, Piers. *Island in the Sky: Building the International Space Station.* London: Aurum Press Limited, 1996. Bizony tells how the International Space Station will be assembled in orbit during an extended sequence of shuttle flights, dockings, and spacewalks. With unrivaled access to NASA and the astronautic sources worldwide, his lively text contains a wealth of information. There are 100 photographs, 60 in color.

Collins, Martin. *Space Race: The U.S.-U.S. S. R. Competition to Reach the Moon.* San Francisco: Pomegranate Communications, 1999. In concise text and abundant illustrations, this book tells the story of the fifty-year struggle for dominion of outer space between the United States and the Soviet Union.

Harvey, Brian. *Russia in Space: The Failed Frontier?* Chichester, England: Springer-Praxis, 2001. This book tells the inside story of the traumatic events that engulfed the once-glorious Soviet space program. It is a story of desperation and decline, but also a tale of heroic efforts to save the Space Station Mir and the construction—along with their old rivals, the Americans—of the new International Space Station.

Haskell, G., and Michael Rycroft. *International Space Station: The Next Space Marketplace.* Boston: Kluwer Academic, 2000. Addresses issues of ISS utilization and operations from all perspectives, especially the commercial viewpoint, as well as scientific research, technological development, and education in the widest sense of the word. Of interest to those working in industry, academia, government, and, in particular, public-private partnerships.

Hudgins, Edward L., ed. *Space: The Free-Market Frontier.* Washington, D.C.: Cato Institute, 2002. Outer space will languish economically until private enterprise can take control, according to this dry and doctrinaire collection of papers from a conference sponsored by the libertarian Cato Institute. The contributors include members of Congress, lawyers, business executives, and an astronaut, and cover such topics as NASA's history, cheaper space travel, opportunities for and barriers to space investment, and legal and property rights in space.

Logsdon, John M. *Together in Orbit: The Origins of International Participation in the Space Station.* Washington, D.C.: National Aeronautics and Space Administration, 1998. Describes the politics and science behind the effort to bring together many nations in the effort to build a space station.

McCurdy, Howard E. *The Space Station Decision: Incremental Politics and Technical Choice.* Baltimore: Johns Hopkins University Press, 1990. The author is a professor of public affairs at American University in Washington, D.C. The events that led up to the decision (1984) to build a permanently occupied space station in low-Earth orbit provide his primary subject matter in the present monograph, but the author's deeper interest has to do with the politics of Big Science. The story is arrestingly told in this nicely produced volume, which provides thirteen pages of plates plus detailed notes and references.

Oberg, James. *Star-Crossed Orbits: Inside the U.S.-Russian Space Alliance.* New York: McGraw-Hill, 2001. Combines personal memoir with investigative journalism to tell the story of the U.S.-Russian space alliance.

Scott, David, and Alexei Leonov. *Two Sides of the Moon: Our Story of the Cold War Space Race.* London: Thomas Dunne Books, 2004. In this unique dual autobiography, Astronaut Scott and Cosmonaut Leonov recount parallel tales of one of the most ambitious contests ever embarked on by humankind.

Siddiqi, Asif A. *The Soviet Space Race with Apollo.* Gainesville: University Press of Florida, 2003.
_____. *Sputnik and the Soviet Space Challenge.* Gainesville: University Press of Florida,

2003. The first comprehensive history of the Soviet piloted space programs, covering a period of thirty years.

Von Bencke, Matthew J. *The Politics of Space: A History of U.S.-Soviet/Russian Competition and Cooperation in Space.* Boulder, Colo.: Westview Press, 1996. This book chronicles the efforts of the United States and the Soviet Union (later Russia) to overcome their political animosities and explore space together. It looks at their respective foreign and domestic policies; military, civil, and commercial influences; and top executive, legislative, and institutional politics. The book examines their separate and joint endeavors from 1945 through 1997.

*Richard A. Sweetsir*

# Dawn Mission

*Date:* Beginning 2006
*Type of program:* Planetary exploration

*The Dawn mission represents the first time that a spacecraft will orbit two separate planetary bodies as part of the same mission. The Dawn spacecraft will examine the two largest asteroids and be the the first fully scientific space mission to use ion propulsion to power the spacecraft throughout its nine-year journey.*

### Key Figures

*Christopher T. Russell*, project director
*A. Coradini*, Institute for Space Astrophysics, Rome
*W. C. Feldman*, Los Alamos National Laboratory
*R. Jaumann*, German Aerospace Center
*A. S. Konopliv*, Jet Propulsion Laboratory
*T. B. McCord*, University of Hawaii
*L. A. McFadden*, University of Maryland
*H. Y. McSween*, University of Tennessee, Knoxville
*C. M. Pieters*, Brown University
*D. E. Smith*, NASA Goddard Space Flight Center
*M. V. Sykes*, University of Arizona
*M. T. Zuber*, Massachusetts Institute of Technology

### Summary of the Mission

The Dawn mission is part of the Discovery Program, which came into being as a result of the November, 1997, Space Science Strategic Plan of the National Aeronautics and Space Administration (NASA). This plan addresses some of the most fundamental questions ever posed to humankind: How did the universe begin, and what evolutionary processes are involved with the planets, stars, and galaxies? How did life originate on Earth, and is life present elsewhere in the universe? What is the cosmic fate of humankind, and can knowledge gained from space exploration improve the quality of life on Earth? The Discovery Program was designed to gain more information about our solar system in order to answer some of these questions.

Discovery missions will systematically explore the planets, many of their moons, and other minor bodies within the solar system. Included in the Discovery Program are the Near Earth Asteroid Rendezvous (NEAR) mission, the Mars Pathfinder mission, the Lunar Prospector mission, the Stardust spacecraft, the Genesis spacecraft, the Comet Nucleus Tour, the MESSENGER (Mercury Surface, Space Environment, Geochemistry, and Ranging) mission, and the Deep Impact mission. Each of these missions was planned to incorporate new technologies into its design to provide the most efficient means to achieve their goals. Additional Discovery missions will search for and focus on planetary systems around other stars.

The objective of the Dawn mission is to investigate the surface features and conditions present on Ceres and Vesta, two of the solar system's largest asteroids. These two main-belt asteroids are believed

267

to be objects that did not complete the process of planetary formation like the more massive and larger Earth or Mars. The reason for this has puzzled scientists for centuries. The two asteroids' relatively small size in relation to Earth and their great distance from Earth have made direct observations very difficult. Observations made from orbit and remote-sensing data will certainly provide scientists with new evidence to understand these objects and the theories of planetary formation better.

The Dawn mission will be constructed using technology and proven hardware that have been used successfully in several other missions, notably the Deep Space 1 spacecraft, which first utilized ion propulsion. Ion propulsion is based on the relatively simple principle that like electrical charges repel each other. To employ this concept for space propulsion, a fluid needs to be electrically charged so that its atoms can be expelled in a specific direction,

which will then produce the effect of pushing the spacecraft in the opposite direction. The fuel used for this propulsion is xenon gas, which interacts with a flow of electrons emitted from a cathode tube. The electron stream enters a magnet-ringed chamber, where the electrons strike the xenon atoms. This impact knocks an electron away from the xenon, creating a positively charged ion. A positively and negatively charged pair of metal grids positioned at the rear of this chamber create a strong electrostatic force on the xenon ions, pulling them out the end of the chamber at a speed of 110,000 kilometers per hour (kph). Once in space, the ion propulsion engine is much more efficient than chemical rocket propulsion. Unlike a chemical rocket, an ion engine provides a nearly continuous rate of propulsion that can last months or even years, giving it the ability to gradually increase its velocity to more than ten times that of a conventional rocket.

The scientific mission of the Dawn spacecraft is to study the physical structure and the geological evolution of Ceres and Vesta. These asteroids are the most massive remnants of the original proto-planets. Ceres, the largest asteroid, at 960 by 932 kilometers in diameter, constitutes more than one-third the estimated total mass of all the asteroids. Vesta, the third largest asteroid, is 530 kilometers in diameter. For some undetermined reason, the accretion process was interrupted and they did not accumulate sufficient mass to qualify as Earth-like planets. To help unravel the mystery, a battery of scientific instruments will be used to gather a variety of data. They include a framing camera, a mapping spectrometer, a laser altimeter, a gamma-ray spectrometer, and a magnetometer.

Both Ceres and Vesta are main-belt asteroids, yet they exhibit very different physical characteristics. It is believed that each followed a very different evolutionary path that can be attributed to

*Workers in a clean room at the Jet Propulsion Laboratory in Pasadena, California, assemble the Dawn spacecraft.* (NASA-JPL)

a variety of different physical and chemical sets of conditions present throughout solar system in its first few million years. As a result, Ceres is believed to be geologically primitive and wet, while Vesta is geologically evolved and dry. This is somewhat disturbing, because present theories suggest that the body with the larger mass (Ceres) should be the one exhibiting the greater diversity in its composition, internal structure, and surface features. Perhaps the fact that Vesta is approximately 62 million kilometers closer to the Sun plays a significant role in the two asteroids' respective evolutionary histories. The asteroids' relative nearness to the Sun may account for the fact that observations made from Earth have also indicated the possibility that frost or water vapor may exist on the surface of Ceres, and perhaps even large quantities of water beneath its surface.

In contrast, Vesta appears to have had a history during which partial melting of its interior eventually gave rise to the extrusion of lavas 4.5 billion years ago. Evidence to support this theory comes from a variety of meteorites called eucrites. These meteorites are very similar to terrestrial basalts in both their texture and basic silicate mineralogy, but they differ significantly in the presence of reduced, metallic nickel-iron alloy metals instead of the oxidized iron minerals found on Earth. Mineralogical analyses of the eucrites compare well to reflectivity studies of the material on the surface of Vesta, but without actual samples collected from its surface no conclusive evidence exists. Observations from the Hubble Space Telescope have detected a huge crater on Vesta estimated to be approximately 430 kilometers across and a billion years old. It is believed that this crater may be the source of the eucrites. Earlier Hubble telescopic observations identified numerous volcanic lava flows that would also support Vesta as the probable parent body for eucrite meteorites.

### Contributions

The knowledge gained from the Dawn mission will be invaluable to both planetary scientists and propulsion engineers. Other spacecraft have made relatively close flybys of several asteroids, including Mathilde, Ida, and Gaspra. Scientists were amazed to see the variety of geological features present on the surfaces of these asteroids. It was readily apparent that each had experienced its own unique history of evolution. In the case of the NEAR mission, the spacecraft actually orbited the asteroid Eros and eventually landed on its surface, providing scientists with their first direct surface contact with an asteroid-sized body.

Each of the above-mentioned asteroids represents a relatively small object when compared to Ceres and Vesta. These two larger bodies represent a stage in planetary development where an object either continues to grow by accretion to true planetary dimensions or completely halts its growth. Close orbital observations of Ceres and Vesta should provide the much-needed data to resolve the main-belt asteroids' position in the evolutionary history of the solar system.

The data collected from orbit around Ceres and Vesta will also provide direct evidence of the chemical and physical nature of the surface conditions present on these miniature planetary bodies. The Dawn mission spacecraft will obtain full surface imagery in seven colors for Vesta and in three colors for Ceres. This will be done in conjunction with a full surface mapping spectrometer. The spacecraft will also acquire neutron and gamma-ray spectra in order to compile an elemental composition surface map for each asteroid. Included in these analyses are the major rock-forming elements (oxygen, magnesium, aluminum, silicon, calcium, titanium, and iron), trace elements (gadolinium and samarium), long-lived radioactive elements (potassium, thorium, and uranium), and light elements (hydrogen, carbon, and nitrogen), which are the major components of many ices.

Such data will provide a comprehensive picture of both asteroids' past evolutionary history and present surface conditions. The knowledge gained from the Dawn mission may also prove valuable for future assessments of the asteroids' resource value. Finally, radio tracking data will be used to determine the basic physical characteristics of these asteroids, such as their mass, gravity field, principal axes, rotational axes, and moments of inertia.

## Context

The Dawn mission plays a very important role in NASA's plan for the exploration of the solar system. It represents a logical step between piloted exploration and robotic missions to the planets. In addition, the potential for an asteroid impact with Earth is real, and such an impact will occur at some time in the future. Present technology is not capable of preventing it from happening, however, and more information about the physical and chemical makeup of an asteroid is needed to help develop adequate defenses against asteroid impacts. The data gained from the Dawn mission and others like it will be of great benefit to the scientists and engineers who will be charged with the task of either altering an asteroid's trajectory or blowing it up far from Earth.

Along with the pure science that will come from the Dawn mission, engineers will better understand ion propulsion and how it may be applied to future missions. The long flight time from Earth to the outer planets is a major problem for mission scientists. New propulsion technologies must be developed to shorten these flights. Cost-effectiveness will always be a concern, and future missions will have to be designed to be more efficient and last longer in order to return as much data as possible from perhaps fewer spacecraft.

**See also:** Asteroid and Comet Exploration; Deep Impact.

## Further Reading

Bottke, William F., Alberto Cellini, Paolo Paolicchi, and Richard P. Binzel. *Asteroids III.* Tucson: University of Arizona Press, 2002. This comprehensive volume continues in the spirit of two previous volumes, *Asteroids* (1979) and *Asteroids II* (1989), with its focus on the latest discoveries dealing with observational telescopic data, spacecraft data, and theoretical modeling of asteroids. Also addresses potential hazards that asteroid impacts present to the Earth. The editors have chosen the contributors well, and this volume makes for a welcome addition to the two previous volumes. Intended for the serious researcher or student of astronomy.

Brandt, John C., and Robert D. Chapman. *Introduction to Comets.* Boston: Cambridge University Press, 2004. Provides detailed discussion of virtually every cometary phenomenon.

Gehrels, Thomas, ed. *Hazards Due to Comets and Asteroids.* Tucson: University of Arizona Press, 1995. A comprehensive collection of papers dealing with virtually all aspects of comets and asteroids, and the potential of impacts with the Earth. The articles reviewing past impact events are very revealing when compared to the subsequent articles, which describe, in detail, the thermal and shock effects associated with these giant impact events. This work is meant for the serious astronomy student or researcher.

Kowal, Charles T. *Asteroids: Their Nature and Utilization.* New York: Halsted Press, 1988. This book was written by the discover of Chiron, one of the first asteroids to be discovered beyond the main belt. Although published in 1988, the volume still provides the reader with a good basis for understanding the physical characteristics and orbital motion of the asteroids, including the Apollos, Amors, and Atens, which cross the Earth's orbit.

Lewis, John S. *Mining the Sky: Untold Riches from the Asteroids, Comets, and Planets.* New York: Perseus Books Group, 1997. Presents a scenario in which the resources of the solar system could be used in order to relieve stresses on Earth's depleted energy reserves.

McSween, Harry Y., Jr. *Meteorites and Their Parent Bodies.* 2d ed. New York: Cambridge University Press, 1999. This second edition builds upon the original 1987 edition by incorporating new information gained from subsequent meteorite studies and asteroid observations. McSween is one of the premier meteorite researchers and proponents of Mars-origin meteorites. He effectively blends his knowledge of meteorites into the question of asteroid parent bodies, especially with eucrites and Vesta.

Verschur, Gerrift L. *Impact: The Threat of Comets and Asteroids.* London: Oxford University Press, 1997. Describes the nature of near-Earth asteroids and comets, and the threat they pose for devastating impacts on Earth.

*Paul P. Sipiera*

# Deep Impact

*Date:* Beginning January, 2005
*Type of program:* Planetary exploration

*Deep Impact is the first space mission designed to study the interior of a primitive celestial body directly. This was accomplished by slamming a heavy impactor into Comet Tempel 1, creating a fresh crater. Never before has any space mission tried to make an impact crater of this size in any object.*

### Key Figures

*Michael F. A'Hearn*, principal investigator
*Michael J. S. Belton*, deputy principal investigator
*James E. Graf*, project manager
*Rick Grammier*, project manager
*Kenneth P. Klassen*, Mission Operations
*Karen J. Meech*, coordinator, Earth-Based Observations
*Joseph Veverka* (b. 1941), Data Processing and Image Analysis
*Donald K. Yeomans*, Dynamics

### Summary of the Mission

In July, 1999, the National Aeronautics and Space Administration (NASA) approved funding under its Discovery Program for a space mission to slam an object into a comet and to study the ejecta from that impact. This proposed mission became the Deep Impact project. By November, 1999, serious work had commenced in design of the project. Except for minor variations, the plans were finalized by May, 2001. These plans called for the development of a flyby spacecraft to study the impact ejecta and for a smart impactor that would guide itself to the target. The comet selected as target was Tempel 1, discovered in 1867 by Ernst Tempel.

Comet Tempel 1 orbits the Sun every 5.5 years. It approached perihelion, the closest point to the Sun in its orbit, in early summer, 2005. A target date for comet rendezvous was selected as July 4, 2005. Earth's orbit placed it in position to launch a mission to arrive at Tempel 1 on this date during a period of only a few weeks in January, 2004, and again for several weeks a year later. These two time periods during which a launch is possible are known as launch windows.

Initially, plans called for the mission to be launched during the first launch window, in January of 2004. However, the spacecraft was not ready for launch at that time, and so launch was postponed until December 30, 2004, the first day of the second launch window, initially reserved as a backup in case the launch could not occur in January, 2004. The launch window was only about three weeks long, and if the spacecraft was not launched in this time, then no rendezvous mission would have been possible. In November, 2004, the launch was further postponed until January 8, 2005, at 19:40 Coordinated Universal Time (UTC). This date was well into the last launch window, and thus it gave very little leeway if problems developed with weather or with the launch vehicle.

The launch vehicle for Deep Impact was a Delta II 2925, a variant of the Delta rocket with nine strap-on solid rockets. As with most launch vehicles, the components were assembled at the launch pad. The first stage of Deep Impact's launch vehicle was lifted into place on November 22, 2004, at Space Launch Complex 17B of the Cape Canaveral Air Force Station, located adjacent to the Kennedy Space Center, in Florida. The solid rocket boosters were attached three at a time over the following eight days.

Meanwhile, the Deep Impact spacecraft was being prepared nearby at the Astrotech Space Operations facility near Titusville, Florida. On October 23, 2004, the spacecraft had arrived from the Ball Aerospace & Technologies factory at Boulder, Colorado, where it had been constructed.

NASA's Deep Impact mission finally launched at 18:47:08 UTC on January 12, 2005, from Cape Canaveral's Complex 17. Six solid rocket boosters fired at liftoff and were jettisoned in flight at a point before the final three solid rocket boosters ignited. Three minutes into ascent, the final solid

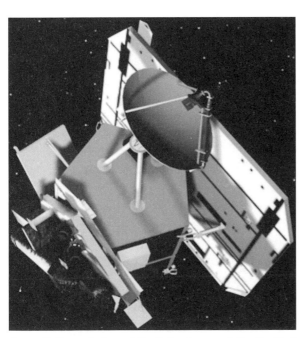

*An artist's rendering of the flyby spacecraft for the Deep Impact mission showing its solar panel (right), high-gain antenna (top), the debris shield (left), and the location of imaging and spectroscopy instruments (box and cylinder, lower left).* (NASA-JPL)

rocket boosters had burned out and dropped away, leaving the main engine to continue to burn. After main engine cutoff and second-stage ignition, the protective fairing covering the spacecraft separated safely, exposing Deep Impact to space conditions. Nine minutes after liftoff, the second stage shut down and left the remaining booster/spacecraft combination in a parking orbit. At 19:11 Coordinated Universal Time (UTC), the second-stage engine reignited for 95 seconds. Third-stage ignition followed second-stage shutdown and separation, and the third-stage engine continued to fire until 19:16 UTC. Spinning motions of the spent third stage were damped out, and at 19:22 UTC the Deep Impact spacecraft separated from the spent booster. Shortly after beginning its journey to the comet, the spacecraft's onboard computer detected an error and put the spacecraft into a safe mode, a shutdown designed to prevent damage to the spacecraft until the problem is corrected.

Only minor course corrections were required over the course of a six-month cruise phase. On July 3 the flyby spacecraft released the impactor so that it could intercept Comet Tempel 1. To prevent collision between the flyby and Tempel 1, the flyby performed an avoidance maneuver, firing thrusters that slowed it down by 120 meters per second (270 miles per hour). Moving down and below the comet, the flyby spacecraft had to turn its cameras back toward Tempel 1 to record the impactor's crash. The impactor's on-board guidance system executed four maneuvers as people all over the world watched on cable news, NASA television, and the Internet. Deep Impact provided tremendous celestial fireworks on the Fourth of July, striking very close to its target at a relative speed of 10.2 kilometers per second (22,800 miles per hour). The time of impact (5:52 UTC) was selected in order for the impact to happen above the horizon simultaneously for two of NASA's Deep Space Network radio telescopes, allowing both receivers to record real-time data transmitted from the spacecraft. The impact excavated a crater in the comet nucleus, allowing the flyby spacecraft to study the material from inside the comet. The flyby spacecraft began imaging one minute before impact, and it was 10,000 kilometers (6,200 miles) away at the moment of impact. The flyby spacecraft continued collecting data as it closed with the comet nucleus, eventually passing within 500 kilometers (310 miles) of it. Selected images were sent back to Earth in near-real time; however, data were collected far more quickly than the communication system could send the information to Earth. Thus, recorded data began to be sent back to Earth starting about fifty minutes after impact, with most of the data returned within a day with provisions to extend data playback for nearly a month if necessary. Shortly after this highly successful encounter, Deep Impact program officials sought funds to continue the mission and redirect the flyby spacecraft toward another comet encounter; preliminary permission was given, but budget consideration held the potential for cessation of the extended mission.

The flyby spacecraft was somewhat irregular in shape, measuring about 1.7 meters wide, 2.3 meters high, and 3.3 meters long and having a mass of 650 kilograms. Power was provided by a 7.5-square-meter solar panel located on one side of the spacecraft bus. The two main science instruments are located on the other side of the main bus. The spacecraft bus housed the propulsion, communication, and computer systems. Mounted to the top of the bus, a 1-meter-diameter dish provided primary communication with Earth. A separate antenna communicated with the impactor. Debris shields, called Whipple shields, protected the spacecraft from damage from impact with high-velocity dust particles shed by the comet.

The principal science instruments on the flyby spacecraft were the High Resolution Instrument (HRI) and the Medium Resolution Instrument (MRI). Both instruments were Cassegrain-type

telescopes, and CCD (charge-coupled device) cameras with filter wheels to permit observations in different colors of light. The HRI was one of the largest instruments dedicated to solar system astronomy ever put aboard a spacecraft. With its telescope having a diameter of 30 centimeters, the HRI had five times the resolution of the MRI and was capable of resolving features smaller than 2 meters when nearest the comet nucleus. The HRI also was able to make measurements in infrared light. The MRI, with its 12-centimeter-diameter telescope, also had a filter wheel to allow observations in different spectral ranges, and it acts as a backup to the HRI. Additionally, at close approach its larger field of view provided images of the entire nucleus, rather than merely a portion like the HRI, which saw only a small portion of the nucleus at a time. Addi-

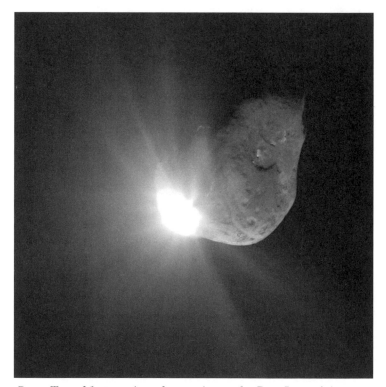

*Comet Tempel 1 approximately one minute after Deep Impact's impactor rammed itself into the comet. The flyby spacecraft's high-resolution camera acquired this image. The bright spot marks the impact area and sunlight can be seen reflecting from the rest of the comet's surface, revealing its pockmarked topography.* (NASA/JPL-Caltech/UMD)

tionally, in the last ten days before the encounter, the MRI acted as a navigation camera for the spacecraft.

The impactor is a 350-kilogram cylindrical device about 1 meter in diameter and 1 meter in length. The impactor was attached to the flyby spacecraft bus until twenty-four hours before impact, and it received all electrical power from the flyby spacecraft's solar panel. After separation, it operated entirely on battery power. The impactor was destroyed in the impact with the comet nucleus, and so it was made of mainly copper, because copper provides the least interference with spectral measurements of the impact ejecta. The impactor contained a small propulsion system to make corrections to its trajectory, in order to impact at the proper point on the comet nucleus, as well as a communication system to send data back to the flyby spacecraft. The main instrument on board the impactor was the Impactor Target Sensor (ITS), which was virtually a duplicate of the MRI, except without the MRI's filter wheel. Thus, in addition to guiding the impactor to the correct target zone, the ITS was capable of taking useful science images on its way to impact with the comet nucleus. Upon impact with the comet, the impactor delivered an energy of 19 gigajoules (equivalent to about 4.8 tons of dynamite) and excavated a sizable crater on the surface of the comet nucleus.

### Contributions

Comets form in the outer solar system of mostly ice, with some dust and rock mixed in, and may be virtually unchanged since the formation of the solar system. For most of the time since their formation, comets have remained either in a zone beyond the planet Pluto called the Kuiper Belt or in a cloud around the solar system called the Oort Cloud. Occasionally the orbits of comets are altered, sending the icy bodies into the inner solar system, where heat from the Sun begins to vaporize the ices that make up a comet's nucleus, shedding dust debris and forming a cloud around the nucleus and the tail of the comet. This cloud prevents astronomers from observing the nucleus of a comet with great detail. The ITS optics aboard Deep Impact were expected to be destroyed prior to impact by collisions with dust particles from the comet, but they still provided the most detailed images of a comet nucleus to date.

Some concerns over the imaging capability of the flyby spacecraft's HRI were reported on March 25, 2005. After a bakeout cycle had been completed to remove residual gases from the optics, mission controllers were unable fully to achieve optimal focus with the HRI because of a slight imperfection in the instrument's mirror. By June, 2005, however, mission scientists were confident that they would be able to produce high-quality images using the HRI, despite its inability to achieve perfect focus. After the images were sent back to Earth, a mathematical process called deconvolution was applied to the images to correct for the focusing problem. This technique should produce images nearly as good as could have been achieved had the HRI mirror been able to provide perfect focus.

By blasting into the comet, Deep Impact permitted the inner layers of a comet to be studied for the first time. The outer portions of a comet in the inner solar system have been altered by the heat from the Sun, but the inner parts may remain as they were when the comet formed. Also, in striking the comet, Deep Impact helped astronomers determine the consistency of the comet nucleus. It is not known if the nucleus is rock-solid, like an iceberg, or is soft and fluffy, like a snowflake. The size of the crater answered some of these questions. Preliminary analysis of images of the comet and data collected by examination of the plume created during crater excavation indicated a surface with much more dust and dirty water ice than had been expected. Scientists would continue to study Deep Impact data for years to come.

### Context

Knowing more about comets permits astronomers to know more about the formation of the solar system. Comet studies also help scientists understand Earth, because some of the water and organic materials needed for life are believed to have been brought to Earth by comets. However, understanding comets better has a more practical application as well, because comets can impact with planets. If a comet were on a collision course with Earth, knowing the structure and consistency of the nucleus of the comet might permit a strategy to be developed to deflect the comet, given sufficient time.

Deep Impact and its sister comet mission, Stardust, were both funded under NASA's very successful Discovery Program. Discovery missions are relatively inexpensive space missions with a narrow focus. An earlier Discovery mission, Deep Space 1 (1998), passed by a comet, but that mission's primary goal was development of a new ion drive for space probes. Other than the Comet Nucleus Tour mission (CONTOUR) of 2002, which lost contact with its spacecraft, Stardust and Deep Impact were the first U.S. missions designed primarily to explore comets. They form part of an international effort to study comets, including Rosetta, launched in 2004 by the European Space Agency (ESA) toward Comet Churyumov-Gerasimenko.

Deep Impact also marks a first in space exploration, in that it attempts to alter the body being studied, in this case by blasting a crater into it. All other space missions have been more passive in their approach, with only minimal impact on the body being studied.

**See also:** Asteroid and Comet Exploration; Cape Canaveral and the Kennedy Space Center; Dawn Mission; Stardust Project.

## Further Reading

Brandt, John C., and Robert D. Chapman. *Introduction to Comets.* Boston: Cambridge University Press, 2004. Provides a detailed examination of virtually every cometary phenomenon.

Crovisier, Jacques, and Thérèse Encrenaz. *Comet Science.* Translated by Stephen Lyle. New York: Cambridge University Press, 2000. An excellent treatise on comet structure and the study of comets, though somewhat technical in places.

Davies, John. *Beyond Pluto.* New York: Cambridge University Press, 2001. A popular book about the Kuiper Belt, a region beyond Pluto from which many comets are believed to originate.

Levy, David. *Comets: Creators and Destroyers.* New York: Touchstone, 1998. Discusses some of the theories of comet impacts bringing life-giving materials to Earth, and perhaps resulting in some of the past mass extinctions.

Sagan, Carl. *Comet.* Rev. ed. New York: Ballantine Books, 1997. Written by a noted popularizer of science, this work describes the pristine remnants left over from the origin of the solar system called comets. Liberally illustrated with photographs, Sagan's book humanizes the study of science and teaches a history of astronomy along the way.

Verschur, Gerrift L. *Impact: The Threat of Comets and Asteroids.* London: Oxford University Press, 1997. Describes the nature of near-Earth asteroids and comets, and their threat of devastating impacts on Earth.

Whipple, Fred L. *The Mystery of Comets.* Washington, D.C.: Smithsonian Institution Press, 1985. An older book, but one that gives an excellent history of comet studies, as well as an easy-to-understand, basic presentation of the standard model for comets as explained by Fred Whipple himself.

*Raymond D. Benge, Jr.*

# Deep Space Network

*Date:* Beginning December 3, 1958
*Type of facility:* Spacecraft tracking network

*The scientific investigation of the solar system by NASA is being carried out largely through the use of unpiloted, automated spacecraft. None of these missions would be possible, however, without the Deep Space Network, which provides the Earth-based radio communications link to all NASA's uncrewed interplanetary spacecraft.*

## Summary of the System

The Deep Space Network (DSN) consists of three deep-space communications complexes located on three continents: at Goldstone in Southern California's Mojave Desert; near Madrid, Spain; and near Canberra, Australia. These locations allow continuous communications with a spacecraft traveling anywhere in the solar system. Each complex comprises four large parabolic dish antennae, one 26 meters in diameter, two 34 meters in diameter, and one 70 meters in diameter. The Network Operations Control Center in Pasadena, California, controls and monitors operations at the three complexes. The Ground Communications Facility provides the circuits that link the complexes to the control center and to the scientists and engineers who build the spacecraft, design its missions, and receive its data.

The functions of the network are to receive the radio signals from the spacecraft, to transmit commands that control the spacecraft operating modes, and to generate the radio navigation data that are used to locate and guide the spacecraft to its destination.

Beginning in 1959 with the Pioneer 3 and 4 missions to the Moon, the network has evolved from communicating with spacecraft at lunar distances (400,000 kilometers) to providing direct radio links to spacecraft that are approaching the edge of the solar system (a distance of 6.4 billion kilometers). The objective has been not only to extend the com-munications link to unexplored regions in space but also to raise the quantity and quality of scientific data returned from each successive mission.

The smallest unit of information in a computer-driven information system is a bit, which represents one of two possible values—for example, yes or no, on or off, 1 or 0. In 1965, the Mariner 4 Mars mission produced the first television pictures of a planet's surface. Because of the distance from Earth (216 million kilometers), the spacecraft data transmission rate was limited to about eight bits per second, requiring about eight hours to return one 240,000-bit picture. Only twenty-two roughly defined digital pictures were obtained, showing a narrow strip of the planet's surface. Nevertheless, it was a scientific feat that was remarkable for its time.

By 1974, during the Mariner 10 flyby mission to Venus and Mercury, the network successfully received transmissions at a data rate of 117,600 bits per second, which provided a clear, sharp pictorial map of nearly all of the sunlit side of Mercury—158 million kilometers away. To achieve this astounding improvement in approximately nine years, both the spacecraft and the network were modified significantly. The main objectives were to improve the quality and quantity of the scientific data returned by increasing the data transmission rate and to improve navigation accuracy.

The primary obstacle to attaining these objectives was the overwhelming distance involved. After traveling across the vastness of interplanetary space, the spacecraft signal reaching the Earth ranges in power from one-billionth of a watt down to one-billionth-trillionth of a watt. This exceedingly small amount of power results from the necessity of minimizing the size and weight of the spacecraft to fit it within the launch vehicle. The spacecraft transmitter power is typically twenty watts, which is the same power used to light an average refrigerator lightbulb. To generate even twenty watts requires a substantial portion of the spacecraft's total power supply. The main technological elements that make it possible to receive, amplify, and extract scientific data from such ultraweak signals are high microwave radio frequencies (2,110 to 2,300 and 8,400 to 8,450 megahertz), an optimum energy-per-bit data transmission scheme, and the state-of-the-art sensitivity and efficiency of the network antennae and low-noise receiving systems.

The first step in the recovery of an ultraweak spacecraft signal is to capture as much of the energy reaching Earth as possible. The network's large parabolic dish antennae are highly efficient in collecting the spacecraft signal and focusing it into a low-noise amplifier system. Unfortunately, the reception process unavoidably includes radio noise, which is naturally generated by nearly all objects in the universe, including the Sun and Earth. In addition to external noise, the operation of the receiver itself contributes internal noise to the process. If there were no noise sources, an amplifier with sufficient gain could increase the weak spacecraft signal to a useful level. Because there will always be noise surrounding the signal, it is the signal-to-noise ratio, not the signal strength, that matters in a communications system. To minimize the addition of noise to the signal, the first amplifier in the receiving system is a maser (a shortened term for microwave amplification by stimulated emission of radiation), which is kept cooled to 4.2 kelvins, the temperature of liquid helium. The maser adds very little internal noise to the amplification process.

NASA's Deep Space Network antenna in Goldstone, California. It obtained the first radar images of Venus. (NASA CORE/Lorain County JVS)

The degree of efficiency with which the signal-to-noise ratio is handled determines the ability of the network to receive spacecraft data at the required data rates. To improve this efficiency, the spacecraft data are coded, which aids in detecting the data bits. Noise interferes with the bit detection process and causes bit errors. Thus, the coding adds bits (additional information) to the data stream that help in detecting errors. Coding allows the communications link to operate at a lower signal-to-noise ratio.

Some of the data returned by the communications link consist of television images of planetary systems and comets. Transmissions of this type of data require high data rates. Other data, including measurements of magnetic fields, gravity fields, temperatures, and electron particle fields, are transmitted at lower data rates. During an encoun-

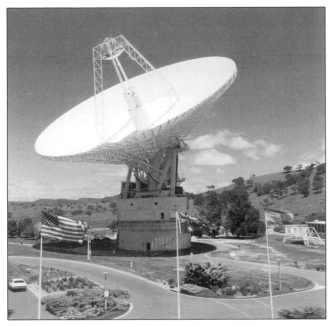

*A view of the Canberra, Australia, 70-meter (230-foot) antenna with flags from the three Deep Space Network sites. The other two sites are in Goldstone, California, and Madrid, Spain. (NASA)*

ter, the high data rates of returned images can range from 21,000 bits per second to 115,000 bits per second, depending on distance. Flyby encounters occur infrequently (every one to three years) and have limited duration (perhaps four months); for most of its lifetime, the spacecraft transmits low-rate data (8 to 7,000 bits per second) and spacecraft-status data (typically 40 bits per second). During low-rate transmissions, which provide more energy per bit, the network receives with its 34-meter-diameter antennae.

The command function of the DSN involves the transmission of information to the spacecraft. Commands implement decisions made by the flight project (the network user) to change the spacecraft operating configuration (its flight path, data rate, instrument activation) either immediately upon receipt or at some definite time in the future. The command link functions at a very low data rate (10 to 1,000 bits per second) and carries a low volume of data that are of extremely high qual-

ity (high energy per bit) to prevent misinterpretation by the spacecraft.

Navigation data are measurements made by the network that are used to determine the spacecraft position and its velocity. The position, range, radial velocity, and acceleration can be determined by extremely precise measurements of the spacecraft radio signal, which require the use of an ultraprecise atomic clock (a hydrogen maser frequency standard). The same radio signal is also a rich source of information about the solar system, because it is affected by magnetic fields, gravity fields, and ionized gases as it travels through space. All three of these functions take place over a single radio link in each direction at the same time.

The design of the communications link for any spacecraft mission must match and balance mission requirements against network capabilities. The scientific objective of the mission requires a certain minimum amount of data to be returned at certain data rates. The bit error rate that will be tolerated in the data stream is another important requirement, and the margin of safety that the network user will accept is another. All these requirements must be balanced with the capacity of the channel to deliver the data, which gradually diminishes as the distance to the spacecraft increases. The Voyager 1 and 2 missions to Jupiter, Saturn, Uranus, and Neptune are prime examples of the use of innovative engineering in the design of a communications link to meet mission requirements.

The Voyager encounters with Jupiter and Saturn were milestones in the evolution of network capabilities. At Jupiter, a data rate of 115,200 bits per second was achieved, which allowed approximately 33,000 very clear pictures of the planet and its five satellites to be received across a distance of 700 million kilometers at the rate of one 5-million-bit picture every 48 seconds. At Saturn, a distance of 1.6 billion kilometers from Earth, the decreasing signal-to-noise ratio reduced the data rate by approximately one-fourth, to 29,000 bits per second.

To capture more spacecraft signal, an experimental antenna array technique, whereby the weak signals received by two or more antennae are combined into a single stronger signal, was successfully employed. The increase in signal power enabled a 50 percent increase in the data rate to 44,800 bits per second, allowing the return of a full-frame picture every 144 seconds. As a result, the network received and delivered thirty thousand high-quality pictures of Saturn, its rings, and its satellites.

After completion of the Saturn encounter, the Voyager 2 mission was extended to Uranus and Neptune. At Uranus, a distance of 3 billion kilometers from Earth, the Voyager signal would drop to less than one-tenth of the signal power received during the Jupiter encounter; at Neptune, 4.6 billion kilometers from Earth, it would drop to less than one-twentieth of the signal power received during the Jupiter encounter.

In order to build on the knowledge of the Uranian system, it would be necessary to obtain scientific data continuously throughout the encounter period and to obtain some 330 pictures during the six-hour near encounter. For several days before and after the encounter, approximately three hundred images per day were requested. Network engineers calculated that the configuration of two 34-meter antennae and one 70-meter antenna would provide a maximum data rate of 19,200 bits per second. This data rate was 10,000 bits short of what was needed to return the desired 330 images during the near encounter.

Looking for an effective but timely and affordable way to increase the data rate, the network sought assistance from the Australian National Radio Astronomy Observatory in Parkes, 225 kilometers from the network complex near Canberra. The Parkes facility is equipped with a 64-meter radio telescope; an analysis of its capability concluded that data rates of 21,600 and 29,200 bits per second could be achieved by adding the telescope to the array at Canberra. Arrangements were made with the Australian radio astronomy community and the Australian government to borrow the telescope for use in tracking the Uranus encounter,

thereby establishing for the first time a cooperative interagency association for the support of a major planetary encounter.

The flight time to Uranus was adjusted in order to have the encounter take place over Australia. The network supplied the necessary receivers, signal combiners, and data recorders and constructed a ground microwave link to relay the Parkes radio signal to Canberra. Starting with initial observations on November 4, 1985, through the six-hour near encounter on January 24, 1986, to the end of the encounter on February 24, 1986, the interagency array returned 2,516 images of Uranus, fulfilling the requirements for high-rate transmission of scientific imaging data.

Neptune is 1.6 billion kilometers beyond Uranus. By adding more interagency antennae and improving the sensitivity of the network antennae and low-noise receiving system, maintaining a 21,600-bit-per-second high-rate capability for the Voyager-Neptune encounter in August, 1989, would be possible. To accomplish this, the network's three 64-meter antennae were extended to 70 meters; the government of Japan agreed to contribute a 64-meter antenna at Usuda to the Parkes-Canberra array; and in the United States, another interagency association was established between the Goldstone complex in California and the National Radio Astronomy Observatory's Very Large Array (VLA) in New Mexico. The VLA consists of twenty-seven 25-meter antennae that can be moved in position along three Y-shaped tracks; the resulting signal capture is the equivalent of that of two and a half 70-meter antennae. The VLA is connected to Goldstone by a domestic communications satellite. With this new capability, the network's status as the most sensitive telecommunications and radio navigation network in the world is ensured.

### Context

NASA was officially established on October 1, 1958. On December 3, 1958, the Jet Propulsion Laboratory (JPL) was transferred to NASA and given responsibility for uncrewed lunar and planetary exploration programs. Eberhardt Rechtin, manager of Telecommunications Development at JPL, was placed in charge of designing and building a network capable of supporting a long-range program of space exploration.

The basic concept was to build a single deep-space communications network to support lunar and planetary missions simultaneously. The network would be a separate entity, responsible for its own research, development, and operation in support of all of its users.

The Deep Space Network has evolved into the largest and most sensitive scientific telecommunications and radio navigation network in the world. Its principal responsibilities are to support uncrewed interplanetary spacecraft missions and to support radio and radar astronomy observations in the exploration of the solar system and the universe. The network is managed by JPL. The antennae, because of their high sensitivity, are also used as radio telescopes to study the structures of stars and are important tools in the search for extraterrestrial intelligence.

The practice of using foreign locations for tracking stations is older than NASA itself. The U.S. armed forces' worldwide Minitrack network was the precursor of both the Deep Space Network and the Spaceflight Tracking and Data Network.

Originally, the DSN had a station in South Africa. This station was closed in 1974 for political reasons, and coverage for that longitude was provided by stations in Ascension Island and Spain.

The DSN has employed indigenous people at the overseas stations as soon as they could be trained. This practice has proved very successful in the operation of the stations and has resulted in deeply rewarding relationships between Americans and the people of the countries involved.

There is another worldwide network that serves the Russian deep-space explorations, but the Russian network does not provide continuous communications, as the Deep Space Network does. Capability in this area has provided the United States with the opportunity to assist other countries engaged in the exploration of the universe. Because of Soviet unwillingness to rely on the goodwill of foreign countries during the Cold War, the Soviet Union's tracking network depended on a fleet of ships outfitted with antennae. Ironically, this independent attitude forced the Soviet Union to ask others for assistance in tracking its spacecraft. The U.S. Deep Space Network provided tracking and data acquisition for some of the Soviet interplanetary missions.

**See also:** Cassini: Saturn; Deep Impact; Galileo: Jupiter; Huygens Lander; International Space Station: U.S. Contributions; Jet Propulsion Laboratory; Magellan: Venus; Mariner 5; Mariner 10; Pioneer Missions 1-5; Pioneer 10; Spaceflight Tracking and Data Network; Spitzer Space Telescope; Ulysses: Solar Polar Mission; Vandenberg Air Force Base.

**Further Reading**

Allday, Jonathan. *Apollo in Perspective: Spaceflight Then and Now.* Philadelphia: Institute of Physics, 2000. This book takes a retrospective look at the Apollo space program and the technology that was used to land humans on the Moon. The author explains the basic physics and technology of spaceflight, and conveys the huge technological strides that were made and the dedication of the people working on the program. All of the major aspects of the Apollo Program are covered, including crews, vehicles, and space suits.

Burrows, William E. *This New Ocean: The Story of the First Space Age.* New York: Random House, 1998. This is a comprehensive history of the human conquest of space, covering everything from the earliest attempts at spaceflight through the voyages near the end of the twentieth century. Burrows is an experienced journalist who has reported for *The New York Times, The Washington Post,* and *The Wall Street Journal.* There are many photographs and an extensive source list. Interviewees in the book include Isaac Asimov, Alexei Leonov, Sally K. Ride, and James A. Van Allen.

Corliss, William R. *A History of the Deep Space Network*. NASA CR-151915. Washington, D.C.: Government Printing Office, 1976. Describes the development of the network from the inception of the technology in the early 1950's to the construction of the stations throughout the world. Describes the early flight projects for which the network provided communications, the contributions made to the disciplines of radio science and radio astronomy, and the state-of-the-art technology that made possible communications to the outer planets.

Davies, John K. *Astronomy from Space: The Design and Operation of Orbiting Observatories*. New York: John Wiley, 1997. This is a comprehensive reference on the satellites that have revolutionized twentieth century astrophysics. It contains in-depth coverage of all space astronomy missions. It includes tables of launch data and orbits for quick reference, as well as photographs of many of the lesser-known satellites. The main body of the book is subdivided according to type of astronomy carried out by each satellite (x ray, gamma ray, ultraviolet, infrared, and radio). It discusses the future of satellite astronomy as well.

Grey, Jerry, ed. *Space Tracking and Data Systems, AAS8*. New York: American Institute of Aeronautics and Astronautics, 1981. Reports on the proceedings of a symposium on space tracking and data systems held in 1981. Provides an overview of activities and plans for the 1980's, supporting developments for the same decade, and new developments extending the network characteristics into the 1990's. Describes programs and operational capabilities of the European, French, German, Japanese, Indian, and Soviet space agencies.

Hey, J. S. *The Radio Universe*. 3d ed. Elmsford, N.Y.: Pergamon Press, 1983. The design of the DSN has its roots in two disciplines. The antennae and their related radio frequency capabilities are based on the technologies developed and used in the field of radio astronomy. This book primarily explains radio astronomy, enabling the reader to understand how it was possible in 1958 to design the network for deep-space telecommunications. Astronomers, using radio techniques in the microwave portion of the electromagnetic spectrum, were studying weak radio signals from the edge of the universe.

Martin, Donald H. *Communication Satellites*. 4th ed. New York: American Institute of Aeronautics and Astronautics, 2000. This work chronicles the development of communications satellites and worldwide networks over the past four decades, from Project Score to modern satellite communication systems.

Poynter, Margaret, and Michael J. Klein. *Cosmic Quest*. New York: Atheneum, 1984. Presents theories about how the universe was formed and how life began in order to understand better how life could have begun in other places and what kinds of "other places" there might be. Discusses how the radio telescope is being used to collect radio wave data and which wavelengths and areas of space seem to offer the best hope for the search.

National Aeronautics and Space Administration. *Space Network Users' Guide (SNUG)*. Washington, D.C.: Government Printing Office, 2002. This users' guide emphasizes the interface between the user ground facilities and the Space Network, providing the radio frequency interface between user spacecraft and NASA's Tracking and Data-Relay Satellite System, and the procedures for working with Goddard Space Flight Center's Space Communication program.

Renzetti, N. A., ed. *A History of the Deep Space Network*. Vol. 1. JPL Technical Report 32-1533. Pasadena, Calif.: Jet Propulsion Laboratory, 1971. Describes the evolution of the technology that made it possible to support both one- and two-way communications between Earth and spacecraft traveling into interplanetary space. Describes each of the deep-space stations of DSN and the communications support provided for the early spacecraft missions to the Moon, Venus, and Mars. Includes photographs of antennae and equipment. Suitable for college students.

*Nicholas A. Renzetti*

# Delta Launch Vehicles

*Date:* Beginning January 25, 1957
*Type of technology:* Expendable launch vehicles

*The Delta launch vehicles became the workhorses of the U.S. space program, playing a crucial role in the history of space exploration since the 1950's.*

## Summary of the Technology

In the mid-1950's, the U.S. Air Force began work on the development of the Thor intermediate range ballistic missile (IRBM). The Thor was to be the Air Force's equivalent of the Army's Jupiter missile. Douglas Aircraft Company of Santa Monica, California, was given the contract for building the airframe. The engine was built by the Rocketdyne Division of North American Aviation, the guidance system by Bell Telephone Laboratories, and the systems engineering by the Aerospace Corporation.

The Thor stood 19.75 meters tall and 266.7 centimeters in diameter (tapering toward the forward end), and at takeoff it developed approximately 667 kilonewtons of thrust. The first attempted test flight of the Thor took place on January 25, 1957. The flight lasted only seconds: The missile exploded after rising only a few centimeters off of the launch pad. After a series of modifications produced successful flights, the Thor was declared operational in June of 1958.

At about that time, it was concluded that the Thor could be used as a first stage of a space vehicle. This first configuration, known as the Thor-Able, was successfully flown in July of 1958. Since then, the Thor has been mated with various upper stages for both National Aeronautics and Space Administration (NASA) and military payloads. The fourth modification of the Thor became known as the Delta (since Δ is the fourth letter of the Greek alphabet). Essentially, the Thor-Delta consisted of modified second and third stages of the Navy's Vanguard satellite launching vehicle united with the Thor as the first stage. This project was initiated as a result of the need for a reliable vehicle to launch civilian payloads into orbit. Because that was the Thor-Delta's function, it was decided to drop the Thor prefix and refer to the rocket as the Delta. A literal or numerical suffix was added to indicate further improvements in the system. Those Thor derivatives that were to be used for military purposes retained the "Thor" title as a prefix.

The first stage of the Delta launch vehicle, a modified Thor, had a length of 18.2 meters and a diameter of 2.44 meters. It produced a liftoff thrust of approximately 667 kilonewtons. The second stage, which was 6.3 meters long and 81.3 centimeters in diameter, produced 34 kilonewtons of thrust at operating altitudes. The third stage was a solid-propellant motor 1.52 meters long and 45.7 centimeters in diameter. It produced a thrust of 12 kilonewtons in space. The guidance system for the rocket was carried in the first two stages. The third stage did not have a guidance system but was spin-stabilized at 140 revolutions per minute by a device located in the second stage.

The first stage had a burning time of about 150 seconds, varying somewhat with the fuel mixture. In a typical flight, this stage would take the vehicle to a height of 66 kilometers. At this point, the first stage would fall away and the second stage would ignite. The 170-second burn of the second stage, plus a few seconds of unpowered flight, would carry the assembly to an altitude of about 260 kilometers. The third stage would then be spun to achieve stability, and the second stage would fall away. The third stage was then ignited and burned for 42 seconds. This burn provided the velocity (approximately 29,000 kilometers per hour) necessary to achieve an orbit. After this velocity was attained, the satellite was separated from the spent third stage. This configuration of the Delta vehicle was capable of sending a 270-kilogram payload into a low-Earth orbit or a 45-kilogram payload into a geosynchronous orbit (an orbit in which the period of revolution is the same as Earth's period of rotation). A satellite in geosynchronous orbit remains over the same position on Earth's surface at an altitude of 35,792 kilometers.

Over the years of its use since 1960, the Delta has had many modifications. The early Delta, as described above, was replaced in October of 1962 by the Delta A. This uprated version featured a more powerful first stage, which improved the payload capacity to a maximum of 320 kilograms.

The Delta B also appeared in 1962. This version, which featured larger second-stage fuel tanks as well as a more powerful motor, increased the payload capacity to 375 kilograms. Later in 1963, improvements in the third stage led to the Delta C. The payload capacity was further increased by this modification to 410 kilograms for low-Earth orbits.

In August of 1964, the first stage of the Delta was augmented with three solid-propellant Castor 1 boosters to form the Delta D. This configuration was also known as the Thrust-Augmented Delta, or TAD. The addition of the solid-propellant boosters uprated the takeoff thrust to nearly 1,500 kilonewtons. The Delta D had the capacity to orbit 580 kilograms. This vehicle, and all future Delta variations, used the strap-on solid-propellant boosters and the uprated main booster engine.

In 1965, NASA introduced the Delta E. The main difference in this vehicle was a virtually redesigned second stage. The original Delta and Delta A rockets had used a 6.3-meter-long, 81.3-centimeter-diameter second stage, which produced a thrust of 34 kilonewtons for a duration of 170 sec-

onds. This stage was increased in length to 7.2 meters for the Delta B, C, and D rockets. The new Delta E second stage was 3.96 meters long and 138.9 centimeters in diameter, and it produced a thrust of 34 kilonewtons for 380 seconds. Modifications were also made to the third stage. The new stage was 1.52 meters long and 50.8 centimeters in diameter, and it developed 12 kilonewtons of thrust. This Delta modification was capable of lifting a payload of 735 kilograms into a low-Earth orbit or could place 550 kilograms into a geosynchronous orbit.

Although the Delta E was used until 1971, in 1968 the Delta L was introduced. It was essentially a lengthened first stage with improved solid-propellant boosters. With the additional length came a new constant diameter of 243.8 centimeters. Prior to the Delta L, the Thor had an inward taper at the forward end. Thus the Delta L was of-

*A Delta launch vehicle ready for liftoff at Cape Canaveral Air Force Station.* (NASA)

ten called the Long-Tank Delta (LTD), or Long-Tank, Thrust-Augmented Delta (LTTAD). All Delta vehicles subsequent to the Delta L would be without the original Thor's taper.

The first-stage main engine of the Delta L produced 765 kilonewtons of thrust. This was supplemented by the three Castor strap-on solid-propellant boosters, which added 680 kilonewtons of thrust. The upper stages of the Delta L were identical to the Delta E. A slight alteration in the third stage produced the Delta M. In the basic two-stage configuration, it was the Delta N. Twenty of the Delta L, M, and N series were launched by NASA between 1968 and 1972.

Four times during the years 1970 and 1971, mission requirements necessitated the addition of more than three strap-on solid-propellant boosters. Six Castor strap-ons were used to produce vehicles known as the Delta M-6 and N-6. These were the same as the Delta M and N, respectively, except for the three additional boosters. As usual, only three solid-propellant boosters ignited with the main engine during liftoff. Thirty-one seconds into the flight, the additional three solid-propellant boosters ignited. This had the effect of increasing the duration of the booster burn. With the addition of the three Castors, payload capacity was increased to 1,300 kilograms.

In 1972, the Delta 900 series was introduced. There were now a total of nine Castor strap-on solid-propellant boosters clustered around the base of the vehicle. There was also an improved second stage, which produced nearly 8,900 newtons more thrust than the second stage of the previous Delta.

At launch, the main engine of the Delta 900 would ignite, as well as six of the Castors. This combination produced a total thrust of 2,135 kilonewtons for a period of 39 seconds. At this time the second group of three Castors would be ignited for an additional 39-second burn. When they burned out and were jettisoned, the main engine continued to burn for an additional 142 seconds.

Eight seconds after the first stage had shut down, a separation occurred. After coasting for

four seconds, the second stage ignited. The duration of the burn time of the second stage depended upon the payload size and the desired orbit. In some cases a second burn, after a period of coasting, was necessary. The second stage had the capability of being restarted in space. A two-stage Delta 900 could place a 1,600-kilogram payload into a low-Earth orbit. A three-stage Delta 904 could place a 635-kilogram payload into a geosynchronous orbit.

A powerful second stage was added to the vehicle in 1974 to form the Delta 2000 series. The rocket motor used to form this second stage was a modified version of the motor that powered the lunar-landing module in the Apollo Program. It was 6.2 meters long and 1.5 meters in diameter, and it weighed more than 6,000 kilograms. The third stage remained the same as that used in the Delta 1000.

The first-stage main engine was changed from the MB-3 engine, which had powered the Delta since 1960, to the Rocketdyne RS-II7, which was developed from the giant Saturn rockets. The new engine produced 900 kilonewtons of thrust for 228 seconds. With its new engine and six Castor solid-propellant boosters firing simultaneously at liftoff, the thrust was 2,300 kilonewtons. At that point in its evolution, the Delta 2000 series rocket bore little resemblance to the original Delta of 1960. The original Delta could place 270 kilograms into a low-Earth orbit. A Delta 2910 could place 2,000 kilograms into a low-Earth orbit and could send 703 kilograms into a geosynchronous orbit.

The Delta 3000 series of launchers was developed to provide a still more powerful vehicle. The 3914 differs from the 2914 in that it has larger and more powerful solid-propellant booster rockets. Because of the increased power from the solids, only five were used on the Delta 3914 at liftoff instead of six, as used on the Delta 2914. After burnout of the five solids, the remaining four would be ignited. The Delta 3914 was capable of placing 930 kilograms into a geosynchronous orbit.

The final version was the Delta 3916/PAM. This configuration stands 35.4 meters tall and develops

3,200 kilonewtons of thrust at launch. Its liquid-fueled first stage is augmented by nine Castor solid-propellant strap-on boosters, six of which fire at lift-off. The other three ignite 58 seconds into the flight. The improved second stage is also liquid-fueled. A Payload Assist Module (PAM) is used on this vehicle instead of the usual solid-propellant third stage. With these modifications, the Delta 3916/PAM can lift 1,270 kilograms into a highly elliptical orbit for transfer into a geosynchronous orbit by the PAM unit.

In February, 1989, the U.S. Air Force launched the first Delta II vehicle. It carried all twenty-one of the operational Global Positioning System satellites. The Delta II 7900-series launch vehicle consists of two liquid-fuel stages with a spin-stabilized solid fuel stage on top. Nine strap-on solid fuel motors augment the thrust of the first stage and per-

mit the lofting of nearly 1,500 kilograms of payload into a geosynchronous orbit. Delta II can carry more than double the weight that the Delta vehicle could manage in the 1980's. Over its forty-year life span, the Delta family of vehicles has racked up what is perhaps the most successful flight record of any rocket currently in service. Of 275 flights, only 15 have been total failures, a success rate of 94.5 percent. The Delta II has had only one total failure (and one partial failure, Koreasat-1)—an incredible 98 percent success rate. The Delta II launched the first eight Globalstar satellites and has lofted fifty-five satellites out of seventy-two in the completed Iridium constellation.

Other successful Delta II launches of note include NEAR Shoemaker (Near Earth Asteroid Rendezvous, February 17, 1996), Mars Global Surveyor (November 7, 1996), Mars Pathfinder (December 4, 1996), Deep Space 1/SEDSAT-1 (October 24, 1998), Mars Climate Orbiter (December 11, 1998), Mars Polar Lander (January 3, 1999), Stardust (February 7, 1999), Landsat 7 (April 15, 1999), IMAGE (Imager for Magnetopause to Aurora Global Exploration—March 25, 2000), Mars Exploration Rovers (June 10 and July 7, 2003), Spitzer Space Telescope (August 25, 2003), and the Swift observatory (November 20, 2004).

The Delta III was developed by Boeing Aircraft Company as a commercial venture to fulfill customer needs for a launch service to accommodate growing satellite sizes. Its first launch was in August, 1998. The RS-27A main engine powers the Delta III first stage. Boeing increased the diameter of the first-stage fuel tank from Delta II to reduce the overall length of the vehicle and improve control margins. To add to Delta III first-stage performance, Boeing uses nine 1.17-meter-diameter solid rocket motors, which are derived from those on Delta II but are larger and produce 25 percent more thrust. Three of the new motors are equipped with thrust-vector control to further improve vehicle maneuverability and control. The Delta III second-stage Pratt & Whitney R110B-2 engine is derived from the R110 power plant flown for more than three decades. This second stage carries

*The Boeing Delta II rocket carried NASA's Far Ultraviolet Spectroscopic Explorer satellite into space.* (NASA)

more propellant than Delta II and burns cryogenic (cold) fuels, which produce more energy, allowing lift of heavier payloads. With a payload delivery capacity to geosynchronous transfer orbit of 3,800 kilograms, Delta III effectively doubles the performance of the Delta II.

The Delta IV launch vehicle uses a new, liquid oxygen/hydrogen "common core" booster powered by a single RS-68 engine. There are several variants of the Delta IV launch vehicle. The Delta IV Medium, designed to replace the Delta II, combines the common core booster with the current Delta II second stage and 3-meter fairing. The larger Delta IV Medium Plus (4, 2), which combines the common core and two solid strap-on motors with a modified Delta III liquid oxygen/hydrogen second stage and 4-meter fairing, is capable of placing 5,760 kilograms into a geosynchronous transfer orbit (GTO). The Delta IV Medium Plus (5, 2), which combines the common core and two solid strap-on motors with a large Delta III-type liquid oxygen/hydrogen second stage and a new 5-meter fairing, can place 4,800 kilograms into a GTO orbit. The Delta IV Medium Plus (5, 4), which combines the common core and four solid motors with a large Delta III-type liquid oxygen/hydrogen second stage and a new 5-meter fairing, can place 6,700 kilograms into a GTO orbit. Finally, the Delta IV heavy combines three common core boosters with a large Delta III-type liquid oxygen/hydrogen second stage and a new 5-meter fairing to place 13,200 kilograms into a GTO orbit. Together, the Delta IV variants are capable of replacing the Delta II and Delta III as well as the heavy Titan IV. In essence, Boeing hopes to use the Delta IV family to address the bulk of the existing and future commercial and government launch market. The initial Delta IV booster was launched on November 20, 2002. The initial Delta IV Heavy booster was launched on December 22, 2004. As the Bush Moon-Mars Initiative develops, the Delta IV Heavy would be under consideration as the launch vehicle for the Crew Exploration Vehicle (CEV), the first new American piloted spacecraft since the space shuttle orbiter. However, in the second half of 2005, NASA studies investigated the advantages of using shuttle-based hardware for a separate heavy-lift booster for cargo and a solid rocket booster-based launch vehicle for the Crew Exploration Vehicle. A decision in 2006 was expected to lead to contracts for the new boosters to support astronauts' continued exploration beyond low-Earth orbit.

**Contributions**

Probably the most significant advance in space methodology occurring during the Delta program is the adaptation of an existing vehicle to accommodate payload requirements, instead of creating another type of booster from scratch. Delta was a composite of already proven components and vehicles. There were spacecraft ready and waiting to be launched, but none of the vehicles under development could have been launched reliably in the near future. The Space Race was already developing; within a few years, both the American and Soviet space programs would send men into space with large launch vehicles. Therefore, various spacecraft containing vital experiments needed to be launched immediately. The composite vehicle, Delta, filled that need.

Delta has undergone fourteen major configuration changes, resulting in a twenty-fold payload capacity increase. The changes varied from thrust augmentation (through the Payload Assist Module) to lighter-weight payload shrouds and an increased volume of the payload area.

Delta enabled scientists to explore and to utilize space for many purposes. The early communications satellites paved the way for an incredible variety of experimentation, including meteorology, solar observation, magnetic field investigations, astronomy, geodesics, biology, and Earth resources mapping. Interaction with other nations has become common as a result of the scientific cooperation made possible by Delta. Delta has chiefly launched experimentation spacecraft for West Germany, France, Japan, Canada, Indonesia, Italy, the United Kingdom, and the European Space Agency.

Another technological advance is the variety of orbits achieved through the Delta program. In addition to the orbital complexity, from low-Earth orbit to high, geosynchronous orbit, orbits around the Moon or the Sun were added to the repertoire.

Delta is well known for its reliability. The use of flight-proven subsystems and components from other launch vehicle systems seems to be a major factor, along with an ongoing attempt to improve the system. The end result is a vehicle with high operability and a relatively low cost per launch, making Delta one of the real bargains for both domestic and governmental use. The first twenty years of Delta launch history produced a 94 percent success rate. In later years, the success rate has climbed to an amazing 98 percent. Trusted to carry extremely valuable payloads whose loss would be catastrophic for many agencies, the Delta launcher is one of the busiest, most successful ever designed.

### Context

In 1957, when the Delta launch vehicle came into use, there was a veritable explosion of activity in the world's space programs. Both the Soviets and the Americans were attempting to create a large vehicle, capable of delivering a nuclear warhead to a desired destination. The American Saturn, designed to launch a lunar mission, was in the conceptual stage. The Atlas Program had its first full-scale launch in 1957. The Soviets were developing their own vehicle, along the same scale as the Atlas. By 1961, they had successfully launched the first human in space. American scientists knew that a smaller payload could be launched in 1957, even though some guidance problems were still to be solved. Early experimentation was extremely important to determine the dangers of launching a human into space. Earth's gravitational fields, weather, and magnetic fields complicated many tests. Tests had been run with smaller rocketry, including the Aerobee, but more power was needed. Therefore, rocket technologists decided to build the composite Delta. One of the first successful launches, that of the Echo communications satellite, by means of which President Dwight D. Eisenhower broadcast his famous Christmas message, is credited both to the Atlas Program and to Delta.

As an "interim" vehicle, the Delta system has been remarkably long-lived. It is probably the busiest, most versatile of all launch systems. Many nations use the Delta for valuable experimentation that has led not only to rapid space technology advances but also to advances in many Earth-related sciences. One of the more recent developments is the use of long-range cameras to record and document the changes occurring on Earth's surface, to manage the world's resources better. International endeavors to improve humankind's situation on Earth, such as search-and-rescue missions, require the complex orbitry possible through Delta's PAM system. Continuing cooperation among nations, especially among nations that do not possess significant launch capabilities, is made not only feasible, but relatively easy, through the use of the Delta launch vehicle.

**See also:** Apollo Program; Atlas Launch Vehicles; Cooperation in Space: U.S. and Russian; Deep Space Network; Environmental Science Services Administration Satellites; Interplanetary Monitoring Platform Satellites; Launch Vehicles; Saturn Launch Vehicles; SMS and GOES Meteorological Satellites; Soyuz Launch Vehicle; Titan Launch Vehicles.

### Further Reading

Baker, David. *The Rocket: The History and Development of Rocket and Missile Technology.* New York: Crown, 1978. A highly detailed, well-illustrated volume telling the story of rocketry from the invention of gunpowder to a human on the Moon. Suitable for a general audience.

Braun, Wernher von, and Frederick I. Ordway III. *Space Travel: A History.* Rev. ed. New York: Harper & Row, 1985. A well-illustrated, chronologically narrated history of rocketry and the theory of spaceflight from the ancient Chinese to the early Apollo Program. Suitable for general readers.

Burrows, William E. *This New Ocean: The Story of the First Space Age.* New York: Random House, 1998. This is a comprehensive history of the human conquest of space, covering everything from the earliest attempts at spaceflight through the voyages near the end of the twentieth century. Burrows is an experienced journalist who has reported for *The New York Times, The Washington Post,* and *The Wall Street Journal.* There are many photographs and an extensive source list. Interviewees in the book include Isaac Asimov, Alexei Leonov, Sally K. Ride, and James A. Van Allen.

Covault, Craig. "Cape Gears for New Vehicles, Launch Surge." *Aviation Week and Space Technology,* March 13, 2000, 54. This article looks at Cape Canaveral and the Kennedy Space Center as the 50th anniversary of the Cape approaches. It discusses some of the Cape's history and looks at its current operations.

Emme, Eugene M., ed. *The History of Rocket Technology: Essays on Research, Development, and Utility.* Detroit: Wayne State University Press, 1964. A collection of fourteen papers written by scientists and historians, covering the development of rocketry from Robert H. Goddard's first liquid-fueled rocket through Project Mercury. Suitable for the layperson.

Heppenheimer, T. A. *Countdown: A History of Space Flight.* New York: John Wiley, 1997. A detailed historical narrative of the human conquest of space. Heppenheimer traces the development of piloted flight through the military rocketry programs of the era preceding World War II. Covers both the American and the Soviet attempts to place vehicles, spacecraft, and humans into the hostile environment of space. More than a dozen pages are devoted to bibliographic references.

Isakowitz, Steven J., Joseph P. Hopkins, Jr., and Joshua B. Hopkins. *International Reference Guide to Space Launch Systems.* 3d ed. Reston, Va.: American Institute of Aeronautics and Astronautics, 1999. This best-selling reference has been updated to include the latest launch vehicles and engines. It is packed with illustrations and figures and offers a quick and easy data retrieval source for policymakers, planners, engineers, launch buyers, and students. New systems included are Angara, Beal's BA-2, Delta 3 and IV, H-2A, VLS, LeoLink, Minotaur, Soyuz 2, Strela, Proton M, Atlas-3 and -5, Dnepr, Kistler's K-1, and Shtil, with details on Sea Launch using the Zenit vehicle.

"Launch Vehicles." *Aviation Week and Space Technology,* January 17, 2000, 144-145. This table details the specifications for each of the current (2000) launch vehicles and spacecraft, as well as the current status.

Ley, Willy. *Rockets, Missiles, and Men in Space.* New York: Viking Press, 1967. A very detailed work starting with the ideas of astronomers such as Galileo Galilei and Johannes Kepler and covering events up to humans in space. The text, suitable for general readers, is supplemented by more technical appendices.

Mari, Christopher, ed. *Space Exploration.* New York: H. W. Wilson, 1999. Twenty-five articles, covering the current state of the space program, reprinted from magazines, are divided into five sections: John H. Glenn, Jr.'s return to space, the exploration of Mars, the International Space Station, recent mining efforts by commercial industries, and new types of space vehicles and propulsion systems.

National Aeronautics and Space Administration. *Countdown! NASA Launch Vehicles and Facilities.* Washington, D.C.: Government Printing Office, 1978. A collection of short articles giving data on various NASA launch vehicles, both active and inactive, and a description of NASA facilities. Suitable for general audiences.

Ordway, Frederick I., III, and Mitchell Sharpe. *The Rocket Team*. Burlington, Ont.: Apogee Books, 2003. A revised edition of the acclaimed thorough history of rocketry from early amateurs to present-day rocket technology. Includes a disc containing videos and images of rocket programs.

Zimmerman, Robert. *The Chronological Encyclopedia of Discoveries in Space*. Westport, Conn.: Oryx Press, 2000. Provides a complete chronological history of all crewed and robotic spacecraft and explains flight events and scientific results. Suitable for all levels of research.

*David W. Maguire*

# Dynamics Explorers

*Date:* February 22, 1978, to February 28, 1991
*Type of spacecraft:* Earth observation satellites

*The Dynamics Explorer spacecraft provided vital data on the energy and momentum of charged particles in Earth's upper atmosphere. They were the first satellites to be placed in Earth orbit for this purpose.*

## Key Figures

*George D. Hogan*, Dynamics Explorer project manager, Goddard Space Flight Center
*Richard A. Hoffman*, Dynamics Explorer project scientist

## Summary of the Satellites

The Dynamics Explorers were designed to investigate the interactions of the region of charged particles surrounding Earth at altitudes above 5,000 kilometers, called the magnetosphere, and the region of charged and neutral particles surrounding Earth below about 2,000 kilometers, called the ionosphere. The charged particles in these regions are often referred to as the magnetospheric and ionospheric plasma.

The proposal to launch two spacecraft into orbits with 90° inclinations and different altitudes arose initially from informal discussions among scientists from many nations in 1974. As a result of the strong case presented by the scientific community, a team was selected by the National Aeronautics and Space Administration (NASA) to design a program to investigate magnetosphere-ionosphere-atmosphere interactions. The initial approach, called the Electrodynamics Explorer, was presented to NASA in the fall of 1976, but budget constraints resulted in approval of a more limited project, the Dynamics Explorer (DE), in the fall of 1977. This mission was to consist of two similarly instrumented satellites able to take measurements at different points in space that are connected by Earth's magnetic field. The instrument payload of each satellite was selected not only to provide the diagnostic measurements, however, but also to perform a variety of experiments; the main object of the DE mission was to provide multiple data sets to advance scientific understanding of coupling processes rather than simply to observe a single geophysical condition.

On August 3, 1981, two satellites were launched into orbits with 90° inclinations on a single Delta 3913 rocket from the Western Test Range in California. The low-altitude satellite, designated DE-2, was placed in an elliptical orbit with a perigee (closest distance to Earth) of 299 kilometers and an apogee (farthest distance from Earth) of 1,003 kilometers. It was stabilized in its orbit by using a momentum wheel to ensure that one axis always pointed toward Earth's center and that the spin axis of the satellite was perpendicular to the orbital plane. The third stage of the Delta rocket was used to boost the high-altitude satellite, designated DE-1, into an elliptical orbit with a perigee of 675 kilometers and an apogee of 24,000 kilometers. DE-1's spin axis also was perpendicular to the orbital plane, and subsequent deployment of its booms and antenna, for magnetic and electric field measurements, stabilized its spin rate to ten revolutions per minute.

Each Explorer was a short cylinder designed and built by RCA's Astro-Electronics Division and powered by efficient solar cells that covered the sides and one end. There was only one failure: A pair of electric field booms on DE-2 could not be completely extended along the spacecraft's spin axis.

As the two spacecraft orbited Earth, the perigee and apogee positions moved around the orbits and gradually decayed in altitude. This allowed a variety of scientific investigations to be carried out. A most significant one took place in the winter of 1981-1982, when the apogee of DE-1 moved over the North Pole, allowing the first global views of the aurora borealis to be recorded with the DE's imaging experiment. Auroras are produced when charged particles in the magnetosphere are accelerated and collide with neutral particles in Earth's upper atmosphere. These pictures provided the first information on the aurora's development over periods as long as five hours, with a time spacing of twelve minutes. As DE-1 was transmitting auroral images, the instruments on DE-2 were measuring neutral and charged particle characteristics at lower altitudes, closer to where the optical emissions originate. Similar instrumentation on DE-1, which was above the emission region, enabled the coupling processes between high and low altitudes to be evaluated.

Approximately three months after launch, the orbital planes of the spacecraft became perpendicular to a line from Earth to the Sun. In this configuration, one end of each spacecraft was pointing at the Sun, and the stabilized DE-2 had to be positioned so that the end containing heat radiators pointed away from the Sun. This maneuver was accomplished by utilizing magnetic torquing coils on the spacecraft that use the force between them and Earth's magnetic field to rotate the spacecraft slowly. Such a procedure was used successfully every six months. The spinning DE-1 satellite did not require such a maneuver to protect it from overheating.

During the first eighteen months of the mission, scientists conducted many investigations by coordinating measurements from each satellite and from ground-based observatories. Because the solar panels around each satellite could not always face the Sun, the amount of power available from the solar batteries varied throughout this period. Careful scheduling of the instruments' operation times was required to accomplish the desired scientific objectives without discharging the batteries and permanently damaging the spacecraft's electrical systems. The successes were the result not only of the DE investigators' participation but also of a vigorous guest investigator program that was instituted one year after the launch of the satellites. This program allowed many scientists other than the principal investigators to have access to the DE data sets and to contribute to their interpretation.

Among the investigations were intense campaigns of coordinated measurements taken during a time of increased solar activity. The associated warming and expansion of Earth's atmosphere subjected the DE-2 satellite to increased frictional drag. Consequently, DE-2's orbit decayed more rapidly than had been expected, and in March, 1983, the satellite reentered the atmosphere and was destroyed. The DE-1 spacecraft was not adversely affected by the high solar activity, thanks to its greater perigee, and it remained operational. It was fortunate that the DE-1 satellite remained active after the demise of DE-2, for it provided valuable images of Earth's auroral emissions through a period of declining solar activity and accumulated measurements of the electric and magnetic fields and plasma around its orbit.

In January, 1986, a unique opportunity arose to view Halley's comet via the imaging experiment on DE-1, because one of the imaging detectors was sensitive to the hydrogen atmosphere that surrounded the comet. Detailed pictures of this atmosphere were transmitted during the month of January as Halley's comet traveled toward the Sun and the hydrogen around the comet expanded. In 1988, the DE-1 satellite continued to operate as the solar activity again began to rise. NASA officially retired the DE-1 satellite, which acquired the first global images of the aurora, on February 28, 1991, after nine years of collecting scientific data. Designed to operate for three years, DE-1 had performed for nearly a decade in space.

The satellite's long life was inevitably accompanied by some subsystem failures. The electrical power system lacked its original flexibility, and continued operation required careful planning and consideration of solar conditions. Despite these difficulties and the shorter lifetime of DE-2, however, the program was an outstanding success. The successes were both scientific and operational and were dependent not only on the U.S. scientific community's skills but also on the dedication of those in the satellite control center and those responsible for receiving the immense amount of data that the DEs collected.

**Contributions**

It should be emphasized that most of the scientific gains from the DE mission have come from the powerful combination of measurements taken simultaneously by a variety of instruments at different altitudes.

The acquisition of multiple-instrument data sets resulted in many advances in scientists' understanding of Earth's plasma environment and its interaction with the Sun's interplanetary magnetic field (IMF) and its charged particle emissions, the solar wind. Measurements taken by the electric field and charged particle velocity instrument on DE-2 illustrated the complexity of the circulation patterns of the charged particles around Earth and their dependence on the magnitude and orientation of the IMF. Similar complexity was seen in the images of the auroral zone obtained from DE-1; the aurora was observed to expand and contract as well as broaden and sometimes bifurcate in response to changes in the IMF magnitude and orientation. Electric field measurements at widely spaced altitudes on DE-1 and DE-2 revealed some of the processes that couple electric fields along Earth's magnetic field lines. These processes involve the acceleration of charged particles in the magnetosphere, which in turn produces the auroral emissions. The relationships between the particle acceleration processes and the generation

and propagation of electric and magnetic field waves throughout the magnetosphere was studied by the plasma wave instrument on DE-1 and by measurement of electromagnetic waves on the ground.

DE-2's neutral particle detectors demonstrated the importance of coupling between the charged and neutral particles in the upper atmosphere by showing that the motion of neutral particles mimics that of the charged particles. These measurements were also used to study the details of energy balance in the upper atmosphere. Measurements of charged particle composition during a period of high solar activity added to scientific knowledge of the ionosphere's behavior. Measurements of low-energy plasma at widely spaced altitudes also allowed the first studies of the processes by which particles are transferred from the ionosphere to the magnetosphere.

**Context**

The Dynamics Explorers were the natural extension of two previous Explorer missions. The Atmosphere Explorers were a series of satellites, the last of which was launched in mid-1970, designed to examine the atmosphere's response to solar radiation. These missions resulted in significant advances in atmospheric chemistry and dynamics, but they also highlighted the need to understand more about the magnetosphere's energy inputs. The International Sun-Earth Explorers were a series of satellites launched in the 1970's to study the interactions of the solar wind and the IMF with Earth's magnetosphere. Again, significant advances were made; this mission also provided the opportunity to divert one of the spacecraft to the Comet Giacobini-Zinner. Nevertheless, it became clear that a complete understanding of solar-terrestrial interaction would require more data on the interactions between the magnetosphere and the ionosphere. The DE mission was undertaken to fulfill this need.

The DEs provided the first demonstration of the optical technology required to image the relatively faint auroral emissions in the presence of sunlit

Earth. They also provided the unique opportunity to examine plasma properties at different locations on a magnetic field line. These two capabilities were essential to the success of the mission. The spacecraft design's soundness and longevity also allowed the United States to participate in the international effort to observe Halley's comet during its 1986 appearance.

The DE project continues to produce vast amounts of data that must be kept accessible to the large number of investigators throughout the United States. Computer facilities have evolved from a central computer and data repository to distributed systems utilizing optical disk storage and high-speed network communications.

By studying Earth's environment and its interaction with the Sun, scientists are better able to assess the properties that make it unique in the solar system. The Dynamics Explorers were devoted to an understanding of the electrodynamics of the upper region of this environment, a region coupled to the lower atmosphere and to the electrical circuits that determine the intensity and location of thunderstorms and weather patterns.

**See also:** Earth Observing System Satellites; Explorers: Air Density; Explorers: Atmosphere; Explorers: Ionosphere; Geodetic Satellites; Global Atmospheric Research Program; Heat Capacity Mapping Mission; Landsat 1, 2, and 3; Landsat 4 and 5; Landsat 7; Mission to Planet Earth; Seasat.

### Further Reading

Burgess, Eric, and Douglas Torr. *Into the Thermosphere.* NASA SP-490. Washington, D.C.: Government Printing Office, 1987. A history of the Atmosphere Explorer program and how it led to the Dynamics Explorers. Suitable for high school and college levels, the book describes how the Atmosphere Explorer mission was conceived and carried out and describes the instrumentation of the later satellites, with numerous illustrations and schematics. It explains how missions are conducted by large teams of investigators, keeps the reader informed of the latest questions being asked, and lists the program's accomplishments.

Deehr, C. S., and J. A. Holtet, eds. *Exploration of the Polar Upper Atmosphere.* Boston: D. Reidel, 1981. This book is a collection of articles resulting from an international meeting of scientists that focused on the interaction of Earth's upper atmosphere with the magnetosphere. Many of the articles are of the tutorial type and are useful to the nonspecialist. Some are more complicated, but these can be easily identified by their elaborate titles. Each of the articles is well illustrated and contains information related to the scientific objectives of the DE mission. Extensive references to detailed research in the field are also provided.

Gavaghan, Helen. *Something New Under the Sun: Satellites and the Beginning of the Space Age.* New York: Copernicus Books, 1998. This book focuses on the history and development of artificial satellites. It centers on three major areas of development—navigational satellites, communications satellites, and weather observation and forecasting satellites.

Hultqvist, Bengt, and Tor Hagfors, eds. *High-Latitude Space Plasma Physics.* New York: Plenum Press, 1983. This book is a collection of review papers designed to keep the nonspecialist apprised of current research achievements in the field. The reader must make a concentrated effort to benefit fully from the information, but the papers are well illustrated and many points can be appreciated even if not fully understood.

Johnson, Francis S., ed. *Satellite Environment Handbook.* 2d ed. Stanford, Calif.: Stanford University Press, 1965. A detailed account, by various contributing authors, of different regions and particle environments that are experienced by Earth-orbiting satellites. The book provides background information on the major processes that take place in the upper atmosphere and is well illustrated with graphs and schematics showing the variations of almost all the geophysical parameters discussed in this article. It not only exposes the reader to the variability of Earth's environment but also provides a useful reference when more quantitative work is required.

Kivelson, Margaret G., and Christopher T. Russell. *Introduction to Space Physics*. New York: Cambridge University Press, 1995. A thorough exploration of space physics. Some aspects are suitable for the general reader. Suitable for an introductory college course on space physics.

Zaehringer, Alfred J., and Steve Whitfield. *Rocket Science: Rocket Science in the Second Millennium*. Burlington, Ont.: Apogee Books, 2004. Written by a soldier who fought in World War II under fire from German V-2 rockets, this book includes a history of the development of rockets as weapons and research tools, and projects where rocket technology may go in the near future.

Zimmerman, Robert. *The Chronological Encyclopedia of Discoveries in Space*. Westport, Conn.: Oryx Press, 2000. Provides a complete chronological history of all piloted and robotic spacecraft and explains flight events and scientific results. Suitable for all levels of research.

*Rod Heelis*

# Early-Warning Satellites

*Date:* Beginning February 26, 1960
*Type of spacecraft:* Military satellites

*Early-warning satellites have orbited the globe since the beginning of the space era. They are designed to detect firings of intercontinental ballistic missiles and to transmit warnings to Earth. They contain sophisticated high-definition cameras and infrared-sensing and scanning equipment.*

## Summary of the Satellites

With the launch of the Soviet satellite Sputnik 1 on October 4, 1957, the "Space Race" between the United States and the Soviet Union began in earnest. Americans suddenly had an urgent need to develop a space program that would counter a perceived threat from the Soviet Union. Shortly after Sputnik 1 was launched and detected over the United States, President Dwight D. Eisenhower was compelled to assure the American people that the satellite did not represent a military threat and that efforts were already under way to launch the first American satellite into space.

In fact, long before the launch of Sputnik 1, the U.S. Armed Forces had been developing space hardware, including rockets and orbital research equipment. Among the promising applications for this equipment was the establishment of a consistent, reliable, instantaneous orbital warning system to detect missile firings in Soviet territory. In the Soviet Union, similar projects were progressing, also under military direction.

The first U.S. early-warning satellite system was organized in 1958 by the Advanced Research Projects Agency (ARPA), which had been created by the Department of Defense. It was called MiDAS (Missile Defense Alarm System). Before its first launch in 1960, however, the MiDAS program was taken over by the Air Force, which by that time had been designated overseer of military space program development in conjunction with the civilian National Aeronautics and Space Administration (NASA).

The first satellite in this series, Midas 1, was launched on February 26, 1960, but it failed to achieve orbit because of a rocket malfunction. Midas 2 was launched successfully on May 24, 1960, aboard a three-stage Atlas-Agena rocket; it went into a circular orbit that took it over southern China and completed one orbit every 94.3 minutes. The satellite weighed 11,000 kilograms and carried instruments weighing 7,200 kilograms. It was cone-shaped, with the nose pointed toward Earth to allow infrared-sensing equipment to detect the firing of missiles. During the orbital life of Midas 2, other U.S. rocket launches were scheduled to allow the Air Force to test the efficiency of the detection equipment aboard the satellite.

Midas 3 was launched on July 12, 1961. It weighed 7,700 kilograms and included a large instrument package. It was launched aboard an Atlas-Agena B rocket designed to boost the satellite into a high polar orbit, which would take it over every point on Earth. Midas 3 went to work almost immediately, traveling over the Soviet Union in its second orbit. It orbited the Earth every 160 minutes at a distance of 2,977 kilometers.

Midas 4 was launched on October 21, 1961, also aboard an Atlas-Agena B rocket. It circled the globe every 2 hours and 52 minutes at an altitude of 3,379 kilometers; its orbit was also polar, and, like Midas 3, it covered every point on Earth.

Midas 4 gained international notoriety as a result of a controversial experiment. Seventy-five pounds of copper needles were dumped into space shortly after orbit was achieved as part of an ill-fated communications experiment designed to determine whether the needles would reflect radio signals transmitted from Earth. An outcry from the international space community failed to persuade President John F. Kennedy to cancel the project, which was conducted by the Massachusetts Institute of Technology for the Air Force.

Shortly after the launch of Midas 4, the Air Force began to cloak the MiDAS program in secrecy. Subsequent launches of similar detection equipment aboard two Discoverer satellites raised concern that the MiDAS system was prone to false alarms. This erraticism was partly the result of the reflection of sunlight off clouds in the upper atmosphere.

In 1963, a new generation of early-warning and reconnaissance satellites began to appear, including some that were recoverable. Approximately twenty were launched per year. With relatively short orbital lives (two weeks or less), these satellites were replaced by a third generation in mid-1966. Aloft for a minimum of fifteen days, these satellites carried high-resolution cameras and equipment that scanned wavelengths in the infrared portion of the electromagnetic spectrum.

Between 1968 and 1971, a generation of geo-synchronous early-warning satellites was launched. ("Geosynchronous" refers to an orbit which is over the equator with the satellite moving in the same direction as the Earth and remaining fixed over a given point on Earth's surface.) These were part of the Ballistic Missile Early-Warning System (BMEWS), the precursor to the Integrated Missile Early-Warning Satellite (IMEWS) system that was launched beginning in November of 1970. These satellites contained infrared telescopes and television cameras that could detect the firing of a Soviet intercontinental ballistic missile (ICBM) within seconds of launch.

In June, 1971, a huge 13,608-kilogram satellite dubbed Big Bird was launched to engage in reconnaissance and, to some extent, in early-warning detection. It was the first in a series of highly sophisticated satellites that contained cameras with the

ability to achieve a resolution of 0.3 meter from an altitude of 161 kilometers. Each also contained six recoverable reentry spacecrafts.

In December, 1976, the Keyhole 11 (KH-11) reconnaissance satellite was launched for the first time. It was similar to Big Bird but transmitted images in digital electronic format rather than via recoverable film spacecrafts. The receiving center for early-warning satellite data in the United States is the North American Air Defense Command (NORAD) headquarters, located 425 meters beneath Cheyenne Mountain in Colorado. Signals from the satellites are picked up at receiving sites in Australia and Guam.

The first Soviet early-warning satellite was Kosmos 169, launched in May, 1967. Early Kosmos satellites were placed in twelve-hour elliptical polar orbits of 500 by 40,000 kilometers. In December, 1968, Kosmos 260 was launched; it appeared to be the last satellite in the Soviet early-warning series until September, 1972, when Kosmos 520 assumed an orbit with the same characteristics. Other, similar, satellites were launched, and in October, 1975, Kosmos 775 became the first Soviet geostationary early-warning satellite, signaling a new era for that country's surveillance of American ICBM sites. Geostationary satellites are placed in orbits at least 35,880 kilometers above the equator and are caused to move in the same direction as the Earth turns on its axis at speeds that make them appear to be stationary.

In November, 1976, U.S. Secretary of Defense Donald Rumsfeld denied reports that the Soviet Union had disabled two U.S. early-warning satellites by pointing high-powered lasers at them. Rumsfeld indicated that the satellites had been damaged by the glare of natural-gas fires along a pipeline in the Soviet Union. The two satellites that were the focus of the reports had been stationed over the Soviet Union to monitor Soviet compliance with the Nuclear Test Ban Treaty, signed in 1963, which limited nuclear testing by the United States and the Soviet Union.

*Newsweek* magazine reported at the same time that the United States was developing a fleet of satellites able to absorb radar, making them undetectable to the Soviets. There has also been speculation since the mid-1960's that both countries have engaged in research that might lead to the deployment of satellites whose mission is to destroy enemy satellites in orbit. Some early Kosmos satellites were observed to maneuver from orbit to orbit and to track and engage in close encounters with other Kosmos satellites. Space observers have noted that such activity might suggest the existence of "killer" satellites that would aggressively seek out and destroy enemy communications and surveillance satellites in wartime.

The Reagan administration announced the beginning of the Strategic Defense Initiative (SDI), a satellite-based, early-warning network armed with laser technology designed to destroy ICBMs before they are able to achieve high altitude. The proposal was controversial from the beginning. Some observers saw it as unworkable, technically unfeasible, and extremely expensive. The Soviet Union was particularly critical of SDI, proclaiming publicly that it represented an expensive and dangerous escalation of the arms race with inherent potential for aggression as well as defense. The Strategic Defense Initiative came to an inauspicious conclusion in 1993 with the inauguration of Bill Clinton as president of the United States. This was made possible by the collapse of the Soviet Union in 1991, a disappointing lack of significant progress in developing SDI technology, and a variety of strategic arms limitations treaties.

### Contributions

Shortly after the Discoverer series of satellites was launched in the early 1960's, American concerns that the Soviet Union was engaged in a massive ICBM buildup were dispelled by high-quality, high-resolution reconnaissance imagery that proved the effectiveness of satellite remote-sensing capabilities. Prior to the satellite era, there was little opportunity to monitor military activities beyond a nation's boundaries. The well-known U-2 incident of 1960 demonstrated the difficulty of using aircraft for surveillance. In addition, even the

sophisticated U-2 aircraft could not fly at altitudes high enough to survey large areas efficiently, thus limiting their potential for monitoring military activity. Satellites, then, because they could be placed in extremely high orbits that took them around Earth in relatively short periods of time, were well suited to the task of early-warning detection of missile launches and other military activity anywhere on Earth and in the atmosphere. Prior to the deployment of these satellites, the early-warning process was more cumbersome, relying on sophisticated networks of land-, sea-, and air-based radar technology that were and continue to be susceptible to various kinds of interference. The U-2 aircraft were able to detect missiles only after such missiles were already in the air, thus costing precious time in an era when intercontinental missiles could travel to targets on other continents in minutes. While these systems were not completely dismantled with the advent of the satellite era, the early-warning satellites offered efficient operation and timely data gathering. They enhanced the defense of host countries and lessened the ability of any nation to launch a first strike without detection.

Another important service provided by early-warning satellites was the detection of nuclear explosions on Earth's surface and in the atmosphere. Originally, satellites in the Vela program, begun with the launch of two satellites on October 16, 1963, were sent aloft to monitor nuclear activity. Eventually, IMEWS satellites took over this task. As nuclear proliferation moved beyond the borders of the United States and the Soviet Union, scientists were able to detect nuclear tests around the world.

In September, 1979, American satellites detected a nuclear explosion in the Southern Hemisphere in the area of South Africa. Scientists determined that the explosion was of low magnitude and that it had occurred in the atmosphere. Some observers speculated that South Africa was engaged in a nuclear testing program, which the South African government quickly denied. Other speculation stemmed from the possibility that a Soviet submarine had been involved in a nuclear accident. The real story behind the explosion was never made public.

Increasingly, sophisticated equipment aboard these satellites began to find other uses, such as the detection of natural resources in regions of the world where they remained undiscovered. Soon, reconnaissance equipment aboard satellites was employed to aid sea navigation and to monitor global weather systems. It was also used for producing land surveys and high-quality mapping of Earth's surface in largely inaccessible regions. It has also been used to track agricultural activity around the world, such as droughts, water resources, and crop yields.

The latest generation of satellites with detection capabilities is able to locate and photograph extremely small objects with remarkable precision. During the 1980's, this technology began to attract the interest of international news-gathering organizations seeking ways to visualize and verify reports of important news events. An example was the 1986 Chernobyl nuclear accident in the Ukraine, an event whose full significance came to light only after high radiation levels were detected in northern Europe. The Soviet Union, unwilling to allow Western reporters to visit the area, was slow to make details public. American satellites were used to locate and identify the source of the radiation, which was determined to be a fire in a nuclear reactor north of Kiev. Reproductions of digital images captured over the site by satellites were among the few visual verifications that such an event had occurred, and they were shown repeatedly on Western television news broadcasts. National security concerns soon placed the issue of news gathering from space before Congress, which held hearings in an effort to determine whether regulation is required in this area.

### Context

That early-warning satellite systems were among the first applications of satellite technology for the United States and the Soviet Union suggests how effective space-based defensive systems are in a nuclear world. In the late 1950's, the United States

was already planning the MiDAS program, and the Soviet Union soon followed with its highly secret Kosmos satellite program. Both countries belonged to military alliances whose members shared information relayed by existing early-warning systems, giving those systems truly global, rather than regional, significance.

The political climate of those early years in the space program was one of American and Soviet distrust and antagonism. The Cold War had not yet begun to give way to the more cooperative era of the mid-1960's that led to joint undertakings such as the Apollo-Soyuz Test Project. The considerable resources allocated to developing early-warning technology reflected that tension. In fact, it was not until shortly before Midas 2 went into orbit that U.S. officials, using data relayed by Discoverer satellites, were able to conclude that fears of a massive Soviet ICBM buildup along its borders were unfounded.

These fears had led to dramatic efforts to prepare defenses against nuclear attack in the United States. Many people built bomb shelters in or near their homes and stocked them with provisions. Schools, hospitals, and other institutions practiced air raid drills. Fears were fueled by the knowledge that nuclear weapons could travel from the Soviet Union and Eastern Europe in minutes and that they would not be detected until already en route. Early-warning satellites, while they did not diminish the perceived threat of a Soviet nuclear attack aimed at the United States, significantly lessened the possibility of a surprise attack. In addition, the constant, instantaneous flow of surveillance data from early-warning satellites increased the risk of immediate reprisal for any aggressor who might launch a first strike.

During the late 1970's and into the 1980's, tensions eased between East and West. Nevertheless, early-warning and surveillance satellites continued to play an important role in tracking military installations and troop maneuvers. In fact, they began to serve in other ways, no longer simply watching for ICBM launches from Soviet and American land bases and warhead-equipped submarines. A newer, more immediate threat was emerging elsewhere: terrorism. Some countries were known to harbor training camps and military installations that were being used as staging areas for terrorists who could and did strike anywhere in the world. Satellites, equipped with sophisticated cameras and sensing devices, could locate such installations, giving military commanders involved in monitoring or intercepting such activity a significant logistical advantage.

As sensing and photographic instruments aboard early-warning satellites have become more sophisticated, their importance to the nations that operate them cannot be overstated. They are the source of highly detailed military and nonmilitary information that would be extremely difficult to acquire in any other way. Now being constructed and launched outside the United States and Russia, early-warning and detection satellites have become important to all nations as a means of deterring aggressive activities by neighboring countries.

**See also:** Apollo-Soyuz Test Project; Applications Technology Satellites; Attack Satellites; Dynamics Explorers; Electronic Intelligence Satellites; Meteorological Satellites: Military; Nuclear Detection Satellites; SMS and GOES Meteorological Satellites; Spy Satellites; Strategic Defense Initiative; Telecommunications Satellites: Military.

## Further Reading

Corliss, William R. *Scientific Satellites*. NASA SP-133. Washington, D.C.: Government Printing Office, 1967. Contains highly technical descriptions of most of the scientific satellites sent aloft during the first decade of the Space Age by all countries. Also included is a chronology of international satellite launches. Contains a chapter describing geophysical instruments and experiments, preceded by an excellent overview of space science in general. Includes index and illustrations.

Daniloff, Nicholas. *The Kremlin and the Cosmos*. New York: Alfred A. Knopf, 1972. An excellent look at the political character of the Soviet space program from the time of Sputnik through the piloted lunar explorations. Includes rare accounts of events that led up to the successful effort to launch the first Sputnik and the surprise within the Soviet scientific community at the dramatic reaction it caused in the United States. Provides the names of many of the key players in the Soviet space program during its most important years and dispels the notion that the Soviets were prone to launch crude, unsophisticated hardware into space to "win the Space Race at all cost." The author contends that among those who oversaw the Soviet program in the early years were many of the world's most talented and skilled space scientists, and he is thorough in his substantiation of that claim.

Day, Dwayne A., John M. Logsdon, and Brian Latell, eds. *Eye in the Sky: The Story of the Corona Spy Satellites*. Washington, D.C.: Smithsonian Institution Press, 1998. The top-secret Corona spy satellites and their photographs were kept out of public sight until 1992. The Corona satellites are believed by many experts to be the most important modern development in intelligence gathering. This book is based upon previously classified documents, interviews, and firsthand accounts from the participants in the program.

Gatland, Kenneth. *The Illustrated Encyclopedia of Space Technology: A Comprehensive History of Space Exploration*. New York: Salamander, 1989. This well-illustrated volume contains a number of chronologies of Soviet, American, and international space programs. Includes a chapter that discusses early-warning satellite systems. Specific examples of successful applications of space technology in this area are cited. Illustrations, photographs, and index.

Gavaghan, Helen. *Something New Under the Sun: Satellites and the Beginning of the Space Age*. New York: Copernicus Books, 1998. This book focuses on the history and development of artificial satellites. It centers on three major areas of development—navigational satellites, communications satellites, and weather observation and forecasting satellites.

Peebles, Curtis L. *The Corona Project: America's First Spy Satellites*. Annapolis, Md.: United States Naval Institute, 1997. This book offers a comprehensive look at America's first spy satellites—both the successes and failures. It discusses Corona photography, as well as the photograph interpreters who were trying to figure out what they were seeing in the pictures.

Ritchie, Eleanor H. *Astronautics and Aeronautics, 1976: A Chronology*. NASA SP-4021. Washington, D.C.: Government Printing Office, 1984. Contains chronological listings of abstracts of press accounts of space activity during 1975 and 1976. Gleaned from popular publications and space agency memoranda released to the general public. Perhaps most useful are summaries of interviews given by Soviet space officials and academicians related to Soviet space activity. Index.

Shelton, William Roy. *Soviet Space Exploration: The First Decade*. New York: Washington Square Press, 1968. Contains a thorough look at the entire Soviet space program up to the lunar and the planetary probe period. Also contains a chapter on automated satellites used for resource exploration. Particularly useful is an excellent chronology of all Soviet spaceflights from Sputnik 1 in 1957 to Molniya 1F in 1967 in the appendix. Index and selected bibliography.

Yenne, Bill. *Secret Weapons of the Cold War: From the H-Bomb to SDI*. New York: Berkley Books, 2005. A contemporary examination of Cold War superweapons and their influence on American-Soviet geopolitics.

Zimmerman, Robert. *The Chronological Encyclopedia of Discoveries in Space*. Westport, Conn.: Oryx Press, 2000. Provides a complete chronological history of all piloted and robotic spacecraft and explains flight events and scientific results. Suitable for all levels of research.

*Michael S. Ameigh*

# Earth Observing System Satellites

*Date:* Beginning 1999
*Type of spacecraft:* Earth observation satellites

*Earth Observing System (EOS) satellites are a constellation of small orbiting platforms equipped with instrumentation designed for long-term observation of Earth's land surface, ice and snow cover, oceans, biosphere, and atmosphere. These satellite missions are part of NASA's Earth Science Enterprise, an integrated effort to study the components of the entire Earth system.*

## Summary of the Satellites

The National Aeronautics and Space Administration (NASA) Earth Observing System (EOS) program is a suite of spacecraft and interdisciplinary investigations dedicated to research on global change. EOS focuses on regional and global climate change, which affects land cover, renewable resources, biodiversity, and human health and quality of life.

The flagship of the EOS series, Terra (originally called EOS AM-1), was launched from Vandenberg Air Force Base in California on December 18, 1999. It follows a Sun-synchronous, polar orbit (that is, one that passes near the Earth's Poles and carries the satellite over any given point on the planet's surface at the same local time; unless otherwise specified, all times mentioned in this article will refer to local time) approximately 705 kilometers above the Earth's surface. The spacecraft descends across the equator at about 10:30 A.M., when obscuring cloud cover is at a minimum. Terra is the first satellite designed for the study of Earth's land, oceans, air, ice, and life as a total global system. It can simultaneously gather data on clouds, water vapor, aerosol particles, trace gases, terrestrial and oceanic properties, the interaction between them, and their effect on atmospheric radiation and climate.

Terra is equipped with five scientific instruments. The Clouds and the Earth's Radiant Energy System (CERES) instrument measures Earth's radiation budget and atmospheric radiation from the top of the atmosphere to the planet's surface. The Multi-Angle Imaging Spectroradiometer (MISR) measures variation in surface and cloud properties and atmospheric particles, using cameras pointed in nine simultaneous different viewing directions. The Moderate-Resolution Imaging Spectroradiometer (MODIS) measures atmospheric, terrestrial, and oceanic processes; surface temperatures of the land and ocean; ocean color; global vegetation; cloud characteristics, temperature, and moisture profiles; and snow cover. The Measurements of Pollution in the Troposphere (MOPITT) instrument maps carbon monoxide and methane concentrations within the lowest region of the atmosphere, enabling researchers to determine the sources and movements of these gases. The Advanced Spaceborne Thermal Emission and Reflection Radiometer (ASTER) provides high-resolution images for detailed study of cloud properties, vegetation, surface mineralogy, soil properties, surface temperature, and surface topography.

The ACRIMSAT spacecraft, launched December 20, 1999, carries the Active Cavity Radiometer Irradiance Monitor III (ACRIM III) instrument. ACRIM III provides precise measurements of the total amount of the Sun's energy that falls on Earth's land surface, oceans, and atmosphere. Previous missions carrying ACRIM instruments have shown that the total radiant energy from the Sun is not a constant, and that additional study is needed to determine what effects a small change in that energy might have on Earth's climate.

The Aqua satellite (formerly EOS PM-1) launched from Vandenberg Air Force Base on May 24, 2002. It was designed to gather data concerning clouds, precipitation, the atmosphere's moisture content and temperature, terrestrial snow, sea ice, and sea-surface temperature. Aqua's orbit is similar to Terra's, with the exception of its equatorial crossing time. Aqua ascends across the equator in the afternoon, to enhance its sensors' collection of meteorological data. Like Terra, Aqua carries CERES and MODIS instrumentation, as well as the Atmospheric Infrared Sounder (AIRS), the Advanced Microwave Sounding Unit (AMSU-A), and the Humidity Sounder for Brazil (HSB). The AIRS instrument measures air temperature, humidity, clouds, and surface temperature. The AMSU-A instrument was designed to profile stratospheric temperatures. The HSB instrument can obtain profiles of atmospheric humidity and detect precipitation under clouds. When analyzed jointly, data from AIRS, AMSU-A, and HSB instrumentation yields highly accurate air-temperature profiles.

The Ice, Cloud, and Land Elevation Satellite (ICESat) mission launched on January 12, 2003, from Vandenberg Air Force Base. The ICESat platform followed a near-polar orbit at an altitude of 600 kilometers, carrying a single scientific instrument: the Geoscience Laser Altimeter System (GLAS). GLAS is designed to measure the elevation of the Earth's ice sheets, clouds, and land. This instrument became particularly useful in assessing the response of ice sheets to climatic change and the relationship between ice sheets and sea-level change. ICESat started its first operational period early in 2004.

Aura (formerly EOS Chem) is the third in the series of EOS missions. Like Terra and Aqua, Aura would follow a polar, Sun-synchronous orbit approximately 705 kilometers above the Earth's surface. The primary objective of the Aura mission is to study the chemistry and dynamics of the Earth's atmosphere, particularly in the upper troposphere and lower stratosphere (5 to 20 kilometers above Earth). Instrumentation included the High Resolution Dynamics Limb Sounder (HIRDLS), the Microwave Limb Sounder (MLS), the Tropospheric Emission Spectrometer (TES), and the Ozone Monitoring Instrument (OMI), all of which measure various chemical species in the atmosphere and other atmospheric properties. Aura launched on July 15, 2004, from the Vandenberg Air Force Base aboard a Delta booster.

Other satellites carrying EOS instrumentation include the Tropical Rainfall Measuring Mission (TRMM, launched November, 1997), which features a CERES instrument and a Lightning Imaging Sensor (LIS); the QuikScat satellite (launched June, 1999), which uses the SeaWinds microwave radar instrument for gathering data on near-surface wind speeds and directions over Earth's oceans;

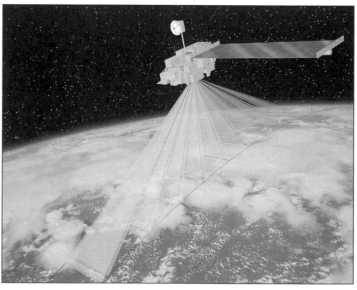

*A computer-generated image of the EOS AM-1 spacecraft orbiting Earth, one in a series of polar-orbiting and low-inclination satellites for long-term global observations of the land surface, biosphere, solid earth, atmosphere, and oceans.* (NASA)

and Landsat 7 (launched December, 1999), which has an Enhanced Thematic Mapper Plus (ETM+) for viewing Earth's terrestrial and coastal areas in multiple spectral bands. Additional EOS instruments were flown on other platforms launched in 2001, 2002, and 2003. The Stratospheric Aerosol and Gas Experiment III (SAGE III), which monitors trace gases and other atmospheric parameters, was aboard Russia's METEOR 3M-1 spacecraft when it launched from the Baikonur Cosmodrome on December 10, 2001. Another flight opportunity for SAGE III remained under discussion. This involved the International Space Station, but a planned 2004 opportunity fell through in the wake of the *Columbia* accident (February 1, 2003). Space station construction and research were expected to resume some time in 2006 with shuttle mission STS-115.

Jason-1, a joint U.S.-France oceanography mission, would carry the Poseidon-2 and Jason Microwave Radiometer (JMR) instruments to study global ocean circulation. This probe launched from the Vandenberg Air Force Base on December 7, 2001.

The Solar Radiation and Climate Experiment (SORCE) included the Solar/Stellar Irradiance Comparison Experiment (SOLSTICE) and Total Irradiance Monitor (TIM) to measure the solar ultraviolet, far ultraviolet, and total irradiance from the Sun. SORCE launched from the Cape Canaveral Air Force Station on January 25, 2003. Two years afterward it continued to record x-ray, ultraviolet, visible, and near-infrared solar radiations.

Data collected by the EOS satellites will be transmitted to various ground stations. NASA has set up the EOS Data and Information System (EOSDIS) to manage, archive, and distribute the data generated during the EOS missions.

### Contributions

The EOS program uses multiple small platforms orbiting in formation to collect a broad range of measurements. This strategy allows collection of near-simultaneous observations, without subjecting the sensitive instrumentation to the magnetic, thermal, and electrical interferences that would arise from densely packing sensors onto a single platform. Also, should a component—or even an entire satellite—fail, only a partial loss in coverage would result. Smaller satellites are less expensive and time-consuming to construct, making it an easier matter to replace a platform in case of failure. NASA contractor TRW has created a modular, standardized platform with subsystems that can readily be adapted for mission-specific requirements. This platform is being used for both the Aqua and Aura satellites.

The various instruments and missions of the EOS program have been designed to reduce key scientific uncertainties and enable reliable modeling and prediction. The program's research efforts can be divided into seven disciplinary areas: radiation, clouds, water vapor, precipitation, and atmospheric circulation; ocean circulation, productivity, and exchange with the atmosphere; atmospheric chemistry and greenhouse gases; land ecosystems and hydrology; cryospheric systems (land ice, sea ice, and snow cover); ozone and stratospheric chemistry; and volcanoes and climate effects of aerosols. EOS research teams will conduct multiple studies in each of these areas.

From EOS satellite data, researchers hope to gain a better understanding of the way that clouds and water vapor affect global climate. The satellites' improved precipitation measurements will contribute to scientists' understanding of the role precipitation plays in connecting atmospheric and surface processes, and to more accurate modeling of precipitation in global climate models. Long-term global measurements of the climate system will facilitate assessment of global change and improve prediction of seasonal and year-to-year climate variability.

Earth's oceans play a major role in modulating global climate, absorbing solar radiation and circulating the heat they gain. Oceans also moderate the effects of increased carbon dioxide levels in the atmosphere by taking up huge quantities of the gas to support the marine food chain. Using data from EOS satellites, oceanographic researchers will gain a greater understanding of the exchange of carbon dioxide, carbon monoxide, and dimethyl sulfide between the atmosphere and oceans. The interplay between the oceans' biological system and the Earth's carbon cycle will also be investigated, along with ocean circulation and productivity.

EOS satellite data will help researchers assess how fluxes of greenhouse gases respond to changes in land cover and land use. The effect of year-to-year climate fluctuation on terrestrial ecology and atmospheric chemistry will be examined, as will the global methane and carbon dioxide budgets' response to changes in global hydrology and patterns of soil moisture in wetlands. EOS data will also be used in studies of the spatial and vertical distributions of ozone in the lower atmosphere.

*The Earth Observing System (EOS) satellites are part of the Earth Science Enterprise, which is an attempt to study the components of the entire Earth system.* (NASA CORE/Lorain County JVS)

EOS data will be key to research efforts examining how changes in land-surface processes and properties interact with and influence climate on a regional and global basis. EOS satellite observations will also contribute to studies of the effect changes in land cover have on land-surface biophysical properties and hydrologic balances. Additional investigations will deal with the effects of changing climate and land-surface characteristics on precipitation patterns, river flows, and surface-water storage; and the impact that land processes such as erosion have on water quality.

EOS data will enhance understanding of cryospheric systems—sea ice, lake ice, river ice, snow cover, glaciers, ice caps, ice sheets, and permafrost. Accurate climate modeling and prediction of global changes must take cryospheric processes into account. An understanding of the cryosphere is also important in modeling the extent of the effects global warming would have on sea level.

Stratospheric ozone both shields the Earth's surface from solar ultraviolet radiation and affects the radiative heating/cooling balance in the atmosphere. Gaining a greater knowledge of ozone's role in climate change is one of EOS's research ob-

jectives, as is learning more about the causes (natural and human-made) of change in stratospheric ozone distribution. Improved accuracy, precision, and resolution of temperature measurements in the lower atmosphere will contribute to ozone research. The SAGE III instrument will provide the first satellite observations measuring the size distribution of polar stratospheric clouds, which can strongly influence the amount of halogen radicals (chemical compounds antagonistic to ozone) in the atmosphere.

Volcanic emissions can cause significant climate variations, with major eruptions leading to measurable climate change within a time scale of a few years. Within the stratosphere, sulfur dioxide released during an eruption can oxidize to sulfuric acid and produce cooling at the Earth's surface. Atmospheric aerosols, including airborne dust and sea salt, particles from fossil-fuel combustion and natural fires, and products from volcanic eruption, contribute to global cooling and interfere with remote sensing of Earth's surface properties. EOS satellite data will make possible the first truly global inventory of volcanic eruptions, and will be used to characterize their number, location, type, and duration. Increased measurements of lava temperature, gas emissions, ground deformation, and eruption plume behavior will contribute greatly to volcanic hazard mitigation. EOS observations will also provide data regarding the effects of natural and human-generated aerosols on atmospheric temperatures and chemical composition.

### Context

The EOS program grew out of the scientific community's recognition that Earth can be understood only as an integrated system, and that human activity can bring about global change. Early planning for the EOS missions began after the establishment in 1979 of the World Climate Research Program, an international effort to gain an understanding of the physical basis of climate. In 1986, the International Geosphere-Biosphere Program (IGBP) was initiated "to describe and understand the interactive physical, chemical, and biological processes that regulate the Earth's unique environment for life, the changes that are occurring in this system, and the manner in which they are influenced by human actions." Satellite systems clearly had an important role to play in such research efforts, with their capability to gather data in different spectral bands over large geographic areas and extended periods of time. In 1990, EOS was made the centerpiece of NASA's Mission to Planet Earth (later renamed the Earth Science Enterprise), an integrated effort to study the components of the entire Earth system—hydrologic, biogeochemical, atmospheric, ecological, and geophysical processes—and the interactions among them. The EOS satellites constitute one of three main components of the Earth Science Enterprise, the other two being an advanced system to handle the satellite data and interdisciplinary teams of scientists to study the data.

The EOS satellites will provide data in support of numerous scientific programs, including the Global Energy and Water Cycle Experiment (GEWEX),the Climate Variability project (CLIVAR) of the World Climate Research Program, and the IGBP's Joint Global Ocean Flux Study (JGOFS). One program depending heavily upon EOS is the U.S. Global Change Research Program, which seeks to quantify and understand seasonal and year-to-year climate variability; climate change over decades and centuries; changes in ozone, ultraviolet radiation, and atmospheric chemistry; changes in land cover, terrestrial systems, and aquatic systems; the impact of human activity on natural systems; the consequences of global change for human and societal health and well-being; and possible responses humankind could take to problems associated with environmental change.

**See also:** Dynamics Explorers; Electronic Intelligence Satellites; Explorers: Air Density; Explorers: Atmosphere; Explorers: Ionosphere; Geodetic Satellites; Global Atmospheric Research Program; Meteorological Satellites: Military; New Millennium Program; Nimbus Meteorological Satellites; Seasat; Stratospheric Aerosol and Gas Experiment.

### Further Reading

Anderson, G. Christopher. "Safety in Numbers for Earth Observing System." *Nature* 346 (August 2, 1990): 399. Discusses the rationale for multiple small platforms versus a few larger platforms.

Golub, Leon, and Jay M. Pasachoff. *Nearest Star: The Surprising Science of Our Sun.* Boston: Harvard University Press, 2002. Although written by two of the most active research astrophysicists, this book is accessible to a general audience. It describes most contemporary advances in solar physics.

King, Michael D., ed. *EOS Science Plan.* Greenbelt, Md.: National Aeronautics and Space Administration, 1999. This document describes how investigators plan to use the data obtained by the EOS program. It includes chapters on radiation, clouds, water vapor, precipitation, and atmospheric circulation; ocean circulation, productivity, and exchange with the atmosphere; atmospheric chemistry and greenhouse gases; land ecosystems and hydrology; cryospheric systems; ozone and stratospheric chemistry; and volcanoes and climate effects of aerosols.

King, Michael D., and Reynold Greenstone, eds. *1999 EOS Reference Handbook*. Greenbelt, Md.: National Aeronautics and Space Administration, 1999. This publication includes an overview of NASA's Earth Science Enterprise, summaries of Earth Observing System satellite platforms and instrumentation, a description of the EOS Data and Information System, and profiles of interdisciplinary science investigations associated with EOS.

Kivelson, Margaret G., and Christopher T. Russell. *Introduction to Space Physics*. New York: Cambridge University Press, 1995. A thorough exploration of space physics. Some aspects are suitable for the general reader. Suitable for an introductory college course on space physics.

Zimmerman, Robert. *The Chronological Encyclopedia of Discoveries in Space*. Westport, Conn.: Oryx Press, 2000. Provides a complete chronological history of all crewed and robotic spacecraft and explains flight events and scientific results. Suitable for all levels of research.

*Karen N. Kähler*

# Electronic Intelligence Satellites

*Date:* Beginning October 11, 1960
*Type of spacecraft:* Military satellites

*Electronic listening satellites, also called electronic intelligence (ELINT) satellites or ferrets, are designed to receive various electronic signals, including radio, radar, teletype, and telemetry signals. These satellites provide a major portion of the intelligence upon which the United States and Russia rely.*

## Summary of the Satellites

During World War II, aircraft, ships, and ground stations monitored enemy radio and radar transmissions. With the start of the Cold War, these systems were again used to gather intelligence; it was very difficult, however, to pick up signals from the interiors of large countries such as the United States or the Soviet Union. An ELINT aircraft flying close to the border could pick up radar sites only within approximately 100 kilometers; the interior of the Soviet Union was impenetrable. One possible solution was to make overflights, as was done with the U-2, but the risk involved with flying over foreign territory limited the number of missions and the amount of data that could be collected. The solution was orbital reconnaissance.

The first orbital ELINTs launched by the United States were with the SAMOS (Satellite and Missile Observation System) photo reconnaissance satellites. ELINT packages were built to be carried aboard the first launches. As it happened, only SAMOS 2 successfully reached orbit—on October 11, 1960. The needs of photo reconnaissance and ELINT satellites are to some extent incompatible. A photo reconnaissance satellite must orbit at a low altitude to maximize resolution. An ELINT satellite should be in a higher orbit to increase the area covered. An altitude of 483 kilometers is normal for ELINT satellites, twice the altitude of photo reconnaissance satellites.

Accordingly, the United States developed specialized ELINT satellites. Two separate types of ELINT satellites were used during the 1960's. The United States used small subsatellites to locate radar sites and areas of coverage. These subsatellites were carried aboard photo reconnaissance satellites. Once in orbit, the subsatellites separated and fired an onboard rocket engine that boosted them into their final orbit. Each subsatellite measured about 0.3 by 0.9 meters and weighed about 57 kilograms. The first ELINT subsatellite was launched on August 29, 1963, aboard a Thor-Agena D from Vandenberg Air Force Base. It finally achieved a 431-by-310-kilometer orbit. As each new generation of photo reconnaissance was introduced, the subsatellites made the transfer. Two different orbits were used—the standard 483-kilometer circular orbit and, starting in 1968, a higher, 1,448-kilometer circular orbit. The subsatellites placed into this higher orbit were used to monitor antiballistic missile radars.

Another type, used by the United States in the 1960's, was the heavy ELINT satellite. This was an Agena upper stage fitted with a 907-kilogram payload; it provided detailed measurements of the radar signals and their location. Unlike the ELINT subsatellites, this type was relatively short-lived. The first was launched from Vandenberg on June 18, 1962, on a Thor-Agena B. It was placed into a 411-by-370-kilometer orbit. Launches of ELINTs continued at a rate of one or two per year until July 16, 1971. After that point, the ELINT satellite program would go in other directions.

By this time, the Soviets had also begun flying ELINT satellites. The first ELINT test satellite was Kosmos 103, launched on December 28, 1965. It used a C-1 booster launched from the Baikonur Cosmodrome. The satellite went into a 600-kilometer orbit, inclined 56°. The first operational Soviet ELINT satellite was Kosmos 189, launched October 30, 1967. The C-1 placed it into a 600-by-535-kilometer orbit—more elliptical than those of the test satellites. Launched from Plesetsk, the Kosmos 189 had an inclination of 74°. The Soviet ELINT satellites are believed to be cylindrical, measuring 2 meters in length and 1 meter in diameter, with paddle-shaped solar panels. Their weight of 875 kilograms is considerably greater than that of the United States' subsatellites.

Introduction of the Soviet-launched ELINT satellite was comparatively slow—only two or three launches were made per year. Then, in 1971, five launches were made, signifying a shift to fully operational status. Beginning in 1972, the satellites were arranged into a network, with each 45° apart. This arrangement allowed for regular coverage of targets. Subsequent years saw a launch rate of three or occasionally four C-1 ELINT satellites.

The Soviets were also flying heavy ELINT satellites. The first of these was Kosmos 389, launched by an A-1 booster from Plesetsk on December 18, 1970. It was placed into a 699-by-655-kilometer orbit. Because this orbit was similar to that of the Soviet Meteor weather satellite, Western space analysts first believed that it was a failed weather mission with the Kosmos label used for cover. It was several years before the satellite's ELINT role was fully realized. It is thought that the Soviet heavy ELINT satellite consists of the Meteor satellite body outfitted with ELINT receivers. Probably a cylinder about 5 meters in height and 1.5 meters in diameter, with two solar panels, this satellite weighed approximately 2,000 kilograms. As with the United States' heavy ELINT satellite, the launch rate of the A-1 ELINT satellite was comparatively low—one or two per year—into the mid-1970's. Starting in 1976, the rate of launches was increased to four to six per year. They were also being placed into a three- and

later six-satellite network. In 1979, the C-1 ELINT satellite was phased out, and the heavy ELINT satellite was shifted to the F-2 booster.

In the first half of the 1970's, the United States introduced one or two new types of ELINT satellites. It has been claimed that some of the Satellite Data System (SDS) military communications satellites were actually ELINT payloads with the code name Jumpseat. The first Jumpseat launch took place on March 10, 1975, with a second following on August 6, 1976. Both the SDS and Jumpseat satellites use the Titan III-B launch vehicle and are placed in highly elliptical 39,300-by-380-kilometer orbits.

The major development in the United States' ELINT satellite program was the move to geosynchronous orbit. A satellite at 35,881 kilometers has an orbital speed that exactly matches the rotation of Earth. Thus, the satellite appears to remain suspended in the same spot in the sky. As a low-orbit ELINT satellite passed over Earth's surface, it could receive signals from a given radar site or radio station for only about ten or fifteen minutes before it passed over the horizon. A geosynchronous ELINT satellite would be able to record data on most of the Soviet landmass constantly.

Since the late 1950's, much of the intelligence the United States had collected on Soviet intercontinental ballistic missile (ICBM) development was based on intercepted telemetry signals from the missiles. These were measurements of the operations of systems aboard the ICBM during the flight. If something should go wrong, these readings would show what changes had to be made. A geosynchronous ELINT satellite could pick up these signals—provided it could carry a large-dish antenna (necessary because the signals are 5,500 times weaker than those received by a low-orbit satellite). The satellite would also have to pick up the desired signals amid all the ones being transmitted from the Soviet Union.

The first geosynchronous ELINT satellite was the Rhyolite. Built by TRW, it was fitted with a 21-meter dish antenna and another smaller array. Its power was supplied by solar panels, and its total weight was 275 kilograms. The first Rhyolite was launched by an Atlas-Agena D from Cape Kennedy on March 6, 1973. It went into a 35,855-by-35,679-kilometer orbit. The Rhyolite was placed over the Horn of Africa, allowing it to receive signals from ICBMs launched from the Baikonur Cosmodrome. The success of the first Rhyolite indicated that a geosynchronous ELINT satellite with the capabilities of ground stations such as the Tacksman 2 facility in northern Iran could be built. TRW began studying this possibility under the code name Argus in the early 1970's. The cost was a problem: Argus would have to duplicate the capabilities the United States already had with its ground stations in Iran, Turkey, and Norway. Also, the Keyhole 11 (KH-11) photo reconnaissance satellite was having problems that required more funding to keep it on schedule. A disagreement developed within the Ford administration over whether to build Argus. Ultimately, the U.S. Congress decided not to fund the project; by late 1975 the project was terminated.

In July of 1976, the Rhyolite and Argus projects were revealed to the Soviets by Christopher John Boyce and Andrew Daulton Lee. Boyce, a TRW employee, photographed documents on reconnaissance satellites, and Lee, a childhood friend and drug dealer, took the film to the Soviet embassy in Mexico City. Their treachery was not discovered until January, 1977. Boyce received a forty-year prison term, and Lee received a life sentence. Soon after, Rhyolite launches resumed. The second was made on May 23, 1977, and was positioned over Borneo to monitor Plesetsk launches. Two more followed on December 11, 1977, and April 7, 1978. They were positioned near the other two and acted as in-orbit spares.

With the loss of the Iranian ELINT ground stations and the signing of the Strategic Arms Limitation Talks (SALT) II Treaty in 1979, the need for an advanced geosynchronous ELINT satellite became greater. According to one account, this need was met by an ELINT satellite code-named Chalet, whose launches may have actually been the launches of early-warning satellites and Defense Satellite Communications System satellites.

The definitive geosynchronous ELINT satellite was launched on January 24, 1985, aboard space shuttle *Discovery* (STS 51-C) from the Kennedy Space Center. According to the Federation of American Scientists, the satellite had two large-dish antennae, was 30.5 meters across when the solar panels were fully extended, and weighed about 2,268 kilograms. The first of a new, improved class of satellite code-named Magnum reportedly is a descendant of the Rhyolite series, designed to intercept Soviet missile test signals (telemetry) and data-links as well as microwaves. The cost ($300 million each, even in 1985) and complexity of the project is such that few have been launched.

The Soviets, in contrast, continued to use low-orbit satellites. On September 28, 1984, they launched a new type, a very heavy ELINT satellite. Kosmos 1603 was placed into a 190-by-180-kilometer orbit, inclined 51.6° by a D-1e booster from the Baikonur Cosmodrome. The satellite maneuvered repeatedly until it ended up in an 856-by-850-kilometer orbit, inclined 71°. These launches continued even after the breakup of the Soviet Union in 1991. Early in the twenty-first century, Russia maintained a constellation of reconnaissance satellites officially called Unmanned Photo Optical Observation Spacecraft.

### Contributions

The information gathered by ELINT satellites covers many fields. The data on Russian radars are used to design electronic countermeasure equipment carried by B-52, B-1B, and B-2 bombers. In the event of war, the countermeasure equipment would jam the radars so that a bomber's location could not be identified. The route a bomber would take as it flew toward its target would also be based on the satellite's information about the radars' coverage. A bomber would fly through gaps in the radars' field.

Monitoring Russian radio communications is another role of the ELINT satellites. By listening to the transmissions between fighter pilots and their ground controllers, for example, it is possible to reconstruct the tactics used, the capabilities of the weapons, and the weaknesses that may be exploited. Radio messages such as these are sent "in the clear," that is, uncoded. More important messages are sent in code. If the codes can be broken, the enemy's most critical secrets are vulnerable. The value of this became evident with what is known as Ultra, the decoding of German messages by the British during World War II. In one case, the Allied commander was able to read the decoded German message before the German commander was able to decode his copy. On several occasions, Ultra was credited with preventing Allied defeats.

### Context

In the broader context of relations between the United States and foreign powers, there are two areas where ELINT satellites have a central role. The first is arms control. Some provisions, such as those on specific categories of weapons, are verified by photo reconnaissance satellites. Others, however, such as the SALT II limits on the increase in the size of ICBMs, are policed through intercepted telemetry.

The second function of the ELINTs is early warning. Before a foreign power could undertake any major military moves, such as invading Western Europe or the oil fields of the Middle East, many preparations would have to be made. Such preparations could be detected in the movement of radar equipment and in an upsurge in radio message traffic between headquarters and the frontline units. Forewarned, the Western allies could undertake diplomatic efforts and prepare their own military forces. The critical task would be to sort out the key information amid the noise and confusion—a task that U.S. intelligence was unable to perform on the eve of Pearl Harbor. Now, as then, the cost of being wrong is high.

**See also:** Applications Technology Satellites; Attack Satellites; Early-Warning Satellites; Earth Observing System Satellites; Meteorological Satellites: Military; Nuclear Detection Satellites; Ocean Surveillance Satellites; Space Shuttle Flights, January-June, 1985; Spy Satellites; Strategic Defense Initiative; Telecommunications Satellites: Military.

### Further Reading

Bamford, James. *The Puzzle Palace.* New York: Penguin Books, 1983. A history of the National Security Agency, which has responsibility for ELINT and code breaking. Contains one chapter on so-called platforms (ground stations, aircraft, ships, and satellites). The primary emphasis is on overall policy. Suitable for high school and college readers.

Clayton, Aileen. *The Enemy Is Listening.* New York: Ballantine Books, 1982. This volume describes ELINT from the working level (the author was a member of Y Service, the British unit that intercepted German communications). The book explains how cryptic radio transmissions were analyzed to reconstruct German tactics. Recommended for general audiences.

Day, Dwayne A., John M. Logsdon, and Brian Latell, eds. *Eye in the Sky: The Story of the Corona Spy Satellites.* Washington, D.C.: Smithsonian Institution Press, 1998. The top-secret Corona spy satellites and their photographs were kept out of public sight until 1992. The Corona satellites are believed by many experts to be the most important modern development in intelligence gathering. The book is based upon previously classified documents, interviews, and firsthand accounts from the participants in the program.

Gavaghan, Helen. *Something New Under the Sun: Satellites and the Beginning of the Space Age.* New York: Copernicus Books, 1998. This book focuses on the history and development of artificial satellites. It centers on three major areas of development—navigational satellites, communications satellites, and weather observation and forecasting satellites.

Klass, Philip J. *Secret Sentries in Space*. New York: Random House, 1971. The first book dealing with military satellites. It covers the historical background and development of reconnaissance satellites. Part of one chapter is devoted to ELINT activities. Recommended for general audiences.

Lindsey, Robert. *The Falcon and the Snowman*. New York: Pocket Books, 1982. The story of the Boyce/Lee espionage case. It covers their childhood, the betrayal of Rhyolite, the discovery of their betrayal, and the trials. Suitable for high school and college audiences.

——————. *The Flight of the Falcon*. New York: Pocket Books, 1985. The continuation of *The Falcon and the Snowman*. This book covers Boyce's escape from federal prison and the manhunt that followed. It details the many false alarms and wild goose chases of the pursuit and the events that led up to Boyce's recapture.

Peebles, Curtis L. *The Corona Project: America's First Spy Satellites*. Annapolis, Md.: United States Naval Institute, 1997. This book offers a comprehensive look at America's first spy satellites—both the successes and failures. It discusses Corona photography, as well as the photograph interpreters who were trying to figure out what they were seeing in the pictures.

——————. *Guardians: Strategic Reconnaissance Satellites*. Novato, Calif.: Presidio Press, 1987. Covers the history and technology of reconnaissance satellites and the profound impact they have had on international relations. The book has chapters on both United States and Soviet ELINT satellites. Suitable for high school and college audiences.

Richelson, Jeffrey. *American Espionage and the Soviet Target*. New York: William Morrow, 1987. This book contains a detailed account of the United States' ELINT activities—ground stations, aircraft, ships, and satellites. It uses the disputed ELINT satellite launch dates.

——————. *America's Space Sentinels: DSP Satellites and National Security*. Lawrence: University Press of Kansas, 1999. This is the story of America's Defense Support Program satellites and their effect on world affairs. Richelson has written a definitive history of the spy satellites and their use throughout the Cold War. He explains how DSP's infrared sensors are used to detect meteorites, monitor forest fires, and even gather industrial intelligence by "seeing" the lights of steel mills.

Yenne, Bill. *Secret Weapons of the Cold War: From the H-Bomb to SDI*. New York: Berkley Books, 2005. A contemporary examination of Cold War superweapons and their influence on American-Soviet geopolitics.

Zimmerman, Robert. *The Chronological Encyclopedia of Discoveries in Space*. Westport, Conn.: Oryx Press, 2000. Provides a complete chronological history of all piloted and robotic spacecraft and explains flight events and scientific results. Suitable for all levels of research.

*Curtis Peebles*

# Environmental Science Services Administration Satellites

*Date:* February 3, 1966, to March 12, 1976
*Type of spacecraft:* Meteorological satellites

*Environmental Science Services Administration (ESSA) satellites provided the first true weather satellite system and helped create a base for meteorological satellites used by scientists on Earth.*

## Key Figures

*Morris Tepper* (b. 1916), director of meteorological programs at NASA
*Michael L. Garbacz,* TIROS flight manager for NASA
*William G. Stroud,* TIROS manager at Goddard Space Flight Center (GSFC)
*Rudolpf A. Stampfl,* TIROS manager at GSFC
*Robert M. Rados,* TIROS manager at GSFC
*William W. Jones,* TIROS manager at GSFC

## Summary of the Satellites

The United States Weather Bureau (later, the Environmental Science Services Administration) participated in the first satellite project to be used to observe the weather, the Television Infrared Observations Satellite (TIROS), from its beginnings in the late 1950's. The bureau, an agency of the Department of Commerce, was responsible for disseminating data returned by satellites. Once the TIROS satellites had become operational in the mid-1960's, the Weather Bureau assumed their management. TIROS Operational Satellite missions were all successful, providing daily information about cloud cover, upper winds, pressure, and precipitation on a global level. This kind of information made possible for the first time daily weather forecasts, storm and marine advisories, gale and hurricane warnings, cloud analysis, and other navigational assistance.

At the headquarters of the National Aeronautics and Space Administration (NASA), Morris Tepper, as director of meteorological programs, headed the TIROS program. In mid-1963, Michael L.

Garbacz was named flight manager and led TIROS for the rest of the 1960's. TIROS was then assigned to NASA's Goddard Space Flight Center and was managed by William G. Stroud, Rudolpf A. Stampfl, Robert M. Rados, and William W. Jones.

All viewers of television are familiar with the satellite maps of Earth shown from space. Early in the twenty-first century, there were four satellites operated by the National Oceanic and Atmospheric Administration (NOAA) to watch the weather over the United States. They covered the West and East Coasts, Alaska, and Hawaii. The first in the series of weather satellites, launched from 1960 to 1965, were called the TIROS. The next nine, from 1966 to 1969, were designated Environmental Science Services Administration (ESSA) satellites for the agency that managed the program. These ESSA meteorological spacecraft, similar in configuration to the TIROS spacecraft, were developed to establish a global weather satellite monitoring system.

Solar cells covered the sides and top of the ESSA

spacecraft, and four whip antennae extended from the baseplates. The ESSA spacecraft was 55 centimeters high and 107 centimeters wide, shaped like an eighteen-sided cylindrical polygon and resembling a hatbox. Its two tape recorders could store up to forty-eight pictures, recorded when ground stations were out of range and transmitted back when they were in sight.

The commitment to provide daily, routine worldwide observations without interruption was fulfilled by ESSA 1 and ESSA 2. Pairs of ESSA satellites enabled full coverage of the globe. Through their onboard data storage systems, the odd-numbered satellites (ESSAs 1, 3, 5, 7, and 9) provided worldwide weather reports to the U.S. Department of Commerce's command and data acquisition stations in Wallops Island, Virginia, and Fairbanks, Alaska, which in turn then relayed them to the National Environmental Satellite Service at Suitland, Maryland, for processing and forwarding to the major forecasting centers in the United States and around the world.

From ESSA 2, with its automatic picture transmission (APT) cameras, came photographs in real time from a 3,000-kilometer-wide area with 3-kilometer resolution. Pictures were taken and transmitted every 352 seconds, allowing a typical APT station to receive eight to ten photographs per day. The first advanced vidicon camera system (AVCS) provided revolutionary information about weather patterns and conditions. These technological marvels featured 800-line cameras with nearly twice the resolution of normal television cameras; they were mounted 180° apart on the sides with 40-centimeter receiving antennae on top.

The ESSA 1 was launched February 3, 1966, on a Thor-Delta 36 launch vehicle from Cape Kennedy and orbited Earth every 99.9 minutes. It weighed 138.3 kilograms and was solar-powered. The satellite had a Sun-synchronous orbit (that is, it remained stationary relative to the Sun), which permitted it to view and record weather in each area of the globe each day. ESSA 1 had an advanced remote sensing and storage device for data acquisition and playback to the two principal United

States command and data acquisition stations. With 14.5 orbits per day, a total of 450 images were available every 24 hours.

The ESSA 2 was launched February 28, 1966, aboard a Delta 37 launch vehicle from Cape Kennedy and rotated around Earth in 113.4 minutes. It weighed 131.5 kilograms. ESSA 2 provided a direct nautical readout of cloud cover photographs to local users. It had a Sun-synchronous orbit that complemented that of ESSA 1, thus completing the first true global weather operational monitoring satellite system (as well as the world's first operational applications satellite). ESSA 1 and ESSA 2 viewed the entire planet on a daily basis.

The ESSA 3 was launched October 2, 1966, from Vandenberg Air Force Base aboard a Delta 41 launch vehicle. It rotated Earth every 114.5 minutes and weighed 147.4 kilograms. This satellite replaced ESSA 1. It had advanced vidicon camera system (AVCS) sensors to provide valuable information about the world's weather patterns and conditions.

The ESSA 4 was launched January 26, 1967, on a Delta 45 launch vehicle, also from Vandenberg, and circled Earth every 113.4 minutes. It weighed 131.5 kilograms. This satellite replaced ESSA 2 and provided daily coverage of local weather systems. A shutter malfunction, however, rendered one camera inoperative soon after launch. ESSA 5 was launched April 20, 1967, on a Delta 48 launch vehicle from Vandenberg and circled Earth every 113.5 minutes. It weighed 147.4 kilograms. It replaced ESSA 3 and furnished daily global coverage of weather systems with an AVCS sensor. Four more ESSA satellites were launched on Delta launch vehicles from Vandenberg, each replacing a previous ESSA satellite and assuming its functions. The last, ESSA 9, was launched February 26, 1969. Circling Earth every 115.2 minutes, it weighed 157.4 kilograms and replaced ESSA 7. It too used an AVCS sensor system.

By the time ESSA 9 was in orbit, there were some four hundred receiving stations in operation around the world, as well as twenty-six universities, up to thirty American television stations, and an unknown number of private citizens receiving the photographs and images.

In 1970, a second generation, called the Improved TIROS Operational System (ITOS), was introduced. The National Oceanic and Atmospheric Administration (NOAA) had taken over the duties of ESSA. NOAA-1 was launched on December 11, 1970, from Vandenberg. Other NOAA satellites began to replace the remaining ESSA satellites, assuming their functions.

### Contributions

The ESSA satellite provided an operational system of considerable value, but it was still a system of limited capability. Weather forecasting is now done numerically, with the aid of computers. Some of the most important data are the vertical temperature structure of the atmosphere, the speed and direction of the winds, and the atmosphere's moisture content. The ESSA system could provide no direct data on these variables. The United States was well aware of these limitations and eventually developed the Nimbus system to improve its method of data acquisition. The Nimbus had remote sensors that could measure temperature and moisture content.

Basically, meteorological satellites are platforms for "topside observation," or electromagnetic scanning, of Earth's atmosphere from above. The scanning is passive in the sense that satellites merely make use of existing radiation emitted or reflected from the atmosphere, without adding to it as radar does. Many experiments with ESSA satellites improved weather forecasting. For example, there have been several crucial experiments measuring the Sun's energy—energy both directly emitted and reflected. Orbiting satellites provide an ideal platform for such measurement. Indeed, the very first successful meteorological satellite experiment, on Explorer 7, launched in 1959, directly measured the Sun's energy for the first time.

Meteorological satellites in general, and the ESSA series in particular, perform several major functions. They record images of clouds and cloud patterns, they track the movement of clouds and weather systems, they measure moisture in the air and the air's temperature, and they help forecast storms, especially severe thunderstorms.

ESSA's radiation sensors may have been its most valuable feature. These instruments measure Earth's infrared and reflected solar radiance; using that information, scientists are able to project cloud formations and other key meteorological features. Other types of satellites use small, selected wavelengths, from ultraviolet to microwave, to measure temperature, ozone, and water vapor in the atmosphere; these spacecraft are sensitive to one or more wavelength bands in the visible and invisible ranges. If visible wavelengths are used, then the satellite's instrumentation detects reflected sunlight, which illuminates cloud vistas not unlike slightly blurred versions of photographs taken directly by astronauts. Those satellites that use invisible wavelengths take advantage of the terrestrial radiation continually emitted by Earth: Higher radiation levels yield better images.

Such sensitive image detectors can measure clouds as small as 1 kilometer. As the satellite spins, they scan Earth's surface. On its Sun-synchronous orbit, the satellite creates a scene to be assembled. The image is formed line-by-line, like a television image, then transmitted to each ground processing station or recorded for later replay. From these data, the ground station can produce a computer-generated map of the area.

### Context

Meteorological satellites have made an enormous impact since the first TIROS satellites were launched in April, 1960. Millions of photographs have been made from space. Nearly all Americans are familiar with them from weather forecasts broadcast on television; they provide cloud patterns to help predict hurricanes, cyclones, and individual thunderstorms.

Meteorological techniques quickly became more sophisticated. Huge panoramic views of the atmosphere directly revealed and confirmed the structure of large, cloudy weather systems that had previously emerged only after painstaking assembly and analysis of synoptic data.

The relatively high resolution of the images (a few kilometers compared with tens of thousands of kilometers of the earlier, coarser maps) revealed a much larger range of structures of cloud systems. Satellites, whether Sun-synchronous or geosynchronous, provide uniform images, covering oceans and landmasses never before observed.

In 1969, an image from ESSA made history by revealing that snow cover over parts of the Midwest was three times the normal level. Measurements showed that it was the equivalent of 15 to 25 centimeters of water covering thousands of square kilometers. A disaster area was declared, flood warnings issued, and before the floods came the area was prepared.

In the longer run, the availability of years of daily coverage of Earth for rain, temperature, clouds, and other meteorological data have enabled researchers to develop climatologies—that is, baselines of what is normal. These baselines are especially important for detecting drought and temperature change over long periods of time. Before ESSA, scientists could make only local observations; they lacked data for the entire planet.

**See also:** Delta Launch Vehicles; Electronic Intelligence Satellites; Explorers: Solar; ITOS and NOAA Meteorological Satellites; Meteorological Satellites; Meteorological Satellites: Military; Nimbus Meteorological Satellites; SMS and GOES Meteorological Satellites; TIROS Meteorological Satellites; Upper Atmosphere Research Satellite.

## Further Reading

Ahrens, C. Donald. *Essentials of Meteorology: An Invitation to the Atmosphere.* 4th ed. Pacific Grove, Calif.: Thomson Brooks/Cole, 2005. This is a text suitable for an introductory course in meteorology. Comes complete with a CD-ROM to help explain concepts and demonstrate the atmosphere's dynamic nature.

_____. *Meteorology Today: An Introduction to Weather, Climate, and the Environment.* 7th ed. Pacific Grove, Calif.: Thomson Brooks/Cole, 2002. A thorough examination of contemporary understanding of meteorology. Includes the contributions made by satellite technology.

Anthes, Richard A. *Meteorology.* 7th ed. Upper Saddle River, N.J.: Prentice Hall, 1997. Volume 8 of the Prentice Hall Earth Sciences Series. Includes illustrations and maps, bibliographical references, and an index.

Bader, M. J., G. S. Forbes, and J. R. Grant, eds. *Images in Weather Forecasting: A Practical Guide for Interpreting Satellite and Radar Imagery.* New York: Cambridge University Press, 1997. The aim of this work is to present the meteorology student and operational forecaster with the current techniques for interpreting satellite and radar images of weather systems in mid-latitudes. The focus of the book is the large number of illustrations.

Grieve, Tim, Finn Lied, and Erik Tandberg, eds. *The Impact of Space Science on Mankind.* New York: Plenum Press, 1976. This book summarizes the uses of space exploration in helping men and women on Earth. The chapter on the environmental satellite discusses the ESSA programs and offers the reader a perspective of where the ESSA satellites fit into worldwide weather satellite programs.

Parkinson, Claire L. *Earth from Above: Using Color-Coded Satellite Images to Examine the Global Environment.* Sausalito, Calif.: University Science Books, 1997. A book for nonspecialists on reading and interpreting satellite images. Explains how satellite data provide information about the atmosphere, the Antarctic ozone hole, and atmospheric temperature effects. The book includes maps, photographs, and fifty color satellite images.

Popkin, Roy. *The Environmental Science Services Administration.* New York: Praeger, 1967. A discussion of the governmental unit that oversaw the ESSA satellites. Places the topic into its contemporary context.

Schnapf, A. "The Development of the Tiros Global Environmental Satellite Program." In *Meteorological Satellites: Past, Present, and Future.* NASA SP-2227. Springfield, Va.: National Technical Information Service, 1982. This fine introduction covers the history of the development of the TIROS system and places the ESSA system in context. A very important and useful document.

Yenne, Bill. *The Encyclopedia of U.S. Spacecraft.* New York: Exeter Books, 1985. Offers an accessible overview of the nature and function of the various ESSA satellites. Intended for the lay reader. Well-illustrated.

*Douglas Gomery*

# Escape from Piloted Spacecraft

*Date:* Beginning 1961
*Type of program:* Piloted spaceflight

*As too often tragically demonstrated, spaceflight is an inherently risky endeavor. Flight engineers for both NASA and the Russian space program have designed a variety of escape mechanisms in an attempt to allow astronauts to escape the spacecraft should a disaster occur during liftoff or later stages of the flight. Escape is not always possible, but the desire to make spaceflight as safe as possible is always present.*

### Key Figures

*Maxime A. Faget* (1921-2004), chief designer for American spacecraft and Mercury launch escape system

*Sergei Pavlovich Korolev* (1907-1966), chief designer of Soviet Vostok, Voskhod, and Soyuz spacecraft

### Summary of the Technology

The first piloted American spaceflights were in the one-person Project Mercury capsules. In the early 1960's, the Mercury design criteria included a system to allow astronauts to escape safely should a mishap occur before or during launch. A nearly 5-meter-tall three-legged tower sat atop the Mercury spacecraft. A small solid-fueled rocket was mounted on the tower. The rocket's three nozzles were pointed slightly outward, so that in the event they were needed they would not fire directly on the piloted spacecraft. An automatic sensing system fired the escape rocket in the event that the mission needed to be aborted during launch. The small rocket was just enough to power the Mercury spacecraft away from the launching rocket and allow it to splash down safely in the Atlantic Ocean. If the launch went well, the astronaut jettisoned the escape tower before reaching orbit. The Mercury craft had no escape system after the initial launch phase of the mission. This Mercury escape system received unplanned tests several times during the development phase when robotic test flights failed during launch, but fortunately it never had to be used during an actual emergency.

The next step in the American piloted spaceflight program was the two-person Gemini spacecraft. The Gemini spacecraft replaced the launch escape tower system with a pilot ejection system similar to that used in jet aircraft. The catapult-rocket-powered ejection seats were capable of blasting the astronauts 150 meters (500 feet) above and 300 meters (1,000 feet) outward from the stricken craft. This system could be used up to an altitude of about 21 kilometers (13 miles) during either launch or reentry. The Gemini system was operated manually rather than automatically, as was the Mercury system. Either the commander or the pilot could pull the rings located near their knees to eject both astronauts. During descent from very high altitudes, a combination balloon-parachute would be used both to slow and to stabilize the astronauts. Otherwise, the falling astronauts would turn and tumble too much. Once lower altitudes were reached, a conventional parachute could take over. The first U.S. spacecraft docking missions occurred during the Gemini Program. One of several reasons for these docking missions was to learn the procedures needed should a space rescue become necessary.

None of the Gemini flights needed to use the ejection system on launch, although the Gemini VI-A crew came closest to ejecting on the pad. Two seconds before liftoff, about a second after main engine ignition, their Titan II first stage aborted the launch. Commander Wally Schirra had the option to eject both himself and pilot Tom Stafford from the potentially dangerous vehicle. If they had ejected, however, the spacecraft would have been useless for their planned rendezvous with Gemini VII. Instead, they waited and were safely removed from the spacecraft.

During the Gemini VIII mission, the spacecraft began to spin uncontrollably during flight. As a result, it made the first U.S. emergency landing of a

piloted space mission on March 16, 1966. However, the ejection seats were not used and a normal landing was made.

The Launch Escape System (LES) for the three-person Apollo spacecraft returned to the concept used by the Mercury program. A four-legged tower and launch escape rocket extended 10 meters (33 feet) from the mounting bolts on the Command Module to the tip of the nose cone. The larger size of the Apollo spacecraft compared to the Mercury spacecraft meant that the Apollo escape system had to be about twice the size of the Mercury escape system. If an emergency occurred during launch, the launch escape rocket could be fired either automatically or manually. The Command

*The X-38 Crew Return Vehicle under the wing of a B-52 in 1997. The X-38 was part of a research project on crew return vehicles for use with the International Space Station.* (NASA)

Module, containing the astronauts, would separate from the vehicle stack. A pitch control motor near the top of the escape system would ensure that the Command Module flew out of the path of the still ascending, perhaps exploding, booster. Once free of danger, the launch escape tower would jettison, allowing the Command Module to parachute into the Atlantic.

These escape systems had no provisions for rescue after the launch phase of the missions. However, during the Skylab program, another Apollo spacecraft was readied for launch after the mission reached orbit. For the first time there was the possibility of rescuing astronauts in space, should an accident occur leaving the astronauts stranded in the Skylab. During this time, an era of U.S.-Soviet cooperation in space first began. A major motivation behind the Apollo-Soyuz docking missions was to learn techniques needed in the event an international space rescue became necessary.

The one-person Vostok craft was the first Soviet spacecraft and the first craft from any nation to carry a man into space. It also carried the first woman into space. The Vostok escape mechanism was a pilot ejection seat that could be used either during launch or after reentry. This simply designed spherical spacecraft had no provisions for a soft landing, and the ejection system was, by design, routinely used prior to landing. The space-suited cosmonaut ejected at an altitude of about 8 to 10 kilometers (5 to 6 miles). Then the spacecraft and the cosmonaut parachuted separately to the ground.

The Soviet two- or three-person Voskhod spacecraft was a modification of the previous Vostok craft. With two or three cosmonauts rather than one, the crew cabin was crowded. To make room, the crew couches were rotated 90° and the ejection seats were removed. A small solid-fueled rocket was added to the landing system to achieve a soft landing. To reduce crowding, the cosmonauts wore space suits only if a spacewalk was planned. The Voskhod therefore had no provisions for the crew to escape in the event of an emergency.

The basic three-person Soyuz spacecraft was originally designed in the 1960's and, with several design updates, is still in use today. Like the Mercury and Apollo craft, the Soyuz has a launch escape system consisting of a tower and a small rocket on top of the spacecraft.

In October of 2003, China became only the third nation to launch a piloted spaceflight program. The Chinese Shenzhou spacecraft is based on the Soyuz design with several modernizations. In particular, the launch escape system has an extra set of motors to allow escape at a higher altitude than the Soyuz.

The first four flights of the U.S. Space Transportation System (the space shuttle), with only two crew members, had ejection seats to allow for the possibility of escape should a mishap occur during launch or reentry up to an altitude of about 30 kilometers (100,000 feet). Starting with the fifth flight of the space shuttle, the larger shuttle crews made ejection seats prohibitively heavy, and managers at the National Aeronautics and Space Administration (NASA) thought shuttle launches were safe enough to remove the ejection systems. After the first four missions, shuttle crews had no launch escape systems, which would prove disastrous five years later during the launch of *Challenger* on STS 51-L.

After the 1986 *Challenger* accident, NASA considered options such as a crew escape spacecraft, small rockets to blast astronauts to safety, and individual ejection seats for the shuttle. However, these systems were deemed too costly, heavy, or bulky and were rejected. The crew bailout system that was finally adopted represented a compromise between crew safety and considerations of weight, cost, and development time. Should the crew need to escape while the shuttle is in a glide (either after an abort or during landing), the side entry hatch is jettisoned. The astronauts, wearing suits partially pressurized against the altitude (up to about 10 kilometers or 6 miles), individually slide out through the door opening and past the left wing on a long escape pole mounted in the cabin. After descent to a reasonable altitude, they parachute to safety.

The International Space Station (ISS) was to have an Assured Crew Rescue Vehicle (ACRV) docked to one of its ports at all times. Development delays and cost overruns resulted in the project's cancellation. Originally planned as backup rescue vehicle, the Soyuz spacecraft, which brings expedition crews and visiting crews to the ISS, now serves as the primary escape system for the station. Because of its limited crew capacity, only two full-time crew members can be on the station.

### Contributions

Each time a space mission ends in disaster, program engineers examine the cause of the failure. They then redesign the spacecraft to prevent the same failure from happening again and to improve astronauts' chances of escape should a similar failure occur. After the 1986 *Challenger* accident, NASA installed a crew bailout system in the shuttle orbiters. Although this would provide no escape during the solid-rocket-burn portion of the launch, it would allow some means of escape after the orbiter separated from the External Tank. It could also be used during the final phases of the landing.

Fortunately, most of the launch escape systems have never actually been tested during a piloted flight. However, on September 27, 1983, the Soyuz launch escape system saved the lives of cosmonauts Vladimir Titov and Gennady Strekalov. The Soyuz T-10-1 launch from the Baikonur Cosmodrome failed when the Soyuz booster caught fire from a fuel spill ignited shortly before liftoff. The cosmonauts used the Soyuz escape rocket system to pull the crew cabin clear of the pad only two seconds before the vehicle exploded. The launch pad was destroyed, but the cosmonauts landed safely 4 kilometers away.

The ejection seat escape systems, such as those used in the Gemini spacecraft, had to be designed for possible use at very low altitudes in the event a failure occurred in the very early launch phase or just before touchdown. The engineering knowledge gained from these designs has been applied to improving the designs of jet ejection seats. So-called zero-zero seats (because they provide safe ejection at zero altitude and zero speed) work well even when the jet is at very low altitude. These seats have saved many pilots from death or serious injury.

### Context

Despite all safety efforts, piloted spaceflight remains a very risky undertaking. Prior to the first piloted spaceflight in 1961, many unmanned test flights failed. Explosions during launch were common. Spacecraft were designed with the assumption that launch failures were a likely possibility and escape systems were needed to ensure crew safety. Over the years, launch vehicles became more reliable through the use of redundant systems. After the space shuttle had been flying for several years, people began to forget about the inherent dangers of launching piloted spacecraft.

After the 1986 *Challenger* accident, people were reminded of the dangers inherent in launching a piloted spacecraft. Engineers renewed the search for viable launch escape systems. Without a major systems redesign, however, only a cursory escape system (using extendable poles and parachutes) would be added to the shuttle orbiters. This required the crew, in the event of an emergency, to be conscious and able to effect the escape.

About a decade and a half later, on February 1, 2003, *Columbia* broke apart during reentry and engineers renewed the quest for escape and rescue systems. Fortunately, these accidents are rare, but they do occur and remind everyone that astronaut escape systems are a necessary feature of all piloted spacecraft. Unfortunately, the cost for providing astronauts and cosmonauts with a viable means of escaping a doomed spacecraft is now considered prohibitive. Unless that line of reasoning is changed, more space travelers will perish in the future.

**See also:** Apollo Program; Apollo 1; Apollo 13; Astronauts and the U.S. Astronaut Program; Gemini Spacecraft; International Space Station: Design and Uses; Mercury Spacecraft; Rocketry: Modern History; Soyuz and Progress Spacecraft; Space Shuttle: Approach and Landing Test Flights; Space Shuttle Activity, 1986-1988; Space Shuttle Mission STS 51-L.

**Further Reading**

Catchpole, John. *Project Mercury: NASA's First Manned Space Programme.* London: Springer-Verlag London, 2001. Highlights the differences in Redstone/Atlas technology and why the need for a launch escape system was vital to the survival of the astronauts.

Hall, Rex, and David J. Shayler. *The Rocket Men: Vostok and Voskhod, the First Soviet Manned Spaceflights.* London: Springer-Verlag London, 2001. The authors detail the Vostok, Voskhod, and Soyuz spacecraft, including the choices for crew safety and escape.

_____. *Soyuz: A Universal Spacecraft.* Chichester, England: Springer-Praxis, 2003. Hall and Shayler review the development and operations of the reliable Soyuz family of spacecraft and discuss the safety features designed into the craft.

Jones, Thomas D. "The View from Here: 'Escape' Velocity." *Aerospace America*, September, 2003. Available online at http://www.aiaa.org. Accessed February 14, 2005. This article describes, from the astronauts' point of view, escape systems that have been used and considered for the space shuttle.

Lovell, Jim, and Jeffrey Kluger. *Apollo 13.* New York: Pocket Books, 1995. This story of the ill-fated Apollo 13 mission clearly illustrates the need for escape and rescue systems during space missions.

Mitchell, Naomi L., and Patricia Q. Roberson. *Aerospace Science: The Exploration of Space.* Maxwell Air Force Base, Ala.: Defense Printing Service Detachment Office, 1994. Chapter 3 of Unit 1 in this Junior ROTC textbook contains a section on space rescue.

Oberg, James. "Max Faget: Master Builder." *Omni* 17, no. 7 (1995): 62. This biographical sketch describes the principal designer behind the launch escape system towers used for the Mercury and Apollo Programs.

Shayler, David J. *Disasters and Accidents in Manned Spaceflight.* Chichester, England: Springer-Praxis, 2000. Shayler examines the challenges that face all crews as they prepare and execute their missions. The book covers all aspects regarding human crews in spaceflight—training, launch to space, survival in space, and return from space—followed by a series of case histories of the major incidents in each of those categories over the past forty years. The sixth section looks at the International Space Station and how it is designed to try and prevent, insofar as possible, major incidents occurring during the lifetime of the Space Station; it also examines the difficulties facing a settlement on the Moon or Mars during the next forty years.

_____. *Gemini: Steps to the Moon.* Chichester, England: Springer-Praxis, 2001. Covers the development of the Gemini Program.

Sullivan, Scott P. *Virtual Apollo: A Pictorial Essay of the Engineering and Construction of the Apollo Command and Service Modules.* Burlington, Ont.: Apogee Books, 2002. Full-color drawings in precise detail provide inside and outside views of the Command and Service Modules, complete with details of construction and fabrication. The launch escape system is likewise detailed.

*Paul A. Heckert and Russell R. Tobias*

# Ethnic and Gender Diversity in the Space Program

*Date:* Beginning March, 1958
*Type of issue:* Sociopolitical

*The American space program was dominated by white, male "flight jocks" from its earliest days in the 1950's. It took twenty years to integrate the program, and several struggles occurred along the way.*

### Key Figures

*Jacqueline "Jackie" Cochran* (1906-1980), pilot and first woman to break the sound barrier

*W. Randolph Lovelace II* (1907-1965), chairperson of the Special Committee for Life Sciences for Project Mercury

*Geraldyn "Jerrie" M. Cobb* (b. 1931), pilot who set world records in aviation for speed, distance, and absolute altitude

*Jane Hart,* member of the Mercury 13

*John H. Glenn, Jr.* (b. 1921), U.S. astronaut

*M. Scott Carpenter* (b. 1925), U.S. astronaut

*Valentina Tereshkova* (b. 1937), Vostok 6 pilot and first woman in space

*Sally K. Ride* (b. 1951), first American woman in space

*Tuân Pham,* Vietnamese citizen who was the first Asian to fly into space

*Arnaldo Tamayo-Mendez,* first space traveler of African descent

*Guion S. Bluford, Jr.* (b. 1942), first African American astronaut

*S. Christa McAuliffe* (1948-1986), schoolteacher and first "citizen in space"

*Eileen M. Collins* (b. 1956), first American woman to command a space shuttle mission, first woman to pilot the shuttle, and first woman to pilot it twice

*Liwei Yang,* first Chinese national and first person to fly in a spacecraft launched by a country other than the United States or Russia

### Summary of the Issue

Before the National Aeronautics and Space Administration (NASA) was created on October 1, 1958, the Air Force was looking to the heavens as an outpost for its eyes in the sky. To that end, it formed the Man-in-Space Task Force and developed a four-phase program to get its crews into space and onto the Moon before anyone else. The first of these programs was called Man-in-Space-Soonest or, in an ironic quirk of acronyms, MISS. If all went according to plan, MISS would put a "man" into orbit by October, 1960.

At the level of government, such "best laid plans" are often changed by congressional committees. As funds for the new civilian space organization began pouring into NASA's coffers, the funding for the MISS project began to diminish. Eventually, MISS was canceled in favor of NASA's own "manned" space program.

Once NASA's plans to put a human into orbit began to gel, the decision as to who would pilot, or occupy, the spacecraft had to be made. Only males between twenty-five and forty years of age would be considered. These men would have to be willing to accept hazards comparable to flying a research aircraft. Parachutists, acrobats, deep-sea divers, and mountain climbers were high on the list of desirable candidates. In an effort to weed out undesirables, NASA required each applicant to be recommended by a responsible organization. Eventually, these criteria were discarded and the search went out to find suitable military personnel with experience as test pilots of jet aircraft. Although race was not mentioned in the specifications, this was the late 1950's, when racism reigned, making it unlikely that NASA would find someone other than a white male to fill all of the requirements.

Shortly after NASA selected the first astronauts in 1959, Dr. W. Randolph Lovelace II, chairperson of the Special Committee for Life Sciences for Project Mercury, met with Jacqueline ("Jackie") Cochran, an experienced and well-known pilot. In 1953, Cochran became the first woman to break the sound barrier. She held more than two hundred flying records. She wanted to fly into space but conceded that she was too old to qualify. Together she and Lovelace decided to test women for the Astronaut Corps. Women were usually smaller and more flexible than their male counterparts, and it was expected that these "astronettes" (a name coined by NASA) would have an easier time fitting into the cramped spacecraft. In addition, Lovelace understood from previous psychological and medical assessments that women could withstand pain, heat and cold, monotony, and loneliness better than men. They contacted twenty-nine-year-old Jerrie Cobb, who had set world records in aviation for speed, distance, and absolute altitude, to see if she would be interested in testing for the astronaut program. She agreed and successfully completed all phases of the testing, including a two-week series of tests at the U.S. Navy School of Aviation Medicine in Pensacola, Florida. Afterward, Lovelace asked Cobb for the names of twenty-five other women aviators for possible further testing.

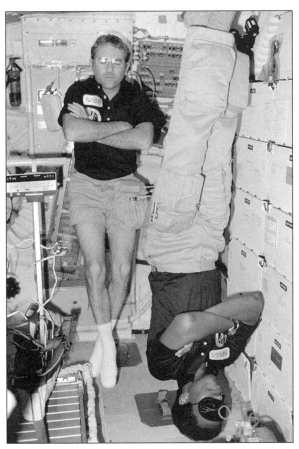

*Guion S. Bluford, the first African American in space on shuttle mission STS-8, "hangs out" with Dick Truly on middeck.* (NASA)

Cochran, a wealthy woman, decided that she and her husband would underwrite the expenses for the selected women. The candidates jumped at the opportunity, and twelve successfully completed the first phase of testing. However, before all phases could be completed, the program was canceled. No one at NASA and no one else connected to the testing has ever publicly explained the reason for the sudden cancellation.

In 1962, Cobb and Jane Hart, a fellow member of the Mercury 13 (as Cobb and the twelve other women had come to be known), lobbied Vice President Lyndon B. Johnson to have Congress review the situation. He agreed, and in July, 1962, the two women testified before the House of Representatives' Space Committee, claiming that NASA had discriminated against women by canceling the program. Two Mercury astronauts, John H. Glenn, Jr., and M. Scott Carpenter, also testified before the committee. They stated that they did not believe there was discrimination in the selection process and that, according to Glenn, "The fact that women are not in this field is a fact of our social order." Cochran then testified at the hearing that both NASA and the military would lose money if they trained women, who would never finish the program because "marriage is the basic objective of all women." She also testified that she did not believe that NASA had discriminated against the women during the testing and selection process. As a result, Congress decided not to force NASA to open the astronaut program to those who were not test pilots.

Project Mercury, America's first small steps into the universe beyond Earth, provided the necessary thrills to encourage President John F. Kennedy to set more heroic goals. As a means to upstage the Soviet Union's cosmonauts, flying in their Vostok spacecraft, Kennedy proposed to send a "man" to the Moon.

Project Gemini, with its two-seat spacecraft, followed Mercury. Bigger crews and an increased number of flights required more astronauts to fill the seats. The "Original Seven" Mercury astronauts could not fly all of the missions. In addition, one of the astronauts, Donald K. "Deke" Slayton, was grounded with a heart murmur; another, M. Scott Carpenter, was not in good standing with NASA because of his apparent inattentiveness during his orbital flight; and Glenn was considered a national treasure whose life could not be risked. The "Second Nine" astronauts, including Neil A. Armstrong, were selected in October, 1962. The following October, fourteen men were added to the roster of astronauts. By 1965, it was clear that the Apollo lunar-landing program would need scientists aboard the lander to conduct geological research. At long last, the ranks of the astronauts would include five nonpilots. It would still, however, lack women and people of color. Nineteen

more pilot astronauts joined the Corps in April, 1966. Eleven scientist-astronauts were chosen the following July. The last group of astronauts, chosen before the close of the Apollo Program, were former pilots of the canceled Manned Orbiting Laboratory program. During the first eighteen years of the American space program, seventy-three white, male astronauts were chosen. That was about to change.

With the advent of the Space Transportation System and the role of the space shuttle in scientific research, a new breed of astronaut was born. Unlike its predecessors, the shuttle did not accelerate rapidly to free itself from the bonds of gravity. Living quarters were roomier and included such amenities as a kitchen and a "restroom." The bone-

*Sally K. Ride was the first American woman in space on shuttle flight STS-7.* (NASA)

rattling water landing was replaced with a gentle glide landing equivalent to that of a commercial jetliner.

In January, 1978, thirty-five astronaut-candidates were selected. They included scientists as well as pilots. They also included women and nonwhite males: Guion S. Bluford, Jr., Ronald E. McNair, Ellison S. Onizuka, Sally K. Ride, and Kathryn D. Sullivan.

### Contributions

Who is best qualified to fly into space? What does it take to acquire the "right stuff"? Since the beginning of the Space Age, these questions have been posed both by those who wanted to fly in space and by those who were going to catapult them into orbit.

In the era before the shuttle, when scientific studies were inadequate to answer the questions, researchers and decision makers had to base their decisions on logic. They reasoned that they were placing the astronauts on top of a very fast-moving guided missile, loaded with extremely explosive propellants, and were accelerating them very rapidly to escape the pull of gravity. Inside their tiny, pressurized spacecraft, the astronauts were protected from the unsurvivable environment of space by thin sheets of metal and multilayered space suits. Hurtling around the globe at speeds of 8 kilometers per second would require lightning-fast reflexes. If the guidance and control systems failed, an astronaut might be subjected to high-speed rotational motion along any of the spacecraft's axes.

As a result, the qualifications of the earliest astronauts were based on these known conditions of the spacecraft and early space travel: The candidate had to be less than forty years of age; NASA did not want someone growing old while the rocket scientists were perfecting the launch vehicle, spacecraft, and other essential hardware. The candidate had to be less than 5 feet, 11 inches tall and in excellent physical condition. Anyone taller than that would be unable to fit inside the tiny spacecraft. The "excellent physical condition" part eliminated those who might have a weight problem or a physi-

cal disability. An astronaut had to have a bachelor's degree or equivalent, although there was no specification for the area of the degree; the educational level, as an indicator of intelligence and learning ability, was what was important. Many thousands of candidates met these requirements, regardless of race or gender.

Each candidate had to be a graduate of a test-pilot school. The astronaut would be flying in a high-performance spacecraft, and at that time it had not yet been determined whether the astronaut would actually pilot the spacecraft. Eligible pilots were required to have accumulated 1,500 hours of flying time and to be qualified as jet pilots. These requirements narrowed the field quite a bit, but it still could include nonwhite and female candidates.

After considering the many possibilities, NASA scientists concluded that seeking candidates among military jet test pilots would be the answer. Jet pilots, it was reasoned, handle very fast-moving aircraft on a daily basis. Test pilots have a great deal of experience handling emergency situations in these fast-moving aircraft. Military test pilots face deadly encounters on a regular basis. Based on all of these criteria, the only candidates remaining, at that time, were white males.

### Context

While America was publicly demonstrating its prowess in space exploration, the Soviets were secretly trying to upstage them in any way they could. Having already beat the United States to orbit with Sputnik in October of 1957, and with a living creature (the dog Laika) the following May, the Soviets wanted to beat the Americans to orbit with a human passenger, too. This they accomplished on April 12, 1961, a scant twenty-three days before Alan Shepard climbed into the heavens aboard *Freedom 7*.

The Soviet Union had bigger boosters, and that gave them the advantage when it came to space firsts. They placed a human, Yuri A. Gagarin, in orbit ten months ahead of John H. Glenn, Jr. A cosmonaut, Gherman Titov, spent a full day in space

*A July, 1979, photograph of astronaut Kathryn D. Sullivan suited to fly the NASA WB-57F reconnaissance aircraft. Sullivan reached a record altitude of 63,300 feet and went on to participate in several shuttle missions and become the first American woman to walk in space.* (NASA)

twenty-one months before Gordon Cooper. They accomplished the first piloted orbital rendezvous mission between two spacecraft three and a half years before Gemini VI-A met Gemini VII in December, 1965. On June 16, 1963, the Soviets launched the first woman into space. Valentina Tereshkova piloted the Vostok 6 spacecraft for three days and amassed more flight time than all of the Mercury astronauts combined. She was not a military test pilot; she was not even a pilot. She was a textile worker and an amateur parachutist. She was also the fiancée of Cosmonaut Andriyan Nicolayev, who

piloted Vostok 3 the previous year. It would be nineteen years before the next woman would fly into space. She was Svetlana Savitskaya, a Soviet test pilot with more than 1,500 hours of flight time in a variety of aircraft. She was the first woman to spend time on a space station. Her flight, like that of her predecessor nearly two decades earlier, served propaganda purposes and came only seven months before the first American woman flew into space. The following year, she returned to space as the first woman to make two spaceflights and the first to perform a spacewalk.

On June 16, 1983, twenty years to the day after Tereshkova's flight, Sally K. Ride became the first American woman in space. The thirty-two-year-old physicist proved, at last, that American women were perfectly capable of surviving a trip into space and performing tasks similar to those performed by their male counterparts. She returned to space the following October, along with Kathryn D. Sullivan, on the first flight to feature two women. Sullivan became the first American woman to perform a spacewalk during the flight aboard *Challenger*.

The first Asian to fly into space was Tuân Pham of Vietnam, a former Vietnamese Air Force pilot. He was part of the sixth international crew flown by the Soviet Union. His Soyuz 37 flight took him for a three-month stay aboard the Salyut 6 space station in July, 1980. The first space traveler of African descent flew on the next Intercosmos mission, Soyuz 38, in September, 1980. Cuban-born Arnaldo Tamayo-Mendez was also a military pilot.

The first African American astronaut was Lieutenant Colonel Guion S. Bluford, Jr., of the U.S. Air Force. He was one of the mission specialists on STS-8, the first night launch in the shuttle program. The six-day scientific mission began with the launch of *Challenger* on August 30, 1983. Two flights later, the STS 41-B mission featured the first untethered spacewalk in history and included in the crew Ronald E. McNair, the second African American astronaut.

The first classified flight in the history of the American space program was STS 51-C in January, 1985. The all-military, five-person crew included the first American of Asian descent, Air Force major Ellison S. Onizuka. The U.S. space program, now integrated, was proceeding at a steady pace toward the day when an ordinary citizen, a nongovernment employee, would fly into space.

On January 28, 1986, the space shuttle *Challenger* was launched on its tenth and final flight. On board was the most integrated flight crew to date. It included three white male crew members, Commander Francis R. "Dick" Scobee, Pilot Michael J. Smith, and Payload Specialist Greg Jarvis. Ellison S.

Onizuka joined Judith A. Resnik, the second woman in space, and Ronald E. McNair, the second African American in space, as mission specialists. The crew was rounded out by the first "citizen in space," schoolteacher S. Christa McAuliffe. For the first time in the history of the American space program, men and women—black, white, and Asian—were equal in space. Seventy-three seconds into the launch of STS 51-L, the vehicle experienced a structural breakup, taking it and its seven-person crew to a watery grave. For the first time in the history of the American space program, a crew was lost during flight.

In March, 1998, NASA announced that Air Force Lieutenant Colonel Eileen M. Collins would become the first American woman to command a space shuttle mission. She had already flown as the pilot on two previous missions, becoming the first woman to pilot the shuttle and the first woman to pilot it twice. The STS-93 launch of *Columbia* to deploy the Chandra X-Ray Observatory took place on July 23, 1999, forty years after W. Randolph Lovelace II had approached Jerrie Cobb to test for his astronette program. The flight was the ninety-fifth in the space shuttle program and the 126th flight piloted by Americans. It took place only thirty-six years after the Soviets sent Valentina Tereshkova into orbit as the pilot of Vostok 6.

In September, 1998, sixty-seven-year-old Jerrie Cobb was honored at the grand reopening of the Pioneer Woman Museum in Ponca City, Oklahoma. The event took place shortly before seventy-seven-year-old John H. Glenn, Jr., made his second trip into orbit. Oklahoma senator James Inhofe addressed the crowd and read a statement from NASA Administrator Daniel S. Goldin that said, in part:

> If everything goes well and the data comes in after the John H. Glenn, Jr., senior citizen spaceflight and the data works out, we will be doing it again. It is logical that a woman would be next and there is no one in America that is more qualified and deserving to be in that space shuttle than Jerrie Cobb.

The first Chinese national and the first person to fly in a spacecraft launched by a country other than the United States or Russia was Liwei Yang, who rocketed into space on October 15, 2003. His spacecraft, Shenzhou 5, was placed into a 200-by-343-kilometer orbit by China's Chang Zheng 2F booster. He spent nearly a day in orbit, completing 14 revolutions of Earth.

**See also:** Cooperation in Space: U.S. and Russian; Gemini Program; Mercury Project; Space Shuttle Flights, 1983; Space Shuttle Flights, 1984; Space Shuttle Flights, July-December, 1985; Space Shuttle Mission STS-63; Space Shuttle Mission STS-95; Space Shuttle Flights, 1999; Space Shuttle Mission STS-93.

### Further Reading

Ackmann, Martha. *The Mercury 13: The Untold Story of Thirteen American Women and the Dream of Space Flight.* New York: Random House, 2003. Tells the story of the thirteen remarkable women who underwent secret testing and personal sacrifice in the hope of becoming America's first female astronauts.

Baker, David, ed. *Jane's Space Directory, 2005-2006.* Alexandria, Va.: Jane's Information Group, 2005. This reference book, updated annually, is invaluable for obtaining general information on virtually any piloted or robotic mission, both those launched by the United States and those launched by other countries. Includes illustrations and a useful index.

Burrows, William E. *This New Ocean: The Story of the First Space Age.* New York: Random House, 1998. A comprehensive history of the human conquest of space, covering everything from the earliest attempts at spaceflight through the voyages near the end of the twentieth century. Burrows is an experienced journalist who has reported for *The New York Times, The Washington Post,* and *The Wall Street Journal.* Interviewees in the book include Isaac Asimov, Alexei Leonov, Sally K. Ride, and James A. Van Allen. Contains many photographs and an extensive source list.

Cobb, Jerrie. *Jerrie Cobb, Solo Pilot.* Tulsa, Okla.: Jerrie Cobb Foundation, 1997. This autobiography tells the fascinating story of this pioneering pilot, detailing her participation in the Mercury 13 project and how she has spent her life in the service of her fellow human beings.

Glennan, T. Keith. *The Birth of NASA: The Diary of T. Keith Glennan.* Edited by J. D. Hunley. NASA SP-4105. Washington, D.C.: Government Printing Office, 1993. A history of NASA's early days and of its first director, the book's author. Details the difficulties, as well as the triumphs, that the space agency experienced during the transitional period of the late 1950's and early 1960's.

Heppenheimer, T. A. *Countdown: A History of Space Flight.* New York: John Wiley, 1997. A detailed historical narrative of the human conquest of space. Heppenheimer traces the development of piloted flight through the military rocketry programs of the era preceding World War II. Covers both the American and the Soviet attempts to place vehicles, spacecraft, and humans into the hostile environment of space. More than a dozen pages are devoted to bibliographic references.

Kevles, Bettyann. *Almost Heaven: The Story of Women in Space.* New York: Basic Books, 2003. The tale of spacefaring women, from Valentina Tereshkova to Kalpana Chawla, including their unique concerns as female astronauts.

Launius, Roger D. *NASA: A History of the U.S. Civil Space Program.* Malabar, Fla.: Krieger, 1994. An in-depth look at the United States' civilian space program and the establishment of the National Aeronautics and Space Administration. Chronicles the agency from its predecessor, the National Advisory Committee for Aeronautics, through the mid-1990's.

Swenson, Loyd S., Jr., James M. Grimwood, and Charles C. Alexander. *This New Ocean: A History of Project Mercury.* Washington, D.C.: National Technical Information Service, 1966. One title in the NASA History series, this book chronicles Project Mercury from its conception during the early days of NASA to its completion following the flight of Gordon Cooper in Mercury-Atlas 9. There are dozens of black-and-white photographs of the piloted as well as robotic flights. Line drawings show the inner workings of much of the equipment related to the missions. An impressive hundred-page appendix lists source notes and bibliographic references. Other appendices include a summary of flight data, functional and workflow organization of Project Mercury, personnel growth, the ground station tracking network, and the cost of the Project.

Von Bencke, Matthew J. *The Politics of Space: A History of U.S.-Soviet/Russian Competition and Cooperation in Space.* Boulder, Colo.: Westview Press, 1996. Chronicles the efforts of the United States and the Soviet Union (later Russia) to overcome their political animosities and explore space together. Examines their respective foreign and domestic policies; military, civil, and commercial influences; and top executive, legislative, and institutional politics. Describes their separate and joint endeavors from 1945 through 1997.

Woodmansee, Laura S. *Women Astronauts.* Burlington, Ont.: Apogee Books, 2002. Includes interviews with many past and current women astronauts.

_____. *Women of Space: Cool Careers on the Final Frontier.* Burlington, Ont.: Apogee Books, 2003. Covers behind-the-scenes space careers from the viewpoint of major figures such as Mars Pathfinder Engineer Donna Shirley, director of the Center for SETI Research Jill Tarter, astrophysicist and celestial musician Fiorella Terenzi, astronomer Sandra Faber, and space artist Lynette Cook.

*Russell R. Tobias*

# Explorers: Air Density

*Date:* February 16, 1961, to December, 1970
*Type of spacecraft:* Earth observation satellites

*Data from the four Air Density Explorers determined the effect upon Earth's upper atmosphere of solar heating over a sunspot cycle, provided the first evidence of the winter helium and summer atomic oxygen bulges, and showed that the atmospheric constituents above 300 kilometers exist in layers and maximize where maximum temperature exists.*

## Key Figures

*William J. O'Sullivan, Jr.,* inventor of the Air Density Explorer concept
*Walter E. Bressette,* designer of Air Density Explorer inflatable sphere
*Claude W. Coffee, Jr.,* project manager
*Gerald M. Keating,* principal investigator

## Summary of the Satellites

In 1956, William J. O'Sullivan, Jr., a member of the National Advisory Committee for Aeronautics, or NACA (the predecessor to the National Aeronautics and Space Administration, NASA), at the Langley Research Center in Virginia, proposed putting a lightweight, inflatable sphere into Earth orbit for the purpose of determining atmospheric density at satellite altitudes. NACA, which was interested in learning more about Earth's atmosphere for the design of space vehicles, approved and funded the proposal. During 1956 and 1957, an inflatable sphere 3.66 meters in diameter was developed. Its inflation system and canister were designed and constructed at Langley for the purpose of providing noninterference scientific payloads on U.S. developmental satellite launching systems.

Between October 23, 1958, and December 4, 1960, three attempts to put the inflatable sphere into Earth orbit were unsuccessful because of launch vehicle failures. At 8:05 A.M. eastern standard time or 13:05 Coordinated Universal Time (UTC) on February 16, 1961, the NASA Scout four-stage, solid-fueled launcher (ST 4) lifted from the NASA Wallops Flight Center, off the coast of Vir-

ginia. Six seconds after the thrust-decay of the fourth-stage rocket motor, a nitrogen gas inflation bottle was opened. The inflation gas pressurized a bellows resulting in bellows elongation and ejection of the compactly folded sphere from an open-ended cylinder.

After ejection, the sphere inflated to its full size in approximately three minutes and was separated from the orbiting rocket motor. Thus, Air Density Explorer 9 achieved orbit at 13:16 UTC at an altitude of 672 kilometers above Earth's surface, traveling at a speed of 7.94 kilometers per second. NASA officials were elated by the successful insertion of Explorer 9 into a satellite orbit by ST 4, because this was the first satellite put into Earth orbit by the Scout launching system at Wallops, and both the Scout development task and Air Density Explorer missions were NASA-Langley projects.

Because its tracking beacon failed soon after the satellite reached orbit, the tracking system for Explorer 9 was based at the Smithsonian Astrophysical Observatory (SAO) Baker-Nunn camera stations. These optical telescope stations could detect the spherical satellite only at sunrise and sunset, and then only when the station was in darkness and the satellite was reflecting sunlight high above Earth. More than one thousand sightings of the satellite were reported during the first 155 days of its orbital lifetime by the Baker-Nunn camera stations and the SAO's ninety-three Project Moonwatch teams (the Moonwatch teams were made up of people worldwide who volunteered to operate satellite optical observing stations early in the morning and late in the evening). The optical teams reported that the satellite was seen as a steady white light as it moved through the star field from one horizon to the other. From the optical sightings it was determined that the initial orbit of Explorer 9 had an apogee (highest altitude in orbit) of 2,586 kilometers and a perigee (lowest altitude in orbit) of 632 kilometers. The inclination of its orbit to Earth's equator was 38.86°, and the satellite required 118.55 minutes to complete one revolution around Earth.

Explorer 9 was constructed from a four-ply laminate consisting of alternate layers of 0.0013-centimeter-thick aluminum foil and 0.0013-centimeter-thick plastic film. The aluminum foil was on the outside surface of the sphere, the plastic film on the inside surface. This arrangement of the laminate material provided the required structural stiffness to maintain the spherical shape of the satellite after inflation in orbit. Without contributing to internal pressure, the laminate made the satellite reflective of sunlight for optical tracking, provided the required electrical conductivity for a tracking beacon antenna, and helped control the temperature of the satellite.

The construction, while containing many more flat gores (forty), was similar to that of a beach ball. The sphere was divided into two hemispheres by a 3.81-centimeter-wide gap of dielectric material so that the two halves of the sphere could be used as the antenna for the tracking beacon. The satellite was often called the "polka-dot satellite" because 17 percent of its outside surface was covered with 5.10-centimeter-diameter white dots. The white dots were required to maintain the temperature of the tracking beacon in orbit between −10° Celsius and +60° Celsius (263 and 333 kelvins). The inflatable

sphere with the attached tracking beacon was folded into a package 21.59 centimeters in diameter and 27.98 centimeters in length so that it would fit into the Scout vehicle. In this folded condition, the thin-walled sphere was able to withstand the severe gravitational force of the launch. The total weight of the Explorer 9 satellite was 6.66 kilograms. When fully inflated, it had a continuous cross-sectional area-to-mass ratio of 15.84 square centimeters per gram.

The total energy of a satellite, which is the sum of its kinetic energy and its potential energy, undergoes change with time in orbit, principally because of solar radiation pressure (photon bombardment) and atmospheric resistance (molecular bombardment). The total energy of a satellite is also related to its mass. In identical orbits, a heavier satellite has more total energy than a lighter one. A satellite remains in orbit because of a balance between its speed around Earth and its height above the ground. Its height determines the amount of pull toward Earth because of Earth's gravitational field, whose force diminishes relative to the force on Earth's surface by the square of the distance from the surface. Its speed determines the centripetal force inward, toward the center of Earth. In a satellite orbit, when the speed of the satellite provides a centripetal force greater than the force of gravity, the satellite gains altitude, and when the speed produces a force less than the pull of gravity, the satellite loses altitude. The time required for a satellite to complete one orbit around Earth (its period) is a measure of its total energy.

When the Sun is behind the satellite in orbit, the Sun's radiation pressure exerts a force on the satellite that adds to the speed of the satellite; when the Sun is in front of the satellite, the radiation force decreases the speed. If the satellite enters Earth's shadow, the satellite gains energy in the shadow when the satellite is moving toward the Sun and loses energy in the shadow when the Sun is behind the satellite. The energy change produced by the lack of the Sun's radiation pressure in Earth's shadow is calculated and subtracted or added to the energy charge observed by the tracking stations in order to deduce the atmospheric density. Unlike solar pressure, atmospheric resistance always acts upon the frontal area of the high-speed satellite; therefore, it always reduces the satellite's speed and its total energy. The atmospheric density is determined in the vicinity of perigee, where the atmospheric resistance is much greater than elsewhere along the orbit.

To deduce atmospheric density over a range of satellite altitudes in a short period of time by a single satellite using the atmospheric drag technique, the satellite experiencing the drag must have a large cross-sectional area-to-mass ratio, because the density at satellite altitude is extremely low (approximately one ten-billionth of a kilogram per cubic meter at 250 kilometers above Earth's surface). The Air Density Explorer satellites, with their large cross-sectional area-to-mass ratios, were designed for this purpose.

To determine atmospheric density in Earth's polar regions, a second inflatable satellite with a diameter of 3.66 meters was inserted into a near-polar orbit on December 19, 1963. The Scout 122 vehicle was launched from the United States Pacific Missile Range (USPMR) at 18:49 UTC, carrying the folded sphere inside its nose cone. At 18:59

*Explorer 24, inflated. The satellite was designed to provide data on relationships between solar radiation and air density.* (NASA)

UTC, the folded sphere was successfully inflated and released in orbit, becoming the Explorer 19 satellite. The initial orbit of the satellite had a perigee of 592 kilometers, an apogee of 2,392 kilometers, an inclination to Earth's equator of 78.62°, and a period of 115.79 minutes. The Explorer 19 satellite was made of the same materials as the Explorer 9 satellite. It weighed 8.10 kilograms, was equipped with a tracking beacon, and had 25 percent of its external aluminum surface uniformly covered with dots of white paint.

The third satellite in the Air Density Explorer series was the inflatable Explorer 24 satellite. Explorer 24 was nearly identical to Explorer 19. It was 3.66 meters in diameter and weighed 8.609 kilograms. Explorer 24 was launched into Earth orbit by the Scout launch vehicle from the USPMR on November 21, 1964, at 14:10 UTC to obtain nighttime atmospheric densities in the North Polar region of Earth. The initial orbit of Explorer 24 had an inclination to Earth's equator of 81.4°, a perigee of 526 kilometers, and an apogee of 2,495 kilometers.

The Explorer 24 satellite was a part of NASA's first dual satellite orbit insertion by the Scout launch vehicle. The Explorer 25 satellite was placed in orbit at the same time as Explorer 24 to complement the atmospheric density information from Explorer 24 by measuring electromagnetic and corpuscular (particle) radiation from the Sun and its relation to atmospheric heating.

The last of the Air Density Explorer satellites was Explorer 39, which was placed in Earth orbit, along with Explorer 40, by the Scout launch vehicle from the USPMR on August 8, 1968. Explorer 39 was put in orbit to determine atmospheric density, while Explorer 40 was to measure the electromagnetic and corpuscular radiation from the Sun during a time of high solar energy output. An inflatable spherical satellite with the same diameter as the others but with a mass of 9.421 kilograms, Explorer 39 was inserted into an orbit that had an inclination to Earth's equator of 80.7°, a perigee altitude of 644 kilometers, and an apogee altitude of 2,525 kilometers.

## Contributions

From the orbital decay of the four Air Density Explorer satellites between February, 1961, and December, 1970, the Earth's atmospheric density was determined at satellite altitudes over a range of latitudes. The Explorer 9 data revealed that atmospheric density at the 675-kilometer altitude was a factor of ten less than the density previously established at that altitude by the U.S. Air Force Research Development Command Model Atmosphere 1959.

The drag measurements of Explorer 19 indicated that a large excess of helium existed above the North Pole in local winter; this phenomenon was called the winter helium bulge. Subsequently, it was discovered from Explorer 39's drag data that a winter helium bulge also existed over the Southern Hemisphere. Thus, the winter helium bulge was found to shift hemispheres every six months. An investigation of the helium distribution over Earth using all the Air Density Explorer data established the existence of a systematic asymmetry in the behavior of the exosphere above both hemispheres. The seasonal variation of helium in the Southern Hemisphere was 80 percent larger than in the Northern Hemisphere.

The latitude and day-night variation of atomic oxygen was established by Explorer 24 data between the altitudes of 500 and 600 kilometers. The atomic oxygen concentrations were found to be greater in the local summer and to coexist with the winter helium bulge.

By comparing the Air Density Explorer's drag determination of helium, at an altitude of nearly 1,000 kilometers, with mass spectrometer measurements of helium from Orbiting Geophysical Observatory 6 (OGO 6), at an altitude of about 500 kilometers, it was determined that the exospheric temperature has two maximums: one at high latitudes in the summer hemisphere and a second near the opposite Pole.

## Context

In 1959, the U.S. Air Force's Cambridge Research Center developed the ARDC Model Atmosphere 1959 for national defense reasons, using orbital decay information from satellites that preceded Explorer 9 into Earth orbit. The orbital decay information from these satellites was obtained in the years 1957 and 1958, during a period of maximum solar activity when the average sunspot number was about 200. (The Sun has many bright spots on its surface that increase and then decrease in number over approximately an eleven-year cycle. It has been determined that these spots contribute appreciably to solar energy output.) The Explorer 9 atmospheric density data, which indicated that the density at an altitude of 675 kilometers was a factor of ten less than the ARDC Model Atmosphere 1959, were obtained during the year 1961, when the sunspot number was less than 80. Thus, the Explorer 9 data clearly showed for the first time the magnitude of the effect upon Earth's upper atmosphere from the level of solar heating. These early Explorer 9 data fueled NASA's desire to determine atmospheric density at satellite altitude with the same technique over a complete solar cycle.

The Air Density Explorer data confirmed the theory postulated prior to 1961 that above 100 kilometers, the density of the lighter atmospheric constituents decreases with altitude less rapidly than the density of the heavier ones—resulting in a layering of the atmosphere, with the heavier constituents, such as nitrogen, forming the lower layers and the lighter constituents, such as helium, forming the upper layers. From the drag data, the variation in altitude of the helium and atomic oxygen layers caused by solar heating was determined. When the helium data were compared to the helium concentrations expected according to the United States Standard Atmosphere Supplements (USSAS) 1966, it was seen that the seasonal varia-

tion relative to the model atmosphere predictions was 80 percent larger in the Southern Hemisphere than in the Northern Hemisphere. This was the first evidence that atmospheric constituents in the upper atmosphere above one hemisphere were distributed differently from those above the other. This information required that the USSAS 1966 be modified.

The discovery of the coexistence of the summer atomic oxygen bulge and the winter helium bulge disproved the idea, used to develop the USSAS 1966, that all atmospheric constituents maximized at Earth's equator regardless of season, and supported the idea that the position of peak upper atmospheric densities coincided with peak exospheric temperatures. From the detection of helium by the Air Density Explorers at 1,000 kilometers and from Orbiting Geophysical Observatory 6 at 500 kilometers, it was discovered that the temperature between these altitudes had a major maximum at high latitudes in the summer hemisphere and a secondary maximum near the opposite Pole.

The information derived from the Air Density Explorer drag data, while at Earth altitudes above 300 kilometers, provided important information for the development of theoretical atmospheric models at all levels of the atmosphere, from which predictions are being made concerning the future influence human-made pollutants will have on Earth's ozone layer and its delicate thermal balance.

**See also:** Dynamics Explorers; Earth Observing System Satellites; Explorers: Atmosphere; Explorers: Ionosphere; Geodetic Satellites; Global Atmospheric Research Program; Heat Capacity Mapping Mission; Landsat 1, 2, and 3; Landsat 4 and 5; Landsat 7; Mission to Planet Earth; Seasat.

### Further Reading

Ahrens, C. Donald. *Essentials of Meteorology: An Invitation to the Atmosphere.* 4th ed. Pacific Grove, Calif.: Thomson Brooks/Cole, 2005. A thorough examination of contemporary understanding of meteorology. Includes the contributions made by satellite technology.

Huggett, Richard J. *Environmental Change: The Evolving Ecosphere.* London: Routledge, 1997. Explores the nature, causes, rates, and directions of environmental change throughout

Earth history. Contains descriptions of cosmic, geological, and ecological environments. Includes bibliographical references and index.

Jacchia, Luigi G. *Electromagnetic and Corpuscular Heating of the Upper Atmosphere.* Amsterdam: North-Holland Publishing, 1963. An analysis of eight satellites in Earth orbit, including Explorer 9, during the years 1958 to 1961. Concludes that atmospheric temperature variations relate to 10.7-centimeter solar flux and indicates that corpuscular radiations have a considerable effect upon the heating of the upper atmosphere. Suitable for advanced high school and college levels.

Jacchia, Luigi G., and J. Slowey. *The Shape and Location of the Diurnal Bulge in the Upper Atmosphere.* Amsterdam: North-Holland Publishing, 1966. An analysis of the drag of Explorers 19 and 24 led to strange results concerning the shape and behavior of the diurnal atmospheric bulge. These findings changed the previous picture of a bulge in which temperature and density decreased nearly symmetrically in all directions to a model of a bulge that follows the latitude migrations of the Sun. Suitable for advanced high school and college levels.

Johnson, R. M., and T. L. Killeen, eds. *The Upper Mesosphere and Lower Thermosphere: A Review of Experiment and Theory.* Washington, D.C.: American Geophysical Union, 1995. This monograph is a review of progress made in recent years in experimental and theoretical investigation of the mesosphere, lower thermosphere, and ionosphere. Includes bibliographical references.

Keating, Gerald M., and Edwin J. Prior. *The Air Density Explorer Satellite Program.* Washington, D.C.: Scientific and Technical Information Program, 1973. This is the report of the International Symposium on Space Technology and Science that met in Tokyo, Japan, in September, 1973. It covers the scientific results of the Air Density Explorers, with emphasis on Explorer 9, Explorer 19, Explorer 24, and Explorer 39.

_____. *Latitudinal and Seasonal Variations During Low Solar Activity by Means of the Inflatable Air Density Satellites.* Amsterdam: North-Holland Publishing, 1967. An analysis of atmospheric densities derived from energy changes of the Explorer 19 and Explorer 24 satellites from December, 1963, through February, 1966, leads to conclusions that winter polar densities above the altitude of 550 kilometers are significantly higher than expected but are consistent with peak temperature. It is suggested that this effect may result from a major heat source in the upper atmosphere in addition to extreme ultraviolet radiation.

_____. *The Winter Helium Bulge.* Amsterdam: North-Holland Publishing, 1967. Analysis of the data from Air Density Explorer satellites 9, 19, and 24 over the altitude range of 390 to 880 kilometers during the years 1961 through 1966 reveals that the diurnal atmospheric density bulge shifts with increasing altitude from the hemisphere on the Sun side of the equator to the winter hemisphere. At altitudes where atomic oxygen is the primary atmospheric constituent, peak density occurs in the summer, and at higher altitudes, where helium predominates, peak density occurs in the winter.

Keating, Gerald M., James A. Mullins, and Edwin J. Prior. *The Polar Exosphere Near Solar Maximum.* Amsterdam: North-Holland Publishing, 1970. Analysis of Air Density Explorer 19 and Explorer 24 drag data between altitudes of 550 and 900 kilometers in the years 1967 and 1968 revealed that with increased heating in the upper atmosphere, resulting from increased solar activity, the global distribution of atomic oxygen at the altitude of 550 kilometers could be determined from the drag data of Explorer 24. Polar atomic oxygen concentrations were found to be 30 percent higher than expected.

Keating, Gerald M., Edwin J. Prior, J. S. Levine, and James A. Mullins. *Seasonal Variations in the Thermosphere and Exosphere, 1968-1970.* Berlin: Akademie Verlag, 1972. From comparison between Explorers 19 and 39 drag measurements and the measurements of the winter helium bulge by the Orbiting Geophysical Observatory 6 mass spectrometer, it was necessary to make major revisions in the generally assumed global distribution of both exospheric heat and atomic oxygen.

Klerkx, Greg. *Lost in Space: The Fall of NASA and the Dream of a New Space Age.* New York: Pantheon Books, 2004. The premise of this work is that NASA has been stuck in Earth's orbit since the Apollo era, and that space exploration has suffered as a result.

Lambright, W. Henry, ed. *Space Policy in the Twenty-first Century.* Baltimore: Johns Hopkins University Press, 2003. This book addresses a number of important questions: What will replace the space shuttle? Can the International Space Station justify its cost? Will Earth be threatened by asteroid impact? When and how will humans explore Mars?

National Aeronautics and Space Administration. *Space Network Users' Guide (SNUG).* Washington, D.C.: Government Printing Office, 2002. This users' guide emphasizes the interface between the user ground facilities and the Space Network, providing the radio frequency interface between user spacecraft and NASA's Tracking and Data-Relay Satellite System, and the procedures for working with Goddard Space Flight Center's Space Communication program.

O'Sullivan, William J., Jr., Claude W. Coffee, Jr., and Gerald M. Keating. *Air Density Measurements from the Explorer IX Satellite.* Amsterdam: North-Holland Publishing, 1963. Describes the Air Density Explorer 9 satellite and the techniques for determining atmospheric density from satellite drag. Illustrates how sunlight affects a satellite orbit, shows a picture of Explorer 9, and presents the Explorer data showing the influence of solar activity on atmospheric density. Suitable for the general reader.

Zimmerman, Robert. *The Chronological Encyclopedia of Discoveries in Space.* Westport, Conn.: Oryx Press, 2000. Provides a complete chronological history of all piloted and robotic spacecraft and explains flight events and scientific results. Suitable for all levels of research.

*Walter E. Bressette*

# Explorers: Astronomy

*Date:* Beginning October 13, 1959
*Type of program:* Scientific platforms

*The Astronomy Explorers collected data on solar radiation, gamma rays, meteoroids, x rays, and radio waves. Knowledge gained from these experiments aided researchers in building technology for space travel and in mapping portions of the universe.*

## Summary of the Satellites

On October 13, 1959, at Cape Canaveral, Florida, the National Aeronautics and Space Administration (NASA) launched Explorer 7, a composite radiation satellite. Explorer 7, constructed in the shape of two cones joined at the bases, weighed 41.5 kilograms and orbited Earth at an altitude of 1,088 kilometers. The launch vehicle was a Juno 2 booster. Explorer 7 carried experiments designed to measure direct radiation from the Sun, particularly that which is transformed into heat by Earth and radiated back into space. Data were relayed to Earth using a transmitter powered by solar batteries.

Explorer 11 was launched from Cape Canaveral on April 27, 1961, using a Juno 2 vehicle. The cylindrical satellite measured 2.26 meters in length and 0.38 meter in diameter, and it weighed 37 kilograms. Explorer 11, which orbited at a distance of 1,779 kilometers from Earth, was a gamma-ray astronomy satellite designed to make the first measurements of gamma rays. Gamma rays are similar to x rays but are nuclear in origin. With greater energy and a shorter wavelength than x rays, they are the most energetic form of electromagnetic radiation.

The experiment package carried by Explorer 11 was powered by twelve nickel-cadmium batteries charged by banks of solar cells; the setup has a life expectancy of 150 years. In order to optimize the satellite's ability to take measurements, the cylinder was programmed to orbit in a controlled tumble at ten revolutions per minute. Crystals on board the satellite emitted an electron and positron when struck by a gamma ray. These impingements were recorded and relayed back to Earth by a four-leaf data transmitter. Explorer 11 was tracked by the Minitrack network, and data were reproduced by NASA's Goddard Space Flight Center. In April, 1961, Explorer 11 became the first satellite to detect gamma rays in space.

Ultimately, scientists hoped that data such as those collected by Explorer 11 would help determine the origin of gamma rays. While some theorists believed that these cosmic rays were emitted from the surface of stars and the Sun, others believed that the rays were created by the cloud of hot gas emitted by an exploding star, or supernova. In order to determine the hazard that meteoroids might pose to space travel, scientists launched several satellites designed to measure the number of micrometeoroids in a given volume of space near Earth. Micrometeoroids are particles of meteoritic dust in space ranging from one to two hundred micrometers in size. Experiments aboard Explorer 23 were similar to those on Explorers 13 and 16, satellites that also were used to measure numbers of micrometeoroids.

Explorer 23 was launched using the fourth-stage motor of a Scout rocket on November 6, 1964, at Wallops Island, Virginia. The satellite was a cylinder 2.34 meters long and 0.63 meter wide; it weighed 134 kilograms. The satellite was fitted with a nose cone and heatshield to protect the experiments during launching. Explorer 23 circled Earth once every 99 minutes at a distance of 977 kilometers while spinning at a rate of 150 times per minute.

Explorer 23 carried four experiments for measuring the number of micrometeoroids in a given volume of space. The most useful data were collected from the 210 stainless steel cells containing helium pressurized at 1,300 millimeters of mercury. The steel cells were 19.05 centimeters long. When struck by a micrometeoroid, a cell leaked gas, which then activated a pressure-sensitive microswitch to open, recording the puncture. The data gathered by Explorer 23 were similar to those obtained by Explorer 16 and were useful in the design of experiments for the Pegasus program.

Experiments aboard Explorer 30, also known as Solar Radiation Satellite 1 (Solrad 1), were part of a study conducted by the Naval Research Laboratory to examine solar radiation. The 24-kilogram, cylindrical satellite was launched into space by a four-stage Scout vehicle on November 19, 1965. Explorer 30 orbited at a range of 704 to 891 kilometers from Earth and was inclined 60° to the equator.

Previous observations had shown that solar x rays caused sudden disturbances in the ionosphere during eruptions, or flares, on the Sun. X rays are a penetrating form of electromagnetic radiation that is nonnuclear in origin. In May, 1966, data transmitted by Explorer 30 showed that a steep rise in solar x rays had taken place, indicating that the upcoming eleven-year sunspot maximum would be very intense. The 1958 sunspot maximum was marked by numerous flares on the Sun. The data from Explorer 30 were also particularly important to the Apollo Program, which required the design of space suits capable of withstanding bursts of x rays.

Explorer 37, also known as Solrad 2, was launched on March 5, 1968. Explorer 37 was another of the Naval Research Laboratory satellites carrying experiments for studying the Sun. The 90-kilogram satellite, shaped like a twelve-sided drum,

*The Advanced Composition Explorer (ACE) is placed on a Delta II rocket launch vehicle at Launch Complex 17A, Cape Canaveral. (NASA)*

was launched into space by a four-stage Scout rocket. Explorer 37 orbited at a distance of 513 kilometers from Earth at an inclination of 59.43°. Attached to the twelve-sided craft were several solar panels and a central band of x-ray detectors and Geiger tubes.

Explorer 37 carried experiments for collecting data on solar ultraviolet radiation and on solar x-ray emissions. Scientists were particularly interested in the solar x-ray data, which would later be used to make short-term solar flare activity forecasts during the U.S. Moon landings. Knowledge of impending solar flares protected astronauts and their communications systems from potential injury.

On July 4, 1968, the cylindrical, 190-kilogram Explorer 38 satellite was launched aboard a three-stage Delta rocket from the Western Test Range in Lompoc, California. Explorer 38, also known as Radio Astronomy Explorer A, would orbit at a distance of 5,862 kilometers from Earth at an inclination of 120.64°. Experiments aboard the satellite were designed to monitor radio waves in the Milky Way.

Once Explorer 38 was in orbit, commands from Earth activated a magnetic coil, which slowed the satellite from ninety-two revolutions per minute to a stop. Following the elimination of spin, six slender metal booms were extended from the craft, where they were stored on motor-driven wheels. These booms were constructed using thin tape designed to form a 1-centimeter-wide tube as it emerged from the satellite. The booms form V-shaped antennae. One was directed away from Earth, and another was directed toward Earth. These antennae were initially extended to a length of 137 meters from the craft, a length that was later increased to 229 meters. The remaining booms formed a 37-meter dipole antenna boom and a 192-meter stabilization boom. Cameras were mounted aboard the satellite to monitor extension of the booms and reveal the position of the antennae so that signals could be associated with their sources in space. The outer V-shaped antenna was designed to sweep the sky as Explorer 38 circled Earth relaying data that would enable radio astronomers to make the first frequency radio map of the Milky Way. The satellite's inner V-shaped antenna was designed to monitor radio waves from Earth, while the dipole antenna would monitor the often-intense radio bursts from the planet Jupiter.

Explorer 42, also known as the X-Ray Explorer, was launched December 12, 1970, by a team of Italian engineers from the University of Rome's Aerospace Research Center. The satellite was launched with the four-stage, solid-fueled Scout rocket at the research center's San Marco platform, located off the coast of Kenya in the Indian Ocean. Using the Italians' platform enabled scientists to place Explorer 42 in orbit around the equator with a smaller and less expensive rocket than would have been required had the satellite been launched from Cape Kennedy. The X-Ray Explorer was the first of three satellites approved under NASA's Small Astronomy Satellite program. Explorer 42 was cylindrical with four paddles; it was 1.6 meters long, 0.56 meter wide, and weighed 143 kilograms.

Aboard Explorer 42 were experiments designed to detect high-energy x-ray sources in space from the Milky Way and other galaxies. By operating above Earth's atmosphere, Explorer 42's instruments could detect x rays from several hundred sources, vastly increasing the knowledge of the universe. In May, 1971, it was reported that Explorer 42 had observed what was believed to be a black hole, the hypothetical remnant of a star that has decayed into an object so dense and small that light no longer escapes from it. Explorer 42's observations were an enormous success, expanding the list of identified x-ray sources from thirty-six to nearly two hundred.

Explorer 44 (Solrad 10), like Explorers 30 and 37, was a joint project prepared by the Naval Research Laboratory and NASA. Launched on July 8, 1971, Explorer 44 was a 118-kilogram, twelve-sided cylinder with four vanes, orbiting Earth at a distance of 632 kilometers. Experiments aboard the satellite were designed to supply data on solar x-ray and ultraviolet emissions. With further information on radiation, scientists hoped to improve their understanding of physical processes such as the potential impact of solar flares on plans for crewed space travel. Designers expanded the data collection capabilities of Explorer 44 by improving the sensitivity and accuracy of its sensors. Furthermore, unlike the sensors aboard Explorers 30 and 37, sensors aboard Explorer 44 were continuously directed toward the Sun.

Explorer 48 was launched from the San Marco platform off the coast of Kenya on November 16, 1972. The cylindrical, 186-kilogram Explorer 48 was the second satellite in NASA's Small Astronomy Satellite program. Unlike Explorer 42, Explorer 48 was designed to detect gamma rays, thus building upon the knowledge gained by Explorer 11.

Explorer 48 carried a gamma-ray telescope, designed to detect gamma rays and determine their intensity, energy, and direction of arrival. Explorer 48's telescope was ten times more powerful than any other detector ever orbited, enabling the satellite to conduct one of the most comprehensive studies of celestial gamma rays ever made. Explorer 48 made observations of gamma rays in the galactic center, galactic plane, and Crab nebula; all of its observations furthered understanding of the dynamics of the Milky Way. Gamma radiation data from the Explorer satellite were also used by scientists studying supernovae with a mass equivalent to the Sun's.

Explorer 49, also known as Radio Astronomy Explorer B (RAE B), was launched on June 10, 1973, aboard a McDonnell Douglas Delta rocket. The cylindrical Radio Astronomy Explorer weighed 66 kilograms. Explorer 49 was the last planned Moon satellite, making a five-day, 400,000-kilometer journey to orbit the Moon. Radio Astronomy Explorer B orbited the Moon at an altitude of 1,098 kilometers.

Experiments aboard Explorer 49 were designed to study low-frequency radio signals emitted from the Milky Way and beyond. Signals from the Sun, Earth, and Jupiter would receive particular attention. Explorer 49 could detect signals from the far side of the Moon, information that was beyond the reach of Earth-based monitors.

Explorer 53, the third in NASA's Small Astronomy Satellite series, was launched from the Italian-owned San Marco platform. Boosted by a Scout rocket, the satellite was cylindrical with four paddles and weighed 193 kilograms. Explorer 53 was placed in orbit around Earth at an altitude of 508 kilometers. Experiments carried aboard Explorer 53 were similar to those carried aboard Explorer 42, although technical improvements in pointing accuracy, spectral range, and time resolution were made in order to increase the chances of identifying x-ray sources in the universe.

The Advanced Composition Explorer (ACE) was conceived at a meeting on June 19, 1983, and launched on a McDonnell Douglas Delta II 7920 launch vehicle on August 25, 1997, from the Kennedy Space Center in Florida. The primary purpose of ACE is to determine and compare the isotopic and elemental composition of several distinct samples of matter, including the solar corona, the interplanetary medium, the local interstellar medium, and galactic matter. It continued its extended mission well beyond the year 2000. The Rossi X-Ray Timing Explorer (RXTE) was launched on December 30, 1995. It is designed to look at cosmic x-ray sources at short time scales over a broad energy range. Astronomers can learn about very fast phenomena such as the flickering of matter as it falls into a black hole or the rotation of disks of matter around a neutron star. They can also look at the longer-term variability of sources, such as the precession of orbiting neutron stars or accretion disks. RXTE was designed for a required lifetime of two years, with a goal of five years. On April 3, 2003, RXTE suffered a corruption of its telemetry interface, but, after rebooting, normal operation was restored. Like so many intrepid robotic craft, RXTE continued well beyond its design lifetime. As of 2005, RXTE-based research continued to routinely provide new and exciting published results concerning such exotic astrophysical objects as magnetars (a rare class of neutron stars with extremely intense magnetic fields), accretion disks illuminated by superbursts, spinning black holes, and the limitation of rotation rates on pulsars. RXTE observations continued to be coordinated with other x-ray telescopes in operation to provide even greater scientific return. Also, objects of interest for their x-ray emissions could also be studied with telescopes viewing those objects in other portions of the electromagnetic spectrum to gain a fuller picture of real-time events.

## Contributions

Experiments aboard the Astronomy Explorers supplied scientists with a wealth of astronomical data. Explorers 7, 30, 37, and 44 collected data on solar radiation, x rays, and solar flares. This information was particularly useful to Apollo scientists

*The Cosmic Background Explorer (COBE) acquired this edge-on infrared image of the Milky Way galaxy. COBE helped prove the Big Bang theory by revealing the existence of the invisible cosmic microwave background radiation.* (NASA-GSFC)

who were designing space suits and spacecraft capable of withstanding bursts of x rays and solar flares. Explorers 11 and 48 measured gamma rays in space, collecting data that improved scientists' understanding of the history of the universe. Explorer 23 collected data on the number of meteoroids in close proximity to Earth. These data helped scientists determine the hazards to which space travelers might be exposed.

Radio Astronomy Explorers A and B (Explorers 38 and 49) carried experiments that aided radio astronomers working on the first frequency radio map of the Milky Way and other galaxies. X-Ray Explorers 42 and 53, both part of the Small Astronomy Satellite series, were designed to detect high-energy x-ray sources in space from the Milky Way and other galaxies.

The Cosmic Background Explorer (COBE) was built and managed by NASA's Goddard Space Flight Center in Greenbelt, Maryland. COBE's primary mission, which began with launch aboard a Delta rocket on November 9, 1989, called for a one-

year observation of the universe at its birth. During its four-year life, COBE addressed such questions as how the universe began, how it evolved to its present state, and what forces govern this evolution. COBE confirmed the Big Bang theory with its precise measurements of the background radiation in the universe.

The Extreme Ultraviolet Explorer (EUVE) was NASA's sixty-seventh Explorer mission. It was launched on June 7, 1992, aboard a Delta II vehicle. The satellite consists of four telescopes, three of which mapped the entire sky in an attempt to determine the existence, direction, brightness, and temperature of sources of extreme ultraviolet radiation. The fourth telescope was used to make spectroscopic observations of the EUV sources to determine composition and temperature.

### Context

The astronomy satellites in the Explorer series made a significant contribution to scientists' understanding of the universe. While many of the data collected by the Astronomy Explorers complemented those collected by other satellites, several of the projects led to discoveries that were significant to space exploration. Explorer 11's detection of gamma rays in space enabled scientists to take a major step toward understanding the history of the universe. Steady state theorists believe that radiant energy in the form of gamma rays may be the key to the creation of matter. Further data on gamma rays were collected by Explorer 48, which carried a gamma-ray telescope capable of determining the rays' intensity, energy, and direction of arrival.

NASA's Small Astronomy Satellite program (Astronomy Explorers 42, 48, and 53) was the first to involve another country, as Italian engineers from the University of Rome's Aerospace Research Center launched the satellites from the San Marco platform off the coast of Kenya.

**See also:** Apollo 16; Explorers: Air Density; Explorers: Radio Astronomy; International Ultraviolet Explorer.

## Further Reading

Davies, John K. *Astronomy from Space: The Design and Operation of Orbiting Observatories.* New York: John Wiley, 1997. This is a comprehensive reference on the satellites that have revolutionized twentieth century astrophysics. It contains in-depth coverage of all space astronomy missions. It includes tables of launch data and orbits for quick reference as well as photographs of many of the lesser-known satellites. The main body of the book is subdivided according to type of astronomy carried out by each satellite (x-ray, gamma-ray, ultraviolet, infrared, and radio). It discusses the future of satellite astronomy as well.

Fabian, A. C., K. A. Pounds, and R. D. Blandford. *Frontiers of X-Ray Astronomy.* London: Cambridge University Press, 2004. Discusses the revolution in research provided by the Chandra and Newton X-Ray Observatories, and those spacecraft that preceded them. Puts the data in cosmological context.

Fink, Donald E. "Explorer 23 Furnishing Data for Pegasus." *Aviation Week and Space Technology* 81 (November 30, 1964): 53-54. This magazine article describes meteoroid detection data gathered by Explorer 23. Provides detailed information on the detectors aboard the Explorer craft and a discussion of the Pegasus spacecraft. College-level material.

Goddard Space Flight Center. *Advanced Composition Explorer (ACE) Lessons Learned and Final Report.* Washington, D.C.: Government Printing Office, 1998. This is the official report of the Advanced Composition Explorer mission. Although it spends much of its space lauding the project's accomplishments, the report provides a very good insight into the satellite and its exploration.

Gregory, William H. "Explorer 53 Seeks Improved X-Ray Data." *Aviation Week and Space Technology* 102 (May 19, 1975): 41-43. This article offers a detailed review of the Explorer 53 project, focusing on the development of improved x-ray detection equipment. Also mentions previous X-Ray Explorers 42 and 48. College-level material.

King-Hele, Desmond G., et al. *The RAE Table of Earth Satellites, 1957-1980.* New York: Facts on File, 1981. Lists the 2,389 launches of satellites and space vehicles from 1957 to 1980. Details include name, national designation, date of launch, lifetime, mass, shape, dimensions, and orbital parameters. An excellent source of basic data on the Astronomy Explorers. Suitable for general reference.

National Aeronautics and Space Administration, Office of Space Science. *Mission Operation Report: Small Astronomy Satellite (SAS-B).* Washington, D.C.: Author, 1972. A detailed description of the Explorer 48 mission and a review and discussion of gamma-ray data collected by Explorer 42. This report gives information on gamma-ray astronomy, launch schedules, spacecraft control, and the gamma-ray experiment aboard Explorer 48. The introductory and review sections are suitable for general reference.

_____. *Mission Operation Report: Solar Radiation Explorer Solrad-C.* Washington, D.C.: Author, 1971. A detailed description of the Explorer 44 mission. Report includes information on mission objectives, spacecraft, launch vehicle, mission support, program management, and project costs. Introductory portions are suitable for general reference.

Schlegel, Eric M. *The Restless Universe: Understanding X-Ray Astronomy in the Age of Chandra and Newton.* London: Oxford University Press, 2002. An accessible introduction to the history, methods of observation, and data of x-ray astronomy. Provides data from the Chandra, Newton, and Astro-E missions as well as from spacecraft that preceded them.

Sullivan, Walter. "An X-Ray Scanning Satellite May Have Discovered a Black Hole." *The New York Times* 120 (April 1, 1971): 20. This article contains a detailed description and evaluation of data collected by Explorer 42 indicating the existence of one or more black holes. Includes commentary by scientists and a description of the spacecraft. Suitable for general audiences.

Watts, Raymond N., Jr. "Explorer 37 Studies the Sun." *Sky and Telescope* 35 (May, 1968): 298-299. Describes data collected by previous solar radiation satellites and gives a detailed description of the Explorer 37 spacecraft.

_____. "The Outlook for Nuclear Spacecraft." *Sky and Telescope* 36 (August, 1968): 88-90. A segment of Watts's article focuses on the Explorer 38 spacecraft, giving a detailed physical description and noting the scientific goals of this Radio Astronomy Explorer.

*Peter Neushul*

# Explorers: Atmosphere

*Date:* April 2, 1963, to June 10, 1981
*Type of spacecraft:* Earth observation satellites

*The Atmosphere Explorer satellites gathered invaluable data on Earth's thermosphere and ionosphere, including information that led to a better understanding of the photochemical processes that occur in the upper atmosphere as a result of the absorption of solar ultraviolet radiation.*

## Key Figures

*David W. Grimes,* project manager, Atmosphere Explorers C, D, and E

*Nelson W. Spencer,* project scientist and principal investigator, neutral temperature experiment

*Erwin R. Schmerling,* program scientist

*Kenneth S. W. Champion,* principal investigator, atmospheric density experiment

*Charles A. Barth,* principal investigator, nitric oxide experiment

*Hans E. Hinteregger,* principal investigator, solar extreme ultraviolet spectrometer experiment

## Summary of the Satellites

Earth's atmosphere is divided in altitude into four major regions, each with its own unique properties: the troposphere, the stratosphere, the mesosphere, and the thermosphere. The thermosphere is a region extending from 90 kilometers out to several hundreds of kilometers and characterized by temperatures that increase with altitude to between 1,000 and 2,000 kelvins, depending on the phase of the solar cycle. The behavior and basic temperature characteristics are dependent on energy from the Sun in the form of extreme ultraviolet photons that are absorbed in this part of the atmosphere.

One of the early pivotal milestones in atmospheric research was the conclusion drawn by Evangelista Torricelli in 1643 that the atmosphere must exert a pressure; he developed the mercury barometer to measure this pressure. It was recognized in early times that the atmosphere at higher latitudes produced spectral displays of colored lights, the aurora. With the advent of spectroscopy,

the details of these and subvisual emissions from high altitudes were observed, providing information on the composition of the upper atmosphere. Following Guglielmo Marconi's transmission of radio waves from England to Newfoundland in 1901, it was realized that there must be a conducting layer surrounding Earth, and it was postulated that this layer must result from the ionization of molecular oxygen and nitrogen by sunlight. Furthermore, it was concluded that solar photons with sufficient energy to ionize molecules would also dissociate molecules into atoms. Thus it was realized that the thermosphere would probably also be composed of chemically active species such as atomic oxygen and nitrogen, and that embedded in the thermosphere was an ionosphere. The latter could be successfully studied from the ground using radio sounding techniques.

With the development of rockets following World War II, primitive instruments were used to probe this region. In the late 1960's, the altitude variation of the concentration of the major ions in the ionosphere was measured. With the launch of the first satellites, an important advance was made: By monitoring the behavior of satellites in orbit compared with their predicted behavior, scientists learned that the atmosphere was exerting a drag on the satellites that could be used to infer the atmospheric concentration. Rocket experiments were improved and extended to satellites such as the Orbiting Geophysical Observatory (OGO) series.

At this point a substantial amount of information concerning the thermosphere had been pieced together from a variety of different measurements often made at different times and under different circumstances. It was known, for example, that the absorption of ultraviolet photons not only causes the major temperature structure but also produces ions and atoms, some of which are in excited energy states. Above about 110 kilometers, the species are no longer well mixed, but their distribution with altitude is one in which the lighter species become progressively more dominant with increasing altitude. In ionizing the neutral species, the solar photons strip electrons from the atoms and molecules, and those photoelectrons have sufficient energy to cause further ionization, dissociation, and excitation. Energetic charged particles (electrons and protons) from the magnetosphere can do the same. The product species then interact chemically to form yet other species. In the course of these photochemical reactions, energy that started as extreme ultraviolet photons is converted into a variety of forms; some of the energy goes into heating the atmosphere, and some is lost in the form of photons radiated by the excited species.

In attempting to understand these phenomena, scientists have had to make many assumptions, including not only many fundamental atmospheric parameters and processes but also the probabilities and efficiencies with which the various chemical reactions occur. Some of these parameters, such as ionization rates and reaction rate coefficients, were computed theoretically, and others were measured in laboratory simulation experiments. The study of the physics and chemistry of the upper atmosphere became known as aeronomy.

The Atmosphere Explorer series of satellites was intended to tie all these pieces together. Explorer 17, which later became known as Atmosphere Explorer A, carried eight experiments to measure the concentration, composition, pressure, and temperature of the thermosphere. The 184-kilogram spherical spacecraft was launched on a Thor-Delta launch vehicle into a 250-by-900-kilometer orbit

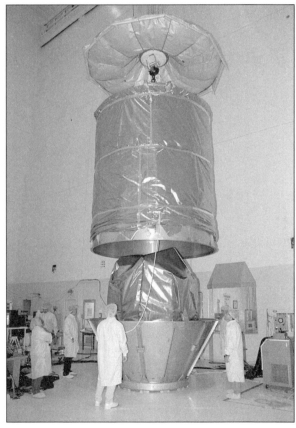

*The Advanced Composition Explorer (ACE) is encapsulated and prepared for launch at the Kennedy Space Center's Spacecraft Assembly and Encapsulation Facility-II. ACE investigated low-energy particles of solar origin and high-energy galactic particles. The collecting power of instruments aboard ACE was 10 to 10,000 times greater than anything previously flown by NASA to collect similar data. (NASA)*

inclined 58° to Earth's equator. The satellite was very simple and provided limited information, but it demonstrated that uncontaminated measurements could be made, and it found neutral helium to be a major species at high altitude.

Explorer 32, which later was called Atmosphere Explorer B, had solar cells to extend its lifetime and included a tape recorder to allow storing of data in between ground station contacts. In addition, the spin axis of the satellite was maintained normal to the plane of the orbit. The payload was made up of eight investigations, and the orbit was chosen to allow separation of local time and altitude effects. The mission provided valuable data on the morphology of the thermosphere.

Atmosphere Explorers C, D, and E (Explorers 51, 54, and 55, respectively) represented sophisticated satellites for the time, including a hydrazine propulsion system to restore apogee altitude in order to compensate for atmospheric drag. A new feature was that perigee (closest orbital approach to Earth) could be lowered much deeper into the atmosphere than had been attempted before and raised again after a chosen period of time. The three satellites were launched into 57°, 90°, and 20° inclination orbits, respectively. With one or two exceptions, the payload was the same on all three missions and included investigations capable of measuring all the major thermospheric parameters identified at the time. The instrument complement included instruments to measure the intensity and spectral distribution of ultraviolet sunlight; the photoelectrons and low-energy charged particles; the neutral and ion composition, temperatures, and concentration; and the visible airglow and emissions from nitric oxide. All these measurements were made simultaneously and, after processing, were entered at fifteen-second intervals in a centralized computer data file. Each investigation team had access to the centralized computer by remote terminals. The team of investigators met regularly to discuss new joint observations and the scientific findings and to plan future measurement sequences. The investigation team included theorists to assist in the interpretation of the multi-

parameter database. This approach, involving multiple instruments, a team of scientists working closely together, and a unified database accessible to all relatively soon after the measurements were made, was new to space science. All three spacecraft were launched on Thor-Delta vehicles. Atmosphere Explorers C and D were launched from the Western Test Range, and Atmosphere Explorer E was launched from Cape Kennedy. Each spacecraft weighed approximately 665 kilograms, of which the scientific instruments made up approximately 95 kilograms.

Atmosphere Explorer C was initially put into an elliptical orbit with perigee at 150 kilometers and apogee at 4,000 kilometers; its perigee was lowered to 129 kilometers on occasion for approximately twenty-four hours. The highly elliptical orbit yielded valuable data on the altitude dependence of the various parameters, allowing considerable progress in modeling and interpretation. One year after launch, the orbit was adjusted until it became circular. The circular phase was maintained for the rest of its useful lifetime, with different altitudes being selected from time to time.

Atmosphere Explorer D, with a 90° inclination orbit, was intended to provide fully global coverage. Unfortunately, after five months, a failure in the satellite's power supply caused the spacecraft to be lost.

Atmosphere Explorer E was initially placed in a 150-to-3,000-kilometer orbit. This spacecraft provided detailed information with extensive local time coverage in the equatorial thermosphere and ionosphere.

**Contributions**

The multiparameter database obtained by Atmosphere Explorers C, D, and E (particularly C) permitted sophisticated photochemical models to be developed. These models were tested for varying conditions in altitude, geographic latitude, local time, season, and magnetic or solar activity. Once tested, the measured parameters were then used as input to models that were in turn used to solve for species or parameters not measured by the instruments in the payload. An example of this was the determination of atomic hydrogen, which is difficult to measure experimentally. The concentration of this species as a function of altitude could be solved for from the relatively simple high-altitude photochemistry, as all the other parameters (oxygen, ionized hydrogen, ionized oxygen) were well measured and the reaction rate coefficients well known.

In other cases, reaction rate coefficients were unknown or still in question. Again, the measured concentrations and temperatures could be used in a chemical balance equation to solve for the rate coefficients. Because the volume of data was so large and spanned a wide range of so many of the important variables, very large statistical samples

could be used to obtain solutions with a high degree of validity. An example of this was the reaction rate coefficient for the recombination of the nitric oxide ion with electrons. This is the principal channel by which the ionosphere is converted back to neutrals. Prior to Atmosphere Explorer C, two attempts to measure this reaction rate coefficient in the laboratory produced results that differed substantially from each other. Using the approach of solving for the reaction rate coefficient as a function of electron temperature, from the photochemical balance equation it was possible to determine very accurately which of the two laboratory measurements was applicable to the thermosphere.

Atmosphere Explorer C made the first measurements of atomic nitrogen in the thermosphere

*The Explorer 17 satellite, a stainless steel sphere that measured the atmosphere's composition, pressure, and temperature in 1963.* (NASA-GSFC)

and allowed some information to be gathered on molecular oxygen concentrations. The latter is particularly difficult to measure in the thermosphere, and while much remains to be done in this area, some measurements were attempted and compared with concentrations inferred from the photochemistry.

The measurements of the extreme ultraviolet solar flux provided the first opportunity to study both short- and long-term responses in the photochemistry. The solar flux measurements permitted improved calculations of the ionization rates of all the major atomic and molecular species. The rate of dissociation of molecular oxygen could be computed over a period of several years, showing the variability over the increasing phase of solar cycle 21 and over the twenty-seven-day solar rotation period. While extreme ultraviolet solar flux measurements remain controversial, Atmosphere Explorers C and E provided an extensive database covering 1974 through 1980.

Atmosphere Explorers C and D led to the first detailed understanding of the high-latitude ion convection patterns. In addition, the distribution of auroral energetic particles was studied in considerable detail.

The mission series allowed the first detailed comparison of measured and calculated photoelectron flux measurements. These calculations can now be done as accurately, as certain cross sections used in the calculations are known. In fact, the calculations can now be used to pin down the cross sections within the uncertainty of the photoelectron measurements.

A storage medium for energy in the thermosphere is in certain excited electronic energy states of atoms and molecules. These are states that, in order to release the energy in the form of a photon, must violate the laws of quantum physics. As a result, this energy is released only after a relatively long period of time. These are known as long-lived, forbidden, or metastable states. The Atmosphere Explorers monitored several of these states and produced detailed studies of their photochemistry and relative importance to the thermosphere.

**Context**

The Atmosphere Explorer mission series, particularly Atmosphere Explorer C, represented the first closely interlinked multi-instrument payloads designed around an investigation team and a centralized database accessible to all investigators and their associates. As such, it represented a transition from single-parameter studies to comprehensive studies of all the principal interacting elements of the thermosphere and ionosphere. The C, D, and E spacecraft were the first to conduct controlled experiments penetrating the atmosphere from distances spanning 130 to 4,000 kilometers in altitude. While many aspects of thermospheric physics and chemistry had been measured or inferred before, the Atmosphere Explorers provided an extensive database of simultaneously acquired parameters from which major advances in the understanding of thermospheric photochemistry were made. Many details that had previously been thought to be understood were found to be different from what had been assumed. Thus, the mission was successful not only in confirming what was known but also in bringing to light what was not known. The photochemical studies also identified those parameters not measured by the Atmosphere Explorer instruments, which would have to be measured in order to advance beyond this point. For example, the doublet D metastable state of the atomic oxygen ion was not measured, yet it appears to play a pivotal role in the photochemistry. In addition, the Atmosphere Explorers lacked the capability to measure the vibration distributions of molecules such as nitrogen, which may be key energy reserves.

The Atmosphere Explorer studies resulted in a detailed assessment of the energetics of the thermosphere, quantifying the flow of solar ultraviolet energy through all the possible energy paths, and finally leading to a computation of the heating and cooling rates. Nothing approaching this level of quantification had been possible previously.

While the series had been specifically designed to study thermospheric aeronomy, important advances were made in other areas also. The mission produced the first detailed study of high-latitude ion convection patterns, a magnetospheric phenomenon.

A substantial database of neutral atmosphere composition was acquired, which allowed empirical models to be developed to the point where the concentrations of the major thermospheric species can be predicted for any specified conditions, with impressive accuracy. The mission was most productive in terms of scientific studies, resulting in well over three hundred papers in the reference literature and hundreds of papers presented at scientific meetings.

In the case of the Atmosphere Explorer series, particularly C, D and E, which focused on the period near the solar cycle minimum (the period of minimum solar activity), the field of thermospheric aeronomy was brought to a state of maturity, and the general understanding of this region of the near-Earth environment was substantially advanced.

**See also:** Dynamics Explorers; Earth Observing System Satellites; Explorers: Air Density; Explorers: Astronomy; Explorers: Ionosphere; Geodetic Satellites; Global Atmospheric Research Program; Heat Capacity Mapping Mission; Landsat 1, 2, and 3; Landsat 4 and 5; Landsat 7; Mission to Planet Earth; Seasat.

## Further Reading

Banks, Peter M., and G. Kockarts. *Aeronomy.* New York: Academic Press, 1973. Two-volume text on the state of knowledge of the chemistry and physics of the upper atmosphere prior to Atmosphere Explorers C, D, and E. Suitable for college-level audiences.

Burgess, Eric, and Douglas G. Torr. *Into the Thermosphere: The Atmosphere Explorers.* NASA SP-490. Washington, D.C.: Government Printing Office, 1987. The official history of the Atmosphere Explorer series. Suitable for high school and college levels, it describes the broad history of atmospheric research, the early rocket and satellite work in this field, and the details of the Atmosphere Explorers and their results. Contains numerous illustrations and photographs of the spacecraft, the instruments, and the scientific objectives, as well as a detailed list of the publications resulting from the mission series.

Dalgarno, A., W. B. Hanson, N. W. Spencer, and E. R. Schmerling. "The Atmosphere Explorer Mission." *Radio Science* 8 (1973): 263. In this special issue of *Radio Science*, dedicated to the Atmosphere Explorer C, D, and E missions, this introductory paper outlines the goals and objectives of the series. (The remainder of the issue consists of papers by each of the investigator teams, describing their investigations.)

Eather, Robert H. *Majestic Lights: The Aurora in Science, History, and Arts.* Washington, D.C.: American Geophysical Union, 1980. This readable history of studies of the aurora is richly illustrated with color and black-and-white photographs. Suitable for high school and college levels.

Gavaghan, Helen. *Something New Under the Sun: Satellites and the Beginning of the Space Age.* New York: Copernicus Books, 1998. This book focuses on the history and development of artificial satellites. It centers on three major areas of development—navigational satellites, communications satellites, and weather observation and forecasting satellites.

Kivelson, Margaret G., and Christopher T. Russell. *Introduction to Space Physics.* New York: Cambridge University Press, 1995. A thorough exploration of space physics. Some aspects are suitable for the general reader. Suitable for an introductory college course on space physics.

Rishbeth, Henry, and Owen K. Garriott. *Introduction to Ionospheric Physics.* New York: Academic Press, 1969. A text on ionospheric physics prior to the Atmosphere Explorer C, D, and E missions. Contains numerous figures and references. Suitable for the college level.

Torr, Douglas G. "The Photochemistry of the Upper Atmosphere." In *The Photochemistry of Atmospheres: Earth, the Other Planets, and Comets,* edited by Joel S. Levine. San Diego, Calif.: Academic Press, 1985. A summary of the state of knowledge of the thermospheric photochemistry. Contains numerous figures, tables, and references. Suitable for the college level.

Zimmerman, Robert. *The Chronological Encyclopedia of Discoveries in Space.* Westport, Conn.: Oryx Press, 2000. Provides a complete chronological history of all crewed and robotic spacecraft and explains flight events and scientific results. Suitable for all levels of research.

*Marsha R. Torr*

# Explorers: Ionosphere

*Date:* Beginning November 3, 1960
*Type of spacecraft:* Earth observation satellites

*Several Explorer satellites carried out the first comprehensive direct measurements of Earth's ionosphere. By collecting data on the temporal and spatial distribution of ionospheric properties, these satellites discovered that the ionosphere is much more intricately structured than previous rocket probes had revealed.*

## Key Figures

*Harry J. Goett* (1910-2000), Goddard Space Flight Center (GSFC) director, 1959-1965
*John W. Townsend, Jr.,* assistant director for Space Science and Satellite Applications at GSFC
*Robert E. Bourdeau,* program manager and project scientist for Explorer 8
*Larry H. Brace,* project scientist for Explorer 22
*R. L. F. Boyd,* principal investigator for Explorer 20 plasma experiment
*John Donley,* principal investigator for Explorer 8 electric field experiment
*James A. Van Allen* (b. 1914), discoverer of the radiation belts that bear his name

## Summary of the Satellites

The Explorer satellites were the most effective among the many tools used by scientists to explore Earth's upper atmosphere. In the nineteenth century, physicists had postulated an electrically conducting region in the upper atmosphere to explain small fluctuations in Earth's magnetic field, but it was not until 1925 that the existence of an ionosphere capable of reflecting radio waves was firmly established. Scientists theorized that solar radiation created these layers of ionized gases extending from 50 to more than 400 kilometers above Earth's surface. For many years the ionosphere could be studied only indirectly, principally through its effects on radio propagation. With the advent of rockets in the late 1940's and satellites in the late 1950's, scientists had a direct means of exploring the ionosphere.

One of the purposes of the International Geophysical Year (July 1, 1957, to December 31, 1958) was to investigate Earth's atmosphere by means of orbiting satellites. Both the United States and the Soviet Union announced that they would launch a series of satellites during this period, and when the Soviet Union successfully orbited Sputnik 1 in the fall of 1957, Americans were shocked at what they saw as a Soviet scientific, technological, and political triumph. Explorer 1, the first American satellite, was launched into orbit at the end of January, 1958. Both the satellite and the network used to track it were developed by the Jet Propulsion Laboratory (JPL) in California, and the Army Ballistic Missile Agency (ABMA) then collaborated on a series of Explorer satellites. The data from Explorers 1 and 3 were particularly significant, for they allowed James A. Van Allen to announce the discovery of highly intense radiation belts around Earth at an altitude of 800 kilometers. He theorized that these phenomena were caused by charged particles trapped in Earth's magnetic field. The ionosphere, which lay far below the Van Allen belts, was not of special concern to these early Explorers.

When the National Aeronautics and Space Administration (NASA) was created in 1958, it acquired the Vanguard project from the Naval Research Laboratory, the Explorer project from ABMA, and the services of JPL. Goddard Space Flight Center, which came into being with the establishment of NASA, became deeply involved with the Explorer spacecraft, because its purpose within NASA was to contribute to the scientific exploration of space. Explorer 6, launched on August 7, 1959, was the first satellite launched by both Goddard and NASA. Before this, the United States had orbited eight satellites and launched a space probe. After the success of Explorer 6, Goddard began a series of satellite launchings that would provide valuable information on Earth's atmosphere—from the lower layers of neutral gases, through the ionosphere of charged gases, to the more distant radiation belts. The first satellite specifically designed for ionospheric studies was Explorer 8, sometimes called the Ionospheric Direct Measurement Satellite, a joint project of Goddard, JPL, and the Marshall Space Flight Center.

On March 23, 1960, the first attempt to orbit Explorer 8 failed when the third stage did not ignite. Seven months later, on November 3, 1960, Explorer 8 was successfully launched at 17:24 Coordinated Universal Time (UTC) from Cape Canaveral, Florida. The launching vehicle was a Juno 2, whose first stage consisted of a Jupiter missile with enlarged fuel tanks and whose three upper stages were powered by Sargeant solid-propellant rockets. The 41-kilogram satellite consisted of two truncated cones joined by a short cylinder. Because scientists wanted Explorer 8 to investigate a wide range of altitudes, the booster's burning time was regulated so that the satellite would achieve a very eccentric orbit. Explorer 8 traveled in a cigar-shaped ellipse with a perigee (the closest approach to Earth) of 420 kilometers, an apogee (the farthest distance from Earth) of 2,290 kilometers, and an inclination of 50°. Its orbital period was close to 113 minutes. Shortly after the payload separated from the rocket, weighted wires slowed the satellite's spin rate to thirty revolutions per minute. Explorer 8 was equipped with eight experiments to study the ionosphere, and its instruments were

powered by eight packs of mercury batteries. Unfortunately, the batteries weakened several weeks earlier than expected, and the last radio signal was received on December 28, 1960, at Quito, Ecuador. Nevertheless, Explorer 8 had transmitted for 15,302 hours, and more than 1,100 kilometers of magnetic tape were needed to record the received data.

The Beacon Explorer A, also known as Explorer 20, was a satellite designed to collect data on upper ionospheric regions inaccessible to "bottomside sounding." Low-frequency radio waves are usually reflected by the ionosphere in ways that depend on the density of charged particles in the lower layers. Because bottomside sounding from Earth cannot get beyond the level of maximum electron density, topside sounding is necessary. Because difficulties with the Scout booster caused launch delays and postponements in 1963, NASA officials decided in 1964 to orbit the topside sounder with a Delta rocket, its most successful booster. At first, the Delta launching from Cape Kennedy on March 19, 1964, appeared to be successful, for two stages of the rocket fired according to schedule; the third stage, however, burned for only half of its programmed time, and the payload did not attain orbital speed. The satellite was destroyed over the South Atlantic.

A topside sounder, also called Explorer 20 and nicknamed Topsi, was launched from Point Arguello, California, on August 25, 1964, by a four-stage solid-fueled Scout rocket. The 44-kilogram conical package was correctly inserted into a near-polar orbit with an inclination of 80°. The orbit had a perigee of 869 kilometers, an apogee of 1,020 kilometers, and a period of 104 minutes. Explorer 20 began transmitting to Earth data on irregularities in the upper ionosphere gathered by its radio frequency sounder. Other instruments began to radio information on electron densities and temperatures in the vicinity of the satellite.

Six weeks later, another satellite was launched by NASA to investigate the ionosphere. This satellite, called Explorer 22, was designed and built by the Applied Physics Laboratory of The Johns

Hopkins University; because of its multiple purposes, it was considerably different in appearance from the earlier ionospheric Explorers. Its top, which resembled the eye of a huge fly, was an eight-sided pyramid covered with 360 fused-silica mirrors to be used in a laser test. A Scout rocket launched the 53-kilogram satellite on October 9, 1964, at 8:00 P.M. Pacific daylight time, which was 03:00 UTC on October 10, 1964. The orbit had an 80° inclination, a perigee of 885 kilometers, an apogee of 1,075 kilometers, and a 105-minute period. Two bar magnets oriented the satellite along the lines of force of Earth's magnetic field to provide stable radio signals for the ionospheric experiments and to keep the reflectors pointing in the direction required for the laser test. Electron density probes were deployed top and bottom, along with two whip antennae and a pair of dipoles that jutted from the ends of four paddles that carried solar cells. These solar panels had twice as many cells as necessary for initial power needs, so that, as cells deteriorated, more could be switched into the circuit to maintain proper total power. When the satellite passed over Wallops Island, Virginia, during the first evening after launching, NASA scientists reflected laser light from the satellite.

Explorer 27, which had essentially the same systems, instruments, and experiments as Explorer 22, was placed into orbit on April 29, 1965, by means of a four-stage Scout rocket. NASA launched this satellite from Wallops Island, Virginia, which had been placed under Goddard's administration. The windmill-shaped satellite achieved an orbit with an apogee of 1,315 kilometers, a perigee of 940 kilometers, and an inclination of 41°. Its orbital period was 108 minutes. As with Explorer 22, NASA scientists used Explorer 27's laser reflectors to determine the precise size and shape of Earth and to locate positions accurately on Earth's surface. Explorer 27's ionospheric data rounded out the information provided by the previous Explorers on the densities, species, and energies of the ionospheric particles as functions of time and of position.

## Contributions

Explorer 8's tasks were numerous and important, and it acquitted itself admirably. Its most important goal was to determine the structure and properties of the ionosphere and to map their daily and seasonal changes. Because these changes can profoundly modify or even interrupt global communications, the knowledge gained by Explorer 8 was important. Its principal measurements were of electron density, positive ion density, and ion masses. One of its interesting findings was that, as its altitude increased, the potential difference between Explorer 8 and its environment switched from negative to positive. Its Langmuir probe measured electron temperatures in the ionosphere (electron temperature is a measure of electron speeds at high altitudes) and found that they show a diurnal variation, with daytime temperatures about 800° Celsius (1,073 kelvins) greater than nighttime temperatures. Explorer 8's ion probe confirmed the predominance of atomic oxygen in the ionosphere, and it also found that the helium ion is an important constituent of the ionosphere above 800 kilometers. Even at an altitude of 2,200 kilometers, it found, helium ions predominate over hydrogen ions.

Explorer 20 probed the electron density of the ionosphere from very high altitudes. Its soundings at six fixed frequencies provided precise electron-density profiles of the ionosphere, but its most significant discovery was of magnetic field-aligned structures and directed propagation effects in the ionosphere. More specifically, Explorer 20 found evidence for the magnetic guidance of ionization along, but not across, field lines. These field-aligned ducts were later shown to be effective in guiding radio energy in the megahertz range.

The chief objective of Explorer 22 was to detect variations and anomalies in the structure of the ionosphere. To accomplish this, NASA scientists modified Explorer 22's instruments so that observers from all over the world, using very simple ground equipment, could make daily measurements of the ionosphere when the satellite was within range. In this way, a comprehensive global picture of the ionosphere was obtained. Although these ionospheric measurements were very important, it was a secondary purpose of Explorer 22 that caught the public's attention. Air Force engineers succeeded in photographing a beam of laser light that had been reflected off the satellite.

The mission of Explorer 27 was similar to that of Explorer 22: to perform a global survey of the ionosphere's electron density. On this occasion, eighty-six ground stations in thirty-six countries were involved, constituting the most extensive international cooperative space program ever undertaken. Explorer 27 also evaluated the use of laser techniques in deriving orbital and geodetic information and, in addition to providing an electron map of the ionosphere, provided data on irregularities in Earth's gravitational field.

### Context

Before Explorers 8, 20, 22, and 27, information about the ionosphere was limited to regions above the more populous parts of the world and to the lower altitudes of the upper atmosphere. In the late 1940's and into the 1950's, scientists used sounding rockets to explore the ionosphere. Rockets such as the Aerobee ascended to great altitudes—though not higher than one Earth radius—and then descended not far from the launch site. These rockets, however, could sample the ionosphere only at selected locations and for very short periods. Scientific satellites were needed for long-duration observations of Earth's ionosphere.

In the early years of the U.S. exploration of space, scientists devised satellites for geophysical research, such as the Orbiting Geophysical Observatory; satellites for micrometeoroid studies; weather satellites, such as the TIROS series; and many others. The simplest of these satellites were the Explorers, whose weights were low and whose instruments were designed for a restricted number of related experiments. Simplicity brought success: During the 1960's, every Explorer that was properly orbited achieved its mission. The average costs

for the Explorers were less than those for any other satellites.

Though they had various shapes, the Explorers were similar in that they all stemmed from a common, straightforward technology. To be sure, they became more complex and specialized as better components became available, but the family relationship continued. Scientists liked the Explorers because they permitted experiments to be carried out expeditiously, and when a discovery was made, an experimenter could follow up quickly with new experiments on another Explorer. This flexibility explains why the scientific community insisted that NASA continue to fund these small research satellites, even in the face of severe competition from the crewed spaceflight program.

The advantages of research satellites were demonstrated by the ionosphere Explorers. These satellites revealed the ionosphere to be a complex, dynamic medium and helped to make aeronomy, the study of the upper atmosphere, a unified field of research. The ionosphere Explorers also illustrated the growing international nature of space science in the 1960's. Initially, only technologically advanced nations such as Great Britain had participated in the Explorer programs, contributing instruments for solar, geophysical, and astronomical satellites, but with Explorers 22 and 27, scientists in more than thirty countries participated in a worldwide study of the ionosphere.

The ionosphere Explorers were part of a successful scientific program. By 1970, Goddard Space Flight Center had launched more than one thousand sounding rockets and more than forty Explorer satellites. The evolution of the Explorer satellites furnishes a good example of the scientific process in operation. Throughout the history of science, new instruments have helped to advance knowledge. Like the telescope and spectroscope, the space satellite gave scientists a powerful new tool to probe nature's secrets. The impact of this new tool was particularly profound on physicists, astronomers, and Earth scientists.

The data from the ionospheric satellites suggested new programs—so many, in fact, that accomplishing all of them simultaneously would have required more resources than NASA could ever have mustered. With the increasing demands of the crewed spaceflight program, NASA was forced to set priorities in space research. The Explorer program had been built around a few key scientists; as NASA's successes engendered a more complex space program, a new consensus had to be found among scientists, politicians, and the American public.

**See also:** Dynamics Explorers; Earth Observing System Satellites; Explorers: Air Density; Explorers: Atmosphere; Geodetic Satellites; Global Atmospheric Research Program; Heat Capacity Mapping Mission; Landsat 1, 2, and 3; Landsat 4 and 5; Landsat 7; Mission to Planet Earth; Seasat.

## Further Reading

Corliss, William R. *Scientific Satellites.* NASA SP-133. Washington, D.C.: Government Printing Office, 1967. More than six hundred satellites were orbited in the first decade of space exploration, and Corliss tries to make sense of them in terms of their equipment and experiments. His study is based on NASA records. Though the book is intended for the layperson, its use of advanced mathematics and technical terms may restrict sections of it to readers with a good scientific background.

Divine, Robert A. *The Sputnik Challenge: Eisenhower's Response to the Soviet Satellite.* New York: Oxford University Press, 1993. A thorough history of the political and scientific decisions made in regard to the International Geophysical Year and the efforts to launch an American Earth-orbiting satellite.

Gavaghan, Helen. *Something New Under the Sun: Satellites and the Beginning of the Space Age.* New York: Copernicus Books, 1998. This book focuses on the history and development of artificial satellites. It centers on three major areas of development—navigational satellites, communications satellites, and weather observation and forecasting satellites.

Kallenrode, May-Britt. *Space Physics: An Introduction to Plasmas and Particles in the Heliosphere and Magnetospheres.* New York: Springer-Verlag, 1998. This illustrated book is an introduction to the physics of space plasmas and its applications to current research into heliospheric and magnetospheric physics. The book uses a new approach, interweaving concepts and observations to give basic explanations of the phenomena, to show limitations in these explanations, and to identify fundamental questions.

Naugle, John E. *Unmanned Space Flight.* New York: Holt, Rinehart and Winston, 1965. Naugle, a director of geophysics and astronomy programs for NASA in the early 1960's, is a physicist who believes that the study of the Sun is the cornerstone of the whole space program. His treatment moves from the Sun through interplanetary space to Earth, which he views as a planet immersed in the atmosphere of the Sun and whose environment is controlled by the Sun's radiations. Naugle makes use of the data collected by Explorers 1 through 19 in his discussion, which is intended for the general reader.

Newell, Homer E. *Beyond the Atmosphere: Early Years of Space Science.* NASA SP-4211. Washington, D.C.: Government Printing Office, 1980. Newell, a former NASA administrator with a background in mathematics and physics, has been with the U.S. space program from its beginnings. He presents the Explorer satellites in several contexts—sociopolitical, scientific, and personal.

Ordway, Frederick, III, Carsbie C. Adams, and Mitchell R. Sharpe. *Dividends from Space.* New York: Thomas Y. Crowell, 1971. The authors explore the space program's contributions to human needs. They analyze the practical benefits to the American space program of the Explorer satellites. The book, which is profusely illustrated, includes a helpful appendix on NASA documents.

Parks, George K. *Physics of Space Plasmas: An Introduction.* 2d ed. Boulder, Colo.: Westview Press, 2004. Provides a scientific examination of the data returned during what might be called the "golden age" of space physics (1990-2002), when more than two dozen satellites were dispatched to investigate space plasma phenomena. Written at the undergraduate level for an introductory course in space plasma, there is also detailed presentation of NASA and ESA spacecraft missions.

Rosenthal, Alfred. *Venture into Space: Early Years of Goddard Space Flight Center.* NASA SP-4301. Washington, D.C.: Government Printing Office, 1968. Rosenthal became Goddard's historian in 1962. The purpose of his historical report is to describe Goddard's origin and its significant early contributions to the American space program. The ionosphere Explorers are competently analyzed in their institutional and scientific contexts. The book has many helpful charts and diagrams.

Zimmerman, Robert. *The Chronological Encyclopedia of Discoveries in Space.* Westport, Conn.: Oryx Press, 2000. Provides a complete chronological history of all crewed and robotic spacecraft and explains flight events and scientific results. Suitable for all levels of research.

*Robert J. Paradowski*

# Explorers: Micrometeoroid

*Date:* June, 1961, to December, 1972
*Type of program:* Scientific platforms

*The Micrometeoroid Explorers, including Explorers S-55, 13, 16, 23, and 45, performed direct measurements of the micrometeoroid environment in near-Earth space, providing important information on an aspect of the space environment that was little understood at the beginning of the age of space exploration.*

## Summary of the Satellites

At the beginning of the Space Age, very little quantitative information was available concerning the threat to satellites and other spacecraft posed by pieces of extraterrestrial matter, collectively called meteoroids. Space engineers and scientists recognized that the success or failure of a mission could rest on their improved ability to define this meteoroid environment in enough detail to allow them to develop appropriate protective measures for spacecraft. Ground-based measurements of the meteoroid environment were essentially limited to meteoroids with masses greater than approximately 0.1 milligram—that is, solid particles with radii greater than approximately 0.01 centimeter. It was suspected that there were many smaller particles that were also traveling at extremely high velocities, some exceeding tens of kilometers per second. The Micrometeoroid Explorers, a group of scientific satellites launched by the United States' National Aeronautics and Space Administration (NASA) between 1961 and 1971, were developed and flown to help resolve questions regarding the nature of such particles.

The terminology associated with meteoroids is sometimes confusing. By generally accepted scientific convention, a meteor is defined as any extraterrestrial particle, originating in the solar system, that can be detected by the human eye, with or without the aid of optical equipment, when it enters Earth's atmosphere. As a meteor enters the atmosphere, it heats up and vaporizes as a result of

frictional effects, emitting visible light as it makes its plunge. Meteors can be detected by radar because of the ionization trail they leave behind in the atmosphere. (In ionization, electrons are stripped from their parent atoms, leaving those atoms with a net positive charge.) If a meteor survives its incandescent descent through Earth's atmosphere and impacts Earth's surface, the remaining solid object is called a meteorite. A micrometeorite is simply a very small meteor (typically with a mass less than one-billionth of a kilogram) that survives its plunge through the atmosphere and is recovered on Earth's surface.

The term "meteoroid" is used in reference to solid objects found in space, ranging from micrometers to kilometers in radius and from fractions of a trillionth of a kilogram to several metric tons in mass. If the meteoroid has a mass of less than 1 gram, the term "micrometeoroid" is generally used.

In the 1960's, scientists were particularly interested in micrometeoroids' flux, directional preference, velocity, size, composition, electrical charge (if any), radioactivity level (if any), momentum, and structure. Aerospace engineers, on the other hand, were principally interested in learning about the micrometeoroid flux, the penetrating ability of these tiny, high-velocity particles, and the average hole size they would create upon impact with spacecraft. In pursuit of information in these areas, NASA engineers designed numerous instruments and experiments to be launched on board various Explorer satellites. Piezoelectric microphones, for example, were to supply information regarding the mass and velocity distributions of micrometeoroids. Detectors with piezoelectric ballistic pendulums measured micrometeoroids' momentum. Pressurized gas can detectors and light-transmission-erosion experiments were used to record micrometeoroid impacts; capacitor detectors recorded the frequency of such impacts and supplied directional information. Light-flash detectors allowed the separation of micrometeoroid mass and velocity data. The wire-grid micrometeoroid detector provided information about the micrometeoroid environment. Finally, time-of-flight experiments were designed to measure micrometeoroids' velocity.

Explorer S-55, also known as Meteoroid Satellite A, had two primary objectives: an evaluation of the Scout launch vehicle and its guidance system and an investigation of the near-Earth micrometeoroid environment and its potential implications for piloted and robotic spaceflight. The Scout launch vehicle for this mission was composed of four solid-propellant stages. The payload was wrapped around the fourth-stage rocket and included an extension on the rocket's nose. The satellite had a diameter of 61 centimeters, a length of 193 centimeters, and a mass of 84.8 kilograms.

The payload consisted of 160 half-cylinder pressurized-cell micrometeoroid detectors, sixty triangular foil gauge micrometeoroid detectors mounted under stainless steel spacecraft skin samples, forty-six copper wire grids wound on melamine cards, two cadmium-sulfide cells mounted in aluminized glass flasks, piezoelectric crystal impact-detecting devices, and five groups of solar cells with various protective coverings. The spacecraft had four whip antennae extending from the nose section and employed two separate telemetry transmitters that were operated by command for a maximum of one minute. Spacecraft power was supplied by solar cells mounted on the sides of the rocket's nose section, as well as by batteries.

Explorer S-55 was launched from NASA's Wallops Flight Center in Virginia on June 30, 1961. After the Scout launch vehicle reached an altitude of approximately 112.7 kilometers, the third-stage engine failed to ignite, and the spacecraft fell into the Atlantic Ocean.

The overall program objectives, launch vehicle, spacecraft design, and payload for Explorer 13 (also known as Explorer S-55A) were similar to those of the unsuccessful Explorer S-55. On August 25, 1961, Explorer 13 was launched from the Wallops facility. The spacecraft attained an orbital inclination (the angle between its orbital plane and the plane of Earth's equator) of 36.4°. Its apogee (farthest distance from Earth during orbit) was 975 kilometers, and its perigee (closest approach to Earth during orbit) was about 125 kilometers. Because the initial orbit was much lower than had been planned, this spacecraft, which was given the international designation 1961-c, reentered Earth's atmosphere on August 28, 1961, after only 2.3 days in space.

Explorer 16 (also called Explorer S-55B) had as its primary objectives the direct measurement of micrometeoroid punctures of structural skin samples, the measurement of micrometeoroid particles with varying momentums, and comparison of the engineering performance of shielded and unshielded solar cells. Explorer 16's launch vehicle was a four-stage, solid-propellant Scout. Payload experiments were mounted around the Scout's fourth stage. The spacecraft had a total mass in orbit of 100.7 kilograms, a diameter of 61 centimeters, and a length of 193 centimeters.

Explorer 16 carried six major experiments: 160 pressurized-cell detectors with varying wall thicknesses, sixty triangular foil gauge micrometeoroid detectors mounted beneath stainless steel skin samples, forty-six copper-wire grid detectors wound on melamine cards, two cadmium-sulfide cells mounted in aluminized glass flasks and covered with a layer of aluminized Mylar, piezoelectric crystal impact detectors, and five groups of solar cells, with samples that were unshielded and others that were shielded by glass or by fused silica.

The spacecraft had four whip antennae mounted on the nose section. There were two independent telemetry transmitters for data storage and command readout. Spacecraft power was supplied by solar cells (mounted on the rocket's nose) and by rechargeable nickel-cadmium batteries.

Explorer 16 was launched on December 16, 1962, from Wallops and was placed in an elliptical orbit with an apogee of 1,177 kilometers, a perigee of 747 kilometers, an inclination of 51.9°, and an orbital period of approximately 104.3 minutes. The last transmission from Explorer 16 (which was given the international designation 1962-bc) was received on July 22, 1963. The dormant spacecraft was expected to continue orbiting Earth for some eight hundred years.

The mission of Explorer 23 (also known as Explorer S-55C) was primarily to provide data on micrometeoroid penetration and the resistance of advanced spacecraft materials to such penetration. Its secondary objective was to test capacitor-type micrometeoroid detectors and to obtain data on the degradation of solar cells. Its scientific payload consisted of four experiments: 216 pressurized stainless steel micrometeoroid detectors with varying wall thicknesses, twenty-four triangular aluminum impact detectors, two cadmium-sulfide cells mounted under aluminized Mylar film, and two capacitor-type micrometeoroid detectors made of a combination of stainless steel, Mylar, and copper. Two groups of solar cell samples were also carried, one set unshielded and the other shielded by a quartz covering.

Power for Explorer 23 was provided by a solar cell and nickel-cadmium battery system. Its telemetry system operated on two frequencies. The spin-stabilized craft carried a timing device designed to terminate data transmissions after one year in space.

Explorer 23 was launched by a four-stage, solid-propellant Scout, whose first three stages were inertially guided, while its fourth stage was spin-stabilized. The satellite was a cylindrical spacecraft 61.0 centimeters in diameter and 233.7 centimeters long, with a mass of 133.8 kilograms. Payload experiments were mounted around the casing of the fourth-stage rocket motor.

Launched from Wallops Island on November 6, 1964, Explorer 23 achieved orbit with an inclination of 51.9°, a perigee of 464 kilometers, an apogee of 982 kilometers, and an orbital period of 98 minutes. This satellite was given the international designation 1964-74A. Except for the cadmium-sulfide cells experiment, which was apparently disabled during launch, all the experiments aboard Explorer 23 functioned properly.

Explorer 46 (also called the Meteoroid Technology Satellite), though not part of the S-55 generation of satellites, is often grouped with those earlier Explorers because of its overall mission objectives. It was designed to provide data on the rates of meteoroid penetration of a bumper-protected target and on overall meteoroid quantities and speeds.

Unlike the cylindrical Explorers 13, 16, and 23, Explorer 46 was a windmill-like spacecraft with meteoroid bumper panels deployed from a polygonal bus. The spacecraft's mass was 136.1 kilograms. One of its main experiments employed a meteoroid bumper target, 27.87 square meters in area, which recorded penetration of the bumper as well as of the main wall of the spacecraft. This target was divided into twelve panels, each with eight meteoroid detectors. Twelve box-shaped meteoroid velocity detectors were mounted around the central hub of the spacecraft. Explorer 46 also carried sixty-four capacitor detectors that provided data on the meteoroid population.

Launched by a four-stage, solid-propellant Scout rocket on August 13, 1972, from Wallops Island, Explorer 46 was given the international designation 1972-61A. It attained an orbit with an orbital inclination of 37.7°, an apogee of 823 kilometers, a perigee of 504 kilometers, and an orbital period of 97.84 minutes. One pair of meteoroid bumpers failed to deploy fully, and secondary meteoroid experiments were suspended in an effort to obtain maximum data from the main bumper experiment. Some twenty meteoroid impacts were recorded through December, 1972. In addition, the capacitor detectors successfully recorded about two thousand micrometeoroid encounters in near-Earth space.

### Contributions

Explorers 13, 16, 23, and 46 were instrumental in providing quantitative data regarding the near-Earth micrometeoroid environment, including information on density, penetration capabilities, range of sizes, velocity, and composition. These data were extremely valuable to both aerospace engineers and space scientists. As a result of these missions (and subsequent missions such as Pegasus 1, 2, and 3), scientists were able to assemble a model of the meteoroid environment that allowed them to develop protective devices—shields, bumpers, or both—for later spacecraft.

Data from these early missions allowed scientists to establish that the meteoroid population originating from comets has a range of mass from 0.000000001 (one-billionth) gram to 1 gram for sporadic meteoroids and from 0.000001 (one-millionth) gram to 1 gram for stream meteoroids. (Stream meteoroids are those that produce the meteor showers seen in Earth's sky during certain seasons of each year.) Meteoroids may be made of stone, iron, or ice.

Earth's gravitational influence has an effect on meteoroids in near-Earth space, increasing the meteoroid flux from certain directions in near-Earth orbit by about 1.7 times the flux values in interplanetary space. This and other factors are considered in estimating meteoroid hazards for spacecraft missions.

### Context

In the early days of the space program, it was impossible to calculate with any certainty the risk to spacecraft of the meteoroid environment. Ground-based observations had made it clear that debris from space is constantly falling into Earth's atmosphere. It is now known that this natural littering from space amounts to about 10 million kilograms per year. Early space planners, however, had little quantitative information regarding the small, high-velocity particles that eventually came to be called micrometeoroids. Data from Explorers 13, 16, 23, and 46 helped establish a good understanding of the actual meteoroid environment surrounding Earth.

Building on the data gathered by the Micrometeoroid Explorers, scientists were able to define in detail the meteoroid environment. Their measurements enabled aerospace engineers to develop techniques for shielding spacecraft against the impact of meteoroid particles at very high velocities. Although special shielding proved necessary for some spacecraft, it was discovered that standard spacecraft "skins" provided adequate protection in most cases. Engineers and scientists now have a reasonably good definition of the meteoroid environment and have developed and tested effective shielding techniques. By the mid-1970's, most major research aimed at defining the meteoroid environment had been terminated. Any remaining findings regarding the hazard of natural space debris will have only marginal implications for spacecraft design and mission planning.

The trend toward larger, more durable space platforms (such as the International Space Station) has necessitated renewed attention from aerospace designers to the possibility of damage from both meteoroids and human-made space debris. Clearly, longer periods in space and larger space structures will increase the likelihood that such structures will experience collisions with meteoroids or other debris. Furthermore, the use of lightweight aerospace materials may make spacecraft vulnerable to damage from smaller particles, so that special meteoroid shields may be required.

Experiments performed in the 1970's (including those carried out by the impact sensors on Explorer 46) indicated that human-made space debris in the form of particles with diameters less than 0.1 millimeter may be more abundant than natural debris of that size. Between 1973 and 1975, the "meteoroid impacts" recorded by instruments on Explorer 46 were especially abundant following launches of solid-fueled rockets. Such a correlation suggests that the near-Earth environment may be increasingly contaminated by plume exhausts from solid-fueled rockets. Fortunately, most of these exhaust particles were too small to damage anything but the most sensitive of spacecraft surfaces, such as coatings on windows or telescopes.

The Micrometeoroid Explorers, then, provided data regarding the micrometeoroid environment, measurements that allowed aerospace engineers to take account of the potential hazard to spacecraft posed by these particles. Data from these Explorers, particularly from Explorer 46, also helped scientists to identify another potential hazard: human-made space debris. The accumulation of such debris represents a serious challenge to those responsible for planning future enterprises in space.

**See also:** Explorers: Ionosphere.

### Further Reading

Brandt, John C., and Robert D. Chapman. *Introduction to Comets*. New York: Cambridge University Press, 2004. Provides detailed coverage of virtually every cometary phenomenon.

Burrows, William E. *This New Ocean: The Story of the First Space Age*. New York: Random House, 1998. This is a comprehensive history of the human conquest of space, covering everything from the earliest attempts at spaceflight through the voyages near the end of the twentieth century. Burrows is an experienced journalist who has reported for *The New York Times*, *The Washington Post*, and *The Wall Street Journal*. There are many photographs and an extensive source list. Interviewees in the book include Isaac Asimov, Alexei Leonov, Sally K. Ride, and James A. Van Allen.

Corliss, William R. *Scientific Satellites*. NASA SP-133. Washington, D.C.: Government Printing Office, 1967. A discussion of the early space missions of the United States, including Explorers 13, 16, and 23. Valuable for its meticulous detail.

Cosby, William A., and Robert G. Lyle. *The Meteoroid Environment and Its Effect on Materials and Equipment*. NASA SP-78. Washington, D.C.: Government Printing Office, 1965. Provides a good technical history of scientists' concerns about the micrometeoroid environment at the beginning of the Space Age and the experiments that were designed to gather information about that environment.

Davies, John K. *Astronomy from Space: The Design and Operation of Orbiting Observatories*. New York: John Wiley, 1997. This is a comprehensive reference on the satellites that revolutionized twentieth century astrophysics. It contains in-depth coverage of all space astronomy missions. It includes tables of launch data and orbits for quick reference as well as photographs of many of the lesser-known satellites. The main body of book is subdivided according to type of astronomy carried out by each satellite (x-ray, gamma-ray, ultraviolet, infrared, and radio). It discusses the future of satellite astronomy as well.

Heppenheimer, T. A. *Countdown: A History of Space Flight*. New York: John Wiley, 1997. A detailed historical narrative of the human conquest of space. Heppenheimer traces the development of piloted flight through the military rocketry programs of the era preceding World War II. Covers both the American and the Soviet attempts to place vehicles, spacecraft, and humans into the hostile environment of space. More than a dozen pages are devoted to bibliographic references.

Johnson, Nicholas L., and Darren S. McKnight. *Artificial Space Debris*. Malabar, Fla.: Orbit, 1987. A comprehensive discussion of the issue of space debris. Detailed and technical, this report is extremely well illustrated and filled with helpful tabulations of data.

Kessler, Donald J., and Shin-Yi Su, comps. *Orbital Debris*. NASA Conference Publication 2360. Washington, D.C.: National Aeronautics and Space Administration, Scientific and

Technical Information Branch, 1985. A set of technical papers, many very well written, concerning the natural and artificial space debris problem. The workshop at which these papers were first presented was held at NASA's Johnson Space Center in 1982.

Launius, Roger D. *NASA: A History of the U.S. Civil Space Program.* Malabar, Fla.: Krieger Publishing Company, 1994. This is an in-depth look at America's civilian space program and the establishment of the National Aeronautics and Space Administration.

Rosenthal, Alfred, ed. *Satellite Handbook: A Record of NASA Space Missions, 1958-1980.* Washington, D.C.: Government Printing Office, 1982. A detailed summary of data pertaining to all early spacecraft and satellites, piloted and robotic, launched by NASA. Includes useful summaries of the Micrometeoroid Explorer missions.

Zimmerman, Robert. *The Chronological Encyclopedia of Discoveries in Space.* Westport, Conn.: Oryx Press, 2000. Provides a complete chronological history of all crewed and robotic spacecraft and explains flight events and scientific results. Suitable for all levels of research.

*Joseph A. Angelo, Jr.*

# Explorers: Radio Astronomy

*Date:* July 4, 1968, to June 3, 1975
*Type of program:* Scientific platforms

*Radio Astronomy Explorers (RAEs) A and B were orbited to detect and measure extraterrestrial radio noise. RAE-A was orbited around Earth, while RAE-B was placed in lunar orbit. They were able to detect very weak solar noise bursts, from as low as 25 kilohertz to as high as 13 megahertz, emanating from outer space.*

### Summary of the Satellites

From the beginning, the United States space program was multifaceted, aimed at exploring the universe beyond as well as at studying Earth's atmosphere, providing remote-sensing and communications capabilities, and putting people on the Moon. For centuries, scientific study of the planets, stars, galaxies, and other extraterrestrial phenomena has been left to astronomers, who have sought to understand the nature of Earth by observing distant worlds and their relationships to one another. The Radio Astronomy Explorer (RAE) series was to take that search beyond the increasingly interference-laden atmosphere of Earth. Each satellite was designed to detect and measure long-wave radio signals emanating from space, within the solar system and beyond.

There were two satellites in the RAE program. RAE-A, also referred to as Explorer 38, was launched from Cape Kennedy on July 4, 1968, aboard a thrust-augmented improved Delta. It was placed in an orbit 6,000 kilometers above Earth. RAE-B, also referred to as Explorer 49, was launched in the same fashion on June 10, 1973.

The RAE series included a jump motor designed to place the satellite into proper orbit. It could not, however, be oriented to a specific target by controllers. Orbital inclination, therefore, was 58°, to allow for frequent sweeps of the Sun, the planets, and the galaxies beyond. The entire celestial sphere could be mapped over the period of one year from this orbital configuration.

Each satellite weighed 125 kilograms. Its 92-by-74-centimeter aluminum cylinder included four 250-meter unidirectional antennae that, when fully extended, made the whole satellite structure as long as the Empire State Building is high. These antennae were made of silver-plated beryllium-copper, 0.005 centimeter thick and 5 centimeters wide, and were arranged in an acute double V. There was also a dipole antenna, 61 meters long, designed to pick up intense bursts of noise from the Sun and the planet Jupiter when the longer antennae were not in position to pick up those signals. The satellites were built for the Goddard Space Flight Center, a National Aeronautics and Space Administration (NASA) facility, by the Applied Physics Laboratory of The Johns Hopkins University.

There were three instruments aboard each satellite. One engaged in antenna impedance measurements using three modes: capacity probe, analog impedance probe, and an electron trap. These modes were attached to the antennae to determine how they varied with antenna distortion and ambient conditions. The second was a cosmic-noise survey instrument with a fast-burst radiometer, which would signal controllers whenever it picked up noise at least ten times as intense as normal background noise. The third, also a cosmic-noise survey instrument, detected and measured extra-long wavelength emissions below the ionospheric cut-off. (Below approximately 15 megahertz, radio waves are unable to penetrate the ionosphere from space and are thus invisible to terrestrial receiving antennae.) Similar instrumentation was carried aboard other scientific satellites launched by the United States, Great Britain, Japan, and the Soviet Union.

The RAE satellites carried six experiments. These included observation of deep-space radiation from ionized hydrogen and synchrotron radiation, detection of low-frequency radiation from the Sun, detection of long-wave radio signals from Jupiter, observation of radiation emanating from Earth's ionosphere, mapping of radio wave sources at low frequencies, and monitoring of radiation caused by oscillations in solar plasma. The satellites were also designed to detect the direction from which measured signals were emanating. This capability allowed experimenters to locate the source of such emissions and, where appropriate, focus in for increased resolution and discrimination.

Ground controllers were able to communicate with the satellites in real time or by recording data for later retrieval on command. The satellites were powered by solar panels.

RAE-A was placed in Earth orbit. RAE-B, launched five years later, was placed in lunar orbit, for scientists had become concerned about terrestrial electromagnetic interference that was significantly reducing the effectiveness of the instrumentation aboard RAE-A. In fact, RAE-A had been able to detect only radio signals emanating from intense solar bursts, and then only in the range of 200 kilohertz to 9 megahertz. RAE-B, from its vantage point in lunar orbit, was able to detect very weak signals over a much broader range, 25 kilohertz to 13 megahertz. One reason for this increased efficiency was that its lunar orbit allowed RAE-B to use the Moon as a shield from electromagnetic radiation, further enhancing the radio-free environment of its position.

### Contributions

The RAE satellite program did not represent the first effort by scientists to determine the strength and source of radio waves reaching Earth's atmosphere from within the solar system and beyond. In fact, Earth-based systems and other satellites had been detecting radio signals from space long before RAE-A was launched in 1968. Radio astronomy had involved the use of huge Earth-based antennae that were designed to pick up random signals over a wide electromagnetic range. To pinpoint the source, two or more of these antennae located at known distances from one another would be oriented to the same signal to form a radio interferometer. This technique allowed for determination of the precise source and wavelength of a particular signal. At the time of the launching of RAE-B, these radio telescopes were located in the United States, Australia, Great Britain, the Netherlands, West Germany, Canada, and the Soviet Union.

Increasingly, however, terrestrial electromagnetic and atmospheric interference has created a cacophony of electronic noise, so that space-generated signals are often lost to Earth-based detection instruments. This interference is being caused by the increased use of the electromagnetic spectrum for everything from broadcasting to cellular phones and microwave ovens.

Another limitation of Earth-based radio telescope technology is the size of Earth's surface, which provides a relatively small base for setting distances. Space-based radio telescopes, perhaps hundreds of thousands of miles apart, would make it possible for the source and wavelength of radio signals from deep space to be precisely identified.

The high-frequency end of the electromagnetic spectrum is the location of the most intense radiation. Known as the gamma-ray portion of the spectrum, this region was investigated by satellite as early as 1961, when Explorer 11 was launched and engaged in an intensive study of that portion of the spectrum. Six years later, interest was rekindled after the United States launched the Vela series of satellites to detect nuclear explosions, as part of the verification process established to monitor adherence by the Soviet Union to the Nuclear Test Ban

Treaty. These satellites unexpectedly detected what appeared to be nuclear explosions in space. Later analysis led to the conclusion that these events occurred naturally.

In 1964, Great Britain experimented with a nondirectional antenna aboard Ariel 2 that picked up signals in the range of 0.75 to 3 megahertz. While the data received from this satellite were useful in determining the frequency and intensity of such radio signals, the source of the signal could not be determined. The Radio Astronomy Explorer program went beyond the nondirectional receiving antenna system to construct large, directional antenna arrays that were, to a limited extent, capable of pinpointing sources of long wavelengths in the solar system and beyond.

RAE-B transmitted data showing that radio noise on Earth tended to relate to the position of Earth and the Sun in relation to each other. For example, it demonstrated that there are two major magnetospheric noise regions in Earth's atmosphere: one in the region of morning, or dawning of the Sun, and the other in the region where the Sun is setting. The most intense electromagnetic noise occurs in late evening, during the four hours leading up to midnight. Late evening noise appears to be generated within 10,000 to 12,000 kilometers of Earth's surface, while daytime noise, the least intense of which occurs around mid-morning, appears to originate within 4,000 kilometers of Earth's surface.

The next step proposed by scientists was to place large radio telescopes or parabolic antenna systems in space to continue the work begun by the RAE satellites. After two years of continuous operation, the Radio Astronomy Explorer program, judged by NASA officials to have been a major success, came to an end. Attention turned from Explorer 38 and Explorer 49 to larger, more complex and sophisticated systems: crewed radio astronomy platforms that would be required if the interference of Earth's atmosphere were to be wholly avoided. It appeared certain that such a move was necessary for the continuation of the quest to understand radiation in the cosmos.

## Context

Radio astronomy has been an important part of the space programs of all nations engaged in space science. In the 1930's, the first directional receiving antennae were pointed toward space; they immediately began to detect a number of radio signals that stood out from the general cosmic noise heard in most regions of the electromagnetic spectrum. This finding led to an increase in scientific interest in radio astronomy and a number of discoveries, among them that objects in the solar system and beyond emit radio waves.

There were limitations, however, to the ability of this land-based technology. Water vapor in the atmosphere and the electrophysical activity in the ionosphere absorb some radio energy, so that some signals are not detectable on Earth's surface. As a result, it was necessary to place radio astronomy instrumentation on satellites, beyond the confines of Earth's atmosphere.

Radiation from space can be categorized as either solar or galactic. In the early days of the space program, most radio wave research was solar in nature, resulting from scientific interest in the Sun and its effect on weather, communications, climate, and atmospherics. Solar cosmic rays are different, however, from galactic rays. They are relatively weak, though they emanate from the high end of the spectrum. They also emit less energy than galactic rays. Solar cosmic rays are directional, while galactic rays are omnidirectional; the latter tend to be located at the lower end of the electromagnetic spectrum. Radio Astronomy Explorer was one of very few satellite programs conducted to continue the search for long waves beyond the Sun, from the planets and the stars.

Astronomers have traditionally classified functions of radio astronomy instrumentation in four categories. The first, observational astronomy, employs spectrographs, photometers, and telescopes to analyze the high end of the spectrum. The second, cosmic-ray astronomy, measures very-high-energy radiation using particle counters and similarly configured telescopes. Active astronomical experiments, which fall into the third category, involve the use of probes to make direct observations of asteroids, comets, and interplanetary gases. An example is the variety of probes sent to meet Halley's comet when it made its appearance in 1987. Artificial comet experiments have been conducted under the aegis of active astronomical experimentation as well. The fourth is what is called cosmology, in which scientists use laboratory instruments to test observations against the general theory of relativity.

Most astronomical research using satellites has fallen into the first two categories, although many of the data retrieved from satellites such as RAE-A and RAE-B are helpful in understanding some of the history of the universe, the Sun, and the planets. Cosmologists are able to extrapolate information useful in the design of instrumentation and space probe trajectories from these data.

Instruments designed to detect solar cosmic rays are very similar to those that pick up galactic cosmic rays. Over the years they have been designed to monitor the intensity, direction, energy level, and position of an incoming radio signal on the electromagnetic spectrum. Since RAE, such instrumentation has been installed on many satellites whose primary missions were to conduct studies totally unrelated to radio astronomy. Interplanetary probes have also carried instrumentation to detect radio waves from space, helping scientists to pinpoint the locations of sources of such radiation with much greater accuracy than was possible from Earth orbit or from Earth's surface.

As the space program progressed, a number of proposals were made to advance radio astronomy using the space shuttle as a staging platform. Efforts were mounted to develop a huge radio telescope to be placed in space for crewed missions aimed at learning more about the nature of the universe and the objects in it. As the Radio Astronomy Explorer program demonstrated, there is much to be learned about the nature of cosmic radiation and the effect it has on Earth and its atmosphere.

**See also:** Apollo 16; Explorers: Astronomy.

### Further Reading

Brun, Nancy L., and Eleanor H. Ritchie. *Astronautics and Aeronautics, 1975.* Washington, D.C.: National Aeronautics and Space Administration, 1979. A chronological compilation of popular-press accounts of space program events during 1975, RAE among them. Includes an index.

Burrows, William E. *This New Ocean: The Story of the First Space Age.* New York: Random House, 1998. This is a comprehensive history of the human conquest of space, covering everything from the earliest attempts at spaceflight through the voyages near the end of the twentieth century. Burrows is an experienced journalist who has reported for *The New York Times, The Washington Post,* and *The Wall Street Journal.* There are many photographs and an extensive source list. Interviewees in the book include Isaac Asimov, Alexei Leonov, Sally K. Ride, and James A. Van Allen.

Corliss, William R. *Scientific Satellites.* NASA SP-133. Washington, D.C.: National Aeronautics and Space Administration, 1987. Contains highly technical descriptions of most of the scientific satellites sent aloft during the first decade of the Space Age. Also included is a chronology of international satellite launches, with comments and technical data. Contains a chapter describing geophysical instruments and experiments, preceded by an excellent overview of space science in general. The RAE satellite program is outlined in detail, and the book contains an excellent overview of the technical nature of radio astronomy. Indexed; includes photographs and illustrations.

Davies, John K. *Astronomy from Space: The Design and Operation of Orbiting Observatories.* New York: John Wiley, 1997. This is a comprehensive reference on the satellites that revolutionized twentieth century astrophysics. It contains in-depth coverage of all space astronomy missions. It includes tables of launch data and orbits for quick reference as well as photographs of many of the lesser-known satellites. The main body of book is subdivided according to type of astronomy carried out by each satellite (x-ray, gamma-ray, ultraviolet, infrared, and radio). It discusses the future of satellite astronomy as well.

Gatland, Kenneth. *The Illustrated Encyclopedia of Space Technology: A Comprehensive History of Space Exploration.* New York: Salamander, 1989. This well-illustrated volume contains chronologies of Soviet, American, and international space programs. It includes a chapter that discusses in detail radio astronomy, and specifically the RAE satellites and their accomplishments. Specific examples of successful applications of space technology in this area are cited. Features illustrations, photographs, and an index.

Gavaghan, Helen. *Something New Under the Sun: Satellites and the Beginning of the Space Age.* New York: Copernicus Books, 1998. This book focuses on the history and development of artificial satellites. It centers on three major areas of development—navigational satellites, communications satellites, and weather observation and forecasting satellites.

Heppenheimer, T. A. *Countdown: A History of Space Flight.* New York: John Wiley, 1997. A detailed historical narrative of the human conquest of space. Heppenheimer traces the development of piloted flight through the military rocketry programs of the era preceding World War II. Covers both the American and the Soviet attempts to place vehicles, spacecraft, and humans into the hostile environment of space. More than a dozen pages are devoted to bibliographic references.

Hirsch, Richard, and Joseph John Trento. *The National Aeronautics and Space Administration.* New York: Praeger, 1973. This excellent chronology of the accomplishments of NASA during the first decade of space exploration describes early plans for the RAE satellite program. Photographs supplement the text. Indexed.

Parks, George K. *Physics of Space Plasmas: An Introduction.* 2d ed. Boulder, Colo.: Westview Press, 2004. Provides a scientific examination of the data returned during what might be called the "golden age" of space physics (1990-2002), when more than two dozen satellites were dispatched to investigate space plasma phenomena. Written at the undergraduate level for an introductory course in space plasma. There is also detailed presentation of NASA and ESA spacecraft missions.

Ritchie, Eleanor H. *Astronautics and Aeronautics, 1976: A Chronology.* NASA SP-4021. Washington, D.C.: National Aeronautics and Space Administration, 1984. This annual, like the Brun and Ritchie volume (see above), contains chronological listings of abstracts of press accounts of space activity. These accounts, gleaned from popular publications and space agency memoranda released to the general public, include the names of hundreds of individuals associated with scores of international space programs. Perhaps most useful are summaries of updates on programs such as RAE, which were ongoing at that time. Includes an index.

Zimmerman, Robert. *The Chronological Encyclopedia of Discoveries in Space.* Westport, Conn.: Oryx Press, 2000. Provides a complete chronological history of all crewed and robotic spacecraft and explains flight events and scientific results. Suitable for all levels of research.

*Michael S. Ameigh*

# Explorers: Solar

*Date:* November 19, 1965, to December 15, 1979
*Type of program:* Scientific platforms

*Explorer probes 30, 37, and 44 monitored x-ray and ultraviolet emissions from the Sun. The data these probes collected improved techniques of predicting solar activity and ionospheric disturbances and increased our understanding of the effects of solar activity on human space missions and radio communications.*

### Key Figures

*M. J. Aucremanne*, program manager on Explorer 30
*R. W. Kreplin*, principal investigator on Explorers 30, 37, and 44
*J. R. Holtz*, Explorer 37 and 44 program manager
*G. K. Oertel*, Explorer 37 program scientist

### Summary of the Satellites

Explorers 30, 37, and 44 were a series of small scientific satellites that primarily monitored x-ray and ultraviolet emissions from the Sun between 1965 and 1979. Each satellite was unique in its design and payload and fulfilled a particular purpose.

It was satellite technology that gave solar scientists and observers their first opportunities to monitor and study solar radiation and emissions from positions and angles high above Earth and its atmosphere. Explorers 30 and 37, also known as Solar or Solrad 1 and Solar or Solrad 2, and Explorer 44 (Solrad 10) made important contributions to the body of solar data collected during the 1960's and 1970's. They were part of a lengthy series of uncrewed satellites that studied near-Earth space, Sun-Earth relationships, interplanetary space, and solar physics, beginning with the launch of Explorer 1 in 1958. (Some sources refer to Explorers 30, 37, and 44 as Solrads 1, 2, and 3.)

The National Aeronautics and Space Administration (NASA) launched Explorer 30 on November 19, 1965, from the Wallops Flight Center at Wallops Island, Virginia. The Wallops center prepares, assembles, launches, and tracks uncrewed space vehicles, ranging from small sounding rockets to larger four-stage rockets. The launch vehicle for Explorer 30 was a four-stage solid-fueled Scout. The rocket weighed 17,463 kilograms and was almost 22 meters tall. NASA's Langley Research Center was in charge of launch vehicle management, and Ling-Temco-Vought (LTV) Aerospace Corporation was the prime launch vehicle contractor.

The satellite's mission was to monitor the Sun's x-ray and ultraviolet emissions for correlation with data from Earth-based optical and radio observations. NASA and the United States Naval Research Laboratory (NRL) coordinated the project. NRL, the spacecraft's prime contractor, was in charge of its solar radiation experiments. The project manager for NASA was M. J. Aucremanne; NRL's R. W. Kreplin served as both project scientist and project investigator.

The spacecraft's design consisted of two 60.9-centimeter hemispheres, separated by an 8.8-centimeter equatorial band. Four telemetry antennae in a "turnstile" arrangement and twelve photometers were attached to the midsection. Six solar cell panels were mounted symmetrically on the satellite. Explorer 30 weighed 56 kilograms at launch.

The payload included ion chambers and a Geiger counter to measure overlapping x-ray bands and ultraviolet rays. The craft stabilized at 60 revolutions per minute. Two low-thrust ammonia vapor jets maintained the spin rate. Input from two Sun sensors controlled the spin axis. The communications system consisted of two command receivers, a digital data storage unit, and a continuously operating analog transmitter. A digital transmitter operated on command. A special coating on the highly polished shell achieved passive thermal control. Solar cells supplied 6 watts of power to the craft. Explorer 30 had an expected life span of one year and an orbital life of two hundred years. Goddard Space Flight Center in Greenbelt, Maryland, provided tracking support for the craft during the mission.

In 1968, NRL and NASA teamed up to launch another satellite that would measure solar radiation, Explorer 37. Kreplin, former project scientist and project investigator for the Explorer 30 experiments, headed the new mission as project manager. He was joined by NASA's J. R. Holtz and G. K. Oertel, program manager and program scientist, respectively. Other investigators for the solar radiation experiment were T. A. Chubb and Herbert Friedman from NRL.

Once again the main objective of the joint project was to measure and monitor selected x-ray and ultraviolet emissions from the Sun. Explorer 37 had a different configuration from and was almost twice as large as its predecessor. The project's scope had also expanded in size and sophistication. As before, NRL was in charge of the solar radiation experiments and served as the prime contractor.

Explorer 37 consisted of a twelve-sided cylinder, 76 centimeters in diameter and about 68 centimeters high. X-ray photometers, Geiger tubes, photomultipliers, a solar aspect system, an attitude control system, and spin nozzles were mounted on a central band around the middle of the cylinder. Solar cells, contained in twenty-four panel arrays, covered the remainder of the vertical sides. Four radio antennae extended from the top and four from the bottom of the spacecraft. The satellite weighed almost 90 kilograms.

The payload reflected the increased requirements of the mission. A scintillation counter, two Geiger tubes, and five x-ray photometers measured x-ray emissions in six bands (compared with Explorer 30's monitoring of only two such bands). Two ultraviolet photometers measured ultraviolet emissions in two bands.

Explorer 37 used a more complex communications system than had Explorer 30. It consisted of two analog transmitters designed for continuous transmission, a digital transmitter that operated on command only, and two four-antenna systems. The satellite also carried two command receivers connected to a decoder system for placing the craft and its experiments into the desired operating mode. A magnetic core memory system collected data from three x-ray detectors for use in timed experiments.

Low-thrust vapor jets maintained the spin rate at about 1 revolution per minute and controlled the spin axis. Two Sun sensors, located 180° apart, generated timed jet pulses to process the spin axis. Silicon solar cells and nickel-cadmium batteries supplied the craft with 27 watts of power.

Explorer 37 was launched on March 5, 1968, from Wallops Flight Center. A malfunction in the first stage of the Scout launch vehicle caused the spacecraft to fail to reach its intended 836.6-kilometer circular orbit. Instead, the satellite went into an 876.9-by-521.3-kilometer orbit at a 59.4° inclination. The flawed orbit did not interfere, however, with the attainment of the mission's scientific objectives. Langley Research Center was in charge of the launch vehicle, and LTV Aerospace Corporation was the prime contractor for the launch vehicle. Explorer 37 had an expected orbital life span of about twenty years. The Goddard Space Flight Center provided tracking services.

The third and final spacecraft in the solar radiation series was Explorer 44, also known as Solrad 10. A joint NASA and NRL project, the mission had two main objectives and involved fourteen different experiments. Its two main experiments entailed continuous monitoring of solar electromagnetic radiation (x-ray and ultraviolet) and mea-

*Explorer 44, also known as Solrad 10, was sent into orbit around the Sun on July 8, 1971, to study the Sun's ultraviolet and x radiation.* (NASA)

The satellite carried equipment for fourteen experiments, including a stellar x-ray telescope. Also on board were eighteen ionization chambers, solar x-ray monitors, a solar electron temperature sensor, a solar Lyman alpha burst monitor, a solar ultraviolet monitor, and a solar ultraviolet continuum flash. A background x-ray level, a solar hard x-ray continuum, a cesium iodide (sodium) scintillating crystal, and a photomultiplier to monitor solar hard x rays were included.

In addition, Explorer 44 carried a lithium fluorine photometer for solar excitation of the F-layer, a thermistor for skin antisolar temperature, and a large-area proportional counter for x-ray variations. Project leaders expected Explorer 44 to function for three years. It reentered Earth's atmosphere on December 15, 1979, after remaining in space for 3,082 days.

E. W. Peterkin of NRL was project manager for the NRL/NASA venture. Kreplin once again served as project scientist and project investigator for both the solar radiation and the stellar x-ray telescope experiments. NASA's J. David Bohlin served as the program scientist. Goddard Space Flight Center handled tracking and data management, and Langley Research Center provided launch vehicle management. Again, LTV Aerospace Corporation was the prime contractor for the launch vehicle.

### Contributions

Data gathered by the Solar Explorers enhanced and complemented the body of solar information gathered in the 1960's and 1970's. Explorer 30 successfully monitored and relayed information about solar x-rays and ultraviolet emissions. Processed data were correlated with observations from ground-based opticals (telescopes) and radio equipment. NRL and NASA had specifically chosen Ex-

suring stellar radiation from other celestial sources. The main purpose was to improve techniques of predicting solar activity and ionospheric disturbances.

Explorer 44 was a twelve-sided structure, 76.2 centimeters in diameter and about 58 centimeters in height. Four symmetrically placed solar cell panels were hinged at the center station. The solar cell panels folded along the length of the structure and opened after the third-stage burnout of the launch vehicle. The panels served as the four elements of a turnstile antenna system.

The electronic subsystems included two nonredundant telemetry transmitters operating on separate frequencies of 137.710 megahertz and 136.380 megahertz. Three spin replenishments and four spin-axis attitude control subsystems maintained the spin rate at 60 revolutions per minute. Solar sensors determined the angle to the Sun and automatically applied control signals to the attitude spin subscriptions. Explorer 44 was the largest of the solar radiation satellites, with a weight of 117 kilograms.

NASA's solid-fueled workhorse, the Scout B, launched the satellite into a 432.8-by-632.3-kilometer orbit on July 8, 1971, from Wallops Flight Center.

plorer 30's launch date to coincide with the final half of the International Quiet Sun Years, a cooperative effort in solar studies of scientists and researchers from all over the world. The 1964-1965 program was similar in scope to the International Geophysical Year of 1957-1958; its purpose was to study the Sun and its terrestrial and planetary effects during the solar minimum, the least active period of the eleven-year cycle of solar activity.

Although Explorer 37 never entered its proper orbit, its scientific objectives were successful. The satellite measured solar x-ray and ultraviolet emissions in various wavelength frequency bands. Scientists also conducted a series of time-limiting observation experiments, using a magnetic core memory system that collected measurements from three x-ray detectors for 14 hours. After the digital data were transmitted to the ground on command, the equipment was reset for another fourteen-hour observation period.

Data gathered by Explorer 37 were distributed to a number of users, including NRL and the Environmental Science Services Administration's Space Disturbances Forecasting Office in Boulder, Colorado.

Explorer 44, the most ambitious of the Solar Explorers, had a twofold mission. Carrying equipment for fourteen separate experiments, it collected specific data on solar activity continuously. The information gathered allowed scientists to understand further how solar flares and ionospheric disturbances might affect future crewed spaceflights and radio communications. In addition, stellar radiation from other celestial objects was measured on command, using the stellar x-ray telescope. The scientific community, under the auspices of the Committee on Space Research (COSPAR), shared the information gathered by Explorer 44. COSPAR was part of the International Committee of Scientific Unions.

Most of the data gathered by Explorers 30, 37, and 44 were not particularly startling or newsworthy, but they did provide a solid foundation of information about solar activity, especially in the areas of x-ray and ultraviolet emissions. The material collected played an important role in the development of more sophisticated solar monitoring activities and projects.

### Context

Throughout history, humans have observed the Sun and pondered its properties with a fascination that, at times, extended to deification and worship. Real understanding and knowledge, however, remained limited by the range of the human eye and Earth-bound technology. It was not until October 10, 1946, that solar observers could look at the Sun from a different perspective. On that date, a German V-2 rocket, captured in World War II, was launched 90 kilometers into space from White Sands Proving Grounds (later named White Sands Missile Range). The instruments it carried recorded the ultraviolet spectrum of the Sun for the first time. It gave a tantalizing glimpse of Earth's mother star from a new point of view. Yet rockets provided only the most hurried of glances. Only when satellite technology became a reality in 1958 could scientists plan and carry out long-term space-based experiments on the Sun and its activities.

The Solar Explorer series, along with other Explorer projects, including the Interplanetary Monitoring Platform and Explorers 18, 21, 28, 33, 34, 35, 41, 43, 47, and 50, extended the knowledge of Sun-Earth relationships and solar activity. During the same time span (the mid-1960's to the mid-1970's), the Orbiting Solar Observatory (OSO) program was developed. OSOs were the first observatory satellites. They were designed to observe, monitor, and record solar data during the Sun's eleven-year cycle.

A decade of almost feverish solar research, 1965 to 1975, culminated in the launch of NASA's Skylab in 1973. Astronauts accumulated enormous amounts of solar data using the Skylab Apollo Telescope Mount. The Apollo Telescope Mount carried an x-ray spectrograph, an x-ray spectroheliograph, an x-ray telescope, a coronagraph, an ultraviolet spectrograph, and an extreme ultraviolet spectrometer-spectroheliometer. This equipment obtained more than 180,000 images and accumulated more than one thousand hours of observation of the Sun.

Other uncrewed solar observers during that time period included Helios 1 and 2. These craft carried probes within the outer edge of the Sun's corona, within 45 million kilometers of the Sun itself.

In 1980, NASA launched the Solar Maximum Mission to coincide with the peak of the eleven-year sunspot cycle. The spacecraft recorded its final data on November 24, 1989, and reentered on December 2, 1989. Other spacecraft have collected valuable solar data in addition to their primary missions. For example, Pioneer 10, launched in 1973, discovered the heliosphere (the Sun's atmosphere) during its journey to Jupiter. Solar radiation research in space has continued through the years. NASA, the European Space Agency (ESA), and the Institute of Space and Astronautical Science (ISAS) of Japan have developed the International Solar-Terrestrial Physics Science Initiative (ISTP), consisting of a set of solar-terrestrial missions that began during the 1990's.

The primary science objectives of the ISTP Science Initiative are to determine the structure and dynamics in the solar interior and their role in driving solar activity; to identify processes responsible for heating the solar corona and its acceleration outward as the solar wind; to determine the flow of mass, momentum, and energy through geospace; to gain a better understanding of the turbulent plasma phenomena that mediate the flow of energy through geospace; and to implement a systematic approach to the development of the first global solar-terrestrial model, which will lead to a better understanding of the chain of cause-effect relationships that begins with solar activity and ends with the deposition of energy in the upper atmosphere. Satellites participating in this program include Wind, Polar, Geotail, SOHO, and Cluster.

**See also:** Explorers: Radio Astronomy.

## Further Reading

Davies, John K. *Astronomy from Space: The Design and Operation of Orbiting Observatories.* New York: John Wiley, 1997. This is a comprehensive reference on the satellites that revolutionized twentieth century astrophysics. It contains in-depth coverage of all space astronomy missions. It includes tables of launch data and orbits for quick reference as well as photographs of many of the lesser-known satellites. The main body of book is subdivided according to type of astronomy carried out by each satellite (x-ray, gamma-ray, ultraviolet, infrared, and radio). It discusses the future of satellite astronomy as well.

Frazier, Kendrick. *Our Turbulent Sun.* Englewood Cliffs, N.J.: Prentice-Hall, 1982. A general study of the Sun, especially in its relationship to Earth. The book contains photographs, charts, graphs, an index, and a chapter-by-chapter selected bibliography. Suitable for general audiences.

Gavaghan, Helen. *Something New Under the Sun: Satellites and the Beginning of the Space Age.* New York: Copernicus Books, 1998. This book focuses on the history and development of artificial satellites. It centers on three major areas of development—navigational satellites, communications satellites, and weather observation and forecasting satellites.

Gibson, Edward G. *The Quiet Sun.* NASA SP-303. Washington, D.C.: Government Printing Office, 1973. Written by American astronaut Edward G. Gibson, in preparation for the Skylab mission, this is a textbook on solar physics. It is written from a physicist's point of view and contains college- and graduate-level material. Topics include the physical properties of the Sun, solar structures, the chromosphere, the corona, and the like.

Giovanelli, Ronald. *Secrets of the Sun.* New York: Cambridge University Press, 1984. Written for the general reader, the book includes information on and a brief history of solar research. Discussions cover sunspots, solar flares, the chromosphere, and solar effects on weather. Contains more than one hundred figures and color plates.

Golub, Leon, and Jay M. Pasachoff. *Nearest Star: The Surprising Science of Our Sun*. Boston: Harvard University Press, 2002. Although written by two of the most active research astrophysicists, this book is accessible to a general audience. It describes most contemporary advances in solar physics.

Heppenheimer, T. A. *Countdown: A History of Space Flight*. New York: John Wiley, 1997. A detailed historical narrative of the human conquest of space. Heppenheimer traces the development of piloted flight through the military rocketry programs of the era preceding World War II. Covers both the American and the Soviet attempts to place vehicles, spacecraft, and humans into the hostile environment of space. More than a dozen pages are devoted to bibliographic references.

Launius, Roger D. *NASA: A History of the U.S. Civil Space Program*. Malabar, Fla.: Krieger Publishing Company, 1994. This is an in-depth look at America's civilian space program and the establishment of the National Aeronautics and Space Administration. It chronicles the agency from its predecessor, the National Advisory Committee for Aeronautics, through the present day.

National Aeronautics and Space Administration. *NASA: The First Twenty-Five Years, 1958-1983*. NASA EP-182. Washington, D.C.: Government Printing Office, 1983. An excellent chronological history of NASA and the United States space program. Designed for classroom use by teachers, the volume contains a wealth of color photographs, charts, graphs, tables, and suggested classroom activities. Topics include piloted and robotic space missions, space science applications satellites and their discoveries, and aeronautics. Suitable for a general audience.

Rosenthal, Alfred, comp. *A Record of NASA Space Missions Since 1958*. Washington, D.C.: Government Printing Office, 1982. A brief summary of all NASA piloted and robotic missions, beginning in 1958. Lists information on spacecraft design, launch vehicle, payload, project objectives and results, and major participants. Some technical data is given. Suitable for general audiences.

Zimmerman, Robert. *The Chronological Encyclopedia of Discoveries in Space*. Westport, Conn.: Oryx Press, 2000. Provides a complete chronological history of all crewed and robotic spacecraft and explains flight events and scientific results. Suitable for all levels of research.

*Ludynne Streeter*

# Explorers 1-7

*Date:* January 31, 1958, to July 24, 1961
*Type of program:* Scientific platforms

*In the aftermath of the Soviet success in launching Sputniks 1 and 2, the first artificial satellites, the United States was compelled to launch a satellite into Earth orbit quickly. After the failures of some early Vanguard launch vehicles and subsequent launch pad explosions, research for what became the Explorer program was approved. Five of the first seven Explorers reached Earth orbit, and the program was considered a major success.*

**Key Figures**

*Wernher von Braun* (1912-1977), technical director, Army Ballistic Missile Agency (ABMA), and director of the Marshall Space Flight Center (MSFC)
*James A. Van Allen* (b. 1914), professor of physics, University of Iowa
*John B. Medaris* (1902-1990), commander, ABMA Ordnance Corps
*William H. Pickering* (1910-2004), director of the Jet Propulsion Laboratory (JPL)
*T. Keith Glennan* (1905-1995), first NASA administrator

**Summary of the Satellites**

The Explorer program began with the United States' first orbital satellite, Explorer 1, launched in the International Geophysical Year (an eighteen-month period during 1957-1958 designated as a time of intense geophysical study). Explorers 1 through 7 were hastily launched by the U.S. Army and the Jet Propulsion Laboratory (JPL) when Project Vanguard showed signs of faltering in the shadow of the Soviet Union's Sputniks 1 and 2.

Technological advances leading up to the launch of Explorer 1 had begun some fifteen years before. In May, 1945, with the sound of Soviet artillery in the distance, German rocket scientists based at Peenemünde, Germany, surrendered to American forces. The effective design and awesome striking power of German offensive missiles had impressed the Allies during the V-2 London bombings. Allied commanders were quick to realize the value of roughly one hundred V-2 rockets for experimentation and the value of securing one hundred German rocket scientists, who were most willing to em-igrate and advance the science of rocketry in the United States. The leader of the German group was Wernher von Braun, who, before his death in 1977, would prove himself invaluable in all phases of U.S. space efforts.

Building on significant advances in rocketry made by the Army, Air Force, Navy, and JPL in the 1940's, the U.S. Army Ordnance Guided Missile Center—later named the Army Ballistic Missile Agency (ABMA)—was formed at the Redstone Arsenal in Huntsville, Alabama, in 1950. There, with the assistance of JPL and, later, von Braun's colleagues, the Redstone rocket was developed from the V-2 for predominantly military purposes. With the Korean War in progress, military ballistic research was well funded and highly successful. By mid-1955, in direct competition with Project Vanguard of the Naval Research Laboratory (NRL), the Army-JPL team submitted a proposal for Project Orbiter to the Department of Defense to fulfill the satellite launch scheduled for the International Geophysical Year.

The Navy plan was ultimately accepted in September, 1955. Morale at the Redstone Arsenal was low after the Navy won its proposal to launch the United States' (and at that time, perhaps the world's) first human-made orbital satellite. Department of Defense funding was not available for non-military research. The Army argued, in vain, for additional research and development support for Project Orbiter, but none was granted. Thus, ABMA and JPL rocket research continued in the shadow of NRL's Project Vanguard.

The Jupiter series evolved from the Redstone ballistic missile, which, in turn, was the direct descendant of the German V-2. Jupiter A was the first modified Redstone, and between September, 1955, and June, 1958, twenty-five Jupiter tests were conducted. Jupiter-C was designed as a vehicle to test nose-cone materials for reentry. Standing 17.7 meters high and measuring 2.7 meters in diameter, the missile weighed roughly 50,000 kilograms. The Jupiter-C was configured with a lengthened Redstone first-stage fueled with hydyne (a mixture of unsymmetrical dimethylhydrazine and diethylene triamine). A cluster of eleven solid-propellant Sargeant rockets made up the second stage, and the third stage was made up of three Sargeants. Different from contemporary rockets, before launch the third stage of Jupiter-C was set into rotation (for stability and added ballistic accuracy), thus creating an unusual "spinning top" effect on the prelaunch rocket.

Various staging combinations and propulsion systems were tested. By September 20, 1956, the ABMA-JPL team had launched a prototype Jupiter-C with a dummy last stage. By the summer of 1957, a suborbital Jupiter-C had reached its target area and was recovered. These launches, and government funding, allowed the Army-JPL team to produce the four-stage Juno 1 (nearly identical to the three-stage Jupiter-C).

Finally, on November 8, 1957, amid the failure of Vanguard missiles and the success of Sputnik 1, the ABMA received authorization from the Department of Defense to revive Project Orbiter and attempt to launch a satellite into low-Earth orbit sometime during the remainder of the International Geophysical Year. The Jupiter-C rocket was reasonably well tested, and confidence in the launch system ran high. Because of previous losses in funding, however, a suitable satellite needed to be designed from scratch. In addition, launch schedules, refitting of launch pads, and tracking systems at Cape Canaveral needed to be organized. The Juno 1 was created from a stock Jupiter-C missile, with an added solid-propellant Sargeant fourth stage as a satellite carrier. On January 31, 1958, some three months after the directive and through an intense concerted effort, the Army-JPL team's Juno 1 rocket successfully launched Explorer 1 from Cape Canaveral, Florida.

The roles of the ABMA and JPL were paramount in the rapid production of the satellite Explorer 1. Acting in concert with physicist James A. Van Allen, researchers had assembled the instrument payload

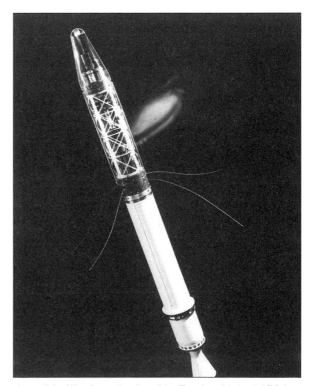

*A model of Explorer 1, placed in Earth orbit in 1958 by a Jupiter-C rocket launched from Cape Canaveral. The satellite discovered the first of two radiation belts surrounding the Earth.* (NASA)

for Explorer 1 and prepared it for launch. Tracking and telemetry systems (known as Microlock) had been tested, and a late January, 1958, launch window at Cape Canaveral had been secured before the next scheduled Vanguard mission (February 5, 1958).

Of the first seven Explorer missions, only five (Explorers 1, 3, 4, 6, and 7) achieved orbit. Explorer 1 was hurled into an orbit with a revolution period of 114 minutes, an initial apogee (farthest distance from Earth) of 2,531 kilometers, and an initial perigee (nearest distance to Earth) of 360 kilometers. Explorer 1, a cylindrical satellite 2 meters tall and 15.2 centimeters in diameter, was mounted atop the fourth-stage Sargeant. With a total weight of 13.97 kilograms, it contained an 8.22-kilogram instrument payload with four antennae and included radio transmitters and power necessary to reach Earth-based monitoring stations. The instruments included a cosmic-ray detector, micrometeorite sensors, and temperature sensors to measure both inside and outside temperature changes.

Explorer 2, a satellite similar to Explorer 1, failed to reach orbit on March 5, 1958, because the fourth stage failed to ignite. Three weeks later, on March 26, Explorer 3 was launched into an orbit with a 2,800-kilometer apogee and a 191.5-kilometer perigee; the orbit decayed and forced reentry after three months, on June 28. The Explorer 3 instrument payload was nearly identical to that of Explorer 1, helping to reproduce and, therefore, verify earlier measurements. On July 26, 1958, Explorer 4 was placed in orbit with an initial apogee of 2,209 kilometers, an initial perigee of 262 kilometers, and a 110-minute orbital period. With a larger payload than its predecessors (11.7 kilograms), Explorer 4 contained radiation and scintillation counters, as well as internal temperature sensors. Explorer 5 failed to reach orbit on August 24, 1958, because of damages incurred during first-stage separation; it plummeted to Earth after eleven minutes.

The National Aeronautics and Space Administration (NASA), created in October, 1958, took control of U.S. space activities, including the Explorer and Vanguard programs. Hence, the next Explorer mission occurred under the new agency's auspices.

Explorer 6 was a 4.20-kilogram, 3.65-meter-diameter inflatable satellite (it resembled a balloon) designed to offer high visibility for Earth-based trackers, allow radar targeting, and provide atmospheric density data. Fired atop a Juno 1 (Jupiter-C) booster with an experimental "apogee kick" fifth stage, Explorer 6 (Beacon 1) failed to reach orbit on October 22, 1958, because of some vibration resulting from third-stage cluster rotation.

During a research and development period at NASA lasting roughly one year, researchers attempted to correct problems found during the previous launch failure. The newer rocket design, named Juno 2, suffered an initial failure on July 16, 1959. In an attempt to launch a multiple-experiment payload (identical to the future Explorer 7), the rocket experienced rapid attitude variations and exploded five seconds after liftoff. A second Explorer 6 attempt, with identical instrument payload, was launched successfully on August 7, 1959, atop a Thor-Able 3 rocket. A Juno 2 did successfully launched the Explorer 7 on October 13, 1959, into an orbit with an initial period of 101.3 minutes, an apogee of 1,096 kilometers, and a perigee of 555 kilometers. Instruments on board the 41.5-kilogram satellite were designed to measure energetic particles and radiation and micrometeorite bombardment. Explorer 7 returned data on Earth's magnetic field and solar flares until July 24, 1961.

**Contributions**

Throughout the last half of the 1940's, many sounding rockets (short-duration subvertical launches to measure physical atmospheric conditions) were fired to perform measurements in the upper atmosphere. By the mid-1950's, prevailing expert opinion supported the necessity of orbital satellites to provide long-duration measurement of atmospheric change rather than the momentary data provided by spot-fired sounding rockets. Destined to set a design precedent for many future classes of spacecraft, the Explorer missions in the late 1950's fulfilled the United States' obligation to participate in the International Geophysical Year, restored pride in the American space effort, and covered altitudes and latitudes far in excess of contemporary Vanguard missions. In combination, the Explorer and Vanguard programs provided new data on Earth's near-space environment, a body of knowledge to be refined and expanded to the solar system and beyond with more sophisticated spacecraft.

Upon orbit, Explorer 1 measured unexpectedly high radiation levels roughly 1,100 kilometers above Earth's surface. Corroborated by Explorer 3 and confirmed by Sputnik 3, the region was named the Van Allen radiation belt, an area high above Earth where cosmic radiation is trapped along the magnetic field lines of Earth.

Micrometeorite sensors detected diurnal variation in meteorite flux and showed that particles between 4 and 9 micrometers in diameter were ten times more likely to strike the sensors than were larger particles. After twelve days of operation, the mass of material striking Earth daily was estimated at 10 million kilograms. Temperature fluctuations were measured both inside and outside the Explorer 1 satellite; they confirmed the success of methods employed to control internal temperature.

Explorer 3, unlike its earlier counterpart, contained a tape recorder that allowed for more complete data transmission. Only 15 percent of the telemetry from Explorer 1 was received by ground stations, while the improved methods of Explorer 3 allowed transmission of 80 percent of its data. The mission added more data to those collected by Explorer 1. Explorer 4 measured, within a wider range of altitude and latitude, cosmic rays and trapped radiation of the Van Allen radiation belt. As confirmed later by Pioneer spacecraft, Explorer 4 discovered two subdivisions within the Van Allen radiation belt.

The Explorer 6 satellite balloon provided atmospheric density distribution data, measured the radiation belt and micrometeorite bombardment, and served as a radar target for ongoing tracking and geodetic research. The instrument package of Explorer 7 suffered from intense saturation of radiation electrons, thus limiting results. One of three micrometeorite sensors was damaged during launch and produced erroneous data. Still, Explorer 7 returned so much useful information that processing it all required more than six years. The spacecraft measured heat and radiation variation in Earth's atmosphere and meteorological changes resulting from such variation. In addition, correlations were documented between auroral activity and trapping of charged particles along Earth's magnetic field lines.

### Context

The early Explorer program, developed during a period of pioneering rocket research, greatly affected the future of uncrewed and crewed spaceflight. The Explorer program advanced the art and science of American rocketry. Expanded growth in the fields of space vehicle engineering, construction, propulsion, and launch technique were predictable outcomes of the Explorer program. It provided a needed boost in morale with the launch of Explorer 1, and, in interfacing with Project Vanguard, added to the strength of both programs. Major advances in electronic circuitry design and miniaturization, missile guidance, tracking, telemetry, and antennae systems were accomplished by the ABMA-JPL team during the formative stages of the Explorer program. The Redstone, Jupiter-C, and Juno 1 rockets were progenitors of the Juno 2, Juno 5, and Saturn missile systems, a series of successful rockets destined to take humankind to the Moon.

In terms of satellite technology, the Explorer 1 led the way for a legacy of long-lasting interplanetary Explorer-class spacecraft and the future development of Ranger, Surveyor, Lunar Orbiter, Mariner, Pioneer, and Voyager spacecraft. The combined data from such missions embraced and developed the science of planetology. Earth-directed investigations heightened environmental awareness and allowed a new view of Earth to be globally shared.

**See also:** Explorers: Air Density; Explorers: Astronomy; Explorers: Atmosphere; Explorers: Ionosphere; Explorers: Micrometeoroid; Explorers: Radio Astronomy; Explorers: Solar; Vanguard Program.

## Further Reading

Bergaust, Erik, and William Beller. *Satellite!* Garden City, N.Y.: Hanover House, 1956. Details the planning for satellite launches during the International Geophysical Year. Now dated by post-IGY satellite development, this volume describes the visionary goals of project scientists for the early Explorer and Vanguard missions.

Bille, Matt, and Erika Lishock. *The First Space Race: Launching the World's First Satellites.* Austin: University of Texas A&M Lightning Source Titles, 2004. A thorough historical perspective on the Army's efforts to launch a satellite (Explorer 1), the Navy's efforts to launch a satellite (the not-so-well-known NOTS), and the Vanguard program.

Braun, Wernher von, and Frederick I. Ordway III. *Space Travel: A History.* Rev. ed. New York: Harper & Row, 1985. Includes a comprehensive, superbly illustrated history of post-World War II rocket research and abundant tables of data on missiles, satellites, and crewed spacecraft. Contains a detailed bibliography. Essential reading for rocketry and space travel enthusiasts.

Burrows, William E. *This New Ocean: The Story of the First Space Age.* New York: Random House, 1998. This is a comprehensive history of the human conquest of space, covering everything from the earliest attempts at spaceflight through the voyages near the end of the twentieth century. Burrows is an experienced journalist who has reported for *The New York Times*, *The Washington Post*, and *The Wall Street Journal.* There are many photographs and an extensive source list. Interviewees in the book include Isaac Asimov, Alexei Leonov, Sally K. Ride, and James A. Van Allen.

Divine, Robert A. *The Sputnik Challenge: Eisenhower's Response to the Soviet Satellite.* New York: Oxford University Press, 1993. A thorough history of the political and scientific decisions made in regard to the International Geophysical Year and the efforts to launch an American Earth-orbiting satellite.

Green, Constance M., and Milton Lomask. *Vanguard: A History.* NASA SP-4202. Washington, D.C.: Government Printing Office, 1970. A history of Project Orbiter, describing the people, agencies, and administrative programs that led to the launch of Explorer and Vanguard satellites. Mission goals and successes are described in detail. The appendices list flight summaries of both the Vanguard and Explorer programs and IGY satellite launches. Contains numerous photographs and diagrams.

Hall, R. Cargill, ed. *Essays on the History of Rocketry and Astronautics.* 2 vols. NASA CP-2014. Washington, D.C.: Government Printing Office, 1976. Printed in two volumes and representing a compilation of papers and memoirs written by active participants, this work traces international efforts in rocketry. Volume 2 concentrates on liquid- and solid-propellant rocket research, including the post-World War II era. Includes good accounts of the early phases of the Vanguard and Explorer projects.

Heppenheimer, T. A. *Countdown: A History of Space Flight.* New York: John Wiley, 1997. A detailed historical narrative of the human conquest of space. Heppenheimer traces the development of piloted flight through the military rocketry programs of the era preceding World War II. Covers both the American and the Soviet attempts to place vehicles, spacecraft, and humans into the hostile environment of space. More than a dozen pages are devoted to bibliographic references.

Lee, Wayne. *To Rise from Earth: An Easy to Understand Guide to Spaceflight.* New York: Checkmark Books, 1996. This is a good introduction to the science of spaceflight. Although written by an engineer with the NASA Jet Propulsion Laboratory, it is presented in easy-to-understand language. In addition to the theory of spaceflight, it gives some of the history of the human endeavor to explore space.

Mari, Christopher, ed. *Space Exploration.* New York: H. W. Wilson, 1999. Twenty-five articles (reprinted from magazines), covering the current state of the space program, are divided into five sections: John H. Glenn, Jr.'s return to space, the exploration of Mars, the International Space Station, recent mining efforts by commercial industries, and new types of space vehicles and propulsion systems.

Neufeld, Michael J. *The Rocket and the Reich: Peenemünde and the Coming of the Ballistic Missile Era.* Cambridge, Mass.: Harvard University Press, 1996. This is the story of the Nazi terror from the sky known as the V-2, tracing its development at Peenemünde, a remote island off the Baltic Coast. In the midst of World War II, von Braun and his team of scientists and engineers developed a weapon of mass destruction that was used by Hitler to wreak havoc on the British Empire. Later, the V-2 was used to develop the ballistic missiles of both the United States and the Soviet Union.

Ordway, Frederick, III, and Mitchell R. Sharpe. *The Rocket Team.* New York: Thomas Y. Crowell, 1979. This work offers a detailed account of the capture, testing, and utilization of the German V-2 rocket cache from Europe. It includes a section on the ABMA launch of Explorer satellites and contains an extensive bibliography on German rocket research.

Piszkiewicz, Dennis. *The Nazi Rocketeers.* Westport, Conn.: Praeger Trade, 1995. Tells the story of Hermann Oberth, Wernher von Braun, and their colleagues as they progressed from the innocent dream of space travel through the development of the V-2 ballistic missile to the transfer of their technological legacy to the Americans. Discusses the somewhat controversial issue of von Braun's involvement with the Nazis and the reasons he and most of his associates came to the United States. References, notes, maps, and photographs.

_____. *Wernher von Braun.* Westport, Conn.: Praeger Trade, 1999. This biographical look at the developer of modern missile technology tells both the public side of von Braun—his contributions to modern rocketry—and his less-well-known activities as a member of the Nazi Party. Fair treatment is given to both his career as a rocket pioneer and his complicity in the Nazi cause.

Stuhlinger, Ernst. *Wernher von Braun, Crusader for Space: A Biographical Memoir.* Melbourne, Fla.: Krieger Publishing, 1996. This is an account of the life and times of Wernher von Braun, focusing on the work he did to get America to the Moon. He also recounts his days developing the launch vehicles for America's first satellites.

Zimmerman, Robert. *The Chronological Encyclopedia of Discoveries in Space.* Westport, Conn.: Oryx Press, 2000. Provides a complete chronological history of all crewed and robotic spacecraft and explains flight events and scientific results. Suitable for all levels of research.

*Charles Merguerian*

# Extravehicular Activity

*Date:* Beginning March 18, 1965
*Type of program:* Piloted spaceflight

*On March 18, 1965, when Russian Cosmonaut Alexei Leonov became the first human to leave a spacecraft and venture into space, humankind moved decisively toward the ability to build space stations and establish colonies in hostile space environments.*

## Key Figures

*Alexei Leonov* (b. 1934), Voskhod 2 pilot who performed the first spacewalk

*Edward H. White II* (1930-1967), Gemini IV pilot who performed the first American spacewalk

*Neil A. Armstrong* (b. 1930), Apollo 11 commander and first lunar walker

*Edwin E. "Buzz" Aldrin, Jr.* (b. 1930), Apollo 11 Lunar Module pilot and second lunar walker

*Bruce M. McCandless II* (b. 1937), STS 41-B mission specialist and first free-flying astronaut

## Summary of the Technology

After humans burst the bonds of Earth and ventured into space in 1961, much work remained to be done. Yuri A. Gagarin's single orbit of Earth on April 12, 1961, was the modest beginning of a space program that would, in time, revise human conceptions of the universe and expand human aspirations and possibilities.

Once it was established that humans could live in space for extended periods and could survive the environment of "weightlessness," or microgravity, to which voyagers in space are subjected, such projects as the physical exploration of the Moon and distant planets and the establishment of orbiting space stations became feasible. In order for people to survive in space, portable environments comparable to Earth's atmosphere had to be created. This hurdle was overcome by the middle of the twentieth century, when airplanes were pressurized so that they could fly at heights where oxygen was insufficient to sustain human respiration. Spacecrafts were similarly pressurized to replicate Earth's atmosphere.

The first piloted spaceflights began in 1961. These flights did not initially enable astronauts to leave their spacecrafts or to engage in the extravehicular activities (EVAs), or spacewalks, that were crucial to achieving the next stage of space exploration and development.

On May 25, 1961, President John F. Kennedy announced the national goal of placing a human on the Moon and returning that human safely to Earth before 1970. As impossible as such a goal first seemed to many people, it was achieved on July 20, 1969, when Neil A. Armstrong became the first human to set foot on the Moon and be returned safely to Earth, along with the spacecraft's crew. President Kennedy's goal could not be achieved until advances had been made that would permit astronauts to leave their spacecrafts and enter the eerie environment of space. In this environment, objects and people are in free fall, traveling at the same speed as their spacecraft, and appear to float. The environment of space cannot support human life. The heat from the Sun reaches temperatures above the boiling point; where the Sun's rays are absent, temperatures fall below −100° Celsius (−212° Fahrenheit).

Before humans could leave the safe, controlled environment of the spacecraft carrying them into orbit, a comparable controlled environment had to be devised in which they could encase themselves. America's seven Mercury astronauts were outfitted with pressurized space suits in 1959, but such space suits were cumbersome and inflexible. They provided safety if pressure inside the cabin failed, but orbiting astronauts wore them uninflated during their trips in space.

The earliest space suits used by American astronauts consisted of an inner layer of rubberized fabric to keep air inside the suit and an outer layer of aluminized nylon fabric that would not stretch.

Space suits were tailor-made for each astronaut. They had metal rings at the elbows and knees to permit bending, but bending caused an uncomfortable increase in pressure.

Spacecraft also had to be designed so that astronauts engaged in EVAs could enter the vacuum outside their spacecraft without compromising the spacecraft's interior atmosphere. In order to exit the spacecraft, astronauts must enter an air lock located between the crew's cabin and the payload bay. Here they don their space suits, after which air is pumped from the air lock prior to their leaving the vehicle. Space suits are cooled by water pumped through them. Most have enough power and oxygen to permit astronauts to work outside for about eight hours.

*On June 3, 1965, Edward H. White II performed the first American "spacewalk" for 23 minutes attached to the spacecraft by a 25-foot umbilical line and a 23-foot tether line. He carried a Hand Held Self Maneuvering Unit (HHSMU), used to move about in weightlessness.* (NASA James McDivitt)

When Alexei Leonov made his first spacewalk in 1965, he was tethered to his spacecraft. He remained outside the vehicle for twelve minutes. When American Astronaut Edward H. White II made his spacewalk ten weeks later, he, too, was tethered to Gemini, the vehicle that brought him to space. Whereas Leonov carried a backpack that provided him with oxygen, White's oxygen came from Gemini through the tether attaching him to the craft. His spacewalk lasted for twenty minutes, during which he tested a handheld maneuvering device whose jets of gas propelled him.

Subsequent astronauts were tethered to the mother ship until the 1969 Apollo mission that put Neil A. Armstrong and Edwin E. "Buzz" Aldrin, Jr., on the Moon. Using backpacks for power and oxygen, they and the Moon walkers who followed them were not tethered. In February, 1984, Bruce M. McCandless II became the first free-flying astronaut who, by using the Manned Maneuvering Unit (MMU), was able to propel himself in space.

In recent years, untethered astronauts have performed complicated chores in space, such as repairing the initially flawed Hubble Space Telescope and making repairs to their own spacecraft. Space suits have become increasingly flexible, permitting considerable movement. They are no longer tailor-made, although the gloves the astronauts wear must be. Free-flying astronauts have made repairs to the Russian space station Mir, which was launched in 1986.

With the construction of the International Space Station, begun in 1999, EVAs have become increasingly commonplace. Untethered EVAs provide the only way to bring to fruition the highly complex activities in space that are now in their infancy. To prevent an astronaut from accidentally drifting off into space, extensible leashes, known as

tethers, are attached to the spacecraft and EVA crew member. While working outside the International Space Station, astronauts wear a self-rescue propulsion device called the Simplified Aid for EVA Rescue (SAFER).

**Contributions**

An incalculable amount of knowledge has been gained through EVAs that have been a fundamental part of the international space program since the late 1950's, when plans were being made by both the United States and the Soviet Union for the first spacewalks. Alexei Leonov's initial excursion outside a space vehicle was of such short duration that it accomplished little except to demonstrate that humans were capable of spacewalks, as they are called, although that designation is somewhat misleading. Humans float rather than walk in space.

*On shuttle mission STS-101 (May, 2000), astronaut James S. Voss handles the main boom of the Russian crane Strela that was connected to its operator post.* (NASA)

Ten weeks after Leonov's first venture into the great vacuum of space, Edward H. White II remained on his tether, floating in space for almost twice as long as Leonov had. White had the added advantage of being able to move about propelled by a maneuvering unit whose oxygen jets moved him in the directions he desired.

This led to the development of the Manned Maneuvering Unit (MMU), which in 1984 enabled Bruce M. McCandless II to float freely in space without a tether and to propel himself as he desired. The MMU made the rescue and on-orbit repair of satellites feasible. Without this achievement, it is doubtful that the development of sophisticated space stations would be possible. The astronauts who construct such composites as the International Space Station must be free to move about and must have the means of controlling where they go in weightless environments.

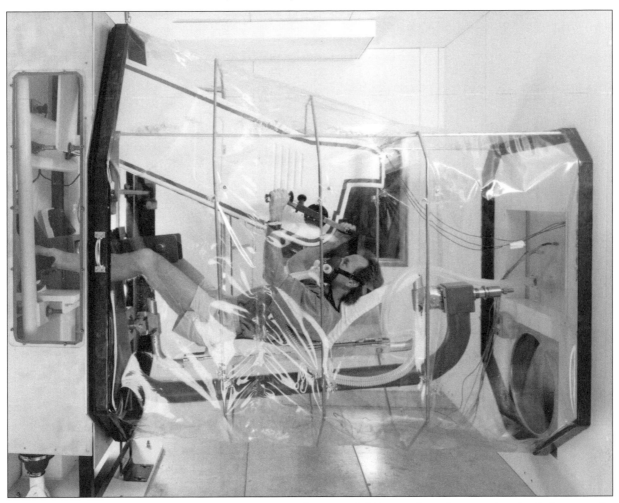

*Astronauts on long-duration spaceflights must exercise during weightlessness in order to maintain skeletomuscular strength. This EVA Exercise Device at Ames Research Center simulates exertion in weightlessness.* (NASA Tom Trower)

The tether that connected White to his spacecraft also brought him the oxygen he needed to live. A tether such as he used not only limits the range of the tethered person but also creates possible hazards that could result in cutting off the oxygen supply.

Space suits had to be redesigned so that they permitted freer motion than the early space suits did but also to be sufficiently self-contained to provide power, insulation, water, and oxygen for the extended periods that astronauts working on complex projects require. As designers worked on making more practical space suits, they found means of making it possible for those who wear them to bend more freely.

Early spacecraft were powered internally by large, heavy batteries that increased their weight to an extent sufficient to reduce the amount of other equipment that could be sent into space at liftoff. Later spacecraft have been powered by light, efficient fuel cells that are fueled by the interaction of oxygen and hydrogen, whose by-product is water that can be used on spacecraft for drinking, bathing, and other purposes. An advance of this sort permits longer, more productive spaceflights than were conceived in the earliest days of the space program.

The operation of the Soviet Union's Mir Space Station demonstrated that humans could exist for extended periods of time in space and that international cooperation could flourish in the space environment. Repairs to Mir by both American astronauts and Soviet cosmonauts showed that EVAs can have extremely practical outcomes.

When Skylab 1 was hurled into space on May 14, 1973, the initial launch went well. An hour into its flight, however, it was learned that during launch, a vital heatshield had been ripped away. This caused the ties that held the two large solar panels needed to power the spacecraft to break, which in turn resulted in one of the panels being ripped off and the other jammed so that it could not unfurl as it needed to in order to function effectively.

Eleven days later, during the Skylab 2 mission, Skylab's first crew established itself in the space station and immediately went to work repairing the damage that had occurred at launch. Astronauts Joseph P. Kerwin and Charles "Pete" Conrad, Jr., spent four arduous hours outside Skylab, working half the time in darkness as the orbiting space station passed from day to night in its hourly rotations around Earth. In a heroic effort, the two astronauts finally snipped the damaged bolts of one solar panel, which unfurled as had been originally intended.

Moon landings and EVAs on the lunar surface have resulted in bringing to Earth the first samples of lunar dust and rocks. About a thousand pounds have been collected and retrieved, offering scientists a rare opportunity to explore the geological constitution and origins of the Moon. The eternal question of whether there is life on other planets will surely be addressed as EVAs explore these distant outposts.

EVAs have not yet been accomplished on the surfaces of other planets, but small vehicles have been placed on Mars and have sent back both a rich photographic record and valuable information about the Martian surface. It is only a matter of time before robotic vehicles will be on other planets to gather information that should enable scientists to gain a broader view of the origins and extent of the universe.

*During extravehicular activities, crews are able to service, repair, and retrieve satellites in orbit.* (NASA CORE/ Lorain County JVS)

### Context

The seemingly outlandish notion of placing an astronaut on the Moon really grew out of the Space Race in which the United States was involved with the Soviet Union during the early 1960's. The Soviet Union's launching of the first artificial satellite, Sputnik 1, in 1957 left Americans embarrassed and bewildered. The Soviets had surged ahead of the United States in conquering space and were forging plans to land a spacecraft and a cosmonaut on the Moon. EVA was fundamental to the achievement of this goal, whether the victor in the Space Race was the Soviet Union or the United States.

The Soviets made the first soft landing on the Moon and accomplished the first flight around the Moon, which produced photographs of the side of the Moon that no one had previously seen. More important was their placing of two Lunokhod roving vehicles on the Moon's surface. These vehicles roved about on the Moon for more than a year and sent more than one hundred thousand photographs back to Earth.

The Space Race diminished with the break-up of the Soviet Union and the accompanying economic crisis that afflicted Russia. The robotic Soviet EVAs that collected Moon samples and returned them to Earth have yielded splendid scientific results and have also caused many Americans to question whether robotic space exploration, whose price tag is lower than that of piloted exploration, is not the best way to learn more about the planets and other celestial bodies in the solar system.

We now know that humans can not only function in space but also survive in it for extended periods of time as long as conditions are provided that are hospitable to life. Certainly, the space environment is fragile, but people in increasingly flexible space suits can work for long periods in the alien environment of space.

One thing is certain: Extravehicular activity, piloted or robotic, is uncovering almost every day new and exciting challenges for the exploration and development of space. The colonization of space, which could never be accomplished without extensive EVAs, is definitely in the future, and work on the International Space Station is under way.

**See also:** Apollo 11; Apollo 15's Lunar Rover; Apollo-Soyuz Test Project; Cooperation in Space: U.S. and Russian; Gemini Program; International Space Station: Living and Working Accommodations; Manned Maneuvering Unit; Manned Orbiting Laboratory; Russia's Mir Space Station; Skylab Program; Space Shuttle; Space Suit Development.

### Further Reading

Abramov, Isaac P., and A. Ingemar Skoog. *Russian Spacesuits.* London: Springer-Verlag London Limited, 2003. The authors, part of the original Zvezda team that manufactured space suits for the first Russian spaceflights, still play an integral role in space suit research and development. The book covers the technical innovations of the past forty years, which enabled Gagarin's first flight in 1961, the first spacewalk in 1965, and the Mir missions of the 1980's and 1990's, culminating in today's International Space Station.

Burrows, William E. *This New Ocean: The Story of the First Space Age.* New York: Random House, 1998. This highly articulate presentation provides a broad background for those interested in knowing how the first four decades of space exploration evolved and what they achieved. Burrows, who writes with vigor, has a complete grasp of his subject matter and presents it with enthusiasm.

Damon, Thomas D. *Introduction to Space: The Science of Spaceflight.* 2d ed. Malabar, Fla.: Krieger Publishing Company, 1995. This comprehensive presentation is rich in illustrative material. It avoids scientific jargon and should be easily accessible to general readers. The chapters titled "Space Shuttle," "Living in Space," "Working in Space," "Space Stations," and "Colonies on Other Worlds" are particularly relevant because they focus considerably on EVAs. The pictures of items such as the air lock and various types of space suits are excellent.

Kozloski, Lillian D. *U.S. Space Gear: Outfitting the Astronauts.* Washington, D.C.: Smithsonian Institution Press, 1999. Chronicles the evolution of the space suit from those developed for high-altitude flight through those used on the space shuttle. There are more than 150 illustrations. Appendices detail the pressure suits in the Preservation/Study Collections and summarize the U.S. piloted spaceflights from Project Mercury through current space shuttle missions.

Lee, Wayne. *To Rise from Earth: An Easy to Understand Guide to Spaceflight.* New York: Checkmark Books, 1996. The format of this extensive presentation is extremely appealing and makes the book easy to understand. Rich in illustrations, it is especially valuable for its chapters titled "Orbital Mechanics Without the Math" and "The Story of the Race to the Moon." Lee has seven specific discussions of EVAs, each clear and concise. No specialized knowledge is required to comprehend the information this book presents.

Mari, Christopher, ed. *Space Exploration.* New York: H. W. Wilson, 1999. Twenty-five articles in this collection are divided into groups, each begun by an introduction by Mari, from sources like *The New York Times, Newsweek,* and *Aviation Week and Space Technology.* The editor devotes an entire section consisting of five articles to the International Space Station; it is particularly useful for the implications it presents about EVA.

Murray, Bruce. *Journey into Space: The First Three Decades of Space Exploration.* New York: W. W. Norton, 1989. Although this volume does not offer much about EVAs, it is valuable for the background material found in it and for its extensive coverage of space exploration to such distant planets as Mars, Saturn, Uranus, and Neptune.

Shayler, David J. *Walking in Space: Development of Space Walking Techniques.* Chichester, England: Springer-Praxis, 2003. An overview of EVA techniques with reference to original documentation and astronaut interviews.

*R. Baird Shuman*

# Extreme Ultraviolet Explorer

*Date:* June 7, 1992, to January 31, 2002
*Type of program:* Scientific platform

*The Extreme Ultraviolet Explorer (EUVE) unveiled the universe in the portion of the electromagnetic spectrum known as the extreme ultraviolet, the shortest ultraviolet wavelengths. EUVE gathered data on a wide range of astronomical objects at these wavelengths—work impossible to conduct from Earth's surface. In fact, prior to the mission, many astronomers thought that extreme ultraviolet astronomy was not possible at all.*

### Key Figures

*Lennard A. Fisk,* associate administrator, Office of Space Science and Applications

*Charles J. Pellerin, Jr.,* director, Astrophysics Division

*Robert Stachnik,* program scientist

*Yoji Kondo,* project scientist

*Robert Spiess,* EUVE spacecraft manager

*Stuart Bowyer,* EUVE science principal investigator

### Summary of the Satellite

Launched into a near-Earth orbit from Cape Canaveral in June, 1992, by a Delta II rocket, the Extreme Ultraviolet Explorer (EUVE) is also known as Explorer 67. Its purpose: to reveal new views of the universe by opening the last remaining window in the electromagnetic spectrum. This window, the extreme ultraviolet range, is thickly curtained by Earth's atmosphere, which is opaque to extreme ultraviolet wavelengths. In addition, it had long been thought that interstellar hydrogen and helium gas would absorb enough extreme ultraviolet radiation that our view of the universe through this window would be too foggy to be useful.

Extreme ultraviolet is the band in the electromagnetic spectrum falling between the ultraviolet light that burns our skin on a sunny day and the x rays that allow physicians to diagnose our broken bones. Blue light has higher energy and shorter wavelengths than red light. Undetectable by our eyes, ultraviolet light has even shorter wavelengths and higher energies than blue or violet light.

X rays are also forms of electromagnetic radiation, but they have even shorter wavelengths and higher energies than ultraviolet radiation. The extreme ultraviolet is the region of the ultraviolet spectrum with the shortest ultraviolet wavelengths (from roughly 10 to 100 nanometers) and the highest energies. Photons of extreme ultraviolet radiation have enough energy to ionize the atom and be absorbed in the process if they were to strike a neutral hydrogen atom. Therefore, if there are enough interstellar hydrogen atoms in a particular direction, the extreme ultraviolet radiation coming from that direction will be absorbed. Our view in that direction will consequently be obscured.

EUVE's two sets of instruments were mounted on a multimission modular spacecraft (MMS) providing the power and support functions for the instruments. The sets of instruments are the sky survey telescopes and the Deep Survey Spectrometer Telescope. Although it was announced on November 17, 2000, that the National Aeronautics and

Space Administration (NASA) would terminate the EUVE mission sometime in late 2001 or early 2002, the spacecraft would continue to orbit Earth until aerodynamic drag would force it to reenter Earth's atmosphere.

Each of the three scanning telescopes on EUVE had an aperture of 40 centimeters (16 inches), and each was mounted so as to allow simultaneous observation of the same part of the sky. They were mounted perpendicular to the rotational axis of the satellite so that, as the satellite spun three times per orbit, these telescopes would scan a band of the sky 2° wide. In six months, these telescopes surveyed the entire sky at four extreme ultraviolet wavelengths.

The Deep Survey Spectrometer Telescope had the same aperture as the scanning telescopes; however, it pointed parallel to the satellite's rotational axis. This pointing strategy allowed it to observe the same location in the sky for a longer time, thereby increasing the sensitivity of the instrument to fainter objects. Half of the radiation striking this telescope was sent to an imaging instrument to detect extreme ultraviolet sources that were too faint for the scanning telescopes. The rest of the radiation was sent to three spectrometers to study the extreme ultraviolet spectra of the detected objects.

To work in the extreme ultraviolet, these telescopes had to have a different design from that of ordinary optical telescopes, which simply absorb extreme ultraviolet light. The telescopes used aluminum very accurately shaped into an inverted cone reflector to focus the extreme ultraviolet light onto the detectors. The reflector shapes were designed so that the extreme ultraviolet radiation struck them nearly parallel to their surface. This grazing incidence reflected extreme ultraviolet radiation rather than absorbing it, as more normal incident angles would do. For the shortest extreme ultraviolet wavelengths, the aluminum was plated with a thin layer of gold. For longer wavelengths, nickel plating was used. Special detectors were also required to detect the extreme ultraviolet radiation.

The first month after launch of the EUVE was reserved for the initial checkout. Spacecraft systems were checked to make sure that they functioned properly. The telescopes were not opened until the end of this phase, because molecules trapped in the spacecraft materials were outgassing into the vacuum of space—a normal process whenever an object is exposed to a vacuum. If the telescopes were opened before the outgassing was finished, then outgassed molecules would contam-

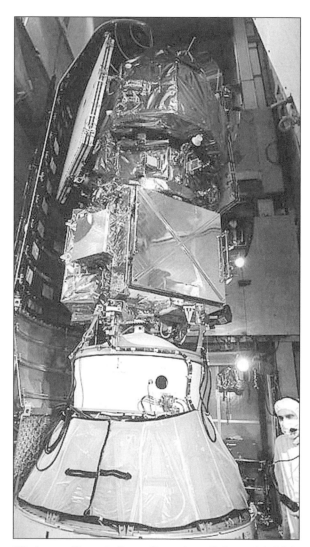

*Workers at Kennedy Space Center install the payload fairing around the Extreme Ultraviolet Explorer (EUVE) mated to a Delta II rocket.* (NASA)

inate the very clean reflective surfaces on the telescopes.

After this initial checkout, six months were devoted to a Full-Sky Survey. The three scanning telescopes surveyed and mapped the entire sky at four separate extreme ultraviolet wavelengths during this time. Simultaneously, a Deep-Sky Survey was conducted in which the Deep Survey Spectrometer Telescope mapped a much smaller portion of the sky, but with greater sensitivity to detect fainter objects. To minimize interference from the Sun, the deep survey telescope was kept pointed in a direction away from the Sun parallel to Earth's shadow. Hence, the portion of the sky observed in the deep survey was a narrow strip directly opposite the Sun during the six-month survey.

Following the initial surveys, EUVE began its continuing mission. Selected objects were studied in detail with the Deep Survey Spectrometer Telescope. These observations were often performed by guest observers selected on the merit of their specific proposals. While these observations were being performed, the scanning telescopes continued to operate and observe whatever crossed their field of view. Initial plans called for a limited lifetime for the EUVE, but it was still operating in the first years of the twenty-first century. An EUVE end-of-mission conference was held in Jennery, California, July 22-24, 2001, to commemorate the accomplishments of this important extreme ultraviolet observatory over the preceding decade. EUVE destructively reentered the Earth's atmosphere on January 31, 2002. It posed no hazard to life or property.

### Contributions

Perhaps the most significant result of EUVE is simply that extreme ultraviolet astronomy is possible. As of 2005, several hundred scientific papers had been published using EUVE data. A 1997 all-sky catalog of objects found by EUVE lists more than five hundred sources. Objects observed by EUVE include more than one hundred cool low-mass stars and white dwarf stars. EUVE has also observed a few very hot and massive stars, active galac-

tic nuclei (AGNs), cataclysmic variables, and solar system objects.

The fact that extreme ultraviolet astronomy is possible reveals something about the Sun's neighborhood. Neutral hydrogen and helium very effectively absorb extreme ultraviolet radiation. Hence, EUVE can detect objects only if the density of hydrogen and helium gas between Earth and the object is greatly reduced. Scientists can observe nearby stars in the extreme ultraviolet because the solar system is located inside a bubble several hundred light-years in diameter, which has a lower-than-normal density of interstellar gas. This bubble is, however, irregularly shaped. It does not extend very far toward the center of the Milky Way, in the direction of the constellations Sagittarius and Scorpius. The bubble extends farther in the direction of the Big Dipper and Orion. The brightest extreme ultraviolet star observed is Epsilon Canis Majoris, which is six hundred light-years distant in the constellation Canis Major, one of Orion's hunting dogs. This observation reveals both that the star is intrinsically very bright in the extreme ultraviolet and that the interstellar bubble extends at least six hundred light-years in that direction.

The stars most commonly observed by EUVE are relatively cool low-mass stars, ranging from a little hotter and more massive than the Sun on down. Cool stars tend to glow at mostly red and infrared wavelengths, in contrast to very hot stars, which glow predominantly at blue and ultraviolet wavelengths. Hence, it would seem that these cool stars should not radiate in the extreme ultraviolet. However, these stars, including the Sun, have an extremely hot but tenuous outer atmosphere (a few million kelvins) called the corona. EUVE allows study of the coronas of these stars, which in some cases are extremely active. On some of these stars, flares have been observed that are thousands of times more intense than solar flares. During a flare, the brightness increases rapidly for a period of several hours.

A large number of white dwarf stars have also been observed by EUVE. White dwarfs are the final corpses of less massive stars. A mass comparable to the Sun's mass is compressed to roughly the size of the Earth. This compression heats the star so that it emits extreme ultraviolet radiation, but the star's small size makes a white dwarf faint and often difficult to detect. EUVE spectra of white dwarfs show a much greater concentration of heavy elements (heavier than hydrogen or helium) near the surface than had previously been expected. If the white dwarf is in a close binary system with another star, material is transferred from the companion onto the white dwarf. This new material on the white dwarf's surface undergoes nuclear fusion reactions that build up until an explosion occurs, creating a cataclysmic variable. In several cases, cataclysmic variables have been studied simultaneously at optical and extreme ultraviolet wavelengths. These studies show how the material accretes onto the white dwarf from the companion.

The irregularities in the interstellar hydrogen gas have even allowed EUVE to observe objects outside the Milky Way galaxy. If an extragalactic object lies in a direction where the interstellar gas is thin all the way to the edge of the Milky Way, then EUVE could observe it. In this way EUVE has been able to take spectra of BL Lac objects. BL Lac objects are very energetic active galactic nuclei that, like quasars, are powered by a supermassive black hole in the nucleus. However, BL Lac objects have featureless spectra that frustrate astronomers' attempts to understand these objects. EUVE has finally obtained spectra of BL Lac objects that show spectral absorption lines. Detailed studies will help us to understand these enigmatic objects.

Closer to home, EUVE has also observed objects within the solar system. EUVE spectra of Mars show helium lines that allow indirect determination of the amount of radioactive decay (which produces helium) in the planet's interior. Jupiter's closest moon, Io, through its volcanic activity, produces a doughnut-shaped ring of gas around Jupiter. EUVE has identified spectral lines of ionized sulfur and oxygen. EUVE ob-

servations of comets, including the well-publicized Hale-Bopp and Hyakutake, have shown that the x-ray emissions from these comets do not originate from the nucleus. Rather, they originate tens of thousands of kilometers closer to the Sun. The Sun is too bright for EUVE to observe it directly, so EUVE observations of the Moon study reflected sunlight.

### Context

Unlike most NASA missions, EUVE was built and operated from a university campus, the University of California at Berkeley. Therefore EUVE has provided unusual educational opportunities relative to other NASA missions. Undergraduate students have been involved in various aspects of the EUVE mission operations and science.

Because it was believed that interstellar hydrogen gas would absorb extreme ultraviolet radiation, most astronomers thought that extreme ultraviolet astronomy would never be possible. A few, however, were not convinced. In the late 1960's, Stuart Bowyer at the Berkeley campus of the University of California learned that the interstellar gas might be distributed unevenly enough to make extreme ultraviolet astronomy possible. He started a research group in extreme ultraviolet astronomy and eventually the Center for Extreme Ultraviolet Astrophysics at Berkeley, which operated EUVE. Before building and launching EUVE, it was necessary to show that such a project would be fruitful. Bowyer's group, with the support of NASA, started small by launching sounding rockets. Each flight allowed five minutes above the atmosphere to test newly developed technology for detecting extreme ultraviolet radiation.

In 1975 the Apollo-Soyuz Test Project carried an early extreme ultraviolet telescope built by Bowyer's group. By observing thirty stars for a total of 20 hours, this mission located four extreme ultraviolet sources. They included a very active corona star, two white dwarfs, and a cataclysmic variable. These Apollo-Soyuz observations were very limited, but they did prove that extreme ultraviolet astronomy was at least possible.

These results provided justification for the initial funding to begin work on EUVE as part of NASA's Explorer program. NASA continued to provide support despite a 1979 recommendation by the National Academy of Sciences to eliminate EUVE. It was incorrectly thought that at most only a dozen EUVE sources would be detected. The Berkeley group persisted, despite the general skepticism about its work. It developed the telescope and detectors needed to detect extreme ultraviolet radiation.

In 1990 ROSAT (the Roentgen X-Ray Satellite) was launched primarily for the purpose of x-ray astronomy. In addition, ROSAT carried a British instrument, the Wide Field Camera, that surveyed the sky at wavelengths on the border between extreme ultraviolet and x rays. This survey identified a few hundred sources.

These efforts were the first, tentative steps toward extreme ultraviolet astronomy. The 1992 launch of EUVE, then, represents the first real concentrated effort at extreme ultraviolet astronomy. Results from the EUVE mission have proved the skeptics wrong: The extreme ultraviolet window provided a new view of many astronomical objects. Opening this new window in the electromagnetic spectrum has revealed fascinating information about the universe.

**See also:** Apollo-Soyuz Test Project; Explorers: Astronomy; Explorers 1-7; Extravehicular Activity; Far Ultraviolet Spectroscopic Explorer.

## Further Reading

Bowyer, Stuart. "Extreme Ultraviolet Astronomy." *Scientific American* 271 (August, 1994): 32-39. This article provides a good summary of the development of EUVE and its early results through 1994. It also has a good inside view: Its author is one of the major pioneers of extreme ultraviolet astronomy and developers of EUVE.

Bowyer, Stuart, Roger Malina, and Bernhard Haisch. "Observing a Partly Cloudy Universe." *Sky and Telescope* 88, no. 6 (December, 1994): 36-40. Also written by EUVE developers, this article provides the same inside view as the article cited above. However, it concentrates less on the history and development of EUVE and places a correspondingly greater emphasis on results from the EUVE mission.

Chien, Philip. "EUVE Probes the Local Bubble." *Sky and Telescope* 83, no. 2 (February, 1992): 161-163. This article, written prior to the launch, describes EUVE, the mission plan, and what astronomers hoped to learn from the mission.

Cowen, R. "Exploring the Extreme Ultraviolet: A NASA Satellite Probes the Local Bubble." *Science News* 141, no. 21 (1992): 344-346. Written just before launch, this article discusses the plan for EUVE and the science behind its mission.

Haisch, Bernhard, Stuart Bowyer, and Roger F. Malina. "The Extreme Ultraviolet Explorer Mission: Overview and Initial Results." *Journal of the British Interplanetary Association* 46, no. 9 (September, 1993). This special issue is entirely devoted to EUVE. This lead article, as its title suggests, provides a good overview. The other articles in the issue, providing more specific details, are in many cases more technical. One of the less technical articles summarizes contributions that undergraduate students made to the EUVE mission.

Leverington, David. *New Cosmic Horizons: Space Astronomy from the V2 to the Hubble Space Telescope.* New York: Cambridge University Press, 2001. Explores the development of space-based astronomical observations from the end of World War II to the Hubble Space Telescope and other major NASA space-based observatories.

Tassoul, Jean-Louis, and Monique Tassoul. *A Concise History of Solar and Stellar Physics.* Princeton, N.J.: Princeton University Press, 2004. A comprehensive study of the historical development of humanity's understanding of the Sun and the cosmos, written in easy-to-understand language by a pair of theoretical astrophysicists. The perspective of the astronomer and physicist are presented.

Zimmerman, Robert. *The Chronological Encyclopedia of Discoveries in Space.* Westport, Conn.: Oryx Press, 2000. Provides a complete chronological history of all piloted and robotic spacecraft and explains flight events and scientific results. Suitable for all levels of research.

*Paul A. Heckert*

# Far Ultraviolet Spectroscopic Explorer

*Date:* Beginning June 24, 1999
*Type of program:* Scientific platform

*The Far Ultraviolet Spectroscopic Explorer (FUSE) is an orbital telescope designed to study wavelengths of ultraviolet light that are inaccessible from either the ground or the Hubble Space Telescope. FUSE is approximately ten thousand times more sensitive than the only previous satellite to study these wavelengths. It investigates questions related to our origins: What were the conditions just after the Big Bang, what are the properties of interstellar clouds that will eventually form solar systems, and how do galaxies evolve?*

## Key Figures

*H. Warren Moos*, principal investigator
*John Hutchings*, Canadian project scientist
*Alfred Vidal-Madjar*, French project scientist

## Summary of the Mission

The Far Ultraviolet Spectroscopic Explorer (FUSE) was launched on June 24, 1999, atop a Boeing Delta II rocket from Cape Canaveral, Florida, to an orbit about 768 kilometers above Earth's surface.

The FUSE mission is part of the National Aeronautics and Space Administration (NASA) Explorer program and was first authorized by NASA on October 5, 1989. FUSE was developed and is primarily operated by The Johns Hopkins University; however, NASA's Goddard Space Flight Center (GSFC), the University of Colorado, the University of California at Berkeley, the Canadian Space Agency (CSA), and France's Centre National d'Études Spatiales (CNES) also collaborated on the project. The team at Johns Hopkins has primary responsibility for managing and planning the science and operations. GSFC provides technical oversight and some administration; it also coordinates the Guest Investigator Program. The CSA contributed the Fine Error Sensor. CNES contributed the telescope diffraction gratings that divide the ultraviolet light into its component wavelengths. The universities in Colorado and Berkeley contributed to the far-ultraviolet instrument development.

FUSE is controlled from an operating station located at the University of Puerto Rico at Mayagüez. Each FUSE orbit takes about 100 minutes. It is in contact with the ground station for about 10 minutes during some of the orbits. Typically this contact will occur during six orbits out of every fourteen orbits. During these brief contact periods, ground controllers must transmit instructions for the spacecraft to operate on its own and receive data collected during previous operations.

The 1,360-kilogram, 5.5-meter-long satellite consists of a 580-kilogram spacecraft component and a 780-kilogram instrument component. The spacecraft allows scientists to point and track the telescope to the desired location in the sky by means of a three-axis inertially stabilized pointing system. Two gyroscopes provide the stability for each axis to allow redundancy in the event of a gyroscope failure. The spacecraft also contains a Fine Error Sensor (FES) to maintain pointing stability to within an accuracy of 0.5 arc second. (An arc second is an angle equivalent to $\frac{1}{3,600}$ of a degree.) The FES consists of a camera that allows selecting a particular star, or other object, and accurately pointing to that object.

The instrument component consists of an ultraviolet telescope. Visible light is the form of electromagnetic radiation to which human eyes are sensitive. Red light has a longer wavelength and lower energy than blue or violet light. Infrared light and all types of radio waves are also forms of electromagnetic waves, with longer wavelengths and lower energies than red light. Toward the other end of the electromagnetic spectrum, ultraviolet light, x rays, and gamma rays have shorter wavelengths and

higher energies than violet light. For comparison, midrange visible light has a wavelength of about 500 nanometers, comparable in size to a typical single biological cell. FUSE is sensitive to the ultraviolet wavelength range from 90 to 120 nanometers. The Hubble Space Telescope is not sensitive to ultraviolet wavelengths shorter than 115 nanometers, and ultraviolet wavelengths shorter than 300 nanometers cannot be observed from the ground.

The telescope optical package consists of four 39-centimeter-by-35-centimeter rectangular mirrors. The light from these mirrors reflects to diffraction gratings that provide high-resolution spectra. These ultraviolet spectra are analogous to rainbows produced by passing visible white light through a prism, but they display much shorter wavelengths and higher resolution. The four diffraction gratings then send the ultraviolet light to two detectors. A detector coated with silicon carbide (SiC) allows scientists to detect the light from 90 to 110 nanometers, and a detector coated with lithium fluoride (LiF) is sensitive from 100 to 120 nanometers. The FUSE telescope does not take pictures. Rather, it measures the ultraviolet spectrum of the object it is observing. The spectral data consist of measurements of the relative brightness of the object at all the ultraviolet wavelengths in the range from 90 to 120 nanometers. The spectral wavelength resolution is about 0.003 nanometer.

The FUSE primary mission lasted for about three and one-half years. During the primary mission phase, only half of the time on FUSE was available to the general astronomical community. The remainder was reserved for the FUSE science team. On April 1, 2003, FUSE entered the extended mission phase, when all of the time became available to the entire astronomical community on the basis of competitive proposals. After completion of the primary mission, the control center in Puerto Rico no longer operated around the clock. Operating costs were reduced by automating some of the control functions.

In addition to the six gyroscopes, the spacecraft guidance and pointing system contains four reac-

*The Far Ultraviolet Explorer in a clean room in June, 1999.* (NASA)

tion wheels. These reaction wheels consist of electric motors spinning flywheels. Accurately accelerating or braking these flywheels causes the spacecraft to spin a small amount, just as a child walking on a playground merry-go-round will cause it to spin in the opposite direction. Hence ground controllers can accurately adjust the direction in which the telescope points without using irreplaceable rocket propellant. The reaction wheels are configured so that any three can point the telescope along the three axes. The fourth is a backup in case of the failure of one of the reaction wheels.

During December, 2001, the second of these four reaction wheels failed. Normally such a failure would doom the rest of the mission; however, mission scientists and engineers worked intensely for two months and devised a substitute procedure for pointing the telescope. Magnetic torquer bars, not originally designed for this purpose, were used with special new software in place of the reaction wheels. Two years later, in July of 2003, mission engineers uploaded new software to the FUSE computers. The new software allows operating and pointing the telescope using the Fine Error Sensor (FES) rather than the normal guidance system. Hence if there are future guidance system failures, the telescope will still be operable. These adjustments represent a major step forward in learning to operate precision-pointed spacecraft without the normal gyroscopic guidance systems. They may therefore prove useful in the event of guidance system failures in future space missions.

In June of 2004, FUSE celebrated its fifth anniversary. During the first five years of operation, FUSE had collected nearly 50 million seconds of scientific data on 2,200 different astronomical objects. These data have led to a few hundred significant scientific papers. On December 27, 2004, FUSE suffered a major malfunction that threatened to halt science operations. One of the two remaining reaction wheels, the one for the roll axis, stalled. FUSE remained in a safe mode for nearly three months, but on March 29, 2005, the control team reopened the telescope's door and the following day achieved first light for the third time. A

*The FUSE ground station in Mayaguez, Puerto Rico. The semispherical object is the protective Radome.* (Tom Ajluni, Swales Aerospace)

spectrum of a previously examined object, the central star in planetary nebula IC2448, was taken to compare present scientific capability with previous observations.

### Contributions

The primary questions to be addressed by FUSE concern our origins. However, the Guest Investigator Program allows any qualified scientist to investigate any other questions that FUSE can help answer.

The ultimate questions about our origins relate to the origin of the entire universe. During the Big Bang, hydrogen, helium, and trace amounts of lithium and beryllium were made; other elements were made later by stars. Normal hydrogen contains just one proton in the nucleus; however, the isotope of hydrogen containing one proton and one neutron, deuterium, was also made during the Big Bang. The exact amount of deuterium made during the Big Bang was very sensitive to conditions such as temperature and pressure a few minutes after the Big Bang. Hence, accurately measuring the amount of deuterium now can tell us about the conditions just after the universe formed.

FUSE scientists have measured the amount of deuterium in the interstellar medium out to about three thousand light-years from the Sun. It is slightly less than previously thought. Eventually these measurements will help cosmologists pin down the conditions shortly after the Big Bang. These measurements, along with FUSE measurements of oxygen and nitrogen, in the interstellar medium will also help astronomers understand the chemical evolution of the Milky Way galaxy.

Also related to our origins, stars, including our own Sun and solar system, form out of the interstellar medium, gas and dust between the stars. Hence, the FUSE measurements of oxygen and nitrogen in the interstellar medium will also help us understand how stars form. FUSE observations made the first discovery of molecular nitrogen in the interstellar medium. Planets, including our own, form from material left over after the star forms. FUSE has found evidence of planets just beginning to form around the star 51 Ophiuchi, which at only 300,000 years old is very young. These observations show that planets can form fairly rapidly. Knowing how fast planets can form will help astronomers estimate the number of planetary systems in the Milky Way galaxy.

FUSE has also discovered a large, hot halo of invisible gas surrounding the Milky Way. It is called the galactic corona, in analogy to the Sun's corona. By observing quasars (the very distant early stages of some galaxies), FUSE has also observed a considerable amount of intergalactic medium—gas between galaxies. One current mystery to modern astronomers is the fact that 90 percent of the mass in the universe is unaccounted for and therefore referred to as the "dark matter." These FUSE observations may account for some of this dark matter.

Other interesting FUSE discoveries include coronal activity around red giant stars and molecular hydrogen in the atmosphere of Mars. FUSE observations also provide or confirm other observations about Mars's atmosphere.

**Context**

The information provided by FUSE was not possible to obtain prior to this mission. Prior to the Space Age, it was not possible to study astronomical objects at ultraviolet wavelengths because Earth's atmosphere is opaque to most ultraviolet wavelengths. FUSE observes a spectral range from 90 to 120 nanometers. The Hubble Space Telescope does not observe ultraviolet wavelengths shorter than about 115 nanometers. The Extreme Ultraviolet Explorer (EUVE) observed ultraviolet wavelengths from 7 to 76 nanometers. The International Ultraviolet Explorer (IUE), during its mission, observed the range from 115 to 320 nanometers. FUSE, therefore, is the only mission observing its particular portion of the ultraviolet spectrum.

The far-ultraviolet portion of the electromagnetic spectrum observed by FUSE was previously observed by Copernicus (the third Orbiting Astronomical Observatory, or OAO 3) in the 1970's. However, FUSE is ten thousand times more sensitive than Copernicus was. The Hopkins Ultraviolet Telescope (HUT) and the Orbiting Retrievable Far and Extreme Ultraviolet Spectrometer (ORFEUS) also observed this spectral regime. However, these missions were flown affixed inside the space shuttle payload bay and therefore provided only brief glimpses. The brief glimpses and lower sensitivity of these missions did not allow the possibility of the discoveries made by FUSE. FUSE will extend our understanding of the universe by greatly enhancing the ability to observe in the ultraviolet.

**See also:** Extreme Ultraviolet Explorer.

**Further Reading**

Chaisson, Eric J., and Steve McMillan. *Astronomy Today*. 5th ed. Upper Saddle River, N.J.: Pearson Prentice Hall, 2004. This introductory college textbook contains information about FUSE and other telescopes in the chapter on telescopes. Other chapters also contain astronomical background information that relates to various FUSE discoveries.

Cowen, R. "Is This Young Star Ready to Form Planets?" *Science News* 160, no. 21 (2001): 326. This brief news article discusses FUSE observations of planets forming around the young star 51 Ophiuchi.

_____. "Milky Way Galaxy: Cloaked in a Hot Shroud?" *Science News* 161, no. 2 (2002): 21. This brief news article describes the discovery of the previously unseen halo surrounding our galaxy.

_____. "UV Telescopes: One Dead, One Revived." *Science News* 161, no. 7 (2002): 110. This article describes the problems with FUSE's guidance system and the steps taken to correct the problems.

Far Ultraviolet Spectroscopic Explorer home page. http://fuse.pha.jhu.edu. Accessed June, 2005. The official FUSE Web site contains a mission overview, science summaries, and public outreach sections as well as information of interest to scientists using FUSE for their research.

Krasnopolsky, V. A., and P. D. Feldman. "Detection of Molecular Hydrogen in the Atmosphere of Mars." *Science* 294, no. 5,548 (2001): 1914. A technical discussion addressed to the scientific community, this article describes one of the discoveries made by FUSE.

*Paul A. Heckert*

# Funding Procedures of Space Programs

*Date:* Beginning 1957
*Type of issue:* Socioeconomic

*Funding of U.S. space programs reached its high point in the 1960's and then declined. Funding depends on the goodwill of Congress, the needs of the current president, and the popularity and visibility of the programs.*

## Key Figures

*Lyndon B. Johnson* (1908-1973), thirty-sixth president of the United States, 1963-1969
*Richard M. Nixon* (1913-1994), thirty-seventh president of the United States, 1969-1974
*T. Keith Glennan* (1905-1995), first NASA administrator
*James E. Webb* (1906-1992), NASA administrator from 1961 through 1968

## Summary of the Issue

The budgetary process for U.S. space programs has accounted for the difference between the funding requests by the National Aeronautics and Space Administration (NASA) and what it ultimately receives. In the strictest sense, the federal budget is the proposed annual financial plan, which Congress considers, approves, and modifies, and the president signs into law. From it, agencies such as NASA request and obtain the financial authority needed to carry out contemplated programs. Ultimately, it acts as the mechanism for control of these operations.

Thus, the federal budget is more than a simple financial document; it helps set goals. It serves as a benchmark for comparing actual and expected accomplishments, it is the basis for authorizing and appropriating legislation, and it is a record of past negotiations and a preview of programs not yet approved. In short, the federal budget translates substantive programs into dollars and cents.

The absence of coordination of legislative review and approval by a single body accounts for the fragmentary nature of the budgetary process. Overseeing House and Senate committees and subcommittees are semiautonomous units that concentrate on limited areas of the budget. Sub-committees usually make only incremental decisions, once a program is in place.

The principal agent in preparing the unified national budget during NASA's early days was the Bureau of the Budget. In July, 1970, this became the Office of Management and Budget (OMB). Both offices set personnel ceilings, established contracting standards and methods for financial reporting, and kept the president informed of the progress and plans of the various agencies, including NASA.

From the beginning, NASA shared several difficulties with other agencies involved in research and development. First, there were no guidelines for a unified national science policy. Thus, it was difficult to evaluate the advancement of the programs, at least in quantitative terms. All too often, policies were subject to political crises. President Lyndon B. Johnson, for example, was so preoccupied with the Vietnam War that NASA funding eventually took a backseat.

In various ways, Congress and the executive branch have tried to systematize the evaluation of governmental programs by quantifying the benefits and the costs; if benefits exceed costs, then the program is funded. Yet NASA programs by their very nature are particularly impervious to cost-benefit analysis. Few NASA programs produce immediate, quantifiable benefits. To study cost effectiveness, NASA must proceed on somewhat doubtful assumptions. Even when the benefits are quantified, other noneconomic considerations too often arise. Thus, critics constantly bring up issues not reducible to quantitative terms. With the space shuttle program, NASA had to take other matters than economics into account: whether the value of technological spin-offs was sufficient to justify the program; whether the shuttle offered enough to the U.S. space program in the scientific/political Space Race with the Soviet Union; whether NASA was prepared to rely on the shuttle for the indefinite future; and finally, whether the United States should make so heavy a commitment to piloted spaceflight when robotic vehicles might reap the same benefits and avoid enormous risks.

In the end, political considerations have dominated the funding process. For example, the Tracking and Data-Relay Communications Satellites were built, among other things, to lessen NASA's dependence on tracking stations located on foreign soil (which could be held hostage by foreign governments). Foreign locations were cheaper, but politics negated their advantage.

Like other agencies, NASA prepares a budget in consultation with the Office of Management and Budget to be included in the president's message to Congress for the following January. Gradually, specific figures are worked out. Nevertheless, NASA has budgetary problems not found in budgets of some other programs. First, current expenditures have to be matched with long-range projects; estimates are constantly made. Second, NASA has to consider current budget needs, the next year's requirements, and preliminary estimates for years beyond that. These interrelations occur because a deficiency in one year might delay a project unnecessarily.

Before 1966, at the height of the Cold War, Congress granted NASA whatever it asked. Later, compromises became necessary. In the late 1960's, the Bureau of the Budget forced the closing of the Electronics Research Center, reduced the number of Surveyor flights from seventeen to ten, and eliminated the Advanced Orbiting Solar Observatory. Nevertheless, NASA tried to keep its programs alive. A project called Nuclear Engine for Rocket Vehicle Application (NERVA) was modestly funded in 1972; during that same year, however, because of expenditures on the Vietnam War, the Office of Management and Budget continued to cut NASA back. Because the NERVA program had no congressional support, request after request was pared down over a three-year span. Because no prospects of an operational nuclear rocket remained, in January, 1972, NASA elected to terminate the project in favor of a smaller nuclear rocket system.

Before 1967, NASA had considerable support in the Congress, especially in the committees that reviewed the agency's budget. All four key chairmen were from districts where NASA had large contracts. Representative George Miller of California was chairperson of the Science and Astronautics Committee from 1961 through 1972. Senator Clinton Anderson of New Mexico chaired the Senate Aeronautical and Space Sciences Committee from 1963 through 1973. Albert Thomas of Texas was the powerful chairperson of the House Independent Offices Appropriations Subcommittee; he was succeeded in 1966 by Joe Evins of Tennessee. Their senate counterpart was Warren Magnuson of Washington State. These men guaranteed strong levels of funding.

Members of Congress from Southern California, Texas (and the rest of the Sun Belt), and the New York and Boston areas saw the great boost to their districts from the billions in governmental spending. When it was decided to build the Manned Spacecraft Center, Vice President Lyndon B. Johnson (who was also chairperson of the Space Council, an advisory committee to President Kennedy), the Speaker of the House, the chair of the House Independent Offices Appropriations subcommittee, and the chair of the House Manned Spaceflight subcommittee were Texans. NASA administrators were careful to keep committee chairs informed, even beyond statutory requirements, to respond to the mood and whims of individual members, and to work within their allocated funds. The Manned Spacecraft Center was renamed the Johnson Space Center in 1973.

Funding for the U.S. space program reached its peak during the Gemini Program in the mid-1960's; it would never again be as great in terms of real, noninflationary dollars. This funding success coincided with the start of Johnson's second term. During nineteen months, Titan II rockets launched ten flights, carrying a total of twenty men. Astronauts walked in space, and the prospects for the future seemed limitless. The combination of social problems at home, the escalation of the Vietnam War, and the beginning of associated inflationary pressures, for which Johnson recommended reduced spending in selected programs, however, caused the beginning of the decline for NASA funding.

Times changed: NASA officials could not come up with any goal as politically attractive as the Apollo Program's race for the Moon. NASA Administrator James E. Webb was extremely reluctant to commit NASA to anything beyond Apollo, because of both the debate in the scientific community over the value of piloted flights and the controversy over the war in Vietnam. Members of Congress wanted programs with an immediate, high-profile payoff, like that of Apollo. Apollo had been important for national prestige and for a vague but strongly felt desire for the long-term benefits for all humanity. In the process of creating the budget, however, it was difficult to establish long-term goals. During most of the 1960's, the budgetary process had worked in NASA's favor because the organization had an overriding mission, few direct competitors, and no opposing vested interests.

After Apollo ended, NASA looked to low-cost, long-term goals. The space shuttle was considered for piloted flight and scientific missions, while much attention shifted to communications satellites, work on meteorology and Earth surveys from space by Landsat, and military spy satellites, which Congress considered more important than more expensive, long-term research and development.

Thus, the budget remained stable at the inflation-adjusted, two-to-three-billion-dollar range. The bottom in real inflation-adjusted dollars would be reached in 1974, the final year of the Nixon administration and the beginning of the Ford administration. In the years that followed, NASA would have to live with a fixed two-billion-dollar share of the budget. In real terms, this amounted to 1.2 percent of the total 1974 federal budget. Today, NASA receives less than 1 percent of the total.

When President Richard M. Nixon took office in early 1969, he commissioned a Space Task Group, led by Vice President Spiro Agnew and Secretary of Defense Melvin Laird, to help further cut NASA's budget in fulfillment of his 1968 campaign promise to curtail NASA operations until the economy could better afford it. Nixon slashed the budget requested, and Congress cut it back further. Critics agreed: With the war in Vietnam and the social problems at home, the United States could not afford to spend so much on NASA.

NASA had to announce major program changes. In 1970, the Apollo Applications Program was renamed Skylab and was pushed back. In addition, instead of seven crews for two space stations, three crews would be sent to one station. Apollo 20 was canceled and the piloted flight to Mars put on hold. By September, 1970, two other Apollo missions were canceled as well.

On March 7, 1970, Nixon made a carefully worded announcement; the space program, NASA in particular, was of low priority to him. Funding continued to fall. Although in 1972 Nixon gave the go-ahead to the space shuttle program, creating some forty thousand new aerospace jobs before an election year, Congress struggled with cost overruns, and NASA could not seem to deal with new technological challenges.

Although precise data are hard to acquire from the four years of President Jimmy Carter's administration, the Soviet Union launched in excess of four hundred spaceflights; Soyuz 32's crew was in space approximately 175 days. In contrast, the United States had only sixty-five launches (excluding classified Department of Defense efforts), mostly robotic communications and meteorological satellites. Funding was low; indeed, NASA simply struggled to hold what it had had in the past in the face of runaway inflation and rising costs. Administrators would shift funds around to keep the shuttle program alive, borrowing production funds for the development program, shifting funds among fiscal years, and continually letting the shuttle schedule slip.

Although President Jimmy Carter announced his policy on space development in June of 1978, little was actually resolved: Budgets remained constant in terms of real purchasing power, and in the best Washington tradition, NASA became like the National Institutes of Health—important, but no one in the general public was sure of the precise benefits. The trend continued with President Ronald W. Reagan's administration. The shuttle remained the top priority, while other NASA funding sputtered. Even after the success of the first shuttle flight on April 12, 1981, support remained lukewarm.

With the *Challenger* accident came no lack of alternatives proposed for NASA. The National Commission on Space in its 1986 report proposed a building-block approach, extending human activities from low-Earth orbit to the Moon and Mars over the next fifty years. A 1987 report by the American Institute of Aeronautics and Astronautics advocated a similar agenda. A NASA report prepared by Sally K. Ride developed a set of options for a more aggressive space program that included a piloted sprint to Mars, a lunar base, more aggressive robotic exploration, and more work on science from satellites, all to be completed by 2010. These ambitious plans, however, would require enormous amounts of money—well in excess of those spent during NASA's heyday in 1965.

As the twentieth century drew to a close, increased pressures on NASA budgets encouraged the well-known and controversial "faster, better, cheaper" approach to mission funding. Although this policy resulted in some successes, particularly at the science end with lunar and interplanetary missions such as the Mars Pathfinder, the notable failures of the Mars Climate Orbiter and the Mars Polar Lander were, in part, attributed to this policy, which some saw as encouraging an overemphasis on budgetary economies.

### Context

NASA's funding depends upon its potential gains, both political and economic. Traditionally, the civilian space program has been justified as a means to realize the international leadership of the United States. If NASA were able to further develop space commercially, the United States would gain greater prestige in the world. This development comes at greater cost, however, as the early, easy gains of the space program have been realized. Moreover, there is competition on the international scene that was not present during the 1960's. Along with Russia, the European nations, Canada, China, and Japan all have serious civilian space programs. Nevertheless, piloted activities aboard a space shuttle provide highly visible symbols of American leadership and competence.

Yet NASA's activities must also yield concrete economic benefits. The civilian space program should provide an economic return, creating knowledge that ultimately improves goods and services. Because of the expense involved, private industry could not have become involved with space exploration and development alone. NASA's activities have continued to help the public, although as always, it is difficult to judge these benefits precisely.

Certainly, NASA programs advance the entire economy; spending on scientific research creates as many jobs as would any program devised for that purpose. Yet how does one measure the effects of the NASA research? Some argue that large development systems, such as the shuttles and space stations, are less likely to aid the public than did earlier Apollo flights. In rebuttal, NASA points to the great advances made because of communications and land-sensing satellites, such as Landsat and TIROS-N.

Regardless of the amount with which Congress and future presidents choose to fund it, NASA will continue to be pressured to increase the effectiveness of its spending. Several proposals include more international cooperation. Thus, the United States participates in the International Space Station (ISS) with other nations.

Another effort will be made to stimulate private investment in space. The NASA budget request for Fiscal Year 2001, for example, reflected four appropriations: Human Space Flight ($5.5 billion); Science, Aeronautics, and Technology ($5.9 billion); Mission Support ($2.6 billion); and Inspector General ($22,000). Human Space Flight was to be dedicated to funding for the ISS and space shuttle programs, including development of research facilities for the ISS and continuing safe, reliable access to space through augmented investments to improve space shuttle safety, support of payload and expendable launch vehicle (ELV) operations, and other investments, including innovative technology development and commercialization. Science, Aeronautics, and Technology was planned for NASA's research and development activities, including all science activities, global change research, aeronautics, technology investments, education programs, space operations, and direct program support. Mission Support was to fund NASA's civil service workforce, safety and quality assurance activities, and facilities construction activities to preserve NASA's core infrastructure. The Inspector General portion of the budget was to be designated for the workforce and support required to perform audits and evaluations of NASA's programs and operations.

In 2004, political bravado and budgetary reality clashed when the administration of President George W. Bush declared a goal of sending humans back to the Moon and on to Mars. Facing the worst federal deficit since the Hoover administration, NASA would have to cancel piloted and robotic missions already in the pipeline in order to fund such a project. In response to the announcement, NASA eliminated all space shuttle flights not directly supporting the ISS, including scheduled servicing missions to the Hubble Space Telescope. Without these essential repairs and upgrades, Hubble would be allowed to wither and die a fiery death. Clearly, realistic goals needed to be set for NASA in both the robotic and piloted exploration of space.

The economic collapse of the Soviet Union removed the political competition that brought a seemingly endless supply of support and funding that NASA enjoyed in the 1960's. NASA and the federal government could not continue to explore space alone. The need for international cooperation and financing had become paramount to extending our knowledge of the universe.

**See also:** Hubble Space Telescope; National Aeronautics and Space Administration; Private Industry and Space Exploration; Space Shuttle Flights, 1999; Viking Program.

### Further Reading

Harvey, Brian. *Russia in Space: The Failed Frontier?* Chichester, England: Springer-Praxis, 2001. This book tells the inside story of the traumatic events that engulfed the once-glorious Soviet space program. It is a story of desperation and decline, but also a tale of heroic efforts to save the Space Station Mir and the construction—along with their old rivals, the Americans—of the new International Space Station.

Hudgins, Edward L., ed. *Space: The Free-Market Frontier.* Washington, D.C.: Cato Institute, 2002. Papers from a conference sponsored by the Cato Institute address NASA's history, cheaper space travel, space investment, and space law.

Johnson, Stephen B. *The Secret of Apollo: Systems Management in American and European Space Programs.* Baltimore: Johns Hopkins University Press, 2002. Examines the challenges of managing space programs.

Levine, Arnold S. *Managing NASA in the Apollo Era.* NASA SP-4102. Washington, D.C.: Government Printing Office, 1982. The best history of the funding of the space agency during its golden age of the 1960's. Although written under NASA sponsorship, it is a well-balanced historical analysis.

United States Congress. Senate. *National Aeronautics and Space Administration Authorization Act for Fiscal Years 2000, 2001, and 2002.* Washington, D.C.: Government Printing Office, 1999. Documentation of NASA's attempts to obtain funding for fiscal years 2000 through 2002. Items requested versus items approved are detailed, as are the reasons NASA has for spending Americans' tax dollars.

*Douglas Gomery*

# Galaxy Evolution Explorer

*Date:* Beginning April 28, 2003
*Type of program:* Scientific platform

*The Galaxy Evolution Explorer (GALEX) was designed to observe ultraviolet emissions and produce a comprehensive survey of galaxies. GALEX data would prove crucial to understanding galaxy formation and evolution.*

## Key Figures

*Kerry Erickson,* GALEX mission manager at the Jet Propulsion Laboratory (JPL)

*James Fanson,* GALEX project manager at JPL

*Peter Friedman,* GALEX project scientist at Caltech

*Anne Kinney,* director of astronomy and physics at NASA Headquarters' Office of Space Science

*Christopher Martin,* GALEX principal investigator at Caltech

*David Schiminovich,* GALEX science operations and data analysis manager at Caltech

## Summary of the Mission

Part of the Structure and Evolution of the Universe program, under the auspices of the National Aeronautics and Space Administration (NASA), GALEX was designed within the constraints of a Small Explorer class mission. Its ultraviolet (UV) detectors were built to provide data complementary to studies under way with the Hubble Space Telescope and the Far Ultraviolet Spectroscopic Explorer. The primary mission of GALEX was to complete an all-sky map of galaxies in the ultraviolet, data which could be compared with all-sky surveys conducted in the infrared.

More specifically, the mission of GALEX was to map the celestial sky in two major portions of the ultraviolet, and in so doing collect data that would determine the history of star formation since the initial formation of galaxies. If present models of the evolution of the universe prove correct, galaxies began forming approximately ten billion years ago. By observing in the ultraviolet, GALEX would find galaxies characterized by young, hot, short-lived stars. Such galaxies display high rates of star

formation and might provide clues to the mechanisms of galaxy formation.

The GALEX mission is led by the California Institute of Technology (Caltech), which manages science operations and data analysis as well. The Jet Propulsion Laboratory (JPL) produced the science instrument. Orbital Sciences Corporation provided spacecraft integration and testing, ground data system and mission operations, and the booster. The Explorer program is managed by NASA's Goddard Space Flight Center (GSFC). Other GALEX participants are the University of California and the Space Telescope Science Institute. GALEX also includes some international cooperation. France's Laboratoire d'Astrophysique de Marseille developed and constructed components of the flight optics, and Yonsei University in Seoul, South Korea, provided test equipment and science operations software.

Compared to NASA's Great Observatories, such as Hubble, Chandra, and Spitzer, GALEX is a rather small spacecraft, which was necessary in order to fit GALEX aboard a Pegasus booster. Constructed out of aluminum, the spacecraft was 1 meter wide at maximum and 2.5 meters high, with a mass of 277 kilograms. GALEX's shape was that of a hexagonal base that housed the scientific instruments and major systems, atop which a telescope and two solar arrays were attached. Those solar arrays generated sufficient energy to charge a 15-ampere-hour nickel-hydrogen battery with 250 watts at 28 volts. A solar battery was necessary because data would be collected while the observatory was in darkness; data would then be replayed to ground stations during the dayside portion of each orbit.

GALEX incorporated a 0.5-meter-diameter telescope outfitted with a pair of ultraviolet detectors, one for detecting emissions in the near-ultraviolet (175 to 280 nanometers) and one for detecting emissions in the far-ultraviolet (135 to 174 nanometers). Of Cassegrain design, this telescope included a primary mirror and a smaller secondary mirror located in front of the primary. Light reflected off the secondary mirror would have to pass through a hole in the primary mirror's central area in order to come to a focus at a detection plane located behind the primary. The nature of the optical system was such that the UV telescope had a field of view more than twice the angular diameter of the full Moon, or five hundred times larger than Hubble's field of view. With a resolution of two million pixels, GALEX's UV detectors could capture hundreds of galaxies in a single image.

Polychromatic light can be dispersed into component wavelengths by refraction using a prism or by diffraction using a grating. The former performs dispersion because the index of refraction of a transparent material depends slightly upon wavelength. The latter performs dispersion by having light pass through a large number of slits or grooves very closely and uniformly spaced in an

*GALEX, mated to the Pegasus vehicle, is rolled out for launch.* (NASA)

otherwise transparent aperture so that different portions of the light travel different distances to the detector. Interference occurs, producing bright lines for component wavelengths well separated from one another.

One unique aspect of the telescope's optics was the inclusion of a "grism," a system combining aspects of both a prism and grating. The grism's grating characteristic incorporated seventy-five grooves per millimeter. The grism's prism characteristic was a dichroic beam splitter that separated out far-UV spectral lines from near-UV ones. Far-UV rays were deflected refractively toward the far-UV detector, while near-UV rays were transmitted to a flat mirror that then reflected them to the near-UV detector. This represented the first use of a UV dichroic beam splitter in space. The grism and an imaging window were mounted on an optical window that could be rotated as needed by a motor. The motor was built to rotate in order to compensate for orbital motion of the spacecraft, thereby preventing blurring of spectra, and to make sufficiently small motions needed to collect spectra for closely spaced objects.

On April 28, 2003, at 11:03 Coordinated Universal Time (UTC), NASA's L-1011 carrier aircraft *Stargazer* departed from the Cape Canaveral Air Force Station. Heading east for a launch zone 160 kilometers off Florida's east coast, *Stargazer* encountered favorable weather. Then, at 12:00 UTC, *Stargazer*'s pilot dropped the Pegasus XL booster while at an altitude of 11.8 kilometers. The booster freely fell for five seconds before its first-stage solid rocket motor ignited. All three stages of the Pegasus properly fired, and after eleven minutes the GALEX spacecraft was released from the booster. This marked the thirty-third consecutive successful Pegasus booster, the nineteenth since 1997. GALEX was dispatched to its operational orbit with an altitude of approximately 688 kilometers in-

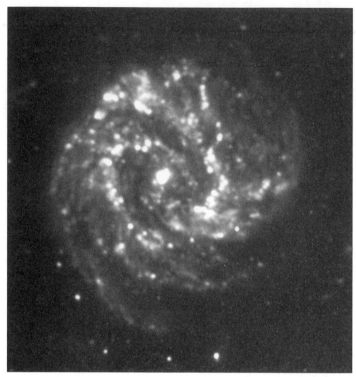

*This image of of the "southern pinwheel galaxy," Messier 83, was taken in July, 2003, during only one orbit of GALEX. Messier 83 is 15 million light-years from Earth.* (NASA)

clined 29° to Earth's equator. In this orbit, GALEX circled the Earth once every 96 minutes.

The GALEX science team honored the *Columbia* space shuttle's final crew by dedicating the spacecraft's first light images, collected on May 21 and 22, to the seven astronauts who died on February 1, 2003. For the initial observations, the team selected an area of the sky inside the constellation Hercules. The pair of short observations demonstrated the optical quality of the telescope and identified more than four hundred stars and star-forming galaxies in the far-UV image and more than fifteen hundred in the near-UV image.

### Contributions

Among the first to be released as part of a large collection of GALEX images was a complete composite image of the Andromeda galaxy, the largest neighboring galaxy in our local group. The image displayed evidence of recent star formation. Other objects studied in this first collection of images were Stephan's Quintet of galaxies, the globular star cluster M2, and a deep-sky image of a portion of the constellation Bootes. These images were chosen as early scientific subjects to demonstrate the power and versatility of GALEX to cover a host of important celestial processes that reveal themselves in the ultraviolet. In regard to the Andromeda galaxy, astrophysicists were interested in determining its brightness, its mass, its age, and how young star clusters were distributed within the spiral arm structures.

On the anniversary of its launch, GALEX imaged a pair of spiral galaxies, M81 and M82, both approximately ten billion light-years away. The former, viewed from above, was similar in size, shape, and luminosity to our own Milky Way galaxy, but M82 was viewed from the side and displayed such a level of violent star formation that gas and dust were ejected perpendicular to its disk.

Late in December, 2004, GALEX produced surprising images of massive newborn galaxies relatively close to the neighborhood of the Milky Way. Models had predicted that formation of large galaxies had drastically diminished since the early era of galaxy formation, and that presently only small ones were forming. This finding by GALEX suggested that galaxy formation remained strong in various portions of the expanding universe. These unexpected images also provided clues to the early stages of galaxy formation, perhaps indicating how the Milky Way may have appeared billions of years ago. Whereas the Milky Way is approximately ten billion years old, these galaxies, each about two to four billion light-years distant, are only between one hundred million and one billion years old. Imaged were about three dozen galaxies identified out of several thousands surveyed thus far. These very young large galaxies appeared ten times brighter in ultraviolet emissions than the present-day Milky Way; they seem typical of galaxies seen at a distance of more than ten billion light-years, which corresponds to galaxies forming very early in the age of the universe.

**Context**

GALEX represents a continuation of a highly successful satellite series going all the way back to the first American Earth-orbital satellite, Explorer 1. The present series of Explorer payloads are meant to be relatively light and inexpensive satellites capable of being produced rapidly and of being sent aloft by smaller boosters. This Small Explorer class has featured satellites with innovative technologies and focused mission objectives.

Acceptance of the notion of island universes or galaxies is less than a century old. Spectroscopic studies by Edwin Hubble of what, previous to his time, astronomers considered nebulae indicated that these objects were moving away from Earth at tremendous speed and had to be located outside the Milky Way. Hubble's observational data played a major role in the development of the "expanding universe" model. Early galaxy studies revolved around categorization of the tremendous number of these objects.

Galaxies are found in a wide range of sizes ranging from dwarf galaxies containing ten million stars or less to massive galaxies encompassing upward of a trillion stars. Galaxies have also been categorized according to structure. There are three main types, with several subcategories in each case: elliptical, spiral, and irregular. Elliptical galaxies are found that are nearly spherical, but most are elongated to one degree or another. Such galaxies are deficient in gas and often do not display significant star formation. Spiral galaxies have central bulges where the density of stars is quite high, spiral arms of varying size and tightness, and the gas and dust necessary for star formation. As the name suggests, irregular galaxies have shapes that are neither elliptical nor spiral; these galaxies may be formed by collisions of galaxies or other catastrophic or unusual processes.

Comprehensive sky surveys were previously conducted in all major portions of the electromagnetic spectrum with the exception of the ultraviolet. GALEX filled in that gap, providing an enormous volume of data from which further studies could be designed using other spacecraft such as Hubble,

Chandra, and Spitzer, and the anticipated James Webb Next Generation Space Telescope. The biggest thing that GALEX would add to the understanding of galactic structure and evolution was that precision measurements of the ultraviolet brightness of galaxies would provide a determination of the rate at which stars form within galaxies, and correlate that activity to the distances of the galaxies. GALEX observations would span approximately 80 percent of the age of the universe.

GALEX data permit scientists to examine galaxies in varying states of formation and at various ages, from very young to very old. The data allow statistical analysis of galactic evolution as a result of three physical properties of the universe. First, the speed of light has a finite value; thus, when looking at an extremely distant galaxy, scientists are seeing that galaxy as it appeared early in the development of the universe. Second, the distribution of galaxies within the universe is essentially uniform; this permits a reliable comparison of galaxies at present with galaxies early in time. Last, because the universe is expanding, light is Doppler-shifted; spectroscopic measurements provide a speed for each galaxy, and the speed is directly proportional to the distance from Earth of these galaxies.

**See also:** Chandra X-Ray Observatory; Explorers 1-7; Extreme Ultraviolet Explorer; Hubble Space Telescope; Small Explorer Program.

**Further Reading**

Binney, James, and Michael Merrifield. *Galactic Astronomy*. Princeton, N.J.: Princeton University Press, 1998. Provides detailed historical background concerning the understanding of galactic structures and formation.

California Institute of Technology. GALEX. http://www.galex.caltech.edu. Accessed January, 2005. Posts general science on the GALEX spacecraft and mission, as well as images.

Waller, William H., and Paul W. Hodge. *Galaxies and the Cosmic Frontier*. Cambridge, Mass.: Harvard University Press, 2003. Provides a great deal of data about galaxies and their cosmological importance.

*David G. Fisher*

# Galileo: Jupiter

*Date:* October 18, 1989, to September 21, 2003
*Type of spacecraft:* Planetary exploration

*Launched from the space shuttle* Atlantis, *the Galileo probe provided more extensive in situ sampling of Jupiter's outer atmosphere, and far closer encounters of its principal satellites, than the previous Pioneer and Voyager missions.*

### Key Figures

*William J. O'Neil,* Galileo project manager at the Jet Propulsion Laboratory (JPL)
*Torrence V. Johnson,* JPL Galileo project scientist
*Neal E. Ausman, Jr.,* Galileo mission director at JPL
*Marcia Smith,* probe manager
*Richard E. Young,* probe scientist
*Wesley T. Huntress, Jr.,* associate administrator, NASA Headquarters Office of Space Science
*Donald Ketterer,* NASA Headquarters program manager for Galileo
*Jay Bergstralh,* NASA Headquarters project scientist for Galileo
*Eugene M. Shoemaker* (1928-1997), astronomer with the U.S. Geological Survey
*Carolyn Shoemaker* (b. 1929), astronomer with the U.S. Geological Survey
*Donald E. Williams* (b. 1942), commander of STS-34

### Summary of the Mission

The Galileo mission to Jupiter was formally approved by the United States Congress in 1977, several years before the space shuttle *Columbia* made its maiden flight into Earth orbit. The mission was a cooperative project involving scientists and engineers from the United States, Germany, Canada, Great Britain, France, Sweden, Spain, and Australia. Even though the uncrewed spacecraft Voyager 1 and Voyager 2 performed flybys of planet Jupiter and its sixteen moons in 1979, the Galileo mission was envisioned to initiate several novel observations of Jupiter, the most massive gas planet of the solar system, and its principal moons, and conduct exclusive, often in situ, experiments on their fascinating environments. The spacecraft was originally scheduled for launch in 1982, but that launch date had to be scrapped due to delays with the space shuttle program. Subsequently, the space

shuttle *Challenger* accident in January, 1986, caused the project to be shelved indefinitely. The Galileo spacecraft was eventually carried aboard the space shuttle *Atlantis* (flight STS-34), and was launched from the shuttle on October 18, 1989. It is estimated that approximately ten thousand people worked directly on Galileo since its inception in 1977.

In order to conserve propellant for orbital activities, Galileo was designed to fly away from Jupiter initially to receive gravity assists (additional gravitational pulls that help accelerate the craft) from Venus and Earth (twice). This maneuver induced an 11.1-kilometer-per-second change in Galileo's velocity—a change that would require 10,900 kilograms of propellant if Galileo were launched directly toward Jupiter. The amount of propellant on board the spacecraft weighed about 900 kilograms, indicating a substantial savings of propellant mass achieved by using the billiards-like gravity-assist technique. The roundabout route taken by Galileo to reach Jupiter covered about 4 billion kilometers. It is worthwhile to note that the Galileo orbiter, which achieved stable Jupiter orbit after December 7, 1995, was designed to use similar close flybys and gravity assists during its complex tour of the Jovian system in order to perform tasks that would require an additional 3,600 kilograms of propellant. Galileo's Earth flybys occurred in December, 1990, and December, 1992. The first flyby brought the spacecraft within 960 kilometers of the Earth, and the second within 303 kilometers. That the spacecraft was within 1 kilometer of its intended path and only 0.1 second early at the time of its closest approach during the second Earth flyby attests the degree of accuracy of spacecraft operational management.

The Galileo spacecraft is named after Galileo Galilei (1564-1642), the great Italian astronomer and physicist who discovered the four large moons of Jupiter (Io, Europa, Ganymede, and Callisto) using handcrafted telescopes. Appropriately, these four moons are called the Galilean satellites. Galileo was the first spacecraft to image the surface of Venus without using radar (a radio-wave pulse detector). Using its near-infrared solid-state imaging camera, it photographed Jupiter's atmospheric banding and its satellites from a half-billion miles away on its way to Venus in December, 1989, and subsequently observed numerous mountain ranges and valleys on Venus's oven-hot surface through its thick atmosphere and clouds on February 10, 1990. The image resolution of Galileo's cameras (the smallest object size that can be detected by them) was around 12 meters, a millionfold improvement over Galileo's original observations.

As a planet, Jupiter is by far the largest in the solar system. It has a volume about

1,400 times that of the Earth; in fact, its volume is 1.5 times the combined volume of all the other planets, moons, asteroids, and comets in the solar system. Jupiter is a "gas giant" planet composed of vast amounts of hydrogen gas. The gas runs thousands of kilometers deep. The gases on Jupiter swirl around in massive hurricanes whose sizes are of the order of the size of the Earth. The famous Great Red Spot on Jupiter is in fact a hurricane three times the diameter of Earth; it has been raging in the Jovian atmosphere for more than three hundred years. It is believed that, given its enormous size, Jupiter would have become a thermonuclear reactor (that is, a star like the Sun), if only it were thirty times heavier. Jupiter rotates about its

*The Galileo spacecraft is prepared for mating with the booster at Kennedy Space Center's Vertical Processing Facility in August, 1989. (NASA)*

axis much faster than the Earth; hence, a Jovian day is only 9 hours and 48 minutes long. This fast rotation causes Jupiter to be somewhat squashed, or oblate: Its equatorial radius is 71,392 kilometers (compared with the Earth's 6,400 kilometers), while its polar radius is about 4,000 kilometers smaller. This causes an object to weigh about 25 percent heavier at Jupiter's poles than at its equator.

In 1994, while the Galileo spacecraft was approaching Jupiter, it became a direct witness to an astounding astrophysical event. The Comet Shoemaker-Levy 9 (SL-9)—discovered on March 24, 1993, by Eugene M. and Carolyn Shoemaker, a husband-and-wife team of astronomers with the U.S. Geological Survey's Lowell Observatory (who have together discovered more impact sites on Earth and comets and asteroids in outer space than anyone in history) and fellow astronomer David H. Levy—had broken up into several small fragments and was expected to plunge directly into Jupiter's atmosphere. The fragmentation of the comet was presumably caused by its having come too close to Jupiter (whose tidal forces, though quite gentle, were enough to break up the weakly built comet) during its closest approach (also called perijove) on July 7, 1992. The collision of Shoemaker-Levy 9's fragments with Jupiter was predicted as early as May, 1993, by Brian Marsden of the Central Bureau for Astronomical Telegrams.

Some astronomers are on the lookout for the doomsday asteroid that one day might impact the Earth. At their velocities of entry, even asteroids not much larger than 1 kilometer in diameter can have a devastating effect on terrestrial life. It is believed that an asteroid colliding with the Earth wiped out the dinosaurs and several other species several million years ago. The possibility of a similar collision occurring again is ever-present; the fact that there may be as many as fifteen hundred to two thousand asteroids or comets in Earth-crossing orbits (of which only a hundred are actually known) makes the possibility of an impact non-negligible. Knowledge of an impending impact may make it vital for humans to look for some means of pre-

vention, or at least to attempt to preserve life on Earth. Fortunately for the Earth (and perhaps for other planets in the solar system), Jupiter, with its powerful gravitational field, has been routinely swallowing runaway comets and asteroids (such as Shoemaker-Levy 9) before they can cause any damage to the other planets.

Project Galileo spent much of the year 1995 preparing for the dual-craft arrival at Jupiter on December 7. In July, 1995, the Galileo probe and the orbiter spacecraft separated to fly their independent missions to Jupiter. After the probe had separated for atmospheric entry, the orbiter's main engine was fired to aim it for going into orbit around Jupiter. During the latter part of 1995, Galileo's engineers had to deal with several spacecraft problems. These included a leaking valve in the main rocket engine, a slipped tape in the tape recorder (which prevented the recording of images of Io and Jupiter during the initial pass), and a malfunctioning high-gain antenna. The problems were either fully or partially resolved during the approach to Jupiter, and the two Galileo craft completed their arrival missions with complete success. The probe entered with an initial velocity of 170,000 kilometers per hour, decelerating for two minutes, then plunging into the wind-torn clouds beneath its Dacron parachute, sending measurements for almost an hour. The orbiter, meanwhile, measured the Jovian environment, received a gravity assist from Io, received and recorded the probe data, then fired its main engine to become the first artificial satellite of Jupiter. The successful arrival was enthusiastically celebrated at NASA Headquarters on December 7, 1995, by William J. O'Neil, project manager; Neal E. Ausman, Jr., mission director; Daniel S. Goldin, NASA administrator; Donald Ketterer, program manager; and Donald E. Williams, commander of STS-34, the shuttle *Atlantis* flight that launched Galileo into space. On December 9, Galileo began relaying the probe data to Earth. The readout was interrupted later in December, when Galileo and Jupiter entered superior conjunction, a position behind the Sun from the Earth's sky, where the Sun's radio noise drowns Ga-

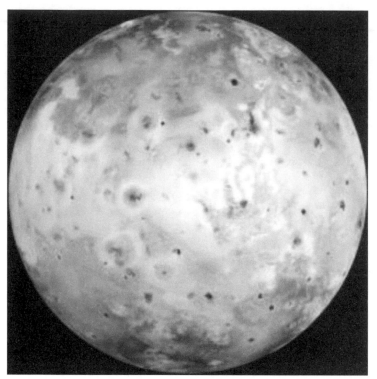

*A high-resolution image of Jupiter's moon Io as Galileo passed approximately 130,000 kilometers above the surface.* (NASA)

lileo's signals for up to two weeks. The data readout resumed in January.

Galileo continued to gather data from the Jovian system nearly five years after its arrival. A large maneuver in March, 1996, raised the inner end of Galileo's orbit away from Jupiter's hazardous radiation belts. The first Ganymede and Io encounters began June 27, 1996, and a second Ganymede encounter on September 6, 1996. The first encounter with Callisto occurred on November 2, 1996, and with Europa on December 19, 1996. Each encounter involved a one-week, high-rate observation of Jupiter and at least one satellite. Each flyby brought Galileo to within a few hundred kilometers of the satellites and gave it a gravity assist into the next orbit. In January, 1997, Galileo and Jupiter entered another superior conjunction, after which the orbiter continued its close flybys for another year. On February 20, 1997, the Galileo orbiter encountered Europa for a second time.

It encountered Ganymede on April 5, 1997, at a distance of only 3,095 kilometers, nineteen times closer than Voyager 2, and, again, on May 7, 1997. This time it got within 1,600 kilometers of the satellite. On June 25, 1997, the probe glided to within 415 kilometers of Callisto. It reencountered it on September 17, 1997. Between November 6, 1997, and February 1, 1999, Galileo played tag with Europa nine times, swooping down on its icy surface. In May 5, 1999, Galileo began another four-visit tour of Callisto, ending on September 16, 1999.

In late 1999 and early 2000, near the end of its two-year mission extension known as the Galileo Europa Mission, the Galileo spacecraft dipped closer to Jupiter than it had been since it first went into orbit around the giant planet in 1995. These maneuvers allowed Galileo to make three flybys of the volcanically active moon Io and also made possible new high-quality images of Thebe, Amalthea, and Metis, which lie very close to Jupiter, inside the orbit of Io. Volcanic calderas, lava flows, and cliffs could be seen in a false-color image of a region near the south pole of Jupiter's volcanic moon, Io. Combining a black-and-white image taken by the Galileo spacecraft on February 22, 2000, with lower-resolution color images taken by Galileo on July 3, 1999, JPL scientists created the image. Included in the image are three small volcanic calderas about 10 to 20 kilometers in diameter.

Galileo's mission was actually extended three times, taking advantage of the unique possibilities of passing very close to several of the Galilean moons. But to preclude the radioactive material in the spacecraft's power generation system from contaminating Europa or any of the other moons, the fourteen-year odyssey of Galileo concluded on September 21, 2003, with a controlled plunge into the outer atmosphere of Jupiter while the spacecraft was on its thirty-fifth orbit. The spacecraft passed into Jupiter's shadow, and the Deep Space Network received its final signal from Galileo at 12:43:14 Pacific daylight time, or 19:43:14 Coordinated Universal Time (UTC). Galileo hit the outer atmosphere just south of the gas giant's equator at a speed of 48.3 kilometers per second. Due to the time delay in receipt of light signals, this message arrived 46 minutes after Galileo was crushed, vaporized, and dispersed into Jupiter's dense atmosphere.

### Contributions

In July, 1994, when the fragments of Comet Shoemaker-Levy 9 were due to collide with Jupiter, more terrestrial telescopes were pointed to Jupiter's region of the sky than ever before in history. In addition, there was the Hubble Space Telescope, which was initially not expected to capture the collision images because the collisions were thought to be obscured by Jupiter's horizon (it was expected to monitor only Jupiter's atmospheric changes due to the impacts). The Galileo spacecraft itself was heading toward Jupiter, ready to record the spectacular astronomical event. As Heidi Hammel of the Massachusetts Institute of Technology described it, Galileo eventually sent back pictures that amounted to "one week of spectacular impacts." As the fragments of Shoemaker-Levy 9 lined up for collision with Jupiter like bomber planes in formation, it took six minutes for the rising plume from each collision to become visible above Jupiter's surface, about thirty minutes for the signal to reach antennae of the three Deep Space Network telescopes (located near Madrid, Spain; Canberra, Australia; and Barstow, California), and about forty minutes for it to be seen by human eyes. In witnessing the Shoemaker-Levy 9 and Jupiter collisions, Galileo achieved a scientific first: It made the first and only documented direct observation of a collision between solar system bodies. Galileo's photopolarimeter radiometer detected the flashes of light caused by the collisions using a 4-inch telescope while it was as far away from Jupiter as the planet Mars is from the Sun. At the time of the colli-

sions, Jupiter appeared at about sixty pixels to Galileo, compared to only about 2.5 pixels to Voyager 2 (which was at a distance of about 41 astronomical units from Jupiter, or about forty-one times the distance from the Earth to the Sun), and two pixels to Voyager 1 (52 astronomical units from Jupiter). At the time of the collisions, Galileo was at a distance of about 240 million kilometers (less than two astronomical units). Hence, Galileo was no doubt at the best possible position to observe the collisions. Galileo's imaging devices recorded six of the impacts (those of fragments G, H, K, L, Q1, and W). Initially, the fragment G was detected to be about 7 kilometers in diameter, with a temperature of 8,000 kelvins, which is hotter than the Sun's surface; less than two minutes later, it had grown to hundreds of kilometers across and had cooled to 400 kelvins. In ultraviolet light, the impacts lasted for about ten seconds; in infrared, they continued up to ninety seconds. The G impact was the largest,

*Earth and the Moon were captured by Galileo in separate images that were combined for this composite view.* (NASA-JPL)

producing a gigantic fireball that shot up as much as 3,200 kilometers above Jupiter's surface, after punching a hole about 80 kilometers below Jupiter's surface.

Despite Galileo's spectacular direct observations of the impacts (the recorded data from which were played back from the spacecraft over a period of several months, extending through January, 1995), one mystery baffled scientists: The Hubble Space Telescope and the Earth-based observatories detected some of the impacts just as soon as Galileo did. This was quite unexpected, because the impacts were known to occur below Jupiter's horizon. Dr. Andrew P. Ingersoll of the California Institute of Technology commented, "In effect, we are apparently seeing something we didn't think we had any right to see." As a possible explanation, Galileo Project Scientist Torrence V. Johnson observed that the Hubble observations of fragments G and W's impacts could conceivably have been due to scattering of light off comet dust or other material at very high altitude. How the material actually found its way to the upper atmosphere of Jupiter, however, is not known.

Around December, 1994, when Galileo was about 175 million kilometers from Jupiter, it encountered the first of several large dust storms. The largest of these, which lasted about three weeks, was encountered around mid- to late August, 1995, at a distance of about 63 million kilometers. Galileo counted up to twenty thousand particles per day during passage through this storm, compared to a normal interplanetary rate of about one particle every three days, according to Dr. Eberhard Grün, principal investigator on the Dust Storm Experiment. Scientists speculate that the dust particles probably emanated from volcanoes on Jupiter's moon Io, or from its faint two-ring system. It is also possible that the particles may be left-over material from the Comet Shoemaker-Levy 9. These dust particles travel through interplanetary space at speeds of between 40 to 200 kilometers per second; however, being microscopic in size, they pose no danger to a spacecraft such as Galileo.

The dust storms encountered by Galileo are made of particles that are less than a tenth the size of typical airborne dust; the normal density of interplanetary dust is about one particle per sixty or so houses on Earth. Compared to particle densities in the Earth's atmosphere, therefore, the interplanetary dust storms encountered by Galileo are quite benign. Galileo took a close-up look at the asteroid Gaspra (roughly a triaxial ellipsoid with dimensions about 19 by 12 by 11 kilometers) during its high-speed flyby, a unique achievement for a spacecraft from Earth. Galileo made the first discovery of an asteroid moon (Ida's moon Dactyl). Because Galileo is actually in Jupiter orbit, it can fly considerably closer to Jupiter and its moons than was possible for the Voyager or Ulysses spacecraft to achieve, because these only flew by the planet. Galileo passed about 100 kilometers closer to the Jovian moon Io than planned; hence, it encountered the moon Ganymede on June 27, 1996, instead of July 4, 1996, as planned. Galileo has already shown that Io has some dramatic volcanoes. Geysers on Io (made mainly of sulfur dioxide and no water) spew out at about 1 kilometer per second, compared to less than 10 meters per second for volcanic geysers on Earth. The volcanic plumes on Io can shoot up to 250 kilometers high, and there are at least two hundred craters on Io with diameters larger than 20 kilometers.

### Context

The investigation of Jupiter by robotic spacecraft began with the Pioneer 10 and 11 spacecraft. Launched on March 2, 1972, Pioneer 10 became the first space probe to attempt passage through the asteroid belt between Mars and Jupiter, beginning its passage on July 15, 1972. It then flew by Jupiter on December 3, 1973, coming within 130,354 kilometers of the gas giant's cloudtops. Having survived both the asteroid belt and Jovian radiation hazards, it took readings and photopolarimeter images as it quickly flew past the planet and headed away from the ecliptic plane and out of the solar system. The Pioneer 11 spacecraft was launched on April 5, 1973, and safely emerged from the asteroid belt on April 19, 1974. Measurements from its sister craft indicated Pioneer 11 could fly closer to Jupiter, and so it was aimed to come within 43,000 kilometers of the cloudtops. This improved Pioneer 11's imaging and made use of a gravitational assist to send it on to Saturn.

The path to Jupiter having been blazed by the Pioneer 10 and 11 spacecraft, the way was clear for Voyager 1 and 2 to make more in-depth investigations with more sophisticated instrumentation and imaging systems. Voyager 1 flew through the Jovian system, achieving closest approach on March 5, 1979, and providing tantalizing views of the Galilean moons before heading out on a trajectory to the Saturn system. Voyager 2 flew through the Jovian system, achieving closest approach on April 9, 1979, and complementing the investigations of moons made by its sister ship before also heading out on a trajectory to the Saturn system. The Voyagers discovered additional moons and a faint ring. Voyager's findings left many unanswered questions, questions that could be better addressed by an orbiting spacecraft. This provided the genesis of the Galileo program. Before Galileo could reach Jupiter, the Ulysses probe, one designed to use Jupiter's gravity to propel it above the ecliptic plane and investigate the Sun at higher latitudes, encountered Jupiter in 1992 and provided additional information and generated new questions about the solar system's largest planet. Ulysses subsequently encountered the Jovian environment again.

Galileo mission data provided answers to many questions regarding Jupiter and its large assembly of satellites (the combination is sometimes compared to a miniature solar system). These include information about their atmospheres, Jupiter's large magnetosphere (and possibly those of its satellites), its unique ring system, the geologic history of the Jovian system, the volcanic characteristics of Io, the possibility of any liquid water under Europa's ice crust, and, perhaps most important, clues to the early history of the solar system, which will help our understanding of our own planet and its relationship to the universe. The Galileo mission required several new technologies to be developed, which are already paying off handsomely in terms of the knowledge gained (and to be gained further) via Galileo's operations.

**See also:** Cassini: Saturn; Deep Space Network; Hubble Space Telescope; Hubble Space Telescope: Science; Magellan: Venus; Mariner 3 and 4; Pioneer 10; Pioneer 11; Planetary Exploration; Space Shuttle Flights, 1989; Voyager Program; Voyager 1: Jupiter; Voyager 1: Saturn; Voyager 2: Jupiter; Voyager 2: Saturn; Voyager 2: Uranus; Voyager 2: Neptune.

### Further Reading

Barbieri, Cesare, Jürgen H. Rahe, Torrence Johnson, and Anita Sohus, eds. *The Three Galileos: The Man, the Spacecraft, the Telescope: Proceedings of the Conference Held in Padova, Italy on January 7-10, 1997.* Boston: Kluwer Academic, 1997. This is a collection of forty-three papers presented at a conference held in honor of three different observers of Jupiter—each named Galileo. Discussed are the man, the spacecraft, and the new Italian National Telescope.

Canning, Thomas N., and Thomas M. Edwards. *Galileo Probe Parachute Test Program: Wake Properties of the Galileo Probe at Mach numbers from 0.25 to 0.95.* NASA SP-1130. Washington, D.C.: Scientific and Technical Information Division, 1988. A space agency document that examines the expected flow turbulence around the Galileo probe's parachute at velocities between 0.25 to 0.95 Mach (where Mach 1 is the velocity of sound in standard Earth atmosphere).

Cole, Michael D. *Galileo Spacecraft: Mission to Jupiter: Countdown to Space.* New York: Enslow, 1999. Suitable for younger audiences. This book explains how space scientists explore the solar system, specifically using the Galileo mission as subject material.

Covault, Craig. "Galileo Jupiter Orbiter/Probe Readied for Launch by Space Shuttle *Atlantis*." *Aviation Week and Space Technology* 131 (September, 1989): 23. An informative report on the preparations for the Galileo orbiter/probe launch from the space shuttle *Atlantis* during the STS-34 shuttle flight.

Fischer, Daniel. *Mission Jupiter: The Spectacular Journey of the Galileo Spacecraft.* New York: Copernicus Books, 2001. Provides a summary of Galileo's extensive findings in the Jovian system as well as the discoveries of the Pioneer and Voyager spacecraft that preceded it.

Hanlon, Michael, and Arthur C. Clarke. *The Worlds of Galileo: The Inside Story of NASA's Mission to Jupiter.* New York: St. Martin's Press, 2001. This text presents the Galileo spacecraft's exploration of asteroids and the Jupiter system while instilling the reader with the excitement of the voyage. The book also discusses the potential for a planetary ocean on the moon Europa.

Harland, David M. *Jupiter Odyssey: The Story of NASA's Galileo Mission.* London: Springer-Praxis, 2000. A detailed scientific and engineering history of the Galileo program, but also includes extensive discussion of Voyager program events and results.

Hartmann, William K. *Moons and Planets.* 5th ed. Belmont, Calif.: Thomson Brooks/Cole Publishing, 2005. Provides detailed information about all objects in the solar system. Suitable on three separate levels: high school student, general reader, and college undergraduate studying planetary geology.

Irwin, Patrick G. J. *Giant Planets of Our Solar System: Atmospheres, Composition, and Structure.* London: Springer-Praxis, 2003. Provides an in-depth comparison of Jupiter, Saturn, Uranus, and Neptune, incorporating data obtained from astronomical observations and planetary spacecraft encounters.

Martin, Donald H. *Communication Satellites.* 4th ed. New York: American Institute of Aeronautics and Astronautics, 2000. This work chronicles the development of communications satellites and worldwide networks over the past four decades, from Project Score to modern satellite communication systems.

Morrison, David, and Tobias Owen. *The Planetary System.* 3d ed. San Francisco: Addison-Wesley, 2003. Organized by planetary object, this work provides contemporary data on all planetary bodies visited by spacecraft since the early days of the Space Age. Suitable for high school and college students and for the general reader.

National Aeronautics and Space Administration. *Space Network Users' Guide (SNUG).* Washington, D.C.: Government Printing Office, 2002. This users' guide emphasizes the interface between the user ground facilities and the Space Network, providing the radio frequency interface between user spacecraft and NASA's Tracking and Data-Relay Satellite System, and the procedures for working with Goddard Space Flight Center's Space Communication program.

Ordway, Frederick I., III, and Mitchell Sharpe. *The Rocket Team.* Burlington, Ont.: Apogee Books, 2003. A revised edition of the acclaimed thorough history of rocketry from early amateurs to present-day rocket technology. Includes a disc containing videos and images of rocket programs.

Tucker, Wallace H., and Karen Tucker. *Revealing the Universe: The Making of the Chandra X-Ray Observatory.* Cambridge, Mass.: Harvard University Press, 2001. In addition to providing a detailed history of the Chandra program, this work goes through the basic physics involved in x-ray astronomy.

Yeates, C. M., et al., eds. *Galileo: Exploration of Jupiter's System.* NASA SP-479. Washington, D.C.: Government Printing Office, 1985. A space agency document outlining the goals of the Galileo Mission.

Yenne, Bill. *Secret Weapons of the Cold War: From the H-Bomb to SDI.* New York: Berkley Books, 2005. A contemporary examination of Cold War superweapons and their influence on American-Soviet geopolitics.

Zaehringer, Alfred J., and Steve Whitfield. *Rocket Science: Rocket Science in the Second Millennium.* Burlington, Ont.: Apogee Books, 2004. Written by a soldier who fought in World War II under fire from German V-2 rockets, this book includes a history of the development of rockets as weapons and research tools, and it projects where rocket technology may go in the near future.

*Monish R. Chatterjee, updated by David G. Fisher*

# Gamma-ray Large Area Space Telescope

*Date:* February, 2007 (projected)
*Type of spacecraft:* Space telescope

*The Gamma-ray Large Area Space Telescope (GLAST) is designed to study the highest-energy gamma rays from celestial objects. It will be about thirty to fifty times more sensitive, detect higher energies, see a larger area of the sky, and provide more accurate positional information than previous missions investigating gamma emissions. By virtue of its sensitivity to higher energies, GLAST will provide new information about the most energetic violent events and objects in the universe, including gamma-ray bursts, the most active galactic nuclei, pulsars, black holes, and the Big Bang itself.*

### Key Figures

*Peter F. Michelson*, principal investigator, Large Area Telescope
*Charles A. Meegan*, principal investigator, GLAST Burst Monitor team
*Giselher Lichti*, co-principal investigator, GLAST Burst Monitor team

### Summary of the Mission

The Gamma-ray Large Area Space Telescope (GLAST) was planned and approved in the late 1990's. GLAST will be launched on a Delta 292H-10 rocket into a circular orbit 550 kilometers above Earth's surface. The orbit will be inclined 28° to Earth's equator. A five-year mission lifetime is planned, but project scientists hope to exceed this goal by operating for at least a decade. The first year of the mission will be devoted to a survey of gamma-ray sources over the entire sky. After this initial survey, guest observers will be able to apply, on a competitive basis, for time to conduct specific observing projects. GLAST will also observe transient targets of opportunity as they arise.

The GLAST project is a team effort that includes the National Aeronautics and Space Administration (NASA) and the Department of Energy in the federal government as well as American universities and research laboratories. The team also includes government and university institutions from France, Italy, Japan, Germany, and Sweden. A few hundred individual scientists and engineers are involved. Spectrum Astro (a division of General Dynamics) is the contractor building the spacecraft.

Visible light is the form of electromagnetic radiation that is detectable by human eyes. Red light has a longer wavelength and lower energy than blue or violet light. Infrared light and all types of radio waves are also forms of electromagnetic waves that have longer wavelengths and lower energies than red light. At the other end of the electromagnetic spectrum, ultraviolet light and x rays have shorter wavelengths and higher energies than violet light. Gamma rays are the form of electromagnetic energy with the shortest wavelengths and highest energies. For comparison, visible light has wavelengths comparable in size to a typical single-celled organism, while gamma rays can have wavelengths comparable in size to the nucleus of an atom. Energies are often measured in electron volts; one electron volt is the energy an electron will have after accelerating across an electrical potential difference of one volt. Because this unit was invented by scientists to measure energies of individual subatomic particles, an electron volt is actually an extremely small amount of energy compared to familiar, everyday macroscopic energies. Visible light has energy on the order of a few electron volts. Gamma rays such as those studied by GLAST can have energies measured in millions of electron volts (megaelectron volts), even billions of electron volts (gigaelectron volts), or more.

GLAST will detect gamma rays ranging in energy from 10 megaelectron volts to 300 gigaelectron volts. Earth's atmosphere is opaque to these gamma rays, so gamma-ray astronomy must be done from space. The GLAST spacecraft will contain two instruments: the Large Area Telescope (LAT) and the GLAST Burst Monitor (GBM). The LAT will detect and measure the energies of gamma rays. It will also make gamma-ray images of objects. The GBM will watch about two-thirds of the sky at a time to look for gamma-ray bursts. These sudden brief flashes of gamma rays occur at random times and locations and are the most energetic events that are observed in the universe. Their nature is still a mystery.

*An artist's rendition of the GLAST spacecraft in orbit, with a cutaway view of the tower array.* (NASA-MSFC)

Because gamma rays are so much more energetic than ordinary visible light, gamma-ray telescopes cannot use designs similar to optical telescopes. They must instead borrow gamma-ray detection methods from nuclear and high-energy physics. The LAT gamma-ray detector has a square array of sixteen towers that alternate thirty-eight silicon strip detectors with thin sheets of lead. The silicon strip detectors were originally designed for high-energy particle physics experiments. A layer of tungsten in the detector creates an electron-positron pair (a positron is the antimatter counterpart to an electron). These particles cascade down the tower, creating additional particles in each detector. Alternate strips give the $x$ and $y$ positional information, so the two-dimensional position where the gamma ray struck the detector is available. Tracking the positional change of the cascading particles through the nineteen pairs of silicon strip detectors allows scientists to reconstruct the original direction of the gamma ray and therefore the position of the gamma-ray source in the sky.

Cosmic rays also trigger the silicon strip detectors, but they will produce a pattern different from that of gamma rays as they pass through the towers. Anticoincidence detectors below the towers will also help differentiate the cosmic rays and gamma rays. Directly below these anticoincidence detectors are cesium-iodide calorimeters to measure the energy of the gamma rays. The LAT also contains a data acquisition system to do preliminary analysis of the detections and to sort out cosmic-ray detections from gamma-ray detections.

The GBM has twelve sodium iodide and two bismuth germanate detectors on opposite sides of the LAT. When gamma rays strike these detectors, they emit flashes of light that in turn are detected by photomultiplier tubes that can detect and measure very faint, brief flashes of light. The sodium iodide detectors are sensitive to low-energy gamma rays from 8,000 electron volts (8 kilo-electron volts) to 1 megaelectron volt. They are designed for greatest sensitivity at the energies where gamma-ray bursts emit the most energy. The bismuth germanate detectors are sensitive to the higher energy range from 150 kiloelectron volts to 30 megaelectron volts. An onboard data-processing unit analyzes the incoming gamma rays from the GBM. When a gamma-ray burst occurs, it computes the position and can allow the GLAST to reposition so that the LAT can observe the burst.

The GBM and LAT will be positioned so that the larger field of view of the GBM encompasses the field of the LAT. Working together, the GBM and the LAT will allow scientists to find and observe gamma-ray bursts as well as other gamma-ray sources. Over the lifetime of the mission, GLAST will provide scientists with considerable new insights into the universe.

### Contributions

The first year of GLAST operations will be devoted to surveying and mapping high-energy gamma-ray emissions across the sky. This survey stage will both find new gamma-ray sources and provide more accurate positions for previously observed sources. Whenever astronomers open up a new spectral window, one of the first steps must be to map the sky in the new window to catalog the objects. After making the catalog, astronomers can study the most interesting objects in detail. This stage always holds the tantalizing possibility that entirely new classes of objects will be discovered.

GLAST will also reveal more about gamma-ray bursts. The first gamma-ray burst was observed on July 2, 1967, by the Department of Defense Vela 4 satellite, which was monitoring for Soviet nuclear tests. The discovery, however, was not announced to the astronomical community until 1973. These bursts last from a fraction of a second to a few minutes. On February 28, 1997, the first afterglow of a gamma-ray burst was discovered. The faint afterglow in longer optical, infrared, or radio wavelengths fades over a period of a few days or weeks

and allows astronomers to measure their distances, which are typically billions of light-years. When the burst fades completely, it reveals a faint, distant galaxy. Hence, these bursts occur in distant galaxies. Gamma-ray bursts seem to be associated with distant supernovae, but there are still many questions about their nature. The GBM will help astronomers identify gamma-ray bursts as they occur and allow them to observe the afterglows with ground-based instruments. The LAT will study the high-energy gamma-ray burst properties and will complement studies by other satellites studying lower-energy gamma rays. The combination of the

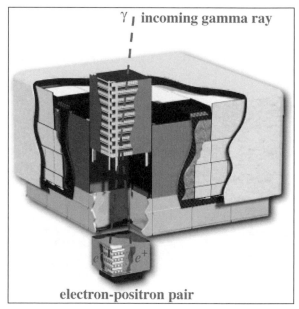

*The LAT gamma-ray detector has a square array of sixteen towers that alternate thirty-eight silicon strip detectors with thin sheets of lead. A layer of tungsten in the detector creates an electron-positron pair (a positron is the antimatter counterpart to an electron). These particles cascade down the tower, creating additional particles in each detector. Tracking the positional change of the cascading particles through the nineteen pairs of silicon strip detectors allows scientists to reconstruct the original direction of the gamma ray and therefore the position of the gamma-ray source in the sky. Shown here is a cutaway of the LAT detector showing an incoming gamma ray producing an electron-positron pair. (NASA/LAT Collaboration)*

GBM and the LAT will provide an unprecedented wide range of gamma-ray energies studied from a burst.

Many distant galaxies produce more energy in their cores than can easily be explained. They go by the generic name active galactic nuclei (AGNs). Most astronomers think that a supermassive black hole in the nucleus provides the energy for an AGN. GLAST will study high-energy gamma-ray emissions coming from AGNs to reveal the nature of their energy source.

As much as 90 percent of the expected matter in the universe has yet to be detected. Astronomers can infer its gravitational effect but have not been able to observe it at any wavelength of electromagnetic radiation—hence the name "dark matter." The nature of dark matter is one of the biggest mysteries of modern astrophysics, but there are several theories. One is that the dark matter is composed of subatomic particles called weakly interacting massive particles (WIMPs). If these WIMPs do indeed make up the dark matter, their interactions should produce high-energy gamma rays that might be detectable from GLAST. GLAST observations may therefore either confirm or refute the WIMP theory.

GLAST is also expected to observe gamma rays produced by other phenomena, such as solar flares and pulsars, to increase our understanding of these phenomena. Previous gamma-ray satellites have identified nearly three hundred gamma-ray sources that are still unidentified. GLAST observations will help scientists identify these sources.

### Context

GLAST will not be the first satellite to explore celestial gamma-ray emissions. Rather, it is part of a series of gamma-ray missions with complementary capabilities. GLAST will improve on the Compton Gamma Ray Observatory (1991-2000), which was one of the most successful gamma-ray missions to date. GLAST will study higher-energy gamma rays than previous missions and will have an improved ability to measure the positions of gamma-ray sources. This improved positional information will allow astronomers to more accurately pinpoint the visible, infrared, or radio source corresponding to a particular gamma-ray source. The LAT will be at least thirty times more sensitive than the corresponding instrument on the Compton Observatory.

The Burst and Transient Source Experiment (BATSE) instrument on the Compton Gamma Ray Observatory detected more than 2,700 gamma-ray bursts. The High Energy Transient Explorer and Swift missions are also designed to catch gamma-ray bursts and quickly relay positional information to ground-based observatories to allow them to study the afterglow. If these missions remain operational when GLAST is launched, GLAST will be able to complement their data by observing the higher-energy gamma rays from the bursts.

Gamma-ray astronomy also has spin-off benefits. Gamma-ray detectors originally designed for gamma-ray astronomy have been applied to breast cancer screening. Preliminary tests using gamma rays rather than traditional x-ray mammograms allow physicians to detect smaller tumors that will be in earlier, more curable stages.

**See also:** Chandra X-Ray Observatory; Compton Gamma Ray Observatory; Hubble Space Telescope; Hubble Space Telescope: Science; Infrared Astronomical Satellite; Orbiting Astronomical Observatories; Orbiting Geophysical Observatories; Orbiting Solar Observatories; Spitzer Space Telescope; Telescopes: Air and Space.

### Further Reading

Chaisson, Eric J. "Astrophysics and Women's Health." *Sensors Magazine* 20, no. 9 (2003): S12. This brief article describes the application of technology developed for gamma-ray astronomy to breast cancer detection.

_____. "Gamma-Ray Large Area Space Telescope." NP-1997 (12)-050-GSFC. Washington, D.C.: Government Printing Office, 1998. This NASA brochure describes the proposed GLAST mission. Despite the fact that some mission details have changed as the mission concept has matured, the article provides a good mission overview.

Chaisson, Eric J., and Steve McMillan. *Astronomy Today.* 5th ed. Upper Saddle River, N.J.: Pearson Prentice Hall, 2004. This introductory college textbook contains information about gamma-ray astronomy in the chapter on telescopes. Other chapters also contain astronomical background information that relates to various GLAST discoveries.

Fabian, A. C., K. A. Pounds, and R. D. Blandford. *Frontiers of X-Ray Astronomy.* London: Cambridge University Press, 2004. Discusses the revolution in research provided by the Chandra and Newton x-ray observatories and those spacecraft that preceded them. Puts the data in a cosmological context.

National Aeronautics and Space Administration, Goddard Space Flight Center. Gamma-Ray Large Area Space Telescope. http://glast.gsfc.nasa.gov and http://www-glast .stanford.edu. Accessed June, 2005. These official GLAST Web sites contain mission overviews, science summaries, mission news, and information of interest to scientists using GLAST. As the mission is launched and progresses, these sites will most likely provide up-to-date information about the mission.

Schilling, G. "Catching Gamma-Ray Bursts on the Wing." *Sky and Telescope* 107, no. 3 (2004): 32. This article is about the Swift mission, but it provides a good and readable summary of scientists' knowledge of gamma-ray bursts.

_____. "Stalking Cosmic Explosions." *Astronomy* 31, no. 2 (2003): 48. This readable article describes a major advance in the understanding of gamma-ray bursts that occurred when the first burst afterglow was observed.

Schlegel, Eric M. *The Restless Universe: Understanding X-Ray Astronomy in the Age of Chandra and Newton.* London: Oxford University Press, 2002. An accessible introduction to the history, methods of observation, and data of x-ray astronomy. Provides data from the Chandra, Newton, and Astro-E missions as well as from spacecraft that preceded them.

Wheeler, Craig J. *Cosmic Catastrophes: Supernovae, Gamma-Ray Bursts, and Adventures in Hyperspace.* Boston: Cambridge University Press, 2000. A complete analysis of high-energy processes at work in the universe. Addresses the contributions of the Compton Gamma Ray Observatory to that understanding.

*Paul A. Heckert*

# Gemini Program

*Date:* December 7, 1961, to November 15, 1966
*Type of program:* Piloted spaceflight

*The Gemini Program placed humans into Earth orbit and taught them how to track, maneuver, and control orbiting spacecraft; how to dock with other orbiting vehicles; and how to reenter Earth's atmosphere and land at specified locations. Later missions in the program were the first to provide astronauts with experience in long-duration flights and extravehicular activity (EVA).*

### Key Figures

*Charles W. Mathews* (1921-2002), Gemini Program manager
*Robert R. Gilruth* (1913-2000), director of the Manned Spacecraft Center, Houston
*George E. Mueller* (1918-2001), associate administrator of NASA for Manned Spaceflight, served as acting director of the Gemini Program
*Christopher C. Kraft, Jr.* (b. 1924), director of Flight Operations, Mission Control
*William C. Schneider* (b. 1923), deputy director of Manned Spaceflight for Mission Operations and mission director on all Gemini flights beginning with Gemini V
*Walter F. Burke*, vice president and spacecraft general manager, McDonnell Aircraft Corporation
*Robert N. Lindley*, engineering manager, McDonnell Aircraft Corporation

### Summary of the Program

Project Gemini was an outgrowth of Project Mercury and its subsequent flights, begun in 1959. Preparing to meet the goal set for the United States by President John F. Kennedy in 1961—to place a human on the Moon before the end of the decade—space experts believed that a transitional program was required to test the systems and maneuvers needed to accomplish such a feat. Earlier Mercury missions had been limited to a single astronaut making at most eighteen orbits, with each orbit lasting approximately ninety minutes. Extending these missions was impossible because of the size of the spacecraft and launch constraints. Orbital maneuverability, the ability of two spacecraft to rendezvous and dock, and prolonged extravehicular activity (EVA) on the part of the astronauts were among the main objectives that would have to be met and perfected before a lunar landing could be attempted.

John C. Houbolt of Langley Research Center (LaRC) had first drawn attention to the fuel-saving advantages of orbital rendezvous in early 1959. As a result of his promotional efforts, which were supported by LaRC personnel, it was agreed almost from the beginning of Project Mercury that an Earth-orbital or Moon-orbital rendezvous would be necessary for a successful lunar-landing mission. In 1961, the Space Task Group (STG), under the direction of Robert R. Gilruth, set about creating the mission requirements of Mercury Mark II; these specifications would demonstrate the feasibility of orbital rendezvous and docking, test whether astronauts could accomplish tasks over extended periods of time while in orbit, and determine the possibility of accurately landing a piloted spacecraft at specific points on Earth.

By October of 1961, the STG had specified the

basic requirements of this enlarged Mercury project, still known as Mercury Mark II. It would consist of a two-person crew, which would be lifted into orbit by a modified Titan II rocket booster. The Atlas-Agena B two-stage rocket would be used to place the Agena B into orbit, and the Mercury Mark II would rendezvous and dock with the Agena B. Because the Agena B had hypergolic fuel (fuel that could be ignited by mixing its two components), the Mercury Mark II, combined with the added fuel source, could then set off on prolonged missions; these missions could last about a week, which would be comparable to the length of a lunar flight.

One of STG's requirements was that existing aerospace technologies be used wherever possible. The U.S. Air Force was therefore assigned the responsibility for meeting Mercury's launch requirements, and McDonnell Aircraft Corporation of St. Louis, Missouri, entered negotiations for a contract for the Mercury Mark II spacecraft. (The Air Force had developed the Atlas booster, and McDonnell at the time was serving as the primary contractor for the Mercury Capsule.)

General Dynamics Astronautics of San Diego, California, which had functioned as the prime contractor for the Air Force, supplying the Atlas launch vehicles that had carried military payloads into space, would provide the Atlas booster, while Lockheed Aircraft Corporation's Missile and Space Division of Sunnyvale, California, would supply the Agena target vehicles. A series of modified Titan II launch vehicles for the Mercury Mark II were to be supplied by Martin Marietta Corporation's Space Systems Division of Denver, Colorado. When the specifications were finalized in October, 1961, STG personnel were in the process of moving into the new Houston headquarters, which was named the Manned Spacecraft Center. At the same time, Gilruth became director of the Center, and James Chamberlin, the engineering chief at STG, was appointed manager of Project Mercury. The Air Force had been developing an advanced and more powerful Agena stage, the Agena D, for its Titan III vehicle. It was decided on June 11, 1961, that Gemini would use the Agena D for its rendezvous missions. On December 7, 1961, Gilruth announced the approval of the Mark II program, which was renamed Gemini for the astrological twins, in view of the two-person spacecraft.

As initially announced by Gilruth, Gemini's budget was slightly in excess of $500 million. A total of twelve spacecraft, two robotic and ten piloted, would be launched beginning early in 1963. The robotic flights would be first, to test overall booster-spacecraft compatibility and systems engineering. Several piloted flights would then follow, culminating in relatively long stays in space and some EVA. The final flights of the Program would perform rendezvous and docking maneuvers with Agena, continuing flight under the power of Agena engines. On November 1, 1963, the Gemini "Project" Office officially became the Gemini "Program" Office. With the completion of Project Mercury, its responsibility was for the Program as a whole and not merely for the spacecraft. So, too, did the name of the endeavor change from Project Gemini to the Gemini Program.

Some delays in the initial launch schedule were caused by the proposed use of a paraglider to make ground landings. Francis M. Rogallo of Langley had designed a collapsible paraglider known as the Rogallo wing; at first every effort was made to have robotic Gemini craft test the device's efficacy in a controlled ground landing. Langley's engineers had difficulty in getting the collapsible paraglider to deploy quickly, however, and many questioned whether the risky procedure should be attempted at all, because the conventional drogue chutes and ocean landings had proved successful. Tests of prototype paragliders were mostly unsuccessful. After more than a year of research and development, Charles W. Mathews put a stop to the paraglider tests. Future piloted flights, including Apollo, were already committed to water landings; thus, the paraglider would have been an unnecessary expense.

The ensuing Gemini flights can be divided into three categories. The first three flights were developmental in nature, and only the last of these,

*The missions of Project Gemini showed that a properly clothed and equipped human could leave the spacecraft and work in space. Medical information gathered on the astronauts increased the knowledge of the effects of weightlessness on human physiology.* (NASA CORE/Lorain County JVS)

Gemini 3, was piloted. Next were the long-duration missions, Gemini IV, V, and VII. These flights culminated in Gemini VII's fourteen-day mission. Last were the rendezvous-docking missions, Gemini VI-A, VIII, IX-A, X, XI, and XII.

Gemini 1 was launched successfully on April 8, 1964. Its mass was about 5,000 kilograms, including the spacecraft, which weighed about 3,500 kilograms, and the adapter section, weighing about 1,500 kilograms. These masses were typical of all Gemini vehicles. The spacecraft itself was 3.35 meters long and 2.3 meters in diameter, in the typical "Coke bottle" shape designed by Maxime A. Faget— a design also used for earlier Mercury spacecraft and modified somewhat for the Apollo spacecraft. The adapter section contained all equipment not required inside the piloted spacecraft, such as fuel cells, oxygen tanks, attitude control thrusters, orbital maneuvering engines, and propellant and reactant tanks.

Gemini 2, the second robotic Gemini, experienced several launch delays; it was even hit by lightning while on the launch pad. It finally got under way on January 19, 1965.

Gemini 3, the last of the developmental flights, was launched on March 23, 1965, and was piloted by astronauts Virgil I. "Gus" Grissom and John W. Young. In the first demonstration of a pilot actively flying a spacecraft, Young made history by changing the orbit of *Molly Brown*. Unfortunately, Young also attracted worldwide attention by presenting Grissom with a corned beef sandwich that the former had smuggled aboard, much to the chagrin of NASA personnel. The flight lasted only 4 hours and 53 minutes.

Gemini IV was launched on June 3, 1965, with Edward H. White II and James McDivitt aboard. (Beginning with Gemini IV, NASA changed to Roman numerals for Gemini mission designations.) It performed a record long-duration flight of 97 hours, 56 minutes. White spent 21 minutes tethered outside the spacecraft performing an EVA. He used a handheld gun powered by compressed gas to maneuver. His statement when ordered back into the Gemini spacecraft by ground control, "This is the saddest moment in my life," attracted worldwide sympathy. The crew exercised with elastic straps, an isometric device used on later spaceflights. Upon the failure of the onboard computer, they had to reenter using a Mercury-type roll, landing 80 kilometers off target.

Gemini V was the first spacecraft to use fuel cells to provide electricity in orbit and performed a series of experiments on visual acuity and radio ranging.

The flexibility of the Gemini Program was perhaps best demonstrated by the flights of Gemini VI-A and Gemini VII. The latter was provisioned for a fourteen-day mission and was used as a rendezvous target by Gemini VI-A. The two craft came within 0.3 meter of each other and maneuvered around each other for 35 minutes—at the time setting an endurance record for astronauts Frank Borman and James A. Lovell.

The failure of the attitude control thruster on Gemini VIII nearly caused disaster for Neil A. Armstrong and David R. Scott, and they returned shortly after docking with the Agena on the seventh orbit.

Gemini IX-A (replacing Gemini IX, which had suffered from several prelaunch problems, including the loss of its Agena) was unable to dock with the replacement Augmented Target Docking Adapter (ATDA), as a result of the nose cone's failure to jettison. It did perform several endurance experiments—Astronaut Eugene A. Cernan performed a two-hour EVA during this mission.

Gemini X, carrying Michael Collins and John W. Young, successfully docked with two different Agena targets. After docking with the first Agena (left over from the ill-fated Gemini VIII flight), the astronauts fired it and reached an apogee (greatest orbital distance from Earth) of 763 kilometers. Then they discarded their first Agena and docked with the second, finally landing on Earth only 8 kilometers from their target.

The last two Gemini missions, XI and XII, focused on Apollo-type objectives; for the astronauts, this meant prolonged flight time as well as docking with Agenas. Edwin E. "Buzz" Aldrin, Jr., of Gemini XII made three EVAs. Both Gemini landed about 2.4 kilometers from their designated ocean sites, which proved that the onboard computer was sufficiently accurate during reentry to enable controlled landings to be planned in the future. Gemini XII returned to Earth on November 15, 1966, and represented a triumph for the Gemini Program.

### Contributions

The knowledge gained from the Gemini Program fell into four areas: rendezvous and docking operations, EVA techniques, operational experiences, and scientific experiments.

Gemini proved conclusively that rendezvous and docking operations were feasible in near-Earth orbits; scientists believed that this would prove true even in lunar orbits. In most flights, the rendezvous

radar functioned reasonably well and was able to provide the flight crew and the onboard computer with the range, range rate (closing speed), and azimuth and elevation angles. When either the radar or the computer failed, alternate systems were employed to complete the rendezvous. The ability of Gemini X to dock with two separate Agena rockets certainly was the culmination of these important program objectives.

EVAs became commonplace in the later Gemini flights. Astronauts proved that they could perform perfunctory repairs to the adapter during flight. Astronauts could work in their shirtsleeves within the spacecraft and readily "suit up" to perform select EVAs.

The operational experience gained—in both program management and launch and flight—was perhaps the most valuable outcome of the Gemini Program as a whole. The spacecraft, modular in design (unlike Mercury), could use conventional components in place of new ones if necessary (as was the case with Gemini V's fuel cells). As a result, the Program could go ahead on schedule. Prior to Gemini, major repairs and the overhaul of entire systems had to be made at or near the launch pad. With Gemini, failure in a system merely resulted in a new module being substituted and activated.

The scientific experiments performed on Gemini flights were broad-ranging and represented a serious effort by NASA to enlist the scientific community in the space agency's efforts. This policy of soliciting experiments in advance was retained in subsequent NASA space programs. Among the more notable science experiments were synoptic weather and terrain photography, which yielded pictures of very high quality; experiments on Gemini 3 with water injected into the plasma sheath during reentry to improve C-band and telemetry signals; measurements of astronauts' bone demineralization, beginning with Gemini IV, which taught astronauts that this deleterious effect of space travel could be controlled by proper diet and sleep; tests of human otolith (inner-ear) stability, which showed how the astronauts could orient themselves in a weightless environment; tests of the

Astronaut Maneuvering Unit (AMU), which functioned adequately; and experiments involving frog eggs, which divide in zero gravity much as they do on Earth.

Gemini also was the first program administered by NASA, at a cost of more than $1 billion and employing more than ten thousand people. Program expertise advanced immeasurably during the four years in which Gemini was active, and much was learned about the handling of budgets, schedules, and personnel that would pave the way for implementation of the Apollo Program.

### Context

With the successful launch of the world's first artificial satellite, Sputnik 1, on October 4, 1957, the Soviet Union achieved a considerable lead in the exploration of space. On April 12, 1961, the Soviets placed the first human in space; Yuri A. Gagarin orbited Earth for 108 minutes and attracted worldwide attention. In June, 1963, Valentina Tereshkova became the first woman in space, completing forty-eight revolutions of Earth.

It was against this background that Gemini entered the public sphere, and it was with Gemini that the United States "caught up" with the Soviets. Each Mercury flight had been preceded by several Soviet flights; after Project Mercury had been announced to the public, there were delays and failures in several of the test launches, causing alarm among a populace that, in the Cold War climate of the time, already felt demoralized by the Soviets' apparent superiority in space technology.

The Soviet Vostok series and the American Mercury flights had occurred during the same time period; the Voskhod series closely paralleled the Gemini flights. In the latter case, Americans learned that the Gemini spacecraft, although not as spacious as its Soviet counterpart, was the technically superior craft. Alexei Leonov, one of the two cosmonauts aboard Voskhod 2, performed the first EVA five days before the launch of Grissom and Young aboard Gemini 3, but by Gemini XII, American EVAs were commonplace events.

Gemini's greatest success, after the Program got under way, was the regularity of its flights. The program averaged one flight every sixty days, proving that the Cape's launch facilities, as well as the Program's management, could handle such intervals comfortably. Gemini paved the way for Apollo, which was scheduled to begin piloted flights shortly after Gemini XII landed.

Despite an eighteen-month delay, caused by the tragic deaths of Grissom, White, and Roger B. Chaffee on January 27, 1967, in a disastrous pad fire, the Apollo Program made progress toward its goal of a lunar landing. The tragedy coincidentally paralleled by the death three months later of Vladimir Komarov aboard a Soyuz flight.

As a transition between Mercury and Apollo, Gemini was a program of firsts: the first EVA, the first long-duration spaceflight, and the first use of an onboard computer to control orbital parameters and the critical reentry attitude of the spacecraft. Perhaps most important, Gemini revitalized the United States' belief in its technological ability and was crucial to the realization of the nation's goal to place the first human being on the Moon.

**See also:** Apollo Program; Atlas Launch Vehicles; Ethnic and Gender Diversity in the Space Program; Extravehicular Activity; Funding Procedures of Space Programs; Gemini Spacecraft; Gemini 3; Gemini IV; Gemini V; Gemini VII and VI-A; Gemini VIII; Gemini IX-A and X; Gemini XI and XII; Lifting Bodies; Manned Maneuvering Unit; Space Suit Development; Space Task Group.

### Further Reading

Borman, Frank, with Robert J. Serling. *Countdown: An Autobiography.* New York: William Morrow, 1988. Borman, who was on the backup crew for Gemini IV, flew on the most spectacular rendezvous mission of the Gemini Program as command pilot on Gemini VI-A. His insights into flight preparations and rendezvous missions are readable and informative.

Cernan, Eugene A., and Don Davis. *The Last Man on the Moon: Astronaut Eugene Cernan and America's Race in Space.* New York: St. Martin's Press, 1999. The story of the last two men to walk on the Moon is told by Cernan, commander of Apollo 17, and Davis, an experienced journalist. Cernan was the first American to spacewalk during Gemini IX-A.

Collins, Michael. *Carrying the Fire: An Astronaut's Journeys.* New York: Farrar, Straus and Giroux, 1974. Collins served on the backup crew for Gemini VII and flew as pilot on Gemini X, which rendezvoused and docked with two different Agena vehicles. Collins is a perceptive and sensitive writer, and his view on the entire space program is extremely valuable.

_____. *Liftoff: The Story of America's Adventure in Space.* New York: Grove Press, 1988. Many books have been written about the Apollo Program, most of them about Apollo 11. Few have given us an inside look at the delicate melding of man and machine. Contributing to this complete history of America's piloted space programs, Collins devotes a large portion to Apollo. He sets the record straight about some of the misconceptions of astronauts and space machines. The book is illustrated with eighty-eight line drawings by James Dean, former NASA art director, that add stark realism to an otherwise unreal world.

Gatland, Kenneth. *The Illustrated Encyclopedia of Space Technology: A Comprehensive History of Space Exploration.* New York: Salamander, 1989. This volume contains several photographs of Gemini craft and gives a brief account of the piloted Gemini missions, putting the Gemini Program into perspective.

Godwin, Robert, ed. *Gemini 6: The NASA Mission Reports.* Burlington, Ont.: Apogee Books, 1999. This collection contains reprints of the Gemini VI press kit, along with pre- and postmission reports. It also includes a CD-ROM featuring more than 50 minutes of 16-millimeter film footage of Gemini VI-A and more than 180 still pictures from the flight. There are two movies about the Gemini Program and an exclusive interview with Gemini VI-A Commander Walter M. Schirra, Jr.

Grimwood, James M., Barton C. Hacker, and Peter J. Vorzimmer. *Project Gemini, Technology and Operations: A Chronology.* NASA SP-4002. Washington, D.C.: Government Printing Office, 1969. An authoritative, day-to-day account of the planning and development of Gemini from its inception as Mercury Mark II to the final flight of Gemini XII. Illustrated.

Grissom, Virgil I. *Gemini: A Personal Account of Man's Venture into Space.* New York: Macmillan, 1968. An experienced astronaut, Grissom flew on Mercury and Gemini missions and was killed in the 1967 Apollo fire. This posthumously published work represents his views on the Gemini Program.

Hacker, Barton C., and James M. Grimwood. *On the Shoulders of Titans: A History of Project Gemini.* NASA SP-4203. Washington, D.C.: Government Printing Office, 1977. This history of the project is more or less the official NASA history of the Gemini Program. Well written and effectively organized, it gives a good picture of the Program's fiscal, political, personnel, technological, and flight problems. Illustrated.

Harland, David. *How NASA Learned to Fly in Space: An Exciting Account of the Gemini Missions.* Burlington, Ont.: Apogee Books, 2004. The nuts and bolts of the Gemini Program are explained in this well-written book. The launch vehicles and spacecraft are detailed, as are the astronauts who flew them and the missions they flew.

Kraft, Christopher C., Jr. *Flight: My Life in Mission Control.* East Rutherford, N.J.: Penguin Putnam, 2002. The first NASA flight director gives an account of his life in Mission Control.

Schefter, James L. *The Race: The Uncensored Story of How America Beat Russia to the Moon.* New York: Doubleday Books, 1999. Journalist Schefter presents the full story of the Space Race beginning in 1963.

Scott, David, and Alexei Leonov. *Two Sides of the Moon: Our Story of the Cold War Space Race.* London: Thomas Dunne Books, 2004. Astronaut Scott and Cosmonaut Leonov recount the drama of the Space Race set in the context of the clash between Russian communism and Western democracy.

Shayler, David J. *Gemini: Steps to the Moon.* Chichester, England: Springer-Praxis, 2001. The development of the Gemini Program from the perspective of the engineers, flight controllers, and astronauts.

_____. *Walking in Space: Development of Space Walking Techniques.* Chichester, England: Springer-Praxis, 2003. A comprehensive overview of EVA techniques.

Shepard, Alan B., Jr., and Donald K. "Deke" Slayton, with Jay Barbree and Howard Benedict. *Moon Shot: The Inside Story of America's Race to the Moon.* Atlanta: Turner, 1994. This is, indeed, the inside story of the Apollo Program as told by two men who actively participated in it. Some of their tales appear here for the first time. Shepard and Slayton discuss the Gemini orbital flights and tell some interesting behind-the-scenes stories of the Program. The book was adapted for a four-hour documentary in 1995.

Siddiqi, Asif A. *The Soviet Space Race with Apollo.* Gainesville: University Press of Florida, 2003.

_____. *Sputnik and the Soviet Space Challenge.* Gainesville: University Press of Florida, 2003. These two volumes constitute the first comprehensive history of the Soviet piloted space programs, from the end of World War II, when the Soviets captured German rocket technology, to the collapse of their Moon program in the mid-1970's.

Slayton, Donald K., with Michael Cassutt. *Deke! U.S. Manned Space: From Mercury to the Shuttle.* New York: Forge, 1995. This is the autobiography of the last of the Mercury astronauts to fly in space. After being grounded from flying in Project Mercury for what turned out to be a minor heart murmur, Slayton was appointed head of the Astronaut Office. Slayton talks of his frustration at being grounded and how he worked to regain flight status. He also discusses the flights of his fellow astronauts during the Gemini Program.

Wendt, Guenter, and Russell Still. *The Unbroken Chain.* Burlington, Ont.: Apogee Books, 2001. The autobiography of the only person who worked with every astronaut bound for space.

*John Kenny*

# Gemini Spacecraft

*Date:* April, 1959, to 1969
*Type of spacecraft:* Piloted spacecraft

*The Gemini spacecraft was developed to bridge the gap between the one-person Mercury vehicle and much more ambitious three-person Apollo spacecraft.*

### Summary of the Program

Originally called "the two-man Mercury" and later "Mercury Mark II," Gemini was first proposed in April of 1959 as a test platform for advanced space technologies. These would include rendezvous and docking, space navigation, new forms of lightweight power supplies, and the study of long-term effects of space travel on the human body.

The prime contractor for the Mercury spacecraft, McDonnell Aircraft Corporation, would also build Gemini. On December 22, 1961, contract NAS 9-170 was awarded for a total of twelve spacecraft, fifteen launch vehicle adapters, and eleven docking adapters for a total cost of $1.290 billion when completed.

Gemini consisted of three modules: the reentry module (or crew cabin), the retrograde section, and the adapter equipment section. The adapter was a truncated cone 228.6 centimeters high, 304.8 centimeters in diameter at the base, and 228.6 centimeters at the upper end, where it attached to the base of the reentry module. The reentry module consisted of a truncated cone, which decreased in diameter from 228.6 centimeters at the base to 98.2 centimeters, topped by a short cylinder of the same diameter and then another truncated cone decreasing to a diameter of 74.6 centimeters at the flat top. The reentry module was 345.0 centimeters high, giving a total height of 573.6 centimeters for the Gemini spacecraft. The entire system weighed in at about 3,851 kilograms.

The titanium crew cabin contained all of the display and control systems, along with life-support and electrical equipment. Additional experimental devices were also supplied, depending on the mission. The two astronauts were seated side by side, angled slightly apart, the commander on the left and pilot on the right. Each crew member had a hatch with an almond-shaped forward-looking window made of two panes of silicate glass and one aluminosilicate panel. The two hatches could swing out from the side of the spacecraft and, unlike the Mercury, could be opened and closed in space for spacewalk activities. The external skin of the cabin was similar to that of Mercury in that it was covered by corrugated "shingles" made out of René 41, a material composed primarily of nickel, chromium, cobalt, and molybdenum.

Also unlike the Mercury, the Gemini was truly maneuverable, thanks to the combination of the Orbital Attitude and Maneuvering System (OAMS) thrusters and a small onboard digital computer. The computer had a clock speed of 8 megahertz and a memory of 4,096 words (39 bits per word). For a basic mission the computer would have four modes: prelaunch, ascent, "catch-up," and reentry.

Gemini had no "launch escape system" tower perched on its nose as Mercury had, instead relying on aircraft-style ejection seats. These seats would be able to rocket a crew member up to 243 meters away in less then 10 seconds, but only after subjecting him to forces up to 24 times gravity in acceleration.

Perched on the top of the cabin section was a "nose" that contained the Rendezvous and Recovery (R&R) section and the Reentry Control System. The R&R section housed both the parachute landing system and the rendezvous radar used in the later flights. Immediately below that was the Reentry Control System: a short cylindrical section that contained 16 small, 111-newton-thrust rockets to orient the spacecraft during the reentry. These engines (and the OAMS engines) used mono-methylhydrazine for fuel and nitrogen tetroxide as an oxidizer. These hypergolic propellants ignited upon contact and need no ignition system. The heatshield was a dish-shaped structure made of a honeycomb mesh filed with ablative material made by Dow Corning.

Immediately behind the reentry module was the adapter section, made up of two main parts. The smaller of the two was the retrograde section: a short conical structure with two crossed-aluminum "I" beams. Mounted in each quadrant was a small solid-propellant retrograde rocket made by Thiokol that supplied 11,070 newtons of thrust. The rockets were fired sequentially at intervals of 5.5 seconds to start reentry.

The adapter's equipment section housed the propellant tanks needed for the OAMS engines, the main oxygen supply, and power systems consisting of either batteries or the more advanced fuel cells (for long-duration missions). Some experiments were also housed there. Scattered about this section were a series of sixteen OAMS rocket engines. Depending upon their location on the adapter, they had a thrust of 111, 378, or 445 newtons. The engines permitted Gemini to change its orientation in space and, unlike the Mer-

*Artist's rendition of the Gemini spacecraft with a cutaway showing the astronauts and structure of the spacecraft.* (NASA)

cury spacecraft, change its orbital plane as well. This capability was necessary for rendezvous and docking.

The Titan II missile supplied by the Air Force would launch Gemini. Measuring roughly 27.5 meters (90 feet) in length, the two-stage Titan weighed about 148,000 kilograms fully fueled. The first stage produced a thrust of 1,900 kilonewtons and the second, 450 kilonewtons.

Compared to the Gemini, the Soviet multicrew spacecraft Voskhod was rather crude. Fashioned out of a hastily modified one-person Vostok spacecraft, the Voskhod would fly only two missions, serving as a short-lived bridge vehicle while the more advanced Soyuz was being built.

A Gemini mission would begin with the "call to stations" at T minus 22 hours, when power was switched on in the launch vehicle. Propellant would be loaded at T minus 12 hours. For rendezvous missions, the Agena target countdown would start at about nine hours before launch. At about T minus 2 hours, the crew would board the spacecraft. The launch vehicle gantry was lowered at T minus 55 minutes. At T minus 4 seconds, the launch panel was armed, and at T minus 0, the twin first-stage engines ignited. The Titan would be held down until full thrust was achieved, at about T plus 3 seconds. Staging occurred at 145 seconds after liftoff, with the second stage dropping off at 345 seconds at about 152 kilometers in altitude and 1,340 kilometers downrange from the launch site.

After the second stage was discarded, the Gemini spacecraft would use the OAMS engines to shape the orbit as needed for the specific flight. When ready for reentry, the crew would orient the spacecraft blunt-end-first and head-down, jettison the adapter equipment section, and fire the four retrograde rockets. Shortly thereafter, the retrograde module would be jettisoned and the spacecraft would be oriented for reentry. The 5.6-meter-diameter conical ribbon drogue chute was deployed at an altitude of about 15 kilometers by a mortar in the nose of the spacecraft. When the spacecraft reached 3.2 kilometers, the 1.6-meter-diameter pilot parachute was deployed, pulling out

the 25.6-meter nylon "ring sail" parachute 6 seconds later.

In the event of a parachute malfunction, the crew would still be able to use the ejection seats. Unlike the Mercury, Gemini had no landing bag to cushion the impact. After main chute deployment, the astronauts would transfer from a single-point suspension to a two-point suspension by depressing the landing attitude switch. On splashdown, the spacecraft would be oriented with the hatches facing skyward, ready for the recovery crew to attach the flotation collar and life raft.

Originally, the Gemini was supposed to use a land recovery system instead of the now familiar water landing. Using a precursor of today's paraglider hang glider wings, the spacecraft would touch down at Edwards Air Force Base on three landing skids. This plan was abandoned in 1964 because of difficulties in deployment of the paraglider. (Early Revell plastic models of the Gemini were, however, sold with optional landing gear.)

Besides the flight spacecraft, several "boilerplate" spacecraft were built. These were little more than crude versions of the vehicle, usually for fit or structural integrity tests. BP-1, for example, was used for testing the paraglider recovery system. BP-2 was used for training the astronauts in ingress and egress during extravehicular activities (EVAs). The third boilerplate was destroyed while testing the ejection seats. BP-4 and BP-5 were used for recovery tests.

Perhaps the most unusual Gemini produced was a full-scale mock-up for the Revell model company. Revell was known for producing a number of high-quality plastic models of the various spacecraft and in 1967 ran a contest in which the grand prize was a full-scale Gemini replica. The lucky winner was a thirteen-year-old boy in Oregon who donated it to the Oregon Museum of Science and Industry, where it can still be seen.

A total of twelve flightworthy spacecraft were built. The first two were used for robotic missions (Gemini 1 and 2), and spacecraft number 3 was used for the first piloted mission in March, 1965. Gemini-Titan 1 (GT-1), launched on April 8, 1964,

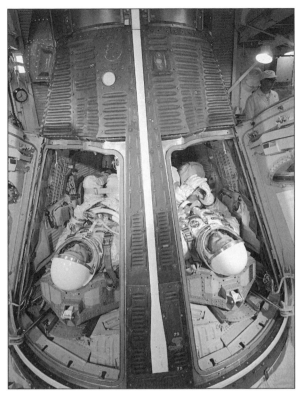

*The two-man Gemini spacecraft in launch orientation with astronauts James McDivitt and Ed White in position.* (NASA)

was a three-orbit, nonrecovery mission to test the basic integrity of both the Titan and the spacecraft. GT-1 was tracked until it reentered on its own after sixty-four orbits. It lacked most of the normal systems, and two equipment pallets were positioned where the crew would eventually be. The second and last robotic flight, GT-2, was launched on January 19, 1965. The objective was to test a full flight profile, focusing on the reentry with a complete spacecraft on an 18-minute suborbital mission. As in the first flight, instruments and cameras took the place of a crew.

Over the next two years, ten piloted missions were flown. Problems with fuel-cell development resulted in the use of batteries for Gemini 3, IV, and VI, limiting the mission durations. Gemini V, Gemini VII, and all subsequent missions would use fuel cells, essential for long-duration flights.

The spacecraft came in two major configurations: the basic version and the more sophisticated rendezvous version. For the latter, additional hardware, such as electrical connections and alignment devices, was added to facilitate docking with the Agena target.

Lockheed Missiles and Space built a number of "target vehicles" to use for rendezvous and docking tests starting with Gemini VI. The Agena with an Atlas first stage would be launched ahead of the Gemini. The 9.4-meter-long Agena had a restartable engine that could be used to boost the twin spacecraft up to record-breaking altitudes. Because of a launch failure with Agena 5002 for Gemini VI, the first docking would not take place until GT-VIII.

In the case of an Agena failure, McDonnell Douglas built a backup docking target called the Augmented Target Docking Adapter (ATDA) out of spare Gemini parts. The ATDA would be used with the Gemini IX-A mission after its own Agena suffered a launch failure. Gemini missions X, XI, and XII would successfully use the Agena for both docking and EVA operations. In addition to docking exercises, the Gemini X and XI missions included tethered experiments to study the dynamics of artificial gravity systems by inducing slow spins in the two attached vehicles.

Even though Gemini XII, launched on November 11, 1966, was the final piloted mission, the program itself was not at an end. The first follow-up program was the Manned Orbiting Laboratory (MOL), sponsored by the Department of Defense. The laboratory would be supplied by a modified "Gemini-B" spacecraft employed as a crew return vehicle. Meant to be used as a platform for long-duration military reconnaissance missions, only a single MOL test flight took place. On October 28, 1966, a Titan III lifted off from the launch pad carrying a refurbished Gemini 2 spacecraft for a 33-minute suborbital flight. The project was canceled in 1969 to avoid unnecessary duplication of effort with the Skylab program, run by the National Aeronautics and Space Administration (NASA).

McDonnell Douglas proposed an expanded Gemini called the Big Gemini (or "Big G") for flying cargo and passengers to the proposed space stations of both NASA and the military. Between the Gemini-like crew cabin and its retrograde module would sit a large cargo container able to hold up to ten passengers or several tons of material. Big G was never officially started, and the project died a quick death when the MOL was canceled.

However, a part of Gemini would live through the 1970's, when a single leftover Gemini hatch would be incorporated into the Skylab space station.

### Contributions

The Gemini development proved the practicality of landing on the ocean versus landing on the earth. The use of ejection seats showed that a costly launch escape tower used in Mercury and later Apollo was not the only means of escaping an aborted launch. As a result, the early shuttle flights used ejection seats. The new advanced onboard computer proved that complicated rendezvous and docking maneuvers could be done without aid from ground control if necessary. In addition, without the new fuel cell technology, Apollo would have been impossible.

The versatile Gemini spacecraft helped NASA gain much-needed flight experience for the upcoming Apollo Program. Specifically, experience was gained in the complicated rendezvous and docking operations that would be used for crew transfers between the Apollo Command Module and the Lunar Module. (Interestingly enough, experience from one docking experiment was used decades later to determine the root cause of a disastrous docking failure with the Russian Mir Space Station.) Additional experience was gained in long-duration spaceflights (those lasting up to two weeks), emergency rescue procedures, and the development of EVA techniques and equipment that would prove extremely useful in both Apollo and post-Apollo programs.

### Context

After Project Mercury ended, it was necessary to gain true operational experience in space before the highly ambitious Apollo Program could commence. Basic understanding of human physiology during long-duration spaceflight was still unknown, and guidance and navigation techniques for sophisticated rendezvous operations were merely the realm of theorists. Moreover, many astronauts needed flight experience before they would be lofted toward the Moon. Building upon many of the technologies learned from Mercury, the Gemini spacecraft was the ideal platform for accomplishing these goals.

**See also:** Tethered Satellite System; Titan Launch Vehicles; Vandenberg Air Force Base; X-20 Dyna-Soar.

### Further Reading

Hacker, Barton C., and James M. Grimwood. *On the Shoulders of Titans: A History of Project Gemini.* NASA SP-4203. Washington, D.C.: Government Printing Office, 1977. The official story from the NASA History Office.

_____. *Project Gemini: A Chronology.* NASA SP-4003. Washington, D.C.: Government Printing Office, 1969. A detailed chronological listing of all important Gemini events from management meetings to tests, decisions, and missions.

Harland, David. *How NASA Learned to Fly in Space.* Burlington, Ont.: Collector's Guide, 2004. A popular account of the Gemini Program.

Shayler, David J. *Gemini: Steps to the Moon.* New York: Springer-Verlag, 2001. Detailed overview of the entire project and missions oriented toward historians or the more technically inclined.

*Michael Smithwick*

# Gemini 3

*Date:* March 23, 1965
*Type of mission:* Piloted spaceflight

*Gemini 3 was the United States' first two-person orbital spaceflight. It tested the spacecraft compatibility and its crew's ability to control a craft in orbital flight. In addition, it performed three science experiments and took some spectacular photographs of Earth. It completed three Earth orbits with a total flight time of 4 hours and 53 minutes.*

### Key Figures

*Virgil I. "Gus" Grissom* (1926-1967), Gemini 3 command pilot
*John W. Young* (b. 1930), Gemini 3 pilot
*Walter M. Schirra, Jr.* (b. 1923), Gemini 3 backup command pilot
*Thomas P. Stafford* (b. 1930), Gemini 3 backup pilot
*Christopher C. Kraft, Jr.* (b. 1924), director of Flight Operations, Mission Control
*Charles W. Mathews* (1921-2002), manager of the Gemini Program
*Robert R. Gilruth* (1913-2000), director of the Manned Spacecraft Center, Houston
*George E. Mueller* (1918-2001), associate administrator for Manned Spaceflight
*L. Gordon Cooper, Jr.* (1927-2004), Gemini 3 capsule communicator

### Summary of the Mission

The Gemini 3 mission carried two American astronauts into Earth orbit for the first time. Its main purpose was to evaluate the Gemini-Titan 2 compatibility during launch and perform a brief series of evaluations of the living quarters and controls available on Gemini. It was the first piloted spacecraft to make an orbital plane change using the Orbital Attitude and Maneuvering System (OAMS).

Prior to Gemini 3, the entire Gemini Program had suffered, particularly with the endemic delays experienced by the robotic Gemini 2. Gemini 1 had been launched on April 8, 1964, and was a considerable success, making sixty-four Earth orbits. The Gemini Program appeared to be progressing smoothly as preparations commenced for the Gemini 2 launch, which was to be the final robotic test flight. The poor summer weather in Florida and several systems failures, however, caused the June launch date to be postponed. When Gemini 2

was finally moved to the launch pad, it was struck by lightning, and after repairs a complete power failure occurred during preliminary launch tests. On January 19, 1965, however, Gemini 2 left Earth to complete its 18-minute suborbital flight. The primary purpose of the flight was to expose Gemini to the most severe heating possible during reentry. Despite total failure of the fuel cells and more than six months' delay in launching, the flight accomplished its purpose in proving that Gemini could withstand severe heating during reentry.

Virgil I. "Gus" Grissom and John W. Young were, respectively, the command pilot and pilot chosen to occupy Gemini 3. Grissom was the oldest astronaut in the Program since John H. Glenn, Jr., had left the Astronaut Corps the previous year and was one of the most popular with training and engineering personnel. At the St. Louis McDonnell facilities, the Gemini spacecraft became known as

*Astronauts Virgil Grissom (left) and John Young preparing for launch in the Gemini capsule. They made America's first two-person flight in the Gemini 3 mission. (NASA)*

Gusmobiles out of respect for the relentless training and fitting efforts in which Grissom was involved. Grissom was also one of the shortest astronauts, and the adjustments made on Gemini seating, originally fixed with him in mind, had to be changed for several of the taller astronauts who flew in Gemini.

The primary mission of Gemini 3 was to test the Orbital Attitude and Maneuvering System (OAMS). Grissom and Young, as well as the backup crew of Walter M. Schirra, Jr., and Thomas P. Stafford, were the best trained of the entire Gemini flight crews; the continual delays of Gemini 2 provided them with additional training time. Grissom named Gemini 3 the *Molly Brown* for a popular musical of the time. (Molly Brown was a wealthy woman from Colorado who survived the tragic sinking of the ocean liner *Titanic* in 1912, and the musical *The Unsinkable Molly Brown* was based on her story.) It was a reminder to him that he had lost his returning Mercury 4 spacecraft in the ocean in 1961. His second choice of a name for the craft was the *Titanic*, so National Aeronautics and Space Administration

(NASA) officials reluctantly went along with the first choice, even though no official NASA literature refers to *Molly Brown* as the designation for Gemini 3.

*Molly Brown* was launched on March 23, 1965, at 14:24 Coordinated Universal Time (UTC, or 9:24 A.M. eastern standard time). The liftoff was so smooth that Young and Grissom later claimed that they hardly knew they were in space. Both were experienced test pilots, but the gravitational pull on their bodies during launch, six times Earth's gravity, would certainly have gotten the attention of most passengers. The Titan II launch vehicle functioned perfectly, however, and within the Gemini spacecraft, both astronauts indicated that it was much quieter than they had anticipated. Gemini 3 entered a nearly perfect orbit, but with a slight over speed it arrived at an apogee (farthest point from Earth) of 175 kilometers and perigee (closest point in orbit to Earth) of 122 kilometers. The programmed parameters had been 182 kilometers and 122 kilometers, respectively.

About twenty minutes into the flight, there was a localized failure in the instrument panel's power supply, and the oxygen pressure gauge constantly monitored by Young showed an abrupt drop. Such a pressure drop would generally be serious, but within 45 seconds both Young and Grissom were convinced that it was a gauge failure and not a real drop in pressure. They then switched to the secondary power supply, and the gauge readings quickly returned to normal.

An hour and a half into orbit, the first OAMS burn occurred and lasted for a carefully planned 75 seconds. This OAMS was the key dynamic feature of the Gemini spacecraft, and its functioning would indeed determine the feasibility of orbital changes and rendezvous in the future. The system worked successfully and reduced the craft's speed by approximately 15 meters per second and placed it in a new orbit with a perigee of 72 kilometers to ensure reentry.

NASA officials had decided prior to the launch of Gemini 3 that the mission would consist of only three orbits. There was some tension between the astronauts as to whether open-ended missions might be allowed. If an open-ended mission were proceeding without any difficulties, it could be extended at the behest of the flight crew or responsible officials such as Christopher C. Kraft, Jr., at Cape Kennedy. At the time, however, NASA was interested in demonstrating successes with the Gemini Program, a conservative attitude that was to dominate the early Gemini flights. Grissom and Young mentioned during flight that they would like to continue past the planned three orbits, but they were called home by L. Gordon Cooper, Jr., the capsule communicator.

The pyrotechnics or explosives used to rid the spacecraft of the adapter package caused the astronauts to experience some shock, but reentry went very well, and the plasma sheath experiment was performed with reasonably good results by Young.

After passing through the plasma sheath phase, Grissom immediately recognized that Gemini 3 would be far off target, by about 70 kilometers. Wind-tunnel tests had indicated that his craft would have greater lift than actually proved to be the case, and his attempts at gliding Gemini closer to the planned landing site failed. Gemini 3 ended up about 85 kilometers short of the planned landing site. When Grissom shifted the spacecraft, the astronauts lurched forward in the cabin and hit the windshield. The impact broke the faceplate on Grissom's space suit, and Young's suit suffered minor damage.

When Gemini 3 landed, all Grissom could see from his window was water because the attitude in landing was about 90° off in roll. The still-attached parachute, dragged by stiff winds, was dragging Gemini's nose under water, but when Grissom released the chute the nose popped up above the water easily.

As a result of their landing so far short of the designated site, the aircraft carrier *Intrepid* was almost 100 kilometers away and would take several hours to reach them. Grissom called for a helicopter to pick them up and refused to open the hatch until Navy swimmers had placed the flotation collar around Gemini 3 to ensure that even if water got inside, it would not sink. Grissom's decision was a result of his Mercury-Redstone 4 experiences; there his actions were alleged to have caused the Capsule to sink.

A public uproar developed when it was discovered that Young had brought a corned beef sandwich on board and offered it to Grissom during the flight. Grissom had taken only a bite or two and had carefully put away the remainder of the sandwich. Several congressmen and NASA officials, however, believed it was a frivolous event that should not be repeated. Most today would agree that, in view of the antics that later provided fodder for the televised appearances of astronauts in space, the case of the corned beef sandwich was a minor infraction of decorum; however, during the Cold War, spaceflight was not taken for granted and was approached with high seriousness by both public officials and private citizens.

### Contributions

Gemini 3 was the qualifying piloted spaceflight for the entire Gemini Program. The evaluations of its design, its compatibility with the Titan II launch vehicle, and its continuous tracking by a reliable worldwide tracking network proved that the practical engineering and technological aspects of the Program were on solid ground.

The successful firing of the Orbital Attitude and Maneuvering System to decrease Gemini 3's velocity by 15 meters per second and reduce its perigee to 72 kilometers caused orbital dynamics groups throughout NASA's network to breathe easier. The OAMS worked very well, and its 75-second application was controlled in the manner that had been accurately planned by dynamics groups. In fact, the launch and tracking techniques that were set in place for Gemini 3 formed the backbone of the NASA network for more than a decade.

The scientific experiments on board produced perhaps the only marked failure. The Sea Urchin Egg Growth Experiment failed as a result of Grissom's breaking a handle which, when released, was supposed to cause sea urchin sperm to mix with sea urchin eggs while in flight. The experiment was meant to determine whether such eggs multiplied or hatched in weightlessness at a rate similar to that at the surface of Earth.

In another experiment, the Reentry Communications Fluid Injection, water was ejected at various rates during reentry. This experiment was designed to determine whether radio frequency blackouts, which occurred while the spacecraft was surrounded by the plasma sheath of hot ions, could be ameliorated. It was successful in that there were increased ultrahigh-frequency signals and C-band (radar or microwave frequencies of 5 gigahertz) signals detected at ground stations. Normally during reentry there were several minutes of radio communications blackout to and from the spacecraft. Most likely, the results of these experiments, which were exclusive to Gemini 3, were used as a basis of subsequent testing done by the military and national security apparatuses of both the United States and the Soviet Union.

Zero-gravity and blood radiation studies were also conducted, using Young's blood. No major biological effects on the blood were noted.

Gemini 3 did not have the lift that the aerodynamics personnel expected it would have. Recognition of this led to more accurate landings in future flights as both the drag and the lift of Gemini were recalculated based on the sinking trajectory of Gemini 3.

### Context

Five days before Gemini 3 was launched, the Soviet Union's Voskhod 2 lifted off from Baikonur

*Mission Control during the flight of Gemini 3.* (NASA)

Cosmodrome with two cosmonauts aboard, Pavel Belyayev and Alexei Leonov. Leonov then left the Voskhod and became the first person to perform an extravehicular activity (EVA), connected to his ship only by telephone and telemetry cables and carrying his required oxygen with him in a backpack tank.

Grissom and Young's flight commenced with the knowledge that, in the popular jargon of the times, "the Russians were ahead." There was pressure on NASA officials to do something spectacular with Gemini 3 in order to prove that the United States was indeed a contender in the Space Race. The continual delays experienced in getting the robotic Gemini 2 off the pad made NASA officials very nervous. With public and congressional clamors for action, it would have been very easy to succumb to pressures and try to bring about some sort of unprecedented event with Gemini 3 and its two-person crew. Merely changing orbits with the OAMS was not an activity that would ignite either public or congressional interest and confidence. There was also continual tension within NASA to try some open-ended missions. The basic premise was that if everything was going well, then the mission could be extended to achieve its technical limits. Gemini 3 could have been extended to at least twenty orbits, and maneuvers more daring than those undergone in shifting the perigee by 50 kilometers could have been tried with a good chance for success.

To its credit, NASA followed the slow-and-steady approach, even though it appeared unduly conservative at times. To the scientists, the changing of orbital parameters for the first time in a well-prescribed and accurate manner was well worth the flight of Gemini 3. It showed that orbital rendezvous and docking were possible. The attainment of this level of control would prove the Gemini and subsequent Apollo Programs successful.

Gemini 3 proved the value of the Gemini spacecraft as a relatively comfortable though small excursion vehicle. Its modular design and the shifting to the secondary power supply early in the flight vouched for the efficacy of redundancy. Launch and astronaut training techniques had indicated that they were functioning smoothly.

After Gemini 3, all the subsequent program launches occurred approximately every sixty days as originally scheduled by the Space Task Group in October, 1962. By the time Gemini XII flew two years later, the public recognized that the United States was at least abreast of the Soviet Union, and the confidence and interest that were necessary to push Apollo along were present. Gemini 3 was a smooth transition between the robotic and the piloted phases of the Gemini Program.

The three scientific experiments had some success. (Grissom later took the blame for breaking the handle on the sea urchin container.) Both astronauts explained that the three experiments had been added rather late, and there was no opportunity to practice doing them during training sessions.

The water injection experiment during reentry did rather well, and both C-band beacons and ultrahigh-frequency signals seemed to be enhanced during periods of maximum water injection. Normally, during reentry, the craft is surrounded by a plasma sheath, actually ionized gases from the heat, and all radio communications to and from the spacecraft go dead for several minutes.

Young's experiment in determining the effects of space radiations and zero gravity on the blood at least established a baseline with which the other Gemini flights could compare similar results. No serious or deleterious effects were noted by either astronaut, even though both experienced nausea in the spacecraft after being buffeted by the ocean waves after landing. Jokes continually dogged Gemini astronauts for getting seasick rather than "space sick."

Overall, Gemini 3 was a great success and spoke well for the Program. The OAMS performed well, and planners became convinced that orbital changes and rendezvous and docking were even easier than some critics had expected.

**See also:** Apollo Program; Apollo Program: Orbital and Lunar Surface Experiments; Extravehicular Activity; Gemini Program; Gemini Spacecraft; Gemini IV; Mercury Project; Mercury-Redstone 4; Space Shuttle Flights, 1983.

## Further Reading

Borman, Frank, with Robert J. Serling. *Countdown: An Autobiography*. New York: William Morrow, 1988. Borman, who was on the backup crew for Gemini IV, flew on the most spectacular rendezvous mission as command pilot on Gemini VI, which orbited and rendezvoused with Gemini VII. His personal insights on flight preparations and undergoing a rendezvous mission are very readable to general audiences.

Cernan, Eugene A., and Don Davis. *The Last Man on the Moon: Astronaut Eugene Cernan and America's Race in Space*. New York: St. Martin's Press, 1999. The story of the last two men to walk on the Moon is told by Cernan, commander of Apollo 17, and Davis, an experienced journalist. Cernan's career began with the Gemini Program.

Grimwood, James M., Barton C. Hacker, and Peter J. Vorzimmer. *Project Gemini, Technology and Operations: A Chronology.* NASA SP-4002. Washington, D.C.: Government Printing Office, 1969. This work is an authoritative day-to-day account of the planning and development of the Gemini Program from inception as Mercury Mark II to the final flight of Gemini XII. Good sketches are included, technical problems are examined, and details are summarized effectively. The book focuses on Gemini throughout its three hundred pages and is quite detailed at times.

Grissom, Virgil I. *Gemini: A Personal Account of Man's Venture into Space.* New York: Macmillan, 1968. Grissom flew on Mercury, on Gemini 3 as command pilot, and was killed in the 1967 Apollo fire. This posthumous work represents the astronaut's views on Gemini. It is suitable for young people and the general public.

Hacker, Barton C., and James M. Grimwood. *On the Shoulders of Titans: A History of Project Gemini.* NASA SP-4203. Washington, D.C.: Government Printing Office, 1977. This 625-page history of the project is more or less the official NASA history of the Gemini Program. It is well written and organized effectively and gives a good picture of the fiscal, political, personnel, technological, and flight problems as they occurred. Illustrated.

Harland, David. *How NASA Learned to Fly in Space: An Exciting Account of the Gemini Missions.* Burlington, Ont.: Apogee Books, 2004. The nuts and bolts of the Gemini Program are explained in this well-written book. The launch vehicles and spacecraft are detailed, as are the astronauts who flew them and the missions they flew.

Kraft, Christopher C., Jr. *Flight: My Life in Mission Control.* East Rutherford, N.J.: Penguin Putnam, 2002. The first NASA flight director gives an account of his life in Mission Control.

Shayler, David J. *Gemini: Steps to the Moon.* Chichester, England: Springer-Praxis, 2001. The story of the development of the Gemini Program from the perspective of the engineers, flight controllers, and astronauts.

Shepard, Alan, Donald K. "Deke" Slayton, with Jay Barbree and Howard Benedict. *Moon Shot: The Inside Story of America's Race to the Moon.* Atlanta: Turner, 1994. This is, indeed, the inside story of the Apollo Program as told by two men who actively participated in it. Some of their tales appear here for the first time. Shepard and Slayton discuss the Gemini orbital flights and tell some interesting behind-the-scenes stories of the Program. The book was adapted for a four-hour documentary in 1995.

Slayton, Donald K., with Michael Cassutt. *Deke! U.S. Manned Space: From Mercury to the Shuttle.* New York: Forge, 1995. This is the autobiography of the last of the Mercury astronauts to fly in space. After being grounded from flying in Project Mercury for what turned out to be a minor heart murmur, Slayton was appointed head of the Astronaut Office. Slayton talks of his frustration at being grounded and how he worked to regain flight status. He also discusses the flights of his fellow astronauts during the Gemini Program.

*John Kenny*

# Gemini IV

*Date:* June 3 to June 7, 1965
*Type of mission:* Piloted spaceflight

*Gemini IV was the second piloted spacecraft in the Gemini series and the first U.S. mission to include an extravehicular activity (EVA). In addition, it established a two-person duration record and set about establishing work-sleep schedules for astronauts during its sixty-two Earth orbits.*

## Key Figures

*James A. McDivitt* (b. 1929), Gemini IV command pilot
*Edward H. White II* (1930-1967), Gemini IV pilot
*Frank Borman* (b. 1928), Gemini IV backup command pilot
*James A. Lovell, Jr.* (b. 1928), Gemini IV backup pilot
*Virgil I. "Gus" Grissom* (1936-1967), Gemini IV capsule communicator
*Charles W. Mathews* (1921-2002), Gemini Program manager
*Kenneth S. Kleinknecht* (b. 1919), Gemini Program deputy manager
*Christopher C. Kraft, Jr.* (b. 1924), director of Flight Operations, Mission Control
*Eugene F. Kranz* (b. 1933), Gemini IV deputy mission controller
*John Hodge,* Gemini IV deputy mission controller
*Charles A. Berry* (b. 1923), Gemini Program physician

## Summary of the Mission

Gemini IV carried two astronauts into orbit; during this time, one of them, Edward H. White II, performed 21 minutes of extravehicular activity (EVA). The flight was the basis for more prolonged orbital flights for astronaut work-sleep cycles; tested the spacecraft's ability to support a long-duration, four-day flight; and for the first time focused on some rendezvous orbital problems by attempting an orbit with the nearby second stage of the launch vehicle.

Gemini 3 had shown that the Gemini craft could perform well for a short-lived mission; it actually completed only three orbits. Besides orbiting and reentering successfully, Gemini 3 was the first spacecraft to change orbital parameters while in flight. (Beginning with Gemini IV, NASA began to use Roman numerals for Gemini mission designations to coordinate with the Roman numeral II, symbolic of the Gemini Program.)

The successful EVA performed by Cosmonaut Alexei Leonov just five days before the launch of Gemini 3 caused considerable turmoil within the National Aeronautics and Space Administration (NASA) hierarchy. Because of the approximately eighteen-month delay in the Gemini Program, it was apparent to all that the United States should have performed the first EVA on either Gemini IV or Gemini V according to the original schedule.

At first, NASA officials downplayed plans for an EVA on Gemini IV. Leonov's Voskhod 2 flight and his EVA had caused dissension as to how to respond to the public and within the Gemini Program. Planning an EVA for Gemini IV would give the appearance of a hasty response to a Russian challenge. Some critics even claimed that the Russians had not planned an EVA until they saw that they could beat the United States to one of its publicly declared goals. Moreover, a failure on Gemini IV would be disastrous to the entire program. Because of the efforts of Kenneth S. Kleinknecht, the deputy manager of the Program, and of the Astronaut Corps, however, it was finally agreed that Gemini IV would make the attempt; White would be the guinea pig.

In addition, Gemini IV had a sixty-two-orbit mission that would take about four days. Charles A. Berry, the NASA physician in charge of space medicine, argued forcefully against such a prolonged mission so early in the Program, remembering the cardiovascular problems experienced by pilots in the last of the Mercury flights, when relatively long missions had been flown. Then, too, Leonov had experienced vision and orientation problems while involved in his EVA on Voskhod 2. Berry, on these grounds, argued for a one-day mission, with longer, four-day flights coming later.

Whatever the ultimate decision about the attempt, White and James A. McDivitt were trained for a possible EVA. A bungee cord with a force constant of 3,000 newtons per meter was incorporated

into their training and required as an in-flight isometric exerciser; it was intended that the cord would ameliorate both the expected cardiovascular problems and the disorientation during EVA. Scientists believed that a good exercise program would cause a good work-sleep cycle and that it was exhaustion that caused disorientation.

On June 3, 1965, at 14:15:59 Coordinated Universal Time (UTC, or 10:15:59 A.M. eastern standard time), McDivitt and White were launched from Cape Kennedy on Gemini IV. It was perhaps the most public of all U.S. launches to date, because rumors of the proposed EVA had circulated in the press both before and after NASA outlined exactly what Gemini IV would do. The Early Bird satellite televised the launch to twelve countries outside the United States, and more than one thousand journalists asked Cape officials for press accreditation and accommodations.

Everything went well during the early launch phase, and McDivitt and White found themselves in a 163-by-282-kilometer orbit, as planned. Their first task, added as a secondary objective of the Gemini IV mission, was to maneuver the craft and attempt a close rendezvous with the second stage of the launch vehicle, which was in an almost identical orbit.

Both McDivitt and White saw the second stage tank at a distance of less than 100 meters and fired the thruster to attempt rendezvous. Both were trained test pilots, and they approached the task very much as they would have while flying jet aircraft. (To catch an object in flight, the craft must speed up and intercept it in the shortest possible spatial interval using the stick for altitude control and direction and the throttle for speed control.)

The two astronauts and many of the engineering personnel watching this rendezvous effort received

*The Gemini IV crew in pressure suits around a model of the spacecraft. From left: Pilot Edward H. White II, Command Pilot James A. McDivitt, and backup crew Frank Borman and James A. Lovell, Jr. (NASA)*

*Helmets on, White and McDivitt take the elevator to the white room to enter the space capsule.* (NASA)

a quick lesson in orbital dynamics. Orbiting objects obey Kepler's laws of motion, and speeding up to catch an object in front causes the spacecraft to fall into an orbit farther from Earth, which in turn lengthens the orbit. To catch an object in orbit, a spacecraft must decrease its speed, thereby decreasing its distance from Earth. When the target vehicle is at the correct position with respect to the pursuing craft, the pursuing craft must then increase its own speed to intercept the target.

The onboard computer that would direct the astronauts in subsequent rendezvous was aboard Gemini IV, but the radar that was to supply the target's coordinates was not yet ready for flight. Besides, Gemini IV had about one-half the amount of fuel for maneuvering that subsequent Gemini craft did, and its fuel level had gotten dangerously low in attempting this secondary mission objective. After more than an hour of attempting rendezvous via astronaut control, the target was still several kilometers away. Because of the planned EVA, it was decided that no more maneuvering thruster fuel should be wasted. Thus, the unaided attempts of the two astronauts were failures, and further rendezvous attempts were called off by ground control.

When the time came for the EVA, both White and McDivitt checked their Gemini IV pressure suits, which were much lighter and more mobile than the earlier Mercury and Gemini space suits. The Gemini IV suits were designed specifically for EVA. When White then tried to open the hatch, it stuck, and only by using brute force, with a wrench as a lever, were the astronauts able to get the hatch open. White then left the Gemini IV spacecraft and spent a total of 21 minutes outside it, controlling his direction with a handheld air gun. After checking some packages on the adapter, he exhausted the gun's air supply. It was time to return to the spacecraft, but White was reluctant to do so. It was the flight director himself, Christopher C. Kraft, Jr., who finally ordered Capsule Communicator Virgil I. "Gus" Grissom to get White back into Gemini IV. White himself seemed enamored of his surroundings and suffered no nausea or any of the other problems that had been feared. White's condition was attributed to his thorough training in finding his bearings on Earth and the Sun relative to his craft; thus, he was always aware of his position relative to Gemini IV.

When he got back inside, White realized how fatigued he was; in the excitement of the EVA, he had not been aware of it. He and McDivitt had to exert a tremendous amount of effort to close the hatch, and both of them were exhausted after the EVA.

During the flight, McDivitt and White measured the radiation present in the South Atlantic anomaly (a pocket of charged particles in the ionosphere), did other tests, and then set about activating their work-sleep cycles as planned. The photographs taken of clouds and large terrain features, particularly those of the Nile Delta and the Sahara Desert, were spectacular.

The onboard computer that was to control reentry attitude and retrofiring sequences failed for the last fourteen orbits, and the astronauts had to land via a rolling, Mercury-type reentry. Their orbital time of 97 hours and 56 minutes established the longest-duration orbital record at the time. They missed their landing point by 80 kilometers, a fact attributed to the failure of the reentry computer system.

McDivitt and White became national celebrities. Their flight proved to the public that the U.S. space program was on a par with that of the Soviet Union; NASA officials and the political leadership were pleased with the successes of Gemini IV. Nevertheless, the computer failure and the difficulties experienced in performing a rendezvous with a target caused some worry. Rendezvous attempts without the aid of computers would be difficult, and the prestige of the onboard computer in planning such orbital changes was greatly enhanced by the failures of Gemini IV.

### Contributions

Gemini IV added to the successes of the Gemini Program in all respects save perhaps for the rendezvous attempts. The mission's rendezvous attempts with a relatively simple target—a vehicle with an orbit almost identical to Gemini's—proved once and for all that an unaided, pilot-controlled rendezvous was virtually impossible. Either the astronauts would have to be experts in orbital dynamics, with its seemingly contradictory flight-path parameters, or the onboard computer would have to work perfectly for the next Gemini craft, which were intended to rendezvous with increasingly difficult targets.

Eleven scientific experiments were performed by the Gemini IV crew, and nearly all were successful. The synoptic terrain and synoptic weather photographs were spectacular. The photographs of the Nile Delta and the Sahara and the overview of giant

*Astronauts Clifford C. Williams, Jr. (left), Frank Borman (center), and Alan Shepard monitor the flight of Gemini IV from their consoles at Mission Control. (NASA)*

hurricanes and pressure centers as seen from Gemini IV convinced meteorologists and geologists that there was considerable knowledge to be gained from such spacecraft.

The three biological experiments—the in-flight exerciser, the in-flight phonocardiogram, and the bone demineralization—also went well. It seemed that there was little or no serious cardiovascular deterioration in either astronaut. Use of the in-flight exerciser, the astronauts' consumption of a considerable amount of food, and the enforced, rigid work-sleep program seemed to alleviate most of the potentially harmful effects of extended spaceflight.

Four space physics experiments were successfully performed. The tri-axis magnetometer monitored Earth's magnetic field by extending a boom out of the adapter and having it change axis to determine the vector components of Earth's magnetic field. Seven sensors placed strategically inside and outside the cabin telemetered radiation levels back to ground stations; this phase of the measurements worked well throughout the four-day mission. The proton-electron measurement determined the radiation level immediately outside the spacecraft; this experiment was controlled by the pilot and telemetered to ground stations successfully when he activated the equipment. The electrostatic charge detector measured a higher-than-expected electrostatic charge on the skin of the spacecraft, but upon evaluation of the data, it was decided that the high readings were the result of calibration problems. In subsequent Gemini missions, the sensors were modified to remove these problems.

The simple navigation experiment using a handheld sextant was a gauge of the fatigued astronaut's ability to navigate and assess his position in space. It worked well, but statistical data were lacking; thus, such measurements cannot be evaluated objectively. Nevertheless, they proved that White was lucid, disciplined, and thinking clearly when he was instructed to make these measurements. All in all, Gemini IV proved that the craft could support two astronauts for prolonged missions without serious effects and that spacewalking with the help of the compressed-air gun was possible.

### Context

News of the EVA of the Cosmonaut Leonov dominated the early days of Gemini IV preflight preparations. The press had dubbed Gemini IV Little Eva, and this name sent shudders through NASA officials, because it seemed that the United States was falling behind the Soviet Union and was hastily adding an abbreviated EVA to the Gemini IV flight plan to win back respect for the American space effort.

Gemini 3 had flown successfully three months previously and had shown that Gemini craft could change orbits effectively under the direction of the two-person crew. Gemini 3 had made only three orbits; thus, the question of whether prolonged spaceflight was feasible in Gemini craft remained unanswered. The success of Gemini IV left little doubt that Gemini craft could eventually perform for a fourteen-day mission if all systems were brought up to program expectations.

Gemini IV successfully performed eleven scientific experiments, while Gemini 3 had only three on board (and these had been added at the last minute). The scientific community, including biologists, space-medicine specialists, meteorologists, and geologists, learned much from these experiments; on subsequent flights, there were requests for hundreds of experiments to be performed. Following Gemini IV, Gemini V carried at least sixteen such scientific experiments, and subsequent Gemini spacecraft would carry between sixteen and twenty experiments each.

Gemini IV convinced the public that the American space program was indeed on a par with Soviet efforts, and McDivitt and White became popular and honored heroes. No other Gemini spacecraft had the press attention of Gemini IV, and only the Apollo lunar-landing craft would attract similar attention. After Gemini IV, the United States launched the subsequent eight Gemini craft at less than sixty-day intervals. Some were long-lived missions of the type made famous by Gemini VII, with its fourteen-day flight.

**See also:** Apollo Program; Apollo Program: Orbital and Lunar Surface Experiments; Extravehicular Activity; Gemini Program; Gemini 3; Gemini V; Gemini VIII; Gemini IX-A and X; Johnson Space Center; Manned Maneuvering Unit; National Aeronautics and Space Administration; Space Suit Development.

### Further Reading

Borman, Frank, with Robert J. Serling. *Countdown: An Autobiography.* New York: William Morrow, 1988. Borman, who was on the backup crew for Gemini IV, flew as command pilot on the Program's most spectacular rendezvous mission, Gemini VI, which orbited and rendezvoused with Gemini VII. His insights on flight preparations and undertaking a rendezvous mission are very readable.

Cernan, Eugene A., and Don Davis. *The Last Man on the Moon: Astronaut Eugene Cernan and America's Race in Space.* New York: St. Martin's Press, 1999. The story of the last two men to walk on the Moon is told by Cernan, commander of Apollo 17, and Davis, an experienced journalist. Cernan, whose spaceflight career spanned both the Gemini and Apollo Programs, fittingly narrates it.

Grimwood, James M., Barton C. Hacker, and Peter J. Vorzimmer. *Project Gemini, Technology and Operations: A Chronology*. NASA SP-4002. Washington, D.C.: Government Printing Office, 1969. An authoritative day-to-day account of the planning and development of Gemini from its inception as Mercury Mark II to the final flight of Gemini XII. Good sketches are included, technical problems are examined, and details are summarized effectively.

Hacker, Barton C., and James M. Grimwood. *On the Shoulders of Titans: A History of Project Gemini*. NASA SP-4203. Washington, D.C.: Government Printing Office, 1977. The official history of the Gemini Program. Well written and organized effectively, it gives a good picture of the fiscal, political, personnel, technological, and flight problems as they occurred. Contains pictures and sketches as well as cogent tables of facts and figures. A bit specialized in parts, but several chapters give a good overview of the Program for the general reader.

Harland, David. *How NASA Learned to Fly in Space: An Exciting Account of the Gemini Missions*. Burlington, Ont.: Apogee Books, 2004. Details launch vehicles, spacecraft, and astronauts.

National Aeronautics and Space Administration, Educational Programs and Services Office. *Manned Space Flight: Projects Mercury and Gemini*. Washington, D.C.: Government Printing Office, 1966. Appearing in the NASA Facts Series, this is a thorough evaluation of both Mercury and Gemini; gives useful illustrations and program features that are unavailable elsewhere. More specialized than the other listed references and a must for devotees of the space program as it was viewed in the 1960's.

Shayler, David J. *Gemini: Steps to the Moon*. Chichester, England: Springer-Praxis, 2001. The story of the Gemini Program from the perspective of the engineers, flight controllers, and astronauts.

—————. *Walking in Space: Development of Space Walking Techniques*. Chichester, England: Springer-Praxis, 2003. Analysis of EVA techniques, drawing on original documentation and interviews with astronauts who conducted EVAs.

Shepard, Alan B., Jr., and Donald K. "Deke" Slayton, with Jay Barbree and Howard Benedict. *Moon Shot: The Inside Story of America's Race to the Moon*. Atlanta: Turner, 1994. This is, indeed, the inside story of the Apollo Program as told by two men who actively participated in it. Some of their tales appear here for the first time. Shepard and Slayton discuss the Gemini orbital flights and tell some interesting behind-the-scenes stories of the Program. The book was adapted for a four-hour documentary in 1995.

Slayton, Donald K., with Michael Cassutt. *Deke! U.S. Manned Space: From Mercury to the Shuttle*. New York: Forge, 1995. This is the autobiography of the last of the Mercury astronauts to fly in space. After being grounded from flying in Project Mercury for what turned out to be a minor heart murmur, Slayton was appointed head of the Astronaut Office. Slayton talks of his frustration at being grounded and how he worked to regain flight status. He also discusses the flights of his fellow astronauts during the Gemini Program.

Wendt, Guenter, and Russell Still. *The Unbroken Chain*. Burlington, Ont.: Apogee Books, 2001. Wendt is the only person who worked with every astronaut bound for space.

*John Kenny*

# Gemini V

*Date:* August 21 to August 29, 1965
*Type of mission:* Piloted spaceflight

*Gemini V advanced confidence that a lunar mission was possible. This flight demonstrated the human ability to survive eight days in space, to conduct rudimentary rendezvous techniques, and to employ fuel-cell electrical power generation systems.*

### Key Figures

*L. Gordon Cooper, Jr.* (1927-2004), Gemini V command pilot

*Charles "Pete" Conrad, Jr.* (1930-1999), Gemini V pilot

*Neil A. Armstrong* (b. 1930), Gemini V backup command pilot

*Elliot M. See, Jr.* (1927-1966), Gemini V backup pilot

*Christopher C. Kraft, Jr.* (b. 1924), director of Flight Operations, Mission Control

*Eugene F. Kranz* (b. 1933), director of Flight Operations, Mission Control

*John Hodge*, director of Flight Operations, Mission Control

*Charles A. Berry* (b. 1923), Gemini Program physician

### Summary of the Mission

As of August, 1965, the National Aeronautics and Space Administration (NASA) had demonstrated that the two-person Gemini spacecraft was capable of performing orbital maneuvers and supporting a space-suited astronaut outside the protective safety of the crew cabin. Gemini astronauts had been in space as long as four days. Gemini-Titan V (GT-V) was to double that endurance record, collecting valuable medical data on the human body's reaction to spaceflight conditions as well as engineering data concerning rendezvous and spacecraft systems. One new system incorporated into the GT-V spacecraft was the fuel cell, a method for generating electrical power from the reaction of liquid hydrogen and liquid oxygen under controlled conditions. As a side benefit of the fuel-cell reaction, pure drinkable water is produced. This in-flight production of water lowers the amount needed to be carried into orbit at launch by a spacecraft.

NASA selected L. Gordon Cooper, Jr., one of the original Mercury astronauts, as command pilot, and Charles "Pete" Conrad, Jr., a member of the second astronaut group, as pilot for GT-V. The backup crew consisted of Neil A. Armstrong (command pilot) and Elliot M. See, Jr. (pilot). Cooper was born March 6, 1927, in Shawnee, Oklahoma. In 1956, he received a Bachelor of Science degree in aeronautical engineering from the Air Force Institute of Technology. Cooper ultimately achieved the rank of colonel in the Air Force. He joined NASA as an astronaut in 1959 and then served as backup pilot for Mercury-Atlas 8 and pilot of the final piloted Mercury flight, Mercury-Atlas 9, during which he spent more than a full day in orbit. Following GT-V, Cooper served as backup command pilot for GT-XII and backup commander of the Apollo 10 lunar flight. Cooper retired from NASA in July, 1970.

Conrad was born June 2, 1930, in Philadelphia, Pennsylvania. In 1953, he received a Bachelor of Science degree in aeronautical engineering from Princeton University. Before retiring from the Navy in 1974, Conrad achieved the rank of captain. This spaceflight was Conrad's first. After GT-V, Conrad served as backup command pilot of GT-VIII, flew as command pilot for GT-XI, served as backup commander of Apollo 9, flew as commander of Apollo 12 (the second piloted lunar landing), and flew as the commander of Skylab 2, during which mission he remained in Earth orbit for twenty-eight days and was instrumental in restoring the damaged Skylab orbital workshop to usefulness. Conrad resigned from NASA early in 1974.

The Gemini V mission was plagued with bad luck. During the first launch attempt, excessive boil-off caused the fuel cell's liquid hydrogen supply to be low. For no apparent reason, the telemetry system malfunctioned. A fire started near Pad 19. Clouds moved overhead, and intense lightning was experienced over the launch site, endangering the astronauts because the erector was down and the launch vehicle stood alone, like a large lightning rod. Filled with volatile fuels, it could have exploded had a lightning bolt struck nearby. Quickly, the erector was set back in place, and the astronauts exited the spacecraft and returned to their living quarters, disappointed.

On August 21, the countdown proceeded more smoothly than in previous piloted launches; there were no unscheduled holds. The liftoff from Pad 19 occurred at 13:00 Coordinated Universal Time (UTC, or 9:00 A.M. eastern standard time). During first-stage ascent, the vehicle subjected the astronauts to excessive longitudinal oscillations (known as "pogo" oscillations), but the ride smoothed out just prior to staging. The spacecraft entered an orbit having an apogee (highest point) of 349 kilometers. Fifty-six minutes after liftoff, Cooper fired thrusters to alter the apogee in preparation for the first major phase of the mission.

The craft carried a special free-flying rendezvous evaluation pod to be ejected from the equipment module. The pod contained a flashing light and radar transponder. After its jettison from the Gemini spacecraft, the astronauts were to seek it out again, using radar in an exercise considered crucial for an attempted docking with an Agena

target vehicle later in the Gemini Program. The rendezvous evaluation pod was ejected at 2 hours, 7 minutes, and 15 seconds ground-elapsed time (GET), and GT-V turned around to face it and lock onto its radar signal. The pod moved away from GT-V at 270°, moving behind and above the spacecraft after reaching a maximum range of 2,354 meters.

Soon, Conrad noticed that the fuel cell's liquid oxygen pressure was low (below the normal reading of 5,500-6,200 kilopascals, or 800-900 pounds per square inch) and falling. Cycling both the heater switch and circuit breaker numerous times failed to improve the situation. A decision would have to be made soon: Either power down the spacecraft, suspending the exercise, or terminate the flight. With GT-V on its second orbit and in trouble, recovery forces steamed into readiness in all primary and contingency recovery zones. Recovery at the end of the sixth orbit would be the first good opportunity if the mission concluded early.

To conserve power, all but the most essential systems were turned off and the spacecraft entered a drifting mode. The troublesome fuel-cell system was producing a mere 13 amperes of current; that was the bare minimum. Mission Control and the astronauts continued to monitor the pressure loss. Eventually, the rate of decrease slowed, but the pressure fell to as low as 500 kilopascals (71 pounds per square inch) before stabilizing. Ground-tests indicated that the fuel cells could generate electricity safely at this low pressure. With at least 13 hours of battery life remaining and a low, but stable, fuel-cell situation, GT-V was cleared to continue for at least one day. Mission Control devised a sequence of maneuvers simulating a rendezvous for the astronauts to perform if the power problem improved substantially. The rendezvous evaluation pod was dead, but rendezvous with an imaginary Agena target could be attempted.

On the second day, the pressure increased gradually and the astronauts were permitted to perform some experiments and systems checks requiring more electrical power. The fuel cells handled the power load adequately, clearing the way for Cooper and Conrad to attempt the rendezvous sequence on the third day of the flight. A combination of ground-based and onboard calculations was used to control the spacecraft maneuvers to a rendezvous with a moving point in space considered to be the location of an imaginary Agena. The first spacecraft maneuver was initiated at 50 hours, 49 minutes, and 57 seconds GET, and the final maneuver at 53 hours, 4 minutes GET, placing the spacecraft in the proper position to attempt a docking had an Agena actually been in space. Following rendezvous completion, the spacecraft was powered down once more.

Later in the mission the spacecraft suffered two other major problems. Six attitude control thrusters failed, forcing the craft to tumble as well as drift through most of the remainder of the flight.

Cooper fired control thrusters only when the tumbling became excessive. Several potential fixes for the thruster problem were attempted, but none restored proper function. In addition, the fuel cells were producing 20 percent more water than anticipated. Because there was no means for dumping excess water overboard, there was concern that the water storage tank would reach capacity and force water back into the fuel cells, with dire consequences. Projections of water buildup indicated that the tank could hold the excess, provided that the rate of production did not increase. More systems were powered down as an extra precaution.

As the mission drew to a close, Hurricane Betsy meandered in the direction of the prime recovery zone, threatening to produce adverse conditions. To avoid these undesirable conditions, the Weather

*Splashdown of Gemini V.* (NASA)

Bureau strongly advised NASA to return GT-V to Earth one orbit earlier than scheduled; Flight Director Eugene F. Kranz agreed and ordered the USS *Lake Champlain*, prime recovery vessel, to steam into a new recovery area farther away from the storm.

One orbit prior to retrofire sequence, the astronauts used one of the two rings of reentry control-system thrusters to position GT-V in the proper attitude. This system was selected because of past difficulties with the orbital attitude and maneuvering system thrusters. Near Hawaii, on the 120th orbit, at 190 hours, 27 minutes, and 43 seconds GET, the retrorockets began to fire, subtracting sufficient forward velocity to drop the spacecraft out of orbit and head it for an Atlantic Ocean splashdown.

Unfortunately, incorrect coordinates had been fed into the spacecraft computer—Earth's rotation rate had been given as 360° per day rather than the more precise 360.98° per day. As a result, onboard calculations steered GT-V away from the planned impact point. Data during reentry indicated that the spacecraft was too high. Cooper responded with maneuvers to generate more drag; this raised gravitational loads from 2.5 to 7.5.

Cooper's actions in response to the "perceived" errors brought GT-V to a splashdown 130 kilometers short of the desired impact point. Recovery helicopters arrived near the spacecraft in short order, dropping a trio of swimmers into the water to secure the spacecraft and extract Cooper and Conrad. The astronauts were then flown back to the carrier by helicopter for an official greeting and extensive medical examinations.

### Contributions

Seventeen experiments were included in Gemini V's flight plan. Only one of these was not conducted—it involved the canceled rendezvous with the pod. The majority of the experiments were either photographic and visual assessments or medical evaluations. Four experiments were sponsored by the Department of Defense (DoD) and another was jointly sponsored by DoD and NASA.

One experiment investigated the electrostatic charge buildup on the spacecraft; GT-IV had carried a similar experiment. Its purpose was to determine if dangerous electrical charges would present a hazard to a docking of a Gemini spacecraft and fuel-laden Agena. Data from GT-V suggested a higher electric potential than Gemini IV data, but this increase posed no threat.

The mission also assessed the ability of astronauts and photographic imaging systems to acquire, track, and photograph selected targets. The four DoD experiments gathered information that was useful to developing equipment for military surveillance and weather satellites. Irradiance measurements, in the 0.2- to 12-micron wavelength band, were taken on selected celestial and terrestrial targets, rocket plumes, and cold objects in space. A pair of missile launches was performed in conjunction with this experiment, the first time astronauts observed such an event from space. The first missile was acquired as it poked through a low cloud deck and was tracked until staging. A second missile was observed at liftoff by Conrad but not by Cooper.

The crew demonstrated extreme visual acuity. Conrad, who had 20/15 vision, claimed to be able to resolve the wakes from vessels at sea, city streets, and airplane contrails.

Cooper and Conrad photographed a number of dim-light phenomena such as zodiacal light, Earth's airglow, and the Gegenschein. Zodiacal light is a faint glow in the sky observed just prior to sunrise or just following sunset. This glow extends along the ecliptic near the rising or setting Sun and is so faint that it is obscured by moonlight. Only in extremely dark skies can the zodiacal light be seen to extend completely around the ecliptic. Gegenschein, photographed from orbit for the first time on GT-V, is a dim glow illuminated in the anti-Sun direction. The mission confirmed the existence of such a dim glow 3° away from the direct anti-Sun direction.

Cooper and Conrad experienced slightly different cardiovascular responses. Cooper pooled 39 percent more blood in his legs than did Conrad.

He also lost 4 percent greater blood plasma volume than Conrad (who lost 4 percent total volume). Conrad's postflight resting heart rate, however, was higher than Cooper's. Two days after splashdown, Conrad's rate returned to normal, and Cooper's did the same two days after that. Neither astronaut suffered any debilitating cardiovascular effects. Bone decalcification began to reverse slowly following the mission for both astronauts.

On board the spacecraft was an in-flight exerciser, which consisted of a pair of elastic cords attached to a foot strap and handle. It required 315 newtons of force to stretch it 30 centimeters. This experiment measured the response of astronauts to given work loads. No evidence of cardiovascular reflex decrement was noted.

### Context

Although Pilot Conrad characterized GT-V as a very boring spaceflight, this mission quietly amassed a number of very important firsts. Admittedly, none of these firsts can be labeled space spectaculars. Nevertheless, GT-V accomplished many tasks necessary to clear a path to the Moon.

Perhaps the most important thing GT-V accomplished was the setting of a new piloted spaceflight endurance record. Cooper, who became the first person to enter space twice, and Conrad accumulated seven days, 22 hours, 55 minutes, and 14 seconds in Earth orbit, beating the previous record set by Soviet cosmonaut Valeri F. Bykovsky on the Vostok 5 mission, traveling a total of 4.8 million kilometers in the process. (The mission goal was eight days; the early conclusion of the mission as a result of poor weather in the original recovery zone prevented achievement of that goal.) This was long enough for an Apollo crew to carry out a conservative lunar-landing mission and doubled the experience gained on GT-IV.

Medical data about human physiological responses to prolonged weightlessness were collected. Although no major medical problem was uncovered that would preclude sending astronauts to the Moon, Cooper and Conrad did experience

difficulty in sleeping, losses of bone calcium, and decreases in blood plasma volume. These three tendencies were first noted on GT-IV. Charles A. Berry, Gemini V flight surgeon, was concerned about this trend and its effect on longer spaceflights. GT-VII, a fourteen-day mission, would further investigate these symptoms. The effects were only temporary; two days after splashdown, both astronauts were physiologically recovered.

This mission also demonstrated the usefulness of fuel cells as a means of generating electrical power and potable water during spaceflight for the first time. For long-duration spaceflight, storage batteries would be impractical, taking up far too much space and weight. (Fuel cells have been used on nearly every American piloted spaceflight since GT-V.) System difficulties experienced on GT-V were not associated with the fuel cells themselves; the pressure decrease was traced to a fault in a heater.

Although the planned rendezvous evaluation pod maneuvers were canceled as a result of powering down the spacecraft after the pod's jettison, GT-V proved that astronauts could perform orbital maneuvers to rendezvous with a target vehicle. Four maneuvers performed by Cooper and Conrad on the third day simulated the type of rendezvous sequence the GT-VI astronauts were expected to perform in order to link up with an actual robotic Agena target vehicle. The four maneuvers were a height adjustment, phase adjustment, out-of-plane burn, and coelliptical adjustment. At the conclusion of these maneuvers, GT-V was precisely at the imaginary target coordinates, and, as a result, the total orbital attitude and maneuvering system had been thoroughly tested. Despite thruster failures later in the flight, GT-V proved the maneuvering capability of the spacecraft and cleared the way for the more complex Gemini missions to follow.

**See also:** Apollo Program; Apollo Program: Lunar Module; Apollo Program: Orbital and Lunar Surface Experiments; Gemini Program; Gemini Spacecraft; Gemini IV; Gemini VIII.

## Further Reading

Cernan, Eugene A., and Don Davis. *The Last Man on the Moon: Astronaut Eugene Cernan and America's Race in Space.* New York: St. Martin's Press, 1999. This autobiography tells the story behind the story of Gemini and Apollo. Cernan, whose spaceflight career spanned both programs, fittingly narrates it.

Cooper, L. Gordon, Jr., with Bruce Henderson. *Leap of Faith: An Astronaut's Journey into the Unknown.* New York: Harper Torch, 2000. Cooper recalls his adventurous life and looks toward the next millennium of space travel.

Hacker, Barton C., and James M. Grimwood. *On the Shoulders of Titans: A History of Project Gemini.* NASA SP-4203. Washington, D.C.: Government Printing Office, 1977. An official NASA history of the Gemini Program. Suitable for general readers, this text includes management and administration history as well as spacecraft development, flight testing, and mission events. Provides appendices and detailed lists of references for locating more information.

Harland, David. *How NASA Learned to Fly in Space: An Exciting Account of the Gemini Missions.* Burlington, Ont.: Apogee Books, 2004. Details the launch vehicles and spacecraft as well as the astronauts and their missions.

Kraft, Christopher C., Jr. *Flight: My Life in Mission Control.* East Rutherford, N.J.: Penguin Putnam, 2002. The first NASA flight director gives an account of his life in Mission Control.

National Aeronautics and Space Administration. *Gemini Midprogram Conference Including Experiment Results.* NASA SP-121. Houston: NASA Manned Spacecraft Center, 1966. Provides a coherent assessment of engineering and experimental data obtained by Gemini flights 3 through 7. Suitable for those with moderate science and engineering backgrounds. References to original publications and reports are included.

Shayler, David J. *Gemini: Steps to the Moon.* Chichester, England: Springer-Praxis, 2001. The development of the Gemini Program from the perspective of the engineers, flight controllers, and astronauts.

Shepard, Alan B., Jr., and Donald K. "Deke" Slayton, with Jay Barbree and Howard Benedict. *Moon Shot: The Inside Story of America's Race to the Moon.* Atlanta: Turner, 1994. This is, indeed, the inside story of the Apollo Program as told by two men who actively participated in it. Some of their tales appear here for the first time. Shepard and Slayton discuss the Gemini orbital flights and tell some interesting behind-the-scenes stories of the program. The book was adapted for a four-hour documentary in 1995.

Slayton, Donald K., with Michael Cassutt. *Deke! U.S. Manned Space: From Mercury to the Shuttle.* New York: Forge, 1995. This is the autobiography of the last of the Mercury astronauts to fly in space. After being grounded from flying in Project Mercury for what turned out to be a minor heart murmur, Slayton was appointed head of the Astronaut Office. Slayton talks of his frustration at being grounded and how he worked to regain flight status. He also discusses the flights of his fellow astronauts during the Gemini Program.

Wendt, Guenter, and Russell Still. *The Unbroken Chain.* Burlington, Ont.: Apogee Books, 2001. Wendt is the only person who worked side by side with every astronaut who left the Cape bound for space.

*David G. Fisher*

# Gemini VII and VI-A

*Date:* December 4 to December 18, 1965
*Type of mission:* Piloted spaceflight

*A lunar mission would entail many significant challenges, two of which would be rendezvous/docking and maintaining human heath and function over the long duration of the flight. In preparation for such a mission, Gemini VI-A tested the ability for rendezvous in space, whereas Gemini VII's flight of fourteen days demonstrated that humans could endure long periods in spaceflight.*

### Key Figures

*Walter M. Schirra, Jr.* (b. 1923), Gemini VI-A command pilot
*Thomas P. Stafford* (b. 1930), Gemini VI-A pilot
*Frank Borman* (b. 1928), Gemini VII command pilot
*James A. Lovell, Jr.* (b. 1928), Gemini VII pilot
*Charles W. Mathews* (1921-2002), Gemini Program director

### Summary of the Missions

The missions of Gemini VI-A and Gemini VII provided necessary experience and information on two important aspects of piloted spaceflight: the effects of long-duration missions on the human body and the procedures necessary for the rendezvous (meeting in space) of two spacecraft. Both of these aspects represented major goals of the National Aeronautics and Space Administration (NASA) for the two-person Gemini Program in preparation for a Moon landing before the end of the 1960's.

The flight of Gemini VI was a planned rendezvous and docking with an robotic Atlas-Agena docking target. The original flight plan called for the liftoff of the Agena docking target on top of an Atlas rocket about an hour before that of the Gemini VI spacecraft. Once the Agena had reached the correct orbit, Gemini VI would be launched. Over a period of time, Gemini VI would catch, rendezvous with, and join or dock with the Agena. Both of these vehicles would be launched from Cape Kennedy, Florida, from separate launch pads.

Rendezvous and docking were seen by NASA as essential to landing man on the Moon, for two spacecraft would be involved in such a mission. One spacecraft would serve as the main vehicle, taking the astronauts to the Moon and returning them to Earth. The second vehicle would land the astronauts on the Moon and return them to the main vehicle, which would remain in lunar orbit.

Walter M. Schirra, Jr., was selected as the command pilot of Gemini VI. Schirra was one of the seven original astronauts and a veteran of spaceflight; he had spent 9.2 hours in orbiting Earth six times during the fifth flight of the one-person Mercury spacecraft. He was to be joined by Thomas P. Stafford. Stafford would be making his first flight, having been selected as an astronaut in the second group of nine astronauts.

The first launch attempt of Gemini VI was made on October 25, 1965. The weather for the dual launches was ideal. The robotic Atlas-Agena docking target left the pad exactly on schedule, at 15:00:04 Coordinated Universal Time (UTC, or 10:00:04 A.M. eastern time). All telemetry indicated that the cutoffs of the main booster and the sustainer engine (a rocket engine that maintains the velocity attained by the rocket's main or booster engine) had occurred as planned. Nevertheless, the docking vehicle apparently exploded when its own engine ignited; the vehicle fell into the ocean, never reaching orbit. The flight of Gemini VI was canceled 42 minutes before its planned launch.

After review of the Gemini flight schedules and major objectives, a daring proposal was made: to launch Gemini VII, make pad repairs (since there was only one Gemini-Titan launch pad available), and then assemble the Gemini VI-Titan launch configuration and launch Gemini VI, redesignated Gemini VI-A, the "A" standing for "augmented." Gemini VII would act as a rendezvous target, carrying the necessary equipment to allow Gemini VI-A to track it.

Gemini VII was a planned fourteen-day mission to test the medical endurance of the astronauts. The command pilot would be Frank Borman, the pilot James A. Lovell, Jr. For both astronauts, selected in the second group of astronauts (along with Stafford), this would be their first spaceflight.

*This photograph of the Gemini VII spacecraft was taken through the hatch window of the Gemini VI spacecraft during rendezvous and station-keeping maneuvers.* (NASA)

Gemini VII was launched right on schedule at 19:30:03 UTC (2:30:03 P.M. eastern time) on December 4, 1965. A maximum perigee (the point in its orbit when an object is closest to Earth) of 161 kilometers and apogee (the point in its orbit when an object is farthest from Earth) of 328 kilometers was achieved by Gemini VII. After assessing pad damage, NASA determined that repair, assembly, and launch of Gemini VI-A could be accomplished during the time constraints of the Gemini VII mission.

Borman and Lovell worked on planned experiments on board Gemini VII while awaiting Gemini VI-A and its astronauts, Schirra and Stafford. Borman and Lovell had new, lightweight space suits (the new suit weighted 7.3 kilograms compared with the older space suit's weight of 10 kilograms). This lighter weight was necessary because of the extended length of their flight and the fact that for the first time in piloted U.S. spaceflights, one astronaut was allowed to take off his space suit while the second remained suited. Only one astronaut was allowed to unsuit because of the possibility of a leak occurring; NASA wanted to make certain that one of the astronauts would be able to handle a possible spacecraft depressurizing emergency at all times.

The second planned launch attempt of Gemini VI-A occurred on December 12, 1965, at 14:54 UTC (9:54 A.M. eastern time). All went according to schedule, including the ignition of the Titan II rocket that was to carry Gemini VI-A, with Schirra and Stafford, into space. Two seconds before liftoff, and only about a second after igniting, however, the Titan II booster shut down. Schirra had the option at that time to eject both himself and Stafford from the potentially dangerous Gemini-Titan. If he had used that option, however, the spacecraft would have been useless for the planned rendezvous.

After close examination of the system, it was discovered that an electri-

cal plug had prematurely disconnected. The plug was intended to
direct electricity away from a master timer aboard the Titan II rocket
until it actually had lifted off. When
the plug disconnected, it sent a signal that the rocket had lifted off
and thus started the master timer.
A blockhouse computer detected
the inconsistency and the fact that
the Titan II rocket had not built up
enough thrust and therefore had
shut down the rocket's engines.
The inspection also yielded another find: a tiny dust cap accidentally left in one of the rocket's
fuel lines. This cap would have restricted fuel flow and would also
have caused an engine shutdown.

Finally, Schirra and Stafford
were launched on December 15,
1965, at 13:32 UTC (8:32 A.M. eastern time). The launch of Gemini

*Another view of Gemini VII in orbit around Earth, from the vantage of the astronauts in Gemini VI.* (NASA)

VI-A placed the spacecraft and astronauts in the orbit necessary to begin maneuvering toward the
Gemini VII spacecraft. Although the initial orbit of
Gemini VI-A was lower than that of Gemini VII, the
lower orbit allowed Gemini VI-A to travel slightly
faster. The relative positions of the two spacecraft
can be likened to cars on a racetrack. A car on the
inside track will circle more quickly than the car on
the outside one. Gemini VI-A was on the "inside
track"—that is, nearer to Earth. The position was
necessary—and planned—because Gemini VI-A
was about 1,900 kilometers behind Gemini VII and
needed to catch up to the "outside" vehicle in order to perform the rendezvous.

At about 18:50 UTC (1:50 P.M. eastern time),
Stafford announced that they had spotted Gemini
VII, even though it was still about 45 kilometers
away. After a series of intricate maneuvers that
lasted nearly six hours, Gemini VI-A and Gemini
VII reached a rendezvous in space. NASA officials
had hoped for a rendezvous distance of 650 meters.
Final analysis of the rendezvous of Gemini VI-A

and Gemini VII revealed that the two spacecraft
had approached a separation of only about one-
third of a meter. Schirra and Stafford could easily
see the twelve-day beards on the faces of Gemini VII
astronauts Borman and Lovell, and Lovell aboard
Gemini VII reported that he could see Schirra's
lips moving—he was chewing gum. The crews also
made a visual and photographic inspection of the
spacecraft. Both Gemini VI-A and Gemini VII had
straps, some as long as 27 meters, trailing off the
rear of the spacecraft; the straps covered an explosive cord that severed all connections between the
spacecraft and the launch vehicle when they separated. These straps were apparently making the
noises reported by the Gemini VII astronauts during the maneuvers.

The two spacecraft continued to fly together for
about five hours, never separated by more than 30
meters. Each of the four astronauts had an opportunity to maneuver his spacecraft. The two spacecraft then separated to a distance of approximately
40 to 77 kilometers so that the crews could sleep.

On December 16, 1965, at 15:29:09 UTC (10:29:09 A.M. eastern time), Gemini VI-A splashed down in the Atlantic Ocean, approximately 19 kilometers from the prime recovery carrier, the *Wasp*. Schirra and Stafford had spent a total of 25.9 hours in space and had completed 17 orbits. Borman and Lovell, in Gemini VII, spent an additional two days in space, landing in the Atlantic Ocean on December 18, 1965, at 14:05:40 UTC (9:05:40 A.M. eastern time). Gemini VII also landed approximately 19 kilometers from its prime recovery carrier, again the *Wasp*. Borman and Lovell had spent a then-record 330 hours, 35 minutes, and 13 seconds in space; their mission lasted a total of 220 orbits, well in excess of 8.5 million kilometers. Their total flight time was more than all Soviet and American total flight time to date.

### Contributions

The flights of Gemini VI-A and VII provided important information necessary for a human being to make a trip to the Moon and return safely to Earth. Foremost was the knowledge and experience gained by the rendezvous of the two Gemini spacecraft. Many questions surfaced prior to the missions about the difficulty of placing two objects in a similar orbit, with one object catching up to the second object in a reasonable amount of time, finding the target object, and approaching it to within a minimal distance without endangering either spacecraft. This, along with docking the two spacecraft, would have to be accomplished for a lunar mission or a permanently piloted space station in Earth orbit to be possible. The successful rendezvous of Gemini VI-A and Gemini VII to within a distance of less than one-third of a meter answered many of these questions.

The effects of the space environment on the human body had been a great unknown. Weightlessness and its effects on the body's systems—respiratory, circulatory, and digestive—were of great concern. How a person would cope with being weightless for a long period of time was another relatively unknown variable. In addition, there was the question of dealing with confinement in a small area for an extended period of time. These were additional questions, questions that Borman and Lovell on Gemini VII would answer. They showed that the human body could withstand spaceflight conditions for up to two weeks with few ill effects. The major problem appeared to be simple boredom during the latter portion of the mission.

Other experiments besides the rendezvous and the study of long-term effects of spaceflight were conducted. Gemini VII performed several experiments. One of the key investigations dealt with stellar navigation. In this experiment, Borman and Lovell were able to determine the position of their spacecraft by observing occultations (an occultation is the interruption of light from a celestial body by the intervention of another body, much like an eclipse) of brighter stars by Earth. They tracked a bright star in the reticle (a series of circles or crosshairs within an optical instrument used for centering) eyepiece of a photoelectric photometer (a scientific instrument used to measure the brightness of an object) and recorded its disappearance behind Earth's limb. The star's disappearance was not sudden; because of Earth's atmosphere, it faded slowly before it eventually disappeared. Another navigational experiment measured the visual contrast between landmasses and water bodies on Earth—helpful in determining the usefulness of Earth in precise navigation.

One of the key experiments performed by the Gemini VII crew did not prove as successful. A test of a two-way laser system for communication was done by pointing a laser at Gemini VII from Earth. The astronauts then were to locate the beam, reorient their spacecraft, and point their own laser beam toward the Earth-based beam. It was found that Borman and Lovell did not have enough time to find the beam, reorient their spacecraft, and aim their own laser during the period that they were over a particular site.

### Context

To explore space and the solar system, people had to be able to adapt to its environment. Earlier piloted flights, both Soviet and American, had demonstrated that an environment could be created in which humans could survive in space for a limited amount of time. These flights had shown that people could control a spacecraft and effect minor repairs, and that many of the ordinary everyday events of living—eating, sleeping, going to the bathroom—could be done in the space environment.

Yet, what about long-term effects? If humanity wanted to go to the Moon—and in the mid-1960's a race for the Moon between the Americans and the Soviets apparently still existed (at least from an American point of view)—then a longer period of time in space than had been demonstrated by the 190.9-hour mission of Gemini V in August, 1965, would be necessary. The major goal of Gemini VII—to demonstrate that people could spend an intermediate length of time in space—provided a major bridge to the Moon and the future of long-term visits to space stations. It would be the longest Gemini mission of the ten to be flown.

The original goal of Gemini VI, to track and locate, rendezvous, and dock with an robotic Agena target, would have to be postponed until the flight of Gemini VIII in March, 1966. Nevertheless, the knowledge gained by the rendezvous maneuvers of Gemini VI-A provided important information about the procedures of such an operation in space, for many had thought rendezvous next to impossible. The future of any piloted long-term mission would depend extensively on rendezvous and docking techniques. Gemini VI-A easily demonstrated—after some early problems and near disasters—that rendezvous and even docking would be relatively easy. This would be later demonstrated and confirmed in other Gemini, Apollo, and Soviet piloted and robotic missions. Gemini VII also provided an opportunity for the study of a number of navigation systems; these systems would be imperative to lunar missions. The data collected during the spaceflight added excellent possibilities.

Earth study from orbit is an important part of space exploration, as had been demonstrated by previous piloted and robotic missions. Photographs and data from the Gemini VII mission added to the newest evaluation of Earth from orbit. Both Gemini VI-A and VII detected, on the night side of Earth, forest fires burning on the island of Madagascar. Additionally, Gemini VII astronauts Borman and Lovell saw the launch of a Polaris missile from a submarine and the reentry vehicle of a Minuteman 2 missile, both visually and instrumentally. The flights of Gemini VI-A and Gemini VII were milestones in the American space program. NASA had demonstrated that it could control two piloted spacecraft in orbit, a major first in the exploration of space by the United States.

**See also:** Apollo Program; Gemini Program; Gemini Spacecraft; Gemini V; Gemini VIII; Manned Maneuvering Unit.

### Further Reading

Borman, Frank, with Robert J. Serling. *Countdown: An Autobiography.* New York: William Morrow, 1988. Borman takes a straightforward look at his Gemini experience with humorous insights into the workings of an astronaut's mind.

Cernan, Eugene A., and Don Davis. *The Last Man on the Moon: Astronaut Eugene Cernan and America's Race in Space.* New York: St. Martin's Press, 1999. Cernan narrates his story of Gemini and Apollo.

Godwin, Robert, ed. *Gemini 6: The NASA Mission Reports.* Burlington, Ont.: Apogee Books, 1999. This collection contains reprints of the Gemini VI press kit, along with pre- and postmission reports. It also includes a CD-ROM featuring more than 50 minutes of 16-millimeter film footage of Gemini VI-A and more than 180 still pictures from the flight. Includes an exclusive interview with Gemini VI-A Commander Walter M. Schirra, Jr.

_____. *Gemini 7: The NASA Mission Reports.* Burlington, Ont.: Apogee Books, 2002. Using copies of NASA documents, this book details the Gemini VII mission. Included are the Gemini VII press kit, which gives insight to the plans for the rendezvous mission, and the crew's technical debriefing that gives the results of the flight. The accompanying CD-ROM contains a recent interview with Lovell, a Gemini VII movie, and Gemini VII Hasselblad mission photographs.

Harland, David. *How NASA Learned to Fly in Space: An Exciting Account of the Gemini Missions.* Burlington, Ont.: Apogee Books, 2004. The nuts and bolts of the Gemini Program are explained in this well-written book. The launch vehicles and spacecraft are detailed, as are the astronauts who flew them and the missions they flew.

Kraft, Christopher C., Jr. *Flight: My Life in Mission Control.* East Rutherford, N.J.: Penguin Putnam, 2002. The first NASA flight director gives an account of his life in Mission Control.

Schirra, Walter M., Jr., with Richard N. Billings. *Schirra's Space.* Annapolis, Md.: Naval Institute Press, 1995. Schirra, the only astronaut to fly in Project Mercury, Gemini, and the Apollo Program, tells of his Gemini VI-A mission. During the flight he, his crewmate Thomas P. Stafford, and the crew of Gemini VII performed the first rendezvous of two orbiting spacecraft. The book is an interesting study of the life of a test pilot—not the loner, but a member of a team, at work and at home. Schirra pulls no punches as he conveys his feelings about NASA, spaceflight, and life in general. There are many black-and-white photographs that add to the text.

Shayler, David J. *Gemini: Steps to the Moon.* Chichester, England: Springer-Praxis, 2001. The development of the Gemini Program and the spacecraft from the perspectives of the engineers, flight controllers, and astronauts involved.

Shepard, Alan B., Jr., and Donald K. "Deke" Slayton, with Jay Barbree and Howard Benedict. *Moon Shot: The Inside Story of America's Race to the Moon.* Atlanta: Turner, 1994. This is, indeed, the inside story of the Apollo Program as told by two men who actively participated in it. Some of their tales appear here for the first time. Shepard and Slayton discuss the Gemini orbital flights and tell some interesting behind-the-scenes stories of the Program. The book was adapted for a four-hour documentary in 1995.

Slayton, Donald K., with Michael Cassutt. *Deke! U.S. Manned Space: From Mercury to the Shuttle.* New York: Forge, 1995. This is the autobiography of the last of the Mercury astronauts to fly in space. After being grounded from flying in Project Mercury for what turned out to be a minor heart murmur, Slayton was appointed head of the Astronaut Office. Slayton talks of his frustration at being grounded and how he worked to regain flight status. He also discusses the flights of his fellow astronauts during the Gemini Program.

Stafford, Thomas P. with Michael Cassutt. *We Have Capture: Tom Stafford and the Space Race.* New York: Smithsonian Institution Press, 2002. Stafford remembers his years as a test pilot, Gemini and Apollo astronaut, Air Force general, president's stand-in at the 1971 Soviet funeral for three cosmonauts, and Apollo-Soyuz astronaut.

Wendt, Guenter, and Russell Still. *The Unbroken Chain.* Burlington, Ont.: Apogee Books, 2001. Wendt is the only person who worked with every astronaut bound for space.

*Mike D. Reynolds*

# Gemini VIII

*Date:* March 16, 1966
*Type of mission:* Piloted spaceflight

*Gemini VIII was the first piloted spacecraft to rendezvous and dock with the Agena, a rocket that could be restarted either from the Gemini spacecraft or by ground control. This accomplishment paved the way for a piloted lunar mission.*

### Key Figures

*Neil A. Armstrong* (b. 1930), Gemini VIII command pilot

*David R. Scott* (b. 1932), Gemini VIII pilot

*Charles "Pete" Conrad, Jr.* (1930-1999), Gemini VIII backup command pilot

*Richard F. Gordon, Jr.* (b. 1929), Gemini VIII backup pilot

*Robert R. Gilruth* (1913-2000), director of the Manned Spacecraft Center, Houston

*George M. Low* (1926-1984), deputy director of the Manned Spacecraft Center

*Charles W. Mathews* (1921-2002), manager of the Gemini Program

*George E. Mueller* (1918-2001), associate administrator for Manned Spaceflight

*John Hodge,* director of Flight Operations, Mission Control

*John F. Yardley* (1925-2001), technical director

*William C. Schneider* (b. 1923), mission director

### Summary of the Mission

The Gemini VIII mission carried two U.S. astronauts into Earth orbit, where they successfully rendezvoused with an already launched Agena rocket "parked" in a circular orbit around Earth. Soon after a successful docking with the Agena, it became obvious to the crew that a serious roll had developed in the Gemini-Agena combination, which was leading to wild and uncontrollable gyrations. Thinking that the fault might lie in the Agena, Gemini VIII's crew detached it from the Agena only to find that one of the craft's own attitude thrusters was stuck in the open position. Gemini VIII was then able to stabilize its attitude briefly and splash down after a flight time of only 10 hours and 41 minutes. The crew was safely recovered in the Pacific.

Before the launching of Gemini VIII, key personnel in the National Aeronautics and Space Administration (NASA) were experiencing some uneasiness regarding the Gemini Program. First, Apollo was scheduled to fly in early 1966, and several key personnel and facilities were dividing their time between the two programs. Prior to Gemini VIII, no successful docking with a fueled rocket booster in near-Earth orbit had occurred, even though it was the prime mission of the entire Gemini Program.

Second, the Agena (Gemini-Agena Target Vehicle, or GATV) was almost a year behind schedule, and some program schedulers doubted that a suitable Agena would be ready before the Apollo Program would dominate NASA's time.

To some of the harder "bottom liners," Gemini had outlived its usefulness in producing seven successful launches, five of which had been piloted by the appropriate crews of two. Unless rendezvous could be accomplished quickly, the entire program would be terminated and personnel and facilities shifted to Apollo to get that program moving with speed at its critical start. In the view of many of its critics, Gemini VIII carried with it the future of the Gemini Program.

With Neil A. Armstrong and David R. Scott as command pilot and pilot respectively, Gemini VIII was launched from Cape Kennedy on March 16, 1966, at 16:41 Coordinated Universal Time (UTC, or 11:41 A.M. eastern standard time). The GATV had been launched at 15:00 UTC (10 A.M.) on the same day and effectively parked in a circular orbit that was approximately 300 kilometers above Earth. Launched with an Atlas rocket, the Agena was thrust initially into a slightly lower orbit, but its engines were fired remotely and the Agena ascended into its proper orbit, much to the relief of ground controllers.

Before the launch of Gemini VIII, one of Scott's parachute harnesses was found to be filled with glue and could not be attached properly. Backup Command Pilot Charles "Pete" Conrad, Jr., and Mc-Donnell Engineer Guenter Wendt had to dig it out with screwdrivers to get Scott properly attached; the launch was delayed as a result.

A Titan II booster launched Gemini VIII into an elliptical orbit of 160 by 272 kilometers. Because of the lateness of the launch, this orbit was not the optimum to bring closest approach to Agena in the least amount of time. One hour and 34 minutes into the launch, therefore, Armstrong adjusted the inertial platform for an altitude change maneuver. To put Gemini VIII in a more circular orbit approaching that of Agena, Armstrong activated the thrusters for five seconds, but upon completing the procedure he noticed a deceleration problem. Varying com-

puter readings at the time made it difficult to tell if the exact deceleration had been obtained. Two hours and 18 minutes into the flight, when Gemini VIII had reached apogee (farthest orbital point from Earth) for the second time, Armstrong made more orbital corrections and noticed again that it was difficult to get coherent computer readings on the orbital changes.

His final maneuver was to place Gemini VIII exactly into the orbital plane of Agena, which he performed, but he received no confirmation from the ground as to how accurate it was. Soon, however, the crew obtained rendezvous radar target acquisition, which showed that Agena was at a range of 332 kilometers from Gemini VIII. Because the rendezvous radar also gave the astronauts the azimuth and elevation angles as well as the closing rate between the two vehicles, the computer was commanded to direct the rendezvous once Armstrong had aligned the inertial platform properly.

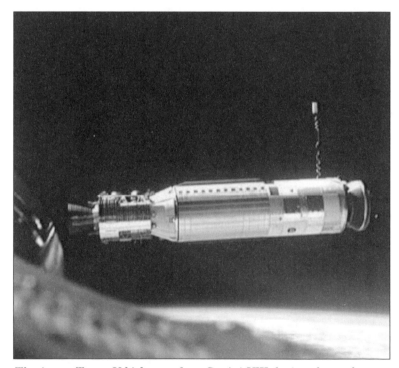

*The Agena Target Vehicle seen from Gemini VIII during the rendezvous.* (NASA)

Five hours and 43 minutes into the flight, Agena could be seen by the astronauts, and soon afterward they closed to an identical orbit to that of Agena, at a distance of about 50 meters. Thus, rendezvous was effectively established for the second time in the Gemini Program. Previously, two Gemini spacecraft, Gemini VI-A and VII, had rendezvoused. Now a Gemini spacecraft had succeeded in docking with its appropriate target vehicle, the Agena, mimicking a docking with a rocket booster that could provide the thrust to boost a spacecraft out of Earth orbit and into a trajectory toward the Moon, in an eventual lunar spaceflight.

Six hours and 32 minutes into the flight, while Gemini was over the Pacific, on the night side of Earth, the order came from a Pacific tracking ship to go ahead and dock. Armstrong then closed on the Agena carefully and finally approached the target adapter at a rate of 8 centimeters per second. The docking was very gentle. Armstrong reported no shocks or oscillations at all, calling the maneuver "a real smoothie."

At last Gemini VIII had accomplished the final and most important mission of the Program, docking in Earth orbit with a vehicle that had a restartable rocket engine. NASA, Lockheed Missiles and Space (the manufacturer of the Agena), Westinghouse (the rendezvous radar builder), and IBM (the maker of the onboard computer) all rejoiced at the success of the rendezvous operation.

Armstrong and Scott then commenced their docked chores—having the combined Gemini-Agena, now tightly linked mechanically, undergo several roll, yaw, and pitch maneuvers. Scott commanded a 90° right turn, which should have been completed in exactly one minute. The maneuver was completed in only 55 seconds, which was strange, but then looking at the "ball" indicator on his instrument panel, Scott noticed that they were in a 30° roll as well.

Scott, who like Armstrong was a veteran test pilot, yelled out to his command pilot, "Neil, we're in a bank." Armstrong's "ball" indicator showed an identical bank that should not have been there. They could not check their attitude, because they

were on the night side of Earth and no accurate radars were tracking them as they sped over the Pacific. Armstrong then tried to correct the roll with attitude thrusters, but the roll worsened. The Gemini-Agena combination was experiencing wild gyrations. The crew tried using Orbital Attitude and Maneuvering System (OAMS) thrusters to correct the gyrations, but again the whirligig motions increased. OAMS was a system designed mainly for changing orbits, but the crew employed it because the attitude thrusters were unable to control the motions.

Noticing that only 30 percent of the OAMS fuel was left, and this would be critical later to reach reentry orbits, Scott and Armstrong became convinced that a spacecraft thruster was causing the problem by being stuck in the open position. They soon realized that they would have to break away from the Agena to analyze their troubles, and on Armstrong's command Scott hit the undocking button. Armstrong then fired the thrusters, and the spacecraft pulled away from Agena.

Detached from Agena, with its large inertial mass, Gemini VIII rolled and gyrated even faster. At that point, both astronauts knew that the problem lay in the Gemini, not in the Agena. Gemini was rolling about one revolution per second, which would have been a dizzying rate in an Earth-gravity environment. Even in the weightlessness of space, the astronauts were becoming dizzy and their vision was blurring. Both vainly tried to correct the spacecraft's attitude with their hand controllers. There was virtually no maneuverability left in the orbital thrusters. The situation looked bleak, because the attitude had to be accurately controlled when the retrorockets were to be fired to commence reentry.

It became clear that one of the sixteen attitude control thrusters was firing continuously. To isolate the faulty thruster, the astronauts turned off all sixteen at once and activated the reentry control system to stabilize the craft. The bad thruster was finally identified, but the activation of the reentry control system forced an early end to the Gemini VIII mission. Over China, Gemini VIII hit the fringes of the atmosphere. Although originally scheduled to touch down in the Atlantic, Armstrong and Scott were glad when they touched down about 500 kilometers south of southern Japan at longitude 136° east. Because they were just west of the International Date Line, they landed on March 17. However, the official landing date—using either eastern standard time or Coordinated Universal Time—was March 16. Their total flight time was 10 hours and 41 minutes.

Planes flying out of Okinawa spotted the spacecraft; the splashdown, fortuitously, had occurred not far from one of the secondary landing sites. Within an hour, the rescue crew had put the flotation collar around Gemini VIII. Armstrong and Scott had landed in relatively rough seas and experienced some seasickness. Three hours after landing, they were brought on board the destroyer *Mason*, and a day later they and the Gemini VIII spacecraft were in Okinawa.

Armstrong and Scott had experienced a most difficult flight and rough seas; they were fortunate to have survived. Luckily, Scott had left the Agena, still in orbit, switched to ground control, so that it could be activated and undergo orbital changes from Earth. Several subsequent orbital and attitude changes were radioed to the Agena, and the rocket performed well beyond design expectations.

McDonnell took Gemini VIII back to St. Louis and diagnosed the thruster problem. Its malfunction had probably been caused by an electrical short circuit, which could have occurred in any of several places in the spacecraft.

### Contributions

The ten scientific experiments aboard Gemini VIII could not be performed adequately, given the abrupt ending of the flight. Nevertheless, several insights gained from this flight were put to good use in later Gemini and Apollo missions.

It was evident to the program managers and technical personnel who evaluated the Gemini VIII flight that the timing of the rendezvous was poor. Armstrong and Scott were unable to receive aid from NASA or McDonnell personnel when the se-

rious roll problems occurred, as they were on the other side of Earth. They were not even able to use Earth's perimeter to gauge how serious or prolonged the roll problem was, because they could not see Earth. Even though their attitude indicator "ball" was correct in describing the roll that had developed, it was difficult for the men to gauge its magnitude. In future programs, all critical maneuvers, such as docking, would be performed when the spacecraft was in direct contact with ground control, and deviations in motions would be readily ascertainable to accurate ground-based radars located in the United States and along its periphery.

Because the rendezvous mode went so well with the combined activity of the rendezvous radar and onboard computer, Gemini VIII proved that the concept worked well. The radar locked onto the Agena at a range of 333 kilometers, which was almost at the maximum design range of 345 kilometers. The final closing rate just before docking, only 8 centimeters per second, was so low and the docking so smooth that the fears of mechanical damage in such maneuvers were alleviated.

Agena did remain in orbit after the return of Gemini VIII and accepted almost five thousand commands, changing its orbit several times on orders from ground control. It was a very successful response vehicle and proved that restartable rocket engines could be used effectively.

Gemini VIII used the onboard computer and had a relatively safe reentry, even though it splashed down at an unscheduled time and place. It was the first reentry totally controlled by the computer.

The failure of the attitude thruster, staying open and causing the serious roll and gyrations, was ascribed to an electrical short circuit. As a result, on subsequent Gemini (and later) craft, the attitude thrusters were isolated from the OAMS thrusters, and each individual thruster was isolated, so that the pilot could turn it off. The thruster problem did not recur on any piloted program.

### Context

Until 1966, the Gemini Program was marked by seven very successful flights, five of which were pi-

*On the deck of the recovery ship USS* Leonard, *Gemini VIII astronauts David R. Scott (left) and Neil A. Armstrong arrive at Nahs, Okinawa.* (NASA)

loted. All seemed to be going well with the Program, and it was suggested by many workers that Gemini could fly every fourteen days, if necessary, without much trouble.

In 1966, with the start-up of the Apollo flights, many key NASA personnel were torn between the two programs. There was even talk of curtailing Gemini after the Gemini VII flight or at least after Gemini VIII. A similar curtailing of Mercury had occurred when Gemini started to fly, as several of the later scheduled flights had been deemed unnecessary.

The problem was the GATV, which was pacing the entire program in 1966. No previous Gemini had docked with a restartable engine such as Agena, and that had been the prime mission of the Gemini Program. Two Gemini craft (Gemini VI-A and VII) had effectively rendezvoused with each

other the previous year but had not actually docked with each other, because no docking adapter was available on either craft.

Unless Agena, manufactured by Lockheed Missiles and Space Division, could be brought on line and readied for flight in early 1966, the entire Gemini Program might be canceled. The Agena development was about a year behind schedule. When the first two Agenas were manufactured, the first was scheduled to be used as an engineering test model and the second to fly with Gemini VIII. When the second exploded accidentally, provisions were made to fly the first manufactured version without adequate engineering evaluation tests.

As it happened, the Agena performed very well during the flight, the rendezvous radar and the onboard computer worked extremely well in the rendezvous mode, and the docking was performed remarkably easily. The failure in the mission was caused by the short circuit of a rocket thruster in the Gemini VIII craft. Luckily, Armstrong and Scott had left the Agena in its ground-controlled mode; after Gemini VIII returned to Earth, the Agena was started and restarted by ground-control personnel. The fact that this orbiting rocket motor accepted and responded to so many ground-control commands subsequent to the flight of Gemini VIII gave proponents of remote-controlled, orbiting rockets considerable confidence for future efforts. Several other orbital corrections could be made remotely in the future in sizable restartable rocket engines; Gemini VIII's GATV was effectively the first.

Apollo would use the lunar-orbital rendezvous between the Apollo craft and the Lunar Modules (LMs) in a manner similar to that established by Gemini VIII. The rendezvous mode used in Gemini was the archetype for all the subsequent piloted lunar missions, and the mechanics of the docking chapters of Apollo were simply scaled-up versions of those of Gemini VIII and its Agena target vehicle.

Gemini VIII also proved that rendezvous could be performed safely without much fuel expenditure. The onboard computer of Gemini VIII that controlled both its rendezvous and reentry proved to be the salvation of subsequent spaceflights; its ef-ficacy was first demonstrated on Gemini VIII. Neil Armstrong's cool behavior under stress may have earned for him the right to be the first human on the Moon on Apollo 11, three years later.

**See also:** Gemini Program; Gemini Spacecraft; Gemini V; Gemini VII and VI-A; Gemini IX-A and X; Gemini XI and XII; Launch Vehicles; Manned Maneuvering Unit.

**Further Reading**

Borman, Frank, with Robert J. Serling. *Countdown: An Autobiography.* New York: William Morrow, 1988. Borman, who was on the backup crew for Gemini IV, flew on the most spectacular rendezvous mission as command pilot on Gemini VI-A, which orbited and rendezvoused with Gemini VII. His personal insights into flight preparations and rendezvous missions are very accessible to general audiences.

Cernan, Eugene A., and Don Davis. *The Last Man on the Moon: Astronaut Eugene Cernan and America's Race in Space.* New York: St. Martin's Press, 1999. Cernan, whose spaceflight career spanned both the Gemini and Apollo programs, was the first person to spacewalk in orbit around the Earth, as pilot of Gemini IX-A, and the last person to leave footprints on the lunar surface.

Grimwood, James M., Barton C. Hacker, and Peter J. Vorzimmer. *Project Gemini, Technology and Operations: A Chronology.* NASA SP-4002. Washington, D.C.: Government Printing Office, 1969. An authoritative day-to-day account of the planning and development of Gemini, from its inception as Mercury Mark II to the final flight of Gemini XII. Good sketches are included, technical problems are examined, and details are summarized effectively.

Hacker, Barton C., and James M. Grimwood. *On the Shoulders of Titans: A History of Project Gemini.* NASA SP-4203. Washington, D.C.: Government Printing Office, 1977. This 625-page history of the project is more or less the official NASA history of the Gemini Program. It is well written and organized effectively, and it provides a good picture of the fiscal, political, personnel, technological, and flight problems as they occurred. Illustrated.

Harland, David. *How NASA Learned to Fly in Space: An Exciting Account of the Gemini Missions.* Burlington, Ont.: Apogee Books, 2004. The nuts and bolts of the Gemini Program are explained in this well-written book. The launch vehicles and spacecraft are detailed, as are the astronauts who flew them and the missions they flew.

Kraft, Christopher C., Jr. *Flight: My Life in Mission Control.* East Rutherford, N.J.: Penguin Putnam, 2002. The first NASA flight director gives an account of his life in Mission Control.

Scott, David, and Alexei Leonov. *Two Sides of the Moon: Our Story of the Cold War Space Race.* London: Thomas Dunne Books, 2004. Astronaut Scott and Cosmonaut Leonov recount the drama of one of the most ambitious contests ever undertaken, set against the clash between Russian communism and Western democracy.

Shepard, Alan B., Jr., and Donald K. "Deke" Slayton, with Jay Barbree and Howard Benedict. *Moon Shot: The Inside Story of America's Race to the Moon.* Atlanta: Turner, 1994. This is, indeed, the inside story of the Apollo Program as told by two men who actively participated in it. Some of their tales appear here for the first time. Shepard and Slayton discuss the Gemini orbital flights and tell some interesting behind-the-scenes stories of the program. The book was adapted for a four-hour documentary in 1995.

Shayler, David J. *Gemini: Steps to the Moon.* Chichester, England: Springer-Praxis, 2001. The story of the Gemini Program and the spacecraft from the perspective of the engineers, flight controllers, and astronauts involved.

Slayton, Donald K., with Michael Cassutt. *Deke! U.S. Manned Space: From Mercury to the Shuttle.* New York: Forge, 1995. This is the autobiography of the last of the Mercury astronauts to fly in space. After being grounded from flying in Project Mercury for what turned out to be a minor heart murmur, Slayton was appointed head of the Astronaut Office. Slayton talks of his frustration at being grounded and how he worked to regain flight status. He also discusses the flights of his fellow astronauts during the Gemini Program.

Wagener, Leon. *One Giant Leap: Neil Armstrong's Stellar American Journey.* New York: Forge Books, 2004. In the first biography of Neil A. Armstrong, Wagener explores the life of one of America's true heroes, based on interviews with Armstrong's family and friends.

*John Kenny*

# Gemini IX-A and X

*Date:* June 3 to July 21, 1966
*Type of mission:* Piloted spaceflight

*Gemini IX-A and X were the seventh and eighth flights of the ten two-person missions of the Gemini Program. Both were three-day flights designed to demonstrate rendezvous and docking techniques and to evaluate extravehicular activity.*

### Key Figures

*Thomas P. Stafford* (b. 1930), Gemini IX command pilot
*Eugene A. Cernan* (b. 1934), Gemini IX pilot
*John W. Young* (b. 1930), Gemini X command pilot
*Michael Collins* (b. 1930), Gemini X pilot
*Elliot M. See, Jr.* (1927-1966), original Gemini IX command pilot
*Charles A. Bassett II* (1931-1966), original Gemini IX pilot
*James A. Lovell, Jr.* (b. 1928), Gemini IX backup command pilot
*Edwin E. "Buzz" Aldrin, Jr.* (b. 1930), Gemini IX backup pilot
*Alan L. Bean* (b. 1932), Gemini X backup command pilot
*Clifton C. Williams* (1932-1967), Gemini X backup pilot
*Robert R. Gilruth* (1913-2000), director of the Manned Spacecraft Center, Houston
*George M. Low* (1926-1984), deputy director of the Manned Spacecraft Center
*Charles W. Mathews* (1921-2002), manager of the Gemini Program
*George E. Mueller* (1918-2001), associate administrator for Manned Spaceflight
*Willis B. Mitchell, Jr.* (b. 1920), manager of the Office of Vehicles and Missions, Gemini Program Office, Manned Spacecraft Center

### Summary of the Missions

Gemini, the second U.S. piloted spaceflight program, was the bridge between the pioneering achievements of Project Mercury and the still-to-be realized lunar missions of the Apollo Program. Gemini missions IX-A and X followed on the heels of the highly successful fourteen-day flight of Gemini VII, a mission that demonstrated that humans could stay weightless for two weeks without serious physical problems, and the prematurely aborted, problem-plagued flight of Gemini VIII. A major objective of the three-day IX-A and X missions was to demonstrate precision spacecraft maneuvering in Earth orbit. This involved rendezvous and docking with an robotic target vehicle which was to be placed into orbit shortly before the launching of the Gemini spacecraft.

Gemini IX suffered from several prelaunch disasters. On February 28, 1966, less than three months before the tentative launch date, astronauts Elliot M. See, Jr., and Charles A. Bassett II were tragically killed when their training plane crashed upon its approach to the St. Louis Municipal Airport. Ironically, their jet hit the roof of a McDonnell Aircraft Corporation building—the very building where Gemini spacecraft IX and X were housed. Moments after the crash, the Gemini IX backup crew, Thomas P. Stafford and Eugene A. Cernan, landed safely in St. Louis. The four had been en route to McDonnell for two weeks of training in the simulator.

Officials at the National Aeronautics and Space Administration (NASA) promptly announced that the mission would not be delayed; it was decided that Stafford and Cernan would fly and that astronauts James A. Lovell, Jr., and Edwin E. "Buzz" Aldrin, Jr.—the backup crew for Gemini X—would be reassigned as their backups. Nevertheless, Gemini IX would not fly on schedule. On May 17, the day of the proposed launch, it was postponed when a booster engine for the Gemini-Agena target vehicle malfunctioned two minutes after liftoff, causing the target vehicle to plunge into the Atlantic Ocean off the coast of Florida.

Following the failure of the target launch vehicle, NASA implemented its contingency plan. After an earlier Atlas failure, NASA ordered prime contractor General Dynamics/Convair to be prepared to furnish a replacement within two weeks of another such disaster. A backup target, the Augmented Target Docking Adapter (ATDA), was successfully launched on June 1 and placed into its proper orbit. One hour and forty minutes later, however, the piloted Gemini flight, now designated IX-A, was again postponed when problems with a data transmitter prevented Mission Control from sending updated information to the spacecraft computer. The ATDA was made from spare parts from the Gemini spacecraft, mated to an Agena docking adapter. Its aerodynamic shell failed to open properly upon reaching orbit, because technicians taped down the separation lanyards. When Stafford saw the ATDA in orbit, he dubbed it the "angry alligator."

Two days later, on June 3, 1966, Stafford and Cernan boarded their conical Gemini spacecraft, which was hoisted atop a Titan II launch vehicle, for a third time. This time the countdown went smoothly, and at 13:39 Coordinated Universal Time (UTC, or 8:39 A.M. eastern standard time), the sev-

enth piloted and third rendezvous mission of the Gemini Program was launched from Cape Kennedy.

Later during the flight, Cernan took a spacewalk that lasted more than three times as long as Edward White's extravehicular activity (EVA) on the Gemini IV mission. Cernan, however, experienced extreme difficulty in donning the Astronaut Maneuvering Unit (AMU), a highly elaborate backpack that he was supposed to strap on and use to propel him through space. The unit, almost a separate spacecraft in itself, with gyroscopes for stabilizing and a host of thrusters for controlling movements, was too large to fit in the cockpit of the spacecraft. Instead, it was housed at the outside rear of the vehicle, in the adapter section. In order to strap it on, Cernan first had to exit the spacecraft, pull himself along the handrails, swing around into the adapter section, and back up against the AMU. In struggling to perform this task in the zero-gravity space environment, Cernan's heart rate soared to 180 beats per minute, and his respiration rate rose to 30 breaths per minute. His exertions so exceeded the limits of his pressure suit's life-support system that his visor completely fogged over. Stafford wisely decided to abort the planned demonstration of the AMU and ordered Cernan to return to the spacecraft.

The 72-hour, 22-minute mission ended successfully after forty-five orbits when Gemini IX-A landed in the Atlantic less than three miles from the carrier Wasp and within one-half mile of the predicted point of splashdown. Both pilots passed the postflight medical exam, although Cernan's exertions in space resulted in a weight loss of 6 kilograms—the most lost by any astronaut on a Gemini flight.

Within a week of Gemini IX-A's splashdown, Gemini X was removed from storage and hoisted to the top of its launch vehicle at Launch Complex 19. Six weeks later, on July 18, both Gemini X and its Agena target vehicle were launched into their designated orbits after a flawless countdown. The 3,600-kilogram spacecraft with its astronauts, John W. Young and Michael Collins, caught the Agena X within five and one-half hours after liftoff and then successfully docked with the target vehicle. This task, however, required the use of more propellant than had been predicted; as a result, constraints were imposed on the remainder of the mission and Mission Control had to develop an alternate flight plan. To conserve fuel, the craft remained docked with the Gemini-augmented target vehicle for 39 hours instead of the planned 16. Docking practice, a secondary objective of the mission, was canceled, as were other secondary mission experiments.

Nevertheless, the mission accomplished a primary objective when it used the propulsion systems of the Agena X to launch the spacecraft toward another target vehicle—the dormant Agena target left from the Gemini VIII mission. To rendezvous with the target, Young fired the Agena X's engine for fourteen seconds, long enough to propel the tandem vehicle into a new orbit 760 kilometers above Earth. After several hours of sleep and a breakfast of toast and bacon, Young fired the Agena's large thruster once more, this time in order to slow his craft and place it inside the circular orbit of the Agena VIII target, some 1,900 kilometers ahead.

While they waited to catch up with the target, Collins conducted a stand-up EVA in which he stuck his head and shoulders out of the spacecraft and exposed a camera lens to an unfiltered view of space. These experiments were cut short, however, when fumes, evidently from the lithium hydroxide in the air-filtering system, temporarily blinded both astronauts. After Collins returned to the spacecraft and the cabin was repressurized, the fumes and the blindness abated. Neither astronaut had problems with the breathing equipment during the remainder of the flight.

On the third day of the mission, the silent Agena VIII target was within range for rendezvous. Young undocked the Gemini X from its Agena workhorse

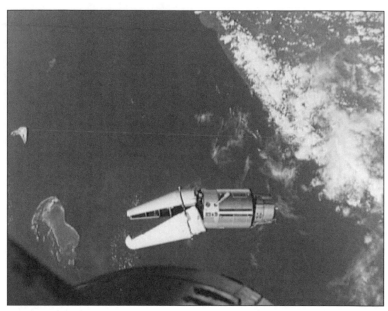

*The Augmented Target Docking Adapter seen from the Gemini IX-A spacecraft.* (NASA)

and skillfully maneuvered the craft close enough for Collins to spacewalk over to the drifting vehicle, where he removed a micrometeorite detection box that had been collecting data for the previous four months. With this EVA task accomplished, ground controllers, still concerned about the dangerously low fuel supply, ordered Collins back into the spacecraft; he did not attempt his remaining EVA maneuvers. Gemini X came to an end on July 21 after a flight of 70 hours and 46 minutes. Splashdown took place less than 10 kilometers from the aircraft carrier *Guadalcanal* in the Atlantic Ocean.

### Contributions

Despite the complications both before and after liftoff and the failure of the mission to dock using the Augmented Target Docking Adapter (ATDA), the Gemini IX-A mission was highly significant. Among other things, it underscored the difficulties humans encounter in working in an alien space environment. Cernan, for example, while unable to test the Astronaut Maneuvering Unit, did demonstrate the human ability to remain outside the spacecraft for a long time without disorientation. Moreover, the difficulties he experienced during his EVA demonstrated that the task of donning the unit was much more arduous than had been anticipated. As a result of his problem, NASA initiated several corrective measures designed to minimize or eliminate these difficulties in future flights. These measures included applying an antifog solution to the space suit's helmet visor, adding foot restraints on the spacecraft adapter section, and replacing zero-gravity aircraft simulations with underwater simulations in an attempt to duplicate the weightless environment of space.

The flight plan for Gemini X included fifteen scientific experiments—more than on any of the other three-day Gemini missions. While the fuel shortage necessitated a departure from the original plan, the crew still completed twelve of the proposed experiments, which ranged from the relatively simple task of photographing colored plates carried into space from Earth (an experiment designed to determine the effect of a vacuum on color photography) to the highly complex assignment of developing a navigation technique that would allow the astronauts to compute their own orbit and calculate for themselves whatever maneuvers were necessary for rendezvous. While the latter experiment was only partially successful, the Gemini X mission achieved its primary objective, rendezvousing and docking with the Agena target vehicle, as well as most of its secondary objectives.

Moreover, during Collins's second EVA—when he removed an experiment package from the passive Agena VIII—he proved that such tasks could be accomplished without excessive workload. In preparing for this EVA, however, Collins felt unduly rushed, receiving less assistance than he had anticipated from Young, who was fully occupied completing the rendezvous operation with the passive target vehicle. In their postflight evaluation, flight planners determined that in future missions the command pilot as well as the extravehicular pilot would be freed from their workload to devote full attention to EVA preparation.

**Context**

Gemini IX-A and X were the seventh and eighth of the ten U.S. Gemini piloted spaceflights. The purpose of the entire program was to build upon the pioneering knowledge of piloted flights gained earlier in Project Mercury in preparation for future lunar landings. These two particular missions achieved successful rendezvous, docking to join two vehicles as a single spacecraft. The missions successfully evaluated the control systems on each vehicle for controlling the combined vehicle, using the target vehicle's propulsion system to maneuver. They performed EVAs successfully in free space and provided for a controlled, precise reentry. Despite some disappointments, Gemini IX-A and X largely achieved their objectives and confirmed much of the technology that subsequently was to be used in future Gemini and Apollo missions.

In Apollo flights, the Lunar Module (LM), on returning from the surface of the Moon, had to rendezvous and dock with the Command and Service Module (CSM). During the Gemini IX-A mission, maneuvers performed during the second rendezvous demonstrated the feasibility of a rendezvous from above—a technique of great importance if the LM should be required to abort a lunar-powered descent. In the Gemini X mission, the spacecraft computer was programmed to use star-horizon sightings for predicting spacecraft orbit. Although the rendezvous for this mission actually was accomplished using ground-computed data, the onboard information was useful in developing space-navigation and orbit-determination techniques—an important step in the development of backup procedures for overcoming equipment failures in future flights. Moreover, the Gemini X crew's ability to dock with the Agena target vehicle and then use the rocket power from that Agena to accomplish a dual-rendezvous operation with the passive Gemini VIII Agena was a major step toward a piloted landing on the Moon.

The crews of Gemini IX-A and X established a number of records. In mission IX-A, the 2-hour, 9-minute spacewalk and the bull's-eye landing less than a mile from the predicted point of splashdown set new EVA endurance and precision-landing records. In mission X, the crew flew higher than any humans had ever flown, spent more time linked with an Agena satellite, and became the first to complete a dual rendezvous in space. In addition, the Gemini X extravehicular pilot became the first human to make physical contact with and work on another orbiting object. Nevertheless, the significance of these missions lay less in the historic "firsts" they achieved than in the role that they played in placing people on the Moon and returning them to Earth before the end of the 1960's. Both the problem-plagued Gemini IX-A and the near-flawless Gemini X made significant contributions to the advancement of the aeronautical knowledge and technology needed to accomplish this national goal.

**See also:** Gemini Program; Gemini Spacecraft; Gemini VIII; Gemini XI and XII; Manned Maneuvering Unit.

### Further Reading

Cernan, Eugene A., and Don Davis. *The Last Man on the Moon: Astronaut Eugene Cernan and America's Race in Space.* New York: St. Martin's Press, 1999. Cernan was the first person to spacewalk in orbit around the Earth, as pilot of Gemini IX-A, and the last person to leave footprints on the lunar surface.

Collins, Michael. "The Bridge." In *Liftoff: The Story of America's Adventure in Space.* New York: Grove Press, 1988. Collins gives a narrative look at the Gemini Program with emphasis on his own Gemini X flight. He takes you inside the confines of the small spacecraft and lets you drift slowly with him toward the Agena target vehicle. The monotony of the voyage and the hectic pace of some of the activities provide contrast in this story of the Program that developed the techniques for rendezvous and docking in space. Illustrated with eighty-eight line drawings that give the reader a view of space unavailable to the camera's eye.

_____. *Carrying the Fire: An Astronaut's Journeys.* New York: Farrar, Straus and Giroux, 1974. Collins's entertaining and inspiring autobiography. Includes a personal account of NASA and the other astronauts involved in Gemini and Apollo; contains numerous photographs, statistical charts, and an outline of the dates, crews, and major accomplishments of all U.S. piloted flights through Skylab.

Hacker, Barton C., and James M. Grimwood. *On the Shoulders of Titans: A History of Project Gemini.* NASA SP-4203. Washington, D.C.: Government Printing Office, 1977. This well-written, official history of the Gemini Program recounts the various problems and successes encountered during development. Contains illustrations and charts. A useful overview.

Harland, David. *How NASA Learned to Fly in Space: An Exciting Account of the Gemini Missions.* Burlington, Ont.: Apogee Books, 2004. The nuts and bolts of the Gemini Program are explained in this well-written book. The launch vehicles and spacecraft are detailed, as are the astronauts who flew them and the missions they flew.

Kraft, Christopher C., Jr. *Flight: My Life in Mission Control.* East Rutherford, N.J.: Penguin Putnam, 2002. The first NASA flight director gives an account of his life in Mission Control.

National Aeronautics and Space Administration. *Gemini Midprogram Conference, Including Experiment Results.* NASA SP-121. Washington, D.C.: Government Printing Office, 1966. This conference presented a summary of the Gemini Program up to February, 1966. A highly informative source for serious students of the early years of the Gemini Program.

National Aeronautics and Space Administration, Educational Programs and Services Office. *Manned Space Flight: Projects Mercury and Gemini.* Washington, D.C.: Government Printing Office, 1966. Appearing in the NASA Facts Series, this reference gives an overview of Projects Gemini and Mercury as they were seen in the 1960's. Includes much information and many illustrations that are not available elsewhere.

Shayler, David J. *Gemini: Steps to the Moon.* Chichester, England: Springer-Praxis, 2001. The development of the Gemini Program from the perspective of the engineers, flight controllers, and astronauts.

——————. *Walking in Space: Development of Space Walking Techniques.* Chichester, England: Springer-Praxis, 2003. This analysis of EVA techniques draws on original documentation and interviews with astronauts.

Shepard, Alan B., Jr., and Donald K. "Deke" Slayton, with Jay Barbree and Howard Benedict. *Moon Shot: The Inside Story of America's Race to the Moon.* Atlanta: Turner, 1994. This is, indeed, the inside story of the Apollo Program as told by two men who actively participated in it. Some of their tales appear here for the first time. Shepard and Slayton discuss the Gemini orbital flights and tell some interesting behind-the-scenes stories of the Program. The book was adapted for a four-hour documentary in 1995.

Slayton, Donald K., with Michael Cassutt. *Deke! U.S. Manned Space: From Mercury to the Shuttle.* New York: Forge, 1995. This is the autobiography of the last of the Mercury astronauts to fly in space. After being grounded from flying in Project Mercury for what turned out to be a minor heart murmur, Slayton was appointed head of the Astronaut Office. Slayton talks of his frustration at being grounded and how he worked to regain flight status. He also discusses the flights of his fellow astronauts during the Gemini Program.

Wendt, Guenter, and Russell Still. *The Unbroken Chain.* Burlington, Ont.: Apogee Books, 2001. Wendt is the only person who worked with every astronaut who left the Cape bound for space.

*Terry D. Bilhartz*

# Gemini XI and XII

*Date:* September 12 to November 15, 1966
*Type of mission:* Piloted spaceflight

*Gemini XI and XII were the final two missions of the Gemini Program, the second U.S. piloted spaceflight program. The primary objectives of both missions included performing rendezvous and docking with the Agena target vehicles and conducting extravehicular activities.*

## Key Figures

*Charles "Pete" Conrad, Jr.* (1930-1999), Gemini XI command pilot

*Richard F. Gordon, Jr.* (b. 1929), Gemini XI pilot

*James A. Lovell, Jr.* (b. 1928), Gemini XII command pilot

*Edwin E. "Buzz" Aldrin, Jr.* (b. 1930), Gemini XII pilot

*Neil A. Armstrong* (b. 1930), Gemini XI backup command pilot

*William A. Anders* (b. 1933), Gemini XI backup pilot

*L. Gordon Cooper, Jr.* (1927-2004), Gemini XII backup command pilot

*Eugene A. Cernan* (b. 1934), Gemini XII backup pilot

*Robert R. Gilruth* (1913-2000), director of the Manned Spacecraft Center, Houston

*George M. Low* (1926-1984), deputy director of the Manned Spacecraft Center

*Charles W. Mathews* (1921-2002), manager of the Gemini Program

*George E. Mueller* (1918-2001), associate administrator for Manned Spaceflight

*Willis B. Mitchell, Jr.* (b. 1920), manager of the Office of Vehicles and Missions, Gemini Program Office, Manned Spacecraft Center

## Summary of the Missions

By the time of the scheduled September, 1966, launch of Gemini XI, the National Aeronautics and Space Administration (NASA) had placed into orbit and safely returned eight Gemini piloted spacecraft. Although one mission (Gemini VIII) had to be aborted after only 10 hours of flight, the astronauts and ground crews of every mission had confronted and overcome all the equipment and human failures that had threatened the success of the flights and the lives of the astronauts. This record of success—sixteen Gemini astronauts in space and sixteen safely returned—inspired high confidence among those at Mission Control, but it did not eliminate the tension that still filled the air at every Gemini liftoff.

On September 9, 1966, after some six months of flight preparation, Command Pilot Charles "Pete" Conrad, Jr., and Pilot Richard F. Gordon, Jr., began their final preflight tasks in anticipation of the scheduled Gemini XI liftoff. According to the flight plan, 97 minutes before a Titan II rocket would send Conrad and Gordon into their orbit, an Atlas booster carrying a robotic Agena target vehicle would be launched from Launch Complex 14. Once both vehicles were in orbit, the Gemini spacecraft would attempt a first-revolution rendezvous and docking with the Agena target.

Shortly after the propellant for Gemini's Titan II rocket had been loaded, a pinhole leak in the stage 1 oxidizer tank of the launch vehicle was discovered. Mission Control's decision to repair the leak required rescheduling the launch for the following day. On September 10, however, the countdown again was halted because of an apparent problem with the autopilot in the Atlas-Agena launch vehicle. While further investigation determined that no hardware replacement was required, the launch again was postponed and was rescheduled for September 12.

At 13:05 Coordinated Universal Time (UTC, or 8:05 A.M. eastern standard time) on September 12, the Atlas-Agena target vehicle was launched. Meanwhile, sitting atop a Titan II carrier in a 3,860-kilogram (8,509-pound) Gemini spacecraft erected on Complex 19, Conrad and Gordon continued their wait, hoping not to face a cancellation for the third time. At 14:42 UTC (9:42 A.M.), less than one second later than programmed and 97 minutes after the Atlas-Agena was launched, the Titan II rocket fired. A primary objective of the mission was achieved 1 hour and 34 minutes later, when the Gemini spacecraft, after five maneuvers, completed a first-orbit rendezvous and docking with the Gemini-Agena target vehicle.

After the rendezvous and docking, the Agena target's propulsion system was ignited to lift the two vehicles to an apogee of 1,369 kilometers, breaking the record set two months earlier by the crew of Gemini X. Later, the Agena was refired, sending the vehicles into a lower orbit of 297 kilometers. During their 71 hours in orbit, Conrad and Gordon completed nearly all the mission's secondary objectives—including docking practice, an extravehicular activity (EVA), scientific experiments, docked maneuvers, a tethered-vehicle test, and a demonstration of automatic reentry.

The only objective not achieved was one experiment—an evaluation of the minimum reaction power tool—which was not performed because Gordon's EVA was terminated prematurely. Gordon's principal EVA was to connect a tether between his Gemini and the Agena to which it was docked so that the two vehicles could later undock and conduct experiments to determine the use of the tether as a fuel-saving device. In training, Gordon had been able to make the necessary attachments easily, but in the weightless environment of space, he found it an extremely difficult task. His body flailed about as he tried to keep from drifting away from the nose of the Gemini while he groped with the connector. Viewing his partner's predicament from the spacecraft window, Conrad could only console his sweating companion with the encouraging words, "Ride 'em, cowboy." The task, which took only twenty-five seconds in simulation, took thirty minutes in flight, and it left Gordon so fatigued that Conrad recalled him to the spacecraft ten minutes early.

After 44 revolutions, Gemini XI returned to Earth in a completely automatic reentry maneuver calculated by the onboard computer. Splashdown

*Gemini XII astronauts James A. Lovell and Edwin E. "Buzz" Aldrin study the flight plans for the final Gemini mission.* (NASA)

occurred less than 5 kilometers from the planned landing point at 71 hours, 17 minutes, after liftoff. The crew was retrieved by helicopter, and the spacecraft was brought aboard the aircraft carrier *Guam*, about an hour after landing.

Like the Gemini XI mission, the twelfth and final flight in the Gemini series experienced preflight difficulties. Before the countdown for the scheduled November 9 launch began, a malfunctioning power supply in the launch vehicle had forced a twenty-four-hour delay. Then, after the deficient hardware was replaced, another malfunction occurred. Following this forty-eight-hour delay, the Gemini Atlas-Agena target vehicle for the Gemini XII mission was launched on November 11 at 19:08 UTC (2:08 P.M. eastern standard time). Shortly thereafter, at 20:37 UTC (3:37 P.M.), the Gemini space vehicle, with Command Pilot James A. Lovell, Jr., and Pilot Edwin E. "Buzz" Aldrin, Jr., was launched from Launch Complex 19.

The major objectives of the mission were to rendezvous and dock (if possible during the third revolution) with the Gemini-Agena target vehicle and to evaluate EVAs. Among the secondary objectives were tethered-vehicle evaluation, fourteen scientific experiments, docking practice, docked maneuvering for a high-apogee excursion, and automatic reentry demonstration. The high-apogee excursion was not attempted, because an anomaly was detected in the Gemini-Agena target vehicle propulsion system, and parking was not attempted, because the target vehicle's attitude control gas was depleted. All other objectives were achieved.

During this four-day flight, Aldrin performed two stand-up EVAs (with his head and shoulders outside the craft) and one umbilical EVA (a spacewalk in which Aldrin was connected to the vehicle with an umbilical cord). While outside the spacecraft, Aldrin performed a number of tasks—cutting wire, hooking and unhooking spring-loaded snap hooks, turning bolts, and plugging electrical connections—with the aid of specially designed space tools. Aldrin also tested a group of restraining devices, handholds, and footrests in-

*Gemini XII propelled into orbit at Kennedy Space Center. The launch vehicle was a Titan II.* (NASA)

stalled to make EVAs less exhausting. In all, Aldrin spent 5.5 hours outside the craft. With a flight plan that included regular rest intervals, Aldrin experienced none of the difficulties suffered by the extravehicular pilots of the Gemini IX-A, X, and XI missions.

Several spacecraft systems experienced technical problems during the flight. Two fuel cell stacks failed and had to be shut down, while two others suffered significant loss of power. Late during the second day of the flight, the crew discovered that little or no thrust was available from two orbital attitude and maneuver thrusters. The spacecraft, however, performed flawlessly during the automatically controlled reentry. Splashdown took place less than 5 kilometers from the planned landing point at 19:21 UTC (2:21 P.M.) on November 15. A helicopter picked up the healthy astronauts and deposited them twenty-eight minutes later on the deck of the aircraft carrier *Wasp*. Gemini XII was a well-executed flight, a splendid ending to a successful program.

### Contributions

The final Gemini missions completed twenty-two of the twenty-five experiments planned by flight controllers. A number were medical experiments designed to monitor the physiological responses of astronauts to heavy work loads. Another experiment studied the effects of weightlessness and radiation on a sample of human blood. The results of these medical tests, coupled with those from other missions, left NASA officials with a generally optimistic view of the human ability to survive and function in space. As far as could be determined, multiday flights did not damage the human body or impair mental acuity or judgment.

An interesting series of experiments designed to increase knowledge of satellites in orbit was conducted by the crew of Gemini XI. In one experiment performed while the spacecraft was docked with its target vehicle, the crew fired boosters to determine whether the mass of an object could be gauged by acceleration. Later, after the two vehicles were attached by a tether, they fired thrusters to send the craft into a slow spin with the Agena in the first attempt to create artificial gravity in space. (This is known as gravity-gradient tethered station-keeping.) Scientists estimated that the gravity created by the rotation of the craft was 0.25 percent of Earth's.

About half of the experiments on the two missions were photographic in technique. Photographs of Earth from space of Earth's terrain provided scientists with a wide range of information. For example, some photographs provided geologists with new evidence to support the theory of continental drift (that is, that the Arabian Peninsula is drifting away from Africa). Others helped geoscientists learn much about the causes and distribution of polluted water; still others provided oceanographers with information about the movement of gulf currents and their impact upon the distribution of shrimp. The astronauts aboard the last Gemini missions also conducted a variety of astronomical photography experiments—such as pointing their camera into space to take ultraviolet photographs of star fields or toward the horizon to obtain measurements of airglow altitude and intensity. Altogether, Gemini astronauts during the ten piloted flights produced some 2,400 photographs, which have proved to be of significant academic value for scientists in disciplines ranging from agriculture and astronomy to urban planning.

Another original focus of the final Gemini missions was upon developing techniques for extravehicular operations in free space. Problems encountered by extravehicular pilots on earlier missions, however, prompted mission designers to forgo the testing of the sophisticated jet-propelled Astronaut Maneuvering Unit (AMU) and to concentrate instead upon the evaluation of body-restraint devices. In all, Gemini XI and XII astronauts spent more than eight hours in EVAs. The results of these exercises demonstrated that all the tasks attempted were feasible when body restraints were used to maintain position. The results also showed that the extravehicular work load could be controlled within the desired limits by the application of proper procedures and indoctrination. Moreover, the results of Gemini XII's EVA confirmed that underwater simulation duplicated the actual space environment with a high degree of accuracy. It was concluded that any task readily accomplished in a valid underwater simulation would have a high probability of success during actual EVA. Thus, the Gemini missions succeeded in demonstrating the basic techniques required for the productive use of extravehicular operations. Nevertheless, evaluations of personal transportation devices—such as the relatively simple handheld maneuvering unit or the highly complex Astronaut Maneuvering Unit—had been too brief to define the full capabilities of the equipment.

### Context

The Gemini Program was designed to extend the experiences gained from Project Mercury to support of the proposed piloted lunar-landing program and other future piloted space programs. The specific objectives of Gemini applicable to the Apollo Program included long-duration flying, rendezvous and docking, postdocking maneuvering, controlled reentry and landing, flight- and ground-crew proficiency, and extravehicular capability. The achievement of these objectives was to provide operational experience and confirm much of the technology that would be utilized in future piloted missions.

The Gemini XI and XII missions were a fitting climax to a well-designed series. While not flawless, in demonstrating first-orbit and third-orbit rendezvous and docking, the feasibility of gravity-gradient tethered station-keeping, and automatic reentry capability, as well as in evaluating a variety of extravehicular operations, these missions added substantially to the body of space knowledge.

Before Gemini, the cumulative total of U.S. human-hours in space was less than fifty-four. By the time Gemini XII was completed, the cumulative human-hours in space had soared to nearly two thousand. During these hours in Earth orbit, Gemini crews completed ten rendezvous, including three re-rendezvous and one dual-rendezvous operation. They computed the maneuvers necessary for rendezvous using ground-based, onboard radar-computer, and manual computations. They developed an optical technique for navigation using information derived from star observations. They demonstrated in the rendezvous operations that the computation and execution of maneuvers for changing or adjusting orbits in space could be performed with considerable precision. Moreover, the nine docking operations achieved in Gemini demonstrated that docking could be accomplished in a routine manner, and that ground-training simulation was adequate for preparing the crews for this operation. Data obtained from flight experiences also established the ideal lighting conditions and preferred approaches and techniques for successful docking operations.

In sum, Gemini XI and XII were the last of a series of missions that together made significant contributions toward the U.S. goal of placing an astronaut on the Moon. The Gemini Program trained flight and ground crews for future space missions. It proved the human ability to function in space for long enough periods to travel and to return from lunar orbit. It demonstrated that humans could perform complicated mechanical and mental tasks with precision while adapting to the spacecraft environment. It advanced spacecraft design practices through the development and testing of systems hardware. Yet, for all of its technological achievements, perhaps the most important legacy Gemini left for future space programs was in demonstrating that successes could be achieved in spite of serious difficulties.

**See also:** Gemini Program; Gemini Spacecraft; Manned Maneuvering Unit; Tethered Satellite System.

### Further Reading

Cernan, Eugene A., and Don Davis. *The Last Man on the Moon: Astronaut Eugene Cernan and America's Race in Space.* New York: St. Martin's Press, 1999. Cernan was the first person to spacewalk in orbit around the Earth, as pilot of Gemini IX-A, and the last person to leave footprints on the lunar surface.

Collins, Michael. *Carrying the Fire: An Astronaut's Journeys.* New York: Farrar, Straus and Giroux, 1974. An entertaining and inspiring autobiography by Astronaut Collins. Includes a personal account of his relationship with NASA and with other astronauts involved in the Gemini and Apollo Programs, as well as numerous photographs, statistical charts, and an overview of the dates, crews, and major accomplishments of all U.S. piloted flights from Mercury through Skylab. Highly informative.

_____. *Liftoff: The Story of America's Adventure in Space.* New York: Grove Press, 1988. Many books have been written about the Apollo Program, most of them about Apollo

11. Few have given us an inside look at the delicate melding of man and machine. Contributing to this complete history of America's piloted space programs, Collins devotes a large portion to Apollo. He sets the record straight about some of the misconceptions of astronauts and space machines. The book is illustrated with eighty-eight line drawings by James Dean, former NASA art director, that add stark realism to an otherwise unreal world.

Godwin, Robert, ed. *Gemini 12: The NASA Mission Reports.* Burlington, Ont.: Apogee Books, 2003. Using copies of NASA documents, this book details the last Gemini flight. Included are the Gemini XII press kit, which gives insight to the plans for the mission, and the crew's technical debriefing that gives the results of the flight. The accompanying CD-ROM contains a recent interview with Lovell and complete 16-mm onboard video from Gemini XII.

Grimwood, James M., Barton C. Hacker, and Peter J. Vorzimmer. *Project Gemini, Technology and Operations: A Chronology.* NASA SP-4002. Washington, D.C.: Government Printing Office, 1969. A chronological account of events between April 24, 1959, and February 2, 1967, relating to NASA in general and the Gemini Program in particular. Each entry is intended to be relatively independent and complete. The chronology is fully documented with sources for each entry. Written for the educated, nontechnical reader. Includes photographs and charts. An excellent sourcebook for any student of the American space program.

Harland, David. *How NASA Learned to Fly in Space: An Exciting Account of the Gemini Missions.* Burlington, Ont.: Apogee Books, 2004. The nuts and bolts of the Gemini Program are explained in this well-written book. The launch vehicles and spacecraft are detailed, as are the astronauts who flew them and the missions they flew.

Kraft, Christopher C., Jr. *Flight: My Life in Mission Control.* East Rutherford, N.J.: Penguin Putnam, 2002. The first NASA flight director gives an account of his life in Mission Control.

National Aeronautics and Space Administration. *Gemini Summary Conference Papers.* NASA SP-138. Washington, D.C.: Government Printing Office, 1967. This report contains twenty-one papers presented at the Gemini Summary Conference held at the Manned Spacecraft Center in Houston in February, 1967. The conference emphasized the highlights of the Gemini Program and especially the flight results of the final five missions. Although some of the reports are highly technical, others are presented in a lively manner without excessive jargon. An excellent source for any serious student of the Gemini Program.

Shayler, David J. *Gemini: Steps to the Moon.* Chichester, England: Springer-Praxis, 2001. The development of the Gemini Program from the perspective of the engineers, flight controllers, and astronauts.

_____. *Walking in Space: Development of Space Walking Techniques.* Chichester, England: Springer-Praxis, 2003. Draws on original documentation and interviews with astronauts with experience in EVAs.

Shepard, Alan B., Jr., and Donald K. "Deke" Slayton, with Jay Barbree and Howard Benedict. *Moon Shot: The Inside Story of America's Race to the Moon.* Atlanta: Turner, 1994. This is, indeed, the inside story of the Apollo Program as told by two men who actively participated in it. Some of their tales appear here for the first time. Shepard and Slayton discuss the Gemini orbital flights and tell some interesting behind-the-scenes stories of the Program. The book was adapted for a four-hour documentary in 1995.

Slayton, Donald K., with Michael Cassutt. *Deke! U.S. Manned Space: From Mercury to the Shuttle.* New York: Forge, 1995. This is the autobiography of the last of the Mercury astronauts to fly in space. After being grounded from flying in Project Mercury for what turned out to be a minor heart murmur, Slayton was appointed head of the Astronaut Office. Slayton talks of his frustration at being grounded and how he worked to regain flight status. He also discusses the flights of his fellow astronauts during the Gemini Program.

Wendt, Guenter, and Russell Still. *The Unbroken Chain.* Burlington, Ont.: Apogee Books, 2001. Wendt is the only person who worked with every astronaut who left the Cape bound for space.

*Terry D. Bilhartz*

# Geodetic Satellites

*Date:* September 9, 1955, to June 1, 1988
*Type of spacecraft:* Earth observation satellites

*The measurement and mapping of the Earth, the science of geodesy, has been revolutionized in coverage, reliability, and precision since 1955 by adaptations of electronic innovations—radar, radio altimeters, magnetometers, return beam vidicon cameras, lasers, and multispectral sensors—to geodetic space satellites linked to extensive and complex land support networks.*

## Key Figures

*Mark Macomber,* commander of the Bureau of Naval Weapons
*H. Arnold Karo,* project director from the Coast and Geodetic Survey
*John D. Nicolaides,* NASA project director
*Jerome D. Rosenberg,* project director from NASA Headquarters
*Richard J. Anderle,* principal investigator at Naval Weapons Laboratory
*Ivan I. Mueller,* principal investigator from Ohio State University
*Lewis Swanson,* principal investigator from the National Geodetic Survey

## Summary of the Satellites

By 1945, the efforts of space theoreticians such as Russia's Konstantin Tsiolkovsky (1857-1935) and the United States' Dr. Robert H. Goddard (1882-1945), coupled with the practical experimentation of Germany's Max Valier, Johannes Winkler, Hermann Oberth, and Wernher von Braun, had produced various rockets capable of extraterrestrial penetration. Some rockets had demonstrated their potentialities during wartime—Germany's famed V-2's (or A-4's) and A-4b's, for example—while others, on academic drawing boards, excited scientific interests.

The possibilities of artificial satellites dedicated primarily to military applications, but not devoid of scientific value, had by the close of World War II already been under study by official and private or semiprivate organizations. Among them were the U.S. Navy's Bureau of Aeronautics and the Douglas Aircraft Company, which had close relationships with the U.S. Army Air Force and with North American Aviation. North American, in turn, was linked through its high-altitude test vehicle with the Navy, which was eager to enhance its surveillance, navigational, and communications capabilities. While the governmental agencies (including the military) and the private and academic organizations participating in the development of space satellites altered—or changed in their importance—during the 1940's and 1950's, their continuing interaction soon characterized emerging satellite geodesy. Project Orbiter, which led to the United States' first satellite, Explorer 1, provided a classic example of the complex collaborative efforts upon which satellite design and deployment were founded. When it gathered in Washington, D.C., on June 25, 1954, the Orbiter team represented the Office of Naval Research, the U.S. Naval Observatory, the Department of Defense, the U.S. Army Ballistic Missile Agency (ABMA), the Harvard College Observatory, and such private concerns as IBM, the Alabama Tool and Die Company, A. D. Little, and the Aerophysics Development Corporation. Thus, very shortly after Goddard's death in 1945, the inspiration and talents of singular individuals engaged in rocket and satellite design had melded into the collective capacities of teams and into the achievements (or the confusion and failure) of projects and programs.

Orbiter never orbited, and its successor, Project Vanguard (which, embarrassingly, was revealed on September 9, 1955, as having been conceived within a laboratory of the Office of Naval Research, unknown to the Orbiter team), suffered numerous failures between 1957 and 1959; nevertheless, their indirect contributions were appreciable. They helped resolve questions about the feasibility of multistage rocket launch vehicles, and although only three of eleven attempts to orbit the Vanguards between December, 1957, and September, 1959, were successful, solar panels proved capable of providing electricity to instrumentation, geophysicists gained new perspectives on the shape of Earth, and opportunities unfolded for more accurate mapping of Pacific islands, long a problem for geodesists. Meanwhile, the United States' response to the Soviets' Sputnik 1—which orbited Earth on October 4, 1957—was to launch Explorer 1, which entered its orbit in January of 1958, triumphing technically because of the miniaturization of its instrumentation and its telemetry system. Though Explorer 1 had no specific geodetic mission, miniaturization freed future satellites to carry more scientific experiments and reduced weights that had to be launched by still-limited rocketry.

As the eighteen-month-long International Geophysical Year approached (July, 1957, through December, 1958), American geodesists had well-established scientific objectives; the Sputniks simply lent them a greater urgency. Indeed, geodesists were already benefiting from applications of electronic distance measuring, which vastly improved the speed and accuracy of their surveys. Thus, after a decade of geodetic advances, they were quite ready for the exploitation of satellites.

Having created the Committee on Applications of Artificial Satellites in December of 1957, geodesists won sponsorship from the American Geophysical Society (AGS). The AGS then presented recommendations to the National Advisory Committee for Aeronautics (NACA), to the new National Aeronautics and Space Administration (NASA), and ultimately, in April, 1959, to Congress. What geodesists wanted was simple: a peaceful utilization of space and an orbiting "flashing light," a satellite that, like a star, would give them a carefully calculated location at regular intervals (an "ephemeris," when locations and times are put in tabular form), allowing Earth observers to fix their positions on its surface (the geoid). Thereafter, known positions could become references for the unknown, and accurate mapping could proceed.

Project ANNA, begun in January, 1961, represented the United States' first attempt to provide such a satellite system. It was a joint effort by the Army, the Navy, NASA, and the Air Force. The project was declassified by the U.S. Department of Defense (DoD), and its overall direction was assigned to Commander Mark Macomber of the Bureau of Naval Weapons, with the collaboration of Admiral H. Arnold Karo of the Coast and Geodetic Survey and NASA's John D. Nicolaides. The ANNAs (1A and 1B) were small, 0.91 meter in length, and almost spherical, weighing 161 kilograms. Launched from Cape Canaveral on May 10, 1962, ANNA 1A failed to orbit. ANNA 1B, however, boosted by a Thor Ablestar rocket, successfully entered a near-circular orbit 1,077 by 1,182 kilometers above Earth. It "fired off" a brilliant flashing light at exact intervals, which ground tracking cameras photographed, permitting measurements with a 10- to 30-meter margin of error. Instrumentation also included a Secor (Sequential Collation of Range) radio transmitter, mobile ground stations used for measurement with the same margin of error, and a Doppler quartz transmitter, which, when timed by ground stations, yielded a 15- to 50-meter margin of error. ANNA 1B's mission was not for specific mapping; rather, it was intended to begin a worldwide triangulation system that would tie the continents together geodetically.

By 1965, data from spreading geodetic satellite networks and from their tracking stations had not only tied the continents together but also caused an information explosion, which, however satisfying, demanded management. NASA consequently created the National Geodetic Satellite Program. Under the overall direction of NASA Headquarters' Jerome D. Rosenberg, a number of participants—NASA's Goddard Space Flight Center, the DoD, the National Geodetic Survey, Ohio State University's Department of Geodetic Science, the Smithsonian Institution's Astrophysical Lab, the U.S. Navy, and the Army Mapping Service—displayed an active interest in converting masses of data into meaningful intelligence. Principal investigators included Richard J. Anderle of the Naval Weapons Laboratory (later of the DoD) and an expert on Doppler satellites, Ivan I. Mueller of Ohio State University, and Lewis Swanson of the National Geodetic Survey.

Most of the geodetic satellite programs were financed by NASA. Indeed, NASA was the most avid consumer of geodetic satellite information. NASA had discovered that tracking its own satellites and spacecraft by conventional means necessitated a pooling of coordinates to reduce errors in determining orbits, in placing tracking stations, and in more accurately describing Earth's magnetic field. Finally, an array of new electronic measuring devices and cameras needed to be calibrated by commonly agreed-upon scientific standards before they could be used in costly programs.

Increasingly, therefore, NASA became a principal in satellite design and mission designations. Although its Beacon Explorers A, B, and C were intended primarily for ionospheric research, they were also employed for geodetic studies. Beacon Explorer A, launched on March 19, 1964, was unsuccessful. Beacon Explorer B, however, orbited by a Delta rocket from Cape Kennedy on October 10, 1964, carried 360 silicon laser reflectors, allowing the first measurements from Earth using a laser beam mirrored back from the satellite. Beacon Explorer C, launched on a Scout rocket from Vandenberg Air Force Base on April 29, 1965, in addition to having laser reflectors, was equipped to measure Earth's gravitational field.

On November 6, 1965, NASA did orbit a purely geodetic satellite from the Cape: GEOS 1/Explorer 29 (Geodetic Earth-Orbiting Satellite). Weighing 175 kilograms, octagonal in shape, with its base constantly facing Earth, GEOS 1 carried the familiar "flashing light" to be photographed from ground tracking stations, along with a quartz laser reflector and a radio transmitter—operating on stable frequencies of 150 to 400 or 162 to 324 megahertz—for measurements based on the Doppler effect. Its orbit was elliptical, ranging from 1,115 to 2,277 kilometers and integrating data into a single, one-world system. GEOS 1 related its data to Earth's center of mass so that site locations could be determined in a three-dimensional coordinate system, and its margin of error was less than 10 meters. Nearly identical in configuration, GEOS 2/Explorer 30 and Explorer 36, launched respectively on November 18, 1965, and January 11, 1968, proceeded with similar global triangulations.

*The LAGEOS 1 satellite, launched in 1976. This passive satellite was designed to measure movements in Earth's crust with a margin of error of 1 to 2 centimeters.* (NASA-MSFC)

Meanwhile, on June 24, 1966, in order to refine the accuracy of a worldwide survey network, PAGEOS (Passive Geodetic Earth-Orbiting Satellite) was launched from Vandenberg. A Mylar- and aluminum-coated balloon measuring 30 meters in diameter and weighing 111 kilograms, PAGEOS was designed to inflate in space. Devoid of instrumentation, it was meant to be a light source, a reflector of the Sun as bright as Polaris, which for five years (it actually lasted for ten) could be photographed from Earth, bringing unprecedented precision to the pinpointing of locations on Earth's surface—as measured from the center of its mass—with a margin of error of 16 to 30 meters.

Few of the many satellites orbiting by the mid-1960's were geodetic, although many could have furnished geodesists with useful information. Declassification of the U.S. Navy's Navigation Satellite System (NNSS) in 1967, therefore, was a boon to geodesists everywhere. Equipped to transmit orbital and timing information, these NNSS satellites allowed portable receivers on the ground to measure Doppler counts over time and compute positions either by reference to a broadcast ephemeris or by using the "precise ephemeris" generated by the U.S. Naval Weapons Laboratory. Thirty satellite passes were recommended before fixing the position of any site, but the original margin of error of 50 to 100 meters was swiftly refined to an accuracy exceeding classical geodetic triangulations. As a result, the exquisite precision necessary for the establishment of geodetic control networks was available to the international geodetic community. NASA's Landsat (for land satellite) series, the first of which orbited on July 23, 1972, and its Seasats (sea satellites), the first of which orbited in June, 1978, made immense contributions to mapping: Landsat scanned nearly every portion of the globe every eighteen days, and Seasat's equally remarkable infrared and multispectral instruments swept the oceans every 36 hours.

Though not developed under the umbrella of NASA's Explorer program, as GEOS 1 and GEOS 2 had been, GEOS 3, launched on April 10, 1975, from Vandenberg Air Force Base, nevertheless had a geodetic mission. It was a Geodynamic Experimental Ocean Satellite, built for the Earth and Ocean Physics Applications Program (EOPAP) for oceanographic research and the improvement of models that ignore heights and depths on the geoid (Earth's surface) and treat the globe as smoothly rounded. Weighing 340 kilograms, GEOS 3 resembled its nominal predecessors, both in its configuration and in much of its instrumentation, but it carried a radar altimeter and was equipped to exchange data with Seasat when it came into service.

That rudimentary geodetic satellite missions were well in hand in the 1970's was manifested on May 4, 1976, when LAGEOS 1 (Laser Geodynamics Satellite) was orbited from Cape Canaveral. A passive satellite, LAGEOS was designed to measure movements in Earth's crust (plate tectonics) with a margin of error of 1 to 2 centimeters. With its 5,900-kilometer orbit, the data it returned aided scientists in understanding continental drift (evidence of plate tectonics). Information from LAGEOS is also useful to earthquake researchers. Unlike satellites that are tracked from Earth, LAGEOS tracks Earth's course, supplying accurate laser measurements to scientists from a dozen countries. Expected to remain functional well into the twenty-first century, LAGEOS will circle Earth for about eight million years carrying various scientists' versions of Earth's history and "messages to the future."

NASA's solar-powered Magsat (Magnetic Field Satellite), orbited from Vandenberg Air Force Base on October 30, 1979. Magsat was designed to help resolve new questions concerning magnetic fields. In the early 1960's, for example, new evidence showed areas of unevenness (anomalies) in Earth's gravitational field. Because such anomalies distorted satellite measurements, sometimes by as much as 50 meters, they greatly concerned geodesists. Yet without reliable descriptions of the gravitational field, the establishment of common, unambiguous coordinate networks was difficult. Weighing only 115 kilograms and placed in a low polar orbit (355 to 562 kilometers), Magsat effectively applied its magnetometers to the reduction of gravity-induced errors of measurement to about 1 meter.

*Explorer 29, or GEOS 1, lifts off from Pad 17A at Kennedy Space Center in December, 1965.* (NASA-KSC)

compare their signals' times of arrival to secure information about the relative motion of the observatories. Pioneered in the 1960's, VLBI was sufficiently refined by 1980 for deployment. Operating within a global network of stations created by NASA's Crustal Dynamics Project, VLBI monitored and measured tectonic plate motion and Earth's "wobble" on its axis. Evidence indicated, too, that large patches of Earth's surface during only a few decades may swell or subside by as much as a centimeter, so that geodesists now are obliged to survey lithospheric motion, if not yet the deeper, underlying mantle.

The *Challenger* accident delayed the planned 1988 launch of the passive, laser-ranging satellite LAGEOS 2. A joint venture of NASA and the Italian Space Plan, LAGEOS 2 was to have been placed in orbit by the United States' space shuttle for study of crustal geodynamics, including measurement of polar movements, movements of the lithospheric plates, and "stretching" of geodetic baselines due to plate tectonics.

The now-elaborate surveying and mapping of Earth's land and oceans continued with Geosat (Geodynamic Earth and Oceans Satellite), orbited from Vandenberg Air Force Base on March 12, 1985. Weighing 636 kilograms, Geosat was powered by the usual solar panels and carried a radio altimeter for measurements. Like Magsat's, Geosat's mission addressed novel geodetic questions that sprang from evidence of Earth's dynamic lithosphere. The importance of these novelties was emphasized by data accumulated through very long-baseline interferometry (VLBI), by which many radio telescopes on different continents observe a quasar or satellite simultaneously and

**Contributions**

The hundreds of artificial satellites and spacecraft, American and foreign, placed in orbit since the 1950's with increasingly sophisticated geodetic missions and equipment, have revolutionized geodesy. Satellite geodesy, unheard of before 1950, is now a flourishing scientific subdiscipline, that in only a few decades has made substantial contributions to a fuller understanding of Earth's surface, its crustal dynamics, its gravitational field, and its very shape and overall movement.

Accurately establishing positions on the planet's surface—the positions of its oceans, continents, islands, cities, towns, farms, and wastelands—and surveying and mapping them are basic geodetic functions. These essential functions proceed by triangulation: the elaboration of measured networks of triangles, large and small. The deployment of artificial satellites—not all of them with primary geodetic missions, and not all of them American—greatly expedited the march of these triangles across the globe.

By 1969, little more than a decade after the successful orbiting of the first American satellite, the missions of ANNA 1B, GEOS (1, 2, and 3), Echo 2, and PAGEOS, among others, had provided an extremely accurate geodetic network stretching from Kwajalein island in the central Pacific to Ascension island, near Africa, in the Atlantic. With 157 U.S. satellites placed in orbit between 1958 and 1963 alone and a flood of data pouring from them to a growing number of ground tracking stations, the North American, European, oceanic, and polar networks were essentially completed by 1988. An international Global Positioning System—a constellation of satellites launched by the DoD—while not error-free, is in place and, according to William Melbourne of the Jet Propulsion Laboratory in Pasadena, California, measures long distances with only centimeters of error for every 1,000 kilometers.

Earth's crustal dynamics, the movements of its lithosphere, have been measured by satellite to within a few centimeters. Such observations confirm the crustal deformation of the United States' West Coast and expansion of the Great Basin and are particularly valuable to seismologists studying areas that are earthquake-prone. In June, 1988, similar measurements were under way along the crustal plates of Central and South America and on a number of small Pacific islands.

Satellite data have also demonstrated anomalies in Earth's gravitational field: Gravity is stronger above seamounts, weaker above the deep ocean trenches. Satellite radio altimeters have measured these anomalies as oceanic swellings or depressions, thereby permitting a partial mapping of the sea floors. Similarly, laser-ranging satellites and VLBI have shown that Earth wobbles on its rota-

tional axis, sometimes for weeks or months, sometimes for years, from 6 to 60 centimeters. Furthermore, an exact measurement has been made of the planet's elliptical shape.

Not least, satellites have carried an evolving array of technologies. ANNA 1B's "flashing light," for example, has become a sophisticated laser beam. Similarly, radio altimeters, magnetometers, laser-ranging devices, Doppler radar, and multi-spectral scanners have supplied mountains of data from which geodesists have refined a richly detailed mapping of Earth.

### Context

Before the development of satellite geodesy, large portions of Africa, Asia, the northernmost region of North America, and the polar and near-polar regions were inaccurately mapped. Tens of thousands of islands had not been brought into geodetic networks, the continents themselves had not been measured in their exact relations to one another, and the greater part of Earth's surface, the oceans and the sea floors, was substantially unsurveyed. Although theories existed before the 1950's about continental drift and sea-floor spreading, extensive, reliable evidence was sparse. Discussions about what later became known as plate tectonics, or crustal dynamics, consequently were conjectural. Even more unfortunate, geodetic investigations tended to be more parochial than international in scope and spirit.

Electronic advances were very important to geodesists. Nevertheless, much of the impetus toward development of artificial satellites derived from Soviet-American rivalries, partially symbolized by the orbiting of the Sputniks. Expensive missiles had to be guided and tracked accurately to targets, targets had to be located, and under the press of international fears, much more had to be swiftly learned about Earth's surface. Moreover, the expenses of satellite research and development were well beyond the means of the geodetic community in the United States. In addition, all the prospective rocket launch vehicles were possessions of the military. Clearly, progress with artificial satellites, whatever their purposes, depended on the financial resources of the nation: that is, on the federal government.

The creation of NASA in 1957, shortly after the success of Sputnik 1, resolved these basic problems. With its resources—not least of which was a public determination to compete with the Soviets—it drew upon the military establishment for rocketry, for its experience with ballistic missiles, for its test ranges, and for its expanding network of tracking stations. At the same time, it was advantageous for geodetic scientists to collaborate with NASA. NASA needed their professional abilities; it had money and access to hardware; and, as Jerome Rosenberg demonstrated, it could manage floods of data and convert them into information. Earth's surface simply had to be better known by both the military and the geodetic communities. Equally for both groups, space was the prestigious and the practical "high ground."

Perspectives of national interest that prevailed from the late 1950's through much of the 1970's changed, of course. First, within the scope of the United States' own worldwide alliance system, NASA's awareness of the international character of space exploration—of which geodetic investigations were a relatively modest part—gradually increased. Second, the enormous expenditures of available scientific talents and of money required by the United States and by the Soviets for space programs proved burdensome. The United States, divided, was becoming more deeply enmeshed in the Vietnam War, the progress of racial relations was proving violent and painful, and inflation worsened. For their part, the Soviets, despite their success in space exploration, were coping with imperial problems in Eastern Europe, with a broken relationship with the People's Republic of China, and with what much of the world increasingly, and accurately, perceived as a laggard political and economic system.

Science, geodesy included, ultimately is international in character. It proceeds by freedom of critical inquiry, free dissemination of intelligence, and cooperation. Heavily burdened nations thus find a measure of collaboration to be economical and in the national interest. Triangulation, surveying, and mapping the recently recognized dynamics of Earth's lithosphere and mantle from geodetic satellites, or from other space vehicles, have become matters of global scientific interest requiring international resources, human and financial. Earth's plates—the collisions they produce, the earthquakes, the faulting, the spreading of sea floors, and the crustal deformations—are not coincident with political boundaries.

**See also:** Dynamics Explorers; Earth Observing System Satellites; Explorers: Air Density; Explorers: Atmosphere; Explorers: Ionosphere; Gemini XI and XII; Global Atmospheric Research Program; Heat Capacity Mapping Mission; Landsat 1, 2, and 3; Landsat 4 and 5; Landsat 7; Mission to Planet Earth; Seasat.

### Further Reading

Burchfiel, B. C. *Continental Tectonics.* J. E. Oliver and L. T. Silver, eds. Washington, D.C.: National Academy of Sciences, 1980. Clear, authoritative writing throughout; despite technical explanations, the subjects are readily understandable. Because much of the work of U.S. geodetic satellites has been focused on plate tectonics, these studies place satellite uses in an overall context. Brief bibliographies.

Davies, John K. *Astronomy from Space: The Design and Operation of Orbiting Observatories.* New York: John Wiley, 1997. This is a comprehensive reference on the satellites that revolutionized twentieth century astrophysics. It contains in-depth coverage of all space astronomy missions. It includes tables of launch data and orbits for quick reference as well as photographs of many of the lesser-known satellites. The main body of book is subdivided according to type of astronomy carried out by each satellite (x-ray, gamma-ray, ultraviolet, infrared, and radio). It discusses the future of satellite astronomy as well.

Hallam, Anthony. *A Revolution in the Earth Sciences: From Continental Drift to Plate Tectonics.* New York: Oxford University Press, 1973. Authoritative and clearly written for intelligent nonspecialists. The role of U.S. satellites is placed in the context of multidisciplinary explorations of Earth's lithosphere and mantle. Adequate bibliography and index.

Henriksen, Soren, Armando Mancini, and Bernard H. Chovitz, eds. *The Use of Artificial Satellites for Geodesy.* Washington, D.C.: American Geophysical Union, 1972. These are symposium papers (Washington, D.C., and Athens, Greece) by specialists. Nevertheless, by selective reading, nonspecialists can readily grasp the excitement, problems, and, above all, the potentialities of U.S. geodetic satellites. The preface, as well as the abstracts and summaries of each paper, are readily understandable. Moreover, plates, photographs, and satellite charts are excellent, and descriptions of the Secor ground equipment indicate the problems and importance of ground tracking equipment. Excellent bibliography with each paper. No index.

Hofmann-Wellenhof, Bernhard, and Helmut Moritz. *Physical Geodesy.* London: Springer-Praxis, 2005. This is an update to a text considered by some to be the introductory book of choice for the field of geodesy. Includes terrestrial methods and discusses contributions made through GPS.

*Institute of Electrical and Electronics Engineers Transactions on Geoscience and Remote Sensing GE-23,* July, 1985, 355-552. This entire issue is devoted to satellite geodynamics, and it is concise, authoritative, and understandable to laypersons. Although there are scores of books and articles from the late 1950's and 1960's dealing with satellite geodesy, there are few recent works, outside of learned journals, covering the last twenty years of progress. Small bibliography; no index.

Kaula, William M. "The Interaction Between Geodesy and the Space Sciences." *Symposium on Geodesy in the Space Age.* Columbus: Ohio State University Press, 1961. This is only one of several papers by distinguished, first-generation experts on satellite geodesy. All deserve a reading, because the basic thrust is clear without the algebra or geometry.

_____. *Theory of Satellite Geodesy: Applications of Satellites to Geodesy.* New York: Dover Publications, 2000. Discusses how Newtonian gravitational theory and Euclidean geometry are used together with satellite orbital dynamic data to determine geodetic information. Requires familiarity with introductory calculus.

Rosholt, Robert L. *An Administrative History of NASA, 1958-1963.* NASA SP-4101. Washington, D.C.: Government Printing Office, 1966. A well-written history that contributes to an understanding of Jerome Rosenberg's tenure during the critical years of NASA's growth. Fine bibliography of original sources; useful index.

*Scientific American* 249 (September, 1983). This is a special issue, bringing summaries to specialists and clear, graphically illustrated explanations to laypersons. Each of these studies incorporates the results of U.S. satellite geodesy into a broader discussion of fresh views of Earth. Articles are authoritative, and brief bibliographical suggestions are appended at the end of the volume.

Smith, James R. *Introduction to Geodesy: The History and Concepts of Modern Geodesy.* New York: John Wiley, 1997. This book provides the professional, as well as student, geodetic investigator with a comprehensive guide to measuring the Earth. The author discusses topics such as traditional survey positioning techniques, geodetic systems, physical and satellite geodesy, the world geodetic system, and the history of geodesy. The book includes lots of illustrations.

Veis, G., ed. *The Use of Artificial Satellites for Geodesy.* New York: Interscience Publications, 1963. Articles have a technical bias but are comprehensible by undergraduates. Each article is preceded by an abstract and complemented with a brief bibliography. The editor, G. Veis, was one of the principal authorities in the 1960's on important aspects of satellite geodesy. The excellent plates, charts, maps, and photographs alone render the volume useful. Each article has a good bibliography.

*Clifton K. Yearley and Kerrie L. MacPherson*

# Get-Away Special Experiments

*Date:* Beginning October 11, 1976
*Type of program:* Scientific platforms

*Engineers of the Get-Away Special (GAS) program developed space technology that allows a wide variety of user groups to have their space experiment packages, or payloads, launched inside small canisters aboard space shuttle missions at relatively low cost.*

## Key Figures

*John F. Yardley* (1925-2001), NASA associate administrator for Space Transportation Systems and originator of the GAS concept

*James S. Barrowman*, flight director of NASA's Special Payloads Division, Goddard Space Flight Center (GSFC)

*R. Gilbert Moore*, the first paying GAS customer and a major shaper of the GAS program's philosophy

*Leonard Arnowitz*, deputy flight director of NASA's Special Payloads Division, GSFC

*James C. Fletcher* (1919-1991), NASA administrator

*Chester M. "Chet" Lee* (1920-2000), NASA director of Shuttle Operations

*Ernest F. Ott*, GAS program manager at NASA Headquarters, Office of Shuttle Operations, and later manager of the GAS Project Office, GSFC

*Donna Skidmore Miller*, second GAS program manager at NASA Headquarters, Office of Shuttle Operations

*L. Rex Megill*, a professor of physics and electrical engineering at Utah State University, faculty adviser for the first GAS payload and an early leader and advocate of GAS-based research

*Dean Zimmerman*, flight director of GAS Payload Processing Operations, Kennedy Space Center

*Lawrence R. Thomas*, manager of the GAS Project Office in the mature phase of the program

## Summary of the Program

The Get-Away Special (GAS) program is a part of the Space Transportation System that addresses the needs of the "small" space experimenter, the experimenter interested in launching a relatively small, simple experiment package (payload) at a relatively low cost in a relatively short time. The word "relative" in this case compares the GAS concept to the major shuttle users, companies and organizations developing large, sophisticated satellites and experiment platforms over many years at a cost of several millions of dollars.

In the summer of 1976, five years before the first shuttle launch, it became clear to John F. Yardley, then-associate administrator for Space Transportation Systems at the National Aeronautics and Space Administration (NASA), that most space shuttle orbiters (the orbiter being the airplane-like winged part of the space shuttle system that carries the payloads and crew to orbit) were going to have a fair amount of residual space left in their payload bays even after the large satellites, laboratories, and experiments were loaded. He proposed to utilize this available capacity by offering small volumes of experiment space to various user groups that might not otherwise get a chance to launch a payload into space: researchers, students, small organizations and companies, and even individual citizens from the United States and abroad. Yardley worked out a concept for such a shuttle service with the chief of NASA's Goddard Space Flight Center (GSFC), Special Payloads Division, James S. Barrowman, and they in turn managed to persuade NASA Administrator James C. Fletcher to support the idea.

The core of the plan was to offer as a standard shuttle service small volumes of experiment space on a space-available basis to any group or individual willing to pay the price, which was deliberately set low to allow a wider base of users to take advantage of the offer. This program—dubbed Get-Away Special after a low-priced travel offer to exotic

places that was being marketed heavily by an American airline that summer—was formally announced by Yardley at an international space conference on October 11, 1976. The first GAS reservation was made by R. Gilbert Moore the next day.

Moore, a seasoned U.S. space industry executive with a broad base of experience working with sounding rockets, was enthusiastic about the idea of providing opportunities for the small user to utilize a system as complex as the shuttle. He became an early, active participant in formulating the GAS program philosophy, working with Yardley, Barrowman, Leonard Arnowitz, Chester M. Lee, Ernest F. Ott, Donna S. Miller, and Dean Zimmerman over the next two years to define the GAS services and technologies. Once the basic details of this concept had been outlined, Barrowman and Arnowitz were given the responsibility of organizing a GAS Project Office at GSFC, and Ott was transferred from NASA Headquarters to GSFC to manage the office. Lawrence R. Thomas inherited this position soon after space shuttle launches started, and his team at GSFC developed the GAS program into a mature set of technologies and capabilities.

The fundamental unit of hardware for the GAS program is the standard cylindrical GAS experiment canister, or "GAS can." Two sizes of these containers are offered: a large cylinder approximately 0.6 meter in diameter and approximately 0.8 meter in length (with a volume of about 0.15 cubic meter) and a smaller cylinder half that long; both resemble thick aluminum garbage cans. GAS payloads are limited in mass to about 91 kilograms for the larger canister and 45 kilograms for the smaller can. The experiments, which must be self-contained (that is, they are not allowed to draw upon any shuttle services beyond three controls operated by an astronaut), are mounted on a circular plate that eventually is bolted to the top end of the cylinder. Cover plates are

*Two Get-Away Special (GAS) canisters installed into the space shuttle* Discovery's *payload bay. The one on the left is the Space Experiment Module (SEM-3) and the one on the right holds commemorative flags to be flown during the STS-91 flight. This mission was the ninth and final docking with the Russian space station Mir.* (NASA)

then bolted to both ends of the cylinder for insulation. Once assembled, the entire GAS package is attached inside the shuttle payload bay in the available open areas around the major payloads.

The number of experiments packaged in each canister is determined by the engineering and experiment control design talents of the user. Experiments may be passive (passive experiments have no active functions and are simply exposed to the space environment) or active (active experiments are designed to respond to one or more of the three astronaut-controlled inputs). Active experiments typically have a control circuit, power subsystem (usually battery-driven), data-recording subsystem (cameras, video, or electronic), and sometimes elaborate thermal control, fluid transfer, mechanical, or atmospheric control subsystems.

One of the primary goals of the GAS program is to provide experimenters, especially novice or unsophisticated experimenters, with the opportunity to send small payloads into orbit at a low cost and without the enormous paperwork and engineering analysis commonly associated with larger space projects. GAS experiment opportunities are reserved for a fee of only hundreds of dollars on a canister-by-canister basis; total fees for the GAS processing and launch services (not including the costs of the experiment and experimenter travel) are generally between $5,000 and $10,000—a very low price by conventional space technology standards. Once a GAS reservation has been recorded at GSFC, various members of the NASA GAS team work closely with the experimenter team to ensure that the experiment design complies with the required shuttle interface requirements (mechanical, electrical, and operational) and safety-related design rules. At various times during the experiment design and construction process (generally requiring between one and three years), NASA and the experiment representatives meet for official design reviews, safety reviews, and prelaunch readiness reviews. A relatively standard sequence of development events has evolved over the lifetime of the program, allowing more experimenter time and effort to be spent on understanding the intri-

cacies of the experiment rather than the intricacies of the shuttle system.

GAS payloads are not launched on all shuttle missions. Some missions contain many major payloads and thus have no available space; others have operational constraints that make GAS payloads impractical. Slots for GAS payloads are typically identified late in the shuttle launch preparation process, usually within the last six months before launch. They are filled on a first-come, first-served basis. It is not uncommon for a GAS payload further down on the reservation list to be launched years ahead of an earlier entry, solely because the payload is prepared for launch.

The first official GAS payload was launched on the fourth shuttle mission on June 27, 1982. R. Gilbert Moore had donated the reservation for this can (a large one) to Utah State University, and faculty adviser Rex Megill managed to integrate eight student-designed experiments into the canister together with two experiments built by Moore's two sons. (Moore, in conjunction with Megill and his Utah State University Student GAS Program, is credited with doing much of the initial experimenter "path finding" in the GAS program: defining interfaces; NASA/user working relationships and procedures; and initial experiment design, control, and mounting concepts.)

The international success of the GAS program is evident when one notes that the second GAS payload launched (aboard STS-5, November 11, 1982) was designed by a West German firm, and the third (aboard STS-6, April 4, 1983) by a Japanese firm. Indeed, in its mature phase, the GAS program maintains a healthy backlog of several hundred reservations from more than two dozen countries. The user community includes individuals and teams of scientific and engineering researchers, students, and faculty from grade school through graduate school; various governmental agencies and laboratories worldwide; public and private companies; hospitals; and even artists and museums.

As the program matured, new technology capabilities were introduced—some by NASA, some by the user community. On the seventh shuttle mission (June 18, 1983), the first GAS can with an opening lid was launched. An experiment was first mounted on the outside of the GAS lid on the eighth mission (August 30, 1983). On April 29, 1985, a small research satellite built by a northern Utah team of students and individuals led by Megill and Moore was ejected from a GAS canister and into its own independent orbit. A more sophisticated GAS package that included an enhanced complement of shuttle interfaces (with more experiment and thermal control functions, electrical power, and a greater mass allowance for the GAS payload) was first offered in 1985 by GSFC as part of a new program called Hitchhiker; the first launch of this system was on the twenty-fourth shuttle mission (January 11, 1986). That same mission included a GAS experiment that relayed its data to amateur radio listeners on Earth via an onboard radio transmitter.

On December 4, 1998, the 151st Get-Away Special payload was launched into space aboard *Endeavour* as part of the STS-88 mission to the International Space Station. VORTEX, the Vortex Ring Transit Experiment (G-093), is an investigation of the propagation of a vortex ring through a liquid-gas interface in microgravity. This process results in the formation of one or more liquid droplets similar to Earth-based liquid atomization systems. In the absence of gravity, surface tension effects dominate the droplet-formation process. The space shuttle's microgravity environment allows the study of the same fluid atomization processes using a larger droplet size than is possible on Earth. This enables detailed experimental studies of the complex flow processes encountered in liquid atomization systems.

### Contributions

About 40 percent of the GAS experiments (official and unofficial) focused on studies of the behavior of fluids and materials in the low-gravity space environment. Another 20 percent were directed at demonstrating various GAS and experiment technologies that were needed to perform additional experiments in the future. About 15

percent dealt with biological investigations, with a like number designed to measure the local environmental conditions around the space shuttle orbiter. The remaining 10 percent of the GAS payloads involved satellite deployments, art projects, and miscellaneous science and engineering support.

A wide variety of results in the fluids and materials area have come from the large number of GAS experiments dedicated to these topics. Several types of metallic substances and alloys were melted, mixed, molded, purified, and solidified. Alloy formation and crystallization properties were studied, as were their heat conduction characteristics. Various organic (that is, carbon-based materials such as oils, plastics, waxes, and proteins) and inorganic materials (minerals, chemicals, and the like) were separated, refined, mixed, and crystallized as well. Fluid transfer and wave phenomena were studied under a variety of conditions. Several new compounds that cannot be made on Earth because of the effects of gravity were made in GAS canisters. A few experiments were designed for forming and curing (hardening) thin films, membranes, and bubbles.

In addition to the expanded GAS technologies mentioned earlier, some interesting technology demonstrations were attempted in GAS cans, though many were not totally successful. A variety of GAS experiment hardware and software concepts were launched, as were many small prototypes of space hardware concepts destined for use on other spacecraft and space platforms. Materials such as electronic components, fabrics, construction materials, lamps, films, and fluids were exposed to the space environment simply to learn how they would react with the space environment. A fair number of mechanical devices, furnaces, and pressure-seal techniques were demonstrated through the GAS experimentation. One experiment even demonstrated that a broad spectrum of soldering techniques was effective in the environment of the space shuttle.

The prospect of biological testing in GAS cans is especially appealing to students. Popular specimens for study include smaller organisms, such as insects, fungi, yeasts, molds, spores, shrimp, bacteria, and worms. Seed and plant growth studies are common. Even human tissue and blood samples were launched in one can.

Environmental measurements were taken inside and outside the GAS canister on several missions. Parameters measured include temperature, pressure, vibration (during launch and in-orbit), noise intensity (during launch and reentry), solar and cosmic radiation levels, local light level intensities, and local concentrations of atoms, molecules, and particulate contamination.

Some of the more unusual applications of the GAS technology included the ejection of two satellites (one for performing radar calibrations, and one for communications), the filming of shuttle payload bay activities for museum shows, and an art project that involved the deposition of a metallic coating on glass spheres. One experiment was designed to bake bread in space.

### Context

The Get-Away Special program is significant within the overall scope of space exploration in two respects. First, it provides professional individuals and small teams of individuals (space scientists, physicists, chemists, materials specialists, biologists, engineers, research and development groups, and the like) with a small-scale opportunity to continue pushing the limits of knowledge about the space environment and how to operate in it. Without the GAS program, many researchers may have been denied access to space as the larger, more expensive shuttle system replaced the smaller rockets used extensively throughout the 1950's, 1960's and 1970's. Indeed, most of the founders of the GAS concept originally got their start in the space business by launching experiments aboard suborbital sounding rockets (small research rockets that attain high altitudes but never enter Earth orbit). These individuals realized the value of this small-scale research and took an active role in preserving such options in the shuttle era.

Perhaps more significant, the GAS program provides an avenue for space-based research to groups and individuals who previously did not have such an opportunity. Guest investigators and student-designed experiments were launched on the U.S. Apollo and Skylab programs, but such arrangements were limited in scope and duration. In the shuttle era, almost anyone able to raise the necessary funds for a GAS launch can launch an experiment. For the first time in history, individuals from all over the planet can individually or collectively design and build a space experiment and have it launched with relative ease. Some student GAS experiments have been launched for a total cost of only $2,000; others have been launched twice within a two-month period. Such low costs and quick turnaround times are unprecedented in the history of space exploration.

In one sense, the GAS program has been more of a success internationally than in the United States. Reservations from dozens of countries have been taken, and several launched payloads have been designed and built by individuals from countries that do not even have a space program. It is surprising that foreign aerospace companies have participated in the program more frequently than their U.S. counterparts have.

An unusual feature of the Get-Away Special program is that the experimenters are not required to furnish NASA with postflight reports on the results of the experiments. Most of the experiments flown have succeeded—at least partially. Experimentation is not without failure. Many of the experiments have flown more than once and have provided data that reflect a variety of in-flight environments.

A GAS canister was aboard *Atlantis* during the STS-106 to the International Space Station (ISS) in the latter half of 2000. G-782, also called Aria-1, was designed to fly forty-seven experiments from students in the St. Louis area. A 70-liter GAS container of less than 30 kilograms, Aria-1 is a joint project between Washington University's School of Engineering and Applied Science and Cooperating School Districts. The primary purposes of this payload are to test the capabilities to develop such a payload and to carry a set of experiments from local-area students.

During the first twenty years of flight aboard the space shuttle, 167 individual GAS missions were flown. Due to the International Space Station assembly sequence, no GAS opportunities will be available for some time. While no shuttle flights were available as of 2005, GAS was considered by NASA to be an acceptable substitute for manifested station components.

**See also:** Geodetic Satellites; Goddard Space Flight Center; Private Industry and Space Exploration; Space Shuttle; Space Shuttle Flights, 1982; Space Shuttle Flights, 1983; Space Shuttle Mission STS 41-B; Space Shuttle Flights, January-June, 1985; Space Shuttle Mission STS 61-C; Space Shuttle Flights, 1998; Spacelab Program.

**Further Reading**

Billings, Linda. "Get-Away Special." *Air and Space*, February/March, 1988. An introductory article about the GAS program in a periodical published by the Smithsonian Institution. The experiences of a New Jersey high school GAS project are described, along with a general summary of the GAS program's history and status. Specifics about other GAS experiments are given, including the first Chinese payload. A mention is made of the contributions of R. Gilbert Moore and Rex Megill. Suitable for all readers.

Burrows, William E. *This New Ocean: The Story of the First Space Age.* New York: Random House, 1998. This is a comprehensive history of the human conquest of space, covering everything from the earliest attempts at spaceflight through the voyages near the end of the twentieth century. Burrows is an experienced journalist who has reported for *The New York Times*, *The Washington Post*, and *The Wall Street Journal*. There are many photographs and an extensive source list. Interviewees in the book include Isaac Asimov, Alexei Leonov, Sally K. Ride, and James A. Van Allen.

Goddard Space Flight Center. Special Payloads Division. *Get-Away Special: The First Ten Years.* Greenbelt, Md.: Author, 1989. An overview of the program by the Get-Away Special team. Readily available by calling or writing the GAS Office at Goddard Space Flight Center.

Harland, David M. *The Space Shuttle: Roles, Missions, and Accomplishments.* New York: John Wiley, 1998. The book details the origins, missions, payloads, and passengers of the Space Transportation System (STS), covering the flights from STS-1 through STS-89 in great detail. This large volume is divided into five sections: "Operations," "Weightlessness," "Exploration," "Outpost," and "Conclusions." "Operations" discusses the origins of the shuttle, test flights, and some of its missions and payloads. "Weightlessness" describes many of the experiments performed aboard the orbiter, including materials processing, electrophoresis, phase partitioning, and combustion. "Exploration" includes the Hubble Space Telescope, Spacelab, Galileo, Magellan, and Ulysses, as well as Earth observation projects. "Outpost" covers the shuttle's role in the Russian Mir program and the International Space Station. Contains numerous illustrations, an index, and bibliographical references.

Jenkins, Dennis R. *Rockwell International Space Shuttle.* Osceola, Wis.: Motorbooks International, 1989. Includes a brief yet concise history of the orbiter and its predecessors. Contains dozens of close-up color and black-and-white photographs, detailing the orbiter's exterior and interior features, including the major subsystems. Some of the text, especially the data tables, is printed in extremely small type.

_____. *Space Shuttle: The History of Developing the National Space Transportation System.* Osceola, Wis.: Motorbooks International, 1996. This is a concisely written technical reference account of the space shuttle and its ancestors, the aerodynamic lifting bodies. It details some of the advantages and inherent disadvantages of using a reusable space vehicle. Each of the vehicles is illustrated by line drawings with important features pointed out with lines and text. The book follows the space shuttle from its original concepts and briefly chronicles its first fifty flights.

National Aeronautics and Space Administration. *The First 100 GAS Payloads.* Greenbelt, Md.: Author, 1995. This pocket-sized book provides brief descriptions of the first one hundred Get-Away Special payloads flown aboard the space shuttle. Most of the descriptions were written by the experimenters. Photographs accompany each description.

_____. *Space Shuttle Mission Press Kits.* http://www-pao.ksc.nasa.gov/kscpao/presskit/presskit.htm. Provides detailed preflight information about each of the space shuttle missions. Get-Away Special packages and experiments are detailed. Accessed March, 2005.

Overbye, Dennis. "The Getaway Kids Shuttle into History." *Discover,* September, 1982. An article focusing on the emotions and excitement experienced by the first group of GAS student experimenters. A description of the pitfalls and failures associated with this GAS payload is included.

Yoel, David, et al. "The First Get-Away Special—How It Was Done." *Space World* T-5-23 (May, 1983): 9-16. An excellent summary of the efforts put into designing, building, testing, and launching the first GAS payload, which contained the GAS student experiments. A good example of how such a project can affect the future career directions of aspiring students.

*Rex Ridenoure*

# Global Atmospheric Research Program

*Date:* 1961 to 1982
*Type of program:* Earth observation

*The Global Atmospheric Research Program (GARP) sought to better scientists' understanding of the atmosphere through space- and land-based observations. Undertaken at the height of the Space Race, it was one of the earliest projects to require the collaboration of scientists and governments worldwide.*

### Summary of the Program

The first meteorological satellite developed in the United States was launched in 1960. During the next few years, most of the development in meteorological remote sensing was carried out by the United States and the Soviet Union, although there were important contributions from other countries, particularly in the development of instruments for use in meteorological satellites.

In 1961, during his State of the Union address, President John F. Kennedy emphasized the importance of the atmospheric sciences and their possible space applications. In response, Special Assistant to the President Jerome Weisner requested that an advisory group headed by Jule Charney prepare a report on the feasibility of a long-term, worldwide study of the atmospheric sciences.

In the fall of 1961, the United States submitted its proposal to the United Nations General Assembly. The proposal resulted in Resolution 1721 (XVI), which recommended that "members and the World Meteorological Organization (WMO) study measures to advance the state of atmospheric sciences and technology in order to improve existing weather forecasting capabilities and to further the study of the basic physical processes that affect climate."

The WMO responded to the U.N. resolution in mid-1962 with an outline plan for an international research program to study the global circulation of the atmosphere, numerical weather prediction (a system whereby mathematical equations are used to describe and predict atmospheric processes), and medium- and long-range forecasting. As a result, the General Assembly in December of 1962 issued Resolution 1802 (XVII) inviting the International Council of Scientific Unions (ICSU) to work with the WMO to develop an expanded program.

Between 1963 and 1964, the ICSU and the International Union of Geodesy and Geophysics (IUGG) worked to find the right organizational forum in which research could proceed at the international level. At its meeting in June, 1964, the ICSU recommended that a program be developed and implemented under the general control of IUGG in collaboration with the Committee on Space Research (COSPAR) and with WMO. Out of this collaboration, the Committee on Atmospheric Sciences (CAS) was formed.

In 1966, at its second meeting, CAS initiated the Global Atmospheric Research Program (GARP). GARP was defined as a "cooperative international meteorological and analytical program with the goal of producing vastly improved understanding of the general circulation of the global atmosphere." The recommendations for GARP also emphasized the "development of improved dynamical models for the general circulation of the global atmosphere."

CAS met again in 1967. It was decided that a number of preliminary studies were necessary before a global experiment could be launched and that GARP subprograms would formulate concrete plans for both the theoretical work and the observational experiments. A full-scale global experiment was planned and designated the First GARP Global Experiment (FGGE), and a joint organizing committee was proposed to guarantee cooperation among the three international organizations concerned: CAS, COSPAR, and WMO.

A study conference was held in Sweden in 1967. Significant progress was made in further defining the plans for a global observational experiment. The project had the following objectives: to obtain a better understanding of atmospheric motion for the development of more realistic models for weather prediction; to assess the ultimate limit of predictability of weather systems; to design an optimum composite meteorological observing system for routine numerical weather prediction of the general circulation; and to investigate, for one year, the physical mechanisms underlying climate fluctuations over periods of a few weeks to a few years and to develop and test appropriate climate models.

In 1972 the planning conference of FGGE was held in Geneva. Further refinements were made in the technology required for FGGE and the various subprograms, many of which had begun to take shape. Among the first was the Complex Atmospheric Energetics Experiment (CAENEX), proposed by the Soviet Union, which sought to "investigate the transfer of radiative energy in the atmosphere and to work out recommendations as to how this energy could be taken into account in the calculations of the dynamics of the atmosphere." A subprogram proposed by Japan was the Air-Mass Transformation Experiment (AMTEX). It was to take place near the southwestern islands of Japan in the winters of 1974 and 1975. Several countries agreed to participate. The Monsoon Experiment (MONEX) was a regional observational subprogram proposed by India and designed to study air-mass transformation in the Arabian Sea during the southwest monsoon season.

Finally, between 1970 and 1974, the first of the GARP subprograms began. They were CAENEX,

the Indo-Soviet Monsoon Experiment, and the GARP Atlantic Tropical Experiment (GATE). Of these, the most notable and ambitious was GATE. The field observations for GATE were held between June 15 and September 30, 1974. Scientific, technical, and support personnel from sixty-nine nations were involved. The area covered by the experiment stretched over one-third of Earth's tropical areas, from the Pacific coast of Central America east across the tropical Atlantic to Africa's east coast.

GATE's principal objectives were to study the structure and evolution of tropical weather systems and to determine how they affect the weather in more temperate latitudes. GATE's observing system included the first Synchronous Meteorological Satellite/Geostationary Operational Environmental Satellite (SMS/GOES), which was used to transmit data. Other observational methods included the deployment of more than thirty ships that took measurements with instruments mounted on tethered balloons and buoys.

In 1975, the WMO Executive Committee approved the final schedule for the FGGE observational period, calling for a preparations phase beginning in September, 1977, followed by an operational phase beginning in September, 1978, and lasting for a year.

Since the very beginning of the planning for FGGE, the project scientists' intention was to observe Earth for a full year. It was recognized, however, that some of the necessary observing systems could not be kept in operation for such a long time. Therefore, the concept of special observing periods was introduced, and two periods—one during January and February, 1979, and another during May and June, 1979—were chosen.

Between 1974 and 1978, three more GARP regional subprograms were undertaken: AMTEX, Monsoon 77, and the Joint Air-Sea Interaction Experiment.

In 1977, preparations for FGGE began. After negotiating with the international satellite agencies concerned, GARP organizers arranged the launch of an almost ideal array of five geostationary satellites (satellites that orbit Earth once every twenty-four hours). The spacecraft were placed at very nearly equidistant intervals around the equator at an altitude of about 36,000 kilometers. In addition to gathering atmospheric and oceanographic data, the geostationary satellites provided communications support for the Aircraft-to-Satellite Data-Relay System.

In addition to those five satellites, four polar-orbiting satellites were deployed. These satellites gathered data in regions, particularly the Poles, that the geostationary satellites were not able to "see." They not only collected important atmospheric and oceanographic data but also aided in the collection of data from instrument platforms on the surface and in the atmosphere.

During the first special observing period, forty ships were used to take wind measurements in the tropics; during the second, there were forty-three. Also deployed during the observing periods were the Tropical Constant Level Balloon System and the Aircraft Dropwindsonde System (ACDWS). The ACDWS consisted of a fleet of nine long-range aircraft flying along six tracks in the equatorial tropics. The Tropical Constant Level Balloon System provided atmospheric data from above the ACDWS. A total of 313 balloons were launched from three sites, two in the Pacific and one in the Atlantic.

Deployed south of 20° south latitude were 301 buoys to make up the Southern Hemisphere Drifting Buoy System. These provided in situ measurements of surface parameters such as pressure and temperature.

During FGGE, in addition to the two special observing periods, there were three regional experiments: the Summer Monsoon Experiment, the Winter Monsoon Experiment, and the West African Monsoon Experiment. The observing phase of FGGE ended successfully in late 1979, and the data analysis phase began.

In 1982 the last of the GARP subprograms, the Alpine Experiment, was completed. It had been a limited-area observational study of air flow over and around mountains.

## Contributions

Because of GARP's length, scope, and ambitious nature, considerable knowledge was gained in three main areas: science, technology, and organization.

The organization of GARP laid the groundwork for the World Weather Watch (WWW) by setting high standards and far-reaching objectives. It demonstrated the benefits of global international cooperation at a level that was unprecedented at the time but that could become common. Also as a result of GARP, much closer working relations between atmospheric and marine scientists developed.

GARP advanced meteorological and oceanographic observing systems and techniques well beyond those envisioned in the 1960's. An international team developed and implemented a worldwide geostationary meteorological satellite system that provided nearly continuous surveillance of weather events and wind patterns around the tropical and subtropical belt. Better satellite systems were developed for data collection and the location of fixed and drifting platforms. FGGE demonstrated the efficiency of a system of drifting buoys, and such systems have become an important source of surface meteorological data in ocean areas. GARP provided the incentive and support for the development of the Navigational Aid (NAVAID) system of upper-air wind soundings, which more than doubled the number of upper-air wind measurements provided by the regular WWW network during the two special observing periods. Finally, the NAVAID system demonstrated the efficiency of new, nearly automatic upper-air sounding equipment.

The scientific accomplishments of GARP include significantly improved understanding of tropical weather systems and the ability to predict weather patterns several days into the future. Overall proficiency in prediction has improved considerably, especially for the tropical regions and the Southern Hemisphere, as a result of the data obtained from better global coverage, improved numerical weather models, and more powerful computers. GARP and FGGE also provided insight into atmospheric processes and helped scientists develop data handling and modeling techniques on which to base even more sophisticated global, regional, and local prediction systems.

### Context

In 1961, when GARP was conceived, numerical weather prediction was only seven or eight years old, and forecasts were typically restricted to one or two days ahead. Observational network requirements for numerical weather prediction were beginning to receive some attention, but a consistent method of prediction was still lacking.

Also at this time, there were only two countries, the United States and the Soviet Union, with the technological capability to launch satellites. These early polar-orbiting meteorological satellites provided limited information—mostly visible images of cloud cover and active weather systems—and had no way of measuring wind speeds, temperature, or moisture content. For information on the latter, meteorologists relied mostly on the inconsistent data from commercial and meteorological ships and remote weather stations. By this time, however, scientists were confident that, if data sources could be improved, programs to forecast long-range global weather systems were within reach.

The ability to understand and predict climate variations would influence governments' decisions concerning the production and distribution of food, water, and energy. It would allow a rational assessment of the possible climatic consequences of the introduction of pollutants, heat, and other substances into the environment. It would aid in the early prediction of catastrophic weather conditions, such as hurricanes, tornadoes, and monsoons, allowing time for mass evacuations.

With the urging of the United Nations and the hard work and dedication of scientific and technical personnel worldwide, GARP provided the impetus for faster development of long-range weather prediction and numerical modeling techniques.

GARP and FGGE had a central role in guiding the design and implementation of the World Weather Watch. GARP produced the resources necessary to establish a global system of geostationary and polar-orbiting satellites that can now be used by all nations in support of their weather and climatic services. It demonstrated the capabilities of several powerful observing techniques that would make valuable additions to future global observing systems. The extensive networks and system planning and the results from the observational phase of GARP will provide a firm foundation for the World Weather Watch into the twenty-first century.

**See also:** Get-Away Special Experiments; Meteorological Satellites; Nimbus Meteorological Satellites; Private Industry and Space Exploration; SMS and GOES Meteorological Satellites; Space Shuttle: Approach and Landing Test Flights; Space Shuttle: Radar Imaging Laboratories; Spy Satellites; Upper Atmosphere Research Satellite.

### Further Reading

Ahrens, C. Donald. *Essentials of Meteorology: An Invitation to the Atmosphere.* 4th ed. Pacific Grove, Calif.: Thomson Brooks/Cole, 2005. A thorough examination of contemporary understanding of meteorology. Includes the contributions made by satellite technology.
*Man and Climate Variability.* Lanham, Md.: Unipub, 1980. Discusses the impact of humans on Earth's climate. Covers topics such as overgrazing, pollution, and deforestation. Also discusses GARP's impact on climatological studies. Suitable for a high school audience.
Möller, Detlev, ed. *Atmospheric Environmental Research: Critical Decisions Between Technological Progress and Preservation of Nature.* Berlin: Springer, 1999. These 181 pages include illustrations, maps, bibliographical references, and an index.

Moran, Joseph M., Michael D. Morgan, and Patricia M. Pauley. *Meteorology: The Atmosphere and the Science of Weather.* 5th ed. Upper Saddle River, N.J.: Prentice Hall, 1997. Comprehensive treatment of the science of weather. Includes colored illustrations and maps, bibliographical references, and an index.

Siedler, Gerold, et al. *GATE: Containing the Results from the GARP Atlantic Tropical Experiment.* Elmsford, N.Y.: Pergamon Press, 1978. Discusses the scientific and technological advancements and achievements of the GARP Atlantic Tropical Experiment (GATE). This text is suitable for a college audience.

Tanczer, T., ed. *First FGGE Results from Satellites.* Elmsford, N.Y.: Pergamon Press, 1981. This book describes the space systems used in FGGE, analyzes the quality of the data retrieved, and details various specific scientific accomplishments of FGGE. Suitable for a college audience.

United States Committee for the Global Atmospheric Research Program. *Understanding Climatic Change: A Program for Action.* Detroit: Grand River Books, 1980. This is a reprint of the 1975 work published by the National Academy of Sciences. It gives a variety of information about the atmospheric research projects, including the Global Atmospheric Research Program. It is fairly detailed and, at times, quite technical.

United States Congress. House Committee on Science. *U.S. Global Change Research Programs.* Washington, D.C.: Government Printing Office, 1996. This congressional report details data collection and scientific priorities used in several terrestrial and space-based studies of Earth's global climate changes. It includes sections on the Global Atmospheric Research Program. There is a bibliographical listing included.

Zimmerman, Robert. *The Chronological Encyclopedia of Discoveries in Space.* Westport, Conn.: Oryx Press, 2000. Provides a complete chronological history of all crewed and robotic spacecraft and explains flight events and scientific results. Suitable for all levels of research.

*Stephanie Gallegos*

# Global Positioning System

*Date:* Beginning 1960
*Type of spacecraft:* Navigational satellites

*The Global Positioning System (GPS), a network of satellites designed originally for military applications, has evolved into a navigation system that allows users to determine their position anywhere in the world, in any weather conditions, twenty-four hours a day.*

## Summary of the System

The Department of Defense describes the Global Positioning System as consisting of three segments: Space, Control, and User. The Space Segment comprises a constellation, or network, of twenty-four satellites orbiting Earth in six orbital planes. The satellites are placed in circular orbits inclined 55° to Earth's equator at a height of 20,200 kilometers above Earth's surface. Each satellite circles the globe once every twelve hours. The satellites are placed in such a way as to ensure that at any time, anywhere on the globe, there are at least five satellites in position to provide location information to anyone equipped with a GPS receiver. The system requires only twenty-one satellites for full operating capability but includes three extra satellites as backups.

The Control Segment is the operational segment of the system. The 2nd Space Operation Squadron of the 50th Space Wing of the United States Air Force, based at Falcon Air Force Base, Colorado, operates the Global Positioning System. A master control station is located in Colorado Springs, Colorado, with five monitor stations and three ground antennae located elsewhere throughout the world. Each monitor station continually tracks each of the GPS satellites it has in view and relays the information it collects back to the master control station. Exact satellite orbits are calculated by the master control station. The master control station then transmits revised navigation messages for each satellite via the ground antennae.

The last of the three segments, the User Segment, consists of the receivers, processors, and antennae that allow land, sea, or airborne operators to receive the GPS satellite broadcasts and compute their precise location, velocity, and time. The location information provided is three-dimensional. That is, in addition to indicating longitude and latitude the receiver provides altitude.

The Global Positioning System works on an extremely simple principle: triangulation. By comparing signals broadcast by three different satellites, a GPS receiver on Earth can determine its own location, altitude, and, if located on a moving object, speed. Each satellite broadcasts its own unique signal on two L-band carrier frequencies, 1575.42 megahertz and 1227.60 megahertz. The system employs signals from clocks to determine the distance from each satellite to a receiver. Each satellite carries four atomic clocks: a working clock, a backup clock, and two other clocks responsible for keeping the first two clocks synchronized.

The satellite broadcasts continuously a pseudo-random noise signal that the receiver decodes to learn the time and the satellite's exact location. The GPS receiver also contains a clock. The receiver calculates the time delay between it and the satellite by comparing its own internal time with the time code broadcast by the satellite. It multiplies the delay by the speed of light to determine how far it is from the satellite, by reference to three different satellites; through triangulation, it then can determine its own location. The GPS satellites broadcast two sets of signals, one for use by civilians and one for the United States military. Civilian GPS signals, constituting the Standard Positioning Service (SPS), can determine location to about 24 meters (80 feet); signals are accurate to within 156 meters for vertical location and to within 340 nanoseconds for time. As a security measure, the military introduced a timing error into the SPS that reduces its accuracy to about 91 meters (300 feet). This feature, known as Selective Availability (SA), was introduced following a 1982 study. At that time an interagency panel consisting of representatives from the office of the secretary of defense, the office of the joint chiefs of staff, the Defense Mapping Agency, branches of the Armed Forces, and U.S. intelligence agencies met to discuss security issues regarding GPS. The panel concluded that the GPS signals had the potential to be used by hostile nations to target locations within the United States or allied nations in time of war. The panel therefore recommended that the SPS be made available to the public with accuracy to within 100 meters horizontally, with an option to degrade the signal even more if necessary for security reasons. The more accurate military signals, forming the Precise Positioning Service (PPS), provide horizontal locations to within 15.25 meters (50 feet).

By the end of 1995, however, innovators in private industry and the Coast Guard had developed systems, known as Differential Global Positioning Systems (DGPS's), that allowed users to pinpoint their location to within about 3 meters (10 feet). A DGPS uses signals from ground stations to provide fixed reference points that correct errors in satellite systems. Because these innovations effectively nullified the effects of SA, the Department of Defense eliminated SA on November 1, 1999.

### Contributions

While the use of GPS technologies was quietly diffusing into a wide community of users in the late 1980's and early 1990's, it was not until January, 1991, and the Persian Gulf War that the general public became fully aware of the accuracy of GPS-guided navigation. The unprecedented extreme accuracy that the allied forces displayed in bombing Iraqi targets generated wide publicity. Using GPS, allied forces were able to locate militarily important targets such as power plants and bridges with precision, while avoiding unnecessary damage to many civilian neighborhoods. Whether this publicity contributed to increased interest in GPS navigation systems for nautical, aviation, and terrestrial vehicles is debatable, however, as GPS was already in wide use by surveyors, archaeologists, mariners, and others. By 1996, GPS receivers that displayed the user's position, velocity, and time were being marketed worldwide by companies such as Rockwell International. Specialized receivers were available to display additional data, such as distance and bearing to selected way points.

Long-haul trucking firms, for example, have found that the use of GPS increases their operating efficiency and safety, reduces costs, and even helps cut the costs of crime. Whereas in the past dispatchers had to rely on truck drivers' telephoning in progress reports to tell the dispatcher that a load had been successfully picked up or delivered, semitractor rigs equipped with GPS receivers allow freight dispatchers to track drivers and loads. After installing GPS receivers, augmented by the use of communications satellite technology to relay specific messages to and from the driver, trucking firms were able to follow individual trucks anywhere. With improved dispatching and communications, trucks began to run with full loads more often, increasing potential earnings for the drivers as well as improving profitability for the freight company. A typical truck GPS system consists of a transceiver unit, an antenna, and a keypad. The antenna is mounted on the roof of the truck cab and is connected to the transceiver with a cable. The keypad, which typically includes a small screen that displays up to four lines of text, allows the driver to send and receive messages. In addition to the messages drivers may send, the transceiver is set to transmit position reports automatically to the dispatcher at defined intervals. Those intervals, determined by the individual trucking companies, can be anywhere from a few minutes to several hours apart.

The dispatcher can request a GPS position for any truck at any time, a feature that, in addition to assisting the dispatcher with making routing decisions, has led to the quick recovery of stolen trucks. While some truck drivers complained about GPS technology contributing to a loss of personal autonomy, most appreciated the improved communications.

By 1995 the cost and availability of GPS receivers had dropped to the point where they were being considered for use in passenger cars. Receivers equipped with map display systems were in wide use in Japan as both taxi drivers and the general public discovered that GPS receivers could help them quickly find their way through Tokyo's confusing streets. In the United States, rental car companies began offering an option they called a satellite guidance system. Systems such as Rockwell's Pathmaster featured a 4-inch color display and a synthesized voice prompt. A user could enter a destination and, using voice prompts such as "left turn ahead," the system would guide the driver through strange cities. GPS systems designed for the general driving public included CD-ROM players for map databases that, in addition to street and highway information, provided listings of hotels, restaurants, and tourist attractions. While the first GPS receivers designed for use in automobiles were stand-alone units, by 1995 electronics firms had begun to market systems that incorporated the GPS receiver into in-dash radio and cassette players.

GPS was also leading to what many analysts termed a revolution in aviation navigation and landing. GPS signals, as broadcast by the satellites alone, were not accurate enough for aircraft navigation over land, but through the implementation of other technology, GPS signals can be augmented to provide greater accuracy and continuity. In the early 1990's the Federal Aviation Administration (FAA) began construction of an aviation navigation system that would combine GPS with a network of ground stations and communication satellites. With the first phase completed in July, 1998, the Wide Area Augmentation System (WAAS) provides navigation coverage over the entire United States as well as U.S.-managed airspace over the Atlantic and Pacific Oceans. WAAS passed its final testing milestone in September, 1999, when the FAA flew successful multiple precision approaches at Morgantown, West Virginia. The system was commissioned early in the twenty-first century, and the FAA began to phase out ground-based radio navigation aids. Engineers in Russia and other countries around the world planned similar systems that would link ground-based air traffic control with GPS.

Aircraft companies such as Boeing were actively involved in the development of GPS navigation systems using DGPS and other technologies. While the FAA concentrated on issues such as the construction of ground stations, Boeing, in partnership with electronics firms such as Rockwell and Raytheon, concentrated on avionics issues. Jim McWha, chief engineer for flight systems at Boeing, said the company focused on problems such as aircraft installation and making GPS approach requirements compatible with existing systems.

GPS was also being used in specialized areas of aviation and agriculture. Crop dusters, for example, now use GPS navigation for spraying herbicides and pesticides on agricultural fields. Navigation systems keep the pilot on track as well as mark the spray activation and cutoff points, replacing the old system of having flagmen in the fields to direct the pilot with hand signals.

Surveyors now rely on GPS to run property lines and find markers. GPS is used in both routine surveying and more unusual applications. After the 1989 earthquake, for example, the city of San Francisco commissioned a survey of the Golden Gate Bridge. Survey points were located down to the centimeter by using differential GPS. In the event of another large earthquake, the points will be resurveyed to learn how far the ground moved.

Archaeologists have used GPS both to locate potential sites for exploration and excavation and to document their discoveries. The lost city of Ubar on the Arabian Peninsula was first noticed in observation satellite images. GPS pinpointed the location. Once archaeologists began excavations, they used GPS receivers to register the location of any artifacts they uncovered. Similarly, paleontologists working in remote locations such as Mongolia's Gobi Desert have relied on GPS. Scientists from the American Museum of Natural History used GPS navigation to find their way across the often untracked and poorly mapped wastelands of the Gobi as well as to mark the location of any fossils they unearthed. Biologists working in the Amazon basin have used GPS to lay out wildlife survey grids and to track animals they have tagged.

At the beginning of the twenty-first century it was clear that GPS receivers would become increasingly commonplace among the general public. Noncommercial use of GPS—for example, in pleasure boating and fishing—continued to climb. Manufacturers had developed GPS receivers with high-resolution display screens that presented data and charts while incorporating a continuous running display of position, speed, and intended course. Having made the transition from commercial trucking fleets to rental cars in only a few years, it seemed likely that GPS receivers would be a common option on new cars being sold to individuals. Similarly, the widespread diffusion of GPS receivers among surveyors, foresters, wildlife biologists, and archaeologists, as well as the proliferation of GPS receivers among outdoor recreationalists, promised that GPS receivers would soon be standard equipment for even casual users. The Global Positioning System, a system developed for use in war, had quickly evolved into a technology with myriad peacetime applications.

**Context**

The Global Positioning System originated as a purely military navigation system. During the Vietnam War, the United States Department of Defense began researching the use of satellites to assist in more accurately bombing enemy targets. Conventional navigation systems that relied on a combination of radar and visual observation often proved inaccurate under combat conditions. Vital targets often were missed, a problem for which the military tried to compensate through the use of saturation bombing, leading, in turn, to increased

damage to nonmilitary targets and the loss of civilian lives. During World War II, for example, whole cities had been leveled in attempts to ensure that one munitions plant was destroyed.

The Navy had begun using satellites for transit navigation with the launch of Transit 1B in 1960. On January 11, 1964, the Army also began launching a series of navigation and positioning satellites. Eight SECOR satellites were placed in orbit in the following twenty-two months. By the late 1960's, three branches of the military—the Air Force, Army, and Navy—had separate navigation satellite systems in place or under development. Following several years of discussion and feasibility studies, in 1973 defense researchers suggested combining the Navy's TIMATION system and the Air Force System 621B three-dimensional navigation system into one Defense Navigation Satellite System (DNSS). The Defense System Acquisition and Review Council rejected the proposed system on the grounds that it failed to address the views and needs of all the military services adequately. The Department of Defense researchers reviewed the existing systems and, after attempting to include the best features of each in a revised proposal for a DNSS, presented their ideas to the council in December, 1993. This time the DNSS passed.

In early 1974, the Department of Defense established a joint program office for a Global Positioning System. The Air Force led the program, with active participation from the other branches of the military services. The first "navigation system using timing and ranging" (Navstar) satellite was launched on February 22, 1978, and became operational on March 29. The satellite remained operational for approximately twenty-two months, until January 25, 1980. At that time its atomic clock failed.

A total of four Navstar satellites were launched in 1978 and two in 1980. By 1985, when the first series of Navstar launches ended, ten satellites had been placed in orbit. Of those ten, only one, Navstar I-10, remained operational a decade later. According to the Air Force, the average operating life of the Navstar I satellites was 7.13 years.

In 1983, a civilian jet liner, Korean Air Lines Flight 007, wandered off course and strayed into Soviet air space. The Soviets erroneously believed the aircraft to be a spy plane and shot it down. In an attempt to prevent any similar tragedies, President Ronald W. Reagan ordered the U.S. military to make GPS navigation signals available for international civilian use. At that time eight Navstar satellites had been launched, but only four were operational. The Air Force launched the last of the first block of Navstar satellites, Navstar I-11, on October 9, 1985.

A series of launch vehicle problems plagued the U.S. space program during the mid-1980's, including the January 28, 1986, *Challenger* accident, so four years were spent refining GPS satellite technology. By 1989, however, the Defense Department was ready to proceed with expanding GPS. The United States Coast Guard, an agency of the Transportation Department during peacetime, was designated to act as the point of contact for civilian information on Navstar. On February 14, 1989, the Air Force began placing the second series, or Block II, of Navstar satellites into orbit. Although the planned GPS satellite constellation was far from complete, civilian entrepreneurs recognized quickly the myriad opportunities that GPS presented and began preparing for use of the planned full constellation. In June, 1993, there were twenty-four operating satellites in place, twenty-one Block II satellites and three Block I satellites, and the Air Force announced that "initial operational capability" had been achieved.

By the time that the last of the twenty-four Navstar II GPS satellites was placed in orbit, in March, 1994, GPS applications could already be found in a wide variety of civilian settings. It was not until July 17, 1995, however, that the military completed its evaluations and declared the Global Positioning System to be fully operational. In a news release, Air Force Vice Chief of Staff General Thomas S. Moorman, Jr., noted that a major milestone had been achieved. GPS, a system originally developed for a very specific military purpose, had been successfully integrated into a wide range of peaceful applications.

Air Force projections had assumed it would be necessary to launch two replacement satellites in 1995, but well after the year 2000, most of the original Navstar II satellites were still operational. Navstar IIR comprises twenty replacement satellites that incorporate autonomous navigation based on crosslink ranging. Lockheed Martin manufactures these satellites. The first launch attempt, in 1997, resulted in a failure. The first IIR satellite to reach orbit was also launched in 1997. The second GPS IIR satellite was successfully launched aboard a Delta II rocket on October 7, 1999.

Russian scientists also developed a satellite navigation system in the 1980's. Like Navstar, the Russian Global Navigation Satellite System (Global'naya Navigatsionnaya Sputnikovaya Sistema or GLONASS) consists of twenty-four satellites in three orbital planes. Each GLONASS satellite operates in a circular 19,100-kilometer orbit at an inclination angle of 64.8°. Each satellite completes an orbit in approximately 11 hours and 15 minutes. In October, 1994, the Global Navigation Satellite System (GNSS) was established, with GPS serving as the first phase. The second phase of GNSS will combine GPS and GLONASS satellites.

The GPS and GLONASS satellites could eventually be supplemented or replaced by purely civilian navigation satellites. The Russian Federation is in the process of implementing the GLONASS to provide signals from space for accurate determination of position, velocity, and time for properly equipped users. GLONASS provides high accuracy and availability to users with continuous, worldwide, all-weather navigation coverage. Three-dimensional position and velocity determinations are based on the measurement of transit time and Doppler shift of RF signals transmitted by GLONASS satellites.

**See also:** Air Traffic Control Satellites; Delta Launch Vehicles; Geodetic Satellites; Global Atmospheric Research Program; Gravity Probe B; Mobile Satellite System; Navigation Satellites; Ocean Surveillance Satellites; Search and Rescue Satellites; Tethered Satellite System; United States Space Command.

**Further Reading**

Andrade, Alessandra A. L. *The Global Navigation Satellite System: Navigating into the New Millennium.* Montreal: Ashgate, 2001. Provides an international view of issues of availability, cooperation, and reliability of air navigation services. Attention is specifically paid to the American GPS and Russian GLONASS systems, although the development of the Galileo civilian system in Europe is also presented.

Bullock, Darcy, and Cesar A. Quiroga. *Development of a Congestion Management System Using GPS Technology.* Baton Rouge: Louisiana Transportation Research Center, 1997. Collaboration between Louisiana State University and the Louisiana Department of Transportation and Development in cooperation with the U.S. Department of Transportation. Includes illustrations and bibliographical references.

Clarke, Charles W. *Aviator's Guide to GPS.* 2d ed. New York: TAB Books, 1996. An explanation of the GPS that focuses on its applications in aviation but is easily understood by the general reader. Includes maps and illustrations.

Dahl, Bonnie. *The User's Guide to GPS: The Global Positioning System.* Evanston, Ill.: Richardsons' Marine Publishing, 1993. An explanation of GPS aimed at recreational boaters. Includes maps and illustrations.

Farrell, Jay A. *The Global Positioning System.* New York: McGraw-Hill, 1998. A concisely written, technical reference on the Global Positioning System for the engineer.

Huang, Jerry. *All About GPS: Sherlock Holmes' Guide to the Global Positioning System.* San Jose, Calif.: Acme Services, 1999. This book is intended to bring the reader's interest toward fundamental sciences and modern technologies. The aspects of concepts, philosophy, methodology, and history are also emphasized.

Kaplan, Elliott D. *Understanding GPS: Principles and Applications.* Boston: Artech House, 1996. An overview of the Global Positioning System accessible to the general reader.

Leick, Alfred. *GPS Satellite Surveying.* New York: Wiley, 1995. A comprehensive text detailing how to use the GPS for land measurements. Contains illustrations and maps.

Letham, Lawrence. *GPS Made Easy: Using Global Positioning Systems in the Outdoors.* Seattle: The Mountaineers, 1995. A guide for the recreational use of GPS technology for camping and hiking.

Logsdon, Tom. *The Navstar Global Positioning System.* New York: Van Nostrand Reinhold, 1992. Overview of the Navstar system. Includes illustrations.

_____. *Understanding the Navstar: GPS, GIS, and IVHS.* New York: Van Nostrand Reinhold, 1995. Comprehensive explanation of the entire Navstar system and its multiple functions.

Simon, Stan. *The Global Positioning System: Markets and Applications.* Burlington, Mass.: Decision Resources, 1991. An entrepreneurial view of GPS.

Zimmerman, Robert. *The Chronological Encyclopedia of Discoveries in Space.* Westport, Conn.: Oryx Press, 2000. Provides a complete chronological history of all piloted and robotic spacecraft and explains flight events and scientific results. Suitable for all levels of research.

*Nancy Farm Mannikko*

# Goddard Space Flight Center

*Date:* Beginning May 1, 1959
*Type of facility:* Space research center

*Operated by NASA, the Goddard Space Flight Center (GSFC) in Greenbelt, Maryland, is the headquarters for the tracking systems that relay voice and data from U.S. vehicles in space to mission managers on the ground. Goddard managers also oversee numerous other NASA satellite and research projects.*

## Key Figures

*Harry J. Goett* (1910-2000), GSFC center director 1959-1965

*John R. Busse*, director of engineering

*John W. Townsend, Jr.*, GSFC center director 1987-1990

*Joseph H. Rothenberg*, GSFC center director 1995-1998

*Alphonso Diaz*, GSFC center director from 1998

## Summary of the Facility

The 1,100-acre Goddard Space Flight Center, located in Greenbelt, Maryland, has been called the intellectual brain trust of the National Aeronautics and Space Administration. Since its founding in 1959, the suburban Washington, D.C., complex has served as the heart of NASA's worldwide tracking and communications network and as one of the U.S. space agency's premier research and development facilities for robotic Earth-orbiting satellites and deep-space probes.

The process by which the Goddard Space Flight Center came to be is as much the story of the United States' early days in space as it is the history of a single facility. On October 4, 1957, the Soviet Union announced the launching into Earth orbit of an 86-kilogram satellite called Sputnik 1 as the Soviet Union's contribution to the International Geophysical Year, a multinational program organized to encourage advances in the Earth sciences. This event, the successful launch of a space satellite by the Soviet Union, led the United States to hasten the progress on its own fledgling space projects.

At that time, the United States had two concurrent space-related research programs. The Naval Research Laboratory was responsible for the development of Project Vanguard, a program to launch at least one scientific satellite into Earth orbit. The U.S. Army sponsored the Explorer project, a program run by Wernher von Braun and other German rocket scientists who had emigrated to the United States after World War II. The Explorer project was organized to develop launch vehicles based on the design of the German V-2 rocket.

The first Vanguard launch, the United States' answer to the Soviet Union's successful Sputnik, ended in flames when the booster rocket exploded on the launch pad. It was not until January 31, 1958, almost four months after Sputnik 1, that the United States officially entered the Space Race, with the launch of Explorer 1 by the Army. The first successful Vanguard satellite was launched less than two months later, in March, 1958.

Shortly after Explorer's launch, President Dwight D. Eisenhower called for the formation of a single civilian federal agency to oversee the nation's peaceful development of space technology. The National Aeronautics and Space Act, signed into law by the president on July 29, 1958, created

NASA and incorporated the civilian National Advisory Committee for Aeronautics (NACA) and both the Naval Research Laboratory and Army satellite programs into a single entity. Shortly after NASA's founding, Project Mercury, a program to put humans into space, was announced as the new space agency's major program.

In early 1959, NASA's management saw the need for the creation of a single facility that would oversee and coordinate advanced theoretical research and practical development of new space projects, a place that would serve as a manufacturing center for new satellites and other space hardware, and as an operations center for the development of new launch vehicles as well as scientific satellites. This new facility would also provide technical support for Project Mercury and would manage the Naval Research Laboratory's Minitrack network of worldwide tracking stations developed originally to support Project Vanguard.

An undeveloped section of the U.S. Department of Agriculture's Beltsville Agricultural Research Center in Greenbelt, Maryland, was selected as the site for the new space center. In May of 1959, NASA's Space Research Center, as it was initially called, was renamed the Goddard Space Flight Center (GSFC), in honor of Robert H. Goddard, the first man to develop a liquid-fueled rocket and the United States' foremost space pioneer.

Once in operation, the Goddard Space Flight Center grew rapidly, as public pressure to counter the Soviets' advances in space research increased. Engineers at Goddard were soon responsible for developing and building American launch vehicles such as the Delta rocket; for managing rocket proving grounds in Wallops Island, Virginia, and at the Atlantic and Pacific Missile Ranges; and for tracking all U.S. satellites in Earth orbit. Within three years, GSFC, one of NASA's fastest-growing facilities, employed more than twenty-eight hundred people on a variety of space projects.

GSFC changed and expanded as the U.S. space program matured. Its Minitrack tracking network went through several technological improvements, finally becoming the Spaceflight Tracking and Data

Network (STDN), which provided tracking and voice, data, and telemetry relay for all the Space Transportation System's space shuttle missions. By the late 1980's, the STDN was being superseded by the Tracking and Data-Relay Satellite System (TDRSS), a series of satellites in geosynchronous orbit around Earth that relay voice and other communications from piloted and robotic vehicles in space to the ground. TDRSS is also managed by GSFC through a ground terminal at White Sands, New Mexico.

The modern-day GSFC, with an annual budget of more than one billion dollars, employs more than three thousand people at its Greenbelt headquarters and is one of the U.S. government's primary scientific research and development facilities. More than nine thousand additional employees work at the Goddard Institute for Space Studies in New York City; the Wallops Flight Facility at Wallops Island, Virginia; the White Sands facility; and the several STDN tracking stations around the world.

At its Greenbelt facility, GSFC operates the NASA Communications Division (NASCOM), which serves as the control center for television, voice, data, and telemetry communications with satellites in Earth orbit and mission controllers on the ground. The control center operates through the STDN stations and the TDRSS satellites in space. NASCOM also provides communications support for space shuttle missions and shuttle payloads and, through the STDN, has supplied launch tracking support for the French Ariane program.

As part of its ground and space networks' communications functions, Goddard also houses the Flight Dynamics Facility (FDF), a data-relay control center responsible for tracking and projecting the orbital paths of both piloted and robotic spacecraft around Earth. Goddard also operates several information and data retrieval and processing centers to augment its primary communications systems.

In the past, before the construction of the White Sands facility, GSFC's communications command and control facilities made it NASA's first choice

*Goddard Space Flight Center oversees Wallops Flight Facility at Wallops Island, Virginia, seen in this aerial view taken in 1982.* (NASA)

as a backup Mission Control center for Mercury, Gemini, Apollo, and Skylab piloted flights. If the primary Mission Control at Johnson Space Center in Houston, Texas, were incapacitated for any reason during a piloted mission, control for the flight would then go to Goddard in Greenbelt. Later, the White Sands facility became the backup Mission Control, and GSFC was designated the interim Mission Control while personnel from Houston were enroute to White Sands.

Goddard still has operational control over the construction and launch of the Delta rocket, one of the workhorse launch vehicles in the U.S. space effort. In the late 1980's, Goddard-operated Delta rockets served as launch vehicles for test projects of the Strategic Defense Initiative, the "Star Wars" missile defense system. GSFC no longer actually manufactures robotic satellites, but it does participate in the design of space hardware and in the management of manufacturing contracts with private companies around the United States.

Many of Goddard's activities center on research and development of the next generation of space hardware and applications. GSFC personnel either manage or otherwise participate in the development of numerous U.S. space projects. These projects include the design, construction, and operation of the Tracking and Data-Relay Satellites; the Active Magnetospheric Particle Tracer Explorers (AMPTEs, scientific satellites launched in 1984 by the United Kingdom, West Germany, and the United States); the Hubble Space Telescope; the Earth Radiation Budget Satellite; and the International Cometary Explorer.

Goddard's research and managerial support for the space shuttle program, apart from its tracking and communications activities during actual missions, has included administering the Get-Away Special, a program that allowed private individuals, companies, and government agencies to create small scientific experiments to be included on shuttle missions. Prior to the tragic explosion of the space shuttle *Challenger* in January of 1986, fifty-three Get-Away Special Experiments had flown on space shuttle missions. GSFC also manages space shuttle satellite servicing and repair missions, such as the Solar Maximum Mission retrieval and repair in 1984.

The Goddard Space Flight Center research activities also involve participation in the design and development of the International Space Station. Other Goddard-supported robotic satellite research projects include a gamma-ray observatory, a cosmic background scientific satellite, an interplanetary monitoring platform, advanced sounding rocket experiments, and satellites designed to study the nature and composition of Earth's crust. Because of its advanced research capability, Goddard has also become involved in several projects unrelated to the space sciences. A variety of medical, computer, and chemical studies are being managed by GSFC.

**Context**

The role of the Goddard Space Flight Center in the broad context of space exploration is significant but difficult to delineate. Goddard has been a principal conduit for a large number of advances made by the U.S. space program since its inception in the late 1950's. Because it operates the primary tracking, communications, and data collection system for a large proportion of NASA's piloted and robotic space missions, GSFC has been at the center of many of the United States' successes and failures in space. Moreover, because Goddard has continually improved the communications and tracking technology it employs, the link between mission managers on Earth and operations in space has matured to a partnership between spacecraft and ground controllers.

As NASA's principal think tank, the Goddard Space Flight Center is the base for more scientists than any other single federal government installation. As a direct result, GSFC has been a vital part of many of the important developments in the Space Age. Many of the most successful scientific satellite programs conducted by NASA have been managed by Goddard.

Originally, Goddard cooperated with other NASA facilities in the design and construction of the vehicles that launched U.S. satellites into Earth orbit and beyond. In the latter half of the 1980's, Goddard began to manage contracts with private companies for the construction of launch vehicles such as the Delta rocket and those used at the Wallops Flight Center for sounding rocket test flights.

The Goddard Space Flight Center supervised the development of the Tracking and Data-Relay Satellites and the TDRSS program that would eventually replace the Spaceflight Tracking and Data Network, also managed by GSFC. The system of geosynchronous satellites affords NASA virtually constant contact with spacecraft in Earth orbit via the Goddard-managed facility in White Sands, New Mexico.

The Goddard Space Flight Center is one of nine major facilities operated by NASA. As one of the space agency's oldest and most complex field centers, it also plays a critical role in managing many of NASA's smaller, more specialized facilities, such as the Wallops Flight Center and the White Sands ground terminal.

Goddard also works closely with major academic institutions, such as the Massachusetts Institute of Technology and the Applied Physics Laboratory of The Johns Hopkins University, to advance space research and to develop new technology for future space missions. In this capacity, GSFC serves as a conduit for advanced research to contribute directly to the U.S. space program.

Goddard's research facilities also cooperate with international studies and programs in specific areas of space exploration. Projects such as a Japanese Earth resources satellite system and the international Active Magnetospheric Particle Tracer

Explorers (AMPTEs) allow NASA, through Goddard, to cooperate with researchers around the world.

As NASA's communications and tracking center, Goddard cooperates with the Johnson Space Center in Houston, Texas, on space shuttle missions and with the Jet Propulsion Laboratory in Pasadena, California, on deep-space missions. The space agency, through its Goddard-managed facilities, has also provided communications support for international space efforts, such as those of France and Japan.

**See also:** Cape Canaveral and the Kennedy Space Center; Deep Space Network; International Space Station: Development; Jet Propulsion Laboratory; Johnson Space Center; Marshall Space Flight Center; Space Centers, Spaceports, and Launch Sites; Space Task Group; Spaceflight Tracking and Data Network; United States Space Command; Vanguard Program.

## Further Reading

Burrows, William E. *This New Ocean: The Story of the First Space Age.* New York: Random House, 1998. This is a comprehensive history of the human conquest of space, covering everything from the earliest attempts at spaceflight through the voyages near the end of the twentieth century. Burrows is an experienced journalist who has reported for *The New York Times, The Washington Post,* and *The Wall Street Journal.* There are many photographs and an extensive source list. Interviewees in the book include Isaac Asimov, Alexei Leonov, Sally K. Ride, and James A. Van Allen.

Davies, John K. *Astronomy from Space: The Design and Operation of Orbiting Observatories.* New York: John Wiley, 1997. This is a comprehensive reference on the satellites that have revolutionized twentieth century astrophysics. It contains in-depth coverage of all space astronomy missions. It includes tables of launch data and orbits for quick reference as well as photographs of many of the lesser-known satellites. The main body of the book is subdivided according to type of astronomy carried out by each satellite (x-ray, gamma-ray, ultraviolet, infrared, and radio). It discusses the future of satellite astronomy as well.

Divine, Robert A. *The Sputnik Challenge: Eisenhower's Response to the Soviet Satellite.* New York: Oxford University Press, 1993. This is a dramatic account of the national hysteria surrounding the Soviet Union's launching of the early Sputniks. It details America's attempts to put its own satellites into orbit and discusses Eisenhower's role in the early exploration of space.

Gavaghan, Helen. *Something New Under the Sun: Satellites and the Beginning of the Space Age.* New York: Copernicus Books, 1998. This book focuses on the history and development of artificial satellites. It centers on three major areas of development—navigational satellites, communications satellites and weather observation, and forecasting satellites.

Isakowitz, Steven J., Joseph P. Hopkins, Jr., and Joshua B. Hopkins. *International Reference Guide to Space Launch Systems.* 3d ed. Reston, Va.: American Institute of Aeronautics and Astronautics, 1999. This best-selling reference has been updated to include the launch vehicles and engines. It is packed with illustrations and figures and offers a quick and easy data retrieval source for policymakers, planners, engineers, launch buyers, and students. New systems included are Angara, Beal's BA-2, Delta III and IV, H-IIA, VLS, LeoLink, Minotaur, Soyuz 2, Strela, Proton M, Atlas-3 and V, Dnepr, Kistler's K-1, Shtil, with details on Sea Launch using the Zenit vehicle.

Lambright, W. Henry, ed. *Space Policy in the Twenty-First Century.* Baltimore: Johns Hopkins University Press, 2003. This book addresses a number of important questions: What will replace the space shuttle? Can the International Space Station justify its cost? Will Earth be threatened by asteroid impact? When and how will humans explore Mars?

Lee, Wayne. *To Rise from Earth: An Easy to Understand Guide to Spaceflight.* New York: Checkmark Books, 1996. This is a good introduction to the science of spaceflight. Although written by an engineer with the NASA Jet Propulsion Laboratory, it is presented in easy-to-understand language. In addition to the theory of spaceflight, it gives some of the history of the human endeavor to explore space.

Levine, Alan J. *The Missile and the Space Race.* Westport, Conn.: Praeger, 1994. This is a well-written look at the early days of missile development and space exploration. The book discusses the Soviet-American race to develop intercontinental ballistic missiles for defense purposes and their subsequent use as satellite launchers.

McAleer, Neil. *The Omni Space Almanac: A Complete Guide to the Space Age.* New York: Ballantine, 1987. A compendium of information on the major developments of the Space Age. Emphasizes the developments of the late 1980's.

Petersen, Carolyn Collins, and John C. Brandt. *Hubble Vision: Further Adventures with the Hubble Space Telescope.* 2d ed. New York: Cambridge University Press, 1998. This picture book of astronomy is both a classroom textbook and an excellent reference. It includes illustrations of exploding stars and colliding galaxies, gravitational lenses, the impact of Comet Shoemaker-Levy on Jupiter, and pictures of other solar systems. It describes celestial objects and the instruments Hubble uses to capture images of them.

Rosenthal, Alfred. *The Early Years, Goddard Space Flight Center: Historical Origins and Activities Through December, 1962.* Washington, D.C.: Government Printing Office, 1964. This commemorative book offers a comprehensive look at the founding of the Goddard Space Flight Center, the Minitrack network, and the beginnings of the Complex Satellite Network.

Wilson, Andrew, ed. *Interavia Space Directory.* 8th ed. Coulsdon, Surrey, England: Jane's Information Group, 1992. This reference book is invaluable for obtaining general information on virtually any crewed or uncrewed mission, both those launched by the United States and those launched by other countries. Includes illustrations and a useful index.

Zimmerman, Robert. *The Chronological Encyclopedia of Discoveries in Space.* Westport, Conn.: Oryx Press, 2000. Provides a complete chronological history of all crewed and robotic spacecraft and explains flight events and scientific results. Suitable for all levels of research.

*Eric Christensen*

# Gravity Probe B

*Date:* Beginning April 20, 2004
*Type of program:* Scientific platform

*The mission of Gravity Probe B (GP-B) is to measure two subtle effects predicted by Albert Einstein's general theory of relativity. If the results support it, GP-B will increase confidence in general relativity; otherwise the results might point the way to a more comprehensive theory.*

## Key Figures

*Leonard Schiff,* physicist
*William Fairbank,* who with Schiff formed a research group at Stanford University
*C. W. Francis Everitt,* principal investigator
*Daniel B. DeBra,* co-principal investigator
*Bradford W. Parkinson,* co-principal investigator
*John P. Turneaure,* co-principal investigator
*William Bencze,* co-investigator
*Robert Brumley,* co-investigator
*Saps "Sasha" Buchman,* co-investigator
*George M. Keiser,* co-investigator
*James M. Lockhart,* co-investigator
*Barry Muhlfelder,* co-investigator
*Michael Taber,* co-investigator

## Summary of the Mission

Consider the space between stars. Does it have properties in addition to being a place in which to put things? People sometimes speak of the "fabric of space" as if space had something in common with a woolen blanket. In a blanket woven on a loom, the vertical threads are called the "warp," while the horizontal threads that weave up and down through the warp are called the "weft" or sometimes the "woof." According to Albert Einstein's general theory of relativity, space has a warp and a woof. In that theory, space and time are welded together into inseparable "spacetime," and gravity, the force flowing from the presence of mass, creates the fabric of spacetime and gives that fabric the properties of curvature and torsion.

For a relatively small mass like Earth's, the twisting of spacetime, called "frame dragging," is very subtle. Consider water in a cup. If a person touched a finger to the water and slowly sketched a circle, a water current would momentarily follow that circle. General relativity predicts that nearby spacetime is drawn along with the spinning Earth, but drawn softly and subtly, more gently than the slightest breeze.

In 1959, Leonard Schiff (and, independently, George E. Pugh) suggested that an orbiting gyroscope could accurately measure two effects predicted by general relativity: the "geodetic effect," manifest when a spinning gyroscope orbits in the gravity well caused by the Earth's mass, and frame dragging, caused by the Earth's spin. By Schiff's cal-

514

culations, the geodetic effect should cause the gyroscope's spin axis to drift by 6.614 arc seconds per year, while frame dragging should cause a drift of 0.0409 arc seconds per year at right angles to the geodetic drift. For comparison, the angle subtended by the thickness of a sheet of paper 20 kilometers away is about 0.001 arc second. To measure these effects to a greater accuracy than 0.001 arc second would require a tracking telescope a thousand times better than those available in 1959 and gyroscopes shielded a million times better from outside forces than the best gyroscopes then used in inertial navigation systems.

Schiff and William Fairbank formed a research group at Stanford University. In 1962 they were joined by C. W.

*An artist's rendition of the Gravity Probe B orbiting Earth.* (NASA-MSFC)

Francis Everitt, who was the chief designer and systems engineer and later became the group leader. Eventually funded by the National Aeronautics and Space Administration (NASA), the group continued to grow. They pushed the frontiers of technology, and then, with help from the Lockheed corporation, the NASA Marshall Space Flight Center (MSFC), and the Jet Propulsion Laboratory (JPL), the hardware was finally built and assembled into GP-B. The project has involved hundreds of people, has taken more than forty years, has survived seven critical reviews that might have terminated it, and has cost $700 million.

GP-B was launched aboard a Boeing Delta II rocket into polar orbit on April 20, 2004, from Vandenberg Air Force Base, California. GP-B orbits 640 kilometers (400 miles) above Earth. Professor Bradford W. Parkinson created the flight program, while Professor John P. Turneaure served as flight hardware manager. Four months were spent in system checkouts, fine-tuning, and calibrations. On August 27 the spacecraft began collecting science data, and apart from some minor problems, GP-B continues to work well. The spacecraft experienced some anomalous behavior when passing over the South Atlantic Anomaly (a pocket of charged particles in the ionosphere), but spacecraft controllers were able to develop means of compensating for the problem. A year after launch the science program continued. Data collection was originally expected to conclude in June, 2005, but the spacecraft continued to provide exciting results well beyond that time. Dr. George M. Keiser is the leader of the data reduction team.

GP-B consists of a spacecraft and a Science Instrument Assembly (SIA). The science assembly is surrounded by a 2,860-liter specially insulated container called a dewar filled with liquid helium. Just as a person is cooled by the evaporation of sweat, when the helium in the dewar absorbs heat, it cools itself by allowing a little helium to evaporate. This process keeps the SIA at about 1.8 kelvins. The dewar is sufficiently insulated to maintain this temperature for up to eighteen months, although this includes time spent waiting on the launch pad. This extreme cold temperature reduces instrument noise to a minimum. Cryogenic helium vapor is used to cool components outside the dewar, and in a stroke of genius, the warmed helium vapor is then vented through the microthrusters that continuously optimize GP-B's orbit. Dr. Michael Taber leads the team that developed and operates the dewar.

The Science Instrument Assembly is made of fused quartz, including the structure, the telescope, and the four gyroscopes—spinning spheres the size of table tennis balls. Fused quartz is made by melting highly purified quartz crystals and allowing the mixture to cool into amorphous glass. Fused quartz can be accurately shaped, is very uniform, and is unusually stable over a wide temperature range.

As GP-B orbits the Earth, electronics continuously monitor the spin directions of its four gyroscopes along with the direction to a star, IM Pegasi. The telescope has a 14.2-centimeter (5.6-inch) aperture and can locate the center of IM Pegasi to within 0.0001 arc second—not a simple task, because the star's image in the telescope is 1.4 arc sec-

onds across. The star's image is first split into two images by a half-silvered mirror, and these images are split again by falling on the peaks of roof-shaped prisms. The roof-ridge of one prism runs horizontally and that of the other runs vertically. Each prism slices the image into two halves and sends these to light detectors. If one of the half-images is brighter, the orientation of the spacecraft is adjusted until the two pairs of half-images are of equal brightness.

Two of the gyroscopes spin clockwise, and two spin counterclockwise, at ten thousand revolutions per minute, but they would soon slow down because of internal friction. To keep friction to a minimum, the gyroscopes spin in a vacuum and are levitated by three pairs of electric coils so that they touch nothing. Their spin axes were all initially aligned in the direction of the guide star, IM Pegasi.

Teams to develop, assemble, and operate the gyroscopes were led by William Bencze, Robert Brumley, and Saps (Sasha) Buchman.

Objects in orbit are weightless because they are in free fall around Earth. This means that one object cannot use its weight to push against another object, because they are both falling at the same rate. Even 640 kilometers above Earth, however, there are traces of atmosphere dragging on the spacecraft. If GP-B slows slightly, the gyroscopes are no longer centered in their housings. One of the gyroscopes is therefore selected as a "proof mass" and the microthrusters adjust the spacecraft to keep that gyroscope centered in its cavity so that it orbits drag-free. Daniel B. DeBra led the work on all control systems, including the drag-free system.

*Gravity Probe B consists of a spacecraft and a Science Instrument Assembly, which is surrounded by a 2,860-liter insulated container called a dewar (shown here) filled with liquid helium. (Lockheed Martin Corporation/ R. Underwood)*

The gyroscopes are coated with superconducting niobium, and as they rotate, electrons just below

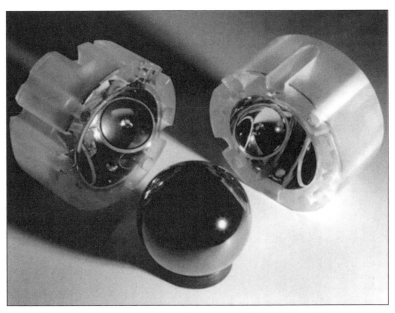

*A coated gyroscope rotor and housings.* (NASA Don Harley)

the surface lag slightly behind the positive niobium ions from which they came. This spinning, separated charge generates a magnetic field called the London moment, named for physicist Fritz London. Since the London moment is exactly aligned with the spin axis, it can be used to indicate the gyroscope's direction. The London moment is monitored by a superconducting coil connected to a superconducting quantum interference device (SQUID). This ultrasensitive combination can detect a gyroscope's tilt of 0.0001 arc second within a few days. To reach this accuracy, Earth's magnetic field is attenuated by a factor of ten trillion by placing the gyroscopes within a Mumetal shield (an alloy composed mostly of nickel with small amounts of copper, iron, and chromium added) and superconducting lead shields. Professor James M. Lockhart and Barry Muhfelder were co-leaders of the gyroscope SQUID readout team.

### Contributions

Although the results of the mission are not yet available, simply building GP-B has been a major accomplishment. It required the invention of new methods for manufacturing and shaping fused

quartz. The four fused-quartz gyroscopes are 3.8 centimeters in diameter, and they are within forty atomic layers of being perfect spheres. If Earth were that smooth, the highest mountains and deepest ocean trenches would be less than 2.4 meters from sea level. GP-B team members were among the first to measure the London moment of a spinning superconductor, the first to use an expanding, superconducting lead bag to exclude magnetic fields, and the first to use a specially designed porous plug to confine superfluid helium without pressure buildup; they also invented and were the first to test the drag-free satellite.

The name Gravity Probe B begs the question, What was Gravity Probe A (GP-A)? GP-A was launched June 18, 1976. Although it was in space less than two hours, it rose 10,000 kilometers above Earth before falling back to the surface. General relativity predicts that the stronger the gravitational field, the more slowly time passes. GP-A carried a very accurate atomic clock that was constantly compared with a second atomic clock on the ground. Earth's field becomes weaker with distance, and as GP-A got farther from Earth, the onboard clock ran slightly faster than its twin back on Earth, just as predicted. GP-A confirmed this prediction of general relativity to within 150 parts per million. The Global Positioning System (GPS) has its roots in GP-A, and if it did not take gravity's effect on time into account, its accuracy would be greatly degraded. It may be that the results of GP-B will become as useful as those of GP-A have become.

### Context

Albert Einstein's theory of general relativity is the most widely accepted theory of gravity, space, and time. Because several of its aspects are difficult to test, it had been one of the least tested of all physical theories. Because it does not incorporate quantum mechanics, the theory, as it stands, cannot be complete. An obvious way to progress is to put the theory's predictions to the most stringent tests possible. GP-B was launched in a quest for fundamental knowledge.

When the GP-B project was first envisioned, few of the predictions of general relativity had been measured with high precision, and none of the predictions about spinning objects had been measured at all. During the ensuing forty years, this has changed. The original tests proposed by Einstein have been repeated with good precision. In ad-dition the geodetic effect has been measured with 2 percent accuracy by laser ranging from the reflector left by the astronauts on the Moon. Frame dragging has been measured with 5 percent to 10 percent accuracy by laser ranging to the Laser Geodynamics Satellites LAGEOS 1 and 2. These effects have also been inferred from analyzing binary pulsars, but in each of these cases assumptions must be made and larger effects must be properly accounted for. The GP-B mission is still considered worthwhile because it will measure the effects more directly and more precisely.

**See also:** Apollo 11; Viking 1 and 2.

**Further Reading**

Everitt, C. W. Francis. "Background to History: The Transition from Little Physics to Big Physics in the Gravity Probe B Relativity Gyroscope Program." In *Big Science: The Growth of Large-Scale Research*, edited by Peter Galison and Bruce Hevly. Stanford, Calif.: Stanford University Press, 1992. An account of GP-B from inception to the beginning of construction. Everitt, the project leader, gives good descriptions of the essential parts of GP-B and the roles played by various team members.

Genz, Henning. *Nothingness: The Science of Empty Space.* Reading, Mass.: Perseus Books, 1999. A fascinating account of ideas about space and time throughout history and into modern times.

Greene, Brian. *The Fabric of the Cosmos.* New York: Knopf, 2004. Writing for the nonspecialist, the author conveys a contagious enthusiasm as he describes insights into space and time gained by Sir Isaac Newton, Albert Einstein, and quantum mechanics.

Hartle, James B. *Gravity: An Introduction to Einstein's General Relativity.* New York: Addison-Wesley, 2002. For the serious reader or the undergraduate physics student, this volume provides an accessible introduction to the esoteric subject of general relativity.

Levi, Barbara G. "Orbiting Gyro Test of General Relativity." *Physics Today* 37, no. 5 (May, 1984): 20-22. A good account of the technological advances (and the people who achieved them) that would make GP-B possible. Includes a brief summary of the history and purpose of the mission.

Stanford University. GP-B home page. http://einstein.stanford.edu. Accessed December 2, 2004. A treasure trove of information and pictures, providing details about all aspects of GP-B, including weekly updates about the mission's status. Contains many helpful links along with a wealth of information on general relativity that is understandable to the layperson.

*Charles W. Rogers*

# Heat Capacity Mapping Mission

*Date:* April 26, 1978, to September 30, 1980
*Type of mission:* Earth observation

*The Heat Capacity Mapping Mission (HCMM) spacecraft surveyed thermal conditions on Earth's surface by measuring the intensity of emitted thermal infrared radiation. HCMM was the first satellite designed to measure thermal inertia, the resistance of substances to changes in temperature.*

## Key Figures

*Dick S. Diller,* HCMM program manager
*Warren Hovis,* HCMM study manager
*Carl L. Wagner,* HCMM project manager
*John C. Price,* HCMM project scientist
*William L. Barnes,* HCMM instrument scientist

## Summary of the Satellite

Since its formation in 1958, the National Aeronautics and Space Administration (NASA) has pioneered the application of remote-sensing technology to the study of Earth. By the mid-1960's, NASA's Earth Resources Survey Program, in cooperation with the Department of the Interior, the Department of Agriculture, and other organizations, had formulated plans for a series of research satellites that would make spaceborne Earth-resources surveys possible. In 1972, the program began with the successful launch of Landsat 1. In 1978, the Heat Capacity Mapping Mission satellite, Seasat, and Nimbus 7 were all deployed.

Early on the morning of April 26, 1978, Applications Explorer Mission A, better known as the Heat Capacity Mapping Mission (HCMM), was launched into orbit by a solid-propellant Scout launch vehicle. The launch and subsequent data gathering culminated an effort that had begun as a study in 1971.

HCMM was a small, inexpensive spacecraft that carried a modified Nimbus 5 surface composition mapping radiometer as its sole experiment. The instrument and spacecraft were assembled over a two-year period ending in mid-1977. A period of calibration and testing followed. The 120-kilogram satellite was a hexagonal prism 64 centimeters wide and, with its antenna, 162 centimeters tall. Two solar cell arrays for converting sunlight to electricity, measuring 86 centimeters by 131 centimeters, were attached on a pivot to the hexagonal structure and moved to face the Sun continually during the daytime portion of the orbit. The instrument package—the Heat Capacity Mapping Radiometer (HCMR)—was mounted on the base of the prism and measured 56 by 30 by 43 centimeters.

After a successful launch, HCMM's controls unfurled the solar paddles and oriented the satellite with the HCMR pointing toward Earth. Ground stations received signals indicating that the craft was in excellent condition. The first images of Earth were transmitted from the radiometer's "visible" channel (which actually was sensitive to the red and near-infrared parts of the spectrum) the day after launch. The thermal infrared (heat-sensing) channel had to cool over a two-week period in order to be sensitive to Earth-radiated heat, and it produced the first thermal images of Earth on

May 11, 1978. Quick-look prints of the two types of data showed excellent data quality. Because of data-processing problems partly attributable to the computer-intensive calibration of the observations, however, it was nearly a year before significant quantities of calibrated data were available for analysis.

Several months after launch, HCMM's batteries, which stored the electrical energy gathered by the solar cells and distributed electrical power to the instrument, control, and transmission systems during the nighttime portion of the orbit, began to degrade. By early February, 1979, HCMM could collect only two scenes per orbit when it was on Earth's night side. By September 30, 1980, after limping along for more than a year at power levels well below design requirements, HCMM ceased functioning altogether. The satellite had worked for a much longer time and had transmitted thousands more scenes than had been expected. It continued to orbit Earth silently, being gradually pulled into the thin outer reaches of the atmosphere, until it was incinerated on reentry early in 1982.

Over its active lifetime, this low-cost spacecraft was immensely successful. It furnished the most useful thermal data on Earth resources that had ever been obtained; it outlasted its expected lifetime of one year; and in nearly six thousand orbits, it transmitted more than twenty-five thousand images of Earth. From the data returned by HCMM, scientists constructed more than one hundred complex thermal images that were "registered," or matched so that the same geographical features carried over from one scene to the next. Extensive scientific results were generated by thirty-four NASA investigations that examined and analyzed the data, and HCMM provided the basis for the first detailed and meaningful predictions of the future course of thermal remote sensing.

HCMM was conceived as a small, low-cost, rapidly developed mission dedicated to studying Earth and its resources. It was to be the first in a series of Applications Explorer Missions; its "younger brother" was known as the Stratospheric Aerosol and Gas Experiment (SAGE).

To be inexpensive, HCMM had to rely on simple systems and existing hardware. Its designers wished to avoid the complexity of onboard tape recorders, so HCMM relayed current images of Earth only when within range of selected NASA tracking stations. To take advantage of existing technology, the Earth-viewing instrument, the HCMR, was nearly the same as a device used on an earlier Nimbus satellite.

The Heat Capacity Mapping Mission concept was elegant: The HCMM satellite would fly over Earth's populated regions in such a way that daytime and nighttime images of the same area could be collected as little as twelve hours apart—in many cases, before the weather over that region had had a chance to change. If the spacecraft passed over the areas at the hottest and coldest times of day, an in-depth study of Earth's thermal responses would be possible.

HCMM's circular orbit was near polar, or inclined nearly 90° relative to the equator, had an altitude of 620 kilometers, and was Sun synchronous (the satellite passed over a certain geographical area at the same local time every day). Because of a slight inclination offset at launch, the nearly ideal 2:00 P.M. to 2:00 A.M. or 19:00 to 07:00 Coordinated Universal Time (UTC) passage times drifted earlier by about forty minutes per year. The importance of obtaining thermal contrast data (data collected at the hottest and coldest times of day) made it desirable to lower the orbit to 540 kilometers nearly two years after launch; this orbital change would torque the plane of orbit so that later passage times would result.

The HCMR instrument—a scanning radiometer—was designed to produce television-like images of Earth. The HCMR consisted of a mirror that rotated with its axis aligned in the direction of spacecraft travel, focusing optics that concentrated the energy received by the mirror, and detectors that converted the energy relayed by the focusing optics into an electrical impulse. At any moment, a 500-meter-wide "spot" on Earth's surface was being scanned by the HCMR. The size of the smallest feature visible to an instrument is known as the instrument's geometric resolution. HCMM could not resolve individual buildings, automobiles, trees, or small streams, but it could view large-area features—such as forests, rivers, cultivated fields, cities, pastures, and geologic formations—at routine intervals. The HCMR's rotating mirror moved the 500-meter spot monitored by the satellite from one side of "ground track" (the satellite's position projected straight down to Earth's surface) to the other. This scanning allowed the detectors to view a swath of Earth's surface 700 kilometers wide. In orbit, the operating sequence was as follows: The rotating mirror scanned one "line" of the surface; the satellite moved along in its orbit while the mirror scanned its own housing, away from Earth, for calibration; a new line on the surface was scanned, adjacent to the previous line; and the process was repeated until an extended image was compiled.

The HCMR detectors, which converted the energy received from Earth into electrical signals, were fabricated from two different substances: silicon, to capture the red and near-infrared wavelengths, and mercury-cadmium-telluride, to capture the infrared wavelengths. This electrical signal was used to control the frequency of the spacecraft's main radio transmitter. On command, this radio transmission could be received with very high-fidelity receivers on Earth. The receiver systems were located at satellite tracking stations at or near Fairbanks, Alaska; Goldstone, California; Greenbelt, Maryland; Merritt Island, Florida; Lannion, France; Madrid, Spain; and Orroral, Australia. Because HCMM had no onboard tape recorder, coverage was limited to areas roughly 2,000 kilometers wide that were centered on each tracking station. Thus, considerable amounts of data were obtained on the coterminous United States, southeastern and northwestern Canada, Alaska, Europe, and eastern Australia. No data were obtained on South America, most of Africa, Asia, or western Australia.

The HCMM data returned in the form of visible and infrared images of Earth were extremely useful. Altogether, more than twenty-five thousand 700-square-kilometer scenes were received, and approximately 10 percent were stored on computer-compatible tapes. Approximately one hundred "elongated" (larger than 700 square kilometers) day and night scenes of the same geographic area were registered, or precisely overlaid by adjusting the data. From these registrations, images showing temperature differences and thermal inertia were created. (Thermal inertia is the resistance of substances to changes in temperature.)

When corrected for atmospheric effects, the infrared data from the HCMR very closely approximated surface temperature. Although some early evidence indicated that the HCMR measurements were more than 5° Celsius too warm, later surface measurements showed the HCMR measurements to be within one degree of the correct temperature.

If one assumes a stable atmosphere and perfect instrument calibration and performance, measurements of a surface's day and night temperature and albedo (reflectance) at visible wavelengths permit determination of the absolute thermal inertia of the surface. There is no such thing as a truly stable atmosphere; air masses are constantly moving. Measurement of day and night temperatures therefore should be taken as close to each other as possible—ideally, twelve hours apart. Instrument performance and calibration are also imperfect. For these reasons, HCMM could not determine the absolute thermal inertia of surface features; rather, it measured "apparent," or relative, thermal inertia.

Although the data returned by HCMM were of high quality, some spacecraft, ground receiving, and processing anomalies should be noted. First, HCMM photographic products exhibited some geometric distortion, amounting to a stretch of approximately 6 percent in the east-west dimension. Additionally, a noticeable skew, or diagonal stretching, of about 9 percent occurred in the photographs taken after the orbit lowering in late February, 1980. Position errors of 50 kilometers or more were revealed when HCMM scenes were compared with appropriate map projections. Second, transmission link and recording medium interference was evident in a few HCMM scenes, particularly the photographs taken near the limits of tracking station reception. Third, some photographically processed scenes appeared inordinately dark or bright, and some visible data exhibited poor contrast. Processing errors were sometimes responsible. Image enhancement, which usually is quite useful, accounted for some of these peculiarities. Finally, HCMM images occasionally exhibited noticeable misregistration near the borders and in or near cloudy areas, and borders were occasionally erratic. These anomalies, however, did not indicate scene distortion.

HCMM's data projects were specifically designed to enable NASA-supported researchers to carry out projects defined by the mission's objectives. These objectives were sixfold: to identify rock types and mineral resource locations, to measure soil moisture effects by observing the temperature cycles of soils, to measure plant canopy temperatures at frequent intervals to determine plant stress and the transpiration of water, to predict the water runoff from snowfields, to measure the effects of urban heat islands, and to map surface thermal gradients on land and in bodies of water.

The attainment of virtually all these objectives depended on the determination of the thermal characteristics of natural materials; even water runoff prediction methods would benefit from knowledge of the surface temperature of snowfields. Knowledge of surfaces' thermal inertia was particularly important.

Just as rocks differ in color, texture, and density, so do they differ in their response to heat. A medium-gray, dense rock, such as granite, tends to heat slowly during the day and to cool slowly at night; it is said to have a high thermal inertia. Conversely, dry, unconsolidated, light sand warms quickly during the day and cools quickly at night; it is said to have a low thermal inertia.

HCMM could measure relative amounts of soil moisture, because water has a high thermal inertia. Moist soil will resist temperature change, and dry soil will heat and cool more rapidly and with greater amplitude.

Vegetation responds to temperature change in a manner that depends on its type and its available supply of water. Generally, vegetation has a fairly high thermal inertia when healthy, with a good water supply and a rapid evapotranspiration rate, and a relatively low thermal inertia when stressed, or lacking sufficient water.

Temperature measurements of a snow-covered surface help to determine the snow's potential rate of melting. Observed fluctuations of the size of snowfields also yield estimates of water runoff.

The heat created by cities is a well-known phenomenon; it can have a significant effect on local weather patterns, human comfort, and moisture retention. Cities are warmer both by day and by night than the surrounding countryside, and the more industrialized the city, the warmer it is. The amplitude, extent, and variations of this heating are of considerable interest to urban planners and to researchers investigating the reasons for local climate change.

Finally, large-scale changes in thermal conditions were examined. Land surface thermal changes on a regional scale may be caused by topography, types of soil and vegetation, weather system routes, and the effects of geographic position. Water surface thermal patterns may provide a key to upwellings, or the movement of colder water to the surface, and may delineate boundaries between different water masses.

Although HCMM ceased operating in 1980, its data still provide valuable information in conjunction with data returned by more recent Earth resources missions.

## Contributions

HCMM was the first satellite specifically designed to observe regions of Earth at their hottest and coldest times. Over two and a half years, HCMM repeatedly imaged significant areas of Earth, both by day and by night.

HCMM was capable of producing interpretable thermal images with good temperature sensitivity.

Thermal inertia images were, at that time, new products requiring careful interpretation. The images did allow identification of some materials and surface conditions, but many factors had to be considered in developing and using the thermal models. Ancillary data from ground and meteorological sources were necessary for quantitative studies, and HCMM was probably most valuable when used in conjunction with other Earth resources data sources, such as Landsat.

Using HCMM data, researchers could distinguish high-density rocks from low-density rocks and unconsolidated materials, such as alluvium, basin fill, and soils. Some dissimilar rock types could be grouped into different classes according to their apparent thermal inertias, and occasionally variations within a rock type could be recognized. The merging of HCMM thermal data with Landsat data helped in the classification. HCMM's 500-meter spatial resolution was observed to be insufficient for the recognition of indicators used in mineral and petroleum exploration. Through superior methods of precision-registering day and night data, certain subsurface geologic features were recognized and subtle topographic formations were enhanced.

Frost-prone and frost-free areas could be recognized through studies of the same area performed on many different occasions. Researchers related pasture surface temperatures to levels of vegetative stress and discriminated some crop types using day-night temperature differences. Given appropriate conditions, they could determine the moisture content of soils and estimate the moisture available to vegetation.

Under certain physiographic circumstances, HCMM could detect ancient subsurface drainage channels. (These channels can be major sources of subsurface moisture.) Near-surface water tables were also recognized under certain conditions. Thermal pollution (temperature changes resulting from human-made effluents) in lakes and rivers was detected when within the satellite's resolution capability. Using HCMM data, scientists observed and mapped the tidal circulation of water masses with different temperatures and traced the extent of major changes in marine currents. Large-scale ocean eddies, or whirlpools, were detected by their temperature differences from the surrounding water.

Investigators used thermal data from HCMM to determine temperature variations within large cities and to relate the variations to the cultural and industrial conditions imposed by humans. The scientists found that removal of vegetation and its replacement with materials that do not retain moisture is the basic cause of urban weather anomalies.

HCMM must be considered a scientific success; it established the value of thermal data for the analysis of planetary surface features and their changes over time.

### Context

HCMM was the first significant effort to study the thermal characteristics of Earth's surface. Before HCMM, thermal experiments by spacecraft were limited to very low-accuracy and low-resolution instruments carried on weather satellites. These satellites proved the possibility of recognizing meteorological conditions through the thermal imaging of clouds and provided broad relative measurements of ocean-surface thermal phenomena.

HCMM introduced improved thermal sensor spatial resolution and greater sensitivity to small temperature differences. The range of HCMM's temperature measurements was adjusted to enhance images of land surface features rather than images of atmospheric phenomena. Just as important, HCMM represented the first deliberate effort to deploy a satellite that would pass over land surfaces at times of thermal extremes.

HCMM data established the extent to which the thermal properties of surface materials could be measured from orbit using existing technology. HCMM's approach was empirical, but because of the lack of adequate models of the thermal properties of land materials, data accumulation was the only feasible course. For this reason, regional patterns in some HCMM data still lack satisfactory explanations.

The thermal results from HCMM had allowed geologists to recognize some features and conditions on Earth's surface not amenable to conventional sampling or remote sensing. Hydrologists and oceanographers found HCMM's thermal sensitivity useful in the delineation of water surface phenomena. Vegetation and soil scientists were introduced to the possibility of mapping plant moisture availability and soil bioproduction potential on a broad regional scale. HCMM showed that useful thermal measurements of Earth's surface could be performed by orbiting satellites.

**See also:** Dynamics Explorers; Earth Observing System Satellites; Explorers: Air Density; Explorers: Atmosphere; Explorers: Ionosphere; Geodetic Satellites; Global Atmospheric Research Program; Gravity Probe B; Landsat 1, 2, and 3; Landsat 4 and 5; Landsat 7; Mission to Planet Earth; Seasat.

## Further Reading

Ahrens, C. Donald. *Essentials of Meteorology: An Invitation to the Atmosphere.* 4th ed. Pacific Grove, Calif.: Thomson Brooks/Cole, 2005. This is a text suitable for an introductory course in meteorology. Comes complete with a CD-ROM to help explain concepts and demonstrate the atmosphere's dynamic nature.

Chen, H. S. *Space Remote Sensing System: An Introduction.* New York: Academic Press, 1985. This description of satellite sensors is written from an engineer's point of view, but it is kept at a basic level and is therefore suitable for the general reader with some technical background. Well illustrated. Includes references.

Harper, Dorothy, ed. *Eye in the Sky: Introduction to Remote Sensing.* 2d ed. Montreal: Multiscience Publications, 1983. A clear explanation of the various uses of satellites. Includes a survey of the Landsat system. Written in clear language aimed at the general reader.

King, Michael D., Yoram J. Kaufman, Didier Tanre, and Teruyuki Nakajima. "Remote Sensing of Tropospheric Aerosols from Space: Past, Present, and Future." *Bulletin of the American Meterological Society* 80, no. 11 (November, 1999). King et al. describe the various satellite sensor systems for tropospheric aerosols from space that are being developed by Europe, Japan, and the United States. They highlight the advantages and disadvantages of each of these systems for aerosol applications.

Ryerson, Robert A., ed. *Manual of Remote Sensing.* 3d ed. 3 vols. New York: John Wiley, 1998. A comprehensive manual covering all aspects of remote sensing, beginning with physical fundamentals and extending to sensors, processing techniques, and applications. Provides explanations of satellite systems and numerous color plates that illustrate satellite applications. Principally intended for professionals in the field of remote sensing, but the second volume is suitable for a broader audience.

Sabins, Floyd F., Jr. *Remote Sensing: Principles and Interpretation.* 3d ed. New York: W. H. Freeman, 1997. An undergraduate-level textbook providing an excellent overview of remote-sensing principles and applications. Illustrated with diagrams, maps, and photographs. Contains much practical information and is highly readable.

Short, Nicholas M., and Robert Blair, Jr., eds. *Geomorphology from Space: A Global Overview of Regional Landforms.* NASA SP-486. Washington, D.C.: Government Printing Office, 1986. This illustrated text addresses Earth's landforms and landscapes as viewed from a number of different spacecraft, including Landsat, HCMM, Seasat, various space shuttles, and the TIROS-N meteorological satellites. Central to the text is a series of annotated plates illustrating tectonic, volcanic, fluvial, deltaic, coastal, karstic, eolian, glacial, and planetary landforms.

Short, Nicholas M., and Locke M. Stuart, Jr. *The Heat Capacity Mapping Mission (HCMM) Anthology*. NASA SP-465. Washington, D.C.: Government Printing Office, 1982. The official history of the Heat Capacity Mapping Mission, this profusely illustrated text describes the principles of thermal remote sensing, the interpretation of HCMM images, and significant results from the NASA-sponsored HCMM investigations. Central to the book is an annotated gallery of one hundred HCMM images.

Zimmerman, Robert. *The Chronological Encyclopedia of Discoveries in Space*. Westport, Conn.: Oryx Press, 2000. Provides a complete chronological history of all crewed and robotic spacecraft and explains flight events and scientific results. Suitable for all levels of research.

*Locke Stuart*

# High-Energy Astronomical Observatories

*Date:* August 12, 1977, to March 25, 1982
*Type of program:* Scientific platforms

*The High-Energy Astronomical Observatories (HEAOs) provided a detailed survey of the celestial sphere, studying x-ray sources and detecting gamma and cosmic radiation.*

### Key Figures

*Richard Halpern,* program manager
*Albert Opp,* program scientist
*Fridtjof A. Speer,* project manager
*Frank McDonald,* project scientist
*Stephen Holt,* project scientist
*Thomas Parnell,* project scientist

### Summary of the Satellites

The High-Energy Astronomical Observatory (HEAO) program was initiated in 1970, following discoveries by other satellites that there was much more x-ray activity in the universe than had been previously supposed. The 12,000-kilogram satellites were to be launched by a pair of Titan III-C boosters. On January 2, 1973, however, soon after contracts had been awarded to TRW Defense and Space Systems Group to build the spacecraft, the program was all but canceled because of budget cuts. Marshall Space Flight Center, the lead center for the HEAO program, and TRW quickly rescaled the project; a series of three smaller satellites, each weighing less than 4,000 kilograms, would be launched by smaller Atlas-Centaurs. The National Aeronautics and Space Administration (NASA) approved the program, and work resumed in July, 1974.

The HEAO program tried a new approach to spacecraft design and construction called "protoflight." Rather than build a prototype followed by a series of flight units, Marshall and TRW instead refurbished the prototype, sending it out into space. This resulted in less waste of materials.

All three HEAO spacecraft used almost identical "buses," or equipment modules to which instrument units were attached. The first and third spacecraft operated similarly, although their scientific objectives were different: The spacecraft rotated, with one side (equipped with solar panels) always facing the Sun and the science instruments directed outward perpendicular to that axis. This allowed every instrument on the spacecraft to scan a thin band of the universe on each orbit and to focus on the "dark" Earth for a brief period to give scientists a zero-level reading on their instruments. The precession of the spacecraft's orbit and Earth's motion around the Sun would shift the scan arc around the celestial sphere so that within six months the entire sky would have been surveyed at least once. The second HEAO spacecraft carried a large telescope that had to be focused on specific targets (discovered largely by HEAO 1) for long periods. The three spacecraft were known as HEAOs A, B, and C during the design phase and were renamed HEAOs 1, 2, and 3 when they achieved orbit.

Development went forward almost without a hitch, although a problem with its rate gyroscopes delayed HEAO 1 for almost six months. Its launch took place at about 2:00 A.M. or 06:00 Coordinated Universal Time (UTC) on August 12, 1977; the spacecraft was placed into an orbit of 424 by 444 kilometers, with an inclination of 22.7°. The observatory began operation a few days later. The satellite was designed for six months of operation, but it functioned for eighteen months. Activity ceased on February 17, 1979, when the craft ran out of attitude control gas.

HEAO 1 carried four instruments designed to study the sky in the "soft" x-ray spectrum covering about 0.2 to 10,000 electron volts. (One thousand electron volts is a kiloelectron volt; visible light is electromagnetic radiation in the energy range between 2.8 and 5.0 electron volts, depending upon the wavelength of the radiation.) The four coarse-resolution instruments had wide fields of view designed for mapping the sky. The large area x-ray survey experiment, sensitive to a wide range of energies, comprised seven modules with fields ranging from 0.5 to 8° (the apparent diameter of the Moon is about 0.5 degree). The cosmic x-ray experiment had collimated detectors for a somewhat narrower range but greater positional accuracy, in the range of 0.2 to 60 kiloelectron volts. The scanning modulation collimator had fine and coarse collimators positioned in front of proportional counters to provide accurate locations for cosmic x-ray sources. The hard x-ray/low gamma-ray experiment sampled the lower end of the gamma-ray spectrum with a small scintillation crystal that could be placed in front of seven detectors designed for different energy levels covering 10 to 10,000 kiloelectron volts. The first three instruments used proportional counters that measured the energy given off by an x ray when it ionizes a gas inside a chamber. Collimators serve to absorb all x rays except those along a narrow axis. The scintillation detectors measure the strength of light flashes caused by scintillation when an x ray is absorbed in a crystal and its energy is re-emitted.

About a month after HEAO 1 was launched, it and a number of other x-ray satellites detected what appeared to be an x-ray nova. By raster scan-

ning—that is, making a series of short scans over a small segment of sky—HEAO 1 was able to locate the source within two days so that optical and radio astronomers could study the phenomenon further.

HEAO 2 was launched on November 13, 1978. The spacecraft was placed in an orbit of 355 by 364 kilometers, with an inclination of 23.5°. The spacecraft was better known as the Einstein Observatory, an informal designation in honor of the one hundredth birthday of Albert Einstein. The craft's general shape was similar to those of the other two HEAOs, but its forward end was slightly pointed where the main telescope and three aspect sensors protruded.

The observatory had a design life of twelve months but operated far longer; on April 25, 1981,

*The first High-Energy Astronomical Observatory (HEAO) was launched August 12, 1977, with the help of an Atlas-Centaur rocket. HEAO 1 carried four instruments designed to study the sky in the "soft" x-ray spectrum. (NASA-MSFC)*

its attitude control gas ran out, and reentry occurred on March 25, 1982. NASA had originally attached a grapple fixture so that the space shuttle's robot arm might retrieve the spacecraft; shuttle delays, however, made that rescue impossible. During its lifetime, the HEAO's attitude control system became erratic for several months, at first rendering the spacecraft useless. Just as engineers were about to try a possible solution, the problem seemed to resolve itself and normal operations resumed. NASA's efforts became useful months later, however, when other portions of the attitude control system started to degrade. HEAO 2 proved a boon to investigating scientists during its operation: For the first six months, the principal investigators were guaranteed 80 percent of its observing time, with the remainder going to guest investigators who proposed targets and investigations. This ratio was gradually reversed.

Where HEAO 1 had coarse-resolution instruments that scanned the sky, HEAO 2 had instruments that were placed on a carousel at the focal plane of a 6-meter telescope. The telescope, a Wolter Type 1, uses parabolic primary and hyperbolic secondary mirrors as do conventional reflector telescopes. Unlike conventional telescopes, however, the Wolter uses a section of these curves where they are almost flat so that an effect known as grazing incidence reflection is achieved (one can see this by viewing a sheet of glass almost edge on), thus reflecting and focusing x rays rather than letting them pass through. While gaining much in image resolution, HEAO 2 was limited to a narrow energy band. With this telescope, HEAO 2 could observe only a 1° patch of sky (about double the Moon's apparent diameter) with a resolution comparable to that of terrestrial optical telescopes.

One of Einstein's instruments was not in the telescope but mounted alongside it. The monitor proportional coun-

ter was an instrument with a wide field of view to correlate observations by the four focal-plane instruments: wide- and narrow-angle imagers and two spectrometers.

The high-resolution imager, which provided pictures with resolutions of 1 to 2 arc seconds (about one two-thousandth the apparent diameter of the Moon), was sensitive to an energy range of 0.15 through 3 kiloelectron volts. It used a microchannel plate that released a cloud of electrons when struck by an x ray. This cascade was then amplified until it triggered an electronic current in a fine wire grid in the back. The imaging proportional counter was similar to the HEAO 1 instruments, but the counter subdivided into a small area to give it 60 arc seconds of resolution over a range of 0.15 to 4 kiloelectron volts. The focal plane crystal spectrometer and the solid-state spectrometer measured the energies of incoming radiation. Both made use of the way in which crystals diffract x rays (much as a prism diffracts light).

*The High-Energy Astronomical Observatories (HEAO) consisted of three satellites designed to study x rays, gamma rays, and cosmic rays. All of the satellites were launched from the Kennedy Space Center aboard Atlas-Centaur rockets. (NASA)*

HEAO 3, the last in the series, was launched on September 29, 1979. It also had a six-month design life, but it operated for twenty-one, until its attitude control gas ran out on May 30, 1981. Reentry occurred on December 7, 1981. While HEAOs 1 and 2 were designed to map the universe using x rays, HEAO 3 was designed to study cosmic rays, either by sampling particulate radiation released by various high-energy sources or by observing the gamma-ray sources in which they may originate. It was placed in an orbit of 424 by 457 kilometers, with an inclination of 43.6°. Most satellites are placed in orbits with an inclination equal to the latitude of the launch site, unless there is a need for a higher inclination. The instruments on HEAO 3 needed to detect cosmic rays (charged atomic nuclei and particles) and thus the satellite was placed in an orbit approaching the cusp of the magnetic poles, where radioactive particles stream in freely. This effectively used Earth's magnetic field as a sort of focusing lens for the two cosmic-ray instruments.

HEAO 3 carried three instruments. The gamma-ray spectrometer was designed to detect electromagnetic radiation in the range of 60,000 to 10,000 kiloelectron volts. It used four high-purity germanium detectors cooled to 90 kelvins by a block of solid methane. Collimators around the crystals fixed the field of view at a few degrees. The other two instruments were designed to sample the stuff of which stars were believed to be made, cosmic rays spewed out by supernovae and altered by their passage through interstellar space. As its name suggests, the "isotopic composition of cosmic rays" experiment measured the composition of cosmic-ray particles, providing important information regarding their life and possible origins. The heavy nuclei experiment instrument was a detector designed to measure the relative abundance of particles heavier than argon and to detect nuclei heavier than uranium.

### Contributions

All three spacecraft considerably extended the edge of the known x-ray universe; thousands of new x-ray sources were discovered. HEAOs 1 and 2 were complementary, the first acting as a guide telescope for the second. HEAO 1 expanded the list of known x-ray sources from the 350 discovered by Uhuru (the first Small Astronomy Satellite) to more than 1,500. It also measured what appeared to be a diffuse x-ray source covering the sky. A "superbubble" of gas with a diameter of 1,200 light-years was found in Cygnus, more than 6,000 light-years away.

HEAO 2 resolved a number of point sources; about 30 percent to 50 percent of them were previously unknown quasars, thus adding to that catalog of hyperactive stellar objects. Two surveys of deep space revealed forty-three x-ray sources. When these "empty" spaces were checked with ground-based photographs, evidence of small quasars was found. A number of x-ray bursters were discovered, which had varying burst rates. HEAO 2 even observed x-ray emissions from Jupiter, apparently caused by intense radiation striking the upper atmosphere, where the aurora would form.

Such discoveries started when the telescope's instruments were activated a few days after launch. The Crab nebula, a remnant of a supernova whose explosion was observed in 1054, was found to be enshrouded in a fog of gas that emitted x rays; its strong x-ray region corresponded to the nebula's brightest visible area. The pulsar whose original explosion produced the nebula was observed winking on and off in the x-ray ranges at the same frequency as in the radio and optical ranges. The Vela supernova remnant, on the other hand, has a steady, nonpulsing x-ray source despite its radio pulsar. A daylong time exposure of Tycho Brahe's supernova remnant shows no discernible trace of its supernova at the center of an expanding gas shell. A new class of Type O stars was discovered in the Eta Carinae constellation, where a bright star flared into notice in 1843, then faded; the star remains the brightest extrasolar infrared source.

Other galaxies, too, were shown to have many x-ray sources. The M31 galaxy in Andromeda revealed seventy-two sources: twenty-one in the core; seven in globular clusters; forty-one with visible objects, dust, and gas; and the rest with uncertain identities. HEAO 1 found the entire Virgo cluster of galaxies to have very strong x-ray emanations, apparently from hot gas diffused throughout the cluster.

Spectral instruments showed emission and absorption lines corresponding to hot gases surrounding a number of supernovae. In particular, the gamma-ray spectrometers on HEAOs 1 and 3 observed variations in the 511-kiloelectron volt emission line from the core of the Milky Way. Cosmic rays generally followed expected distributions, although transuranic elements (elements having an atomic number greater than that of uranium) were not found in the quantities expected; indeed, they were hardly observed.

### Context

Until the discovery of x rays in the 1950's, it was thought that stars would be visible only in light and radio waves. The discovery in 1962 of two x-ray sources—first the Sun, then SCO X-1 (in Scorpius)—opened up the field of x-ray astronomy. Instruments carried by sounding rockets, balloons, and finally the satellite Uhuru produced a map of more than three hundred confirmed sources. Scientists suspected that these were merely the strongest sources and that many more would be found as detectors became more sensitive.

In that respect, HEAO represented a considerable gamble for NASA. Even though the HEAO program actually flown was much smaller than the recommended program, it produced a substantially broader understanding of the energy cycles involved in the universe and of the origin and fate of matter and energy.

The abundances of cosmic-ray particles were found to be a close match for mass ratios in the solar system, although theory had predicted that cosmic rays would be richer in heavy elements. One theory that has emerged from this finding is that the heavy elements are not created directly in the supernova blast but are produced when lighter cosmic rays, normally emitted by stars, are accelerated by magnetic fields and collide with cosmic dust countless times.

Theories about the formation of stars, the solar system, and the Milky Way itself are being restudied as well. The Cygnus superbubble, a ring of hot gas some one thousand light-years wide, apparently generated a number of stars as a shock wave swept through space and compressed gas enough to form new stars. What was the original force that would have had to release the energy of twenty supernovae? Several superbubbles filled with tunnels of hot, thin gas pepper the galaxy. One theory suggests that the Milky Way's spiral arms are the result of such a shock wave sweeping around the galaxy.

Before the Einstein Observatory was launched, it was expected that only supernovae, their remnants, and other oddities would have detectable x rays, and that the x rays produced by the Sun are a special phenomenon. Observations, however, indicate that virtually every star emits x rays. Further, these studies have provided vital clues as to why the tenuous corona enveloping the Sun is hotter (by a million kelvins) than the visible surface of the Sun. Observations by Skylab's x-ray telescopes lent support to a theory that acoustical shock waves carried heat from the interiors of stars to the corona. Observations of Type O stars, which are not believed to have convection zones, showed them to have strong x-ray emanations. Unexpectedly, cooler Type K stars were found also to be extremely strong in x rays. Thus, scientists have hypothesized the presence of extremely strong magnetic fields; otherwise, the corona would be blown away by the high temperatures.

Black holes, where physical theories simply do not work, probably were observed by the HEAO spacecraft, although few astronomers will identify an object as a black hole. Data from HEAO 1 observations of Cygnus X-1, a blue supergiant orbiting an invisible companion nine times as massive as the Sun, seem to confirm the theory that a black hole is surrounded by a large accretion disk of matter waiting to be swallowed. The crowding in this "traffic jam" creates temperatures ranging from 30 million to 1 billion kelvins.

Evidence for a black hole at the center of the Milky Way was gained by HEAO 3's gamma-ray spectrometer. The 511-kiloelectron volt emissions match exactly those produced by the annihilation of electrons and antielectrons (positrons). The current theory holds that antimatter is not naturally present but is created by the intense magnetic field of a black hole rending gamma rays into matter. Because the brightness of the source varies over a six-month period, the object can be no more than one-half light-year across, suggesting that the black hole is about one hundred to one million times as massive as the Sun. X-ray emissions from the nuclei of other galaxies indicate that this may be a normal condition. Indeed, in the cases of a quasar and two highly active galaxies, x-ray-emitting gas jets may have been blown off at right angles to the accretion disk by intense radiation pressure and twisted magnetic fields within the rapidly rotating object.

In many cases the data from HEAO have raised more questions than they answered. The lack of x-ray sources within many known pulsars and supernova remnants is forcing scientists to rework their theories. Observations of pulsars that do emit, however, have revealed that the magnetic fields at the surface of these collapsed objects is several trillion times stronger than that of Earth. In sum, the data from the HEAO spacecraft showed that the universe must be studied in all portions of the electromagnetic spectrum, not merely in a few narrow slices.

**See also:** Cape Canaveral and the Kennedy Space Center; Compton Gamma Ray Observatory; Heat Capacity Mapping Mission; Marshall Space Flight Center; Orbiting Astronomical Observatories; Swift Gamma Ray Burst Mission; Telescopes: Air and Space.

### Further Reading

*Astronomy and Astrophysics for the 1980's: Reports of the Astronomy Survey Committee.* 2 vols. Washington, D.C.: National Academy Press, 1983. The space science community's unresolved astrophysical questions and proposed major missions are discussed in these volumes. College-level reading.

Burrows, William E. *This New Ocean: The Story of the First Space Age.* New York: Random House, 1998. This is a comprehensive history of the human conquest of space, covering everything from the earliest attempts at spaceflight through the voyages near the end of the twentieth century. Burrows is an experienced journalist who has reported for *The New York Times, The Washington Post,* and *The Wall Street Journal.* There are many photographs and an extensive source list. Interviewees in the book include Isaac Asimov, Alexei Leonov, Sally K. Ride, and James A. Van Allen.

Cornell, James, Wendell Johnson, and Carroll Dailey, eds. *High Energy Astronomy Observatory.* NASA EP-167. Washington, D.C.: Government Printing Office, 1980. This booklet, largely devoted to the discoveries of HEAO 2, is especially helpful for its comparisons of optical, radio, and x-ray images. Illustrated.

Davies, John K. *Astronomy from Space: The Design and Operation of Orbiting Observatories.* New York: John Wiley, 1997. This is a comprehensive reference on the satellites that revolutionized twentieth century astrophysics. It contains in-depth coverage of all space astronomy missions. It includes tables of launch data and orbits for quick reference as well as photographs of many of the lesser-known satellites. The main body of book is subdivided according to type of astronomy carried out by each satellite (x-ray, gamma-ray, ultraviolet, infrared, and millimeter, and radio). It discusses the future of satellite astronomy as well.

Dooling, Dave. "Window on Violent, Cataclysmic Universe Closes." *Space World* S-5-221 (May, 1982): 16-17. This reprinted newspaper article surveys the results of the HEAO program as they were seen after the last observatory's reentry.

Fabian, A. C., K. A. Pounds, and R. D. Blandford. *Frontiers of X-Ray Astronomy*. London: Cambridge University Press, 2004. Discusses the revolution in research provided by the Chandra and Newton X-Ray Observatories and by their predecessors. Puts the data in cosmological context.

Gavaghan, Helen. *Something New Under the Sun: Satellites and the Beginning of the Space Age*. New York: Copernicus Books, 1998. This book focuses on the history and development of artificial satellites. It centers on three major areas of development—navigational satellites, communications satellites, and weather observation and forecasting satellites.

Heppenheimer, T. A. *Countdown: A History of Space Flight*. New York: John Wiley, 1997. A detailed historical narrative of the human conquest of space. Heppenheimer traces the development of piloted flight through the military rocketry programs of the era preceding World War II. Covers both the American and the Soviet attempts to place vehicles, spacecraft, and humans into the hostile environment of space. More than a dozen pages are devoted to bibliographic references.

Launius, Roger D. *NASA: A History of the U.S. Civil Space Program*. Malabar, Fla.: Krieger Publishing Company, 1994. This is an in-depth look at America's civilian space program and the establishment of the National Aeronautics and Space Administration. It chronicles the agency from its predecessor, the National Advisory Committee for Aeronautics, through the present day.

Schlegel, Eric M. *The Restless Universe: Understanding X-Ray Astronomy in the Age of Chandra and Newton*. London: Oxford University Press, 2002. An accessible introduction to the history, methods of observation, and data of x-ray astronomy. Provides data from the Chandra, Newton, and Astro-E missions as well as from spacecraft that preceded them.

Tucker, Wallace H. *The Star Splitters: The High Energy Astronomy Observatories*. NASA SP-466. Washington, D.C.: Government Printing Office, 1984. This well-written account narrates the history of the HEAO program and then describes modern astrophysics as it relates to x-ray observations. Illustrated.

Wheeler, Craig J. *Cosmic Catastrophes: Supernovae, Gamma-Ray Bursts, and Adventures in Hyperspace*. New York: Cambridge University Press, 2000. A complete exposé of high-energy processes at work in the universe. Includes the Compton Gamma Ray Observatory's contributions to that understanding.

*Dave Dooling*

# Hubble Space Telescope

*Date:* Beginning April 24, 1990
*Type of spacecraft:* Space telescope

*The Hubble Space Telescope (HST) is the largest optical astronomical observatory to orbit the Earth. Situated above the Earth's atmosphere, it has relayed the clearest images of distant objects in the universe.*

## Key Figures

*Galileo Galilei* (1564-1642), Italian astronomer and physicist
*Hermann J. Oberth* (1894-1989), German rocket scientist
*Edwin P. Hubble* (1889-1953), American astronomer
*Steven A. Hawley* (b. 1951), STS-31 mission specialist
*Richard O. Covey* (b. 1946), STS-61 commander
*Kenneth D. Bowersox* (b. 1956), STS-61 pilot
*Kathryn C. Thornton* (b. 1952), STS-61 mission specialist
*Claude Nicollier* (b. 1944), STS-61 mission specialist
*Jeffrey A. Hoffman* (b. 1944), STS-61 mission specialist
*F. Story Musgrave* (b. 1935), STS-61 mission specialist
*Thomas D. "Tom" Akers* (b. 1951), STS-61 mission specialist

## Summary of the Technology

The Hubble Space Telescope (HST) has provided the most detailed images of distant objects in the universe. Although solar system probes transmitted images of certain planets and moons, they were limited to what they could see within our solar system. In contrast, the HST has been able to image both solar system objects and deep space objects from its location in Earth orbit.

The science of optical astronomy began in 1609 when the Italian scientist Galileo Galilei first pointed his telescope toward the night sky. He made many observations of craters on the Moon, spots on the Sun, and moons around Jupiter that proved the Earth was not unique. When Galileo aimed his telescope at the heavens, he observed faint stars that could not be seen with the unaided eye. He concluded that these stars must be more distant than the planets.

In 1919, the American astronomer Edwin P. Hubble began a study of star systems, called galaxies. Using the Hooker telescope at the Mount Wilson Observatory in California, Hubble measured the distance to the Andromeda galaxy, a neighbor of our own Milky Way galaxy. Hubble also made measurements of the distances and motions of other galaxies, from which he concluded that the universe was expanding.

In 1957, the Soviet Union initiated the Space Age with the launch of the first artificial satellite, Sputnik 1. The United States also launched Earth-orbiting satellites for communication and for studying many parts of the electromagnetic spectrum: gamma rays, x rays, ultraviolet light, infrared and radio waves. They had names like Uhuru, Einstein, the Infrared Astronomical Satellite (IRAS), and the Cosmic Background Explorer (COBE).

The National Aeronautics and Space Administration (NASA) launched robotic probes from

Earth to specific areas in the solar system: Mariner 10 to Venus and Mercury; Magellan to Venus; Viking to Mars; Galileo to Jupiter; Pioneers 10 and 11 and Voyager 1 to Jupiter and Saturn; and Voyager 2 to Jupiter, Saturn, Uranus and Neptune. These craft relayed the first detailed views of these planets.

Astronomers on Earth built observatories with larger mirrors in order to see fainter and more distant celestial objects. However, they continued to be hindered by what they could see through the haze and turbulence of the Earth's atmosphere. Placing an optical telescope above the atmosphere in orbit about Earth originally was suggested in the 1920's by the German rocket scientist Hermann J. Oberth. Oberth's suggestion was not acted upon until NASA began a study in 1962. Fifteen years later, Congress approved the plan of a space tele-scope and it became an official NASA project. The proposed space telescope was named the Hubble Space Telescope in honor of Edwin P. Hubble's important contributions to astronomy.

The HST was designed by astronomers and engineers to function somewhat like an Earth-based reflecting telescope. While the telescope orbited Earth, two solar panels on HST pointed toward the Sun to provide energy for scientific instruments. A door at one end of the telescope opened and light struck the larger (primary) mirror, which was 2.4 meters in diameter. This mirror reflected light toward the smaller (secondary) mirror, which was 0.3 meter in diameter. From there, the light was again reflected and passed through a hole in the primary mirror. The focused light was converted into an electrical signal that was transmitted by satellite to White Sands, New Mexico, and then to the Goddard Space Flight Center and Space Telescope Science Institute, both in Maryland.

The original instruments of HST included the Wide Field/Planetary Camera 1, the Faint Object Spectrograph, the High Resolution Spectrograph, the High Speed Photometer, the Faint Object Camera, and Fine Guidance Sensors. All of the instruments were built to be modular in design so that they could be replaced in case of a system failure or during routine upgrading of equipment.

HST was designed as part of a series of orbiting observatories dedicated to measuring many parts of the electromagnetic spectrum. Instruments on HST allow measurements of infrared and ultraviolet radiation as well as visible light. Other orbiting observatories in the series are the Compton Gamma Ray Observatory (GRO), launched on April 5, 1991, aboard the space shuttle *Atlantis*; the Chandra X-Ray Observatory, deployed from the space shuttle *Columbia* in July, 1999; and the Space

*The Hubble Space Telescope, held in space by the Remote Manipulator System on board the space shuttle* Discovery *during the STS-31 mission.* (NASA)

Infrared Telescope Facility (SIRTF), launched atop a Delta vehicle in August, 2003.

HST weighed approximately 11,000 kilograms when it was loaded into the cargo bay of the space shuttle *Discovery*. It was launched into orbit from Cape Canaveral on April 24, 1990. On the following day, the telescope was deployed by American Astronaut Steven A. Hawley, using the 15-meter (50-foot) mechanical arm of the shuttle.

The primary mirror on the HST was designed at the Perkin-Elmer Corporation (later renamed Hughes Danbury Optical Systems) in Danbury, Connecticut. Several months after deployment, astronomers on Earth discovered that technicians had made it slightly too curved, which caused much of the starlight to be out of focus. Once the defect was identified, specialists planned a mission during which astronauts would replace an optical component in HST and allow light to be focused as originally planned.

The flaw in the mirror was only one of several problems with HST. The devices that help align the observatory with targets, the gyroscopes, were not reliable; one failed in 1990 and two failed in 1992. Astronomers would not be able to point the telescope at any object if another gyroscope failed. These units would either have to be replaced or repaired.

The two solar arrays, which provided energy to power the telescope's instruments, also had a problem. The material used in the arrays expanded when HST was facing the Sun and contracted when the telescope was entering darkness. The temperature reaction of the panels caused HST to vibrate uncontrollably, which further blurred the images. Engineers on Earth had to develop new software to try to compensate for the jitter. Even with the software, there were certain times when observa-

*A long view of the Hubble Space Telescope flying over Shark Bay, Australia. The photo was taken from the space shuttle* Discovery. *(NASA)*

tions could not be made because of the erratic motion.

Engineers from Ball Aerospace and Communication Group in Colorado developed a solution for the problem with Hubble's primary mirror. They studied technical drawings of the telescope and determined that a set of corrective lenses could compensate for the flaw. The lenses would have to be placed along the internal path of light. Their solution was to replace Hubble's photoelectric photometer with the Corrective Optics Space Telescope Axial Replacement (COSTAR) optics units. Once in place, small robotic arms would move the mirrors into position to refocus the fuzzy light.

A space shuttle mission was designed in which astronauts would rendezvous with HST, install Wide Field/Planetary Camera (WFPC) 2 and COSTAR and replace the solar panels and gyroscopes. The crew of STS-61 trained in large swimming pools with instruments and equipment similar to those needed to repair HST. They used a model of HST and practiced removing and installing components, a dress rehearsal for the real mission in space.

On December 2, 1993, the space shuttle *Endeavour* and its STS-61 crew lifted off from Cape Canaveral. Upon achieving orbit, American astronauts Kenneth D. Bowersox and Richard O. Covey maneuvered *Endeavour* so that its path would intersect the orbit of HST. Two days later, Swiss Astronaut Claude Nicollier used the shuttle's robotic arm to reach out and grapple HST. The telescope was then moved onto a special turntable inside the shuttle's cargo bay so that the astronauts could repair or replace the malfunctioning instruments.

To repair HST, astronauts had to wear space suits during extravehicular activity (EVA). While outside the shuttle, the astronauts moved using a combination of tethers (rope) and holding on to the railing on HST. An astronaut inside the shuttle operated a mechanical arm that was used to move some of the EVA astronauts toward HST.

During the first day of repair, astronauts restored the malfunctioning gyroscopes to operational condition. They also inspected the solar panels and discovered that one panel was severely bent. The other panel was mechanically rolled up, leaving the bent panel still unfurled against HST.

The solar panels are very important because it is these arrays that take energy from the Sun and convert it into energy to power equipment on HST. With the gyroscopes functioning, a good power supply was needed and the focus was turned to the solar panels. During the second EVA, Astronaut Kathryn C. Thornton successfully detached the bent solar panel from HST. She held the broken panel above her head. As the shuttle backed away, Thornton released the panel and it fluttered and fell into orbit around Earth like a bird in flight.

On the third EVA, astronauts removed the HST's original Wide Field/Planetary Camera 1, WFPC1. In its place, they installed the new state-of-the-art WFPC2.

The fourth EVA was dedicated to correcting the flawed optics of HST. To make room for COSTAR,

astronauts removed the photoelectric photometer, the least-used scientific device on HST. With the photometer out, COSTAR fit like a glove into HST. Once in place, scientists on Earth positioned a series of internal mechanical arms that moved the corrective lenses in front of the mirror, like a pair of eyeglasses.

The HST was released back into orbit on December 10, 1993. Astronomers on Earth were able to confirm the success of the repair mission a few weeks later. The new optics worked perfectly. In February, 1997, the space shuttle *Discovery* returned to Hubble for a second servicing mission during STS-82. Two advanced instruments—the Near Infrared Camera and Multi-Object Spectrometer (NICMOS) and the Space Telescope Imaging Spectrograph— were swapped out with the two first-generation spectrographs. The astronauts also replaced or enhanced several electronic subsystems and patched unexpected tears in the telescope's shiny, aluminized thermal insulation blankets, which gives the telescope its distinctive foil-wrapped appearance.

Tasks planned for the third servicing mission grew into a lengthy list of repair jobs in addition to the installation of new instruments—so lengthy, in fact, that the work was split into two missions, 3A and 3B. In December, 1999, the STS-103 mission brought *Discovery* to rendezvous with Hubble for the 3A servicing mission. Astronauts replaced faulty gyroscopes, which had suspended science observations for nearly a month. The telescope also got a new high-tech computer and a data recorder. The astronauts left the telescope in "better-than-new" condition.

In March, 2002, the STS-109 mission brought *Columbia* to a rendezvous with the Hubble Telescope for the 3B servicing mission. This proved to be *Columbia*'s final successful mission, and astronauts performed five spacewalks to install new solar arrays, the Advanced Camera for Surveys (ACS), a new power control unit, and a new cryocooler for the NICMOS instrument to replace one that had failed in the intervening years since its installation on Hubble. The latter action restored NICMOS to scientific function.

A fourth servicing mission had been planned for late 2003 or early 2004, but those plans fell through when *Columbia* was lost during reentry on February 1, 2003, and shuttle flights were grounded pending resolution of the accident. During this hiatus, NASA Administrator Sean O'Keefe decided that, for safety reasons, all future shuttle flights should be restricted to orbital inclinations similar to that of the International Space Station. Hubble was too far afield for a shuttle to rendezvous with it and then change its orbit to reach the Space Station in the event that the astronauts need a safe haven (in the event, for example, that the orbiter were damaged too severely to attempt reentry). That left Hubble in a precarious situation: Its batteries or gyroscopes would fail by 2008 without further intervention, and the Cosmic Origins Spectrograph and Wide Field Camera 3 were stuck in final preparations at the Goddard Space Flight Center with no flight opportunity.

NASA considered robotic efforts to fix Hubble, but these proved too challenging and costly given the time constraint, so such plans were soon abandoned. When O'Keefe was replaced by Michael Griffin, the question of adding another servicing mission to the shuttle program manifest was revisited. The possibility of shuttle astronauts replacing critical equipment and installing new, highly anticipated scientific instruments capable of expanding our understanding of the early universe was again considered seriously. Flight-hardware preparations for a fourth servicing mission resumed in 2005 under orders from Administrator Griffin even though no decision was made to adopt a shuttle flight to accomplish that servicing. When *Discovery* flew the Return to Flight mission (STS-114) in late July and early August, 2005, Administrator Griffin mentioned several times that the shuttle would be used to finish construction of the International Space Station and repair the HST before being retired in 2010. Even at this point, no shuttle-based Hubble servicing mission was officially scheduled, but the likelihood of one being adopted for 2007 or 2008 had increased significantly. Griffin and the scientific community truly wanted Hubble to be operational when the James E. Webb Space Telescope, Hubble's next-generation replacement, was launched in 2011.

### Contributions

The HST has provided new views of objects within our solar system. Storms on the gas giant planets (Jupiter, Saturn, Uranus, and Neptune) have been observed. Huge storms were seen on Jupiter and Saturn.

In July, 1994, HST observed the fragments of Comet Shoemaker-Levy 9 slam into the atmosphere of Jupiter. Images were available hours later to both professional astronomers and science enthusiasts through various Internet sites. The pictures showed a series of dark blotches that appeared on Jupiter and could be seen for months following the event.

In 1995, HST took images of Saturn as its rings appeared "edge on" from our perspective on Earth. S1995S3, a newly discovered moon orbiting Saturn, was viewed.

Peering outside our solar system, the HST has viewed stars at various stages of their lives. The Orion nebula, a huge cloud of dust and gas, is a nursery of young stars. It is one of the most active regions of star birth in the Milky Way galaxy. Within the center of the nebula, HST imaged four stars surrounded by gas and dust. Scientists have named these new objects proplyds, or protoplanetary disks.

The HST has observed M16, the Eagle nebula, in the constellation of Serpens. The images show columns of cool interstellar dust and gas. Astronomers have counted about fifty young stars within these pillarlike structures. An elusive brown dwarf, believed to be a star with not quite enough mass in its core to shine like other stars, has been seen by HST. The star, known as GL229B, orbits a red dwarf star and is the faintest object ever seen around a star beyond the Sun. HST has observed groups of stars known as star clusters. Images of the globular star cluster M4 show rarely seen white dwarf stars, remnants of stars that blew off their outer envelopes of gas.

Observations of unstable stars have been made by HST. The star Eta Carinae has been seen as a luminous star that is both unstable and prone to violent outbursts. The star shines with three million times the energy of the Sun as it emits streamers of material into space.

Remnants of violent star deaths have been seen. The debris of Supernova 1987A shows three rings of expanding and glowing gas, surrounding the area of the star explosion.

Moving out of the Milky Way galaxy, HST has imaged other galaxies. The elliptical galaxy in Virgo named M87 appears as an elongated ball of stars from ground-based telescopes. However, images taken by HST show that the center of the galaxy is spinning very rapidly and is ejecting a jet of material into space.

Cepheid variable stars, which vary in brightness on a periodic basis, have also been studied with HST. A mathematical relationship between the period of brightness and its luminosity helps astronomers measure distances to galaxies. The observations of Cepheids taken by HST are being used by cosmologists to estimate more accurately the rate at which the universe is expanding.

According to the Space Telescope Science Institute in Baltimore, Maryland (as of 2000):

> In its first ten years of surveying the heavens, the Hubble Space Telescope has made 330,000 exposures and probed 14,000 celestial targets. It has whirled around Earth 58,400 times, racking up 2.4 billion kilometers, approximately equal to making eight round trips to the Sun. The orbiting observatory's observations have amounted to 3.5 terabytes of data. Each day the telescope generates enough data—3 to 5 gigabytes—to fill a typical home computer. Hubble's archive delivers between 10 and 15 gigabytes of data a day to astronomers all over the world. Astronomers have published 2,651 scientific papers on Hubble results.

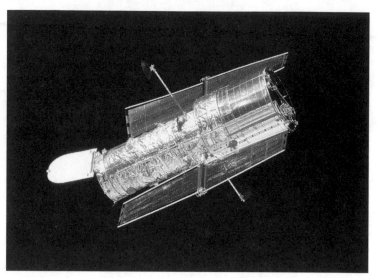

*A close view of the Hubble Space Telescope during the second servicing mission, STS-82.* (NASA)

### Context

Observations made by HST have advanced current understanding of astronomy and objects in the universe. HST provided the first conclusive evidence for the existence of supermassive black holes. Its observations suggest that the object at the center of the Galaxy NGC 4261 is 1.2 billion times the mass of the Sun, yet occupying an area of space not much larger than the solar system. The elliptical Galaxy M87 appears to conceal a black hole with a mass of 2.6 billion suns.

Observations of Cepheid stars have yielded new distances to galaxies and to their rate of movement. According to data taken with HST, the universe is younger than had been previously suggested from observations on Earth.

The origin of the universe has been a subject of much debate. The most current theory, known as the Big Bang, suggests that the universe derived from an explosion about fifteen billion years ago. By resolving the shapes of very distant galaxies, HST has provided the first direct visual evidence that the universe is evolving and expanding, as predicted by theory.

Astronomers have not been able to account for 90 percent of the matter in the universe. This missing mass was suggested as being faint red dwarf stars, which could not be seen from Earth. However, observations taken by HST show that there are not enough small red dwarf stars to account for the missing mass.

Observations of protoplanetary disks, like those seen in the Orion nebula, suggest that Earth and neighboring planets are not the only solar systems in the universe. The protoplanetary disks are believed to be embryonic planetary systems around young stars.

Within our solar system, HST provided detailed views of planets and moons once possible only when robotic spacecraft were sent to each destination. However, Hubble's observations from Earth orbit show that the atmospheres of the gas giants are more dynamic than previously believed. Huge storms were seen as they developed on Saturn and Jupiter. The Great Dark Spot on Neptune has disappeared and has been replaced by a larger storm near the planet's pole, one example of the changing nature of the outer planets.

By observing objects both within the solar system and near the edge of the observable cosmos, scientists have updated their understanding of astronomy. In doing so, they continue to learn more about Earth and its place in the universe.

In 2004, political bravado and budgetary reality clashed when the White House declared a goal of sending humans back to the Moon and on to Mars. Facing the worst federal deficit since the Hoover administration, funding for such projects would require NASA to cancel piloted and robotic missions already in the pipeline. In response to the announcement, NASA eliminated all space shuttle flights not directly supporting the International Space Station. These included scheduled servicing missions to the Hubble Space Telescope. However, after the success of STS-114, NASA planned to visit Hubble one last time in 2007 or 2008 to change out instruments and replace its gyroscopes with the intent of keeping the telescope in service until at least 2011, when its heir apparent, the James E. Webb Space Telescope, is expected to launch. Scrapping the final servicing mission raises the

likelihood that Hubble will fail before Webb is on orbit.

Because the planning for and development of a fourth servicing mission was very far along and the scientific promise of such a mission is extraordinarily high, NASA has decided to consider a novel idea to service Hubble without astronauts— a completely robotic servicing mission. Not only would the mission benefit Hubble by extending its life and enhancing it scientifically, but also such a mission, which advances robotic techniques in space, would have far-reaching implications for NASA's ever-expanding frontiers of exploration.

It has not yet been decided whether NASA will pursue a robotic servicing mission to Hubble. NASA wishes to develop the still-new mission concepts and designs further before making a final decision about whether the mission is feasible or advisable. Should it be decided to proceed with the implementation of such a mission, the targeted launch date would be at the end of 2007 or early 2008.

The James E. Webb Space Telescope (JWST) is a large, infrared-optimized space telescope scheduled for launch in August, 2011, by a launch vehicle not yet indicated. JWST is designed to study the earliest galaxies and some of the first stars formed after the Big Bang. These early objects have a high red shift from our vantage-point, meaning that the best observations for these objects are available in the infrared. JWST's instruments will be designed to work primarily in the infrared range of the electromagnetic spectrum, with some capability in the visible range.

JWST will have a large mirror, 6.5 meters (20 feet) in diameter, and a sunshield the size of a tennis court. Neither the mirror nor the sunshade can fit onto the rocket fully open, so both will fold up and open only once JWST is in outer space. JWST will reside in an L2 Lissajous orbit, about 1.5 million kilometers (1 million miles) from the Earth. L2 is one of five Lagrangian points where the pulls of the Earth and Sun combine to form a point at which a third body of negligible mass—a satellite—

would be stationary relative to the two bodies. Achieving this point precisely is difficult, so a special Lissajous orbit (a periodic orbit in which there is a combination of planar and vertical components) called a "halo" orbit will be used. A halo orbit is one in which a spacecraft will remain in the vicinity of a Lagrangian point, following a circular or elliptical loop around that point.

Mission goals are to determine the shape of the universe, explain galaxy evolution, understand the birth and formation of stars, determine how planetary systems form and interact, determine how the universe built up its present chemical/elemental composition, and probe the nature and abundance of dark matter.

**See also:** Chandra X-Ray Observatory; Compton Gamma Ray Observatory; Extravehicular Activity; Extreme Ultraviolet Explorer; Funding Procedures of Space Programs; Gamma-ray Large Area Space Telescope; High-Energy Astronomical Observatories; Hubble Space Telescope: Science; Hubble Space Telescope: Servicing Missions; Space Shuttle Flights, 1990; Space Shuttle Flights, 1991; Space Shuttle Flights, 1992; Space Shuttle Flights, 1993; Space Shuttle Mission STS-61; Spitzer Space Telescope; Telescopes: Air and Space.

### Further Reading

Bahcall, John, and Lyman Spitzer, Jr. "The Space Telescope." *Scientific American* 247 (July, 1982). The article describes the early history of the space telescope and its evolution from committee to construction. It details original plans for use and includes photographs of the primary mirror, sketches of the telescope, and the specific components inside. This reference is useful as an early view of the project and its goals, and it details procedures on the use of the HST.

Chaisson, Eric J. "Early Results from the HST." *Scientific American* 266 (June, 1992). This article was written by the former director of public affairs for the Space Telescope Science Institute. It details some of the unique observations made by HST before the repair mission. Photographs illustrate the clarity of HST compared with ground-based images. A scale illustrates the time light takes to travel great distances in the universe. The article is suitable for general audiences.

_____. *The Hubble Wars.* New York: HarperCollins, 1994. This text begins with an account of the deployment of the HST. It follows with the identification of the flawed optics and the work of scientists and engineers to develop a solution for the telescope. Early science results are displayed in text and through real and false-color photographs. This reference is suitable for general audiences.

Davies, John K. *Astronomy from Space: The Design and Operation of Orbiting Observatories.* New York: John Wiley, 1997. This is a comprehensive reference on the satellites that revolutionized twentieth century astrophysics. It contains in-depth coverage of all space astronomy missions. It includes tables of launch data and orbits for quick reference as well as photographs of many of the lesser-known satellites. The main body of the book is sub-divided according to type of astronomy carried out by each satellite (x-ray, gamma-ray, ultraviolet, infrared, millimeter, and radio). It also discusses the future of satellite astronomy.

Field, George, and Donald Goldsmith. *The Space Telescope.* Chicago: Contemporary Books, 1989. This was the first comprehensive book about the HST. It is a good introductory text. Because it was written before launch, some of the information is not currently accurate, particularly information on the instruments and their use. This book is well suited for the general reader who is looking for an overview of the original space telescope project.

Fischer, Daniel, and Hilmar W. Duerbeck. *Hubble Revisited: New Images from the Discovery Machine.* New York: Copernicus Books, 1998. This book concentrates on the discoveries of the Hubble Telescope from 1992 through 1997. Containing more than 140 spectacular images, the text explores a wide range of astronomical topics—from the births and deaths of stars to quasars and black holes—and includes self-contained portraits of astronomers as well as explanations of astronomical topics and instruments.

Jones, Brian. "The Legacy of Edwin Hubble." *Astronomy* 17 (December, 1989). The HST was named in honor of the accomplishments of the American astronomer Edwin P. Hubble. This article provides an overview of Hubble's accomplishments, from his observations of galaxies and Cepheid stars to building a distance scale of the universe. Article includes a photograph of Hubble, a page from his observing log, and a graph of his velocity-distance relationship.

Kerrod, Robin. *Hubble: The Mirror on the Universe.* Richmond Hill, Ont.: Firefly Books, 2003. The book covers the universe in six sections: "Stars in the Firmament," "Stellar Death and Destruction," "Gregarious Galaxies," "The Expansive Universe," "Solar Systems," and "The Heavenly Wanderers."

Maran, Stephen. "Hubble Illuminates the Universe." *Sky and Telescope* 83 (June, 1992). This article summarizes the initial problems with the primary mirror in the HST. It outlines the repair mission and its goals and objectives. The majority of the text is devoted to the observations made by HST before the repair. Although not in perfect focus, the images were better than could be achieved from ground-based telescopes on Earth.

Neal, Valerie. *Exploring the Universe with the Hubble Space Telescope.* NASA NP-126. Washington, D.C.: Government Printing Office, 1989. This book, prepared by NASA, is geared toward general audiences and educators. It details the observation strategy of HST, its operation, and opportunities for making discoveries using the specialized equipment. Each instrument is outlined in text and illustration. Although the book was printed before the telescope was launched, it provides a very comprehensive guide for the HST.

Petersen, Carolyn Collins, and John C. Brandt. *Hubble Vision: Further Adventures with the Hubble Space Telescope.* 2d ed. New York: Cambridge University Press, 1998. This picture book of astronomy is both a classroom textbook and an excellent reference. It includes illustrations of exploding stars and colliding galaxies, gravitational lenses, the impact of Comet Shoemaker-Levy on Jupiter, and pictures of other solar systems. It describes celestial objects and the instruments Hubble uses to capture images of them.

Scott, Elaine, and Margaret Miller. *Adventure in Space: The Flight to Fix the Hubble.* New York: Hyperion Books for Children, 1995. This children's book is dedicated to the story of the HST repair mission. Each astronaut is shown in photographs and described in the text. They are shown rehearsing for the repair mission by working in large swimming pools, to simulate the conditions of working in space. Spectacular pictures taken from the space shuttle show the astronauts working on the HST.

Smith, Robert W. *The Space Telescope: A Study of NASA, Science, Technology and Politics.* New York: Cambridge University Press, 1989. This text illustrates the scholarly history of the early years of the space telescope project in the 1970's and 1980's. It is based on interviews with those people directly involved in the program: scientists, astronomers, and engineers. Although many references focus intentionally on the telescope's specialized instruments, this book brings the people behind the telescope into the spotlight.

Wilkie, Tom, and Mark Rosselli. *Visions of Heaven: The Mysteries of the Universe Revealed by the Hubble Space Telescope.* London: Hodder & Stoughton, 1999. This is an excellent, well-written look at some of the most incredible images taken by the Hubble Space Telescope. The images, most of which are in color, are sharp and clear and cover a wide range of celestial objects. *Visions of Heaven* presents these remarkable pictures with a concise narrative that reveals the thrilling and moving stories behind them.

*Noreen A. Grice, updated by Russell R. Tobias*

# Hubble Space Telescope: Science

*Date:* Beginning April 24, 1990
*Type of spacecraft:* Space telescope

*Since ancient astronomers first tracked the night sky, the mapping of the universe has been envisioned. Today, because of what the Hubble Space Telescope (HST) and other instruments have "seen," a picture of our cosmic history is slowly unfolding. The HST has sent back terabytes of incredible pictures that have amazed not only scientists but also the public at large.*

## Summary of the Program

On April 24, 1990, in the cargo bay of the space shuttle *Discovery*, the Hubble Space Telescope (HST) was launched with an array of glitches. It was the size of a bus, cost more than $1.5 billion, was powered by solar panels, and was to orbit at 600 kilometers above the Earth. Within a month of launch, scientists became conscious that the telescope was not focused properly and that it had a bad case of the shakes. The National Aeronautics and Space Administration (NASA) eventually admitted that the primary mirror, which cost so much money, was the wrong shape. Prior to launch, two telescope mirrors had been tested separately, but, because of expense, not together. Finally, engineers found that the grinding of the mirror was off by a minute fraction of a millimeter. The mirror was precise, but not accurate.

To correct the telescope, scientists were able to make a "contact lens," repair the shakes, and fix the fluttering solar panels. Two repair teams were assembled, and three years later, on December 2, 1993, with much rehearsal and a series of incredible spacewalks, the telescope was finally fixed. The HST was now able to see a 100-watt bulb on the Moon.

The HST does not take actual pictures, but beams back digital information in bits and bytes. These are reassembled by computer to make the images that have graced magazines and newspapers and have kept the astronomers in awe of the amount of data of light and radiation intensity. This magnificent telescope has made more than 271,000 observations and has studied more than 13,670 objects. The HST also works in conjunction with other radio telescopes to gain expansive knowledge that continues to unlock the secrets of the universe.

Servicing missions were planned to maintain and upgrade this extremely valuable tool. One, STS-109 (March, 2002), installed a new Advanced Camera for Surveys (ACS) system, replacing the Faint Object Camera. ACS is ten times more effective than the previous system. The ACS is expected to complete a map of the universe, which will help scientists understand what dark matter is, how it is distributed, and its relationship to the life cycles of galaxies and stars. This same service mission revitalized the Near Infrared Camera and Multi-Object Spectrometer (NICMOS), which will pick out the brightness of supernovae and their red shifts, and will "see" more distant galaxies.

## Contributions

HST has provided insights into the solar system, how it was formed and the morphology of the planets. Some of these insights have come from its observations outside the solar system. The HST also observed, for the first time, disk-shaped clouds in the Orion nebula. Until data from HST were available, there was no evidence of flat, spinning disks

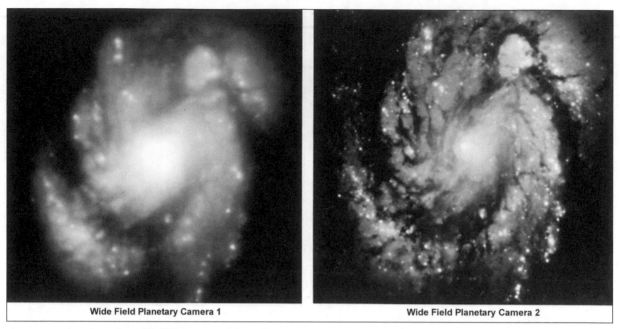

**Wide Field Planetary Camera 1**                    **Wide Field Planetary Camera 2**

*A comparative view of two Hubble images of the galaxy M100: before (left) and after the first Hubble servicing mission to install the Wide Field and Planetary Camera 2.* (NASA)

of gas accreting that pointed to proof of the theory that the origin of planets was from flat disks. A warp in the Beta Pictoria system, indicating strong gravitational pull from an object such as a planet, was the first indirect evidence of a planet outside our solar system. Speculation ended when the HST discovered forty to fifty Jupiter-sized masses around other stars.

The telescope, a stationary observatory, was directed at the planets in the solar system. Myriad data were returned to Earth. Venus, the planet closest to Earth, is shrouded in mystery, with thick, heavy clouds. The HST mapped the cloud tops and may have discovered high and low pressure cells, indicative of a stormy atmosphere. It also identified sulfuric acid clouds and rain triggered by a volcanic eruption in the 1970's.

Since the time the first telescopes studied Mars, the Red Planet has fascinated people. In 1976 the Viking 2 Lander returned much data about the Martian atmosphere and surface, reporting a very large impact basin, a massive volcano, and huge dust storms. Scientists studying the deuterium

(heavy hydrogen) in the Martian atmosphere estimate that Mars may have had a layer of water from 3 to 27 meters deep. More studies are planned to learn when and how Mars lost its water. HST supplied weather data for the Mars Pathfinder mission in 1997, providing statistics that showed Mars to be cooler, drier, and cloudier than it was during the 1970's and 1980's, when Viking first touched down. Clouds and seasonal polar changes were also observed. The HST found only small dust storms and is still looking for the huge dust storms reported by Viking that would be detrimental to piloted landings.

With its onboard instrumentation, HST could measure the effects of the magnetic fields on Jupiter and take sharp ultraviolet pictures of its gases. Data about a three-hundred-year-old ammonia storm was updated daily. The data allowed scientists to map the most intense magnetic field (with a current of 1 million amperes) in the solar system. In 1994 the world watched as the comet-string of Shoemaker-Levy 9 crashed pieces onto the clouds and thus affected the aurora of Jupiter and the magnetosphere. The aurora constantly changed and mutated, and storms and wind shears were recorded. Gas in the aurora was heated to 927° Celsius. Scientists noted that the Great Red Spot constantly changed color, as did the cloud belts. A Jovian ring was found to contain oxygen.

The improved HST revealed ozone on Ganymede and oxygen on Europa. The telescope mapped the surface features of two of Jupiter's moons, Callisto and Ganymede. In 1996 HST caught the largest volcano on Io, another of Jupiter's moons, in the process of erupting. Io, a silicate-rich moon, has the widest temperature range in the solar system. The huge gravitational field of Jupiter is believed to cause the volcanoes by flexing and melting material, resulting in sulfur flows and plumes of silicon dioxide. This gas appears to feed a ring around Jupiter called the Io plasma torus.

HST has returned clear images of Saturn's rings and its upper cloud decks. It also mapped a huge storm, possibly an eruption of ammonia ice crystals, beneath the clouds. This storm, which had winds clocked at 500 meters per second, was a short-lived event—unlike the Great Red Spot of Jupiter.

Titan, a moon of Saturn that is a little smaller than Mars, was mapped using near-infrared radiation. The solid surface is obstructed by photochemical smog, but the HST instrumentation was able to "see" beneath the cover. Bright and dark regions that might be continents, oceans, or craters were photographed, indicating Titan may have rain and oceans of methane and ethane. Four new Saturnian moons were discovered in 1995.

In 1781, when William Herschel discovered Uranus, he noted clouds on the surface. The next detailed look at Uranus in 1986, by Voyager 2, showed it a featureless blue ball with a hydrogen atmosphere and no clouds. This feature was not detected again until HST saw "spring" on the planet after an eighty-four-year winter. The first images returned in 1994 were sharp details of clouds, rings, and moons.

HST saw distant Neptune with incredible clarity. Dark storms raced across the planet, and high-altitude cirrus clouds were visible. Because HST was able to track storms on the outer gas planets, new theories, such as an internal heat source in Neptune, are now being discussed. Neptune had changed in appearance since the Voyager 2 flyby, as evidenced by the discovery of the Great Dark Spot and Great Dark Spot 2, which have since disappeared.

Pluto, an ice ball covered in frozen methane, was discovered in 1930. Its companion, Charon, was located in 1978. HST pictures show that Pluto has ragged ice polar caps and bright and dark features that might be caused by impact craters or by the cold, long winters creating icy patterns. The pictures sent back pose more questions about the remotest planet than they answer. The HST mapped Pluto and Charon together in 1994, revealing incredible differences. Pluto is mostly rock with smooth ice on its surface. Charon, one-half its size, is mostly ice with a blue cast. Though they are in close proximity, they are very different in structure. Pluto may not actually be a planet but rather a super-comet in the Kuiper Belt.

One of the most important functions of the telescope is to track down "killer" meteors and comets. In 1995 HST found evidence of a primordial reservoir of 200 million comets at the edge of the solar system, proving the existence of the Kuiper Belt, which was proposed in the early 1960's. Here, astronomers saw forty large-sized protocomets that were not previously visible. This region will help reveal how the solar system was formed.

In 1995 HST mapped Vesta, one of the biggest asteroids found between Mars and Jupiter. Surprisingly, the surface showed a huge impact crater and traces of lava flows—implying that the interior is molten. Also helpful to scientists was a piece of Vesta that crashed into Australia in 1960. Having an actual sample from the same asteroid as HST was observing was a scientific first, and the data corresponded. Vesta had yet more to offer astronomers, because part of it had broken off, leaving an exposed mantle. No other object in the solar system has an exposed fresh interior. A company called SpaceDev plans to use the information gathered to stake claims and eventually survey or mine the asteroids.

The HST has helped scientists investigate stars, galaxies, and the universe. With incredible clarity, it has caught stars at various stages of their life cycles. Humans, despite their short life spans, have now observed the births, lives, and deaths of stars. The Orion nebula is a stellar nursery with many gases only 300,000 years old and consisting primarily of hydrogen, oxygen, and nitrogen. HST has photographed many plates that, when put together, show protostars, the birth of a star, and its protoplanets. Ground-based instruments see space in wide angle, but the HST can zero in on the fields, so the picture of star formation is becoming more precise and much better understood.

The HST can sense infrared light that is too cool to "see." It can look through dust clouds. Material in the Orion nebula was in chaos—shooting out hydrogen gas at 1.6 million kilometers per hour. Yet, star creation was detected within these gases. In the Eagle nebula astronomers were able to see beautiful towers of gas compacting under its own gravity, the earliest stages of star formation. HST also captured later stages of star birth, evidenced by pulsating jets of gas moving at speeds of 800,000 kilometers per hour. Astronomers' dreams came true when HST actually witnessed Supernova 1987A, the first supernova observed since 1604. Though it was 160,000 light-years away, two luminous rings were visible. New stellar theories posit that the rings were not ejected but were formed at the same time as the explosion and were a proto-stellar disk. Pulsars, expected in the center of this supernova, were not observed; however, as predicted, a pulsar was seen in the Crab nebula. In February, 1994, HST obtained a spectacular image of three rings around this explosion, and later oxygen, nitrogen, and sulfur were detected. The spectrograph also picked up hydrogen traveling at the astounding rate of 52.8 million kilometers per hour. Eta Carinae appears to be on the verge of a supernova and has already blown off an irregular-shaped nebula of gas at 160,000 kilometers per hour. Melnick 42, a star 2.5 million times brighter than the Sun and a

short-lived supernova, is blowing away material at 2,900 kilometers per second, which equates to losing the mass of four Suns every million years. The brighter the star, the faster it burns out. In 1997, hidden by dust clouds, HST found the brightest star in the Milky Way. It is 25,000 light-years away and is as bright as 10 million Suns.

In April, 2000, HST's WFPC2 witnessed an unusual planetary nebula, NGC, 5,400 light-years from Earth. This nebula, which glows like a big eye, was ejected several thousand years ago from a hot star in the constellation Aquila. The hottest area is a circular ring around the center of the stellar remnant; cool gas tends to lie in long streamers pointing away from the central star with a tattered-looking ring at the outer edge. The origin of the cooler clouds is uncertain, but the shapes are affected by the radiation and stellar winds that emanate from the hot (140,000° Celsius) star. The nebula is expanding at 25 miles per second and its diameter is 600 times the diameter of the solar system. HST also found that Betelgeuse, a variable star thought to be a binary system, has a hot spot, which could cause the brightness to change during rotation. The telescope is separating true binary stars from those that, by their interactions, dump material from one to the other. HST has been able to distinguish individual stars in global clusters of 300,000 stars. Once thought to be only old star groups, these clusters were revealed to harbor hot young stars. The space telescope surprisingly observed that evolution of stars could be changed by close interactions with other stars.

Many of the theories of the life and death of stars have been observed, analyzed, and proven by HST. Other observations continue to lay new groundwork. For instance, scientists, who thought red giant stars were round, were surprised to see photographs showing myriad shapes from symmetrical to chaotic. Other discoveries include information that red dwarfs are not as numerous as once thought and that ice bodies like comet heads the size of Earth are found in planetary nebulae.

HST found that black holes, merely a theory in the 1980's, are quite common. In 1994 HST mea-

*One of the many dramatic and detailed images taken by Hubble, this of the spiral galaxy NGC 4414. (NASA)*

sured the velocity of gas surrounding a black hole and discovered a supermassive disk of 3 billion suns concentrated to the size of the solar system. One sun that is 500 million times the mass of the Earth's Sun has been found, and many galaxies may have one or more at their central cores. Some black holes are dormant, and some may swallow up star after star or whole galaxies. Theories suggest that the larger the galaxy, the more massive the central black hole.

Theories regarding the origin of quasars, very distant and very bright radio sources, are up for scrutiny. The latest theories suggest their derivation may be a black hole. HST has gathered data that have stumped the astronomers because these very strong radio sources can be quite different from each other, tend to evolve, and were much brighter in the past. Visible light from a neutron star was seen for the first time by HST.

Some stars are chemically peculiar, showing spectra of mercury 204, while others show very high values for gold and platinum in their top layers. HST found heavy metals such as krypton and lead in the spaces between nearby stars. HST has also observed white dwarfs at 620,000 kelvins, some of which have highly ionized metals like iron on the surface. All this gives a surprising picture of the past but poses many questions concerning the formation of the heavier elements. Because the heavy metals should sink, new ideas on the formation of stars need to be considered. Data from the Andromeda galaxy, the nearest major spiral galaxy, showed helium-burning stars, not just hydrogen-fusing ones.

Many scientists believe that at least 90 percent of the matter in the universe is hidden in an exotic "dark" form, which has not yet been detected by Earth instrumentation. This "dark matter," if sufficient, could cause the universe to collapse instead of allowing it to expand forever. In the 1970's astronomers determined that stars in the middle of a galaxy moved as fast as those on the edges, suggesting something massive, possibly dark matter, must be changing their expected motion. One theory is that some protostars do not quite make it to incandescence and, though the star eventually collapses, not enough energy is present to radiate sufficient light to be seen by normal means. These cool small stars, called brown dwarfs, have been observed by HST, proving their existence. In 1994 HST ruled out red dwarfs as dark matter because it could resolve the images to a clear point.

HST captured galactic collisions between supernovae and several billion new stars. The resulting huge bursts of fireball-like gamma radiation are yet to be explained. HST found that star-forming spiral galaxies were more prevalent in the early universe, and interacting galaxies that merge and disrupt each other are more important than previously thought. HST detected ordinary changes in the formation of galaxies by gravitational attraction and flattening due to rotation. Explosions in galaxies, the normal formation and evolution of stars, and the short lives of massive stars were witnessed. The telescope saw stars recycle their particles into the interstellar dust, where new stars will form.

HST took a census of galaxies in a small region of the sky near the Big Dipper for ten days. The field of view contained three thousand galaxies; a magnitude of thirty was the faintest. Astronomers determined that the peak period of star formation was eleven billion years ago and was twelve times the present rate.

Clusters of galaxies appear to form where hydrogen filaments intersect. In recent years supercomputer models predicted an intricate web of gas filaments where hydrogen is concentrated along vast, chainlike structures. Hydrogen clouds flowing along these chains should collide and heat up—squelching the formation of more galaxies in the hottest regions of space. If this is true, then star birth was more abundant in the early universe, when the hydrogen was cool enough to coalesce. An oxygen "tracer," probably created in the core of an exploding star through nuclear fusion, was seen by the HST's Imaging Spectrograph. HST found the spectral "fingerprints" of intervening oxygen superimposed on a quasar's light. (Astronomers detected the highly ionized oxygen by using the light of a distant quasar to probe the invisible space between the galaxies, not unlike shining a flashlight beam through a fog.) HST was able to see across billions of light-years of space and observed four separate filaments of the invisible hydrogen laced with the oxygen.

Scientists wondered what happened to the huge amount of hydrogen that was formed in the first few minutes after the Big Bang and which should make up half of the matter in the universe. HST came to the rescue by finding the presence of highly ionized oxygen between the galax-

ies, implying that huge quantities of hydrogen in the universe are so hot that they escaped detection by normal observational techniques. HST has determined an accurate deuterium-to-hydrogen ratio in interstellar space used to determine the density of the universe and has detected twelve intergalactic hydrogen clouds near the Milky Way. HST caused astronomers to upgrade their estimate of the number of galaxies in the night sky from 50 billion to a staggering 200 billion. In fact, there may be more than 20,000 billion galaxies in all. In December, 1996, the Deep Field Imaging Telescope took pictures of galaxies that had never been seen before and that had needed 5 billion years for their light to reach Earth. Some galaxies "seen" may be between ten and fourteen billion years old, around the commonly accepted time that the universe began. Astronomers were shocked that galaxies were formed so early in the history of the universe.

*Gaseous pillars made of dense intersteller hydrogen gas and dust, remnants of hot, massive stars that are being created in the Eagle nebula.* (NASA CORE/Lorain County JVS)

HST has detected Cepheid variables in other galaxies and uses them to determine distances. By counting white dwarfs and examining their color, astronomers can tell the temperature of the stars and their rate of cooling. Thus, they can calculate the age of the universe, now estimated at 14.8 billion years. Yet, additional information gathered by the orbiting HST is mystifying scientists. The data seem to show that our universe may be only eight billion years old, rather than fifteen billion years, as commonly thought. This is a huge discrepancy, which would make some stars and galaxies older than the universe itself. Much more research is needed to begin to understand this observation.

One of four science instruments aboard NASA's Hubble Space Telescope suspended operations in August, 2004. The instrument, called the Space Telescope Imaging Spectrograph (STIS), was installed during the second Hubble servicing mission in 1997 and was designed to operate for five years. It either met or exceeded all its scientific requirements. Hubble's other instruments—the Near Infrared Camera and Multi-Object Spectrometer (NICMOS), the Advanced Camera for Surveys, and the Wide Field/Planetary Camera 2—were all operating normally.

Hubble teamed with ground-based telescopes to observe exploding stars in galaxies whose light was emitted when the universe was half its present age. The preliminary result, if confirmed, will be one of the most important scientific discoveries of our time—that the expansion of the universe is accelerating, driven by an unknown force. The telescope's exquisite images of dying stars help scientists understand the death process and how it is influenced by each star's specific circumstances. Only Hubble can chronicle the spectacular changes as the blast debris expands over time.

Hubble revealed stunning views of the northern and southern lights on Jupiter, Saturn, and Jupiter's moon Ganymede, as well as imagery of the dynamic electrical interactions between Jupiter and its satellite Io. Until Hubble, scientists could not determine if mysterious, intense bursts of gamma rays originated in our own galaxy, far across the universe, or somewhere in between. The telescope traced these bursts to the outskirts of faint, distant galaxies in the early universe.

April 24, 2005, marked the fifteenth anniversary of the Hubble Space Telescope's launch. No other telescope in space has had a similar impact on the field of astronomy.

**Context**

Galileo constructed his first telescope in the seventeenth century, opening up a whole world of controversy and creating the science of astronomy. Sixty years later, Sir Isaac Newton invented the first reflecting telescope. In the 1920's Edwin P. Hubble used a 100-inch reflector on Mount Wilson to study the scale of the universe, to show that the universe was expanding and to prove that there are other galaxies beyond our own. Karl G. Jansky of Bell Labs, who in 1931 detected radio waves emanating from the Milky Way, was largely ignored.

A space telescope was first suggested in 1923 by Hermann J. Oberth, the German pioneer in rocketry, and again in 1946 by astronomer Lyman Spitzer, Jr. A space telescope project was envisioned in 1952 and 1966 based on the facts that in outer space stars do not twinkle and radiation is minimal with no interference from pollution and city lights. By 1968 an ad hoc committee was formed by NASA to determine how a telescope could be used in astronomical research. A more complete discussion took place in 1974. The first plan sent to Congress by NASA was rejected as too expensive, but the United States managed to get its European counterparts interested. The project was downscaled, and in 1977 an enthusiastic astronomical committee gave full support to the HST project. Preliminary design and scientific proposals were completed in 1979, the critical design review was in 1980, and scientific contracts were awarded. Planetary scientists soon realized that the Venus-orbiting Magellan and the flybys, Pioneer and Voyager, were sending only snapshots. They wanted constant observation of the planets to map surface changes and were excited by the prospects of the HST.

HST was to be the first of four orbiting observatories, of which three have been launched. It is expected to last until 2011. The Compton Gamma Ray Observatory was launched in 1991, and the Chandra X-Ray Observatory was launched on board *Columbia* during the STS-93 mission in July, 1999.

The fourth, the Spitzer Space Telescope (formerly SIRTF, the Space Infrared Telescope Facility), was launched into space by a Delta rocket from Cape Canaveral, Florida, on August 25, 2003. During its 2.5-year mission, Spitzer will obtain images and spectra by detecting the infrared energy, or heat, radiated by objects in space between wavelengths of 3 and 180 microns (1 micron is one millionth of a meter). Most of this infrared radiation is blocked by the Earth's atmosphere and cannot be observed from the ground.

In 2004, NASA eliminated all space shuttle flights not directly supporting the International Space Station as a result of budgetary cutbacks. These missions included scheduled servicing missions to the Hubble Space Telescope.

The James E. Webb Space Telescope (JWST) will be the successor to Hubble. It is a large, infrared-optimized space telescope scheduled for launch in August, 2011. JWST is designed to study the earliest galaxies and some of the first stars formed after the Big Bang. These early objects have a high red shift from Earth's vantage point, meaning that the best observations for these objects are available in the infrared range of the electromagnetic spectrum. JWST's instruments will be designed to work primarily in the infrared, with some capability in the visible range.

JWST will have a large mirror, 6.5 meters (20 feet) in diameter, and a sunshield the size of a tennis court. Both the mirror and sunshade are unable to fit onto the rocket fully open, so both will fold up and open only when JWST is in outer space. JWST will reside in an L2 Lissajous orbit, about 1.5 million kilometers (1 million miles) from Earth. L2 is one of five Lagrangian points, locations where the pulls of the Earth and Sun combine to form a point at which a third body of negligible mass—a satellite—would be stationary relative to the two bodies. Precisely achieving this point is difficult, so a special Lissajous orbit (a periodic orbit in which there is a combination of planar and vertical components) called a "halo" orbit will be used. A halo orbit is one in which a spacecraft will remain in the vicinity of a Lagrangian point, following a circular or elliptical loop around that point.

Mission goals are for it to determine the shape of the universe, explain galaxy evolution, understand the birth and formation of stars, determine how planetary systems form and interact, determine how the universe built up its present chemical/elemental composition, and probe the nature and abundance of dark matter.

**See also:** Asteroid and Comet Exploration; Cassini: Saturn; Funding Procedures of Space Programs; Galileo: Jupiter; Hubble Space Telescope; Hubble Space Telescope: Servicing Missions; Planetary Exploration; Telescopes: Air and Space; Viking Program.

### Further Reading

International Astronomical Union Colloquium. *Scientific Research with the Space Telescope 54.* Cambridge, England, and Marshall Space Flight Center, Ala.: Author, 1979. This book summarizes the work of fourteen invited lecturers of observations with the space telescope and includes comments as given by the lecturers and discussion of all nontechnical questions. The discussion is somewhat technical but a good resource for the early space telescope research projects and planning.

Kerrod, Robin. *Hubble: The Mirror on the Universe.* Richmond Hill, Ont.: Firefly Books, 2003. The book covers the universe in six sections: "Stars in the Firmament," "Stellar Death and Destruction," "Gregarious Galaxies," "The Expansive Universe," "Solar Systems," and "The Heavenly Wanderers."

Leverington, David. *New Cosmic Horizons: Space Astronomy from the V2 to the Hubble Space Telescope*. New York: Cambridge University Press, 2001. *New Cosmic Horizons* tells the story of space-based astronomy since World War II. Leverington examines how politics in the United States, the Soviet Union and Russia, and Europe modified their space astronomy programs.

Livio, Mario, Keith Noll, Massimo Stiavelli, and Michael Fall, eds. *A Decade of Hubble Space Telescope Science*. Space Telescope Science Institute Symposium series. New York: Cambridge University Press, 2003. Proceedings of the Space Telescope Science Institute Symposium, held in Baltimore, Maryland, from April 11-14, 2000. Representing some of the most important scientific achievements of the Hubble Space Telescope in its first decade of operation, this collection of review articles is intended for researchers and graduate students.

Petersen, Carolyn Collins, and John C. Brandt. *Hubble Vision: Further Adventures with the Hubble Space Telescope*. 2d ed. New York: Cambridge University Press, 1998. Contains numerous excellent pictures of the latest findings of the space telescope. It fills in the scientific background that many readers might lack—those interested in amateur astronomy will gain knowledge from the remarkable discoveries. The authors explain the background leading up to each series of celestial observations and what was learned from previous missions. They explain what each study was designed to look for, how the data were interpreted, what the data mean, and the resultant knowledge. The authors present the information gathered by HST in the proper perspective of what is currently known about astronomy.

Wilkie, Tom, and Mark Rosselli. *Visions of Heaven: The Mysteries of the Universe Revealed by the Hubble Space Telescope*. London: Hodder & Stoughton, 1998. An easy-to-read and enjoyable book about some past history of telescopes, their observations, the history of HST, and its recent discoveries. Much information about the planets and the universe are tracked from the very first telescope. The book is a fine overview of astronomy and how the HST has helped to prove some theories and dispel others. Contains incredible pictures from the space telescope with excellent color plates.

*Judith Belsky Farrin, updated by Russell R. Tobias*

# Hubble Space Telescope: Servicing Missions

*Date:* Beginning 1993
*Type of program:* Piloted spaceflight

*The Hubble Space Telescope (HST) was designed to be serviced routinely by spacewalking astronauts. Shuttle missions to maintain and repair the telescope, referred to as Hubble servicing missions, became imperative to correct unexpected problems with the telescope's optical prescription and malfunctions of key systems. HST servicing missions clearly demonstrated the utility of trained humans in space to further the cause of basic science.*

## Summary of the Missions

The Hubble Space Telescope (HST) was deployed in Earth orbit by mission STS-31 on April 24, 1990. The following day, Mission Specialist Steven A. Hawley released the telescope from *Discovery*'s Remote Manipulator System (RMS) arm after minor problems with the Observatory's solar arrays were fixed. The first images taken by HST came in early May. Initially scientific reaction was enthusiastic, but it soon became painfully obvious that the telescope's primary mirror had been ground precisely but incorrectly, rendering the Observatory incapable of performing the frontline research for which it had been designed.

An intense characterization of HST's spherical aberration was started, and it was determined that corrective optics could be incorporated into a second-generation Wide Field/Planetary Camera 2 (WFPC2) and a unit called the Corrective Optics Space Telescope Axial Replacement (COSTAR). Astronauts could exchange COSTAR for the High Speed Photometer (HSP), a scientific package that had to be sacrificed to insert COSTAR into HST's light path and properly restore focus for the remaining scientific packages. WFPC2 would replace the original instrument. HST's Goddard High Resolution Spectrograph (GHRS) needed some repairs but would remain in the telescope after the

first servicing mission. The task of repairing HST fell to STS-61 Commander Richard O. Covey, Pilot Kenneth D. Bowersox, and Mission Specialists Thomas D. "Tom" Akers, Jeffrey A. Hoffman, F. Story Musgrave, Kathryn C. Thornton, and Claude Nicollier.

STS-61, the first servicing mission (SM1), got under way on December 2, 1993, when *Endeavour* lifted off at 09:26:59.983 Coordinated Universal Time (UTC, or 4:27 A.M. eastern standard time). On December 4, Commander Covey parked *Endeavour* just 11 meters from HST so that Mission Specialist Nicollier could grapple the telescope with *Endeavour*'s RMS arm and then berth it upon a special restraint within the payload bay for subsequent repair work.

Extravehicular activity (EVA) teams alternated consecutive days spent out in *Endeavour*'s payload bay, with Musgrave and Hoffman performing spacewalks one, three, and five, and Thornton and Akers performing spacewalks two and four. During EVA 1, new rate sensor units were installed inside HST's aft shroud, and electronic control units and fuses were changed. During EVA 2, HST's solar arrays were replaced with new units. During EVA 3, the old WFPC was exchanged for the new one, and new magnetometers were installed atop the telescope's barrel. During EVA 4, HSP was removed and replaced by COSTAR. During EVA 5, solar array drive electronics units were replaced and repairs were made to GHRS.

HST's new solar arrays unfurled at the end of EVA 5, and then on December 10 the astronauts released the telescope back into its own independent orbit. *Endeavour* touched down at the Kennedy Space Center at 05:25:37 UTC (12:26 A.M. eastern standard time) on December 13.

Five weeks after STS-61, HST took its first post-repair images, and the results proved that corrective designs and astronauts' efforts had overcome the telescope's unfortunate spherical aberration. HST was now capable of fulfilling its original scientific goals.

Whereas SM1's purpose was essentially one of repair, plans for the second servicing mission (SM2) included scientific equipment change-outs and systems upgrades as well as repairs. COSTAR and WFPC2 had to remain, but it was time to provide HST with infrared detection systems in exchange for two experiment payloads that had fulfilled their scientific objectives. On STS-82 the Faint Object Spectrograph (FOS) would be replaced by the Near Infrared Camera and Multi-Object Spectrometer (NICMOS), and the Space Telescope Imaging Spectrograph (STIS) would be substituted in place of GHRS. Also since STS-61, HST developed a few problems needing attention. A Fine Guidance Sensor suffered mechanical wear and needed replacement. Magnetometers installed on STS-61 required new thermal covers, and solar array drive electronics needed to be replaced.

STS-82, the second servicing mission (SM2), got under way on February 11, 1997, at 08:55:17.017 UTC (3:55 A.M. eastern standard time) when *Discovery* lifted off into clear skies carrying Commander Kenneth D. Bowersox, Pilot Scott J. Horowitz, and Mission Specialists Mark Lee, Gregory J. Harbaugh, Steven L. Smith, Joseph R. Tanner, and Steven A. Hawley into orbit. Two days into the flight, Hawley used the RMS arm to grab what he once had deployed, putting it down on the same payload fixture upon which it had been serviced more than three years earlier.

EVA teams alternated consecutive days spent out in *Discovery*'s payload bay, with Smith and Lee performing spacewalks one and three, as well as a contingency fifth EVA, and Tanner and Harbaugh performing spacewalks two and four. EVA work began just prior to midnight on February 13 and continued with one each day over the subsequent four mission shifts.

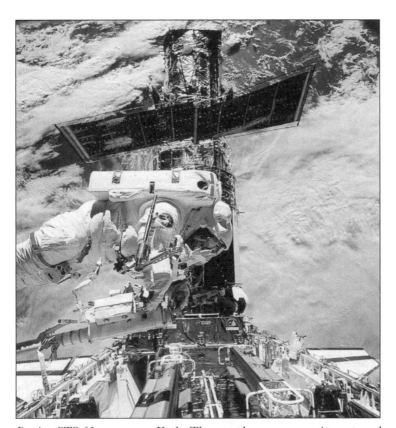

*During STS-61, astronaut Kathy Thornton hovers near equipment used to service the Hubble Space Telescope.* (NASA)

During EVA 1, GHRS was removed from HST and, after it was stowed for the return to Earth, the available space in the telescope was outfitted with STIS. Also FOS was removed and in its place NICMOS was inserted into HST.

During EVA 2, Tanner and Harbaugh observed cracks in some of HST's multiple layer insulation (MLI) and discolorations on EVA handrails. Fearing that flaking of insulation and paint could produce particulates potentially harmful should they infiltrate into the telescope's delicate optics, they devised temporary repairs to some of these blankets. The main activities, however, included Fine Guidance Sensor (FGS) replacement, engineering, and science tape recorder replacement, photograph documentation of present solar-array conditions, and Optical Control Electronics (OCE) change out. While EVA 2 was under way, Bowersox and Horowitz performed a maneuver to gently boost HST's orbital altitude.

During EVA 3, one data interface unit was exchanged for one that had failed shortly after STS-61, a new solid-state recorder was installed, one of four Reaction Wheel Assemblies (RWAs) was exchanged, and more MLI inspections were made in preparation for creating makeshift patches for some of the worst areas.

The MLI situation warranted a fifth contingency EVA so that repairs could be attempted. Meanwhile the astronauts began constructing patches during periods between spacewalks.

During EVA 4, a solar-array drive electronics unit was replaced, covers were placed atop the magnetometers installed on STS-61, and some initial MLI patches were put in place. Virtually the entire EVA 5 time line, apart from cleanup and storing of tools and equipment, was devoted to MLI patch attachment. The STS-82 EVA team tied STS-61's record for most spacewalks on a single shuttle mission.

These four spacewalking repairmen worked a total of 33 hours and 11 minutes spacewalking in the payload bay, leaving HST ready to perform research that would revolutionize infrared astronomy textbooks.

Hawley released HST back into its own independent orbit on February 19 at a record shuttle altitude: 619 kilometers. Two days later, *Discovery* touched down at Kennedy Space Center (KSC), completing STS-82 with a night landing.

In the weeks following STS-82, a complete checkout of HST's new equipment proceeded without major problems. Discrepancies in sensitivities of the new infrared detectors developed, but fortunately, everything could be addressed and they could still accomplish their scientific goals. Initial NICMOS and STIS science operations began in March, 1997. Orbital verification periods concluded by summer and intensive infrared studies commenced.

NICMOS incorporated a new cryogenic cooling design. Rather than liquid helium, NICMOS employed a solid block of nitrogen. Unfortunately a thermal pathway developed, permitting heat to slowly consume that solid nitrogen and degrade NICMOS's cooling system. With the nitrogen gone, NICMOS's sensor could no longer collect useful data, so an accelerated NICMOS program was devised to maximize its use at the expense of WFPC2 while coolant still remained. Nitrogen ran out well before the scheduled third servicing mission (SM3), shutting down NICMOS. Fortunately, realizing that that eventuality had been inevitable, HST managers had already initiated an engineering project to devise an alternate cooling system that astronauts could attach to NICMOS to restore its scientific capability.

In 1998 it became apparent that HST's gyroscopes were wearing out faster than expected. The telescope was outfitted with six units, three of which were necessary to provide the stable attitude control needed for scientific work. Two had failed since STS-82, and in early 1999 a third began displaying abnormalities. NASA split the original Servicing Mission 3 into two parts because of the large number of activities that needed to be accomplished. The scale of the third servicing mission (now called "SM3A") was cut back, with some originally scheduled tasks such as NICMOS repair delayed to a fourth visit (SM3B). Nevertheless, the SM3A schedule was advanced. That became fortuitous: in late April a third unit failed, eliminating any gyroscopic redundancy. STS-103 was scheduled for an October launch, but shuttle safety concerns after STS-93 grounded the fleet until December. On November 13 a fourth gyroscope went down, forcing HST into a deep safe mode in which science could not be performed.

SM3A, STS-103, got under way at 00:49:59.986 UTC on December 20, 1999 (7:50 P.M. eastern standard time on December 19), when *Discovery* lifted off, carrying Commander Curtis L. Brown, Jr., Pilot Scott J. Kelly, and Mission Specialists C. Michael Foale, John M. Grunsfeld, Claude Nicollier, Steve Smith, and Jean-François Clervoy into orbit. Two days later Commander Brown parked *Discovery* within 11 meters of HST, and Clervoy grappled it with the RMS arm and berthed it in the bay for the EVA work.

EVA teams alternated consecutive days spent out in *Discovery*'s payload bay, with Smith and Grunsfeld performing spacewalks one and three and Foale and Nicollier performing spacewalk two. Originally, four spacewalks had been incorporated into STS-103's flight plan, but in order to have the mission end prior to year's end and avoid "Year 2000" (Y2K) issues, NASA had to drop one that had largely involved MLI repairs.

During EVA 1, the payload bay was configured for STS-103's repair efforts, new gyroscopic packages called rate sensor units were exchanged for the old defective gyroscopes, some get-ahead work on NICMOS was performed, and voltage/temperature improvement kits were installed on the telescope. During EVA 2, HST's old computer was replaced by a 486-based microprocessor, and a new FGS was installed. EVA 3 was performed on Christmas Eve, and during this spacewalk optical control enhancement units were installed to improve the newly installed FGS's effectiveness, an old solid-state recorder was exchanged for a new one, and some new MLI blankets were attached to problem areas. STS-103's spacewalks entered the history books as NASA's second, third, and fourth longest. Total EVA time on this mission amounted to 24 hours and 33 minutes.

HST was returned to its own independent orbit on Christmas Day. NICMOS had to wait for a fourth servicing mission, but the rest of HST was now back in business. Several weeks of verification work under ground control remained before science could resume, but in the meanwhile, *Discovery* returned to Earth on December 28, landing in darkness at KSC at 00:00:47 UTC.

During Servicing Mission 3B (SM3B), a new science instrument was installed: the Advanced Camera for Surveys (ACS). *Columbia* and its crew of seven began the twelve-day STS-109 mission at 11:22:02.021 UTC on March 1, 2002. The STS-109 astronauts performed five spacewalks in five consecutive days to service and upgrade Hubble. STS-109 mission specialists John M. Grunsfeld and Richard M. Linnehan conducted the mission's first, third, and fifth EVAs. Mission specialists James H. Newman and Michael J. Massimino performed the second and fourth spacewalks. Accomplishments of the spacewalks included the installation of new solar arrays, a new camera, a new power control unit, a reaction wheel assembly, and an experimental cooling system for the NICMOS unit.

**Contributions**

Although HST servicing missions were themselves not scientific research missions, the target of their repair work proved to be one of the most productive space-based research tools ever devised and deployed in orbit. Its unfortunate initial spherical aberration provided a dramatic opportunity to prove the worth of humans in space. Had HST not been designed for intervention and periodic change out of individual scientific packages, it would have been cheaper, but it would also have turned out to be a total failure. Because of talented, dedicated individuals, HST was transformed from the butt of jokes to an awe-inspiring observatory producing scientific data that revolutionized our understanding of the universe.

The HST was designed in a modular fashion to allow servicing in the sense that astronauts could repair certain malfunctions and could exchange scientific equipment as needed to support the overall scientific objectives well beyond the telescope's deployment on shuttle mission STS-31. As the telescope was originally slated to support science for fifteen years, it was anticipated that several dedicated space shuttle servicing missions would be required over HST's orbital lifetime. However, what was never anticipated was the need to dispatch a shuttle mission to salvage the telescope's initial mission. Having bypassed a ground-based optical test that would have discovered the telescope's precise but incorrect mirror curvature, it was not until HST started collecting data that it became apparent that HST, as launched, could not fulfill the tremendously ambitious scientific goals for which it had been designed.

### Context

STS-61, the first servicing mission (SM1), essentially provided a salvage mission rather than a servicing mission as such. That mission had to install a sophisticated optical fix to provide HST with the necessary resolution for its scientific harvest to in-

clude landmark observations and discoveries. In essence, the telescope's very future hinged upon the success of STS-61. Over the course of five spacewalks, rotating teams of two spacewalking astronauts carefully inserted new instruments that restored HST's scientific potential.

STS-82, the second servicing mission, represented the first HST-dedicated mission of the servicing variety that program managers had originally anticipated. Scientific equipment that had gathered highly valuable data since installation either preflight or on STS-61 was replaced with new detectors. One particularly was meant to expand HST's investigations into the infrared region of the electromagnetic spectrum. Also some HST systems were upgraded, and some temporary patchwork repairs made to the telescope's thermal blankets.

STS-103, the third servicing mission, represented a blending of the identities of purpose displayed by the first two servicing missions. It had been placed in the shuttle schedule as an opportunity to change out scientific equipment and make some repairs, but its immediacy increased when HST suffered unexpected failures in its gyroscope pointing system. Indeed, for a few weeks while the shuttle fleet was grounded for safety inspections, HST entered a safe mode wherein no scientific data could be collected, because too many gyroscopes had failed to support accurate observations. In that sense, STS-103 shared aspects of STS-61 in that it had to restore HST to scientific capability.

*On shuttle mission STS-82, astronauts Gregory Harbaugh and Joseph Tanner replace the Hubble Space Telescope's Magnetic Sensing System (MSS) protective caps with new, permanent covers. (NASA)*

During STS-109, the fourth servicing mission, the Advanced Camera for Surveys (ACS) was installed on Hubble in March, 2002, replacing the Faint Object Camera. Consisting of three cameras, the ACS is providing an unparalleled view of the universe. The most widely used camera consists of two charge-coupled devices

(CCDs) with a total of 16 million pixels, which can be used with more than a dozen different filters to isolate different colors ranging from the blue through optical to the near infrared regions of the electromagnetic spectrum. A second camera with 1 million pixels provides the highest spatial resolution currently available on HST, while the third camera operates in the far ultraviolet. The ACS has twice the field of view and a higher sensitivity than the older main camera, the Wide Field/Planetary Camera 2, which is still in operation.

STS-109 accumulated a total of 35 hours and 55 minutes of EVA time. Through STS-109, eighteen spacewalks by fourteen different astronauts were conducted during four servicing missions for a total of 129 hours, 10 minutes.

In response to budgetary cuts in the wake of the STS-107 accident, NASA eliminated all space shuttle flights not directly supporting the International Space Station. These included scheduled servicing missions to the Hubble Space Telescope. NASA had planned to visit Hubble one last time in 2006 to change out instruments and replace its gyroscopes with the intent of keeping the telescope in service until at least 2011, when its heir apparent, the James E. Webb Space Telescope, is expected to launch. Scrapping the final servicing mission raised the likelihood that Hubble would fail before Webb was in orbit.

Because the planning for and development of the fourth servicing mission was very far along and the scientific promise of such a mission was extraordinarily high, NASA decided to consider a novel idea to service Hubble without astronauts—a completely robotic servicing mission. Not only would the mission benefit Hubble by extending its life and enhancing it scientifically, but also such a mission would advance robotic techniques in space and would also have far-reaching implications for NASA's ever-expanding frontiers of exploration.

NASA has not yet decided whether to pursue a robotic servicing mission to Hubble. NASA wishes to develop the still-new mission concepts and designs further before making a final decision about whether the mission is feasible or advisable. Should it be decided to proceed with the implementation of such a mission, the targeted launch date would be at the end of 2007 or early in 2008.

**See also:** Hubble Space Telescope; Hubble Space Telescope: Science; Space Shuttle Flights, 1992; Space Shuttle Flights, 1993; Space Shuttle Mission STS-61; Space Shuttle Flights, 1997; Telescopes: Air and Space.

### Further Reading

Chaisson, Eric J. *The Hubble Wars: Astrophysics Meets Astropolitics in the Two-Billion-Dollar Struggle over the Hubble Space Telescope.* New York: HarperCollins, 1994. This book provides thorough insight into the management of HST program development; the political aspects of such a large scientific endeavor; and the task of identifying the nature of, cause of, and blame for the incorrect optics HST had when launched. Includes a brief description of STS-61's repair efforts in space.

Grunsfeld, John M. "Healing Hubble." *Sky and Telescope,* April, 2000, 36-42. Written by one of the STS-103 spacewalkers, this article describes Grunsfeld's personal experiences on the third servicing mission to the HST. Includes numerous photographs taken during the mission.

Kerrod, Robin. *Hubble: The Mirror on the Universe.* Richmond Hill, Ont.: Firefly Books, 2003. The book covers the observable universe in six sections: "Stars in the Firmament," "Stellar Death and Destruction," "Gregarious Galaxies," "The Expansive Universe," "Solar Systems," and "The Heavenly Wanderers."

Smith, Robert W. "HST in Orbit." *Sky and Telescope*, April, 2000, 28-34. Describes HST's first decade of science and the data's impact upon astronomical advancement. Provides numerous color photographs taken by HST.

_____. *The Space Telescope: A Study of NASA, Science, Technology, and Politics.* New York: Cambridge University Press, 1989. Written prior to HST's launch, this book provides a scientific and management history of the development of this large orbiting astronomical observatory. Missing is any discussion of the problems encountered shortly after HST's orbital deployment, as this work was published prior to STS-31.

*David G. Fisher, updated by Russell R. Tobias*

# Huygens Lander

*Date:* October 15, 1997, to January 14, 2005
*Type of spacecraft:* Planetary exploration

*Titan, the largest moon in the Saturn system and the second largest moon in the solar system, is shrouded from telescopic observation in electromagnetic wavelengths visible to the human eye. The only ways to determine surface information are to use imaging radar and a probe that enters the atmosphere and proceeds to the surface. The former is the job of the Cassini orbiter, and the latter is the job of the Huygens probe, which Cassini transported to the Saturn system.*

### Key Figures

*Robert Mitchell*, Cassini program manager at the Jet Propulsion Laboratory (JPL)

*Earl H. Maize*, Cassini deputy program manager

*David Southwood*, science director, European Space Agency (ESA)

*Jean-Jacques Dordain* (b. 1946), ESA director general

*Jean-Pierre Lebreton*, ESA Huygens mission manager

*Alphonso Diaz*, associate administrator of science at NASA

*Claudio Sollazzo*, ESA Huygens operations manager

*Martin Tomasko*, principal investigator for the Descent Imager and Spectral Radiometer

*Marcello Fulchignoni*, principal investigator for the Huygens Atmospheric Structure Instrument

*Hasso Niemann*, principal investigator for the Gas Chromatograph Mass Spectrometer

*Guy Israel*, principal investigator for the Aerosol Collector and Pyrolyser

*Michael Bird*, principal investigator for the Doppler Wind Experiment

*J. C. Zarnecki*, principal investigator for the Surface Science Package

### Summary of the Mission

Shaped like a flattened gumdrop, the Huygens probe was outfitted with 48 kilograms of science instruments housed within a protective thermal shield. These included the Huygens Atmospheric Structure Instrument (HASI), the Gas Chromatograph Mass Spectrometer (GCMS), the Aerosol Collector and Pyrolyser (ACP), the Descent Imager and Spectral Radiometer (DISR), the Doppler Wind Experiment (DWE), the Surface Science Package (SSP), and the Accelerometer subsystem (ACC).

HASI included sensors to record the electrical properties of Titan's atmosphere. Accelerometers would measure movements along three indepen-dent orthogonal directions during descent. HASI data would provide information about wind speed and atmospheric density, and it would measure electrical permittivity and conductivity in the atmosphere and near the surface. DWE was designed to provide data about wind speed during atmospheric entry. DISR incorporated imaging and radiation sensing equipment. This experiment would produce the visual and infrared pictures of the surface scientists so greatly anticipated. A calibrated light source would provide illumination down close to the surface. Scattering of sunlight by aerosols in the atmosphere would be measured. The imaging system was designed to provide horizontal views at surface level and also to photograph the underside of the cloud deck. The radiometer portion of the DISR instrument would measure the upward and downward radiation budget in the atmosphere.

GCMS was a device capable of determining chemical composition of the atmosphere. Samples were to be collected at different altitudes. This experiment was also designed to provide compositional information about the surface atmospheric environment in the event of a safe landing. During entry, samples altered by heating would be passed through the Aerosol Collector and Pyrolyser. Samples sent to the ACP would be vaporized and complex organic molecules would be decomposed so that these pyrolyzed products could be sent to the GCMS for analysis. The SSP incorporated sensors to measure a variety of physical properties of the surface. It would record the probe's orientation at impact, note the deceleration profile leading up to impact, and ascertain whether the surface was liquid or solid. If liquid, the surface would have waves, and the movement of the probe would provide information about wave speed and wave height.

Structurally, the Huygens probe consists of a spin-eject device, a back cover, an aft cone, a top platform, an experiment platform, a fore dome, and a front shield. Attached to the side of the Cassini orbiter by the Probe Support Equipment ring, the Huygens probe remained inactive during the trip from Earth to Saturn except for biannual functionality checks. The probe was released on December 24, 2004; shortly thereafter, the Deep Space Network picked up a signal indicating that all active onboard systems were in proper condition. At the time, Cassini was on a collision course with Titan. The orbiter had to maneuver out of that trajectory on December 27 so that it would be in position to collect the data streaming from Huygens when it entered Titan's atmosphere. Actually, the maneuver set up an encounter with Iapetus that would then fine-tune Cassini's orbit so that it was in proper position for the Huygens encounter.

The National Radio Astronomy Observatories' Robert Byrd Green Bank Telescope (GBT), in Green Bank, Pocahontas County, West Virginia—the largest steerable radio telescope on Earth—picked up a signal early in the morning local time on January 14, 2005, which indicated that Huygens' onboard timer had activated properly as the probe neared Titan's outer atmosphere. No science data were included in the signal beamed directly to Earth; science data would be transmitted to the Cassini orbiter for relay back to Earth after the entry and landing probe had stopped transmitting. The data would have to be recorded on board Cassini's solid-state recorder because the orbiter had to be turned away from Earth in order to receive the data from Huygens.

Special ceramics on the front shield, materials similar to those on a space shuttle orbiter, managed the thermal load as the probe slammed into Titan's thick atmosphere and decelerated. A pilot parachute deployed, followed one minute later by initiation of transmissions to Cassini. The drogue parachute deployed just as the 100-meter GBT received its first signal from the probe. It took Huygens nearly 148 minutes to pass through Titan's atmosphere and impact upon the surface at 16.2 kilome-

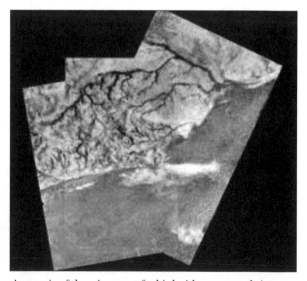

*A mosaic of three images of a high ridge area and river system on Saturn's moon Titan, taken by the Huygens Descent Imager and Spectral Radiometer.* (NASA/JPL/ESA/University of Arizona)

ters per hour. Loading upon the probe reached fifteen times Earth-normal gravity. A penetrometer on the base of the probe deployed fifteen centimeters into some rather unusual surface material. Scientists initially appraised the soil's mechanical consistency as similar to wet sand or claylike materials. Signals from the probe continued to be received well after the time it was anticipated to be able to last, either afloat or on a surface.

### Contributions

Cassini entered Saturn orbit on July 1, 2004. An initial close approach to Titan occurred within 36 hours of orbital insertion. Titan remained a mystery despite the fact that Cassini had produced many images of the large moon during its early close encounters. It had been anticipated by many that liquids would be found on the surface of Titan, but Cassini data thus far had yet to provide definitive proof of liquid hydrocarbons.

An initial press conference held at the Space Operations Control Center of the European Space Agency (ESA) only a few hours after the Huygens encounter revealed that only channel B data had been received in total from the probe. Fortunately, the probe's two data channels had been built to be essentially duplicates of each other. Thus, if all went according to the nominal plan, Cassini would have relayed essentially identical data sets generated by Huygens' channels A and B. If one failed, then the other was a redundant system.

That foresight had paid off during the encounter, because channel A's data were not received. The loss of channel A data was traced to a software commanding error. The only scientific loss was data concerning wind speed, which had been designated to be determined only through channel A. Data concerning wind speeds encountered during descent eventually were obtained in a different fashion using other aspects of space probe signals; in brief, this involved Doppler wind data coming from examining characteristics of the probe's signals relayed to Cassini. Wind motions could be inferred from small changes in signal frequency.

Initial images of the unusual surface Huygens encountered revealed ice blocks strewn around, features potentially indicative of fluid flow, and indications of recent activity. Missing was a landscape of impact craters, ridges, and hills. In short order, DISR images were processed and pieced together to provide a panoramic view that one would see if standing above the probe.

Data analysis continued in concentrated fashion over the next few hours. After further analysis, ESA scientists proclaimed that Huygens had landed in Titanian mud and was resting frozen and powerless at 93 kelvins. The probe's batteries were now dead, but they had lasted far longer than anticipated. Huygens had slowed to a gentle 5.4 meters per second before impact and had picked up a sideways drift of 1.5 meters per second. One scientist joked that Huygens had not landed with a splash or thud, but with a splat in the cryogenic mud. Huygens provided data that produced geological evidence for precipitation, erosion, mechanical abrasion, and other fluvial activity. Apparently physical processes at work on Earth were also at work on Titan, only at a greatly reduced temperature and in a very different atmosphere. Planetary scientist Toby Owens joked that scientists had long been searching for liquid on Mars, but Huygens found them, not buried beneath a solid surface but right there on Titan. He predicted the weather report for Huygens would likely be rain, but precipitation of liquid methane.

### Context

Four decades after the initial successful interplanetary probe, the Mariner 2 flyby of Venus, it was rather rare for a probe to be making an initial examination of a major object in the solar system. Huygens was one of the exceptions. Although the Pioneer 11 probe flew past Saturn in 1979 and was followed in 1980 and 1981 by the Voyager 1 and 2 spacecraft, respectively, no high-resolution images of the surface of Titan, the largest moon in the Saturn system, had been taken prior to Huygens. The Voyagers had taken images of Titan, but all that was revealed was a thick atmosphere and a haze layer. Subsequent ground-based images, as well as images taken by the Hubble Space Telescope, did record differences in albedo and the possibility of surface irregularities; these were not images taken in the visible portion of the electromagnetic spectrum.

With Cassini still in orbit, gathering tremendous amounts of data and transmitting thousands of images, there were no immediate plans for a subsequent mission to the Saturn system. It is likely that the data returned by the Huygens probe will provide the definitive picture of this tantalizing, large moon for some time to come. The data had tremendous implications for comparative planetology. For example, Huygens data, along with information from Cassini, provided an explanation for the large quantity of nitrogen retained by Titan's atmosphere. This would be compared with Galileo data about large moons orbiting Jupiter that did not retain any original nitrogen. Scientists hoped that Huygens data might provide a basis for understanding the origin of the methane found in Titan's atmosphere, something the large moon should have lost if it were primordial. Was the methane produced only recently, and is it in the process of being lost to space? Or is the methane being somehow replenished? Answers to these and many other questions could provide a hint as to the nature of the early Earth long before life began.

**See also:** Cassini: Saturn; Jet Propulsion Laboratory; Pioneer 11; Planetary Exploration; Voyager Program; Voyager 1: Jupiter; Voyager 1: Saturn; Voyager 2: Jupiter; Voyager 2: Saturn.

### Further Reading

Coustenis, Athena, and Fred Taylor. *Titan: The Earth-Like Moon.* New York: World Scientific Publishing, 1999. An overview of Titan compiled prior to the Cassini mission, stressing the expectation that the large moon shares aspects of an early Earth.

Encrenas, Thérèse, R. Kallenbach, T. Owen, and C. Sotin. *The Outer Planets: A Comparative Study Before the Exploration of Saturn by Cassini-Huygens.* New York: Springer, 2005. A comparative study of the formation and evolution of the outer planets, the atmospheres of the outer planets and Titan, and moons and ring systems. Includes early Cassini data.

Harland, David M. *Mission to Saturn: Cassini and the Huygens Probe.* London: Springer-Praxis, 2002. A detailed scientific and engineering history of the Cassini program, but also includes extensive discussion of the events and results from the Voyager program.

Hartmann, William K. *Moons and Planets.* 5th ed. Belmont, Calif.: Thomson Brooks/Cole, 2005. Provides detailed information about all objects in the solar system. Suitable for high school students, general readers, and undergraduates studying planetary geology.

Irwin, Patrick G. J. *Giant Planets of Our Solar System: Atmospheres, Composition, and Structure.* London: Springer-Praxis, 2003. Provides an in-depth comparison of Jupiter, Saturn, Uranus, and Neptune, incorporating data obtained from astronomical observations and planetary spacecraft encounters.

Lorenz, Ralph, and Jacqueline Mitton. *Lifting Titan's Veil: Exploring the Giant Moon of Saturn.* London: Cambridge University Press, 2002. Written prior to the Huygens landing, this text provides the state of understanding about Titan prior to the initiation of Cassini's scientific returns. Anticipates the excitement of discovery that Huygens would bring.

Morrison, David, and Tobias Owen. *The Planetary System.* 3d ed. San Francisco: Addison-Wesley, 2003. Organized by planetary object, this work provides contemporary data on all planetary bodies visited by spacecraft since the early days of the Space Age. Suitable for high school, college, and general readers.

*David G. Fisher*

# Infrared Astronomical Satellite

*Date:* January 25 to November 21, 1983
*Type of spacecraft:* Space telescope

*The atmosphere is substantially opaque to infrared radiation, which occupies the spectral range between visible light and shortwave radio emissions. From its vantage point in Earth orbit, the Infrared Astronomical Satellite (IRAS) enabled scientists to study in detail, and without atmospheric interference, the heat emitted by astronomical sources.*

## Key Figures

*Walker E. "Gene" Giberson*, NASA program manager
*Peter F. Linssen*, Netherlands Agency of Aerospace Programs

## Summary of the Satellite

On January 25, 1983, the Infrared Astronomical Satellite (IRAS) was successfully launched into a polar orbit 900 kilometers high by a Delta 310 rocket. The launch took place at the Western Test Range at Vandenberg Air Force Base in California. Designed for a useful life of two hundred days, IRAS exceeded its life expectancy by 50 percent and continued to collect and relay high-quality data to Earth until November 21, 1983, when its supply of liquid helium, necessary for cooling its telescope, was exhausted.

During its ten-month mission, IRAS monitored the infrared radiation, or heat, given off by astronomical objects. The infrared survey of the universe that IRAS successfully completed from above Earth's obscuring atmosphere was the first of its kind to be attempted. The program was an international effort, with organizations from the Netherlands, the United States, and the United Kingdom participating in design, construction, testing, operation, data collection, and data analysis. The project cost $189 million: The United States spent $119 million, and the Netherlands and the United Kingdom spent $45 million and $25 million, respectively.

Primary responsibility for building the spacecraft was given to two Dutch companies, Fokker and Hollandse Signaalapparaten, with the Netherlands Agency for Aerospace Programs and the United States' National Aeronautics and Space Administration (NASA) providing guidance. The design and development of the infrared telescope was entrusted to NASA's Jet Propulsion Laboratory (JPL). Ball Aerospace Division, under contract to JPL and Ames Research Center, built the telescope and cryogenic dewar, a large container designed to enclose the entire telescope system and maintain its temperature below 10 kelvins (−263° Celsius) indefinitely. The telescope mirrors were fabricated by the Perkin-Elmer Corporation, of Connecticut. The United Kingdom, through its Science and Engineering Research Council (SRC), handled satellite tracking, data acquisition, and preliminary data analysis. JPL was charged with processing the masses of data returned by IRAS to produce an infrared catalog of the sky.

The basic telescope design was a Ritchey-Chrétien Cassegrain. The mirrors were made of beryllium metal, ground and polished into ellipsoids. Beryllium was chosen because it would be compatible with the extremely low temperatures at which the telescope would operate. The primary mirror, 0.57 meter in diameter, was configured to collect the infrared radiation from an astronomical source and reflect it back toward a small secondary mirror, which returned the rays through a hole in the center of the primary mirror and thence to a focus. The focal length of the system was 5.5 meters, giving the telescope a focal ratio (the ratio of a refracting device's length to its diameter) of f/9.6. The infrared detector array, located in the focal plane (at the telescope focus), gave the telescope a 1.1° field of view.

At infrared wavelengths, objects glow with the electromagnetic radiation signature of the infrared light they emit. Infrared radiation appears in different wavelengths just as visible radiation does. For example, green light has a wavelength of about 0.5 micrometer, or one-half of a millionth of a meter, while the wavelength of red light is closer to 0.7 micrometer. In contrast, the infrared radiation detected by IRAS had wavelengths ranging from almost ten to more than one hundred micrometers.

There were sixty-two individual infrared sensing elements in the detector array, separated into four groups, with each group designed specifically to monitor a different band of the infrared spectrum. Fifteen silicon-arsenide detectors monitored wavelengths near 12 micrometers and were sensitive to the warmest astronomical sources. Sixteen silicon-antimonide detectors monitored wavelengths of 25 micrometers. For observations of the coldest sources, germanium-gallium sensors were used; sixteen elements detected wavelengths of 60 micrometers, and fifteen detected wavelengths of 100 micrometers.

The entire telescope system was cooled to below 10 kelvins and maintained at that temperature throughout the mission. Because infrared radiation is heat, the telescope itself, if not properly cooled, would become the predominant source of heat detected by the heat sensors. Consequently, the system was enclosed in a double-walled container, the dewar, and the space between the walls was filled with 72.3 kilograms of liquid helium at launch time. The initial supply of helium slowly boiled away at a temperature of 2.55 kelvins during the satellite's three hundred-day lifetime. In addi-

tion to keeping the telescope temperature below 10 kelvins, the dewar kept crucial elements of the system—principally the focal plane assembly, consisting of the detector array and electronic instruments—even colder, at the boiling point of liquid helium. The sensitivity thereby attained would have allowed the telescope to detect the heat generated by a flashlight bulb at a distance of 2,800 kilometers.

The IRAS satellite weighed 1,020 kilograms and was neatly packaged as a cylinder 3.5 meters long and 1.7 meters in diameter. At the time of launch, a protective cover was in place that served to protect the telescope from contamination. This cover was ejected six days after launch, and on February 9, 1983, after a period of in-orbit testing and calibration, the satellite began scientific observations. In its polar orbit, IRAS circled Earth in the "twilight zone," following the terminator, the imaginary line separating Earth's sunlit side from its night side. IRAS could thus draw uninterruptedly on the power supplied by the Sun's rays and simultaneously make observations in all directions.

During its three hundred days of operation, IRAS continuously scanned the sky for infrared radiation from astronomical objects. A full 95 percent of the sky was surveyed at least twice, and 72 percent was covered three times; only 2 percent was not scanned at all. The satellite's orientation around all three of its axes was under computer guidance, which guaranteed a pointing accuracy of better than 20 arc seconds. The powerful onboard computer controlled all satellite functions and was capable of being reprogrammed in orbit, providing great flexibility in the mission profile.

Three modes of observation were available. One, known as slow-scan, was used for the standard survey operation, in which the telescope swept the sky at a constant angular velocity, allowing complete sky coverage in approximately thirty days. A faster scanning mode allowed mapping of a small region of sky several times in succession in order to enhance faint details in that area. A pointing mode was used to examine details of faint, small objects of interest.

Data were collected continuously and stored on board the satellite. Twice daily, all stored data were transmitted back to Earth over seven and a half minutes. The amount of data thus stored and transmitted every twelve hours was said to equal the amount of information contained in the Bible. IRAS was the most sophisticated infrared telescope ever designed and built. It was the first cryogenically cooled telescope launched by NASA.

### Contributions

While completing its planned general survey of 95 percent of the sky, IRAS detected more than two hundred thousand starlike heat sources and made a number of other interesting discoveries.

A previously unknown ring of dust was detected near the asteroid belt between the orbits of Mars and Jupiter. Dust appears to be present in the universe wherever there is low-density or low-temperature gas. These ubiquitous dust grains are excellent emitters of infrared radiation and are responsible for much of the infrared energy emitted by a variety of sources.

IRAS detected several new asteroids and five new comets. Comets were found to be much dustier than had been supposed. The well-known Comet P/Tempel 2, it was discovered, has a 32-million-kilometer dust tail; during the comet's sixteen previously recorded passages, the tail had never been observed. A previously unknown object, dubbed 1983TB, was found orbiting in the path of the Geminid meteor shower; evidently, it is the worn-out hulk of a comet, the parent of the Geminid meteor stream. Its path carries it within 19 million kilometers of the Sun, closer than any other known body.

IRAS surprised astronomers by recording the presence of wispy clouds of cold dust farther out in space. Because of their appearance, the clouds were immediately labeled "infrared cirrus. " Distances to the patches of infrared cirrus had not been determined by the late 1980's, and it was not known whether the clouds are a local phenomenon or a general characteristic of the galaxy. IRAS also detected a more uniform distribution of cold dust that pervades the galaxy in all directions.

Perhaps the most celebrated result from IRAS was the discovery of a cold (168 kelvins) cloud of solid particles around the bright star Vega. Theoretical analysis showed that each of the particles must be at least 1 millimeter in diameter, one thousand times larger than ordinary dust grains. The amount of material in the cloud is indeterminate, but it could be as much as three hundred Earth masses, which would be comparable to the total mass in the Sun's planetary system. Vega is relatively young, less than a billion years old, and the observed particle cloud extends into space roughly twice as far from Vega as Pluto's orbit is from the Sun.

Infrared images returned by IRAS of a large region of space in the constellation Orion showed for the first time the true extent of the vast clouds of dust and gas in this star-forming neighborhood. The Orion nebula, M42, is the middle "star" in Orion's sword, hanging from his belt, and is easily visible to the naked eye. In telescopic photographs it appeared as a beautiful, multicolored expanse of glowing gas, energized by hot, young stars buried in its depths. The very cold dust detected by IRAS pervades the entire constellation, however, and M42 and the Horsehead nebula appeared in some images as small, bright areas against the expansive background of the dust.

New infrared sources in the vicinity of several dark clouds were discovered. These may be protostars, new stars in the very early stages of formation. The infrared emissions of eighty-six galaxies were measured; the Crab nebula's magnetic field was measured for the first time and found to be about one hundred times stronger than the galaxy's general interstellar magnetic field; and the recording of infrared emissions from a sampling of quasars added to the pool of information that will eventually unlock the secrets of those enigmatic objects.

### Context

IRAS opened a new window on the universe. Galileo initiated the era of instrumental astronomy when he built the first astronomical telescope and gazed through it at the heavens in 1609. In the early nineteenth century, Sir William Herschel studied the refraction of sunlight through a prism. He discovered that a thermometer placed just outside the red end of the solar spectrum experienced a rise in temperature, an effect of invisible solar infrared radiation.

It was another 150 years before sufficiently sensitive infrared detection systems enabled astronomers to peer through Earth's atmosphere and routinely observe extraterrestrial sources of infrared energy. By 1975, approximately two thousand infrared sources had been identified and cataloged.

IRAS's enormous sensitivity compared with previous ground-based observing systems could be attributed to its operation at cryogenic temperatures and to its location above the atmosphere. Not only does Earth's atmosphere provide just a few spectral "windows" through which infrared observations may be made, but even in the windows, the atmosphere itself is a strong emitter of infrared radiation, and ground-based systems must be designed specifically to reject the unwanted atmospheric radiation.

The engineering accomplishments represented by the fully functioning IRAS spacecraft established a new level of sophistication in infrared technology. This technological knowledge would be used and expanded in the next generation of infrared instrumentation. Known as the Space Infrared Telescope Facility (SIRTF), that advancement was the last of the series referred to as NASA's Great Observatories. Once SIRTF was commissioned in orbit after launch on August 25, 2003, it was renamed the Spitzer Space Telescope in honor of Lyman Spitzer, Jr., an astronomer who spent many years advocating the placement of large telescopes in orbit, where they would be free from the obscuring effect of light absorption by gases in the Earth's atmosphere.

The IRAS data have increased a hundredfold the inventory of known infrared astronomical sources. Morever, interpretation of some of the observations mentioned earlier may change some theories about the astrophysical environment. For years, scientists have been aware of the presence of dust in the ecliptic (the plane of Earth's orbit around the Sun); the dust is the source of the zodiacal light and the Gegenschein, phenomena in which sunlight scattered by this dust can be seen by the naked eye on Earth under the right conditions. The newly discovered dust ring between Mars and Jupiter is in a different location and is something of a puzzle. The ring lies at a small angle to the ecliptic, and the grains are small enough to absorb and emit the energy of sunlight in such a way as to cause themselves to spiral slowly into the Sun. This process should result in elimination of the dust ring in forty or fifty thousand years. The fact that the ring exists means either that it is of relatively recent origin or that the dust is being continually replaced, perhaps by the debris from asteroid collisions.

The cold dust found in the galaxy may produce a hitherto unsuspected dimming of light from large distances, meaning that the astronomical distance scale, which depends on the apparent brightness of objects, might have to be revised and that the universe may be smaller than has been thought.

The available evidence concerning the particle cloud surrounding Vega has fueled speculation that it represents an early stage in the formation of a planetary system. If this were the case, however, Vega's planets would not develop in the same way as the solar system did; such a system around Vega would have been undetectable by IRAS. Nevertheless, the detection of the cloud was the first evidence ever found for the existence of solid bodies in orbit around any star other than the Sun.

**See also:** Chandra X-Ray Observatory; Compton Gamma Ray Observatory; Gamma-ray Large Area Space Telescope; Hubble Space Telescope; Huygens Lander; International Ultraviolet Explorer; Orbiting Astronomical Observatories; Orbiting Geophysical Observatories; Orbiting Solar Observatories; Spitzer Space Telescope.

**Further Reading**

Burnham, Robert. "IRAS and the Infrared Universe." *Astronomy* 12 (March, 1984): 6-22. An excellent discussion for the layperson of the important early IRAS results. Features an image, returned by IRAS, of infrared energy sources in the Milky Way.

Burrows, William E. *This New Ocean: The Story of the First Space Age.* New York: Random House, 1998. This is a comprehensive history of the human conquest of space, covering everything from the earliest attempts at spaceflight through the voyages near the end of the twentieth century. Burrows is an experienced journalist who has reported for *The New York Times, The Washington Post,* and *The Wall Street Journal.* There are many photographs and an extensive source list. Interviewees in the book include Isaac Asimov, Alexei Leonov, Sally K. Ride, and James A. Van Allen.

Davies, John K. *Astronomy from Space: The Design and Operation of Orbiting Observatories.* New York: John Wiley, 1997. This is a comprehensive reference on the satellites that have revolutionized twentieth century astrophysics. It contains in-depth coverage of all space astronomy missions. It includes tables of launch data and orbits for quick reference as well as photographs of many of the lesser-known satellites. The main body of the book is subdivided according to type of astronomy carried out by each satellite (x-ray, gamma-ray, ultraviolet, infrared and millimeter, and radio). It discusses the future of satellite astronomy as well.

Glass, I. S., Richard Ellis, John Huchra, Steven Kahn, George Rieke, and Peter B. Stetson. *Handbook of Infrared Astronomy.* London: Cambridge University Press, 1999. This book provides a wealth of information about the revolution in infrared astronomy in a way that the nonspecialist can appreciate.

Heppenheimer, T. A. *Countdown: A History of Space Flight.* New York: John Wiley, 1997. A detailed historical narrative of the human conquest of space. Heppenheimer traces the development of piloted flight through the military rocketry programs of the era preceding World War II. Covers both the American and the Soviet attempts to place vehicles, spacecraft, and humans into the hostile environment of space. More than a dozen pages are devoted to bibliographic references.

Launius, Roger D. *NASA: A History of the U.S. Civil Space Program.* Malabar, Fla.: Krieger Publishing Company, 1994. This is an in-depth look at America's civilian space program and the establishment of the National Aeronautics and Space Administration. It chronicles the agency from its predecessor, the National Advisory Committee for Aeronautics, through the late twentieth century.

Leverington, David. *New Cosmic Horizons: Space Astronomy from the V2 to the Hubble Space Telescope.* New York: Cambridge University Press, 2001. This is a broad treatise exploring the development of space-based astronomical observations from the end of World War II to the Hubble Space Telescope and other major NASA space-based observatories.

Mampaso, A., M. Prieto, and F. Sanchez, eds. *Infrared Astronomy.* London: Cambridge University Press, 2004. This text provides a series of lectures from nine infrared astronomy researchers that cover the birth and death of stars as well as the nature of the interstellar medium, intergalactic medium, brown dwarfs, and extrasolar planets.

Robinson, Leif J. "The Frigid World of IRAS, Part 1." *Sky and Telescope* 67 (January, 1984): 4-8. This article is a compilation of early results from the IRAS mission. It includes a magnificent infrared image of Orion. Suitable for general audiences.

Schorn, Ronald A. "The Frigid World of IRAS, Part 2." *Sky and Telescope* 67 (February, 1984): 119-124. A discussion of more of the main results from the IRAS mission. Topics include the study of infrared radiation from young and old stars, the Crab nebula, and external galaxies. Suitable for a general audience.

Sterrenburg, Frithjof. "IRAS: Mission Invisible." *Astronomy* 11 (April, 1983): 66-70. This article describes the IRAS project at time of launch. Special emphasis is given to a description of the Dutch Additional Experiment (DAX). Includes a line drawing showing the details of IRAS and a large color photograph of IRAS in the laboratory before launch. Suitable for a general audience.

*Alan F. Bentley*

# Intelsat Communications Satellites

*Date:* Beginning April 6, 1965
*Type of spacecraft:* Communications satellites

*Intelsat was created by international agreement to develop and operate a global satellite communications system that would guarantee access to international satellite communications for all member nations of the International Telecommunication Union, an agency of the United Nations.*

### Summary of the Satellites

The International Telecommunications Satellite Consortium (Intelsat) was created on August 20, 1964, as a direct result of a call by the United States for the establishment of an international commercial satellite communications system. Presidents Dwight D. Eisenhower and John F. Kennedy both proposed the undertaking publicly in 1961, and the United Nations issued a formal endorsement in December of that same year. Its purpose would be to provide nondiscriminatory satellite access to all nations in a timely manner. The satellites were to be used for international communications traffic only and would not compete with private companies engaged in developing, deploying, and operating domestic communications satellites. Over the years, however, exceptions have been made for member countries that do not have domestic satellite systems of their own.

The United States Congress passed the Communications Satellite Act of 1962 on August 31, 1962, formally delineating a purpose for such a system: to serve all nations in the interest of world peace and international understanding. The act also created the Communications Satellite Corporation (COMSAT) to serve as the representative of the United States in the proposed organization. COMSAT became a legal entity on February 1, 1963. It is privately owned and financed by publicly held stock. Its original offering of two hundred million dollars sold quickly to approximately one hundred thousand investors.

Negotiations among nations interested in forming a communications satellite consortium began in early 1964, and on August 20, 1964, an interim agreement creating Intelsat was approved by nineteen governments, including that of the United States. In the original interim agreement, COMSAT was designated manager of operations and maintenance. It was also charged with the design, development, construction, and deployment of the system's satellites and related equipment. After 1979, however, full managerial responsibility rested with an executive staff that answered to a twenty-eight-member board of governors composed of representatives of member countries.

In 1998, Intelsat's Assembly of Parties voted to restructure and eventually commercialize Intelsat, a key indicator of the organization's commitment to change in this vibrant industry to ensure Intelsat's continued vitality and resilience as the organization moved into the twenty-first century. Governments around the world agreed with Intelsat and their customers that the organization should privatize. Privatization was expected to allow Intelsat to meet new competitive challenges while continuing to ensure global satellite connectivity and coverage for users in all countries.

Intelsat is a commercial enterprise and, as such, seeks to make a profit. It leases half-circuits, or one-way lines of communication, from either Earth station to satellite or satellite to Earth station. Lease prices set by the governing board take the form of annual fees intended to recoup investment capital and cost of operation and to return a modest annual profit. The consortium has a network of Earth stations, or downlinks, worldwide. Some are owned by Intelsat, though most are owned by host countries. In 1969 there were 24 commercial Earth stations operated by Intelsat. By 1980 that number had grown to 263 in 134 countries. By 1999, Intelsat had 143 member countries and signatories.

Intelsat also operates geosynchronous satellites and related ground control equipment for spacecraft tracking, maneuvering, and operation. Geosynchronous satellites are those with orbits directly above the equator and whose direction and velocity cause them to move in concert with the turning of Earth on its axis. They are placed at least 35,680 kilometers above the equator and directed by ground-controlled onboard maneuvering rockets into orbits that appear to put them in a stationary position relative to Earth. Periodic adjustments are required to maintain their proper position, and they can be moved to different locations at the discretion of the operator until the fuel in the maneuvering rockets is exhausted.

The first generation of Intelsat satellites operated on extremely low power, which required the construction of elaborate and expensive Earth stations. To many small and underdeveloped nations the cost of these multimillion dollar installations was prohibitive, one reason the Intelsat charter membership included fewer than twenty countries.

The first satellite in the Intelsat series was Intelsat I, better known as Early Bird, launched by COMSAT aboard a Thrust-Augmented Delta booster rocket from Cape Kennedy on April 6, 1965. Built by the Hughes Aircraft Company, it weighed 39 kilograms and provided 240 telephone circuits, or one television circuit. That meant that when it was used to relay a television broadcast, all 240 telephone circuits were preempted. The estimated investment cost per circuit over the course of its projected life span was $32,500. The first publicly viewed television transmission over Intelsat I occurred on May 2, 1965. Early Bird was turned over to Intelsat on June 28, 1965. It was designed to

*The first Intelsat, Intelsat I or Early Bird, was launched in April, 1965. Built by Hughes Aircraft Company, it had 240 telephone circuits that cost about $32,500 per circuit.* (NASA CORE/Lorain County JVS)

remain in service for under two years but continued to function for several years beyond its expected life span.

Intelsat II F-1, also referred to as Intelsat II-A, was launched October 26, 1966. The first in the Intelsat II series manufactured by Hughes Aircraft Company, it also contained 240 telephone circuits, or one television circuit. It weighed 87 kilograms and had a life expectancy of three years. It was placed in a twelve-hour orbit, despite original plans that called for a twenty-four-hour orbit. On January 11, 1967, Intelsat II-B, Pacific 1, was launched to provide coverage in the Pacific Basin. The second in the Intelsat II series, it was identical to Intelsat II-A. Transpacific service was initiated on the day of launch. The Intelsat II series continued with the launch of Intelsat II-C, Atlantic 2, placed in geostationary orbit on March 22, 1967, close to the then still-operational Early Bird. Intelsat II-D, the last in the Intelsat II series, was launched on September 28, 1967.

The Intelsat III series, manufactured by TRW, had a false start when Intelsat III F-1 was destroyed in an unsuccessful launch on September 18, 1968. The first operational satellite in the series was Intelsat III F-2, launched on December 18, 1968, weighing 145 kilograms and containing 1,500 telephone circuits. Intelsat III F-3, launched on February 6, 1969, followed it less than two months later. These satellites replaced the transatlantic and transpacific satellites in the Intelsat II series, and on May 22, 1969, a truly global network was created for the first time, with the successful placement of a third Intelsat III series satellite over the Indian Ocean, thus providing coverage to more than 90 percent of Earth's population. That year, Intelsat provided global coverage of the first Apollo Moon landing. Four more satellites in the Intelsat III series were launched, of which two were failures.

The first of eight satellites in the Intelsat IV series was launched on January 26, 1971. Intelsat IV was bigger and more powerful with four thousand telephone circuits plus two television circuits. It was five times as powerful as Intelsat III and was designed for a service life of seven years. All but one in this series were successfully launched and placed into operation.

Intelsat IV-A, first placed in operation with the launch of Intelsat IV-A F-1 on September 26, 1975, contained six thousand telephone circuits and twenty transponders for television transmission. It measured 5.9 meters top to bottom and weighed 790 kilograms. This satellite was built at a cost of $23 million and cost about the same amount to launch into orbit. It also had a useful life expectancy of seven years, with an investment cost per circuit year of $1,100 in 1977.

Eleven Intelsat V and V-A series satellites were placed in operation. Built by Ford Aerospace and Communications Corporation, Intelsat V was a high-powered Ku-band series launched between 1980 and 1985. It contained twelve thousand voice circuits on each satellite and provided different communication capabilities to different regions of the world with global, spot, and zone relay capabilities on the four- to six-gigahertz band and the 11- to 14-gigahertz band. Six Intelsat V-A satellites had fifteen thousand two-way voice circuits each. As of 2005, only one of the Intelsat V/V-A series satellites was still in operation. Launched in June, 1985, Intelsat VII hovers over the Atlantic Ocean Region (AOR), serving the Americas, the Caribbean, Europe, the Middle East, India, and Africa.

The Intelsat VI design features thirty thousand two-way circuits with fifty transponders, at a cost of $785 million. There are five Intelsat VI series spacecraft, and all five are in operation. The prime contractor for the VI series was Hughes Aircraft Company. Three Intelsat VI satellites are in service in the AOR, and two are in service in the Indian Ocean Region (IOR). The Intelsat VI series satellites are the largest commercial spacecraft ever built and launched. Their high-powered Ku-band spot beams provide a variety of voice, video, and data services to a wide range of Earth station antenna sizes, including very small Earth stations located on customer premises.

The most noted of the Series VI spacecraft is Intelsat VI-4 (known as Intelsat 603 on orbit). After being launched atop a Titan II vehicle on March 14, 1990, it was stranded in a useless low orbit when the Titan's second stage failed to separate. Two years later, during the maiden flight of the space shuttle *Endeavour*, four spacewalking astronauts successfully retrieved it and attached a new second-stage motor to its base. Later, it was blasted into its designated geosynchronous orbit over the Atlantic Ocean Region.

Four of the eight Intelsat VII/VII-A satellites are in service in the Atlantic Ocean Region, one in the Indian Ocean Region (IOR), and two in the Pacific Ocean Region (POR). The Intelsat VII/VII-A series of satellites are a powerful, versatile fleet of spacecraft that are deployed in all Intelsat service areas. Design of the series focused on the special requirements of the Pacific Ocean Region, but the satellites have unique features facilitating operation in the other ocean regions as well. Optimization for operation with smaller Earth stations, efficient use of capacity, and digital operation were key goals in development of the series. Other enhance-

*Intelsat VI from the vantage of space shuttle orbiter* Endeavour *during STS-49 in May, 1992. During the mission four astronauts successfully retrieved the satellite and attached a new second-stage motor to its base.* (NASA)

ments incorporated in the design include command security, improved reliability, and improved ability to operate in inclined orbit.

Four of the four Intelsat VIII satellites are in service in the Atlantic, Indian, and Pacific Ocean Regions. The last Intelsat VIII/A satellite (Intelsat 805) was launched successfully on June 18, 1998, for deployment in the Atlantic Ocean Region (AOR) at 304.5° east longitude. The Intelsat VIII/ VIII-A series has been designed to meet the needs of Intelsat users throughout the system for improved C-band coverage and service. These spacecraft incorporate six-fold C-band frequency reuse, two-fold frequency reuse of expanded C-band capacity, and the highest C-band power level ever for an Intelsat satellite.

### Contributions

Experiments by the National Aeronautics and Space Administration (NASA) demonstrated the potential for satellite communication on a global scale only a few years before Intelsat was created. In 1960, aluminized balloons were launched into orbit at 1,000 miles above Earth as part of an experiment to test passive reflection of radio signals. The experiments were a success, demonstrating that bouncing radio signals off satellites was possible. The American Telephone and Telegraph Company had launched its domestic communications satellite, Telstar, in 1962, and NASA had launched its communications satellite, Syncom 1, in 1964. Both were successful.

With each new series, Intelsat was able to enhance the transcontinental communications capabilities of its satellites, significantly increasing the amount of transoceanic voice, image, and data traffic. In addition, the increased circuitry and cost efficiencies that resulted from the longer life expectancies of later satellites significantly reduced the cost of these communications. For example, the cost for one hour of prime time color television transmission on Intelsat I was $22,350 in 1965; by February of 1977, that cost had been reduced 80 percent, to $5,100, and continued to drop.

Intelsat was also created to compete with transoceanic cable, which was inadequate to meet the demand in the early 1960's. In 1965, there were 317 cable circuits available. With the launch of Intelsat I, the number of circuits almost doubled despite the fact that Intelsat I was the weakest satellite with the least capacity of any launched during the Intelsat program.

In 1974, Intelsat implemented the world's first international digital voice communications service, a precursor of today's advanced digital networks. Later that year, Intelsat activated a direct hot line between the White House and the Kremlin. Intelsat also provided facilities to make communication available to population centers that were difficult to reach because of rough terrain or unpredictable weather conditions. In 1978, an estimated one billion people in 42 countries watched World Cup Soccer matches via the Intelsat system. With the launch of the Intelsat V spacecraft series in the 1980's, the first use of the dual-polarization technique was implemented. During the same decade, the Intelsat VI series allowed broadcasters to transmit news feeds via the Intelsat system using very small, easily transportable Earth stations. In 1994, Intelsat transmitted the historic South African elections via seven Intelsat satellites in three ocean regions.

By 2005, seventeen Intelsat operational satellites carried the bulk of international telephone traffic and virtually all intercontinental television traffic. More than 140 member countries and more than forty investing entities contributed to Intelsat's cooperative of networks linked via seventeen geostationary satellites, bringing global access to more than two hundred nations and territories around the world.

The seven newest Intelsat IX series spacecraft would deliver public and private voice/data networks, Internet and Intranet, broadcast video, and broadband (high-data-rate telemedecine and tele-education, interactive video, and multimedia). Built by Space Systems/Loral and designed to replace the Intelsat VI satellites, the Intelsat IX series reflects a total investment of approximately one billion dollars.

**Context**

Politically, Intelsat has always been a victim of the divergent interests of its members. Through COMSAT, the United States took a controlling hand in the consortium's affairs. Having emerged during the most productive years of space systems development, COMSAT was alone in its ability to provide the necessary equipment and operational expertise required to operate such a system. Voting rights, ownership interest, and circuit access were doled out according to formulas based on early financial support and developmental and scientific contributions by member nations. COMSAT received the lion's share.

This arrangement quickly became controversial. Membership had grown quickly in the early years and began to include many small countries that had great need for satellite-assisted communications facilities mostly because of inadequate telecommunications infrastructures at home. Many of these nations were unable to make significant operational contributions to Intelsat, financial or otherwise, yet they wanted a greater say in the planning and operation of the system.

Also, the original estimated cost of developing and orbiting Intelsat I, two hundred million dollars, proved to be much higher than the actual cost of eighty million dollars. Demand for satellite time had been overwhelming from the start, and Intelsat quickly began to take in large revenues. It made plans to build and deploy bigger satellites with more circuit capacity. With less need for financial support from the United States and other large nations, pressure began to mount for greater independence in the operation and control of the organization and the satellite system.

The extent to which COMSAT had assumed control over Intelsat during the early years is perhaps best illustrated in the way it handled the construction of the Intelsat III series of satellites. In February of 1966, COMSAT petitioned the Federal Communications Commission (FCC) for permission to participate in the construction of the satellites after negotiating with an American company, TRW, which it favored to build the series. In effect, COMSAT, as the manager of Intelsat and agent for the United States with its controlling interest in the consortium, was in a position to benefit from the partnership with TRW by authorizing itself to proceed. The FCC was unsure of its jurisdiction over Intelsat and was reluctant to act.

Meanwhile, Hughes Aircraft Company, which had built Intelsat I and the Intelsat II series, claimed that it could construct a series of satellites even more advanced than the planned Intelsat III series and have them ready by the launch dates set by Intelsat. Before the FCC acted, however, COMSAT received authorization from the Intelsat governing body to conclude an agreement with TRW for the construction of the Intelsat III satellites. In June of 1966, the FCC concurred but told COMSAT that it must request FCC approval before it leased half-circuits from Intelsat or offered them to other American international carriers. The Intelsat Interim Committee viewed this rulemaking by the United States government as direct interference in its operations. The United States, on the other hand, saw a potential conflict of interest on the part of COMSAT.

From February 24 to March 21, 1969, a conference was held in Washington, D.C., to discuss formalizing a structure that would include an independent governing body and weaken the control of COMSAT. The meeting ended in a stalemate after heated debate. Issues included whether the system should place greater emphasis on servicing the needs of industrial countries. Some original members were utilizing the system less than they had originally anticipated and stood to lose favored ownership status to others who had used it more during the first few years of operation.

Almost four years later, on February 12, 1973, an agreement was finally reached that significantly restricted the power of COMSAT. A governing board was established that was to work in concert with COMSAT in anticipation of the election of a director general by December 31, 1976. With that election, COMSAT's management duties would come to an end, and Intelsat would achieve legal status for the first time.

Another development that affected the future of Intelsat in the eyes of its members was a decision by the FCC to allow private companies to construct and deploy communications satellites and operate them under the aegis of free enterprise. On November 28, 1984, United States president Ronald W. Reagan issued a statement calling for the creation of new global satellite systems that would compete with Intelsat. He suggested that having alternate systems available would be in the national interest. The FCC had authorized the construction of five such systems by July of 1988. Other countries had for some time been creating domestic satellite systems, including Canada's Anik/Telesat, established in 1969, Symphonie, operated by West Germany and France, Indonesia's Palapa, and Italy's Sirio.

The Soviet Union, which had not joined Intelsat despite being qualified to do so, had extended a proposal for its own international satellite network to be called Intersputnik. Later, its Interkosmos program provided launch facilities and support for a number of Eastern Bloc countries. In 1988, one international communications satellite system, PanAmSat, entered into an agreement with Peru to provide an international communications service. It was the first system to enter into direct competition with Intelsat.

**See also:** Amateur Radio Satellites; Applications Technology Satellites; Mobile Satellite System; Private Industry and Space Exploration; Space Shuttle Mission STS 51-A; Space Shuttle Flights, 1992; Space Shuttle Mission STS-49; Telecommunications Satellites: Private and Commercial; Tracking and Data-Relay Communications Satellites.

**Further Reading**

Bilstein, Roger E. *Orders of Magnitude: A History of the NACA and NASA, 1915-1990.* 2d ed. NASA SP-4406. Washington, D.C.: Government Printing Office, 1989. This heavily illustrated history of the United States space program is written for the layperson. A small volume, it takes a broad approach with a focus on the political climate during the height of the space program. Particular emphasis is placed on the crewed programs, but Intelsat is described in the context of the exploitation of space science that proved beneficial to large segments of Earth's population. Included is a discussion of the events that led to what is described as "The New Space Program." Contains a bibliography and an index.

Butrica, Andrew J. *Beyond the Ionosphere: Fifty Years of Satellite Communications.* NASA SP-4217. Washington, D.C.: Government Printing Office, 1997. Part of the NASA History series, this book looks into the realm of satellite communications. It also delves into the technology that enabled the growth of satellite communications. The book includes many tables, charts, photographs, and illustrations, a detailed bibliography, and reference notes.

Elbert, Bruce R. *Introduction to Satellite Communication.* Cambridge, Mass.: Artech House, 1999. This is a comprehensive overview of the satellite communication industry. It discusses the satellites and the ground equipment necessary to both the originating source and the end user.

Gavaghan, Helen. *Something New Under the Sun: Satellites and the Beginning of the Space Age.* New York: Copernicus Books, 1998. This book focuses on the history and development of artificial satellites. It centers on three major areas of development—navigational satellites, communications satellites, and weather observation and forecasting satellites.

Hirsch, Richard, and Joseph John Trento. *The National Aeronautics and Space Administration.* New York: Praeger, 1973. Describes the agency, its organization, its programs, and its relationship to other government agencies. Also contains an interesting account of the relationship between NASA and the press during the most active years of the crewed lunar exploration program. COMSAT's plans to launch satellites on the space shuttle as a way of cutting costs is described. Appendices include a particularly useful list of U.S. and Soviet crewed spaceflights launched from 1961 to 1971.

Ley, Willy. *Events in Space.* New York: David McKay, 1969. This volume contains summaries of all satellite programs, national and international, that were carried out during the 1960's. Particularly useful is a series of tables and glossaries that define space jargon and describe many of the satellite and rocket series of the decade. Also included are lists of satellite launches complete with launch dates and other information relative to the satellites and the programs with which they are associated.

National Aeronautics and Space Administration. *NASA, 1958-1983: Remembered Images.* NASA EP-200. Washington, D.C.: Government Printing Office, 1983. This oversized paperback traces various programs of the space agency from their beginnings. Included is a chapter on space sciences and another on Earth orbit applications, including communications. Illustrated with color photographs.

_____. *Space Network Users' Guide (SNUG).* Washington, D.C.: Government Printing Office, 2002. This users' guide emphasizes the interface between the user ground facilities and the Space Network, providing the radio frequency interface between user spacecraft and NASA's Tracking and Data-Relay Satellite System, and the procedures for working with Goddard Space Flight Center's Space Communication program.

Paul, Günter. *The Satellite Spin-Off: The Achievements of Space Flight.* Translated by Alan Lacy and Barbara Lacy. Washington, D.C.: Robert B. Luce, 1975. A survey of the commercial, scientific, and communications applications that developed from the space research of the 1960's and early 1970's. Contains a comprehensive account of the early, politically charged days of Intelsat. This book is written from the perspective of the European community, which makes it necessary reading for those seeking a broad understanding of Intelsat. In addition to communications applications, space medicine, meteorology, cartography, agriculture, and oceanography are discussed.

Shelton, William Roy. *American Space Exploration: The First Decade.* Boston: Little, Brown, 1967. Another historical account of the space program, with comprehensive listings of every American spaceflight launched during the period from 1957 to 1967.

Zimmerman, Robert. *The Chronological Encyclopedia of Discoveries in Space.* Westport, Conn.: Oryx Press, 2000. Provides a complete chronological history of all crewed and robotic spacecraft and explains flight events and scientific results. Suitable for all levels of research.

*Michael S. Ameigh*

# International Space Station: Crew Return Vehicles

*Date:* Beginning 1995
*Type of spacecraft:* Piloted spacecraft

*A lifeboat in space, the Soyuz TMA spacecraft affords astronauts aboard the International Space Station (ISS) an escape route in case an accident renders the station uninhabitable. A Soyuz taxi crew normally delivers a new Soyuz spacecraft to the station every six months. The taxi crew then returns to Earth in the older Soyuz spacecraft.*

### Key Figures

*Alfred J. Eggers, Jr.* (b. 1922), assistant director of Research and Development Analysis and Planning at what later became the NASA Ames Research Center
*John Muratore,* X-38 program manager at Johnson Space Center
*Bob Baron,* X-38 program manager at the Dryden Flight Research Center

### Summary of the Program

In 1957, while studying nose cone aerodynamics with his engineering team at the Ames Research Center in California, Alfred J. Eggers, Jr., had a brilliant insight. Blunt, symmetrical nose cones, the team found, could survive the intense heat created by friction during reentry into the atmosphere. Eggers reasoned that flattening its shape would give the nose cone aerodynamic lift; it could glide to Earth instead of simply plummeting. The lifting body was born, and with it the possibility of a spacecraft capable of a crew-controlled landing on Earth. Eggers's later work for the National Aeronautics and Space Administration (NASA) included the first crewed vehicle based on the concept of the M2-F1, an unpowered lifting body, tested in 1963. As well as an ancestor to the space shuttles, the M2-F1 led directly to X-38, an experimental vehicle that is to develop into the Crew Return Vehicle (CRV) for the International Space Station (ISS).

The M2-F1 proved difficult to control, but later designs based upon the SV5 lifting body shape established that these squat, blunt-nosed craft, powered or unpowered, could be landed safely. Of particular importance was the X-24A, flown twenty-eight times from 1969 to 1971. More than 7.3 meters long and 4.3 meters wide, it looked like a baking potato with a tapered nose at one end and three stubby fins at the other, but it withstood the fierce stresses of supersonic speed easily, reaching Mach 1.6, and was flown at altitudes up to 22 kilometers. It was both sturdier than the space planes envisioned by crewed spaceflight enthusiasts and simpler than rocket planes like the X-15.

As plans for a space station took shape during the early 1990's, NASA officials considered possible designs for a CRV, a critical safety feature. A tight budget limited their search to vehicles whose hull shape and internal subsystems could be developed from existing technology. After rejecting capsule and innovative lifting body candidates, NASA settled on an enlarged version of the X-24A and christened it the X-38. In 1995 the X-38 program was launched at the Johnson Space Center in Texas with a small budget and a small development crew. They first built a model that was 1.2 meters long and weighed 68 kilograms. R. Dale Reed, a retired NASA engineer, carried it to 3 kilometers in a Cessna airplane and dropped it to test its stability. The model glided for 3 kilometers and then landed under a parachute. On the basis of this success, NASA contracted with Sealed Composites, a California company, to build three full-size hulls to test handling capabilities in the atmosphere. Engineers at the Johnson Space Center installed the internal hardware and software, including computers and flight control systems. The majority of the equipment, about 80 percent, was off-the-shelf, although some of it had never before been applied to a crewed spacecraft.

The first completed X-38, known as Vehicle 131, was moved to the Dryden Flight Research Center in California in May, 1997, and the first test flight took place in March, 1998, when the vehicle was dropped from a B-52. Because of delays, a revised four-phase testing schedule called for continued atmospheric flight testing of the vehicles through 2001, followed in 2002 by a test return from low-Earth orbit after Vehicle 201 was lifted into space aboard a space shuttle. The final CRV vehicles based on the X-38 tests were slated for delivery to the ISS in 2006. Because the ISS had a crew before the CRV could become operational, they used Russian-built Soyuz spacecraft as lifeboats.

On April 29, 2002, NASA announced the cancellation of the X-38 program due to budget pressures associated with the International Space Station (ISS). The X-38 was two years short of completing its flight test phase. This cancellation was part of a larger plan that emerged late in 2002 wherein NASA would proceed with an Orbital Space Plane capable of both crew transport and crew re-turn missions from the ISS. The X-38 was developed to provide only a crew return capability from the ISS.

The Soyuz (Russian for "union") is a thoroughly tested, reliable vehicle, the workhorse of Russian piloted spacecraft since 1967. It was the primary ferry for the Salyut and Mir space stations. Weighing 6,000 kilograms, the spacecraft has a conical descent module, in which the crew rides; a cylindrical equipment module, which contains electrical power equipment and rocket engines and is attached aft of the Command Module; and forward of the descent module, a spherical orbital module with docking equipment. Unfortunately, the crew compartment holds only three people, and this limits the ISS crew.

The current Soyuz spacecraft (Soyuz-TMA) incorporates several changes to accommodate NASA requirements, including more latitude in the height and weight of the crew and improved parachute systems. During the two-year period following the STS-107 accident, the Soyuz-TMA delivered crews and up to 100 kilograms of payload to the station, and returned crews and up to 50 kilograms.

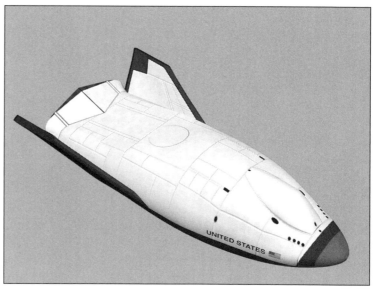

*An artist's concept of the X-38 Crew Return Vehicle (CRV). It was designed to stay in orbit for three years and to take the place of the Russian Soyuz capsule. (NASA)*

The Soyuz TMA craft is designed to operate for fourteen days and can be stored on orbit for 200 days. The vehicle is replaced by another either bringing a new permanent station crew or a short-stay visiting crew. The craft's length is 6.98 meters and has a maximum diameter of 2.72 meters. Its total habitable volume is 9.00 cubic meters ($m^3$). Of its 7,250-kilogram total mass, 900 kilograms are its nitrogen tetroxide and unsymmetrical dimethyl hydrazine propellants. It uses solar panels that span 10.60 meters and have an area 10.00 square meters for its electrical system.

Soyuz's orbital module is 3.0 meters long and has a diameter of 2.3 meters. It houses all the equipment that not needed for reentry, such as experiments, cameras, or cargo. It also contains the docking port and can be isolated from the descent module to act as an air lock if needed.

The descent module can safely carry a crew of three. It is 2.2 meters long and has a diameter of 2.2 meters. Used for launch and reentry, it is covered by a heat-resistant coating. After falling through the atmosphere, a braking parachute is deployed, followed by the main parachute, which slows the craft for landing. At 1 meter above the ground, solid-fuel braking engines mounted behind the heatshield fire to provide a soft landing at a speed of 2.6 meters per second. Its headlight shape—a hemispherical forward area joined by a barely angled cone (7°) to a classic spherical section heatshield—allows a small amount of lift to be generated due to the unequal weight distribution. This lift allows the module to be guided with a landing accuracy of 30 kilometers.

The service module is 2.3 meters long with a diameter of 2.1 meters. It has a pressurized container shaped like a bulging can that contains thermal control systems, the electric power supply, radio communications and telemetry, and instruments for orientation and control. A non-pressurized part of the service module contains the main engine, the liquid-fuel maneuvering engines, low-thrust attitude control engines, and the retrorockets for deorbit. Outside the service module are the sensors for the orientation system and the solar array.

### Contributions

The X-38 program was an exercise both in technology and in project management. Although four-fifths of the technology was off-the-shelf, the heat-resistant coating for the vehicle's thermal tiles was new and crucial, in that it was expected to toughen the tiles and make them last through far more flights than the tiles of the space shuttles, thereby reducing maintenance costs. Moreover, just because hardware comes off the shelf does not mean it is outdated; more to the point, it all must be integrated into a novel configuration for novel purposes, then tested and modified. For instance, the parafoil, developed by the U.S. Army, gives NASA extensive experience in testing a new landing system. The testing led to a new "zero-stage reefing" system to stabilize the parafoil and a control package to modify its shape in flight.

Nothing is a certainty when traveling into the void of space. The development of new technology, based upon tried and true hardware, can prove to be impossible. Reality rears its ugly head when ideas compete with budgetary constraints and schedule deadlines. The X-38 could not get past the developmental stage and Soyuz—the true workhorse of piloted spaceflight—came to its rescue.

### Context

According to Muratore, the purpose of the X-38 program was not just to build a new piloted spacecraft. It was to prove that NASA could do it for far less money than ever before. The agency estimated a new spacecraft-type CRV would cost about $2 billion to make flight-ready. It expected to build four operational CRVs based on the X-38 for about $500 million, one-quarter that amount, including the $80 million spent on building and testing the X-38 itself.

Neither supporters nor critics want to see a disaster aboard the Space Station, especially one that leads to the crew's death. Safety is paramount in design considerations, and the CRV was the last resort to keep crew members alive. Fortunately, the Russians had Soyuz to step in when the CRV project was canceled. This means that the crew still has a vehicle in which to return to Earth if a catastrophic failure leaves the station uninhabitable. There are many possible causes for this failure—destructive radiation from solar storms, hull breaches from impacts by meteoroids or space junk, and systems failures that shut down power or poison the air. Accidents aboard Russia's Mir Space Station during the late 1990's demonstrated chillingly the need to plan for such contingencies.

The X-38's could have served an even broader function if they were converted into utility vehicles to transfer crew and cargo between vehicles in or-

bit—in other words, space taxis. They could have become the first reusable, multipurpose craft dedicated to work primarily in space, even if in emergencies they could also land.

Once again, the reliability of the Russian Soyuz spacecraft has stepped in to save the International Space Station. Originally designated as a backup to the CRV, the Soyuz TMA continues to serve in the role of shuttlecraft and lifeboat.

**See also:** International Space Station: Living and Working Accommodations; International Space Station: Modules and Nodes; International Space Station: 2002; Lifting Bodies; Space Shuttle: Ancestors; Space Stations: Origins and Development.

## Further Reading

Crouch, Tom D. *Aiming for the Star: The Dreamers and Doers of the Space Age.* Washington, D.C.: Smithsonian Institution Press, 1999. A readable general history of spaceflight. The final chapter discusses NASA research efforts and goals, mentioning such reusable launch vehicles as the DC-XA, X-33, X-34, and X-38.

Hall, Rex, and David J. Shayler. *Soyuz: A Universal Spacecraft.* Chichester, England: Springer-Praxis, 2003. The authors review the development and operations of the reliable Soyuz family of spacecraft, including lesser-known military and unmanned versions. Using authentic Soviet and Russian sources, this book is the first-known work in the West dedicated to revealing the full story of the Soyuz series, including a complete listing of vehicle production numbers.

Messerschmid, Ernst, and Reinhold Bertrand. *Space Stations: Systems and Utilization.* Berlin: Springer-Verlag, 1999. This technically detailed review of space station designs and the science of the systems includes a history of space station concepts and a section about Crew Return Vehicles.

Shayler, David J. *Disasters and Accidents in Manned Spaceflight.* Chichester, England: Springer-Praxis, 2000. Covers all of spaceflight by a human crew—training, launch to space, survival in space, and return from space—followed by a series of case histories and a section on the International Space Station.

Thompson, Milton O. *At the Edge of Space: The X-15 Flight Program.* Washington, D.C.: Smithsonian Books, 2003. Thompson, a former test pilot, tells the story of the X-15 program.

Thompson, Milton O., and Curtis Peebles. *Flying Without Wings: NASA Lifting Bodies and the Birth of the Space Shuttle.* Washington, D.C.: Smithsonian Institution Press, 1999. Thompson, a test pilot of the vehicles mentioned in the book, and Peebles describe the design and development of the precursors to the X-38, including the X-24A, its direct ancestor. They also discuss its use as a CRV. For general readers.

Zubrin, Robert. *Entering Space: Creating a Spacefaring Civilization*. New York: Jeremy P. Tarcher/Putnam, 1999. An extended argument by an aerospace engineer for developing space travel, this book's vision is largely based upon contemporary science and technology. Two chapters discuss the crucial role of reusable vehicles, including the X-38 and the prototypes under development.

*Roger Smith and Russell R. Tobias*

# International Space Station: Design and Uses

*Date:* Beginning 1984
*Type of program:* Space station, piloted spaceflight

*The International Space Station (ISS) is an Earth-orbiting facility designed to house experimental payloads, distribute resource utilities, and support permanent human habitation for conducting research in the microgravity environment of space.*

### Key Figures

*Richard Kohrs,* director of NASA's Space Station Freedom program

*Eugene F. Kranz* (b. 1933), NASA Mission Operations director at the Johnson Space Center

*David C. Leestma* (b. 1949), Flight Crew Operations director at the NASA Johnson Space Center

*Bryan D. O'Connor* (b. 1946), Space Station Redesign director, responsible for transforming Space Station Freedom into the International Space Station

*Robert Phillips,* flight director scientist of NASA's Space Station Freedom project

*Wilber Trafton,* head of the U.S. portion of the International Space Station effort

### Summary of the Facility's Design

At the beginning of the twenty-first century, a permanently inhabited scientific laboratory accommodates space-based research for extended periods. The International Space Station (ISS) provides scientists in universities, industry, and government with new and exciting research opportunities and capabilities that greatly exceed those available from any other space program. It furnishes researchers with unprecedented opportunities to conduct scientific, technological, and commercial research in space to benefit humanity, foster international cooperation in the post-Cold War world, and inspire other such endeavors with its spirit of exploration and achievement. Upon its completion, ISS will have an on-orbit mass of approximately 420 metric tons (924,000 pounds). The assembly's complete configuration will be 88 meters long and 37 meters high, with an end-to-end wingspan width of 110 meters. It will orbit Earth at low altitude, ranging from 350 to 450 kilometers above the planet's surface, nominally inclined at 51.6° to Earth's equator. The American National Space Transportation System (NSTS) and Russian launch vehicles service it. Humankind's first permanent outpost in space, ISS pursues cutting-edge science, as well as augmenting ground-based research, to accelerate scientific discovery and spur practical applications.

ISS is designed and developed by a consortium of six space organizations: Agenzia Spaziale Italiana (ASI) of Italy, the Canadian Space Agency (CSA), the European Space Agency (ESA, including Belgium, Denmark, France, Germany, Italy, the Netherlands, Norway, Spain, and the United Kingdom), the United States' National Aeronautics and Space Administration (NASA), the National Space Development Agency (NASDA) of Japan, and the Russian Federal Space Agency (RSA). The facility has been designed to include seven laboratories for use by its member nations. Three of these laboratories are designated for Russian research modules. Two laboratories will be operated by the United States; one is the Centrifuge Accommodation Module (CAM) and the other, the U.S. Laboratory Module (Lab), a life science laboratory. One laboratory, the Columbus Attached Pressurized Module, will be used by the ESA. Finally, Japan will operate the Japanese Experiment Module (JEM), to which is attached an exposed platform, or "back porch," ten mounting spaces for experiments that require direct contact with the space environment, and a small robotics arm for payload operations on the exposed platform. In April, 1999, the JEM was named Kibo, which means "hope" in Japanese.

When it is fully assembled, ISS will house a crew of six members at any given time. Three primary

systems will support this crew: Flight Crew Integration (FCI), which provides accommodations necessary to ensure proper integration of human beings into the on-orbit space station and will maximize crew productivity; Crew Health Care System (CHeCS), a facility to maintain crew health, in part by monitoring the crew's environment; and the Flight Crew Equipment (FCE), hardware to support the crew's day-to-day living and working activities.

To ensure the safety of the crew and the successful completion of the experiments, ISS and its elements must be carefully designed, manufactured, assembled, and operated. Design considerations are largely dictated by the natural and induced environments that will affect ISS during the various mission phases: ground operation, ascent, orbital cargo transfer, and on-orbit operations. The envi-

*This drawing shows special equipment called a Flight Telerobotic System that astronauts could use to build and repair the space station.* (NASA CORE/Lorain County JVS)

ronmental elements critical to ISS design are neutral atmosphere, thermal conditions, plasma, ionizing radiation, meteoroids and other orbital debris, electromagnetic interference, contamination, acoustics, and microgravity. ISS must be engineered to withstand the effects of all these conditions.

First, ISS is designed to orbit through Earth's upper, tenuous atmosphere, at altitudes ranging between 300 and 500 kilometers above Earth's surface. Here, oxygen molecules are broken into atomic and ionic oxygen, which affects exposed surfaces severely. Although atmospheric density and pressure are considerably reduced, there is enough to generate drag, which results in orbit decay and aerodynamic torque. The attitude control and stabilization system is designed to handle the latter effects. For the former effects, protective coverings and proper materials are selected.

On orbit, ISS is exposed to intense thermal environments caused by the direct solar constant, solar radiation reflected from Earth, emission of thermal radiation from Earth, and "space sink" (deep-space temperature). All these thermal conditions have an effect on ISS, creating surface-temperature variations, thermal stress, possible damage to the vehicle's capacity for heat reflection, and variations in the efficiency of the solar arrays' ability to generate power. In designing the thermal control system, engineers considered materials and coatings that will maximize such properties as absorptivity, emissivity, and transmissivity.

Ionospheric plasma can have several adverse effects on the vehicle, creating plasma waves, arcing and sputtering at significant negative potential relative to the plasma, spacecraft charging at high inclinations, electrostatic discharge (a corona-like phenomenon), current imbalance between ISS and the ambient plasma, and geomagnetic field effects. To mitigate the potential difference between ISS structure and the plasma, a plasma contractor is directed to control this difference to within q40 volts of the ionospheric plasma potential.

ISS is continuously exposed to ionizing radiation, which has its greatest effect over a region of Earth called the South Atlantic Anomaly. The anomaly is an area in the Southern Hemisphere lying between the southern tip of Africa and South America, which can cause severe electromagnetic disturbances on spacecraft. This ionizing radiation consists of galactic cosmic rays and secondary radiation such as heavy ions from the inner trapped radiation belt. Ionizing waves cause both chemical and molecular bonds to break, thereby affecting living organisms, chemical processes, materials, and electronics. Such effects could jeopardize life, let alone disturb scientific experiments aboard ISS. Proper shielding from these forms of radiation is essential to mitigate those radiation effects.

Meteoroids and other orbital debris may range in size from approximately one micron (one one-thousandth of one millimeter) to several centimeters in diameter. Impact from debris at the lower end of this range can, among other effects, degrade performance of solar arrays. To overcome this problem, engineers use multilayer shielding to limit the effects of small particle impacts. At the upper end of the range, centimeter-sized debris can defeat ISS's protection system, causing potentially catastrophic loss of equipment, crew, or the station itself. Characterization and mitigation of the effects of such large debris are extremely difficult. The best solution, if a collision cannot be avoided, is to provide redundant paths for critical electrical, fluid, and gaseous lines.

Another environmental hazard, electromagnetic interference (EMI), can produce equipment malfunctions, noise, and crosstalk between wires. Electromagnetic compatibility is necessary to help the system function as intended without degradation and malfunction. This compatibility can be achieved through the use of shielding and proper grounding of equipment and wiring.

Contamination generally consists of neutral molecules and particulates released through outgassing from the ISS and spacecraft. Such contaminants, in combination with solar ultraviolet radiation and atomic radiation, can alter the thermo-optical properties of the vehicle's surface, in turn degrading systems designed to maintain thermal control, field of view of the payload, optical telescopes, signal intensity, and solar arrays; increased background noise and interference patterns may also result. In-flight contamination effects are minimized in part by choosing proper materials.

Acoustical conditions on ISS, produced by the combined noise from all operation systems, can cause human fatigue, thereby degrading human-machine system effectiveness. Therefore, the facility is designed so that total noise does not exceed that of the typical office environment.

A critical aspect of the ISS environment, finally, is its prolonged access to an extremely low level of net gravity acceleration, or microgravity. This "weightlessness" affects the crew's activities, noise levels, equipment vibrations, transient disturbances, payload locations, station configurations, and activity schedules. In response to microgravity, for example, designers must locate payloads near the station's center of gravity, implement vibration abatement designs, and minimize equipment noise.

### Contributions

All activities on Earth, including human life, have evolved under the direct influence of gravity, one of the four fundamental forces of nature. Given the microgravity environment of space, ISS provides scientists with a laboratory in which to explore the role that gravity plays in the fundamental principles that govern basic physical and biological processes—such as the burning of fuels, the solidification of metals, the growth of crystals, the life cycle from conception to old age, and the systems of the body, ranging from the musculoskeletal system to the immune system.

Scientists can use low gravity to test fundamental theories of physics with degrees of accuracy that far exceed the capability of the Earth-bound science. Observing physical processes in low gravity will expand our understanding of the structure of matter, as well as changes in states of matter (including those changes responsible for the long-sought high-temperature superconductivity), properties and behaviors of fluids (liquids and gases), and complex combustion processes. Findings in mate-

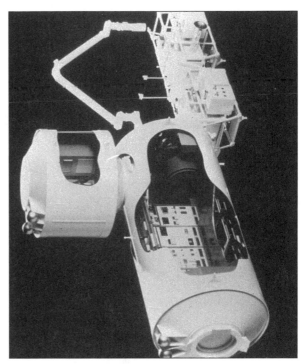

*This drawing shows Japan's design for an "outdoor" platform with a remotely controlled arm to do the work. This allows the experiments to be exposed to space while the astronauts remain inside.* (NASA CORE/Lorain County JVS)

rial science alone will have very broad applications in industrial processes, including the production of semiconductors, glasses, metal alloys, polymers, and ceramics. Fundamental knowledge of fluid behavior is essential to industrial activities, ranging from energy production to material engineering. Combustion is used to produce 85 percent of Earth's energy; even small improvements in combustion efficiency will have large economic and environmental benefits.

In the low-gravity environment, the biological sciences—including the study of the development, growth, and internal processes of plants and animals—will benefit medical, agricultural, pharmaceutical, and other bioindustries. Orbital research enhances the ability to describe proteins, enzymes, and viruses at the molecular level, enabling scientists to develop new drugs and vaccines much more quickly and effectively. Study of the processes controlling the growth of human tissue outside the body (tissue culturing) may lead to an improved understanding of normal and abnormal (cancerous) tissue development, with important implications for the development of new drug therapies and applications for transplant research. New insights into physiology—such as how the heart and lungs function, the growth and maintenance of muscle and bone, perception, cognition, balance, and the regulation of the body's many systems—will be gained in the low-gravity environment.

Such groundbreaking new insights have been achieved already from the experiments flown aboard the space shuttle. It has been demonstrated that when gravity's effects are substantially reduced, secondary forces, such as surface tension, can predominate and become more significant than previously observed on Earth. Metals, for example, solidify through the growth of microscopic, tree-shaped "dendrites," which develop an intricate structure in the low-gravity environment of space. Based on orbital research, new mathematical techniques have been developed to model the behavior of molten metals. The flame of a candle in low gravity forms a spherical shape, unlike the conical shape it takes on Earth. Under normal Earth gravity, gas occupies the top of the pipe when flowed with liquid; in reduced gravity, the gas is surrounded by the liquid, forming a core down the middle of the pipe.

Such low-gravity behaviors offer seemingly endless opportunities for discovery and use of new properties and new materials. Certain proteins grown in space, for example, form superior, regularly shaped crystals that provide the precise structural development data needed for the design of new drugs and chemicals; Earth-grown protein crystals are irregular in shape, thus making them unusable for gathering such precise data. Data from space on human insulin will lead to a drug that will bind insulin, thereby improving treatments for diabetes. A colon cancer tumor grown in a rotating-wall bioreactor shows cell groupings similar to those found in tumors in the human body;

conventionally grown cells form smaller and less developed groups, which are much less valuable for research. The bone loss observed in space bears strong similarities to bone loss associated with aging on Earth. Astronauts lose the same percentage of their total bone mass over a period of eight months in space as the average person loses in ten years between the ages of fifty and sixty. Drugs and physical therapies are being developed to combat and possibly reverse some of these effects in space, and their outcomes will have significant implications for aging on Earth.

Systems of "telemedicine" (the use of telecommunications to exchange medical data and images) that have been developed to maintain astronauts' health have the potential to reduce health care costs and improve the quality of health care on Earth. Human occupation of ISS inevitably necessitates such advanced life-support technologies as crop growth research capable of improving hydroponics and other controlled production systems; improved air and water quality sensors and analyzers and air revitalization systems; automatic systems for identifying microbes to detect a broad range of infectious diseases; advanced robotics and remote operation systems; improved power generation and storage systems that include particularly flexible thin-film solar arrays; and advanced waste-processing and recycling techniques to reduce pollution. For example, ISS is designed with closed-loop systems for water use and conservation of air in which about 30 liters (8 gallons) of water per astronaut will be used for hygiene and cooking, compared with an average of 606 liters (160 gallons) per person on Earth. Applied on Earth, this technology could benefit water-starved regions of the planet. Another example of potential Earth applications from the design of ISS is a more efficient burner: If engineers succeed in building a burner that is only 2 percent more efficient, and if such a burner were used routinely on Earth, the savings alone—approximately $7.9 billion per year—would pay the United States' costs for ISS in less than two years.

Knowledge gained from ISS will therefore exceed the by-products of its scientific experiments:

Even the basic life-support and health-maintenance systems developed for this facility will both improve life on Earth and make possible the next generation of long-term habitation of space.

### Context

ISS is the largest internationally cooperative scientific program in history, drawing on the resources and expertise of thirteen nations. It is making good use of existing member nations' space technology, capabilities, and hardware to design and build a better space station for less money in a relatively short period of time. Compared with Freedom (the space station proposed by the United States in January, 1984), ISS has nearly twice the power (110 versus 56 kilowatts), almost double the pressurized volume (1.2 cubic meters versus 0.65 thousand cubic meter), twice the number of laboratories (six versus three), and a larger full-time crew (six versus four).

ISS has benefited from the post-Cold War atmosphere of U.S.-Russian scientific cooperation and, in turn, may help to ensure that reductions in defense and foreign aid can continue by providing meaningful jobs for aerospace workers and a peaceful outlet for nuclear and missile expertise. Such cooperative efforts can reduce the risk of nuclear proliferation and slow traffic in high-technology weaponry to developing nations, all while developing space technology at a fraction of the cost of the arms buildup during the Cold War era.

Finally, ISS provides scientists and engineers with their first opportunity to design, develop, operate, and study the systems required to maintain a habitable, long-term spacecraft cabin environment. As a result, a new generation may be inspired to excel in science, mathematics, and engineering, just as, in 1969, the children of an earlier generation were inspired by the first view of Earth from the Moon.

**See also:** International Space Station: Development; International Space Station: Living and Working Accommodations; International Space Station: Modules and Nodes; International Space Station: U.S. Contributions.

### Further Reading

Bizony, Piers. *Island in the Sky: Building the International Space Station.* London: Aurum Press, 1996. Bizony's text and 160 photographs describe the plans for building the International Space Station.

Boeing Missiles and Space Division, Defense and Space Group. *International Space Station Alpha: Reference Guide.* Washington, D.C.: National Aeronautics and Space Administration, 1995. This handbook provides a quick reference for Space Station personnel and top-level sources of ISS data in a convenient size. References to applicable documents are provided for the user to obtain controlled data.

Fawcett, Michael K. *From Centralized to Distributed: The Evolution of Space Station Command and Control.* Houston: NASA Space Station Program Office, 1995. This paper documents the evolution in the command and control concept for the core systems of the Space Station.

Harland, David M. *The Space Shuttle: Roles, Missions, and Accomplishments.* Hoboken, N.J.: John Wiley, 1998. *The Space Shuttle* is written thematically, rather than purely chronologically. Topics include shuttle operations and payloads, weightlessness, materials processing, exploration, Spacelabs and free-flyers, and the shuttle's role in the International Space Station.

Haskell, G., and Michael Rycroft, eds. *International Space Station: The Next Space Marketplace.* Boston: Kluwer Academic, 2000. Examines uses of ISS from all quarters: commercial, scientific, technological, and educational.

Lindmoyer, A., J. Theall, and J. Williamsen. *Meteoroid and Orbital Debris Risk Mitigation for the International Space Station, 1995.* Proceedings of the forty-sixth International Astronautical Congress, Oslo, Norway, 3-5, Rue Mario-Nikis, 751015 Paris, France. This paper examines the threat to ISS imposed by the meteorid/orbital debris and risk management approaches implemented to mitigate the threat.

Logsdon, John M. *Together in Orbit: The Origins of International Participation in the Space Station.* Washington, D.C.: National Aeronautics and Space Administration, 1998. Describes the politics and science behind the effort to bring together many nations in the building of a space station.

McCurdy, Howard E. *The Space Station Decision: Incremental Politics and Technical Choice.* Baltimore: Johns Hopkins University Press, 1990. The author is a professor of public affairs at American University in Washington, D.C. The events that led up to the decision (1984) to build a permanently occupied space station in low-Earth orbit provide his primary subject matter in the present monograph, but the author's deeper interest has to do with the politics of Big Science. The story is arrestingly told in this nicely produced volume, which provides thirteen pages of plates plus detailed notes and references.

National Aeronautics and Space Administration, Johnson Space Center. *Space Station White Papers.* Part 1. NASA OA311. Houston: Author, 1994. This pamphlet describes the needs and uses of the International Space Station.

National Aeronautics and Space Administration, Space Station Program Office. *A Science and Technology Institute in Space: The Promise of Research on the International Space Station.* Houston: Author, 1995. This fact book gives an overview of science, technology, and research facilities on the Space Station, as well as a complete assembly diagram.

Shayler, David J. *Disasters and Accidents in Manned Spaceflight.* Chichester, England: Springer-Praxis, 2000. Covers all aspects of spaceflight: training, launch to space, survival in space, and return from space.

Spencer, Ron. *International Space Station Assembly Sequence.* Houston: NASA Space Station Program Office, 1995. This paper outlines the assembly of ISS in a sequence of phases.

Von Bencke, Matthew J. *The Politics of Space: A History of U.S.-Soviet/Russian Competition and Cooperation in Space.* Boulder, Colo.: Westview Press, 1996. This book chronicles the efforts of the United States and the Soviet Union (later Russia) to overcome their political animosities and explore space together. It looks at their respective foreign and domestic policies; military, civil, and commercial influences; and top executive, legislative, and institutional politics. The book examines their separate and joint endeavors from 1945 through 1997.

*M. A. K. Lodhi*

# International Space Station: Development

*Date:* Beginning 1984
*Type of program:* Space station, piloted spaceflight

*The International Space Station (ISS)—an Earth-orbiting facility designed to house experimental payloads, distribute resource utilities, and support permanent human habitation in space—was designed to be completed in three phases. Its first phase began in 1993.*

### Key Figures

*Richard Kohrs,* director of NASA's Space Station Freedom program
*Eugene F. Kranz* (b. 1933), NASA Mission Operations director at the Johnson Space Center
*David C. Leestma* (b. 1949), Flight Crew Operations director at the NASA Johnson Space Center
*Bryan D. O'Connor* (b. 1946), Space Station Redesign director, responsible for transforming Space Station Freedom into the International Space Station
*Robert Phillips,* flight director scientist of NASA's Space Station Freedom project
*Wilber Trafton,* head of the U.S. portion of the International Space Station effort

### Summary of the Facility's Development

The Soviet Union launched the world's first space station, Salyut 1, in 1971—a decade after launching the first human into space. The United States sent its first space station, the larger Skylab, into orbit in 1973, and it hosted three crews before being abandoned in 1974. Russia continued to focus on long-duration space missions and in 1986 launched the first modules of the Mir Space Station.

In 1993, U.S. president Bill Clinton called for a redesign of the Space Station program in order to reduce costs and include more international involvement. The National Aeronautics and Space Administration (NASA) presented three different options, of which the White House selected the one dubbed Alpha. The project thus became known as International Space Station Alpha. The name "International Space Station" (abbreviated MKS in Russian) represents a neutral compromise that ended a disagreement about a proper name for the station. The initially proposed name, "Space Sta-

tion Alpha," was rejected by Russia because it would have implied that the station was something fundamentally new, whereas the Soviet Union already had operated eight orbital stations long before the ISS launch. The Russian proposal to name the Space Station "Atlant" was in turn rejected by the U.S., which was worried about that name's similarity to *Atlantis,* the name of a shuttle orbiter.

The redesigning effort simplified many subsystems, resulting in a planned completion of the station assembly thirteen months earlier than originally scheduled, a permanent crew presence starting five years sooner than previously anticipated, and savings of $2 billion. A single management team replaced previously dispersed management, and an Integrated Product Team (IPT) became responsible for making decisions affecting product suitability, quality, and cost.

When the assembly is completed, a permanent on-orbit crew of six, who will conduct research, will inhabit the ISS. It will have an on-orbit mass of approximately 420 metric tons. The completed truss, measuring 108.4 meters in length, will hold systems requiring exposure to space, such as communication antennae, external cameras, mounts for external payloads, equipment for temperature control, transport around the station's exterior during spacewalks, robotic servicing, stabilizing, and attitude control. The truss will also support eight Suntracking solar array pairs, providing the station with 110 kilowatts of electrical power—twice as much as the canceled Freedom space station was designed to have and more than ten times as much as the U.S. Skylab or the Russian Mir Complex. The total pressurized volume upon completion of the assembly will be 1,200 cubic meters, as opposed to Freedom's 651 cubic meters, with five Laboratory Modules (versus three) and six crew members (versus four). The station will have a total cost of $25.1 billion and annual operating costs of $1.3 billion (versus $2.6 billion).

The ISS is divided into segments defined according to the international partners' levels of responsibility. Segments are constructed from one or more functional elements that include modules, nodes, truss structures, solar arrays, and thermal radiators.

Modules are pressurized cylinders of the habitable space on board the station. They may contain research facilities, living quarters, and any vehicle operational systems and equipment the astronauts may need to access. Nodes connect the modules and offer external station access for purposes such as docking, extravehicular activity (EVA) access, and unpressurized payload access. Trusses are erector-set-like girders that link the modules with the main solar power arrays and thermal radiators. Together, the truss elements form the Integrated Truss Structure. Solar Arrays collect and convert solar energy into electricity for the station and its payloads. Thermal radiators emit excess thermal energy into space.

An assembly sequence for the ISS was developed that requires fifty assembly and utili-

zation flights (including resupply missions) on five different launch vehicles from three countries—Russia, Japan, and the United States. The facility's development has been broken into three phases. Phase I, which operated from March, 1995, through May, 1998, was a joint program between the existing space facilities of both NASA and the Russian Space Agency (RSA), during which operational concepts required for ISS assembly were tested several times in a sort of dress rehearsal. Phase II, spanning November, 1998, to August, 2001, began the actual construction of ISS and depended on the successful endeavor of Phase I. The hardware must be flown in a particular order as the functionality of ISS is incrementally increased. Phase II concluded after ten assembly flights, resulting in a sustainable station permanently occupied by humans and capable of continuous on-orbit payload operations. Phase III, which began in September, 2001 and will end in 2010, will see growth of the configuration as power is increased and science platforms are added.

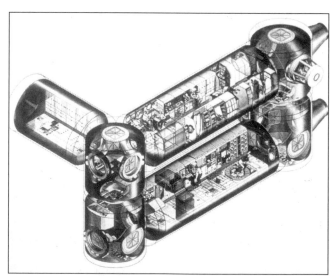

*This drawing shows how the large modules can be stacked on the International Space Station. The smaller modules would be used for storage and equipment. Because the space station needs to be as self-sufficient as possible, the equipment for recycling air and water is in these modules.* (NASA CORE/Lorain County JVS)

Phase I used existing assets—primarily U.S. shuttle orbiters and the Russian Mir Complex—to build joint space experience, start joint scientific research, and test and validate hardware. This phase began with Russian cosmonaut Sergei Konstantinovich Krikalev's flight aboard space shuttle *Discovery* during February 3-11, 1994. In March, 1995, U.S. Astronaut Norman E. Thagard lifted off in the Russian Soyuz TM-21 spacecraft with two Russian cosmonauts for a three-month stay on Mir. On June 29, 1995, space shuttle *Atlantis* docked with the Mir station for the first time and picked up three cosmonauts, as well as experiment samples and other items from the station, for return to Earth.

Astronaut Shannon W. Lucid, from STS-76, was on board the Mir Space Station from March, 1996, until September, 1996, when John E. Blaha, on STS-79, replaced her. Astronaut Jerry M. Linenger replaced Blaha in January, 1997, on mission STS-81. STS-84, launched on May 15, 1997, carried C. Michael Foale to replace Linenger. STS-86 was launched on time on September 26, 1997, with David A. Wolf taking over the responsibilities of Foale on board Mir. A nighttime launch of STS-89 took place on January 23, 1998. This mission returned Wolf to Earth; Andrew Thomas took over his duties. The final shuttle-Mir mission launch took place on June 2, 1998. STS-91 rendezvoused with Mir and returned Andrew Thomas to Earth on June 12. Several concepts were tested for possible use on the ISS, including a new kind of solar power system in which the Sun heats a working fluid to drive a turbine, generating more electricity than the photovoltaic solar arrays currently used.

Phase II consisted of twelve U.S. and four Russian assembly flights. The United States committed to funding spacecraft development and the Russians to funding booster and launch services. The first flight of this phase took off from the Baikonur Cosmodrome atop a Russian Proton launch vehicle in November, 1998. The unpiloted rocket's cargo was the U.S.-purchased space tug, known by the Russian acronym FGB (*funktsionalya-gruzovod blokor,* or Functional Cargo Block). The twenty-ton

FGB, called Zarya, which means "sunrise" in Russian, provides its own communications, power systems, attitude control, and propulsion during early assembly of ISS, as well as berthing ports for additional modules. Zarya provided temporary control of ISS until the Service Module (SM) reached the Space Station in July, 2000.

The first assembly space shuttle flight to ISS was launched in December, 1998. The STS-88 *Endeavour* mission delivered Unity Node 1, two storage racks, and two Pressurized Mating Adapters (PMAs) to the front end of the FGB. Unity Node 1 is a pressurized module that serves as a structural core for future additions to the U.S. segments and provides a pressurized tunnel between the U.S. and Russian modules. It carries the first U.S. power system, provides storage space for supplies and equipment, has berthing ports and attachment points for modules and the station's large truss, and includes a docking port for shuttle orbiters. PMA-1 is a pressurized tunnel that allows the crew to move between the U.S. and Russian segments of the ISS. PMA-2 is a shuttle docking port that will be used for most of the subsequent shuttle missions.

In July, 2000, the Zvezda Service Module, with living and working space for three crew members, was launched from Baikonur and docked with FGB to assume the permanent duty of primary attitude control, communications, and life support for the ISS. Russian Progress Spacecraft—automated freighters—will periodically dock at the SM to deliver supplies, fuel, and equipment. Two Soyuz spacecraft provide assured crew-return capability without the space shuttle present. The Soyuz craft can operate for fourteen days and can be stored on orbit for two hundred days.

Prior to the arrival of first permanent crew (Expedition One), STS-92 delivered the Integrated Truss Structure (ITS) Z1, Pressurized Mating Adapter 3, and four Control Moment Gyros (CMGs) in October. ITS Z1 was an early exterior framework to allow solar arrays to be temporarily installed on Unity for power. PMA-3 provided a shuttle docking port for installation of permanent solar arrays on STS-97 in November and for the Destiny Laboratory Module, which arrived at the station in February, 2001, on STS-98 (ISS-5A).

Soyuz TM-31 was launched from the Baikonur Cosmodrome on October 31, 2000, at 07:52:47 UTC. It carried the three-person Expedition One crew. They remained on board for five months. The second truss piece launched, the P6 Truss, contains a solar array, radiators, and plumbing. It was mounted to the Z1 Truss by the STS-97 crew but will be moved to the port side main truss by the STS-119 crew after the five port truss pieces have been assembled by other crews.

STS-98 delivered and mated the U.S. Destiny Laboratory Module to the forward port of the Unity Node in February, 2001. This module allows astronauts to work inside the pressurized facility and conduct research in numerous scientific fields. Destiny comprises three cylindrical sections and two end cones that contain the hatch openings through which astronauts enter and exit the module.

Launched on STS-100 in April, 2001, the Canadarm2 is a bigger, better, smarter version of the space shuttle's original robotic arm. Canadarm2 is capable of handling large payloads of up to 116,000 kilograms and can assist with docking the space shuttle. Canadarm2 can move end over end to reach many parts of the ISS in an inchworm-like movement.

In July, 2001, the STS-104 crew delivered the Quest Airlock and installed it on the station's Unity Node. Quest is designed to be the primary air lock for the station, designed to be able to host spacewalks with both American and Russian space suits. It consists of two segments: the Equipment Lock, which stores space suits and equipment, and the Crew Lock, from which astronauts can exit into space. The Quest Airlock was necessary because American space suits will not fit through a Russian air-lock hatch and have different components, fittings, and connections. The Quest Airlock is designed to contain equipment that can work with both types of suit. The addition of the air lock signaled the completion of Phase II of ISS, meaning the orbiting station had taken on a degree of self-sufficiency and capabilities for full-fledged research in the attached Laboratory Module.

In September, 2001, Progress M-SO1 (ISS-4) delivered the Pirs (Russian for "pier") Docking Compartment and docked it to the nadir port on Zvezda. The Docking Compartment provides docking ports for the Soyuz and Progress (resupply) spacecraft. It also has an air lock to accommodate spacewalks by Russian cosmonauts wearing Orlan-M space suits.

Phase III of ISS includes increased power supply. This will be followed by increased science capability and an increase in the permanent on-orbit crew size to six upon completion of the assembly. This phase will include thirteen U.S., eight Russian, two European, one Japanese, and two Japanese/American assembly flights. Most of the Phase III assembly will constitute the addition of new international laboratories.

The ISS Truss forms the backbone of the Space Station, with mountings for unpressurized logistics carriers, radiators, solar arrays, and other equipment. The S0 (Segment Zero) Truss, the center segment of eleven integrated trusses, was attached to the top of the Destiny Laboratory on April 11, 2002, by the STS-110 crew. The S0 Truss acts as the junction from which external utilities are routed to the station's pressurized modules. The port and starboard trusses attach to the S0 Truss. The P1 and S1 Trusses connect to the port and starboard sides of the S0 Truss respectively. The STS-112 crew attached the S1 Truss on October 10, 2002, to begin Phase III. The STS-113 crew attached the P1 Truss on November 26, 2002.

On February 1, 2003, NASA suffered the loss of *Columbia* and its crew of seven. The sixteen-day mission of STS-107 was dedicated to research in physical, life, and space sciences. *Columbia* broke up approximately 16 minutes before landing, during reentry over Texas, en route to the Kennedy Space Center. After a more than a two-year hiatus, the shuttle returned to the ISS in July, 2005. The assembly flights would not resume until after a second logistics flight, however. The first four Russian solar arrays were planned for delivery in 2006 to

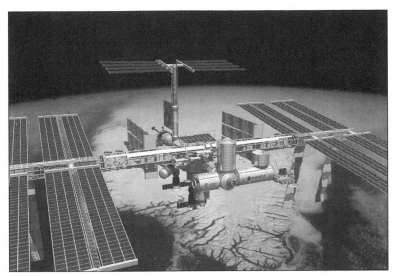

*A 1996 concept drawing for the International Space Station at completion.* (NASA-MSFC)

maximize available power for science. Truss Segments Port 3 and 4 (P3/P4) were to be delivered and installed on the first port truss segment (P1), directly opposite the S1 Truss, which provides the same structural support. The remaining starboard and port truss segments will be delivered thereafter.

The final four Russian solar arrays are to be delivered in July, 2006, marking the independence of the Russian and U.S. assemblies. Russia will modify the backup Functional Cargo Block (FGB-2) into a Multipurpose Laboratory Module (MLM). It will replace the canceled Universal Docking Module (UDM). The MLM should be ready for launch in November, 2006. The MLM will be used for experiments, docking, and cargo, serve as a work and rest area for the crew, and be equipped with an attitude control system that can be used as a backup by the ISS. It will be docked onto the Zarya Control Module side docking port.

Node 2 is the Space Station's "utility hub," containing eight racks that provide air, electrical power, water, and other systems essential to support life on the spacecraft. Node 2 is the second of three connectors between the ISS modules and is set for launch in January, 2007, aboard STS-120.

Columbus is a science laboratory contracted by ESA. The laboratory is a cylindrical module very similar in shape to the Multi-Purpose Logistics Module. Once launched on STS-123 (scheduled for April, 2007), it will be attached at Node 2's starboard side, with the cylinder pointing outward. The module contains ten International Standard Payload Racks (ISPRs). Four racks are located on the forward side, four on the aft side, and two in overhead locations. Three deck racks are filled with life-support and cooling systems; the remaining deck rack and two overhead racks are storage racks. An additional four payloads can be attached as external payloads outside the port cone.

The Special Purpose Dexterous Manipulator, or Canada Hand, is a smaller, two-armed robot capable of handling the delicate assembly tasks currently managed by astronauts during spacewalks. It is scheduled to be attached to the ISS during STS-125 in November, 2007.

The Japanese Experiment Module (JEM), named Kibo (Japanese for "hope"), consists of four components: the Pressurized Module (PM), the Exposed Facility (EF), the Experiment Logistics Module (ELM), and the Remote Manipulator System (RMS). The Pressurized Module is the core component. It is of cylindrical shape, 11.2 meters long and 4.4 meters in diameter. It contains ten International Standard Payload Racks (ISPRs). The Exposed Facility, also known as the "terrace," is located outside the port cone of the PM (which is equipped with an air lock hatch). Experiments are fully exposed to the space environment here. The Experiment Logistics Module contains a pressurized section to serve the PM and an unpressurized section to serve the EF. It will be located atop the port side of the PM and is highly movable. It is intended as a storage and transportation module.

The Remote Manipulator System is a robotic arm, mounted at the port cone of the PM, intended to service the EF and to move equipment from and to the ELM, which will be launched in October, 2008, aboard STS-129. The Pressurized Module will arrive in January, 2009, aboard STS-130.

The Science Power Platform (SPP) is a Russian element of the ISS that will be delivered by the space shuttle in July, 2009, on STS-132. It will provide additional power for the ISS as well as roll-axis control capability for the orbital facility. The SPP will have four solar arrays and a robotic arm provided by the ESA dedicated to maintaining the SPP.

The Centrifuge Accommodations Module (CAM) will provide controlled gravity for experiments and the capability to expose a variety of biological specimens to artificial gravity levels between 0.01g and 2g. The CAM will be attached to Node 2 of the ISS in October, 2009, during the STS-133 mission.

The Cupola is a U.S. element of the ISS that will provide direct viewing of robotic operations and the space shuttle payload bay, as well as a spectacular observation point for the astronauts to view Earth. It is approximately 2 meters in diameter and 1.5 meters tall. It has six side windows and a top window, all of which are equipped with shutters to protect them from micro meteorite damage. The Cupola is designed to be attached to the Unity Module and is scheduled for launch in April, 2010.

Although the assembly portion of the ISS is scheduled to end in 2010, subsequent flights to the International Space Station will support it for the next ten years or more.

### Context

The concept of an outpost orbiting Earth and occupied by human beings dates back to 1869, when an American novelist, Edward Everett Hale, published a science-fiction tale called "The Brick Moon," in which he envisioned a satellite as a navigational aid for ships at sea. In 1903, a Russian schoolteacher, Konstantin Tsiolkovsky, wrote a novel, *Beyond the Planet Earth,* based on sound science, in which he described orbiting space stations. Hermann Oberth, a Romanian, coined the term "space station" in 1923.

The American Robert H. Goddard launched the first liquid-fueled rocket in 1927. In 1928, Herman Noordung, an Austrian, published the first blueprint for a space station. The ideas of Oberth and Noordung greatly influenced the rapid advancement of rocketry during World War II, especially in Germany, where the V-2 rocket emerged. The range of this missile was about 500 kilometers, and it became a prototype for both U.S. and Soviet rockets after the war. In May, 1955, the Soviets began working on the launch site Baikonur, from where the world's first intercontinental ballistic missile was test-flown in August, 1957.

On October 4, 1957, the Soviets launched the first artificial satellite to orbit Earth, Sputnik 1, which triggered further Cold War competition between the United States and the Soviet Union. The United States established the National Aeronautics and Space Administration (NASA) in 1958. In 1959, NASA recommended that a space station be established. The Soviet Union launched the first human being, Yuri A. Gagarin, into space on April 12, 1961.

In January, 1969, Soyuz 4 and Soyuz 5 docked in Earth orbit—the first time that two piloted spacecraft came together in space—and the Soviet Union declared that they constituted the first space station. The same year, the U.S. spacecraft Apollo smoothly landed on the Moon, and NASA proposed a one-hundred-person permanent space station called Space Base. On April 19, 1971, the Soviet Union launched the first piloted space station, Salyut. The United States launched the Skylab space station in May, 1973, hosting three different three-person crews for twenty-eight, fifty-six, and eighty-four days, respectively. Together, the Skylab crews conducted more than one hundred experiments and observations and performed an unscheduled spacewalk to correct a solar array. Skylab reentered Earth's atmosphere in 1979 and was destroyed during reentry.

International cooperation in the endeavor to establish a space station may be considered to have begun when the European Space Agency (ESA) formally agreed to participate in the space program jointly with the United States in 1973. The first international spacecraft docking, the Apollo-Soyuz Test Project (ASTP), took place in 1975. In April, 1981, the United States launched the first space shuttle and in May, 1982, founded the Space Station Task Force, which proposed international participation in the Space Station's development, construction, and operation. In January, 1984, President Ronald W. Reagan called for participation by U.S. allies and gave the space station the name Freedom.

By the spring of 1985, Japan, Canada, and the member nations of ESA (Belgium, Denmark, France, Germany, Italy, the Netherlands, Norway, Spain, and the United Kingdom) each had signed a bilateral memorandum of understanding with the United States for participation in the station project; formal agreements were signed in September, 1988. In 1992, the United States agreed to buy Soyuz vehicles from Russia, and the joint program between the U.S. space shuttle and Russian space station Mir began. Freedom underwent six major redesignings until 1994, when the task of ISS's development was broken out into its current three phases.

When assembling the Space Station, many factors must be considered: not only technical but also economic and political. In incrementally building from one stage of assembly to another, an active, controllable spacecraft must be maintained at all times. Hardware added from each assembly flight significantly changes the mass properties and functionality of orbiting vehicles, resulting in a unique spacecraft at each stage. The assembly sequence must therefore be designed so that sufficient docking ports for supply and future assembly are preserved at every step along the way. System functionality must periodically be increased to meet the needs of the growing configuration. Interdependence and development schedules of international elements form another driving factor in the developmental sequence. Assembly elements for each stage are also constrained by the available volume and payload mass capabilities of applicable launch vehicles. Program requirements, such as early science capabilities, also must be factored in when determining the developmental sequence.

To maximize the return on investment from the station's finite on-orbit lifetime of fifteen years, it is desirable to complete ISS as fast as possible. However, the availability of applicable launch vehicles, funding constraints, and hardware development have to be phased over a period of time.

**See also:** Cooperation in Space: U.S. and Russian; Funding Procedures of Space Programs; International Space Station: Crew Return Vehicles; International Space Station: Design and Uses; International Space Station: Living and Working Accommodations; International Space Station: Modules and Nodes; International Space Station: U.S. Contributions; Space Stations: Origins and Development.

**Further Reading**

Bizony, Piers. *Island in the Sky: Building the International Space Station.* London: Aurum Press, 1996. Bizony tells how the International Space Station will be assembled in orbit during an extended sequence of shuttle flights, dockings, and spacewalks over a period of five years. With unrivaled access to NASA and the astronautic sources worldwide, he provides a lively text that contains a wealth of information. There are one hundred photographs, sixty in color.

Boeing Missiles and Space Division, Defense and Space Group. *International Space Station Alpha Reference Guide.* Washington, D.C.: National Aeronautics and Space Administration, 1995. This document provides generic background information on the design and assembly sequence of the International Space Station. It is illustrated with computer images showing the station in its various configurations.

Bond, Peter. *The Continuing Story of the International Space Station.* Chichester, England: Springer-Praxis, 2002. Bond describes the development and evolution of space stations, with particular emphasis on the International Space Station, beginning with the revolution that began in 1970, when Salyut 1, the world's first space station, was sent into orbit by the Soviet Union.

Fawcett, Michael K. *From Centralized to Distributed: The Evolution of Space Station Command and Control.* Houston: NASA Space Station Program Office, 1995. This paper documents the evolution in the Command and Control concept for the core systems of the Space Station since 1986.

Harland, David M., and John E. Catchpole. *Creating the International Space Station.* London: Springer-Verlag London, 2002. An overview of the events leading to the construction and commissioning of the ISS.

Haskell, G., and Michael Rycroft, eds. *International Space Station: The Next Space Marketplace.* Boston: Kluwer Academic, 2000. Examines applications of the ISS: commercial, scientific, technological, and educational.

Logsdon, John M. *Together in Orbit: The Origins of International Participation in the Space Station.* Washington, D.C.: National Aeronautics and Space Administration, 1998. Describes the politics and science behind the effort to bring together many nations in the building of a space station.

McCurdy, Howard E. *The Space Station Decision: Incremental Politics and Technical Choice.* Baltimore: Johns Hopkins University Press, 1990. The author is a professor of public affairs at American University in Washington, D.C. The events that led up to the decision (1984) to build a permanently occupied space station in low-Earth orbit provide his primary subject matter in the present monograph, but the author's deeper interest has to do with the politics of Big Science. The story is arrestingly told in this nicely produced volume, which provides thirteen pages of plates plus detailed notes and references.

National Aeronautics and Space Administration. *Future Space Shuttle Missions.* Washington, D.C.: National Aeronautics and Space Administration, 1996. This paper lists the planned space shuttle flights through STS-113 in May, 2001, and includes all of the Phase II and most of the Phase III flights.

_____. *International Space Station Assembly Sequence: Revision E (March, 2000 Planning Reference).* Washington, D.C.: Government Printing Office, 2000. This is the official assembly sequence planning reference for the International Space Station. It details the assembly schedule and the various components of the Station.

_____. *International Space Station (ISS): Phase I-III Overview.* Houston: Author, 1995. A brief overview of the three phases of ISS's development.

_____. *International Space Station: Russian Space Stations.* Houston: Author, 1995. An outline of the historical development of the Russian space program.

_____. *International Space Station: The International Space Station Program Is Underway.* Houston: Author, 1995. A brief description of three phases of development of the ISS, with a detailed summary of Phase I.

_____. *International Space Station: U.S. Space Station History.* Houston: Author, 1995. Examines the historical development of the International Space Station.

Spencer, Ron. *International Space Station Assembly Sequence.* Houston: NASA Space Station Program Office, 1995. Offers a brief account of the three phases of the assembly.

Von Bencke, Matthew J. *The Politics of Space: A History of U.S.-Soviet/Russian Competition and Cooperation in Space.* Boulder, Colo.: Westview Press, 1996. This book chronicles the efforts of the United States and the Soviet Union (later Russia) to overcome their political animosities and explore space together. It looks at their respective foreign and domestic policies; military, civil, and commercial influences; and top executive, legislative, and institutional politics. The book examines their separate and joint endeavors from 1945 through 1997.

*M. A. K. Lodhi and Russell R. Tobias*

# International Space Station: Living and Working Accommodations

*Date:* Beginning 1984
*Type of spacecraft:* Space station, piloted spacecraft

*A crew of six persons is planned to live and work both in and out of the International Space Station (ISS). Among the earliest goals in building the ISS was the completion of basic systems to create an environment that would accommodate long-term human habitation.*

### Key Figures

*Richard Kohrs,* director of NASA's Space Station Freedom program
*Eugene F. Kranz* (b. 1933), NASA Mission Operations director at the Johnson Space Center
*David C. Leestma* (b. 1949), Flight Crew Operations director at the NASA Johnson Space Center
*Bryan D. O'Connor* (b. 1946), Space Station Redesign director, responsible for transforming Space Station Freedom into the International Space Station
*Robert Phillips,* flight director scientist of NASA's Space Station Freedom project
*Wilber Trafton,* head of the U.S. portion of the International Space Station effort

### Summary of the Facility's Accommodations

The International Space Station, or ISS, represents a global partnership of sixteen nations. When completed, the 450-metric-ton space station will include six laboratories and provide more space for research than any spacecraft ever built. Internal volume of the Space Station will be roughly equal to the passenger cabin volume of a 747 jumbo jet.

More than forty spaceflights over five years and at least three space vehicles—the space shuttle, the Russian Soyuz rocket, and the Russian Proton rocket—will deliver the various space station components to Earth orbit. Assembly of the more than one hundred components will require a combination of human spacewalks and robot technologies.

The International Space Station has been permanently occupied since November, 2000. Until the STS-107 accident grounded the shuttle fleet in February, 2003, three crew members spent tours on the station every six months. During the hiatus

occasioned by the accident, the crew was reduced to two for safety considerations. Once the station is completed, an international crew of up to seven will live and work in space for periods of between three and six months. Crew Return Vehicles will always be attached to the station to ensure the safe return of all crew members in the event of an emergency.

Through November, 2002, when the STS-113 crew delivered the P1 Truss and the Expedition Six crew, the station had twelve major components in place. These included Zarya, Unity, Zvezda, the Z1 Truss, the P6 Integrated Truss, the Destiny Laboratory Module, the Space Station Remote Manipulator System (SSRMS, or Canadarm2), the Joint Airlock, Pirs, the S0 Truss, the S1 Truss, and the P1 Truss.

The 19,323-kilogram pressurized module Zarya ("sunrise" in Russian) was launched atop a Russian

Proton 8K82K vehicle on November 20, 1998. The flight is designated as ISS-1 A/R in the assembly sequence. The "1" indicates the first major component launch. The "A" refers to American participation (construction or launch) and the "R" refers to Russian participation. The Zarya Control Module, known in Russia as the Functional Cargo Block (*funktsionalya-gruzovod blokor,* or FGB), was funded by NASA and built by Khrunichev in Moscow under subcontract from Boeing Aircraft Company for NASA. The Khrunichev Space Center, home of Rosaviakosmos (Russian Aviation and Space Agency), is the Russian state space research and production center. Zarya's design is taken from the TKS military station resupply spacecraft of the 1970's and the later 77KS Mir modules. Zarya is a self-supporting active vehicle. It provides propulsive control capability and power through the early assembly stages, fuel storage, and rendezvous and docking capability to the service module.

In early December, 1998, the STS-88 (ISS-2A) mission saw space shuttle *Endeavour* attach the Unity Module to Zarya, initiating the first ISS assembly sequence. Unity was launched passive with two Pressurized Mating Adapters (PMAs) attached and one stowage rack installed inside. PMA-1 connects U.S. and Russian elements. PMA-2 provides a shuttle docking location. PMA-1 was fitted to Unity's aft port. The crew conducted three spacewalks to attach PMA-1 to Zarya. In addition to connecting to Zarya, Unity serves as a passageway to the U.S. Destiny Laboratory Module and the Joint Airlock. Unity is 5.5 meters long, 4.6 meters in diameter, and fabricated of aluminum. It has six hatches that serve as docking ports for the other modules.

Two shuttle missions—STS-96 (ISS-2A.1) in June, 1999, and STS-101 (ISS-2A.2a) in May, 2000—supplied the two modules with tools and cranes and delivered provisions in preparation for the arrival of the Zvezda Service Module and the station's first permanent crew. Zvezda (Russian for "star"), built and financed by Russia, docked with the ISS on July 26 at 00:45 Coordinated Universal Time (UTC) and became the third major component of the station. Zvezda provides the early station living quarters, life-support system, electrical power distribution, data-processing system, flight control system, and propulsion system. It also provides a communications system that includes remote command capabilities from ground flight controllers. Although many of these systems have been or will be replaced by later station components, Zvezda will always remain the structural and functional center of the Russian segment of the International Space Station. Zvezda has a mass of 19,051 kilograms, a wingspan of 29.7 meters, and a length of 13.1 meters.

STS-106 (ISS-2A.2b) and the crew of *Atlantis* visited the station in September, 2000, to deliver supplies and outfit Zvezda in preparation for the first permanent crew, Expedition One. Prior to that crew's arrival, STS-92 (ISS-3A) delivered the Integrated Truss Structure (ITS) Z1, Pressurized Mating Adapter 3 (PMA-3), and four Control Moment Gyros (CMGs) in October. ITS Z1 was an early exterior framework to allow solar arrays to be temporarily installed on Unity for power. PMA-3 provided a shuttle docking port for installation of permanent solar array on STS-97 (ISS-4A) in November and for the Destiny Laboratory Module, which arrived at the station in February, 2001, on STS-98 (ISS-5A). The CMGs provide nonpropulsive (electrically powered) attitude control by means of a constant-rate flywheel mounted on a set of gimbals. The flywheel orientation can be changed by torquing the gimbals, which redirects the rotor's angular momentum. In accordance with conservation of angular momentum, any change in the CMG momentum is transferred to the ISS, producing a change in the station's pitch (up or down motion), yaw (left or right motion), and roll (clockwise or anticlockwise motion).

Soyuz TM-31 (ISS-2R) was launched from the Baikonur Cosmodrome on October 31, 2000, at 07:52:47 UTC. When it docked with the ISS on November 2, the first human presence on the Space Station was established. The three-person crew included Commander William M. Shepherd, Soyuz Commander Yuri Pavlovich Gidzenko, and Flight

Engineer Sergei Konstantinovich Krikalev. The Soyuz spacecraft, which remained docked to the station after the crew returned to Earth, provides assured crew return capability without the space shuttle present.

The Soyuz craft can operate for fourteen days and can be stored on orbit for two hundred days. The vehicle is replaced by another, either bringing a new permanent station crew or a short-stay visiting crew. Soyuz's orbital module houses all the equipment that is not needed for reentry, such as experiments, cameras, and cargo. It also contains the docking port and can be isolated from the descent module to act as an air lock if needed. The descent module can safely carry a crew of three and is used for launch and reentry. The service module contains thermal control systems, the electric power supply, radio communications and telemetry, and instruments for orientation and control. A nonpressurized part of the service module contains the main engine, the liquid-fuel maneuvering engines, low-thrust attitude control engines, and the retrorockets for deorbit.

The second truss piece launched was the P6 Integrated Truss, which contains a solar array, radiators, and plumbing. It was mounted to the Z1 Truss by the STS-97 (ISS-4A) crew but will be moved to the port side main truss by the STS-119 (ISS-15A) crew after the five port truss pieces have been assembled by other crews.

STS-98 (ISS-5A) delivered and mated the U.S. Destiny Laboratory Module to the forward port of the Unity Module in February, 2001. Astronauts work inside the pressurized facility to conduct research in numerous scientific fields. Scientists throughout the world will use the results to enhance their studies in medicine, engineering, biotechnology, physics, materials science, and Earth science. The aluminum U.S. Laboratory Module is 8.5 meters long and 4.3 meters wide. It comprises three cylindrical sections and two end cones that contain the hatch openings through which astronauts enter and exit the module. In Destiny are five life-support system racks that provide electrical power, cooling water, air revitalization, and temperature and hu-

midity control. In March, 2001, six additional racks were flown to Destiny on STS-102. Four standoffs provide raceways for module utilities—interfaces for ducting, piping, and wiring to be run to and from the individual racks and throughout the lab. Twelve racks that will provide platforms for a variety of scientific experiments will follow on subsequent missions. In total, Destiny can hold twenty-three racks—six each on the port and starboard sides and overhead, and five on the deck.

Launched on STS-100 (ISS-6A) in April, 2001, the Canadarm2 is a bigger, better, smarter version of the space shuttle's original robotic arm: 17.6 meters long when fully extended with seven motorized joints, a mass of 1,800 kilograms, and a diameter of 35 centimeters. The arm is capable of handling large payloads of up to 116,000 kilograms and can assist with docking the space shuttle. Canadarm2 can move end over end to reach many parts of the Space Station in an inchworm-like movement, limited only by the number of Power Data Grapple Fixtures (PDGFs) on the station. PDGFs located around the station provide power, data, and video to the arm through its Latching End Effectors (LEEs). The arm can also travel the entire length of the Space Station using the Mobile Base System (MBS). The MBS is a work platform that moves along rails covering the length of the space station and provides lateral mobility for the Canadarm2 as it traverses the main trusses. It was added to the station during STS-111 (ISS-UF-2, Utility Flight 2) in June, 2002.

In July, 2001, the STS-104 (ISS-7A) crew delivered the Quest Joint Airlock Module and installed it on the station's Unity Node. Expedition Two Flight Engineer Susan J. Helms used the Canadarm2 to lift Quest from the orbiter's payload bay. At the same time, STS-104 mission specialists Michael L. Gernhardt and James F. Reilly conducted a spacewalk, guiding Helms's movements as she installed Quest onto the station. The Quest Airlock is designed to be the primary air lock for the International Space Station, able to host spacewalks with both the American and Russian space suits. Before Quest was attached, Russian spacewalks were required to be done from the service module and American spacewalks were possible only when the space shuttle was docked. The Quest Airlock consists of two segments, the Equipment Lock, which stores space suits and equipment, and the Crew Lock, from which astronauts can exit into space. It was derived from the space shuttle air lock, although it was significantly modified to waste less atmospheric gas when used. It was attached to the starboard side of the Unity Module during STS-104. It has mountings for four high-pressure gas tanks, two containing oxygen and two containing nitrogen, which provides for atmospheric replenishment to the American side of the Space Station, most specifically for the gas lost after a hatch opening during a spacewalk.

The Quest Airlock was necessary because American space suits will not fit through a Russian airlock hatch and have different components, fittings, and connections. The air lock is designed to contain equipment that can work with both types of suits. Quest is designed to provide an environment where astronauts can "camp out" before a spacewalk in a reduced-nitrogen atmosphere to purge nitrogen from their bloodstreams and avoid decompression sickness in the pure-oxygen atmosphere of the space suit. The Boeing Aircraft Company built the air-lock and tank systems out of aluminum. The air lock is 5.5 meters long, has a diameter of 4 meters, weighs 6,064 kilograms, and encompasses 34 cubic meters of volume.

In September, 2001, Progress M-SO1 (ISS-4) delivered the Pirs ("pier" in Russian) Docking Compartment and docked it to the nadir port on Zvezda. RKK Energia manufactured the Russian Pirs. The Docking Compartment is similar to the ones used on the Mir Space Station. It provides docking ports for the Soyuz and Progress (resupply) spacecraft. It also has an air lock to accommodate spacewalks by Russian cosmonauts wearinCompartment g Orlan-M space suits. The Docking is 5 meters long, 2.2 meters in diameter, and has a mass of 3,630 kilograms.

Progress M-SO1 was the designation given to the service module section of a Progress M supply craft. The Pirs docking and air lock module for the ISS replaced the standard cargo and fuel sections of Progress. It also carried an astronaut chair, a space suit, a small crane, and some equipment for the Zvezda Module of the ISS. The Progress M-SO1 later undocked from the Pirs nadir port to leave it free for future dockings. Pirs gave extra clearance from the station for spacecraft docking underneath Zvezda. Pirs is scheduled to be moved to the Science Power Platform, side docking port, above the Zvezda Module in 2009.

The ISS Truss forms the backbone of the Space Station, with mountings for unpressurized logistics carriers, radiators, solar arrays, and other equipment. The S0 (Segment Zero) Truss, the center segment of eleven integrated trusses, was attached to the top of the Destiny Laboratory Module on April 11, 2002, by the STS-110 (ISS-8A) crew. The S0 Truss acts as the junction from which external utilities are routed to the station's pressurized modules. These utilities include power, data, video, and ammonia for the Active Thermal Control System. The truss also provides a mounting point for electronic equipment such as the Main Bus Switching Units, four of the DC-to-DC Converter Units, and four Secondary Power Distribution Assemblies. Also mounted on S0 are the station's four GPS (Global Positioning System) antennae and two rate gyroscopes.

The port and starboard trusses attach to the S0 Truss. The P1 and S1 Trusses connect to the port and starboard sides of the S0 Truss respectively. Each contains carts to transport the Canadarm2 and astronauts to worksites, as well as radiators to dissipate heat. Going out to the port side, the P3 and P4 Trusses together contain a pair of solar arrays, a radiator, and a rotary joint to aim the solar arrays. The P5 Truss provides a mounting for the P6 Truss, once it has been moved from the Z1 Truss. Similarly, the S3, S4, S5, and S6 Trusses provide two more pairs of solar arrays, more radiators, and another rotary joint. The STS-112 (ISS-9A) crew attached the S1 Truss on October 10, 2002. The STS-113 (ISS-11A) crew attached the P1 Truss on November 26, 2002.

On February 1, 2003, NASA suffered the loss of *Columbia* and its crew of seven. The sixteen-day mission of STS-107 was dedicated to research in physical, life, and space sciences. *Columbia* broke up approximately 16 minutes before landing, during reentry over Texas, en route to the Kennedy Space Center.

For more than two years, the remaining shuttle fleet—*Atlantis, Discovery,* and *Endeavour*—was grounded while improvements were made to the External Tank. With only the Soyuz spacecraft to act as a transport and rescue craft, the permanent crew was reduced to two persons. No construction on the station was attempted during this period. Shuttle flights to the Station resumed in July 2005. However, loss of foam insulation from *Discovery*'s external tank during the STS-114 launch resulted in further delays in the schedule. The launch of the second logistics mission of 2005, STS-121, was delayed until the foam issue could be addressed. Station construction, planned for early 2006, nevertheless would not resume until these issues were satisfactorily resolved.

### Contributions

In general, the scientific utilization of ISS will inject a fresh vigor into core sciences, from fundamental physics to biotechnology. Of particular value will be the data gathered on the humans who live and work on ISS. ISS will offer an unprecedented and unparalleled opportunity to researchers from universities, industry, and governments to expand the promising research begun on Russia's Mir Space Station and the United States' space shuttle by conducting hundreds of high-quality science and technology experiments year-round. Living on ISS will also support global environmental observations and high-energy astrophysics research.

Technology to ensure the crew members' safety against the hostile space environment and to maximize their productivity will, in turn, have important benefits for life on Earth. Advanced life-support technology has multiple applications to health care, pollution control, environmental cleanup, agriculture, and various industries.

Especially important, aging due to living in the "gravity-free" environment of ISS will provide insights into osteoporosis, muscle atrophy, the neurovestibular system, and cardiovascular function that will help medical researchers understand the role of mechanical forces and countermeasures in human health as people age. By studying the rapid bone loss, molecular and cellular mechanisms governing muscle loss, adaptation of the nervous system to sensory and motor input, and blood circulation of astronauts and PIs as they live and work in space, scientists will be better able to provide countermeasures to overcome those threats, not only in space but probably on Earth as well. Nutrition plans, drugs, and exercise regimens such as strength-training techniques developed for living in space will be used to improve the quality of life for the Earth-bound bedridden and as a preventive measure to delay the consequences of aging. Knowledge from living and working in space will be applied to treating cardiovascular and cardiopulmonary diseases, such as heart attacks and strokes, which affect millions of people on Earth.

ISS's value is expected to extend beyond the pure sciences into the social and political sciences: As the world redefines itself in the wake of the Cold War, the international crew who work and live together in ISS (and other people involved in the program on the ground) are expected to act as a catalyst for international cooperation. ISS may serve as a "sociopolitical laboratory," in which the power of nations to live peacefully together on Earth and to work together to achieve shared goals is tested and developed.

### Context

Since the time when long-term occupation of space was considered a possibility, a major concern has been the effects of microgravity on humans living and working in space. Some of the effects of "weightlessness" are amusing and fun. Astronauts in space do not need to exert as much force in doing work or physical exercise. They can move from one place to another while lying flat. They can leave an object anywhere in space, and it will stay there in relation to them. While shaving, for example, an astronaut can leave his shaving stick in front of his face and it will hang there while he applies shaving cream.

More serious effects of microgravity, however, pose concerns for those planning the future of space travel. Chief among these is a phenomenon called "space aging," which manifests itself in several forms. Studies and observations show that osteoporosis—a debilitating disease, occurring mostly in elderly people, that leads to bone fragility due to loss of bone mass—is a serious health threat to those working and living in space. Astronauts' and PIs' weight-bearing bones become less dense during spaceflight, mimicking the effects of osteoporosis. Most significant, the rate of bone loss in space is ten times as fast as that for people between the ages of fifty and sixty on Earth. Patients on Earth—the elderly and the bedridden—are less physically active than average. The space crew, although physically active, are not subjected to the same mechanical forces because they are not working against gravity's pull. Scientists believe that their decreased mechanical loading—the stress on bones caused by everyday activity—may contribute to their bone loss.

Space research has also shown that space travelers lose muscle mass and strength during spaceflight at a very rapid rate—again, because of the effects of microgravity. During two weeks of spaceflight, astronauts lose as much muscle mass as people on Earth lose during decades of aging. In space, muscles do not have to bear the weight of the crew member's body. This loss is similar to that experienced on Earth when a person is wheelchair-bound or is undergoing bed rest. Muscle loss is a particular risk to space crews, for it can hinder their ability to evacuate the living/working module safely during emergencies.

Cardiovascular functions also undergo different stresses in microgravity. Heart rate, blood pressure, and lung function all change during spaceflight. Over time, interactions between the kidneys and the endocrine system alter the balance of fluids in space crews' bodies. The function of baroreceptors—pressure sensors in the neck arteries that play a role in controlling blood pressure—degrades during spaceflight. Baroreflex abnormalities seen in space crews are similar to those found in the elderly or in patients who have had heart attacks.

During spaceflights, crew members' blood and heart volumes decrease, and they are less able to conduct vigorous activities. This "decoordinating" is similar to the effects of prolonged bedrest after injury or severe infection. During spaceflight, crew members experience changes in the neurovestibular system—the sensory and motor network that includes the inner ear and provides sense of balance—which lead to disorientation, dizziness, motion sickness, instability, and falls. The incidence and prevalence of vestibular disorders are known to increase with age.

Countermeasures for overcoming or minimizing the threat to space crews' health are being evaluated. To some extent space crews are being trained to live with some of these effects. Exercise, strength-training techniques, nutrition plans, and use of pharmaceutical drugs to simulate bone formation limit bone loss and muscle atrophy.

**See also:** Cooperation in Space: U.S. and Russian; Ethnic and Gender Diversity in the Space Program; International Space Station: Design and Uses; International Space Station: Development; International Space Station: U.S. Contributions; Space Shuttle: Life Science Laboratories; Space Stations: Origins and Development.

## Further Reading

Bizony, Piers. *Island in the Sky: Building the International Space Station.* London: Aurum Press, 1996. Bizony tells how the International Space Station will be assembled in orbit during an extended sequence of shuttle flights, dockings, and spacewalks over a period of five years. With unrivaled access to NASA and the astronautic sources worldwide, he provides a lively text that contains a wealth of information. There are one hundred photographs, sixty in color.

Bond, Peter. *The Continuing Story of the International Space Station*. Chichester, England: Springer-Praxis, 2002. Bond describes the development and evolution of space stations, with particular emphasis on the International Space Station, beginning with the revolution that began in 1970, when Salyut 1, the world's first space station, was sent into orbit by the Soviet Union.

Harland, David M., and John E. Catchpole. *Creating the International Space Station*. London: Springer-Verlag London, 2002. Overview of the events leading to the construction and commissioning of the ISS.

Haskell, G., and Michael Rycroft, eds. *International Space Station: The Next Space Marketplace*. Boston: Kluwer Academic, 2000. Examines the ISS from all perspectives: commercial, scientific, technological, and educational.

Logsdon, John M. *Together in Orbit: The Origins of International Participation in the Space Station*. Washington, D.C.: National Aeronautics and Space Administration, 1998. Describes the politics and science behind the effort to bring together many nations in the building of a space station.

McCurdy, Howard E. *The Space Station Decision: Incremental Politics and Technical Choice*. Baltimore: Johns Hopkins University Press, 1990. The author is a professor of public affairs at American University in Washington, D.C. The events that led up to the decision (1984) to build a permanently occupied space station in low-Earth orbit provide his primary subject matter in the present monograph, but the author's deeper interest has to do with the politics of Big Science. The story is arrestingly told in this nicely produced volume, which provides thirteen pages of plates plus detailed notes and references.

National Aeronautics and Space Administration. *Extravehicular Activity*. NASA Mail Code HS-30. Houston: Author, 1995. Contains information on the ISS's capability to support tasks such as assembly, maintenance, servicing, and repair of the Space Station and other large space structures.

_____. *Extravehicular Robotics*. NASA Mail Code HS-30. Houston: Author, 1995. Information on the ways in which the EVR system facilitates what EVA does.

_____. *Flight Crew Systems ISS*. NASA Mail Code HS-30. Houston: Author, 1995. This is ISS information on flight crew support and facilities.

_____. *International Space Station Assembly Sequence: Revision E (March, 2000 Planning Reference)*. Washington, D.C.: Government Printing Office, 2000. This is the official assembly sequence planning reference for the International Space Station. It details the assembly schedule and the various components of the Station.

_____. *A Science and Technology Institute in Space: The Promise of Research on the International Space Station*. Houston: Author, 1995. This fact book gives an overview of science and technology on the Space Station and research facilities on it. Contains a complete assembly diagram.

National Aeronautics and Space Administration, Johnson Space Center. *Environmental Control and Life Support Systems*. NASA Mail Code HS-30. Houston: Author, 1995. This technical data manual contains information on a safe, habitable, shirt-sleeve environment within ISS pressurized areas for the on-orbit crew.

National Aeronautics and Space Administration, Space Station Program Office. *International Space Station Fact Book*. Houston: Author, 1995. Provides some statistics of ISS and its assembly schedule.

_____. *Space Station White Papers*. Part 7. NASA Mail Code OA311. Houston: Author, 1994. This white paper discusses the issues of aging in space and provides responses to commonly asked questions about the Space Station.

Shayler, David J. *Disasters and Accidents in Manned Spaceflight*. Chichester, England: Springer-Praxis, 2000. Covers human spaceflight from training and launch to survival in the space environment and return to Earth.

Von Bencke, Matthew J. *The Politics of Space: A History of U.S.-Soviet/Russian Competition and Cooperation in Space*. Boulder, Colo.: Westview Press, 1996. This book chronicles the efforts of the United States and the Soviet Union (later Russia) to overcome their political animosities and explore space together. It looks at their respective foreign and domestic policies; military, civil and commercial influences; and top executive, legislative and institutional politics. The book examines their separate and joint endeavors from 1945 through 1997.

*M. A. K. Lodhi and Russell R. Tobias*

# International Space Station: Modules and Nodes

*Date:* Beginning January 25, 1984
*Type of spacecraft:* Space station, piloted spacecraft

*When the International Space Station (ISS) is completed, a crew of up to seven astronauts and mission specialists from various nations will live and work in space for periods of between three and six months, during which they will perform research in materials science, microgravity, space medicine, and Earth observation. Crew Return Vehicles will always be attached to the ISS to ensure the safe return of all crew members in the event of an emergency.*

### Key Figures

*Richard Kohrs*, director of NASA's Space Station Freedom program

*Eugene F. Kranz* (b. 1933), NASA Mission Operations director at the Johnson Space Center

*David C. Leestma* (b. 1949), Flight Crew Operations director at the NASA Johnson Space Center

*Bryan D. O'Connor* (b. 1946), Space Station Redesign director, responsible for transforming Space Station Freedom into the International Space Station

*Robert Phillips*, flight director scientist of NASA's Space Station Freedom project

*Wilber Trafton*, head of the U.S. portion of the International Space Station effort

### Summary of the Facility's Modules and Nodes

The National Aeronautics and Space Administration (NASA) proposed a space station to President Richard M. Nixon, as part of a combined space shuttle/space station program, in the early 1970's. However, budget restrictions allowed only the space shuttle to be developed at that time. President Ronald W. Reagan gave his support for a space station on January 25, 1984, when he announced in his State of the Union Address: "I am directing NASA to develop a permanently piloted space station, and to do it within a decade."

NASA developed plans for this space station, named Freedom, but increasing costs forced a series of redesigns. International partners, including the European Space Agency (ESA), Canada, and Japan, were brought into the Space Station project to assume some of the financial burden. By the early 1990's this had evolved into a project in which the United States, Europe, and Japan would each provide pressurized laboratory facilities, linked together to form the Space Station.

On June 17, 1993, President Bill Clinton called for the United States to "work with our international partners to develop reduced-cost, scaled-down version of the original Space Station [and] seek to enhance the opportunities for international participation in the Space Station." Russia, which has extensive experience in operating Space Station Mir, launched in 1986, was brought into the group of international partners. A major purpose of adding Russia to the Space Station team was to persuade Russia to comply with international restrictions on the export of missile technology by giving Russian missile design and production facilities work to do. Russia was expected to receive $400 million in aid for the launch services and hardware it contributes to the Space Station.

In 1994 and 1995, the fifth major redesign since 1984 transformed the project into the International Space Station. This new design made significant use of Russian space technology, much of it developed for Russia's Mir Space Station. Some of the Space Station hardware, originally scheduled to be launched on the U.S. space shuttle, would be carried into space on less costly, robotic Russian Proton boosters. In addition, the international partners would have access to the Russian Mir Space Station in the period before the International Space Station is completed, providing them with opportunities to begin research experiments using space station facilities.

The International Space Station uses the modular construction technique initially developed for Space Station Freedom. Each module is about 4 meters in diameter and about 10 meters long, similar in size to a school bus. Individual modules are assembled on the ground and launched into space in the cargo bay of the U.S. space shuttle.

Because of the increasing amount of debris orbiting the Earth, the modules are shielded to protect them from puncture by small orbiting fragments. Objects larger than a few centimeters in size are tracked by ground-based radar, and the Space Station can use the shuttle's propulsion system to alter its orbit, if necessary, to avoid close encounters with larger orbiting debris.

The modules are pressurized to one atmosphere, the pressure experienced at Earth's surface, and the crew is able to work inside the Space Station in a "shirt-sleeve environment," without the use of space suits. Each module has an air lock at one or both ends, allowing an individual module to main-tain air pressure if a leak should develop somewhere else in the station.

The modules are linked together using nodes. Box-shaped tunnels with connectors mating to the air locks on the modules provide the connections between the modules. The module and node design of the International Space Station provides flexibility as well as the capability for expansion, because additional modules can be added to unused air locks on the nodes. One such expansion would allow an increase in the number of crew members by adding an additional habitation module. Additional research space could be made available by adding laboratory modules, and these modules could be outfitted with newly developed research equipment, not available at the time of the original design.

When completed, the International Space Station will include major elements constructed by the United States, Europe, Canada, Japan, and Russia. Individual modules provided by the United States, Europe and Japan would be linked together, drawing their electrical power from a common source.

*The first Multi-Purpose Logistics Module (MPLM), built by Alenia Aerospazio of Italy, designed to provide storage and additional work space for up to two astronauts when docked to the International Space Station. (NASA)*

Canada has provided the 17-meter-long robotic Canadarm2 used for assembly and maintenance tasks on the Space Station. The European Space Agency is building a pressurized laboratory to be launched on the space shuttle and logistics transport vehicles to be launched on the Ariane 5 launch vehicle. Japan is building a laboratory with an attached exposed exterior platform for experiments as well as logistics transport vehicles. Russia is providing two research modules. The first, the service module, has its own life-support and habitation systems, a science power platform of solar arrays that can supply about 20 kilowatts of electrical power for logistics transport vehicles. The second is a Soyuz spacecraft, which will aid in crew return and transfer. In addition, Brazil and Italy are contributing some equipment to the station through agreements with the United States.

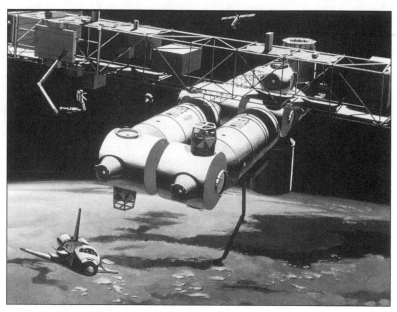

*An artist's rendition of the laboratory, habitation, and logistics modules docked to the space station. They are designed to fit in the cargo bay of the shuttle used to transport them into orbit.* (NASA CORE/Lorain County JVS)

The United States provided the Destiny Laboratory Module, where experiments are conducted. Destiny uses a rack and tray mounting system for experiments and research equipment, allowing considerable flexibility. Equipment for a particular experiment is flown to the Space Station preassembled into trays that can be inserted directly into the racks mounted on the walls of the module. Following completion of the experiment, if the equipment is no longer needed, the tray can be flown back to Earth, and a new tray containing equipment for a different experiment can be placed into the rack.

The ESA will develop the Columbus Orbital Facility, to be built jointly by Germany and Italy. The Columbus Orbital Facility is scheduled for launching in 2007 on STS-122. The Columbus module will also use the rack-mounting system for experiments. The experiments on the Destiny Laboratory and

on the Columbus Orbital Facility are expected to focus on materials processing, microgravity, and space medicine.

The Japanese National Space Development Agency will provide the Japanese Experiment Module (JEM), scheduled for a September, 2007 launching. The JEM, named Kibo ("hope" in Japanese), is significantly different in design from the U.S. and European laboratory modules. The pressurized cylinder of the Japanese module is only half the length of the Columbus module. There are air locks on both ends of the JEM. The air lock on the end opposite the node connecting the JEM to the rest of the Space Station will open onto an "exposed facility," a box-shaped platform exposed to the vacuum of space. The exposed facility is referred to as the "back porch" of the JEM. Crew members can easily reach the exposed facility through the air lock to deploy, retrieve, or activate experiments. The exposed facility will provide convenient attachment points for experiments that must be done in a vacuum, instruments to observe Earth, and certain types of materials processing which, for safety reasons, cannot be done within the station.

Node 2, built for NASA under a barter agreement with the European Space Agency, is set for launch aboard STS-120 in December, 2006. In exchange for Node 2, NASA will launch the European Columbus Laboratory on board a space shuttle mission the following March. The aluminum node is 7.2 meters long and 4.4 meters in diameter. Its pressurized volume is 70 cubic meters, and its launch mass is approximately 13,600 kilograms. The Japanese Experiment Module will be linked to Node 2 on the station when it arrives in September, 2007. Node 2 controls and distributes resources from the Integrated Truss Structure and Destiny Laboratory Module to the connected elements: the European Columbus Laboratory, Centrifuge Accommodation Module, Japanese Experiment Module, Multi-Purpose Logistics Module, and H II Transfer Vehicle. It also provides a working base point for the Space Station Remote Manipulator System.

The Centrifuge Accommodations Module (CAM) will provide controlled gravity for experiments and the capability to expose a variety of biological specimens to artificial gravity levels between 0.01g and 2g. It will simultaneously provide two different artificial gravity levels and provide partial g and hyper-g environments for specimens to investigate altered gravity effects and g thresholds. It will provide short-duration and partial g and hyper-g environments for specimens to investigate temporal effects of gravity exposure and provide Earth simulation environment on ISS to isolate microgravity effects on specimens. Most important, the CAM will provide Earth simulation environment on ISS to allow specimens to recover from microgravity effects and in situ 1g controls for specimens in microgravity. The CAM will attach to Node 2 of the ISS.

The Cupola is a U.S. element of the ISS that will provide direct viewing of robotic operations and the space shuttle payload bay, as well as an observation point from which the astronauts can view Earth. Designed by Alenia from Italy, it is approximately 2 meters in diameter and 1.5 meters tall. It has 6 side windows and a top window, all of which are equipped with shutters to protect them from micrometeorite damage. When the Cupola was designed, it was intended that one of the two identical robotic workstations to control the Canadarm2 would be eventually mounted in the Cupola. The Cupola is designed to be attached to the Unity Module and is scheduled for launch in April of 2010.

The Canadians, who developed the robotic arm used on the space shuttle, built the Space Station Remote Manipulator System (SSRMS), or Canadarm2. The SSRMS is used to move components into position during the assembly of the Space Station. Following assembly, the SSRMS will be used for maintenance, to move components and equipment from one part of the station to another, and possibly for the deployment of satellites. Although the Canadians will not provide a separate module for the Space Station, a certain percentage of the experimental area of the Space Station is allocated to Canadian researchers because of the Canadian contribution of the SSRMS.

Two Assured Crew Return Vehicles are attached to the Space Station at all times. These Russian Soyuz spacecraft, provide a rescue capability, allowing the crew to abandon the Space Station, should a massive failure occur, and return safely to the Earth. The Soyuz is a well-proven spacecraft, which has been used to carry cosmonauts into space and return them to Earth since its first launch in 1967. The Soyuz craft can operate for fourteen days and can be stored on orbit for 200 days. The vehicle is replaced by another either bringing a new permanent station crew or a short-stay visiting crew.

The first launch of hardware for the International Space Station occurred in November, 1998, with the first crew of three astronauts occupying the partially constructed assembly in October, 2000. The assembly of the Space Station is expected to be completed by 2010, and a crew of up to seven astronauts is expected to occupy the station at that time. The plan is to continuously occupy the International Space Station, with individual crew members

*A wide view of workers in the Space Station Processing Facility building a reusable logistics carrier, the Multi-Purpose Logistics Module. It is the primary delivery system used for International Space Station cargo requiring a pressurized environment.* (NASA)

typically spending several months on the station before being returned to Earth.

### Contributions

Research on the International Space Station (ISS) will focus on three areas: space medicine, materials processing, and Earth observation. NASA's flight director Scientist for Space Station Freedom, predecessor to the International Space Station, sees the Space Station "as the next logical platform in the continuing exploration of our solar system. We need to live and work [in space] for long periods of time, continuously, before we can move to a lunar outpost, and then on to Mars."

The ISS provides an environment of microgravity, where the force that gives weight to objects is reduced to less than one-millionth its value on the surface of the Earth. Because of this near-weightless environment, experiments and manufacturing processes that are impossible on Earth can be performed in space. Crystals can be grown in more perfect shapes and with higher purity than on Earth. Perfectly spherical objects, such as ball bearings, can be formed in space by allowing drops of liquid to cool in microgravity.

The U.S. Destiny Laboratory Module houses a large centrifuge, which will rotate to produce an acceleration similar to gravity. The centrifuge is the central instrument in a series of experiments in space medicine, designed to study the adaptation to weightlessness and to determine the minimum acceleration needed to inhibit the negative effects of weightlessness, including fluid loss, increase in the heart rate, loss of calcium from the bones, and muscle deterioration. A thorough understanding of the long-term effects of weightlessness is required before human missions to Mars can be undertaken. The Space Station will serve as a permanently piloted orbiting platform from which the Earth can be observed and monitored. NASA expects the Space Station to play an important role in the Mission to Planet Earth, an effort to monitor global change and to distinguish changes caused by human activity from those resulting from natural processes. The Space Station will fly over most of the world's rain forests, allowing long-term monitoring of deforestation and its effects, and it is in a position to monitor atmospheric chemistry, important to understanding global warming.

### Context

The United States launched its first space station, Skylab, on May 14, 1973. Over the next year Skylab was piloted by three three-person crews, the first occupying the station for twenty-eight days, the second for fifty-nine days, and the third for nearly three months. These Skylab astronauts provided U.S. medical experts with their best physiological data on the effects of prolonged exposure to weightlessness. The Skylab crews also operated a telescope for astronomical and solar observations.

The Soviet Union launched its first of its seven Salyut space stations on April 19, 1971. The Salyut program continued through Salyut 7, launched on April 19, 1982. Crews aboard the Salyut space stations conducted experiments in Earth observation, materials processing, astronomical observation, and space medicine. Salyut 7 carried a furnace, weighing 135 kilograms, to allow high-temperature materials processing in space. On February 19, 1986, the Soviet Union launched its Mir Space Station, which was operated by Russia until it reentered the atmosphere in March, 2001.

Experiments on the space shuttle, and the Salyut and Mir space stations, have demonstrated that crystals can be grown to larger sizes and with fewer defects in the weightless environment than can be produced on Earth. Large, defect-free crystals could be useful in developing faster computer chips. The microgravity environment also allows high-temperature materials to be produced in a purer form than is possible on Earth. On Earth, liquids are confined in a container, and at high temperatures, material from the container walls frequently ends up in the liquid. On the Space Station, where there is no gravity, engineers are able to suspend liquids in space and avoid contamination from the container walls. Materials processing experiments on the International Space Station will focus on the production of materials that can be produced only in space. The Space Station is regarded as a first step toward space industrialization.

Experiments in space manufacturing on the International Space Station may have immediate practical benefits on Earth. In the 1980's a collaboration between the pharmaceutical manufacturer Johnson & Johnson and the aerospace manufacturer McDonnell Douglas produced a miniature space manufacturing facility that flew on the U.S. space shuttle. They were able to demonstrate that some chemicals, which are extremely difficult to produce in pure form on Earth, can be produced efficiently in space. Although the high cost of access to space precluded cost-effective production of these chemicals in space in the 1980's, the pro-

duction of medicines in space is expected to be one of the goals of early space station research.

If any of these space manufacturing efforts proves successful in developing economically viable products, the flexibility of the modular design would allow new modules containing manufacturing equipment to be attached to the basic structure.

A satellite repair facility was incorporated into the preliminary design of Space Station Freedom. The intention was that malfunctioning satellites in orbit around the Earth could be retrieved using a space tug. Astronauts on the Space Station could perform repairs, and the satellites could then be redeployed. Although this repair facility was eliminated from the International Space Station, the flexibility of the Space Station design would allow its incorporation at a later date.

The International Space Station provides an isolated environment, where experiments with the potential for contamination of Earth can be conducted in safety. The Space Station can also serve as a departure and return point for piloted missions to the Moon and Mars.

The involvement of the Russians in the International Space Station provides the opportunity for U.S. and European astronauts to conduct long-term experiments in space prior to, and in preparation for, the International Space Station.

In March, 1995, Astronaut Norman E. Thagard rode in the Russian Soyuz TM-21 spacecraft with two Russian cosmonauts for a three-month stay on Mir. On June 29, 1995, the shuttle *Atlantis* docked with the Mir station for the first time and picked up three cosmonauts, as well as experiment samples and other items from the station, for return to Earth. Astronaut Shannon W. Lucid from STS-76 was on board the Mir Space Station from March to September of 1996, when John E. Blaha on STS-79 replaced her. Astronaut Jerry M. Linenger replaced Blaha in January, 1997, on mission STS-81. STS-84 launched on May 15, 1997, carrying C. Michael Foale to replaced Linenger. STS-86 was launched on time on September 26, 1997, with David A. Wolf taking over the responsibilities of Foale on board

Mir. A nighttime launch of STS-89 took place on January 23, 1998. This mission returned Wolf to Earth, and Andrew S. W. Thomas took over his duties. The final shuttle-Mir mission launch took place on June 2, 1998. STS-91 rendezvoused with Mir and returned Thomas to Earth on June 12.

**See also:** Cooperation in Space: U.S. and Russian; International Space Station: Development; International Space Station: Living and Working Accommodations; International Space Station: U.S. Contributions; Space Stations: Origins and Development.

**Further Reading**

Bizony, Piers. *Island in the Sky: Building the International Space Station.* London: Aurum Press, 1996. Bizony tells how the International Space Station is assembled in orbit during an extended sequence of shuttle flights, dockings, and spacewalks over a period of five years. With unrivaled access to NASA and the astronautic sources worldwide, he provides a lively text that contains a wealth of information. There are one hundred photographs, sixty in color.

Bond, Peter. *The Continuing Story of the International Space Station.* Chichester, England: Springer-Praxis, 2002. Bond describes the development and evolution of space stations, with particular emphasis on the International Space Station, beginning with the revolution that began in 1970, when Salyut 1, the world's first space station, was sent into orbit by the Soviet Union.

David, Leonard. "Countdown to *Freedom.*" *Final Frontier,* May/June, 1992, 44-47. A well-illustrated account of NASA's plans for Space Station Freedom, including photographs of mock-ups of the design and a discussion of the design, the costs, and the objectives of the project.

Foley, Theresa M. "Space Station: The Next Iteration." *Aerospace America,* January, 1995, 22-27. An illustrated, in-depth description of the International Space Station, focusing on its mission and the roles of the individual international partners. This article includes descriptions of the individual modules.

Harland, David M., and John E. Catchpole. *Creating the International Space Station.* London: Springer-Verlag London, 2002. Covers events leading to the construction and commissioning of the ISS.

Haskell, G., and Michael Rycroft, eds. *International Space Station: The Next Space Marketplace.* Boston: Kluwer Academic, 2000. Examines uses of the ISS: commercial, scientific, technological, and educational.

Johnson-Freese, Joan, and George M. Moore. "Space Station Reconceptulized: An Apollo-Soyuz Project for the 1990's." *Spaceflight* 36 (February, 1994): 40-43. An illustrated account of the new International Space Station after the addition of Russia as a partner. Focuses on design, including descriptions of the individual modules, and costs.

Kluger, Jeffrey. "NASA's Orbiting Dream House: Space Station Freedom." *Discover,* May, 1989, 68-72. Discusses NASA's plans for Space Station Freedom, describing the individual modules and their roles in the overall station. Although outdated in their design, the discussion of the objectives of the Space Station is equally relevant to the new design.

Logsdon, John M. *Together in Orbit: The Origins of International Participation in the Space Station.* Washington, D.C.: National Aeronautics and Space Administration, 1998. Describes the politics and science behind the effort to bring together many nations in the building of a space station.

McCurdy, Howard E. *The Space Station Decision: Incremental Politics and Technical Choice.* Baltimore: Johns Hopkins University Press, 1990. The author is a professor of public affairs at American University in Washington, D.C. The events that led up to the decision (1984) to build a permanently occupied space station in low-Earth orbit provide his primary subject matter in the present monograph, but the author's deeper interest has to do with the politics of Big Science. The story is arrestingly told in this nicely produced volume, which provides thirteen pages of plates plus detailed notes and references.

National Aeronautics and Space Administration. *International Space Station Assembly Sequence: Revision E (March, 2000 Planning Reference).* Washington, D.C.: Government Printing Office, 2000. This is the official assembly sequence planning reference for the International Space Station. It details the assembly schedule and the various components of the station.

Strode, Scott. "Space Station Freedom." *Aerospace America,* April, 1993, 38-40. An illustrated account of Space Station Freedom, including descriptions of the individual modules, their functions, and the roles of each international partner.

Von Bencke, Matthew J. *The Politics of Space: A History of U.S.-Soviet/Russian Competition and Cooperation in Space.* Boulder, Colo.: Westview Press, 1996. This book chronicles the efforts of the United States and the Soviet Union (later Russia) to overcome their political animosities and explore space together. It looks at their respective foreign and domestic policies; military, civil, and commercial influences; and top executive, legislative, and institutional politics. The book examines their separate and joint endeavors from 1945 through 1997.

*George J. Flynn and Russell R. Tobias*

# International Space Station: U.S. Contributions

*Date:* Beginning 1984
*Type of program:* Space station, piloted spaceflight

*The International Space Station (ISS) program is the largest international scientific project in history, drawing on the resources and experience of sixteen nations, led by the United States.*

### Key Figures

*Richard Kohrs*, director of NASA's Space Station Freedom program

*Eugene F. Kranz* (b. 1933), NASA Mission Operations director at the Johnson Space Center

*David C. Leestma* (b. 1949), Flight Crew Operations director at the NASA Johnson Space Center

*Bryan D. O'Connor* (b. 1946), Space Station Redesign director, responsible for transforming Space Station Freedom into the International Space Station

*Robert Phillips*, flight director scientist of NASA's Space Station Freedom project

*Wilber Trafton*, head of the U.S. portion of the International Space Station effort

### Summary of U.S. Contributions to the Facility

In an era of constrained budgets throughout the world, no single nation can realistically pursue ongoing human spaceflight. If humanity is to reap the great potential from human spaceflight and experimentation in microgravity, it is essential that international cooperation be successful. Perhaps the most significant U.S. contribution to the International Space Station (ISS), therefore, is the bringing together of sixteen nations to form the largest international scientific cooperative program in history. To complete the program on schedule is a big challenge by itself. To do so by marshaling the efforts of nations from around the world with different backgrounds, languages, cultures, educational systems, technical standards, and experience to create a unified station in space—with all of its associated supporting systems, facilities, and personnel—is the most complicated and challenging international peacetime effort ever undertaken.

In order to deal with a new concept of this complexity, new systems of management, new international relationships, new types of partnerships, and new funding mechanisms had to be developed. The ISS is a joint project of sixteen countries: the United States, Russia, Japan, Canada, Brazil, and eleven countries of the European Union (EU). Their respective space agencies are the U.S. National Aeronautics and Space Administration (NASA), the Russian Federal Space Agency (RSA), the Japan Aerospace Exploration Agency (JAXA), the Canadian Space Agency (CSA), the Brazilian Space Agency (Agência Espacial Brasileira, or AEB), and the European Space Agency (ESA). Participating ESA members include Belgium, Denmark, France, Germany, Italy, Netherlands, Norway, Spain, Sweden, Switzerland, and the United Kingdom.

The United States has been pivotal in organizing and developing ground facilities for the program, providing two of its finest existing facilities for international use. Johnson Space Center in

Houston, Texas, is the site of the Space Station Control Center (SSCC), which will integrate the Space Station mission. For payload control, the United States has established the Payload Operation Integration Center (POIC) at Marshall Space Flight Center in Huntsville, Alabama. All six ISS space agencies, representing the sixteen nations as partners, have the mandate to retain liaison personnel at the SSCC and at the POIC. These two central facilities, in addition to their central international role, will perform research functions, as will other U.S. space centers: Jet Propulsion Laboratory in Pasadena, California; Langley Research Center near Hampton, Virginia; Lewis Research Center in Cleveland, Ohio; and Goddard Space Flight Center in Greenbelt, Maryland. The other twelve nations will develop Payload Operations Control Centers at their own space centers, which will be networked into the integrated control and research operations of the Space Station through SSCC and POIC.

As a precursor to the construction of the ISS, the United States has conducted joint projects with international partners. The United States and Russia, in particular, conducted the NASA/Mir program as Phase I of the Space Station program. The NASA/Mir activities included (1) cosmonaut flights on board the U.S. shuttle, (2) U.S. shuttle rendezvous and close approach to Mir, (3) up to two years of astronaut-stay time on board Mir, (4) up to ten shuttle-Mir docking missions, (5) joint research activities, and (6) joint development of new technology and spacecraft improvements. In order to cover RSA's additional cost of the expanded Phase I and selected Phase II activities, NASA is providing $100 million per year for four years.

NASA has a bilateral agreement with the Italian Space Agency (ASI), apart from ASI's role as an ESA member, to build three Multi-Purpose Logistics Modules (MPLMs). The MPLMs are launched aboard the shuttle, and each carry resupply items, such as food and scientific equipment, to ISS, returning to Earth with items such as experiment samples. The MPLMs can also carry and return International Standard Payload Racks (ISPRs), the basic internal building blocks for station research facilities.

In collaboration with Japan, the United States is contributing some Phase III joint space shuttle flights. Phase III began in 2001 and will continue through 2010. It marks the beginning of utilization flights and the completion, in 2010, of the Space Station's assembly. Two joint Japanese/American flights are scheduled and will deliver the Japanese Experiment Module (JEM), Experimental Logistics Module Exposed Section (ELMES), Exposed Facilities (EF), and Pressurized Module (PM). The JEM complements the U.S. laboratory with significant pressurized space.

Noncollaborative U.S. contributions include habitation, laboratory, and centrifuge accommodation modules. Each of these modules encloses about 110 cubic meters of space and is designed to have the same air pressure as that on Earth's surface.

ISS will carry a total of six laboratories, including the U.S.-supplied Centrifuge Accommodations Module (to be launched in 2009) and Destiny Laboratory Module (launched in February, 2001). With the exception of the centrifuge, all laboratory facilities will be designed to fit into the ISPRs. This modular approach allows facilities to be developed by one partner to fit into rack space supplied by another, and to be upgraded and modified as needed.

The U.S. Destiny Laboratory Module provides a shirt-sleeve environment for research, technology development, and repairs by the on-orbit crew. Systems include life-support, electrical power, command and data handling, thermal control, communication, flight crew support systems, and a vacuum system. There are a total of twenty-four racks in this laboratory.

The 4,000-kilogram Centrifuge Accommodations Module (CAM) will enable life scientists aboard ISS to conduct research on the long-term effects of gravity on living plants and animals. The CAM comprises a 2.5-meter-diameter Centrifuge Rotor, two microgravity Habitat Holding Unit Racks, Plant and Rodent Habitats, and a Glovebox. Gravity levels will be selectable, from the microgravity level of the Space Station to 2g. The CAM will support ongoing investigations with statistically significant sample sizes for extended durations, including multigenerational studies, and will collect biological samples on-orbit in the microgravity environment. CAM data are required to be displayed on-orbit and downlinked. The data include rodent biotelemetry data (such as body temperature and heart rate), video images of plants and animals, specimen environmental conditions, data from science protocols conducted in the Glovebox, and hardware engineering parameters. The CAM will provide the means to gain an understanding of the effects of gravity on both the structure and the function of plants and animals, as well as to investigate potential countermeasures for the changes observed in microgravity.

The United States will contribute additional elements to the ISS. Among the important ones is the Russian Functional Cargo Block (*funktsionalya-gruzovod blokor,* or FGB) module, built in Russia and purchased from it for $190 million. This 20,000-kilogram element includes the energy block, contingency fuel storage, propulsion, and multiple docking points. The FGB or Zarya (Russian for "sunrise") is a pressurized module with large-propellant storage tanks and interfaces to other elements at its forward, aft, and nadir ends. It was launched on a Russian Proton rocket in November, 1998, and was the first ISS element launched. The forward end of the FGB contains a U.S.-compatible docking mechanism that is matched with the U.S. Pressurized Mating Adapter (PMA) 1. PMA-1 is the sole interface between Russian and U.S. seg-

*The U.S. Laboratory Module on a workstand in the Space Station Processing Facility at the Kennedy Space Center. The laboratory comprises three cylindrical sections and two end cones. It provides a shirtsleeve environment for research in the life sciences, microgravity sciences, Earth sciences, and space sciences.* (NASA)

ments and is connected to the aft end of the Unity Node (launched December, 1998). The FGB is a self-sufficient spacecraft with its own communication, attitude control, and power systems. Until the SM arrives, the FGB will provide all critical functions for the ISS and will serve as a structural spacer between the Russian and U.S. segments. It is the building block for future additions to the ISS. However, the FGB has limited resources and provides only temporary control of the ISS; it is not capable of being refueled directly from its Progress Supply Vehicle. In July, 2000, the Russian Zvezda Service Module, with living and working space for three crew members, was launched from Baikonur and docked with FGB to assume the permanent duty of primary attitude control, communications, and life support for the ISS.

Two U.S. nodes—Node 1 (Unity) for storage space and Node 2, with racks of equipment to convert electrical power for use by the international partners—constitute another U.S. contribution, carrying the first U.S. power system and providing storage space for supplies and equipment, with berthing ports and attachment points for modules and the station's large truss, and including a docking port for shuttle orbiters. PMA-2, attached to Unity, is a shuttle docking port that will be used for most of the shuttle missions.

An especially important U.S. contribution, the Quest Joint Airlock will make it possible for the crew to transfer from the pressurized modules into the space vacuum. Another element contributed by the United States is the Cupola, which provides direct viewing capability for robotics operations and payload viewing. Other elements include the photovoltaic power arrays (which provide the power to the station), the power storage system, the main 95-meter truss, and a boom that facilitates the addition of modules to the station. The Mobile Transporter (MT), is a cart that is the U.S. portion of the Mobile Service System, which rolls up and down the station's main truss boom to get the robots to their work locations.

A total of thirty-nine U.S. space shuttle flights are planned to ferry cargo and the crew between the Space Station and Earth from 1998 until completion. Of these flights, thirty-five were dedicated to assembly and four to utilization and outfitting. The United States will build and maintain integrated station systems, including electrical power, data, thermal control, the crew health system, environmental control and life support, control moment gyroscopes for attitude control, and the communication system. Part of the communication system is the U.S. Deep Space Network, on the ground, and the Tracking and Data-Relay Satellite System (TDRSS), an array of telecommunications satellites already in space.

More than five hundred U.S. companies are working as contractors, associates, and suppliers to build ISS's hardware and module facilities, of which three are prime: The Boeing Aircraft Company is responsible for building the pressurized U.S. laboratory, the habitat module, connecting nodes, the environmental control system, and the life-support system. McDonnell Douglas is to manufacture the truss, data-handling systems and hardware, and crew health care and monitoring equipment. Rockwell Aerospace, Rocketdyne is responsible for building end-to-end electrical power system architecture for the ISS, which consists of power generation and energy storage subsystems. In addition to these corporate contractors, a large number of principal investigators, technical personnel, and students from U.S. universities and research institutions have been contributing their expertise to ISS.

In terms of cash, as the leader of the ISS effort, the United States is contributing $17.4 billion to the program—an amount that does not include prior development costs. Because this commitment was made before Russia joined the program, some expect that the Russian contribution will save the United States approximately $2 billion.

### Context

Philosophically speaking, the greatest contribution of the United States to the ISS is to inspire a new generation around the world to explore and achieve, while pioneering new methods of education to teach and motivate the next generation of scientists, engineers, entrepreneurs, and explorers. The United States has thus provided the driving force for emerging technologies. By promoting the idea of internationalizing the Space Station, the United States has transformed the dream of living and working in space into the International Space Station, a tangible symbol of the power of nations to work together on peaceful initiatives and a test for building mutual trust and shared goals.

Russia's vast experience in long-duration spaceflight will provide immense benefits to this international partnership. ISS is using Russian space technology, capability, expertise, and hardware to build a better space station, both cost-effectively and sooner than anticipated in the old days of the United States' plans for Space Station Freedom. By channeling the aerospace industry of Russia and other countries into nonmilitary pursuits, the United States has helped reduce the risk of nuclear proliferation and slow traffic in high-technology weaponry to developing countries. As a result of the U.S. effort, the international partners' commitments will total over $9 billion.

The per-capita contribution of each U.S. citizen to the ISS amounts to $9 per year for the 1998-2005 assembly period and will constitute one-seventh of one percent of the U.S. federal budget. The ISS constitutes less than 15 percent of NASA's total budget. The returns on this investment, by comparison, are expected to be vast in knowledge gained and practical applications. For example, combustion has been a subject of vigorous scientific research for more than a century and accounts for approximately 85 percent of the world's energy production as well as a significant fraction of the world's atmospheric pollution. By conducting research on ISS, scientists will be able to study subtle aspects of combustion normally masked by fluid flows caused by Earth's gravity. Breakthroughs in combustion science could have far-reaching effects for the world economy and the environment. A 2 percent increase in burner efficiency would save the United States $8 billion per year. Such savings would recuperate the U.S. investment in ISS within only two years.

A less tangible but potentially far more valuable return on the ISS investment concerns future generations: Across the nation, those university teachers and laboratory researchers who are contributing to ISS are at the same time sparking further interest among their students. If even a small percentage of these young minds are inspired to pursue science and technology, a better quality of life may be secured for years to come.

**See also:** Cooperation in Space: U.S. and Russian; Funding Procedures of Space Programs; International Space Station: Crew Return Vehicles; International Space Station: Design and Uses; International Space Station: Development; International Space Station: Living and Working Accommodations; International Space Station: Modules and Nodes; International Space Station: 1998; International Space Station: 1999; International Space Station: 2000; International Space Station: 2001; International Space Station: 2002; International Space Station: 2003; International Space Station: 2004; Materials Processing in Space; National Aeronautics and Space Administration; Space Stations: Origins and Development.

## Further Reading

Bizony, Piers. *Island in the Sky: Building the International Space Station.* London: Aurum Press, 1996. Bizony tells how the International Space Station will be assembled in orbit during an extended sequence of shuttle flights, dockings, and spacewalks over a period of five years. With unrivaled access to NASA and the astronautic sources worldwide, he provides a lively text that contains a wealth of information. There are one hundred photographs, sixty in color.

Boeing Missles and Space Division, Defense and Space Group. *International Space Station Alpha: Reference Guide*. Washington, D.C.: National Aeronautics and Space Administration, 1995. Ready reference for Space Station personnel and top-level sources of ISS data in a convenient size. References to applicable documents are provided for the user to obtain controlled data. An electronic version is available via MOSAIC; the address is http://issa/www.jsc.nasa.gov.html. Accessed March, 2005.

Bond, Peter. *The Continuing Story of the International Space Station*. Chichester, England: Springer-Praxis, 2002. Bond describes the development and evolution of space stations, with particular emphasis on the International Space Station, beginning with the revolution that began in 1970, when Salyut 1, the world's first space station, was sent into orbit by the Soviet Union.

Harland, David M., and John E. Catchpole. *Creating the International Space Station*. London: Springer-Verlag London, 2002. Covers events leading to the construction and commissioning of the ISS.

Haskell, G., and Michael Rycroft, eds. *International Space Station: The Next Space Marketplace*. Boston: Kluwer Academic, 2000. Examines the uses of the ISS, from commercial to scientific.

Logsdon, John M. *Together in Orbit: The Origins of International Participation in the Space Station*. Washington, D.C.: National Aeronautics and Space Administration, 1998. Describes the politics and science behind the effort to bring together many nations in the building of a space station.

McCurdy, Howard E. *The Space Station Decision: Incremental Politics and Technical Choice*. Baltimore: Johns Hopkins University Press, 1990. The author is a professor of public affairs at American University in Washington, D.C. The events that led up to the decision (1984) to build a permanently occupied space station in low-Earth orbit provide his primary subject matter in the present monograph, but the author's deeper interest has to do with the politics of Big Science. The story is arrestingly told in this nicely produced volume, which provides thirteen pages of plates plus detailed notes and references.

National Aeronautics and Space Administration. *International Space Station Assembly Sequence: Revision E (March, 2000 Planning Reference)*. Washington, D.C.: Government Printing Office, 2000. This is the official assembly sequence planning reference for the International Space Station. It details the assembly schedule and the various components of the station.

_____. *International Space Station Fact Book*. Houston: Author, 1995. Contains ISS statistics and the assembly schedule.

_____. *Science and Technology Institute in Space: The Promise of Research on the International Space Station*. Houston: Author, 1995. This fact book gives an overview of science and technology on the Space Station, including its research facilities. Contains an assembly diagram.

_____. *Space Station White Papers, Part 1*. NASA Mail Code OA311. Houston: Author, 1994. Describes the need for and uses of the International Space Station.

National Aeronautics and Space Administration, Johnson Space Center. *A New Era of Discovery: Plans for Research on Space Station*. NASA Mail Code OA311. Houston: Author, 1995. Provides a summary of the Space Station, its research capabilities, science research, technology, and the commercial development of space.

National Aeronautics and Space Administration, Space Station Program Office. *Global Participation in the International Space Station.* Houston: Author, 1995. A brief account of the participation and contribution of international partners to the Space Station.

Von Bencke, Matthew J. *The Politics of Space: A History of U.S.-Soviet/Russian Competition and Cooperation in Space.* Boulder, Colo.: Westview Press, 1996. This book chronicles the efforts of the United States and the Soviet Union (later Russia) to overcome their political animosities and explore space together. It looks at their respective foreign and domestic policies; military, civil and commercial influences; and top executive, legislative and institutional politics. The book examines their separate and joint endeavors from 1945 through 1997.

*M. A. K. Lodhi, updated by Russell R. Tobias*

# International Space Station: 1998

*Date:* January 1 to December 31, 1998
*Type of mission:* Space station, piloted spaceflight

*Despite financial problems, in 1998 the Russian Space Agency launched the Zarya Control Module, the first component of the International Space Station (ISS), and soon afterward NASA orbited the Unity connecting module. Space shuttle* Endeavour *crew members joined the two components.*

### Key Figures

*Robert D. Cabana* (b. 1949), STS-88 commander
*Frederick W. Sturckow* (b. 1961), STS-88 pilot
*Jerry L. Ross* (b. 1948), STS-88 mission specialist
*Nancy J. Currie* (b. 1958), STS-88 mission specialist
*James H. Newman* (b. 1956), STS-88 mission specialist
*Sergei Konstantinovich Krikalev* (b. 1958), STS-88 mission specialist, Russian Space Agency

### Summary of the Mission

After years of design revisions and political maneuvering, the National Aeronautics and Space Administration (NASA) and the Russian Space Agency (RSA) launched, connected, and tested the first components of the International Space Station (ISS). Almost to the very last moment, however, a near-catastrophic depression of the Russian economy delayed ISS construction and even jeopardized the existence of the sixteen-nation collaborative space venture. Nevertheless, the RSA successfully launched the Zarya Control Module, and the space shuttle *Endeavour* later ferried to it the Unity Module, a storage compartment and passageway to other components to be added during future space shuttle flights.

The Russian economy suffered shortages and inflation throughout the 1990's, but the problems became particularly acute during 1998. Because of the resultant scarcity of government revenues, RSA could not meet its assembly schedule, much less promise to fulfill its original contract with NASA and partners from Brazil, Belgium, Canada, Denmark, France, Germany, Italy, Japan, the Nether-

lands, Norway, Spain, Sweden, Switzerland, and Great Britain. In late May the ISS consortium approved a revised schedule to ease the pressure on Russia. Construction would begin the following November, a year's delay, and was planned for completion in 2004.

RSA canceled construction of two life-support modules and one storage chamber, which also saved the agency the cost of three launches. However, the savings were not enough. For $60 million in ready cash, RSA sold NASA some of the research space aboard the ISS allotted to cosmonauts during the assembly phase. The transaction meant that until the Space Station was completed Russia would be a builder but not a user of the facility. In one last-minute cost-cutting attempt, RSA requested a change in orbit for ISS so that it would be closer to the Russian Mir Space Station and therefore easier to transfer equipment from the old space station to the new station. NASA refused the alteration.

Meanwhile, Yevgeni Shaposhnikov, adviser on space to Russian president Boris Yeltsin, announced that Russia would try to keep the Mir Space Station

621

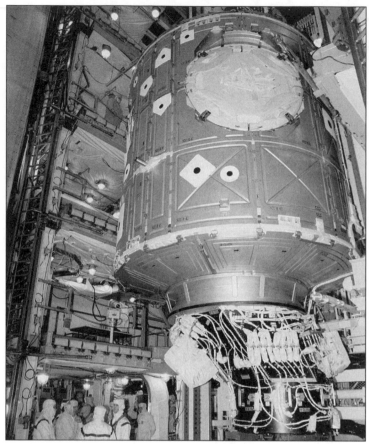

*The Unity Module was prepared for launch on mission STS-88. It is a passageway that connects to the living and working areas of the space station. As part of the International Space Station it was attached to the Russian-built Zarya Control Module. (NASA)*

State Research and Production Space Center, the 19-metric tonne (21-ton) Zarya, a cylinder 12.5 meters (41 feet) long and 4 meters (13.5 feet) in diameter, contains attitude and orbit control systems, solar generators, and docking ports for additional modules. The solar panels charge six batteries, which together provide three kilowatts of power. Its two rocket engines and thirty-six maneuvering thrusters, fed by a six-ton propellant reservoir, are to keep ISS in position over the Earth. The control rockets are needed to keep the station aloft because it orbits close enough to Earth for atmospheric particles to hit it and gradually slow it down. Without periodic boosts, ISS would sink deep into the atmosphere and burn up. "Zarya" was the radio call sign adopted by Soviet Mission Control when Yuri A. Gagarin went into orbit. The Russians adopted Zarya and the image of the Sun rising over the Earth as a symbol of the new era in space exploration made available by the ISS.

The launch and orbital insertion went perfectly, and it was a heady moment both for Zarya mission controllers at the Russian Mission Control Center in Korolev, Russia, and for their American counterparts at the ISS Flight Control Room of the Johnson Space Center in Houston, Texas. The Proton left Zarya in an orbit with a 354-kilometer (220-mile) apogee or high point and a 185 kilometers (115-mile) perigee or low point. Communications antennae and 24-meter-long (80-foot-long) solar panels unfurled as programmed. Tests during the following days revealed that the rockets and thrusters operated normally, and controllers successfully used the engines to raise Zarya's orbit. There were malfunctions, but all were minor. A software glitch made it appear that the module was too humid, one of the battery rechargers was not working

operational longer than previously planned. NASA officials worried that the move would further distract RSA and impede its work on ISS modules. The economic and political maneuvering notwithstanding, on November 20 a Proton booster lifted Zarya ("sunrise" in Russian) into orbit from the Baikonur Cosmodrome in Kazakhstan. NASA Administrator Daniel S. Goldin and RSA General-Director Yuri Koptev were on hand to watch the launch and proclaimed it a vindication of the Russian space program and the international partnership.

Developed from the Functional Cargo Block design used for Mir and built by the Khrunichev properly, and two small antennae failed to deploy fully for the Telerobotically Operated Rendezvous (TORU) system, a manual backup for the station's automated docking system.

On December 4 the space shuttle *Endeavour* blasted off from the Kennedy Space Center (KSC) in Florida to carry the Unity storage and connecting module aloft to Zarya. Commanded by Robert D. Cabana and piloted by Frederick W. Sturckow, Space Transportation System mission STS-88 had four mission specialists to mate the two modules and perform other construction and installation: Nancy J. Currie, Jerry L. Ross, James H. Newman, and Cosmonaut Sergei Konstantinovich Krikalev, who had much experience on the Mir Space Station. The Unity Module, 6.7 meters (22 feet) long and 5.5 meters (18 feet) in diameter, has six hatches, some of which serve as passageways from Zarya to other modules, including the U.S. laboratory and habitation modules; others are for docking. It also is an equipment storage site, but despite these modest functions, it is a complex unit and required many hours of labor to make operational. Because it connects work spaces and living areas, all fluids, electrical and data systems, environmental controls, and life-support systems pass through Unity, involving 50,000 mechanical parts, 216 fluid and gas lines, and 121 electrical cables.

*Endeavour* caught up with Zarya on its 222nd orbit, 386 kilometers (240 miles) above Earth. On December 7, Ross and Newman conducted the first of an estimated 160 spacewalks necessary to put together the entire ISS. During it and their subsequent two walks for STS-88, they connected Unity to

Zarya and attached cables while Currie held Zarya in place with the shuttle's 15-meter (50-foot) robotic arm. At other times she shifted Ross and Newman around as if the arm were a cherry picker. The pair also freed the jammed antennae of TORU. After the two modules had been slowly pressurized, Cabana and Krikalev entered Unity side by side, symbolizing the solidarity of NASA and RSA. Then the rest of the crew followed and began unscrewing the approximately seven hundred bolts holding down protective panels aboard the two modules and transferred equipment from the shuttle. Besides bringing tools, computers, and clothing for

*The Unity Module sits "atop" the International Space Station against a backdrop of Earth, sea, and clouds in 1999.* (NASA)

ISS's permanent crew, they installed a direct communications link between Zarya and ISS Mission Control at the Johnson Space Center, and Krikalev replaced the malfunctioning battery charger. For him the work was a taste of things to come; he was scheduled to be the flight engineer for the first permanent crew.

While still docked, *Endeavour* fired its main engines and carried the inchoate ISS into a higher orbit to avoid using up the propellant stored in it. The shuttle undocked on the twelfth day of its mission and returned to KSC. By the end of the year, flight controllers had used the station's main rockets in a series of boosting maneuvers that lifted it into a nearly circular orbit—an apogee of 412 kilometers (256 miles) and a perigee of 399 kilometers (248 miles)—at an inclination of 51.6° to the equator. It completed an orbit every ninety-two minutes. To stabilize the station and moderate temperatures aboard, it was sent into an "X-nadir" spin, which points Unity Earthward and Zarya toward deep space while they spin along their long axis at 1° per second.

### Contributions

The year's work on ISS brought no new scientific knowledge, because no research was done in space, but it did afford practical experience in microgravity construction and in testing and repairing equipment. The experience was thought to be sorely needed. NASA considered STS-88 one of the most complex shuttle missions ever flown; nevertheless, this first stage of work was expected to be among the simplest in the assembly of the Space Station. The year's delay in launching Zarya gave the shuttle crew extra time to practice for their work. Future crews could not look forward to so much practice, planners thought, and would have to draw upon the experience of *Endeavour*'s crew.

The construction proved to be especially arduous, even for Ross, NASA's most experienced spacewalker. The intense cold made cables stiff and difficult to attach. The space suit's gloves, which must be thick to protect the hands from numbing with cold, were awkward and slowed assembly. The energy spent in just keeping their bodies in the proper position exhausted the spacewalkers. Tools and a thermal cover got away from them and floated off into space. A jet backpack that was specially designed to bring a spacewalker back to the station if the safety tether broke used up its fuel too quickly to permit maneuvering.

The mission was a clear success nevertheless. Not only did the astronauts and cosmonaut work flawlessly together to assemble the two modules; they also showed that repairs could be made efficiently. As important, space agency leaders in Russia and the United States, along with their fourteen other partner nations, learned to negotiate economic and political problems and to endure delays without scuttling the ISS program.

### Context

As costs for the International Space Station climbed well beyond initial estimates, NASA faced widespread criticism and political opposition. Having already paid Russia $240 million to help pay for Zarya, Congress balked at NASA's request for $660 million more to get Russia through the following five years of construction and launches. When NASA bought research time aboard ISS from Russia for $60 million, some members of Congress angrily complained of deception, calling it a "smoke-and-mirrors" ploy. Others, while dubious of RSA's ability to deliver hardware on time because of Russia's economic crisis, wanted to keep the Russians closely involved, even if it meant extra money from the United States. Otherwise, they feared, unemployed Russian aerospace engineers might offer their services to rogue nations, such as Libya and Iraq, to help them build guided missiles and other weapons.

NASA also faced divided popular opinion about ISS. In 1997 the agency opened an ISS assembly line at the KSC to tourists, hoping it and walk-through models would inspire political support from the public. Whatever effect the exhibits had on vacationers, editorials in the United States and overseas were often still highly critical. England's *Economist* magazine called for cancellation of the project as wasteful and without a clear mission, while *The New Scientist* characterized it as an expensive nightmare. In the United States, many scientists, even those who supported space exploration, were incensed because ISS construction used up government money that could finance scientific research. In an influential *New York Times* editorial, Timothy Ferris, a prominent science writer, scoffed at ISS as "little more than a Motel 6 in low-Earth orbit" and without scientific value. Even though supporters insisted that ISS research facilities would make possible new discoveries in medicine and electronics, creating high-tech jobs on Earth to convert the discoveries into consumer goods, critics were little mollified.

Broad political support was needed if NASA, RSA, and its partners were to complete the construction. The launch of Zarya was only the first of nine Proton launches scheduled for RSA. NASA looked forward to thirty-five more shuttle missions following STS-88, during which astronauts would log more than 1,100 hours of spacewalks. Exasperated at congressional criticism, NASA Administrator Goldin asked Congress to commit itself to financing ISS throughout its five years of construction or cancel it immediately. Congress chose not to cancel.

**See also:** Cooperation in Space: U.S. and Russian; Funding Procedures of Space Programs; International Space Station: Crew Return Vehicles; International Space Station: Design and Uses; International Space Station: Development; International Space Station: Living and Working Accommodations; International Space Station: Modules and Nodes; International Space Station: 1999; Materials Processing in Space; National Aeronautics and Space Administration.

**Further Reading**

Bizony, Piers. *Island in the Sky: Building the International Space Station.* London: Aurum Press, 1996. Intended for a general audience, this book introduces the idea of the Space Station to general readers, both in its historical development and in its technology. Bizony is especially trenchant in explaining the Byzantine domestic and international politics behind ISS. Lovely photographs and illustrations supplement the text.

Bond, Peter. *The Continuing Story of the International Space Station.* Chichester, England: Springer-Praxis, 2002. Bond describes the development and evolution of space stations, with particular emphasis on the International Space Station, beginning with the revolution that began in 1970, when Salyut 1, the world's first space station, was sent into orbit by the Soviet Union.

Ferris, Timothy. "NASA's Mission to Nowhere." *The New York Times*, November 29, 1998. Ferris deftly argues the position of many scientists: support for the American space program but sharp criticism of ISS because it takes money away from other science projects and lacks a well-defined purpose.

Harland, David M., and John E. Catchpole. *Creating the International Space Station.* London: Springer-Verlag London, 2002. Overview of events leading to the construction and commissioning of the ISS.

Messerschmid, Ernst, and Reinhold Bertrand. *Space Stations: Systems and Utilization.* Berlin: Springer, 1999. Thorough and technically sophisticated, this volume reviews space station designs in general and that of ISS specifically. It also discusses the science of the subsystems, the support vehicles, logistics and communications, the orbital environment, and factors affecting the crew. A wealth of photographs, graphs, tables, and illustrations accompanies the discussions.

National Aeronautics and Space Administration. *Space Shuttle Mission Press Kits.* http://www.shuttlepresskit.com/index.html. Provides detailed preflight information about each of the space shuttle missions. Accessed March, 2005.

Osterwalder, Anja. *Space Manual.* Stuttgart, Germany: Die Gestalten Verlag, 1999. Produced in the form of a wire-bound manual, this booklet uses minimal text interspersed among collages of graphics, photographs, and illustrations to describe ISS, its research, the space environment, and the physical effects of microgravity on the crew. Useful as an introduction to space station science, especially for young readers.

*Roger Smith*

# International Space Station: 1999

*Date:* January 1 to December 31, 1999
*Type of mission:* Space station, piloted spaceflight

*Delays in assembling and launching components disrupted the construction schedule for the International Space Station (ISS) and heightened the controversy surrounding it. Nevertheless, crew members of the space shuttle* Discovery *added new equipment, repaired malfunctions, and resupplied the station, while ground controllers resolved a series of minor threats.*

## Key Figures

*Kent V. Rominger* (b. 1956), STS-96 commander
*Rick D. Husband* (1957-2003), STS-96 pilot
*Tamara E. Jernigan* (b. 1959), STS-96 mission specialist
*Ellen Ochoa* (b. 1958), STS-96 mission specialist
*Daniel T. Barry* (b. 1953), STS-96 mission specialist
*Julie Payette* (b. 1963), STS-96 mission specialist, Canadian Space Agency
*Valery Ivanovich Tokarev* (b. 1952), STS-96 mission specialist, Russian Federal Space Agency, RSA

## Summary of the Mission

The beginning of 1999 found the International Space Station (ISS), launched late the previous year, in good shape. It orbited Earth with an apogee (high point) of 412 kilometers (256 miles) and a perigee (low point) of 399 kilometers (248 miles), and its systems for the most part functioned normally. The year's schedule for construction and supply, however, was not in good shape. Short of funds, the Russian Space Agency (RSA) slowed completion of the service module Zvezda, due to be launched and mated to the Zarya and Unity Modules of the ISS. RSA promised a launch, but the year dragged on without it, disrupting other events planned by the National Aeronautics and Space Administration (NASA) and its fourteen other ISS partners. The completion of Space Transportation System mission 96 (STS-96), during which the space shuttle *Discovery*'s crew transferred supplies and attached essential construction equipment, provided a much-needed success for the program.

The addition of the Zvezda (Russian for "star") was to be the program's highlight for the year. It contained laboratory facilities and crew living space and was needed if the first permanent crew was to occupy ISS in 2000, as scheduled. RSA, however, was unable to build the module on time, and was almost a year and a half behind schedule in January. Subsequent problems with testing the systems and computer software pushed back the launch date for Zvezda from July to September, even though RSA threw a celebration for its official completion on April 26. To make matters worse, RSA further delayed assigning a launch date after one of its Proton boosters crashed. Because Protons lift Russia's ISS components into orbit, RSA began a thorough investigation of the boosters in December to avoid a disaster with Zvezda. A supply flight to ISS by the space shuttle *Atlantis*, scheduled for December 2, had to be put on hold because it required the Russian service module to be successful. By then,

627

however, NASA had its own reasons for keeping *Atlantis* on the ground. Short circuits that occurred during the launch of *Columbia* the previous month induced the agency to ground the entire shuttle fleet until their electrical wiring could be inspected.

Meanwhile, ISS had a series of minor malfunctions. Early in the year it was discovered that batteries aboard Zarya discharged faster than expected. Controllers had to shut down some heaters and smoke detectors to conserve power. The problem turned out to be not with the batteries themselves but with the eighteen charge-discharge integrated circuit units, known by their Russian acronym MIRTs. Plans were laid to replace them during the next space shuttle mission. One antenna of a U.S. communications system did not function in certain positions, and, late in the year, tests revealed that the primary automatic docking system, called by its Russian name Kurs, produced inaccurate velocity readings because of electromagnetic interference from other electronics systems on the station.

Three problems caused outright alarm. Crew members of the space shuttle *Endeavour* mission to ISS the previous year reported that stale air on the station—the buildup of carbon dioxide—gave them nausea, headaches, and dry, itchy eyes. Although they insisted the problem was fleeting, NASA managers ordered that air samples be taken by the next shuttle's crew. More serious, on two occasions space junk looked as if it might collide with ISS. The first, in June, was the more threatening. United States Space Command, which monitors human-made objects in orbit, warned that a piece of a Russian rocket could come within two-thirds of a mile of the station, too close for comfort. Russian flight controllers near Moscow radioed instructions to ISS to activate its engines and get out of the way, but the instructions were flawed, and the sta-

*The Canadian Space Agency's first contribution to the International Space Agency, the Space Station Remote Manipulator System (SSRMS), is the primary means of transferring payloads between the orbiter payload bay and the International Space Station for assembly.* (NASA)

tion's computers rejected them. The botched attempt left the station vulnerable, but the space junk ended up missing the station by 4.5 miles. Then, in October, Space Command warned that a wayward Pegasus rocket hulk could pass within a mile of the station. This time flight controllers succeeded in firing ISS's engines so that it had a 24-kilometer (15-mile) safety margin.

Otherwise, the ISS program proceeded satisfactorily. Controllers managed the cyclic charging and discharging of the batteries to keep them in optimal condition. Tests of communications equipment and automated rendezvous systems took place regularly. Members of *Discovery*'s crew trained at the Gagarin Cosmonaut Training Center in Star City, Russia, on how to replace the faulty MIRTs on Zarya, and astronauts slated for other shuttle missions went to the Cosmodrome in Baikonur, Kazakhstan, to learn about Zvezda.

On May 27, space shuttle *Discovery* lifted from Kennedy Space Center (KSC) in Florida and docked with ISS late the following day. Aboard were STS-96

Commander Kent V. Rominger, Pilot Rick D. Husband, and Mission Specialists Tamara E. Jernigan, Ellen Ochoa, Daniel T. Barry, Julie Payette (a Canadian astronaut), and Valery Ivanovich Tokarev (a Russian cosmonaut). On May 30, Jernigan and Barry went for a spacewalk, during which they installed equipment outside ISS that would be needed for future construction: sections of cranes, foot restraints designed to fit both American and Russian space boots, and three bags of tools and handrails. They also attached insulation and inspected the paint on the Zarya and Unity modules, assisted by Ochoa, who operated the shuttle's robot arm. Lasting 7 hours and 55 minutes, it was the second longest spacewalk ever.

On the following day Jernigan and Tokarev opened the hatch connecting Unity to *Discovery*. Over the next three days, Payette and Tokarev replaced the eighteen MIRTs and took air samples, Barry and Husband replaced a power distribution unit and transceiver for the disabled communication equipment aboard Unity, and Barry and Tokarev installed muffling panels over fans inside Zarya to reduce the noise level. Meanwhile, Ochoa directed crew members in transferring cargo to ISS from the SPACEHAB Logistic Double Module, which *Discovery* carried in its cargo bay. Altogether, they moved 1,618 kilograms (3,567 pounds) of goods intended for use by the first permanent ISS crew, including seventy-five gallons of water, laptop computers, clothing, sleeping bags, spare parts, and medical equipment. During rest periods crew members spoke with television and radio reporters in the United States, Canada, and Russia.

After nearly eighty hours aboard ISS, the crew returned to *Discovery*, and Rominger and Husband used the shuttle to boost the station slowly into a higher orbit. They then undocked and flew two laps around ISS in order to perform a final inspection. *Discovery* returned to KSC on June 6. STS-96 was proclaimed a critical milestone and a complete success, the crew having completed all assigned tasks, often ahead of schedule.

After STS-96, Boeing Aircraft Company, the prime contractor for ISS, delivered to KSC the backbone of the long truss structure, called Z1, which will support solar panel arrays. NASA also finished tests on the Control Moment Gyros, which, when installed, would make it possible to control the position of the Space Station without use of thruster propulsion. The gyros would conserve fuel and move the station more smoothly than do thrusters, thereby disrupting scientific experiments less. At the close of 1999, ISS circled the Earth with an apogee of 396 kilometers (246 miles) and a perigee of 377 kilometers (234 miles), slowly spinning to distribute the heat from sunlight evenly. It had completed more than 6,300 orbits. When it would receive its next module and further supplies was yet to be determined.

### Contributions

The second year of space station operations and the second shuttle flight to service ISS afforded astronauts and cosmonauts more experience in operational methods. Chief among these was the first large cargo transfer from SPACEHAB. It was a difficult procedure. *Discovery*'s crew had limited time, approximately 750 items to move, and narrow space to pass through. Accordingly, logistics experts planned the operation so that supplies were shifted in eight stages in order of their importance to ISS function and safety: ingress and safety (for example, air sampling bottles and sound mufflers), critical spares (MIRTs, communications processors), incremental assembly (tools, tethers, cameras, space suits), crew health maintenance (medical supplies and equipment), prepositioned equipment and provisions for future missions, resupply (lubricant, drill charger), spares and nonassembly equipment (IMAX 3D movie hardware, clothing, sleeping bags), and detailed test objectives (ties, tape, cables). According to the crew, their movements in and out of SPACEHAB and through Unity and Zarya, passing each other in the process, turned into an intricate ballet. It worked. They fulfilled all objectives on time.

Ground controllers learned, somewhat to their dismay, that maneuvering ISS to avoid collisions with space junk could be more difficult than expected. Incorrect instructions sent to ISS resulted in failure on one attempt, and although the margin of safety was nearly five times larger than Space Command had originally calculated, the incident inspired a review of procedures. NASA scientists estimated that ISS stands a 6 percent chance of being penetrated by space junk, and at least twice a year an orbital adjustment will be necessary. Accordingly, prompt, successful maneuvering is critical.

### Context

While productive, 1999 was a troubled year for the Space Station program. NASA and RSA found yet again that politics, distinct operational traditions, and differing equipment standards could produce obstacles and consternation. The Russian-made Zarya did not meet NASA noise level standards, forcing NASA to install mufflers, an unwelcome extra expense. Moreover, the Russian government's opposition to the brief air war between the North Atlantic Treaty Organization (NATO, led by the United States) and Yugoslavia imperiled cooperation between NASA and RSA, even though the leaders of the two agencies, Daniel S. Goldin and Yuri Koptev, reportedly had a good working relationship. Repeated assurances from RSA that Zvezda would be completed, tested, and placed in orbit during the year—assurances that were not met—frustrated the other members of the ISS consortium, and none more so than the United States.

In large part, the difficulty for RSA was financial. The Russian government had undergone a fiscal crisis for years, exacerbated by expenditures for a military campaign in Chechnya, and could not pay the space agency's expenses. Goldin wanted to give RSA as much as $600 million to help it through lean times, but Congress and President Bill Clinton balked at the idea. Eventually, the United States did offer Russia $100 million to buy Soyuz spacecraft, which would be used as lifeboats for ISS, but Goldin was forced to shift $1.2 billion from other projects to make up for Russian lapses, and a backlog of components built in the United States and other countries accumulated, awaiting launch.

Yet another point of contention was the thirteen-year-old Russian space station Mir. NASA wanted RSA to abandon Mir in order to concentrate its resources on ISS, but Russian national pride produced a backlash of public opinion in favor of keeping it operational. Russia agreed not to send crews up to Mir unless outside sources paid all expenses, but this compromise did not please NASA.

As the year passed, the estimated completion date for ISS slipped from 2004 to 2006. Meanwhile, the estimated total costs, as reported in the media, rose from $40 billion at the beginning of the year to $100 billion by the end. Although the variation in part depended upon whether operating expenses were included, the size of the amount worried politicians and taxpayers alike. Complicating matters somewhat, two high-ranking NASA officials in charge of ISS resigned during the year, Randy Brinkley, the program manager, and Gretchen McClain, the deputy associate administrator in charge of managing the station's annual budget and providing liaison to the White House and Congress. Brinkley was replaced by Tommy W. Holloway, formerly a space shuttle administrator, and McClain by Michael Hawes, the project's chief engineer. NASA denied that either resignation was a sign of upheaval in the ISS program.

Controversy over how science projects would be selected for ISS prompted NASA to seek advice from the National Research Council (NRC). In mid-December, the NRC panel studying the question, chaired by Cornelius Pings, a former head of the Association of American Universities, suggested that NASA create an extragovernmental institute to schedule projects and manage scientific experimentation aboard the station. The recommendation got a mixed reception from NASA officials.

**See also:** Cooperation in Space: U.S. and Russian; Funding Procedures of Space Programs; International Space Station: Crew Return Vehicles; International Space Station: Design and Uses; International Space Station: Living and Working Accommodations; International Space Station: Modules and Nodes; International Space Station: 1998; International Space Station: 2000; Materials Processing in Space.

**Further Reading**

Bizony, Piers. *Island in the Sky: Building the International Space Station.* London: Aurum Press, 1996. Intended for a general audience, this book introduces the idea of the Space Station, both in its historical development and in its technology. Bizony is especially trenchant in explaining the complex domestic and international politics behind ISS. Lovely photographs and illustrations supplement the text.

Bond, Peter. *The Continuing Story of the International Space Station.* Chichester, England: Springer-Praxis, 2002. Bond describes the development and evolution of space stations, with particular emphasis on the International Space Station, beginning with the revolution that began in 1970, when Salyut 1, the world's first space station, was sent into orbit by the Soviet Union.

Harland, David M., and John E. Catchpole. *Creating the International Space Station.* London: Springer-Verlag London, 2002. An overview of events leading to the construction and commissioning of the ISS.

Lawler, Andrew. "NRC Panel to Propose Station Institute." *Science,* December 17, 1999, 2251. Explains the controversy over control of science projects aboard ISS and an advisory panel's recommendation that a commission independent of NASA be given control to avoid the agency's red tape.

Messerschmid, Ernst, and Reinhold Bertrand. *Space Stations: Systems and Utilization.* Berlin: Springer, 1999. Thorough and technically sophisticated, this volume reviews space station designs in general and that of ISS specifically. It also discusses the science of the subsystems, the support vehicles, logistics and communications, the orbital environment, and factors affecting the crew. Photos, graphs, tables, and illustrations clarify the discussions.

Osterwalder, Anja. *Space Manual.* Stuttgart, Germany: Die Gestalten Verlag, 1999. Published in the form of a wire-bound manual, this booklet uses minimal text interspersed among collages of graphics, photographs, and diagrams to describe ISS, its research, the space environment, and the physical effects of microgravity on the crew. Useful as an introduction to space station science, especially for young readers.

Stine, G. Harry. *Living in Space.* New York: M. Evans, 1997. A lucidly written handbook for general readers describing the conditions under which humans live and work in space and the systems aboard spacecraft and space stations that make piloted spaceflight possible.

Vizard, Frank. "ISS Isn't About Science." *Popular Science,* February, 1999, 73. An aerospace veteran argues that ISS has a larger purpose than either science or national prestige: that it is an essential stepping stone for space exploration. Vizard's comments eloquently summarize the position of the program's defenders.

*Roger Smith*

# International Space Station: 2000

*Date:* January 1 to December 31, 2000
*Type of mission:* Space station, piloted spaceflight

*Although behind in its construction schedule and pestered by technical problems, the International Space Station (ISS) grew with the addition of a major module and solar panels and housed its first crew, who spent most of their time turning it into a home and workplace.*

## Key Figures

*William M. Shepherd* (b. 1949), U.S. Navy captain and Expedition One ISS commander

*Yuri Pavlovich Gidzenko* (b. 1962), Russian Air Force lieutenant colonel, Soyuz commander, and Expedition One ISS crew member

*Sergei Konstantinovich Krikalev* (b. 1958), Russian Expedition One flight engineer

*James D. Halsell, Jr.* (b. 1956), STS-101 mission commander

*Scott J. Horowitz* (b. 1957), STS-101 pilot

*Susan J. Helms* (b. 1958), STS-101 mission specialist

*Yuri Vladimirovich Usachev* (b. 1957), STS-101 mission specialist

*James S. Voss* (b. 1949), STS-101 mission specialist

*Mary Ellen Weber* (b. 1962), STS-101 mission specialist

*Jeffrey N. Williams* (b. 1958), STS-101 mission specialist

## Summary of the Mission

The first crew for the International Space Station (ISS) took up residence in November of 2000, a milestone for the massive program. Yet the year proved to be contentious for its partners. The Russian Space Agency (RSA) delayed launch of the Zvezda Service Module, throwing off the schedule for building and staffing the station, much to the chagrin of the National Aeronautics and Space Administration (NASA), Japan, Canada, Brazil, and the European Space Agency (ESA). The consortium revised the schedule, intensifying the preparatory work in the second half of the year, and successfully readied the station, but the year ended with project managers facing further delays and mounting criticism, especially related to cost overruns.

ISS had minor but nagging malfunctions during the year. The batteries in Zarya—the control module funded by the United States, built by Russia, and launched in 1998—failed to charge properly, as had been the case in 1999. There was no immediate impairment for the station because it is designed to operate on three batteries. Eventually, a fourth battery also showed signs of trouble. In late January, one of two remote power controller modules, which route electricity to the American-built Unity Module, failed to work smoothly. Flight controllers also found that a crane on ISS's hull was not latched down properly. However, none of these malfunctions stood in the way of docking the long-awaited Zvezda module (*zvezda* means "star" in Russian).

ISS was ready for Zvezda, but RSA was not ready to launch it. Zvezda is a key component of the station and had to be in place before the first permanent crew could take occupancy, because it contains their quarters and laboratory facilities.

Equally important, it has rocket engines that are the primary means for ISS to maintain its orbit, which, if left untended, degenerates as atmospheric particles hit the station and gradually slow it down. Two crashes of Russia's massive Proton boosters in 1999 led RSA to cancel further launches pending a review of the booster's design. Originally scheduled to be launched by Proton in 1998, Zvezda, RSA promised, would go into orbit in May. Meanwhile, ISS managers grew increasingly worried: By mid-May, the perigee (low point) of the ISS had dropped to 344 kilometers, and the station was losing 2.5 kilometers weekly. If it descended too far into the atmosphere, it would tumble out of control and tear apart. At last, on July 12 RSA launched Zvezda, and it docked with the Zarya module on July 25 after a flawless maneuver by RSA ground control.

*An artist's rendering of the solar arrays of the Zvezda Service Module in the process of deployment.* (NASA)

NASA had decided not to wait for Zvezda, however. It quickly readied the space shuttle *Atlantis* on a mission of "home improvement work," and Space Transportation System 101 (STS-101) launched on May 19, commanded by James D. Halsell, Jr., and piloted by Scott J. Horowitz. Aboard were mission specialists Susan J. Helms, Yuri Vladimirovich Usachev, James S. Voss, Mary Ellen Weber, and Jeffrey N. Williams. For Voss, Helms, and Usachev, the visit was a preview: They were to be the Space Station's second permanent crew.

First, Voss and Williams went on a spacewalk, during which they battened down the incorrectly mounted crane, attached a second heavy-duty crane built by Russia, replaced a faulty antenna, and installed handrails and a camera cable. Later the same day Usachev and Helms opened the hatches between the shuttle and ISS. Over the next three days, the crew replaced the defective batteries, installed new storage compartments and sound-insulation panels to dampen deafening noise from

the circulation system, emplaced three fire extinguishers and ten smoke detectors, and replaced a radio telemetry system. The crew then transferred more than 1,500 kilograms of new equipment and supplies from the SPACEHAB Logistics Double Module in *Atlantis*'s cargo bay, including exercise equipment. Halsell and Horowitz fired *Atlantis*'s engines twenty-seven times to lift ISS into an orbit with an apogee (high point) of 408 kilometers and perigee of 383 kilometers.

On August 8, another 616 kilograms of supplies and fuel arrived at ISS with the first mission of Progress, a fully automated, Russian-built resupply spacecraft of the Soyuz family. On September 10, space shuttle *Atlantis* again docked at the station for a seven-day visit during STS-106. This time, crew members conducted a spacewalk in order to connect power, data, and communications cables to Zvezda while other astronauts and cosmonauts unloaded 3,000 kilograms of supplies and equipment and prepared ISS modules for the first permanent crew. One month later, space shuttle *Discovery* arrived during STS-92 for further construction. During four spacewalks, its crew installed the 13,971-kilogram Z1 Integrated Truss Structure, an exterior framework to support solar panels; a communications system to support science projects and television; electrically powered gyroscopes for attitude control; and the Pressurized Mating Adapter 3, which provided a shuttle docking port for later installation projects.

Finally, on November 2, the first ISS crew docked at their newly prepared home in space aboard a Russian Soyuz space capsule commanded by RSA's Yuri Gidzenko, a lieutenant colonel in the Russian Air Force. At 10:31 Coordinated Universal Time (UTC), he and fellow cosmonaut Sergei Krikalev opened the hatch between Soyuz and ISS and floated into the station, followed by the Space Station commander, U.S. Navy Captain William M. Shepherd. The Soyuz remained attached to the station to serve as an emergency lifeboat. Once aboard, the crew began bringing ISS fully to life, checking out communications, activating computer systems and food warmers, charging batteries, starting water processors and the toilet, and restoring one of two defective batteries. It was an experienced crew. Gidzenko had already spent 179 days in space as commander of Russia's Mir Space Station. Krikalev had accumulated 484 days in space aboard Mir and space shuttle flights, including one aboard *Endeavour* in December, 1998, for construction work on ISS (he was the first person to visit the station twice). On November 6, they began the first of their exercise sessions, pedaling on an ergometer-equipped bicycle in Zvezda in order to minimize bone and muscle loss during their stay in space.

After spending two days largely in the heavily shielded section of Zvezda to avoid radiation exposure during a solar storm, the crew received more supplies upon arrival of the second Progress resupply craft on November 17. On December 2 space shuttle *Endeavour* docked at the station as part of STS-97. In the course of the three spacewalks of this crucial mission, the shuttle crew, aided by the ISS crew, removed a half-ton solar array structure from the shuttle's cargo bay with its Canadian-built robot arm and attached it to the ISS's Unity Module. Spanning 73 meters (240 feet), like giant wings spread perpendicular to the living modules, the solar arrays generate about 50 kilowatts of power, a fivefold boost for the station. This increase was important because before it the Elektron oxygen generator, a heavy electricity user, could be operated for only short periods, and the crew otherwise had to rely on bottled oxygen. The shuttle also brought supplies and computer hardware.

At year's end, the station was in good shape, orbiting approximately every 90 minutes at an altitude of 383 kilometers. The crew had already begun the station's first science projects. These included experiments in crystal growth, methods to decrease the effects of vibration on payloads in space, Earth observations, and tomato seed germination.

**Contributions**

The space shuttle and Progress missions to ISS accomplished all of their goals and often finished ahead of schedule, giving NASA officials reason to boast. In fact, they compared the space shuttle crews' efficiency to that of the pit crews working on racecars. Thus, much of the benefit from the ISS program during the year came from the practical experience gained by the crews, especially in spacewalks, equipment installation and start-up, and docking maneuvers.

Not all the experience was wholly advantageous, however. NASA learned to its dismay that RSA, its principal partner, was seriously underfunded and that accordingly its contributions to ISS would be chronically delayed. Furthermore, a cultural difference between astronauts and cosmonauts became apparent, although not divisive. NASA trained astronauts for missions by use of precisely scripted practice sessions for each major objective, such as installation of solar arrays, and expected the astronauts to follow the mission plan punctiliously in space. By contrast, RSA gave cosmonauts more general preparations for missions and allowed them greater flexibility in adapting their work to suit specific objectives. Rigid NASA mission schedules annoyed cosmonauts, and the Russians' penchant for individual initiative frustrated NASA.

### Context

Serious trouble loomed for the ISS program during 2000. On February 16, NASA Administrator Daniel S. Goldin assured a congressional committee that it would be a landmark year for the station despite unforeseen challenges. Even so, he admitted to being disappointed with RSA, particularly over the repeated delays in launching Zvezda. Largely because of these delays, ISS construction was at least twenty-seven months behind schedule and unforeseen costs for his agency were mounting. Worse, unsubstantiated rumors circulated that money given to RSA by NASA was being misused, although not by RSA itself. The rumor made ISS look bad to Congress and the political masters of NASA's partner agencies in other countries. The issue of Russia's Mir Space Station further troubled the political atmosphere. After agreeing to abandon Mir so it could concentrate its resources on ISS, RSA sent cosmonauts to reopen the old station in April. Even though a Netherlands-based corporation paid for the trip, NASA officials were upset. American public opinion also displayed some anti-Russian tendencies, and critics urged NASA to end the partnership.

NASA's reports of Russian safety violations in the construction of Zvezda exacerbated the controversy. Particularly at issue was the noise level aboard the Space Station, which could damage the crews' hearing and interfere with operations.

There was also worry that ISS was inadequately protected from space debris and vulnerable to explosive decompression. NASA had to fix such problems, incurring further cost overruns. Meanwhile, the United States' own commitment to the Space Station had shrunk to providing a single laboratory module, a single living module, a crew lifeboat, the truss, and the solar panels, and support in the scientific community was waning. Some scientists questioned the cost-effectiveness of ISS-based research, which had always been a primary purpose for the program. One ISS partner, British National Space Center Administrator Paul Murdin, called ISS a white elephant. Despite the controversies, NASA and RSA pressed ahead in their partnership, and the first crew ended the year with high-minded optimism. For the Space Station's log entry on New Year's Eve, the commander wrote a poem that placed ISS in the grand tradition of exploration and discovery:

> Though star trackers mark Altair and Vega
> Same as mariners eyed long ago
> We are still as wayfinders of knowledge
> Seeking new things that mankind shall know.

**See also:** Cooperation in Space: U.S. and Russian; Funding Procedures of Space Programs; International Space Station: Crew Return Vehicles; International Space Station: Design and Uses; International Space Station: Living and Working Accommodations; International Space Station: Modules and Nodes; International Space Station: 1999; International Space Station: 2001; International Space Station: 2002; Materials Processing in Space.

### Further Reading

Bond, Peter. *The Continuing Story of the International Space Station.* Chichester, England: Springer-Praxis, 2002. Bond describes the development and evolution of space stations, with particular emphasis on the International Space Station, beginning with the revolution that began in 1970, when Salyut 1, the world's first space station, was sent into orbit by the Soviet Union.

Harland, David M., and John E. Catchpole. *Creating the International Space Station.* London: Springer-Verlag London, 2002. A comprehensive review of the historical background, rationale behind, and events leading to the construction and commissioning of the International Space Station.

Messerschmid, Ernst, and Reinhold Bertrand. *Space Station: Systems and Utilization*. Berlin: Springer, 1999. Thorough and technically sophisticated, this volume reviews space station designs in general and that of the International Space Station specifically as it was originally envisioned. It also discusses the science of the subsystems, support vehicles, logistics and communications, the orbital environment, and factors affecting the crews. With photos, graphics, and tables clarifying the text.

Oberg, James. *Star-Crossed Orbits: Inside the U.S.-Russian Space Alliance*. New York: McGraw-Hill, 2002. A former NASA engineer, Oberg draws on contacts within the agency, and so his unstinting critique of U.S.-Russian misunderstanding and mutual manipulation during construction of the International Space Station has an insider's authority.

Zimmerman, Robert. *Leaving Earth: Space Stations, Rival Superpowers, and the Quest for Interplanetary Travel*. Washington, D.C.: Joseph Henry Press, 2003. This history of space stations concludes with a chapter about the International Space Station. Basing his discussion on interviews with former ISS crew members, the author reviews the difficulties in U.S.-Russian cooperation, the controversy over the station's purpose, and the technical challenges. He suggests that ISS provides training for the expertise needed to send spacecraft to the planets.

*Roger Smith*

# International Space Station: 2001

*Date:* January 1 to December 31, 2001
*Type of mission:* Space station, piloted spaceflight

*During 2001, the International Space Station (ISS) was a busy construction site and science platform, but the program nevertheless faced increasing criticism over its budgeting and scientific justification, and a decline in confidence in NASA forced a reduction in the long-term goals for the station.*

## Key Figures

*William M. Shepherd* (b. 1949), U.S. Navy captain and Expedition One ISS commander

*Yuri Pavlovich Gidzenko* (b. 1962), Russian Air Force lieutenant colonel, Soyuz commander, and Expedition One ISS crew member

*Sergei Konstantinovich Krikalev* (b. 1958), Russian Expedition One flight engineer

*Yuri Vladimirovich Usachev* (b. 1957), Russian ISS commander for Expedition Two

*James S. Voss* (b. 1949), U.S. Army colonel and Expedition Two flight engineer

*Susan J. Helms* (b. 1958), U.S. Air Force colonel and Expedition Two flight engineer

*Frank L. Culbertson, Jr.* (b. 1949), U.S. Navy captain and ISS commander for Expedition Three

*Vladimir Nikolaevich Dezhurov* (b. 1962), Russian Air Force lieutenant colonel, Soyuz commander, and Expedition Three crew member

*Mikhail Tyurin* (b. 1960), Russian Expedition Three flight engineer

*Yuri Onufrienko* (b. 1961), Russian Air Force colonel and ISS commander for Expedition Four

*Daniel W. Bursch* (b. 1957), U.S. Navy captain and Expedition Four flight engineer

*Carl E. Walz* (b. 1955), U.S. Air Force colonel and Expedition Four flight engineer

## Summary of the Mission

The International Space Station (ISS) expanded in size and capability during a troubling year for the National Aeronautics and Space Administration (NASA). The station underwent three crew changes, saw six visits from space shuttles and two Soyuz taxi flights, received three Progress resupply craft payloads, and had a new docking module sent up on its own automated rocket. With the addition of the new modules and a robotic arm, it became the largest, most sophisticated space station ever, surpassing Russia's Mir. Despite these achievements, cost overruns and delays caused by faulty equipment clouded the program's reputation and threatened its future.

The Expedition One crew began the year maintaining scientific experiments in crystal growth, vibration control, and seed germination while preparing ISS for further construction with the arrival of the next shuttle. During the Space Transportation System mission 98 (STS-98), space shuttle *Atlantis* docked with ISS on February 9 for a seven-day visit. In its cargo bay, the shuttle brought the American-built Destiny Laboratory Module, which was intended to become the center of experimentation in space once its twenty-four laboratory racks were delivered during subsequent shuttle missions. As a colleague inside *Atlantis* moved Destiny into position with the spacecraft's Canadian-built robotic arm, shuttle astronauts conducted spacewalks to attach the new module to ISS and then connected cables and other equipment to it. Destiny is 8.5 meters long and 4.3 meters in diameter; its addition brought the mass of ISS to 122 tons and increased its habitable volume by 41 percent. It has a special window that allows crew members to take high-resolution photographs and videos as part of the station's Earth observation mission.

In an unplanned maneuver, shuttle commander Kenneth D. Cockrell used *Atlantis*'s boosters to raise the orbit of ISS by 2 kilometers in order to avoid a piece of space junk.

A Progress cargo ship brought up supplies for ISS in late February (additional Progress missions took place in May and November), and then on March 10 space shuttle *Discovery* arrived for an eight-day visit during STS-102. Aboard was the Expedition Two crew, ISS commander Yuri Usachev of the Russian Space Agency (RSA) and flight engineers James S. Voss and Susan J. Helms from NASA. The crew exchange entailed a procedure in which Expedition Two members one by one moved their custom seat-liners from the space shuttle to the Soyuz capsule docked at ISS, while the Expedition One crew did the reverse: Yuri Gidzenko changed places with Usachev, Sergei Krikalev with Voss, and Expedition One commander William M. Shepherd with Helms. During spacewalks, astronauts moved the Italian-built Leonardo Multi-Purpose Logistics Module into docking position; the ISS crew then transferred supplies from the module into the station, after which Leonardo was returned to *Discovery*'s cargo bay.

A spacewalk by Helms and Voss turned into the longest in history, lasting 8 hours and 29 minutes, as small problems put them behind

schedule. While the new crew settled into their routine, further problems surfaced. The Ku-Band communications system for television had software glitches that misaligned its antenna, the exercise treadmill required repair, and Destiny's carbon dioxide removal system was not working.

Space shuttle *Endeavour* (STS-100) arrived on April 21 for eight days of repair, construction, and supply transfer, during which American and Canadian shuttle astronauts delivered and installed the Canadarm2. The 17.6-meter robot arm was a key component needed to move heavy loads into position during future construction and was the first of three pieces of the Space Station's Mobile Service System; a mobile work platform and mechanical hand were to arrive later. Canadarm2 has greater range of motion than a human arm and is guided with the help of four color cameras.

*Endeavour*'s visit turned out to be timely, as ISS's three command and control computers malfunctioned after the hard drive in one of them failed. Until the problem was fixed, station communications were routed through *Endeavour.*

No sooner had the shuttle departed than the first Soyuz taxi crew (designated by the Russians as

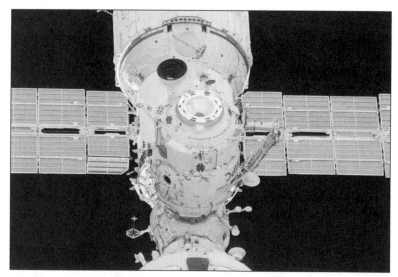

*The Russian-built Pirs docking compartment, which was mated to the International Space Station's Zvezda Service Module in September, 2001.* (NASA)

ISS Visiting Crew 1) arrived April 30 for a six-day stay, and it caused a media sensation. In accordance with RSA regulations, no Soyuz can remain in space longer than six months, a measure to prevent fuel degradation, so the taxi crew flew up to exchange their craft with the Soyuz that had been docked at the station, serving as a lifeboat, since it ferried up Expedition One in November, 2000. The new Soyuz also brought supplies for ISS. However, what captured all the attention was the arrival of American businessman Dennis Tito along with Soyuz commander Talgat Musabayev and flight engineer Yuri Baturin. Tito had paid RSA $20 million for the trip into space, and his entry into ISS made him the first space tourist aboard a space station.

In June, the ISS crew was preparing for another visit from space shuttle *Atlantis* when problems developed in Canadarm2. The robot arm had to be working properly in order to lift the new Quest Airlock from the shuttle's cargo bay and steady it for attachment to the station's Unity Module. Accordingly, STS-104 was postponed two weeks while NASA engineers and MD Robotics of Canada identified a computer malfunction in the arm's control system and developed a software patch to correct it. *Atlantis* docked on July 13 and Quest was installed. A week later, astronauts used the air lock for the first spacewalk to egress from ISS rather than from a shuttle.

Expedition Two was relieved by Expedition Three after the new crew rode to ISS aboard space shuttle *Discovery* during STS-105, beginning August 12: ISS commander Frank Culbertson of NASA and pilot Vladimir Dezhurov and flight engineer Mikhail Tyurin of RSA. With *Discovery* came the Leonardo Multi-Purpose Logistics Module full of supplies and equipment; upon being emptied, it was then filled with trash for return to Earth. Spacewalks by shuttle crew members attached the Materials International Space Station Experiment (MISSE) and a spare part for the station's cooling system. On September 16 the station was again expanded when the Russian-built Pirs ("pier" in Russian) docking compartment arrived by automated

rocket and was mated to Zvezda. The 4.9-meter, 3,000-kilogram module was designed to dock Russian spacecraft and act as an air lock and stowage area.

Another taxi crew brought up a fresh Soyuz spacecraft for exchange on October 23 and spent eight days helping Expedition Three with operations and research. On December 7, the space shuttle *Endeavour* carried Expedition Four to ISS during a ten-day stopover as part of STS-108. With it was the Italian-built Raffaello Multi-Purpose Logistics Module, carrying six tons of equipment and supplies. The new crew included commander Yuri Onufrienko of RSA and flight engineers Carl E. Walz and Dan Bursch of NASA.

**Contributions**

As in previous years, astronauts improved their skills while assembling the Space Station, learning to use new equipment, and making repairs, but during a year dominated by construction and repair, scientific investigations nevertheless started in earnest. Nearly fifty thousand hours of varied research occurred aboard ISS during its first full year of operation. Some of the work came from efforts to take advantage of ISS to develop products with commercial potentiality, but much of it sought to understand how the body reacts to low gravity (space medicine) or involved other fundamental biological research. Astronauts measured their rate of bone loss and studied methods to protect people from space radiation; they conducted research in the growth of kidney stones and in lung function. They used the microgravity environment of low-Earth orbit to track tiny disturbances caused by aerodynamic drag within the station as it moved; such disturbances affect both research and the crew. They germinated seeds and nurtured the plants, monitored crew interactions, and at the end of the year started cultures of human colon, kidney, and ovarian cells intended for medical research later on Earth.

Materials science also occupied the ISS crew. They grew crystals and examined their structures. Outside ISS, MISSE began an eighteen-month study in which 750 material samples were exposed to radiation in space for later analysis by ground-based scientists. The Expedition Three crew started a projected fifteen-year program to work with middle school students in acquiring knowledge from Earth observation.

### Context

A budgetary audit of ISS by the presidential administration of George W. Bush uncovered $4.8 billion dollars in cost overruns, outraging congressional representatives and prompting the NASA Advisory Council to warn that the entire international spaceflight enterprise was being undermined by a loss in confidence in NASA management. NASA's principal partner, Russia, was also having trouble financing its share of ISS. In an attempt to acquire ready cash, RSA accepted Dennis Tito's $20-million-dollar bid for a ride to ISS as part of a Soyuz taxi crew and trained him despite NASA's objections. A showdown between RSA and NASA took place when NASA initially refused to allow Tito to complete his training at American facilities. Russia countered that if Tito was rejected there would be no Soyuz flight, threatening the partnership's future. Eventually, NASA relented and Tito had his space vacation, but the incident soured American-Russian cooperation on ISS.

Even as scientific research got under way, criticism of it mounted in tandem with calls to review the entire space station program. The U.S. National Academy of Sciences concluded that ISS's protein crystal growth experiments—a keynote space commerce program—had little merit and called upon NASA to assure that the effort would produce significant results or drop it. The Advisory Council also complained of a lack of clarity in ISS science projects. Meanwhile, the Bush administration ordered NASA to shelve plans for expanding ISS to accommodate seven crew members and ended the program to develop a new crew lifeboat to replace Soyuz spacecraft. The outlook for ISS began to shrink.

**See also:** Cooperation in Space: U.S. and Russian; Funding Procedures of Space Programs; International Space Station: Crew Return Vehicles; International Space Station: Design and Uses; International Space Station: Living and Working Accommodations; International Space Station: Modules and Nodes; International Space Station: 2000; International Space Station: 2002; Materials Processing in Space.

### Further Reading

Bond, Peter. *The Continuing Story of the International Space Station.* Chichester, England: Springer-Praxis, 2002. Bond describes the development and evolution of space stations, with particular emphasis on the International Space Station, beginning with the revolution that began in 1970, when Salyut 1, the world's first space station, was sent into orbit by the Soviet Union.

Harland, David M., and John E. Catchpole. *Creating the International Space Station.* London: Springer-Verlag London, 2002. Examines the events leading to the construction and commissioning of the International Space Station.

Messerschmid, Ernst, and Reinhold Bertrand. *Space Station: Systems and Utilization.* Berlin: Springer, 1999. Thorough and technically sophisticated, this volume reviews space station designs in general and that of the International Space Station specifically as it was originally envisioned. It also discusses the science of the subsystems, support vehicles, logistics and communications, the orbital environment, and factors affecting the crews. With photos, graphics, and tables clarifying the text.

Oberg, James. *Star-Crossed Orbits: Inside the U.S.-Russian Space Alliance.* New York: McGraw-Hill, 2002. A former NASA engineer, Oberg draws on contacts within the agency, and so his unstinting critique of U.S.-Russian misunderstanding and mutual manipulation during construction of the International Space Station has an insider's authority.

Zimmerman, Robert. *Leaving Earth: Space Stations, Rival Superpowers, and the Quest for Interplanetary Travel.* Washington, D.C.: Joseph Henry Press, 2003. This history of space stations concludes with a chapter about the International Space Station, based on interviews with former ISS crew members.

*Roger Smith*

# International Space Station: 2002

*Date:* January 1 to December 31, 2002
*Type of mission:* Space station, piloted spaceflight

*The International Space Station (ISS) continued to grow in size and sophistication during a year that also was crowded with scientific research, but budgeting problems and scaled-back plans for its future created conflict among the sixteen nations involved in the program.*

## Key Figures

*Yuri Onufrienko* (b. 1961), Russian Air Force colonel and ISS commander for Expedition Four

*Daniel W. Bursch* (b. 1957), U.S. Navy captain and Expedition Four flight engineer

*Carl E. Walz* (b. 1955), U.S. Air Force colonel and Expedition Four flight engineer

*Valery Grigorievich Korzun* (b. 1953), Russian Air Force colonel and ISS commander for Expedition Five

*Peggy A. Whitson* (b. 1960), American Expedition Five flight engineer

*Sergei Yevgenyevich Treschev* (b. 1958), Russian Expedition Five flight engineer

*Kenneth D. Bowersox* (b. 1956), U.S. Navy captain and ISS commander for Expedition Six

*Donald R. Pettit* (b. 1955), American Expedition Six flight engineer

*Nikolai Mikhailovich Budarin* (b. 1953), Russian Expedition Six flight engineer

*Yuri Pavlovich Gidzenko* (b. 1962), Soyuz 4 commander

*Roberto Vittori* (b. 1964), Italian flight engineer

*Mark Shuttleworth*, a South African who was the second space "tourist" to ISS

## Summary of the Mission

Crews of the three expeditions aboard the International Space Station (ISS) during 2002 increased the number and complexity of scientific research while construction advanced with the help of four space shuttle missions. By the end of the year, about 34,000 kilograms of hardware had been delivered to ISS, bringing its total mass to nearly 150,000 kilograms. During eighteen spacewalks, ISS and shuttle crews installed trusses lengthening the station to 52 meters with a 27.4-meter mast supporting solar panels spanning 73 meters. The station also experienced troublesome equipment malfunctions, but as in previous years, the greatest danger to ISS came from below on Earth, where scientific panels challenged the program and plans

for it by the National Aeronautics and Space Administration (NASA), and where conflicts with international partners clouded the sixteen-member coalition's future.

The malfunctions, although few and relatively minor, were potentially serious. The crew of Expedition Four—ISS commander Yuri Onufrienko of the Russian Space Agency (RSA) and flight engineers Daniel Bursch and Carl E. Walz of NASA—were busy with the mission's fourteen new experiments and fifteen ongoing experiments on February 4, when an accidental computer shutdown interrupted the station's ability to keep the solar panels pointed at the Sun. As a precaution, flight controllers shut down many station systems, including the Payload Operations Center, where some experiments are conducted. No damage ensued, and after six hours, power was fully restored.

In May, the Elektron oxygen generator failed. The generator, which produces oxygen from water by electrolysis, had given the crew intermittent problems. The complete failure forced the crew to use solid-fueled canisters to produce oxygen until they repaired Elektron under the guidance of Russian flight controllers three days later. They were in no danger of running out of oxygen—the backup oxygen supply was good for three months—but the canisters' flammability made them risky to use. Also in May, a cooling system failure in the Zvezda Service Module caused some systems to shut down in the Destiny Laboratory Module. Flight controllers restarted all systems within forty minutes, and neither Destiny nor its scientific payload was harmed.

On June 8, a much more serious malfunction occurred when debris in its bearings caused one of the station's four 363-kilogram control gyroscopes to grind to a halt. Spinning at 6,600 revolutions per minute, the gyroscopes create torque that can be used to change the attitude of ISS without expending rocket fuel. The malfunctioning gyroscope, a major component, did no damage to the station, but its replacement posed a vexing problem for NASA. ISS administrators, however, were pleased with how quickly and successfully crews and flight controllers responded to these and other, less serious problems during 2002.

The first of the year's construction projects arrived with space shuttle *Atlantis* during Space Transportation System mission 110 (STS-110). In the course of a week of joint operations, the ISS and shuttle crews took four spacewalks and used

*A Soyuz spacecraft docked to the Pirs compartment at the International Space Station in November, 2002. (NASA)*

both the shuttle and the station robotic arms to install the 13.4-meter-long Starboard-Zero (S0) Truss, part of the station's eleven-piece Integrated Truss Structure, which forms the external framework. The full truss is designed to hold solar panels and convey cooling and electrical systems. It also supports the Mobile Transporter, a 2.7-by-2.6-meter platform on a 110-meter rail track that can move up to 20,900 kilograms of cargo. Initial tests of the transporter were successful despite a minor problem with its position sensors.

Only ten days later, the Soyuz 4 taxi crew arrived to rotate ISS's Soyuz lifeboat. With Soyuz commander Yuri Gidzenko of RSA and his Italian flight engineer, Roberto Vittori, came South African businessperson Mark Shuttleworth, the station's second space tourist. This international Soyuz crew helped the ISS crew carry out experiments during their weeklong visit.

After a delay of five weeks resulting from a jammed docking mechanism at the Space Station and bad weather on the ground, space shuttle *Endeavour* finally got airborne on June 5. The STS-111 mission docked to the ISS two days later to begin a nine-day stay. It carried the Expedition Five crew: ISS commander Valery Korzun, flight engineer Sergei Treschev (both of RSA), and flight engineer Peggy Whitson of NASA. The shuttle and ISS crews transferred cargo from the Italian-built Leonardo Multi-Purpose Logistics Module and prepared for further construction by affixing attachment mechanisms outside the hull during three spacewalks. (ISS also received cargo from three Progress automated resupply spacecraft during the year.)

In early September, the Expedition Five crew completed an important materials science study that used the antivibration Microwave Science Glovebox, which had been brought to the station in June, to solidify semiconductor crystals from molten indium antimonide doped with tellurium and zinc. A month later, they achieved another scientific first by producing a crop of "space beans": forty-two soybean seeds germinated out of the eighty-three planted. These results came as part of

the average of twenty-seven experiments running during the expedition.

Space shuttle *Atlantis* docked on October 9 during STS-112 with the 13.7-meter Starboard 1 (S1) Truss. During three spacewalks, crew members attached it to the S0 Truss. Meanwhile, other crew members transferred cargo and used the shuttle's thrusters to raise ISS's orbit. *Atlantis* undocked on October 16, after installing new experiments in place of completed experiments, which were returned to Earth for analysis.

On October 31, the fifth Soyuz taxi crew docked to the station in a new, upgraded spacecraft to serve as the station's lifeboat. The new Soyuz, Soyuz Transport Modification Anthropometric 1 (TMA-1), contained redesigned seats and suspension to accommodate American astronauts, who on average are taller than cosmonauts, and a new set of computer displays. The Soyuz spacecraft has made three major changes since the first one appeared in 1967. The Soyuz Transport (Soyuz T), introduced in 1979, featured larger solar panels for longer independent flights to the Salyut space stations and carried a crew of two. Soyuz Transport Modification (TM) began service in 1986. It featured multiple improvements in the design, including the introduction of a new weight-saving computerized flight-control system and an improved emergency escape system. These upgrades allowed for three crew members, while they could be still protected with pressure suits.

Space shuttle *Endeavour* docked as part of STS-113 on November 25 for the last shuttle mission of the year. Aboard was the Expedition Six crew: ISS commander Kenneth Bowersox (NASA) and flight engineers Donald R. Pettit (NASA) and Nikolai Budarin (RSA). *Endeavour* also brought the Port 1 (P1) Truss, which is 13.7 meters long. During three spacewalks, crew attached it perpendicular to the left side of the S0 Truss. Following the shuttle's departure on December 2, the Expedition Six members began work setting up their nineteen new experiments. At year's end, ISS was stationed at an altitude of 386 kilometers, having completed orbit number 60,000 earlier in the year.

## Contributions

By the end of 2002, ISS had logged more than ninety thousand hours of science operations—sixty-five NASA-sponsored projects alone—even though the station's science facilities were still being assembled. This research reflected eight categories of descending NASA priority, Associate Administrator Mary Kicza told a congressional committee in June, and each produced results during the year.

All the research took advantage of the microgravity environment, either as a variable in an experiment or as a means to produce a result. Studies in bioastronautics, concerned with human health in space, defined the relation of skin exposure from radiation to organ exposure (much of the data were taken following spacewalks), discovered that in space spinal cord reflexes decline and then after several months rebound, and examined lung function. Fundamental space biology experiments grew liver cells to learn about cell behavior and produced space beans as a possible food and oxygen source for future spacefarers.

Biotechnology efforts examined the molecular structure of proteins and nucleic acid for possible pharmaceutical applications and grew three-dimensional tissue for applications in tissue technology and disease research.

Combustion research experimented with conditions of ignition in the convection-free environment with the aim to produce more economic and environmentally benign techniques.

Fluid physics experiments sought clues to improve spacecraft system designs by modeling fluid behavior. In fundamental physics, experiments tested basic theories with greater precision than possible on Earth and in one series of experiments involving colloids observed a new phenomenon in how matter spontaneously organizes itself as it changes states.

Materials science projects developed methods for semiconductor, metal, ceramic, and polymer manufacturing, as well as biomaterials fabrication; of particular note was the growth of large, pure zeolite crystals, which have wide application in chemical and petroleum processing.

Finally, space product-development efforts sought to improve crop development and sensor technology, drawing from $50 million in commercial investments for the projects.

In addition, regular Earth observations by ISS crew produced data on air quality, urban spread, water control, and fire control worldwide. Crew members were able to watch the spread of massive blazes in Arizona and Colorado, downloading information about hotspots and fire movement for possible use by firefighters.

## Context

A successful year of construction and scientific research did not silence critics of the Space Station program. To control cost overruns, NASA's new director, Sean O'Keefe, approved a reduction in ISS construction called "Core Complete." It tabled plans for the crew return vehicle needed as a lifeboat for a seven-person crew, restricted the ISS crew to three, and indefinitely postponed adding some new American-built modules. Furious, both the European Space Agency and Russia accused NASA of unilaterally breaching its contract in the partnership. Without a new lifeboat to replace Soyuz, RSA would have to spend its budget on resupplying ISS rather than on research. Meanwhile, budgetary constraints caused Japan to delay launch of its science module for two years, and Brazil dropped plans to build its own research pallet.

Domestic critics were likewise upset with the Core Complete statement. They pointed out that with a crew of three there would be only eight hours per week of work time available for research after station housekeeping chores were finished. That was not enough to make ISS a major research facility, the National Academy of Sciences warned. In response, NASA proposed yet another plan for ISS: increased shuttle flights, new pressurized modules, revised research priorities, shuttles specially outfitted for visits as much as a month long to increase personnel strength for short intervals, and possible ways to maintain a larger permanent crew. The agency also promoted the commercialization of ISS resupply operations and considered sifting some research operations from the crew to ground-based controllers through telepresence.

NASA tolerated the presence of ISS's second space tourist, Mark Shuttleworth, but O'Keefe squelched as frivolous RSA's bid to fly the winners of a proposed U.S. reality television show to the station. However, NASA was not opposed to mass media exposure in itself: ISS's public approval increased following the April release of *Space Station,* a three-dimensional IMAX documentary filmed on board ISS in 2000.

**See also:** Cooperation in Space: U.S. and Russian; Funding Procedures of Space Programs; International Space Station: Crew Return Vehicles; International Space Station: Design and Uses; International Space Station: Living and Working Accommodations; International Space Station: Modules and Nodes; International Space Station: 2001; International Space Station: 2003; Materials Processing in Space.

### Further Reading

Bond, Peter. *The Continuing Story of the International Space Station.* Chichester, England: Springer-Praxis, 2002. Bond describes the development and evolution of space stations, with particular emphasis on the International Space Station, beginning with the revolution that began in 1970, when Salyut 1, the world's first space station, was sent into orbit by the Soviet Union.

Harland, David M., and John E. Catchpole. *Creating the International Space Station.* London: Springer-Verlag London, 2002. Looks at the historical background leading to the construction and commissioning of the International Space Station.

Messerschmid, Ernst, and Reinhold Bertrand. *Space Station: Systems and Utilization.* Berlin: Springer, 1999. Thorough and technically sophisticated, this volume reviews space station designs in general and that of the International Space Station specifically as it was originally envisioned. It also discusses the science of the subsystems, support vehicles, logistics and communications, the orbital environment, and factors affecting the crews. With photos, graphics, and tables clarifying the text.

Oberg, James. *Star-Crossed Orbits: Inside the U.S.-Russian Space Alliance.* New York: McGraw-Hill, 2002. A former NASA engineer, Oberg draws on contacts within the agency, and so his unstinting critique of U.S.-Russian misunderstanding and mutual manipulation during construction of the International Space Station has an insider's authority.

Zimmerman, Robert. *Leaving Earth: Space Stations, Rival Superpowers, and the Quest for Interplanetary Travel.* Washington, D.C.: Joseph Henry Press, 2003. This history of space stations concludes with a chapter about the International Space Station. Basing his discussion on interviews with former ISS crew members, the author reviews the difficulties in U.S.-Russian cooperation, the controversy over the station's purpose, and the technical challenges. He suggests that ISS provides training for the expertise needed to send spacecraft to the planets.

*Roger Smith*

# International Space Station: 2003

*Date:* January 1 to December 31, 2003
*Type of mission:* Space station, piloted spaceflight

*Loss of the space shuttle* Columbia *in February, 2003, caused NASA to suspend construction of the International Space Station (ISS) and reduce the permanent crew from three to two, yet crews were still able to conduct some scientific research.*

## Key Figures

*Kenneth D. Bowersox* (b. 1956), U.S. Navy captain and ISS commander for Expedition Six
*Donald R. Pettit* (b. 1955), American Expedition Six flight engineer
*Nikolai Mikhailovich Budarin* (b. 1953), Russian Expedition Six flight engineer
*Yuri Ivanovich Malenchenko* (b. 1961), Russian Air Force colonel and ISS commander for Expedition Seven
*Edward T. Lu* (b. 1963), American Expedition Seven flight engineer
*C. Michael Foale* (b. 1957), American ISS commander for Expedition Eight
*Alexander Yurievich Kaleri* (b. 1956), Russian Expedition Eight flight engineer
*Pedro Duque* (b. 1963), Spanish flight engineer, Visiting Crew 5

## Summary of the Mission

At year's opening, the National Aeronautics and Space Administration (NASA) and its international partners were anticipating a busy schedule of construction for the International Space Station (ISS) with the hope of finishing the "Core Complete" version of it a little more than a year later. The breakup of space shuttle *Columbia* as it reentered the atmosphere on February 1 threw all planning for ISS into turmoil. Pending an investigation of the accident, NASA grounded all three remaining shuttles, so the principal ferry and cargo vehicles for ISS were no longer available. Since most major components for ISS, such as trusses and habitat modules, were designed for shuttles to carry into orbit, construction halted. The ISS would continue to be crewed and resupplied by the Soyuz and Progress spacecraft. However, this limited the permanent crew to two and the experiments to a bare minimum.

Expedition Six crew members—ISS commander Kenneth Bowersox of NASA and flight engineers Donald R. Pettit (NASA) and Nikolai Budarin (RSA)—were beginning the third month of their mission when the *Columbia* accident occurred. They were as shocked by the loss of their comrades as the rest of the astronaut-cosmonaut community. Moreover, they had another source of distress: How would the ISS be resupplied and the crew rotated back to Earth? A Russian-built Progress resupply spacecraft docked on February 4 with more than two tons of food, water, scientific equipment, and housekeeping supplies. There was a Soyuz spacecraft parked at another port as a lifeboat in case an emergency forced the crew to abandon the station. Nevertheless, Expedition Six had been scheduled to leave aboard space shuttle *Atlantis* in March, and that was no longer possible.

After taking time to hold a memorial ceremony for *Columbia* on February 12 in conjunction with the larger one taking place on Earth, Expedition Six focused on work while awaiting NASA's decision on the future of ISS. The crew had a packed schedule overseeing an average of twenty-seven research projects per week, some new and some ongoing, as well as station-keeping chores. Additionally, Pettit spent his modicum of free time hosting his Internet-based "Saturday Morning Science," a popular program of experiments for young people.

In late February, the ISS program's Multilateral Coordination Board agreed to an interim operations plan. The station would remain crewed, but the number would be reduced from three to two, and crews would rotate every six months when the Soyuz spacecraft are exchanged; scientific research would continue, albeit at a greatly reduced pace. After the shuttles resumed flight, ISS construction would recommence, and the crew level would return to three. In the meantime, expedition members were to concern themselves with maintenance primarily, and there was usually plenty of it to do. Since November, the Expedition Six crew had had electrical problems with the Microgravity Science Glovebox, used for experiments that have to be free of vibrations, and they were unable to return it to full operation until March.

The Expedition Six mission was extended one month to permit preparation of their taxi home. On April 28, a Soyuz spacecraft with Expedition Seven aboard docked at the Zarya Control Module. Bowersox, Budarin, and Pettit introduced ISS to their replacements—the new ISS commander Yuri Malenchenko of RSA and flight engineer Edward T. Lu of NASA—and helped transfer cargo until they left on May 3. Normally, Soyuz spacecraft transport three persons, but with a reduction in crew size, Malenchenko and Lu were able to bring a specially

*Expedition Six flight engineer Nikolai Budarin in the Soyuz spacecraft docked to the International Space Station, January, 2003. ISS Commander Ken Bowersox can be seen lower right. (NASA)*

designed cargo container in the third seat. For a while there were three Russian-built vehicles parked at ISS, the first time this had occurred. Progress remained docked in case ISS needed to use its remaining fuel to raise its orbit and to collect additional rubbish before the next resupply visit.

In addition to doing maintenance, the new crew upgraded computer software in anticipation of resumed construction and used the station's Canadarm2 to inspect the outside hull, a task previously performed by shuttle and Soyuz crews. They modified procedures for scientific research, gradually increasing the amount of work done per week from six to fifteen hours. Additionally, Lu took over as host of "Saturday Morning Science." On June 11, another Progress craft brought 1,600 kilograms of fuel, food, supplies, and scientific gear. It was modified to carry extra water as well, which was needed. Previously, visiting shuttles had delivered large quantities of water to ISS; without these visits a shortage developed on the station. On August 31, after much rescheduling, another Progress carried up cargo. On August 10, Malenchenko married Ekaterina Dmitriev during the first "space wedding." The bride was at the Johnson Space Center in Houston, 386 kilometers below, and the ceremony took place via video. The honeymoon was delayed until Malenchenko returned to Earth.

Expedition Eight arrived at ISS aboard a Soyuz spacecraft on October 20 to relieve Expedition Seven. The new crew had a temporary member. Along with new ISS commander C. Michael Foale of NASA and flight engineer Alexander Kaleri of RSA came Pedro Duque of Spain under a contract between the European Space Agency and its Russian counterpart. Until he departed with the Expedition Seven crew on October 27, Duque performed a series of experiments in the Microgravity Science Glovebox using materials brought during the latest Progress cargo mission. For their part, Foale and Kaleri expected to match the pace of research that Expedition Seven had achieved and intended to complete 360 hours of science work.

However, an increasing number of equipment problems affected conditions on board. Their ex- ercise equipment, important to maintaining the crew's health, was not working properly; they had to nurse a balky air-conditioning system; and sensors to monitor air and water quality had failed. The prospect of undetected water- or air-borne toxins endangering the crew prompted the RSA to test samples brought back with Expedition Seven. The results revealed the station's environment to be safe, but Soyuz or Progress missions could not deliver spare parts for such damaged systems readily. This difficulty became particularly pressing in mid-December, when one of the control moment gyroscopes began showing signs of trouble. Each of the four gyroscopes weighs 363 kilograms, and they are used to change the station's attitude in space without expending valuable propellants. One had already broken down in 2002, and if a second failed, the crew would have to use thrusters for attitude control. Accordingly, fuel for the station would take up more of their weight allotment on subsequent Progress missions.

### Contributions

ISS managers and NASA scientists were pleased that the reduced crew was able to accomplish more scientific research than anticipated. By the time Expedition Seven departed, astronauts and cosmonauts had completed 1,513 hours of work involving seventy-four investigations, as well as experiments during science programs for primary and middle school students. Revised priorities decided upon even before the *Columbia* accident had refocused the research primarily on the long-term effects of life in space upon human health (bioastronautics) and commercial experiments, although work in basic biology and physics was done too.

Particularly of note were pulmonary function tests performed on ISS crew members (especially following spacewalks), radiation monitoring, bone-density tests, and muscle and joint reaction tests on the legs. Commercial experiments with controllable fluids (also called "smart materials") and formation of crystals in plasma produced results that were promising for practical applications in automotive and aerospace industries. Other experiments concerned plant growth, pharmaceutical synthesis, and petroleum refining. Additionally, the crews observed environmental phenomena on Earth, such as glaciers, wild fires, and conditions affecting air and water quality.

### Context

NASA's decision to ground the space shuttle in the wake of the *Columbia* accident left ISS only 40 percent complete and already $22 billion over budget. The agency also had critical short-term and long-term decisions to make. First was how to keep ISS supplied and crewed. There were both internal and external recommendations, as well as some political pressure, to end the ISS program or at least mothball the station, but the international ISS consortium rejected them. NASA's solution was to ask RSA to produce more Progress and Soyuz spacecraft to replace the shuttles until they could fly again. That created further problems. RSA agreed but, badly in need of money, at first demanded $100 million in advance, funds not budgeted for NASA. Even when a satisfactory agreement was reached, it was uncertain whether Progress missions, which hold one-tenth as much cargo as a shuttle, would suffice. As it turned out, consumption of supplies aboard the station was slower than expected. For the long term, NASA needed a replacement for the aging space shuttles, but no definite plan existed. The agency discussed using an entirely new Orbital Space Plane, albeit without deciding on a basic design, or possibly a redesigned Apollo spacecraft.

NASA officials insisted that ISS was a sturdy platform that could survive the temporarily reduced level of resupply and repair. They emphasized that ISS crew were getting much more research done than expected. Yet criticism of that science persisted. The president of the prestigious American Physical Society told a U.S. Senate committee that there was no scientific justification for a space station. Even NASA's own experts were disapproving. One scientist called the medical research aboard ISS poorly planned and wishful rather than rigorous. Meanwhile, a NASA-appointed task force concluded that the "Core Complete" version of the Space Station, even when construction was finished, would not be large enough to justify calling it a research center. They urged that the full original design for a seven-person crew be built. Given past cost overruns, that possibility seemed slight. Members of the U.S. Congress, responsible for most of the program's financing, viewed NASA's handling of ISS warily. "We are one problem away from abandoning the Space Station," warned Representative Bart Gordon (Democrat of Tennessee), the Ranking Member of the House Science Committee.

**See also:** Cooperation in Space: U.S. and Russian; Funding Procedures of Space Programs; International Space Station: Crew Return Vehicles; International Space Station: Design and Uses; International Space Station: Living and Working Accommodations; International Space Station: Modules and Nodes; International Space Station: 2002; International Space Station: 2004; Materials Processing in Space.

### Further Reading

Bond, Peter. *The Continuing Story of the International Space Station.* Chichester, England: Springer-Praxis, 2002. Bond describes the development and evolution of space stations, with particular emphasis on the International Space Station, beginning with the revolution that began in 1970, when Salyut 1, the world's first space station, was sent into orbit by the Soviet Union.

Harland, David M., and John E. Catchpole. *Creating the International Space Station.* London: Springer-Verlag London, 2002. A comprehensive review of the historical background, rationale behind, and events leading to the construction and commissioning of the International Space Station.

Messerschmid, Ernst, and Reinhold Bertrand. *Space Station: Systems and Utilization*. Berlin: Springer, 1999. Thorough and technically sophisticated, this volume reviews space station designs in general and that of the International Space Station specifically as it was originally envisioned. It also discusses the science of the subsystems, support vehicles, logistics and communications, the orbital environment, and factors affecting the crews. With photos, graphics, and tables clarifying the text.

Oberg, James. *Star-Crossed Orbits: Inside the U.S.-Russian Space Alliance*. New York: McGraw-Hill, 2002. A former NASA engineer, Oberg draws on contacts within the agency, and so his unstinting critique of U.S.-Russian misunderstanding and mutual manipulation during construction of the International Space Station has an insider's authority.

Zimmerman, Robert. *Leaving Earth: Space Stations, Rival Superpowers, and the Quest for Interplanetary Travel*. Washington, D.C.: Joseph Henry Press, 2003. This history of space stations concludes with a chapter about the International Space Station. Basing his discussion on interviews with former ISS crew members, the author reviews the difficulties in U.S.-Russian cooperation, the controversy over the station's purpose, and the technical challenges. He suggests that ISS provides training for the expertise needed to send spacecraft to the planets.

*Roger Smith*

# International Space Station: 2004

*Date:* January 1 to December 31, 2004
*Type of mission:* Space station, piloted spaceflight

*Because the space shuttle fleet remained grounded throughout the year in the wake of the February, 2003, shuttle accident, the International Space Station (ISS) saw no significant construction, and the two-person crews spent most of their time in maintenance, although they did conduct some scientific research.*

### Key Figures

*C. Michael Foale* (b. 1957), English ISS commander for Expedition Eight

*Alexander Yurievich Kaleri* (b. 1956), Russian Expedition Eight flight engineer

*Gennady Ivanovich Padalka* (b. 1958), Russian Air Force colonel and ISS commander for Expedition Nine

*Edward Michael "Mike" Fincke* (b. 1967), U.S. Air Force lieutenant colonel and Expedition Nine flight engineer

*André Kuipers,* Dutch flight engineer, Visiting Crew 6

*Leroy Chiao* (b. 1960), American ISS commander for Expedition Ten

*Salizhan Shakirovich Sharipov* (b. 1964), Russian Expedition Ten flight engineer

*Yuri Georgievich Shargin,* Russian Space Forces lieutenant colonel and Visiting Crew 7 flight engineer

### Summary of the Mission

Through 2004, the space shuttle fleet of the National Aeronautics and Space Administration (NASA) stayed on Earth while a panel investigated the loss of space shuttle *Columbia* in 2003. This meant that the International Space Station (ISS) lacked its primary personnel and cargo transfer vehicles. The three expeditions at the station were therefore limited to crews of two instead of three, as before the *Columbia* accident, and no major construction took place.

Those crews had to spend an increasing amount of time repairing equipment and depended upon Russian-built Progress automated cargo spacecraft, which were not intended to assume the full resupply burden for their food and water. The crews experienced several trying episodes because of equipment and supply problems, yet they were able to carry out the station's original primary mission to some extent: scientific research. However, even while crew performance exceeded the expectations of NASA and its fifteen partner nations, U.S. long-term space policy began to change in ways that narrowed ISS's mission.

Above all else, the ISS crews were to keep the Space Station in good running order and prepare it for the resumption of space shuttle flights in 2005. In addition to routine maintenance chores, there was frequent need for challenging repairs. ISS commander C. Michael Foale of NASA and flight engineer Alexander Kaleri of the Russian Space Agency (RSA), the Expedition Eight crew, began the year nursing a malfunctioning oxygen generator, Elektron, which persisted in being troublesome. It was not approved for around-the-clock use until October, forcing crews to supplement their air supplies from backup oxygen canisters.

Problems with the air quality monitoring system also were not solved until October.

Early in January, flight controllers had detected a drop in pressure that the crew traced (only after much searching) to a condensation-venting hose for the large viewing window in the Destiny Laboratory Module. Foale and Kaleri noted materials drifting away from the station and strange noises coming from the hull (eventually thought to be from metal sheets flexing in the sunlight). In April, one of the control moment gyroscopes lost power; the gyroscopes are needed to control the direction that the station faces, and one of the four units had already malfunctioned. Not until July 2 was the problem corrected, and the gyroscope returned to operation.

The long delay in repairing the gyroscope occurred because of yet another persistent mainte-nance headache, the crews' space suits. On February 26, Foale and Kaleri had just begun a spacewalk to install equipment that had been brought the month before by a Progress resupply craft, when Kaleri reported a problem with his space suit's environmental system. "I have rain inside my helmet," he told flight controllers in Moscow. Both were ordered back inside after fourteen minutes—a quick end to the first spacewalk to occur without a crew member inside ISS. The Expedition Nine crew—ISS commander Gennady Padalka of RSA and flight engineer Mike Fincke of NASA—had to abort a spacewalk in June after ten minutes when a manual switch for the oxygen system froze in Fincke's space suit. Theirs was the spacewalk to fix the gyroscope, as well as to install new equipment and experiment packages on the hull. Crew members spent long periods troubleshooting space-suit problems, espe-

*The crew of the International Space Station in October, 2004 (from left): Yuri Shargin, Gennady Padalka, Leroy Chiao, Mike Fincke (seated), and Salizhan S. Sharipov.* (NASA)

cially the cooling systems, and the American-built space suits eventually were set aside as unusable, leaving only the Russian-built models.

Expedition Nine arrived by a Soyuz spacecraft on April 21. With Padalka and Fincke was a temporary crew member, André Kuipers, a Dutch astronaut from the European Space Agency (ESA). He spent nine days aboard ISS conducting research in life sciences and returned to Earth with the Expedition Eight crew. Padalka and Fincke earned much praise from ISS managers during their six-month mission because despite space-suit problems they were able to repair the gyroscope and carry out more scientific work than the scheduled twenty-four projects; the crew often spent their free time on science. Nine's crew also continued the Internet-based "Saturday Morning Science" program for students and spoke to conferences of science teachers. Padalka and Fincke unwittingly created a serious problem for their successors when, in an attempt to vary their diets, they consumed too many rations.

Expedition Ten ISS commander Leroy Chiao of NASA and flight engineers Salizhan Sharipov of RSA and Yuri Shargin of the Russian Space Forces docked with ISS on October 16. They narrowly avoided an accident when the autopilot brought the Soyuz spacecraft to the station too fast. A visiting crew member (officially, the sole member of Visiting Crew 7), Shargin completed a program of scientific experiments and then returned to Earth with Expedition Nine eight days later. Chiao and Sharipov looked forward to a busy schedule of spacewalks to prepare for the first ATV and later the return of the shuttles and hoped to work on fourteen science projects. By mid-December, however, they realized that less food was aboard than the inventory system had led them to believe, and no automated Progress spacecraft would bring supplies to make up for the shortfall before the end of December. To make supplies last, they decreased their intake from 3,000 calories to 2,700 calories per day. Undaunted, the pair said that they looked upon the challenge as a kind of camping adventure. Both lost weight waiting for the Progress re-

supply craft, which arrived on December 26 with 2.5 tons of food, water, and other supplies, including Christmas gifts from their families and a special holiday meal. They spent New Year's Eve at an altitude of 375 kilometers watching fireworks near cities under their orbital path.

### Contributions

Although expedition crews carried out research in some other scientific areas, NASA began to confine ISS science increasingly to those projects that studied some aspect of living and working in space. To improve construction techniques, for example, Expedition Nine experimented with different materials for soldering. Much of the work, however, was devoted to bioastronautics. Crews monitored the effects of radiation and microgravity on their hearts, lungs, and muscles; NASA was particularly interested in the chronic calcium loss from bones that spacefarers experience, which can result in the formation of kidney stones as well as weakening of the skeletal system. The Expedition Nine crew also conducted the first ultrasound body scan in space to practice the technique and determine its accuracy for future use by spacefaring medics.

Other scientific work aimed for practical applications. Crews studied the behavior of particles suspended in liquids (colloids) and the formation of bubbles in metals, both to improve materials processing, and they grew yeast cells in culture for possible use in the pharmaceutical industry. The Expedition Ten crew photographed the aftermath of the devastating tsunami of December 26 in the Indian Ocean to identify coastal changes, and all crews sought to inspire future space scientists through programs intended for students and their teachers.

### Context

On January 14, the Bush administration announced a series of new goals in outer space for the United States. NASA was to begin work on returning Americans to the Moon in order to build a base there for a crewed voyage to Mars. As part of the new policy, ISS was to be completed in 2010, at

*The International Space Station as of October 15, 2004, in a computer-generated image showing Soyuz 9 docked to the Pirs aft port, Soyuz 8 docked at the Zarya nadir port, and the Progress resupply vehicle at the aft end of the Zvezda Service Module.* (NASA)

which time the aging space shuttle fleet would be retired. NASA's scientific mission for ISS narrowed to studying the effects of long-duration spaceflight on humans and equipment in preparation for the Mars trip. In order to fund the lunar and Mars expeditions, robotic missions would be curtailed and the Hubble Space Telescope would be permitted to die a fiery death before its work could be completed.

NASA's partners in the ISS program—RSA, ESA, Brazil, and Japan—were already displeased with earlier reductions in the station's design and crew and suspected that their principal reason for joining the program, scientific research, would be sacrificed to the new policy. In part to reassure its

partners, NASA agreed to increase the crew to six in 2009, using two Soyuz spacecraft as lifeboats, and replace the shuttles with a new Orbital Space Plane able to keep the station supplied and working to its capability. However, the space plane was not yet planned or funded.

American critics continued to attack the ISS program because of its cost overruns and modest scientific achievements. The right-wing newspaper *The Washington Times*, for instance, called ISS a financial black hole. Meanwhile, more than ninety tons of components accumulated at the Kennedy Space Center in Florida awaiting delivery to ISS by space shuttle, including new science and habitat modules. This hardware, built in thirty-seven states and fifteen countries, cost $60 million per year to keep ready for space.

**See also:** Cooperation in Space: U.S. and Russian; Funding Procedures of Space Programs; International Space Station: Crew Return Vehicles; International Space Station: Design and Uses; International Space Station: Living and Working Accommodations; International Space Station: Modules and Nodes; International Space Station: 2003; Materials Processing in Space.

**Further Reading**

Bond, Peter. *The Continuing Story of the International Space Station.* Chichester, England: Springer-Praxis, 2002. Bond describes the development and evolution of space stations, with particular emphasis on the International Space Station, beginning with the revolution that began in 1970, when Salyut 1, the world's first space station, was sent into orbit by the Soviet Union.

Harland, David M., and John E. Catchpole. *Creating the International Space Station.* London: Springer-Verlag London, 2002. A comprehensive review of the historical background, rationale behind, and events leading to the construction and commissioning of the International Space Station.

Messerschmid, Ernst, and Reinhold Bertrand. *Space Station: Systems and Utilization.* Berlin: Springer, 1999. Thorough and technically sophisticated, this volume reviews space station designs in general and that of the International Space Station specifically as it was originally envisioned. It also discusses the science of the subsystems, support vehicles, logistics and communications, the orbital environment, and factors affecting the crews. With photos, graphics, and tables clarifying the text.

Oberg, James. *Star-Crossed Orbits: Inside the U.S.-Russian Space Alliance.* New York: McGraw-Hill, 2002. A former NASA engineer, Oberg draws on contacts within the agency, and so his unstinting critique of U.S.-Russian misunderstanding and mutual manipulation during construction of the International Space Station has an insider's authority.

Zimmerman, Robert. *Leaving Earth: Space Stations, Rival Superpowers, and the Quest for Interplanetary Travel.* Washington, D.C.: Joseph Henry Press, 2003. This history of space stations concludes with a chapter about the International Space Station. Basing his discussion on interviews with former ISS crew members, the author reviews the difficulties in U.S.-Russian cooperation, the controversy over the station's purpose, and the technical challenges. He suggests that ISS provides training for the expertise needed to send spacecraft to the planets.

*Roger Smith*

# International Sun-Earth Explorers

*Date:* October 22, 1977, to September 11, 1985
*Type of program:* Scientific platforms

*The three International Sun-Earth Explorers (ISEEs) performed observations and experiments in deep space and Earth's magnetosphere. After the original set of experiments was completed, ISEE 3 was sent around the Moon to perform the first close encounter between a spacecraft and a comet.*

### Key Figures

*Martin Titland*, ISEE program manager
*Robert Farquhar*, team leader for International Cometary Explorer

### Summary of the Satellites

The three International Sun-Earth Explorer (ISEE) satellites were the vehicles of a series of complex passive experiments designed to discover the nature, composition, and behavior of Earth's magnetosphere, the layer of highly charged subatomic particles that makes up the planet's upper atmosphere. ISEE 3, after the completion of its original mission, was moved from its "halo" orbit between Earth and the Sun and put to further use as the International Cometary Explorer (ICE). In 1985, ICE became the first human-made satellite to have close contact with the tail of a comet, the Comet Giacobini-Zinner. The International Sun-Earth Explorer missions were among the first joint deep-space missions between the United States and the European Space Agency (ESA). As such, the ISEE program was a valuable step in the building of cooperation between the United States and its European allies.

ISEEs 1, 2, and 3 were conceived as part of the International Magnetosphere Study, a mid-1970's multinational program studying the structure of Earth's magnetosphere. This area of energetic particles around most of Earth (which includes the Van Allen belts of radioactive particles) extends nearly 64,000 kilometers from Earth in the direction of the Sun and many millions of kilometers behind Earth,

forming an invisible "tail," much like that of a comet. This magnetotail was one of the principal phenomena to be examined by the ISEE satellites.

In addition, the ISEE satellites returned data on related phenomena, including the solar wind (waves of energized particles released by the Sun), the "bow shock" created by the supersonic impact of the solar wind against the magnetosheath (the blunted layer of the magnetosphere that faces the Sun), and the magnetopause (the brief area between the bow shock and the magnetosphere).

The ISEE satellites were preceded into space by earlier Explorer spacecraft that made up the Interplanetary Monitoring Platform program (notably Explorers 18, 21, 28, 33, 34, 35, 41, 43, 47, and 50, along with ESA's Geodetic Earth-Orbiting Satellites, or GEOS). These probes had returned data on a variety of atmospheric and deep-space phenomena.

The impact of charged ions from the magnetosphere has been known to affect and sometimes damage electronic and computerized instrumentation on piloted and robotic spacecraft. It was, therefore, deemed essential to the success of future Earth-orbital and deep-space missions to gain a better understanding of the component parts of this natural phenomenon.

ISEE 1 contained thirteen experiments to measure the magnetosphere, magnetosheath, and magnetotail; the craft was designed and managed by Goddard Space Flight Center, located in Greenbelt, Maryland. The satellite was a 1.73-by-1.61-meter cylinder with sixteen sides. At launch it weighed 340 kilograms.

ISEE 2 contained several experiments similar to those on ISEE 1 and was manufactured by the Dornier Company of West Germany through the European Star Consortium and managed by the European Space Technology Center. The uncrewed spacecraft was a 1.27-by-1.1-meter cylinder and weighed 157.3 kilograms at the time of launch.

ISEE 3, built by Fairchild Industries and also managed by Goddard Space Flight Center, was similar in size, shape, and weight to ISEE 1. ISEE 3 also contained instruments to measure magnetic fields and the energized particles in the solar wind and the magnetosphere, including an electrostatic analyzer, a magnetometer, plasma and radio wave receivers, an energetic proton particle detector, a scintillation detector, and particle detector telescopes. Once in space, all three satellites could be moved around in orbit through the use of small hydrazine thrusters.

The first two satellites of the ISEE program were launched atop a Delta rocket booster from Cape Canaveral, Florida, on October 22, 1977, and placed in orbits of approximately 340 by 137,847 kilometers around Earth. In these positions, ISEEs 1 and 2 conducted measurements of the magnetosphere via radio command from Earth. ISEE 1 was used as the control satellite, the stationary point of measurement. Through a series of firings of the satellite's small engines, ISEE 2 was moved away from ISEE 1 to perform additional measurements.

ISEE 3, launched from Cape Canaveral by a Delta rocket on August 12, 1978, was the first satellite to be placed in

a "halo" orbit around a point in space between Earth and the Sun. This point, known as a libration point, allowed the satellite to conduct solar wind and magnetotail measurements at far greater distances from Earth than had ever been done before. It also gave investigators an hour's advance warning of solar wind disturbances that might be measured by the near-Earth ISEEs 1 and 2. ISEEs 1 and 2 provided ample data about the energized protons, electrons, and ions that make up the magnetosphere and the magnetotail during their three-year life spans. ISEE 3, with an expected ten-year lifetime, gave valuable insights into the composition of the solar wind and Earth's upper atmosphere, in conjunction with the other ISEE spacecraft.

*The International Sun-Earth Explorer C undergoing tests at Goddard Space Flight Center in November, 1976.* (NASA)

By 1981, when it was determined by NASA that the United States would not have the opportunity to launch a probe to study the return of Comet Halley in 1985-1986, a team of investigators at Goddard Space Flight Center, led by Robert Farquhar, determined that ISEE 3 could be moved from its position in orbit between Earth and the Sun and placed on a flight plan to encounter Comet Halley. Further study showed that, while the craft could encounter the comet's tail, it could not do so at a distance that would allow its communications equipment to stay in contact with Earth. An alternative plan, however, was soon developed.

The Comet Giacobini-Zinner is visible from Earth every six and a half years. Through a complex, three-year-long series of maneuvers beginning in 1982, ISEE 3 was set on a course that would take it through the comet's tail, within 10,000 kilometers of its nucleus. This event took on added significance because it was due to occur six months before spacecraft from the USSR, Japan, and ESA were to pass through the tail of Comet Halley. ISEE 3, renamed the International Cometary Explorer (ICE), would be the first spacecraft to have direct contact with a comet.

In order to attain the velocity and propulsion necessary to move from its libration-point orbit to encounter the comet, ICE was placed on one of the most complicated interplanetary trajectories ever devised. The satellite had to loop around the Moon five times and fire its small engines fifteen different times over three years to gain the momentum to intersect with Giacobini-Zinner's tail. Finally, on September 11, 1985, at 7:00 A.M. eastern daylight time or 11:00 Coordinated Universal Time (UTC), ICE passed through the comet's tail at its closest point. Data from ICE indicated that the comet's tail is composed of slow-moving, cold plasma, water, and carbon monoxide ions. The comet is also known to leave a trail of debris in the wake of its orbit around the Sun. This trail resulted in a spectacular meteor shower in 1946, when Earth itself passed through Giacobini-Zinner's tail. This fact and close measurements by ICE have led scientists to conclude that comets are loose collections of primordial dirt and ice passing through space, possibly even remnants of the materials from which the solar system's planets originated. The Sun's heat as it melts the ice is believed to cause the tail effect visible to observers on Earth.

Less than one year after its encounter with Giacobini-Zinner, in March of 1986, ICE also made a distant pass by Comet Halley. In 1988, ICE finished its work when its solar orbit took it out of contact with Earth. The satellite will return to the vicinity of the Earth in the year 2012. Prior to the *Columbia* accident there had been some rather speculative discussions about mounting a retrieval mission so that the plucky spacecraft could be placed on display in the National Air and Space Museum in Washington, D.C. In the wake of the restriction of space shuttle flights to missions sharing the same orbital inclination as the International Space Station, this idea had to be discarded.

### Contributions

The International Sun-Earth Explorers gave scientists their first prolonged opportunity to obtain data pertaining to the magnetosphere, the invisible shroud of energy that surrounds Earth. As such, the ISEE series represented a milestone in human progress toward understanding our home planet and the solar system in which it resides. The ISEEs' three-pronged coverage of deep-space events created a "road map" of the magnetosphere and its interaction with the constant stream of solar wind. The movements of charged particles through space and around the planet were more closely charted than ever before.

ISEE 3, as ICE, gave scientists their first look at the tail of a comet. It was learned that a comet is colder and much broader than originally expected. For example, ICE encountered atmospheric ions more than 1.6 million kilometers from the comet. This fact and other data retrieved from the ICE mission are still being studied and analyzed by scientists around the world.

The ISEE missions also produced intangible benefits on Earth, beyond the data they sent back from space. As one of the first international projects conducted jointly by NASA and ESA, ISEE paved the way for future cooperation on the Spacelab module carried on board the American space shuttles and on other uncrewed programs such as the Active Magnetospheric Particle Tracer Explorers (AMPTEs), which were launched in 1984. ISEE 3/ICE proved the benefits of putting a satellite in a libration-point orbit, at the point where the gravitational fields of Earth and the Sun intersect. This distant "outpost" gave scientists at Goddard Space Flight Center and at facilities around the world one of the first in-depth, long-distance views of magnetic phenomena in Earth's upper atmosphere.

ICE also showed that a satellite, even though its design did not call for it, could be put to multiple uses and become a versatile tool in space for scientists. Its lunar-assisted trajectory to the Comet Giacobini-Zinner proved the efficacy of a method of propulsion that could be used for crewed interplanetary missions in future years.

### Context

The ISEE/ICE satellites were three in a long series of spacecraft designed to study phenomena in near-Earth space. Explorers 18, 21, 28, 33, 34, 35, 41, 43, 47, and 50 and ESA's GEOS satellites, all coming before or during the ISEE/ICE operational lifetime, helped scientists paint a multidimensional picture of humans' environment on Earth and in space.

With the exception of single passes made by Pioneer 8 in the late 1960's, ISEE 3 made the farthest and the first sustained observations of solar wind phenomena and of Earth's magnetotail. The satellite's instruments also produced images of the bow shock created by the impact of the solar wind on the magnetosphere.

Several satellites studying the same phenomena have followed the ISEEs. Most notable of these projects is the Active Magnetospheric Particle Tracer Explorer (AMPTE) series, managed by the Goddard Space Flight Center. These satellites injected barium and lithium atoms into the magnetosphere to permit measurements of the flow of particles through the different densities and bands of the upper atmosphere.

While ICE was the first satellite to have a close encounter with a comet, it preceded by only six months visits by spacecraft from the Soviet Union, Japan, and the ESA to Comet Halley, perhaps the most famous of celestial visitors to Earth's neighborhood. If it returns to Earth orbit in the year 2012, ICE will be the first deep-space, long-distance satellite to return to Earth after nearly a quarter-century in space. What it will be able to reveal to investigators on Earth remains until that time a matter of conjecture.

**See also:** Compton Gamma Ray Observatory; Dynamics Explorers; International Space Station: 2004; Pioneer Missions 6-E; Solar Maximum Mission; Stardust Project.

### Further Reading

Baker, David, ed. *Jane's Space Directory, 2005-2006.* Alexandria, Va.: Jane's Information Group, 2005. The definitive reference manual on all types of space exploration. Each listing provides a concise yet informative examination of its topic. An excellent starting point for most laypersons' investigations into a space-related subject.

Brandt, John C., and Robert D. Chapman. *Introduction to Comets.* New York: Cambridge University Press, 2004. Provides a detailed exposé about virtually every cometary phenomenon.

Chaikin, Andrew. *Space.* London: Carlton Books, 2002. A large-image picture book spanning crewed and robotic exploration of space. Provides pictures of Earth and special resources as well.

Davies, John K. *Astronomy from Space: The Design and Operation of Orbiting Observatories.* New York: John Wiley, 1997. This is a comprehensive reference on the satellites that have revolutionized twentieth century astrophysics. It contains in-depth coverage of all space astronomy missions. It includes tables of launch data and orbits for quick reference as well as photographs of many of the lesser-known satellites. The main body of the book is subdivided according to type of astronomy carried out by each satellite (x-ray, gamma-ray, ultraviolet, infrared and millimeter, and radio). It discusses the future of satellite astronomy as well.

Gavaghan, Helen. *Something New Under the Sun: Satellites and the Beginning of the Space Age.* New York: Copernicus Books, 1998. This book focuses on the history and development of artificial satellites. It centers on three major areas of development—navigational satellites, communications satellites, and weather observation and forecasting satellites.

Launius, Roger D. *NASA: A History of the U.S. Civil Space Program.* Malabar, Fla.: Krieger Publishing Company, 1994. This is an in-depth look at America's civilian space program and the establishment of the National Aeronautics and Space Administration. It chronicles the agency from its predecessor, the National Advisory Committee for Aeronautics, through the present day.

McAleer, Neil. *The Omni Space Almanac: A Complete Guide to the Space Age.* New York: Ballantine Books, 1987. This compendium of information about the major developments of the Space Age puts emphasis on recent developments and their import for the future.

National Aeronautics and Space Administration. *International Cometary Explorer Satellite Media Kit.* Washington, D.C.: Goddard Space Flight Center, 1985. This series of news releases and background articles produced for public consumption during the ICE mission provides an excellent comprehensive overview of the ISEE/ICE missions. Information ranging from mission and communications schedules to a longer history of the missions is included in this package.

Yenne, Bill. *The Encyclopedia of U.S. Spacecraft.* New York: Exeter Books, 1985. This book summarizes design information on all American spacecraft, both piloted and robotic. With minimal substantive information in each listing, this manual is best used for brief, peripheral research.

Zimmerman, Robert. *The Chronological Encyclopedia of Discoveries in Space.* Westport, Conn.: Oryx Press, 2000. Provides a complete chronological history of all crewed and robotic spacecraft and explains flight events and scientific results. Suitable for all levels of research.

*Eric Christensen*

# International Ultraviolet Explorer

*Date:* January 26, 1978, to September 30, 1996
*Type of program:* Scientific platform

*The International Ultraviolet Explorer (IUE) was conceived as a next step in increasingly sophisticated orbiting observatories designed to observe the ultraviolet portion of the spectrum. Launched in 1978, it has proved to be the most productive and oldest functioning satellite in the history of the space program.*

## Key Figures

*Yoji Kondo,* NASA project scientist
*Leon Dondey,* NASA Astronomy Explorers manager
*Albert Boggess, Jr.,* IUE project scientist
*Robert Wilson,* European IUE project director
*George Sonneborn,* IUE resident astronomer

## Summary of the Satellite

The International Ultraviolet Explorer (IUE) was a joint project between the National Aeronautics and Space Administration (NASA), the European Space Agency (ESA), and the British Science and Engineering Research Council (SRC). Each agency contributed financially as well as scientifically to the project. NASA provided the spacecraft and scientific instruments. It launched the IUE and set up a ground station at Goddard Space Flight Center in Greenbelt, Maryland. The ESA provided the solar arrays, which supply the spacecraft with power, and built a second ground station in Villafranca del Castillo, Spain. The SRC took responsibility for the satellite's four television cameras.

IUE carried a 45-centimeter telescope equipped with two spectrographs. A spectrograph records information by passing light or other radiation through a narrow slit and then through a prism, which separates the radiation into its component wavelengths. The result, a spectrogram, was then recorded on film.

The two spectrographs on board IUE were designed to study short and long ultraviolet wavelengths in the electromagnetic spectrum. Light waves are measured in angstroms, which are equal to one one-hundred-thousandth of a centimeter. Ultraviolet wavelengths range from 100 to 4,000 angstroms. Between its two spectrographs, IUE was able to measure ultraviolet radiation from 1,150 to 3,200 angstroms. Along with the spectrographs, two fine error sensors tracked the stars and allowed ground observers to aim the telescope at specific stars.

The satellite, weighing 671 kilograms, measuring 4.3 meters long, octagonal in shape, and with a rocket engine and a telescope tube extending from opposite ends, sported two large arrays of solar cells extending to either side, looking somewhat like a pair of wings. It was launched by a Delta rocket from Kennedy Space Center on January 26, 1978. IUE was placed into an elliptical orbit inclined 34.5° with respect to the equator, having a closest approach distance of 30,252 kilometers and a maximum distance of 41,315 kilometers. The orbit was such that IUE was in communication range with the ESA facility for only ten hours daily. NASA used the satellite sixteen hours a day, and the European agencies operated it for the remaining eight hours.

A sophisticated computer console called the experiment display system was used from the ground for operations coordination and image processing. All spacecraft systems, including those that control orientation, focusing, telescope temperature, and the status of the spectrograph cameras, were monitored from this console. A television monitor displayed spectra from the objects under observation, and a built-in computer allowed a quick preliminary analysis of the images.

The IUE was the first space mission designed to be used by visiting scientists rather than by a select group of researchers. Astronomers from around the world used the satellite if their observing proposals were accepted. In its first year of operation alone, more than two hundred scientists from seventeen countries participated in research using the IUE.

The International Ultraviolet Explorer was run much like a traditional observatory. The visiting astronomers directed the operations in real time. That is, the telescope responded immediately to the operators' instructions, and raw data were displayed directly after exposure, allowing the astronomer unprecedented flexibility. Research plans could be modified as the session proceeded, based on the results as they were obtained. The scientists went to the ground facility, either at Goddard or in Spain, spent several days making observations, and then returned home to analyze their data.

The IUE could be oriented to point anywhere in space, but it had to be at least 43° away from the Sun. If it was closer than that, the Sun's heat would disturb the temperature balance of the mechanism that focused the telescope. Although each day one-quarter of the sky could not be observed by IUE, over the course of a year the entire sky could be scanned.

To make observations, the satellite was skewed to point at the target area, and a television image of that field of view was transmitted to the ground. The exact target was then identified, and the light from the target was directed into one of the spec-

trographs. The spectrum was then recorded and transmitted to the ground, where it was reconstructed on the screen of the monitor in the control room. Once the spectrum had been recorded, the image was processed by the computer to provide the type of information requested by the observer.

To help the visiting researchers, the IUE ground facilities had a full-time staff of experts who supervised the scientific operations, conducted image processing, and maintained the calibration of the spacecraft. For the first six months, the original observer had exclusive rights to his or her data. After that, all data passed into the public domain. All the observations made using IUE are collected in the IUE Archive at Goddard Space Flight Center. The archive continues to be a tremendous resource. Astronomers can request data about whatever they are studying and perhaps gain information in areas that the original observer did not consider.

The IUE was designed to last for three years, but it exceeded all expectations, providing more than ten years of service. In large part, the ingenuity of its operators accounts for its longevity. In August, 1985, NASA engineers and scientists rescued the spacecraft from certain death after one of its three remaining gyroscopes had failed. Three gyroscopes are necessary to position the spacecraft to point at different targets and to maintain contact with Earth. By using other instruments aboard the satellite in conjunction with the two working gyroscopes, engineers solved the problem and IUE continued to function. After that, NASA developed a plan to control the telescope with only one working gyroscope.

In addition to its longevity, IUE proved to be a source of other pleasant surprises to scientists and engineers working with it. When the satellite was launched, it was expected to be capable of exposure times of fifteen to twenty minutes. By testing its capabilities, researchers were able to make exposures of up to fifteen hours, with a record exposure time of twenty-four hours. Such long exposures allowed study of very faint objects outside the

Milky Way, as well as cool stars, which do not radiate strongly in the ultraviolet. In many such instances, IUE proved itself capable of making observations that its designers never intended it to make. In one case, in order to observe Halley's comet close to the Sun, the telescope was manipulated so that it could be pointed toward the Sun, violating the 43° limit.

IUE was shut down on September 30, 1996. It produced more published scientific papers than any previous astronomical satellite. It provided information about physical conditions in the central regions of distant galaxies that may contain black holes. It also provided scientists with more knowledge of the physical conditions in very hot stars, the effect of solar winds on the atmospheres of the planets in our solar system, and the loss of mass from stars when stellar winds and flares occur.

### Contributions

The International Ultraviolet Explorer revolutionized almost every area of astronomy. It was instrumental in gathering research on hot and cool stars, energetic galaxies, x-ray sources, solar system objects, and quasars, and was particularly useful to scientists studying binary star systems.

The IUE was especially suited to the study of young, hot stars, because the bulk of their radiation is emitted in the ultraviolet range. With information provided by the IUE, researchers were able to map the regions of star formation in the Milky Way and other galaxies. A major breakthrough in the study of stars revealed that many stars have hot outer atmospheres similar to that of Earth's Sun. Cooler stars were also studied extensively. The Sun is a cool star, and data on cool stars help scientists to learn not only about other stars but also about processes occurring in the Sun itself.

One of the landmarks of ultraviolet astronomy has been the discovery of the nature and character of the dust and gas between the stars, the interstellar medium. It was found to be quite different from what was expected. One of the major findings was

the discovery of hot gases as well as cool material between the stars.

IUE research confirmed some standard theories that had been hypothesized for some time. One was the presence of a hot halo of gas around the Milky Way, first proposed by the astrophysicist Lyman Spitzer, Jr., in 1956. IUE has also discovered evidence of a "gravitational lens," in which a double image of an object is created when the gravitational field of a very massive body acts as a lens, splitting the light from the more distant object. The idea of a gravitational lens was intimated by Albert Einstein's theory of general relativity but until the IUE had never been supported by solid evidence.

Another area where IUE has contributed significantly is in the study of novae, events in which a star suddenly brightens thousands or millions of times. IUE has observed more than one dozen novae in its lifetime. It was also instrumental in the study of Supernova 1987A, in which a star in a galaxy located fairly close to the Milky Way exploded. IUE helped researchers determine which star had exploded.

IUE has also pointed its ultraviolet telescopes toward comets. It observed Comet Kohoutek in 1976, finding a spectacular object in the ultraviolet spectrum (the visible display in the night sky had been disappointing). It also observed Comet IRAS-Iraki-Alcock in 1983 and Comet Halley in 1986. It was found that the compositions of comets are similar, suggesting a common origin.

Besides being an important tool of research on its own, IUE was extremely useful when used in conjunction with other space missions. When several instruments sensitive to different wavelengths are used simultaneously, information on many levels can be obtained. The satellite was also used successfully during the Voyager flybys of the outer planets. The resolution of the instrument and its ability to observe over long periods of time allowed discoveries that would not otherwise have been possible.

In addition to solving many mysteries and casting light on others, IUE created a few mysteries of its own. For example, a number of objects have been detected for which there is no known object in the visible spectrum. It has also been found that other, previously known stars have very odd ultraviolet spectra that do not agree with current theories.

**Context**

Scientists can discover the nature of celestial objects only by the way those bodies emit, absorb, or alter electromagnetic radiation. The celestial objects that are most familiar emit light in the visible spectrum. Yet visible light is only one portion of the electromagnetic spectrum, which can be considered a range of wavelengths of energy. The longest wavelengths have the lowest energy. These are the radio waves and infrared radiation (heat). Visible light is approximately in the middle of the spectrum. Radiation of progressively shorter wavelengths than that of visible light moves into the ultraviolet and finally into the x-ray and gamma-ray ranges.

The more information one can uncover about a celestial object in each of these ranges, the more completely one will understand its nature. For example, a star may be quite dim in the visible portion of the spectrum but at the same time be a source of intense radio or x-ray emissions. Because the human eye cannot record information from the whole range of electromagnetic radiation, instruments must be designed to be sensitive to and record these wavelengths.

The ultraviolet spectrum is particularly difficult to observe from Earth, because Earth's atmosphere filters out most of the ultraviolet radiation coming from space. Many of the most abundant elements are detected only in the ultraviolet range. It is therefore important to the study of the universe to place instruments outside Earth's atmosphere in order to glean information from this important range of the spectrum.

Because of the necessity of observing from outside Earth's atmosphere, ultraviolet astronomy did not begin until the 1950's, the dawn of the Space Age. At that time, scientists were dependent on rockets and balloons to carry ultraviolet experiments. In the late 1960's, orbiting satellites began to carry ultraviolet telescopes. Most notable of these was the third Orbiting Astronomical Observatory, also known as Copernicus. Starting from virtual ignorance about the ultraviolet, researchers used the early satellites to pave the way for this area of astronomy. Yet none of these instruments had the resolution to detect information from very dim or distant sources. Because of a relatively low orbit, Earth blocked out much of the satellites' range, and they could not remain in continuous communication with receiving stations on Earth.

With the launch of the IUE, ultraviolet astronomy took a giant leap forward. There are few areas of astronomy to which IUE observations have not made significant contributions. The IUE satellite was a telescope in the groundbreaking stages of a new science. Although it had unprecedented resolution for an ultraviolet instrument and had observed objects fainter than those observed previously, its capability allowed only the strongest ultraviolet sources to be observed.

At the time that the IUE was conceived, in the late 1960's, the space program enjoyed a level of government and public support that was destined to diminish in the subsequent decade. The IUE was considered another step along the way to the Hubble Space Telescope, a 2.4-meter orbiting telescope that would be the first permanent optical and ultraviolet space observatory. The Hubble Space Telescope was launched aboard *Discovery* in April, 1990. Until then, IUE remained the sole orbiting observatory surveying the ultraviolet sky. In spite of the tremendous progress made by the IUE, the future of ultraviolet astronomy lies in the future of the space program, when launches of new and more sophisticated instruments can be made.

**See also:** Far Ultraviolet Spectroscopic Explorer; International Sun-Earth Explorers; Orbiting Solar Observatories; Telescopes: Air and Space.

**Further Reading**

Ahrens, C. Donald. *Essentials of Meteorology: An Invitation to the Atmosphere.* 4th ed. Pacific Grove, Calif.: Thomson Brooks/Cole, 2005. This is a text suitable for an introductory course in meteorology. Comes complete with a CD-ROM to help explain concepts and demonstrate the atmosphere's dynamic nature.

Cornell, James, and Paul Gorenstein, eds. *Astronomy from Space: Sputnik to Space Telescope.* Cambridge, Mass.: MIT Press, 1985. An overview of twenty-five years of astronomical research from space. Summarizes what has been learned in different areas of astronomy in the form of short articles written by experts in each field. Suitable for those with some science background.

Henbest, Nigel. *Mysteries of the Universe.* New York: Van Nostrand Reinhold, 1981. Explores the limits of what is known about the universe. Ranges from current theories about the origin of the solar system and the universe to exotic astronomy and astronomy at invisible wavelengths. Contains information about ultraviolet astronomy and some results of IUE observations.

Hofmann-Wellenhof, Bernhard, and Helmut Moritz. *Physical Geodesy.* London: Springer-Praxis, 2005. This is an update to a text considered by some to be the introductory book of choice for the field of geodesy. Includes terrestrial methods and discusses contributions made through the Global Positioning System (GPS).

"Infrared Cirrus (Infrared Astronomical Satellite)." *Sky and Telescope* 73 (June, 1987): 601-602. An article similar to the one in *Astronomy* magazine cited below, but generally more in-depth and slightly more technical. Geared toward the student with some background in astronomy and spaceflight.

Kaula, William M. *Theory of Satellite Geodesy: Applications of Satellites to Geodesy.* New York: Dover Publications, 2000. Discusses how Newtonian gravitational theory and Euclidean geometry are used together with satellite orbital dynamic data to determine geodetic information. Requires familiarity with introductory calculus.

Kivelson, Margaret G., and Christopher T. Russell. *Introduction to Space Physics.* New York: Cambridge University Press, 1995. A thorough exploration of space physics. Some aspects are suitable for the general reader. Suitable for an introductory college course on space physics.

Leverington, David. *New Cosmic Horizons: Space Astronomy from the V2 to the Hubble Space Telescope.* New York: Cambridge University Press, 2001. This is a broad treatise exploring the development of space-based astronomical observations from the end of World War II to the Hubble Space Telescope and other major NASA space-based observatories.

Parks, George K. *Physics of Space Plasmas: An Introduction.* 2d ed. Boulder, Colo.: Westview Press, 2004. Provides a scientific examination of the data returned during what might be called the "golden age" of space physics (1990-2002) when over two dozen satellites were dispatched to investigate space plasma phenomena. Written at the undergraduate level for an introductory course in space plasma, there is also detailed presentation of NASA and ESA spacecraft missions.

Shore, L. A. "I.U.E.: Nine Years of Astronomy." *Astronomy* 15 (April, 1987): 14-22. This article describes the satellite, giving information on its status as well as the results of observations. Details about working with the telescope from ground stations are provided.

Directed toward general audiences, although the language is sometimes technical. Illustrated.

Wheeler, J. Craig. *Cosmic Catastrophes: Supernovae, Gamma-Ray Bursts, and Adventures in Hyperspace.* New York: Cambridge University Press, 2000. A complete exposé of high-energy processes at work in the universe. Includes contributions made by the Compton Gamma Ray Observatory and other space-based observatories to that understanding.

Zaehringer, Alfred J., and Steve Whitfield. *Rocket Science: Rocket Science in the Second Millennium.* Burlington, Ont.: Apogee Books, 2004. Written by a soldier who fought in World War II under fire from German V-2 rockets, this book includes a history of the development of rockets as weapons and research tools, and projects where rocket technology may go in the near future.

*Divonna Ogier*

# Interplanetary Monitoring Platform Satellites

*Date:* November 26, 1963, to October 25, 1973
*Type of program:* Scientific platforms

*The ten Interplanetary Monitoring Platform satellites (IMPs), part of the Explorer program, measured cosmic radiation levels, magnetic field intensities, and solar wind properties in the near-Earth and interplanetary environment.*

### Summary of the Satellites

The Interplanetary Monitoring Platform (IMP) program included ten missions and formed part of the Explorer program. A primary scientific goal was to collect data from a particular region of interplanetary space for a significant portion of the solar cycle (an interval of approximately eleven years during which sunspot activity reaches a peak, then diminishes); as a secondary goal, the IMPs were intended to help provide some operational support for the crewed Apollo missions by sensing radiation levels in space between Earth and the Moon.

Each of the IMP spacecraft is known by at least two labels: An IMP letter designation was assigned prior to launch, and an Explorer number designation was given afterward. Some of the satellites were also known by an IMP number. IMP A (also known as IMP 1) was built by the Goddard Space Flight Center, part of the National Aeronautics and Space Administration (NASA), which also contributed some scientific experiments on board the spacecraft. Other experimenters included groups from the University of California at Berkeley, the University of Chicago, the Massachusetts Institute of Technology, and the Ames Research Center.

The first seven IMPs had designs similar to Explorers 12, 14, and 15. Each had a main structure with a flat, octagonal shape 71 centimeters across and 20 to 30 centimeters deep, and an average mass of 76 kilograms. The octagonal structure allowed easier access for testing and replacement of components. Each spacecraft used four solar panels as well as rechargeable silver-cadmium batteries to provide power for the experiments and telemetry transmitters. The average data rate for the transmitters was 10 to 100 bits per second. Other prominent features included the magnetometers. Most IMP spacecraft had two, mounted on 2-meter (6- to 7-foot) booms extending from the main body; on some IMPs, a third magnetometer was mounted on top of the spacecraft on a telescopic boom. The first generation of IMPs (IMP 1 through IMP 5) was spin-stabilized at 20 to 28 revolutions per minute about an axis perpendicular to the plane of the octagon.

IMP 1 (Explorer 18) was launched from Cape Canaveral on November 26, 1963, using a specially modified Delta launch vehicle (the third stage provided an extra 1,227 newtons of thrust). The spacecraft achieved an elliptical orbit with a perigee (the closest distance to Earth's surface) of 192 kilometers and an apogee (the farthest distance from Earth's surface) of 197,585 kilometers, giving it an orbital period of approximately 2,294 hours. The orbital inclination (the angle between the orbit's geometric plane and Earth's equatorial plane) was 33°. IMP 1 carried seven experiments, including a rubidium-vapor magnetometer (housed in a 33-centimeter-diameter sphere on the telescopic boom); two flux gate magnetometers (for measuring magnetic field intensity and direction); a low-energy charged particle detector and a grid device for separating electrons and low-energy positive particles, particle telescopes (Geiger-Müller counters), and an ion chamber (all for detecting and measuring the direction, intensities, and compositions of cosmic rays); a curved-plate electrostatic analyzer; and a thermal ion electron experiment.

IMP 2 (Explorer 21) was launched from Cape Kennedy on October 4, 1964, but because of a failure in the Delta-Thor launch vehicle, it achieved an elliptical orbit with an apogee of only 95,575 kilometers instead of the intended 203,539 kilometers. IMP 2 was inclined 37.5° to the equator and had an orbital period of approximately thirty-five hours. The principal consequence of the failure was that the spacecraft could provide data on the magnetic field only from within the magnetosphere (the intense portion of Earth's magnetic field that deflects much of the highly energetic stream of charged particles from the Sun) rather than from the transition region (essentially, the boundary of the magnetosphere). IMPs 2, 3, D, and E carried sets of experiments similar to those on IMP 1.

IMP 3 (Explorer 28) was launched from Cape Kennedy on May 29, 1965, into an elliptical orbit of 195 by 263,604 kilometers (this was higher than the intended apogee of 209,170 kilometers because of an overextended propellant burn in the launch vehicle's third stage). The spacecraft's orbital inclination was 34° and the orbital period was approximately 142 hours.

IMP D (Explorer 33) was launched on July 1, 1966, from Cape Kennedy, using a three-stage, Thrust-Augmented Delta launch vehicle. Because the second and third stages of the Delta provided too much thrust, the retromotor (which was to have slowed the spacecraft for injection into lunar orbit, "anchoring" it there) was able only to inject the spacecraft into an elliptical Earth orbit of 15,897 by 435,331 kilometers, with an inclination of 29° and an orbital period of approximately 309 hours.

IMP E (Explorer 35) was launched from Cape Kennedy on July 19, 1967, and achieved an orbit about the Moon with perilune (the closest distance to the Moon's surface) of 805 kilometers and apolune (the farthest distance from the Moon's surface) of 7,401 kilometers, at an inclination of 147°. Its lunar-orbit period was approximately eleven hours.

IMPs 4 and 5 departed significantly from the preceding satellites in the series by carrying eleven and twelve experiments, respectively. These experiments included a three-axis flux gate magnetometer, a range-versus-energy-loss detector, an energy-versus-energy-loss detector, a low-energy proton and alpha particle detector, an ion chamber, a low-energy solar flare electron detector, a solar proton monitoring experiment, a cosmic-ray angular distortion analyzer, a low-energy telescope, a plasma experiment, and two low-energy proton and electron differential energy analyzers.

IMP 4 (Explorer 34) was launched on May 24, 1967, from Vandenberg Air Force Base in California into an elliptical orbit of 248 by 211,080 kilometers, with an inclination of 67° and an orbital period of approximately 104 hours. IMP 5 (Explorer 41) was launched on June 21, 1969, from Vandenberg Air Force Base into an elliptical orbit of 338 by 213,812 kilometers, with an inclination of 83.8° and an orbital period of approximately 106 hours.

The second generation of IMP spacecraft used a considerably larger structure to support a greater number of experiments. Each vehicle consisted of a sixteen-sided drum, approximately 1.35 meters in diameter and 1.83 meters high. Most of the electronics and experiments were mounted on an aluminum honeycomb shelf in the upper part of the drum. Electrical power was provided by three solar arrays attached to the outside of the drum, with silver cadmium batteries for storage; the average power consumption was 110 watts. Because of their average mass of 278 kilograms, the spacecraft required larger launch vehicles; IMP 6 (Explorer 43) used a Delta M-6 (a Thor liquid first stage with six solid-fueled rocket boosters). Data were sent back to Earth at rates of 1,000 to 1,600 bits per second.

IMP 6 (Explorer 43) was launched on March 13, 1971, from Cape Kennedy into an elliptical orbit of 235 by 196,533 kilometers, with an inclination of 28.8° and an orbital period of approximately 94 hours. IMP 7 (Explorer 47) was launched on September 22, 1972, from Cape Kennedy into an elliptical orbit of 249,395 by 397,423 kilometers, with an inclination of 17.2° and an orbital period of approximately 21 days, 20 hours. IMP 8 (Explorer 50) was launched on October 25, 1973, from the Cape into an elliptical orbit of 226,869 by 465,001 kilometers, with an inclination of 27.8° and an orbital period of approximately 24.1 days. IMPs 7 and 8 were positioned so that during a solar flare one satellite could observe the effects on the dark side of Earth, while the other was making the same observations on the sunlit side.

### Contributions

Collectively, the ten IMP spacecraft contributed significantly to knowledge of space physics in the near-Earth, Earth-Moon, and interplanetary environments. Like other long-term programs, such as the Orbiting Geophysical Observatory satellites, the IMP series provided an extended examination of the interactions of the solar wind with Earth's magnetic field over a substantial portion of one solar cycle. A large part of the data collected pertains to specialized areas in space physics; only those findings of general interest from particular satellites are mentioned here.

IMP 1 discovered a region of high-energy radiation beyond the Van Allen radiation belts (the bands of energetic, charged particles trapped in the geomagnetic field). In addition, the satellite provided data indicating that the solar wind has a spiral character in interplanetary space, resulting from the rotation of the Sun as it emits the streams of particles. Most notable was the discovery of a stationary shock wave in the solar wind. This is a region where the flow properties of the wind change drastically; these changes are caused by interaction of the geomagnetic field and the magnetic field generated by the wind. The shock wave precedes Earth in its motion through the solar wind by 86,250 kilometers.

IMP E in its lunar orbit recorded no such shock wave in the solar wind, indicating no lunar magnetic field. Consequently, the Moon has no radiation belts, ionosphere, or magnetosphere as Earth does. The probe also encountered a wake in the region "behind" the Moon in its motion through the solar wind (a region shielded from the solar wind by the Moon) extending more than 160,000 kilometers from the Moon. One result of this absence of charged-particle flow is a distortion in the interplanetary magnetic field in the wake region.

### Context

The IMP missions played a significant role in extending knowledge of the near-Earth, Earth-Moon, and interplanetary magnetic fields and their interactions with the solar wind. This information was used to construct new maps of the field in these regions and to create more accurate models of the dynamic magnetic field-solar wind interactions. In addition, the satellites helped to provide operational support for several of the piloted Apollo missions to the Moon, giving real-time information to Apollo mission controllers on solar flares and the attendant increased radiation levels.

The second group of IMPs, starting with IMP 6, were part of a new generation of satellites in the U.S. space program; they were larger and were capable of supporting greater numbers of even more complex scientific experiments.

**See also:** Explorers: Solar; Goddard Space Flight Center; International Sun-Earth Explorers; International Ultraviolet Explorer; Orbiting Geophysical Observatories.

### Further Reading

Baker, David, ed. *Jane's Space Directory, 2005-2006.* Alexandria, Va.: Jane's Information Group, 2005. This reference work, devoted to the various international space programs, gives considerable background on the technology used for space exploration. Its section on the IMP satellites is easy to read and informative.

Corliss, William R. *Scientific Satellites.* NASA SP-133. Washington, D.C.: Government Printing Office, 1967. Gives a history of scientific satellite missions from 1958 to 1967. Describes major subsystems common to all satellites, devoting significant coverage to scientific instrumentation for spacecraft use. Includes an appendix of all U.S. scientific missions (with descriptions of the satellites and their experiments) flown through early 1967. For technical audiences.

Davies, John K. *Astronomy from Space: The Design and Operation of Orbiting Observatories.* New York: John Wiley, 1997. This is a comprehensive reference on the satellites that have revolutionized twentieth century astrophysics. It contains in-depth coverage of all space astronomy missions. It includes tables of launch data and orbits for quick reference as well as photographs of many of the lesser-known satellites. The main body of the book is subdivided according to type of astronomy carried out by each satellite (x-ray, gamma-ray, ultraviolet, infrared and millimeter, and radio). It discusses the future of satellite astronomy as well.

Gavaghan, Helen. *Something New Under the Sun: Satellites and the Beginning of the Space Age.* New York: Copernicus Books, 1998. This book focuses on the history and development of artificial satellites. It centers on three major areas of development—navigational satellites, communications satellites, and weather observation and forecasting satellites.

Heppenheimer, T. A. *Countdown: A History of Space Flight.* New York: John Wiley, 1997. A detailed historical narrative of the human conquest of space. Heppenheimer traces the development of piloted flight through the military rocketry programs of the era preceding World War II. Covers both the American and the Soviet attempts to place vehicles, spacecraft, and humans into the hostile environment of space. More than a dozen pages are devoted to bibliographic references.

Johnson, Francis S., ed. *Satellite Environment Handbook.* 2d ed. Stanford, Calif.: Stanford University Press, 1965. This excellent technical overview covers all near-Earth environments, including the magnetosphere. Includes graphs, illustrations, and references.

King, J. H. *IMP I, H, and J: Final Report.* Springfield, Va.: National Technical Information Service, 1973. Gives detailed engineering descriptions of Explorers 43, 47, and 50, including accounts of the construction and testing of these satellites. For technical audiences.

_____. *IMP Series Report/Bibliography.* NASA TM-X-68817. Springfield, Va.: National Technical Information Service, 1971. Includes descriptions of the engineering subsystems of Explorers 18, 21, 28, 33, 34, 35, 41, and 43, as well as summary logs of the activities involved in construction and testing of the spacecraft. Contains an extensive bibliography of NASA publications related to these missions. For technical audiences.

Kivelson, Margaret G., and Christopher T. Russell. *Introduction to Space Physics.* New York: Cambridge University Press, 1995. A thorough exploration of space physics. Some aspects are suitable for the general reader. Suitable for an introductory college course on space physics.

Launius, Roger D. *NASA: A History of the U.S. Civil Space Program.* Malabar, Fla.: Krieger Publishing Company, 1994. This is an in-depth look at America's civilian space program and the establishment of the National Aeronautics and Space Administration. It chronicles the agency from its predecessor, the National Advisory Committee for Aeronautics, through the present day.

Leverington, David. *New Cosmic Horizons: Space Astronomy from the V2 to the Hubble Space Telescope.* New York: Cambridge University Press, 2001. This is a broad treatise exploring the development of space-based astronomical observations from the end of World War II to the Hubble Space Telescope and other major NASA space-based observatories.

Yenne, Bill. *The Encyclopedia of U.S. Spacecraft.* New York: Exeter Books, 1985. This well-illustrated volume provides an overview of all craft, including the IMPs, used in the U.S. exploration of space. Includes several appendices and a helpful table of abbreviations and acronyms.

Zimmerman, Robert. *The Chronological Encyclopedia of Discoveries in Space.* Westport, Conn.: Oryx Press, 2000. Provides a complete chronological history of all crewed and robotic spacecraft and explains flight events and scientific results. Suitable for all levels of research.

*Robert G. Melton*

# ITOS and NOAA Meteorological Satellites

*Date:* January 23, 1970, to May 30, 1979
*Type of spacecraft:* Meteorological satellites

*ITOS is an acronym for Improved TIROS Operational System, a series of meteorological satellites developed by the United States to maintain constant surveillance of weather conditions around the world. The National Oceanic and Atmospheric Administration operated the ITOS satellites during the 1970's.*

## Summary of the Satellites

Among the first applications devised for artificial satellite technology was the monitoring and analysis of global weather conditions. Scientists quickly recognized the value of satellites for the tracking and reporting of weather systems and for predicting weather patterns.

The surveillance technology aboard these satellites improved quickly over the first twenty years of the space era, resulting in increased efficiency and capabilities. Early experiments with balloons and relay satellites led to the discovery of a variety of atmospheric characteristics previously unknown to scientists who studied the atmosphere. These discoveries included the existence of a band of radiation around earth (the van allen radiation belts) and other phenomena that scientists began to suspect were key elements in Earth's ever-changing weather patterns.

At the time, the only way to study these phenomena was to place satellites containing specialized sensors in orbits that would make possible the identification of relationships between changes in activity such as solar radiation or cosmic-ray bombardment of Earth's surface and weather change. The necessity for such satellites led to the development of sophisticated television cameras and infrared sensing instruments that, when placed aboard orbiting satellites, proved to be even more useful as tools of meteorology than had been expected. As a result, weather-sensing technology was among the first to be applied to artificial satellites placed in orbit by the United States.

The first weather satellite, launched in 1960, was TIROS 1. (TIROS is an acronym for Television Infrared Observations Satellite.) In 1966, the TIROS Operational System (TOS) was established to provide daily observations of global weather without interruption. It included a number of satellites of similar configuration and was operated by the Environmental Science Services Administration (ESSA), which was succeeded by the National Oceanic and Atmospheric Administration (NOAA) in 1970. Thus, TIROS satellites were succeeded by other first-generation satellites known as the ESSA series, which was followed by ITOS/NOAA. Once operational, NOAA weather satellites became part of that agency's National Operational Meteorological Satellite System (NOMSS).

ITOS, an acronym for Improved TIROS Operational System, was the second generation of weather satellites. It included six satellites launched between January 23, 1970, and July 29, 1976. The ITOS satellites were built by RCA, under the direction of engineers at the Goddard Space Flight Center, an arm of the National Aeronautics and Space Administration (NASA). The ITOS series was designed to provide daily real-time coverage of Earth's cloud cover and other atmospheric conditions, using daytime and nighttime instrumentation.

The first of these satellites, ITOS-1, provided meteorological data that were dramatically improved over the data that had been relayed to Earth by the earlier TIROS and ESSA satellites. It included equipment that could provide direct automatic picture transmission and that could gather and store data for relay at a later time. It was the first satellite to provide around-the-clock radiometric data, and its imagery was significantly better than that of TIROS. The satellite weighed 313 kilograms and was 1.24 meters high and 1.02 by 1.02 meters wide. With shields and antennae deployed, it had a span of 4.3 meters. On board were two advanced 2.54-centimeter vidicon television cameras and two automatic picture transmission devices. Each of the satellite's three winglike panels contained solar cells. There were four communications antennae, two vertical-temperature profile radiometers, two high-resolution radiometers, and two solar proton monitoring systems. Visible channel resolution was 3.7 kilometers, and infrared resolution was 7.4 kilometers. The second ITOS satellite was NOAA-1 (or ITOS-A), launched on December 11, 1970.

Within ITOS, which represented the second generation of weather satellites, was a second generation of ITOS satellites, ITOS-D. Similar to ITOS-1 and NOAA-1, this series of satellites was more sophisticated. They were also a bit larger; each weighed 340 kilograms and measured 1.016 by 1.016 by 1.219 meters. Rectangular in shape, the main structure of the satellite was a three-axis stabilized despun platform, a major improvement in design that facilitated continuous orientation toward Earth's surface. New onboard equipment included very high resolution radiometers and scanning radiometers, along with vertical temperature profile radiometers for keeping track of atmospheric temperatures. It also contained equipment for moni-

*The meteorological satellite NOAA-K at Vandenberg Air Force Base, almost ready for launch.* (NASA)

toring proton and electron flux, phenomena related to solar activity that are thought to affect weather.

Six ITOS-6 satellites had been planned, but only four were launched, because of the unexpected longevity of the individual satellites. The ITOS-D series included NOAA-2 (ITOS-D), launched on October 15, 1972, NOAA-3 (ITOS-F), launched in 1973, NOAA-4 (ITOS-G), launched in 1974, and NOAA-5 (ITOS-H), launched on July 29, 1976.

The ITOS satellites were placed in Sun-synchronous orbits. These orbits were near-polar, the most efficient for the observation of Earth's surface and its near atmosphere. They were placed at an altitude of 1,463 kilometers, and the duration of each orbit was 115 minutes. Sensors aboard the satellite were able to view a track of such width that Earth's entire surface was covered in only 12.5 orbits each day.

Data gathered by the ITOS satellites were continuously fed to automatic receiving stations around the world. These stations were referred to as APT stations, for automatic picture transmission. Additional data stored by the satellite were periodically fed to receiving stations in the United States, which then relayed them to the National Environmental Satellite Service facilities at Suitland, Maryland. There the information was processed before being distributed around the world.

The ITOS program was superseded by the third-generation operational polar-orbiting environmental satellite system, with the launch of TIROS-N in 1978. The last of the ITOS satellites, NOAA-5 (ITOS-H), remained in orbit until the summer of 1979.

### Contributions

The ITOS satellites enabled a significant increase in the amount and quality of meteorological data sent back by artificial satellites. Their increased sophistication meant that more area could be covered in less time with more accuracy than had been the case with the first series, the TIROS and ESSA satellites. Ten TIROS satellites had been launched between 1960 and 1966 in the first effort to use space technology for weather surveillance. Each carried two miniature video cameras, and half carried infrared sensors and radiation sensors that could determine how much radiation entered the atmosphere and how much was reflected by Earth's surface. The ITOS series cameras were equipped with vidicons that were larger than those installed in the TIROS cameras and thus could cover more area.

The TIROS satellites did not provide constant, real-time weather information on demand. The imagery they sent back was sporadic, as their mission was to determine the feasibility of using satellites for meteorological purposes rather than to function as a full-fledged monitoring operation. It

was not until the TIROS Operational System was established that a satellite was used as a constant, twenty-four-hours-a-day source of weather information. Nine satellites in this, the ESSA series, were launched between 1966 and 1969. With this series came the first daily photography of Earth's entire surface.

It was not until the launch of the ITOS series, however, that consistent, reliable data became available instantaneously, as well as from images stored on board the satellites. For the first time, meteorologists were able to gather weather data from parts of the world that contained no reporting stations. These regions included much of the ocean areas in both hemispheres, as well as largely uninhabited deserts and inaccessible or inhospitable regions, such as the North and South Poles.

The range of information collected by the ITOS satellites was enormous. They could detect tropical storms that often turn into hurricanes or typhoons, other less violent storm systems, jet-stream variations, storm fronts, upper-level disturbances, snow cover, even fog. These satellites have enhanced the ability of meteorologists to forecast the weather, often on a long-term basis, and to track dangerous storms on a minute-by-minute basis. The ITOS satellites were also the first to be equipped with instrumentation designed to sense changes in atmospheric conditions caused by solar activity.

Another important function of the ITOS satellites was to provide infrared images that could be used to prepare special Earth surface temperature charts. These charts, which were regularly updated, included surface temperatures of the oceans by regions, which made them useful to the captains of oceangoing vessels, commercial fishermen, and others in the maritime industry. These infrared data were also extremely important to meteorologists, who, through analysis, were able to locate and document conditions that might lead to tropical storms on the world's oceans.

Before the age of satellite imagery, weather conditions had been determined by the amassing of data from observation points around the world.

Often, these data were incomplete or outdated because of the difficulty of obtaining accurate readings or updates, particularly in areas that were inaccessible or inhospitable. During the first half of the twentieth century, the United States began to employ aircraft to gather meteorological data. This development was a major step forward, but it proved to be a technique that had significant shortcomings, because it was impossible to get the global weather picture quickly and accurately. When satellites did come on the scene, they first served to corroborate or support the information coming in from weather observation posts and aircraft and helped in the analysis of those data. After the satellites took over the task of daily, continuous monitoring of the weather, the roles were reversed, with aircraft used to enhance and support satellite-generated meteorological data.

### Context

Since the earliest days of the space era, artificial satellites have been invaluable in keeping track of global weather conditions. The impact of ITOS and other meteorological satellites has been felt in agriculture, maritime shipping, air transportation, and weather-disaster warning and control efforts worldwide. They have also been used in conjunction with Earth resources satellites and other land-sensitive satellites to track ice floes in the polar regions, in an effort to determine the effects of such phenomena on global weather. Later generations of weather satellites would be equipped with search-and-rescue instrumentation that would help locate downed or lost aircraft and oceangoing vessels anywhere on Earth's surface.

This technology also has military and defense applications, as demonstrated by the United States Department of Defense weather satellite program that began operation during the mid-1960's. Known as the Defense Meteorological Satellite Program (DMSP), it was a polar-orbiting satellite system, operated by the United States Air Force. Its purpose was to support military operations around the world and to engage in research that would help engineers design more sophisticated and effi-

cient sensing technology for installation in future meteorological satellites.

The United States was not alone in the use of satellites to keep track of Earth's weather. In fact, a working cooperation among several nations existed under the auspices of an international agency, the World Meteorological Organization, which included Japan, the United States, the Soviet Union, and the European Space Agency (ESA) nations. The United States also entered into a cooperative venture with France, which led to the establishment of a program known as EOLE. EOLE was an 84-kilogram satellite launched in August, 1971, to receive and retransmit data from five hundred weather balloons that contained instrumentation designed to monitor the upper atmosphere.

In 1977, a Japanese satellite, Himawari 1, was launched from Cape Canaveral and positioned over the Pacific Ocean. During that same year, Meteosat, a satellite constructed for ESA, was launched from Cape Canaveral and positioned over the eastern Atlantic Ocean, where it worked in concert with Himawari, ITOS, and other international satellites to acquire and transmit global weather data. This cooperation led to increased use of satellite data by scores of countries, among them several in the Third World.

The Soviet Union was also an early player. Beginning in the early 1960's, it began to experiment with weather-sensitive technology as part of the Kosmos series of satellites. Weather satellites were placed in polar orbits at altitudes of approximately 900 kilometers. For the Soviet Union, polar orbits are required because of its geographical position, which includes territory too far north to be photographed from an equatorial orbit. The Kosmos series eventually led to the Meteor series of meteorological satellites, each of which provided continuous, uninterrupted data in conjunction with others to provide wide-ranging weather coverage. Meteor 1 was launched from Plesetsk on March 29, 1969.

Early Soviet satellites carried video cameras and infrared scanners. The Soviets indicated that they were instrumental in predicting snowmelt from mountain ranges, an important element in the evaluation of prospects for spring flooding, irrigation, and seasonal crop yields. Since the 1970's, the Soviets developed geostationary weather-monitoring satellite technology to support the polar-orbiting satellites.

International space cooperation has been consistent and productive from the earliest days of the space era. The ITOS program, which was developed and operated in conjunction with other programs around the world, remains a monument to such cooperation. The real-time capabilities of its technology resulted in many useful real-world applications that have made satellite meteorological imagery accessible to ordinary people. Today, daily newspapers and radio and television broadcasts include up-to-the-second satellite weather data as part of their weather coverage. Tropical storms, hurricanes, tornadoes, thunderstorms, and snowstorms can be viewed by the public as they are developing as a result of technology advanced through the research and development in the ITOS program. While that technology has advanced significantly since the 1970's, the ITOS satellites were the first to prove that a consistent, reliable, dependable, twenty-four-hours-a-day weather-monitoring system could be established using space hardware. Indeed, it set the standard for technology that would follow.

**See also:** Environmental Science Services Administration Satellites; Interplanetary Monitoring Platform Satellites; Meteorological Satellites; Meteorological Satellites: Military; Seasat; SMS and GOES Meteorological Satellites; TIROS Meteorological Satellites.

## Further Reading

Ahrens, C. Donald. *Essentials of Meteorology: An Invitation to the Atmosphere.* 4th ed. Pacific Grove, Calif.: Thomson Brooks/Cole, 2005. A thorough examination of contemporary understanding of meteorology. Includes the contributions made by satellite technology.

_____. *Meteorology Today: An Introduction to Weather, Climate, and the Environment.* 7th ed. Pacific Grove, Calif.: Thomson Brooks/Cole, 2002. A thorough examination of contemporary understanding of meteorology. Includes the contributions made by satellite technology.

Bader, M. J., G. S. Forbes, and J. R. Grant, eds. *Images in Weather Forecasting: A Practical Guide for Interpreting Satellite and Radar Imagery.* New York: Cambridge University Press, 1997. Offers meteorologists and forecasters an overview of the current techniques for interpreting satellite and radar images of weather systems in mid-latitudes. Heavily illustrated.

Gavaghan, Helen. *Something New Under the Sun: Satellites and the Beginning of the Space Age.* New York: Copernicus Books, 1998. This book focuses on the history and development of artificial satellites. It centers on three major areas of development—navigational satellites, communications satellites, and weather observation and forecasting satellites.

Heppenheimer, T. A. *Countdown: A History of Space Flight.* New York: John Wiley, 1997. A detailed historical narrative of the human conquest of space. Heppenheimer traces the development of piloted flight through the military rocketry programs of the era preceding World War II. Covers both the American and the Soviet attempts to place vehicles, spacecraft, and humans into the hostile environment of space. More than a dozen pages are devoted to bibliographic references.

Hoyt, Douglas V., and Kenneth H. Shatten. *The Role of the Sun in Climate Change.* Oxford, England: Oxford University Press, 1997. This book discusses the interaction between the

Sun and the Earth's atmosphere and how the latter is shaped by solar activity. It describes many of the different cyclic events that affect our climate and how they can be used or abused. It contains an extensive bibliography.

Ley, Willy. *Events in Space*. New York: Van Rees Press, 1969. Features summaries of all satellite programs, national and international, that were carried out during the 1960's. Particularly useful is a series of tables and glossaries that define space jargon and describe many of the satellite and rocket series of the decade. Also included are lists of satellite launches, complete with launch dates and other information relative to the satellites and the programs with which they were (or are) associated.

National Aeronautics and Space Administration. *NASA, 1958-1983: Remembered Images*. NASA EP-200. Washington, D.C.: Government Printing Office, 1983. This paperback book traces various programs of the space agency from its beginnings. Included is a chapter on space sciences as well as one on Earth-orbit applications, including meteorology. Heavily illustrated with color photographs.

Parkinson, Claire L. *Earth from Above: Using Color-Coded Satellite Images to Examine the Global Environment*. Sausalito, Calif.: University Science Books, 1997. A book for nonspecialists on reading and interpreting satellite images. Explains how satellite data provide information about the atmosphere, the Antarctic ozone hole, and atmospheric temperature effects. The book includes maps, photographs, and fifty color satellite images.

Paul, Günter. *The Satellite Spin-Off: The Achievements of Space Flight*. Translated by Alan Lacy and Barbara Lacy. New York: Robert B. Luce, 1975. A survey of the commercial, scientific, and communications applications that developed from the space research of the 1960's and early 1970's. Contains a comprehensive account of the early developments in scientific research that led to the evolution of weather satellites. This book is written from the perspective of the European community. In addition to meteorological applications, space medicine, communications, cartography, agriculture, and oceanography are discussed.

Seinfeld, John H., and Spyros Pandis. *Atmospheric Chemistry and Physics: Air Pollution to Climate*. New York: John Wiley, 1997. This is an extensive reference on atmospheric chemistry, aerosols, and atmospheric models. While the book may be too complex for the average reader, it is extremely useful as a research tool on the science of atmospheric phenomena.

Shelton, William Roy. *American Space Exploration: The First Decade*. Boston: Little, Brown, 1967. Another historical account of the space program, with comprehensive listings of every American spaceflight launched during the period between 1957 and 1967.

Zimmerman, Robert. *The Chronological Encyclopedia of Discoveries in Space*. Westport, Conn.: Oryx Press, 2000. Provides a complete chronological history of all crewed and robotic spacecraft and explains flight events and scientific results. Suitable for all levels of research.

*Michael S. Ameigh*